LOW TEMPERATURE PHYSICS - LT 13

Volume 4: Electronic Properties, Instrumentation, and Measurement

LOW TEMPERATURE PHYSICS - LT 13

Volume 1: Quantum Fluids

Volume 2: Quantum Crystals and Magnetism

Volume 3: Superconductivity

Volume 4: Electronic Properties, Instrumentation, and Measurement

LOW TEMPERATURE PHYSICS - LT 13

Edited by

K. D. Timmerhaus
University of Colorado
Boulder, Colorado and
National Science Foundation
Washington, D.C.

W. J. O'Sullivan
University of Colorado
Boulder, Colorado

and

E. F. Hammel
Los Alamos Scientific Laboratory
University of California
Los Alamos, New Mexico

Volume 4: Electronic Properties, Instrumentation, and Measurement

PLENUM PRESS • NEW YORK-LONDON

Library of Congress Cataloging in Publication Data

International Conference on Low Temperature Physics, 13th, University of Colorado, 1972
Low Temperature physics—LT 13; [proceedings]

Includes bibliographical references.
CONTENTS: v. 1. Quantum fluids.—v. 2. Quantum crystals and magnetism.—v. 3. Superconductivity.—v. 4. Electronic properties, instrumentation, and measurement.
1. Low Temperatures—Congresses. 2. Free electron theory of metals—Congresses. 3. Energy-band theory of solids—Congresses. I. Timmerhaus, Klaus D., ed. II. O'Sullivan, William John, ed. III. Hammel, E. F., 1918- ed. IV. Title.
QC278.1512 1972 536'.56 73-81092

ISBN 978-1-4684-2693-9 ISBN 978-1-4684-2691-5 (eBook)
DOI 10.1007/978-1-4684-2691-5

The proceedings of the XIIIth International Conference on Low Temperature Physics, University of Colorado, Boulder, Colorado, August 21-25, 1972, will be published in four volumes, of which this is volume four.

softcover reprint of the hardcover 1st edition 1974

A Division of Plenum Publishing Corporation
227 West 17th Street, New York, N.Y. 10011

United Kingdom edition published by Plenum Press, London
A Division of Plenum Publishing Company, Ltd.
4a Lower John Street, London W1R 3PD, England

Contents*

Electronic Properties

1. Plenary Topics

2. Electron–Electron Interactions

3. Cyclotron Resonance

* Tables of contents for Volumes 1, 2, and 3 and an index to contributors appear at the back of this volume.

4. Fermi Surfaces

5. Fermi Surfaces—Stress Effects

6. Lifetimes

7. Alloys

8. Electron–Phonon Interaction

9. Electronic Properties

10. Transport—Electrical

11. Transport—Thermal

12. Energy States and Heat

Instrumentation and Measurement

13. Plenary Topics

14. Thermometry

15. Superconducting Instruments

16. Cryogenic Instrumentation and Measurement Techniques

ELECTRONIC PROPERTIES

1

Plenary Topics

Measurement of Relaxation Times and Mean Free Paths in Metals

R. G. Chambers

H. H. Wills Physics Laboratory
University of Bristol, England

The area of interest in Fermi surface studies has changed appreciably over the past few years. The shapes of the Fermi surfaces of most simple metals and of many transition metals are now fairly well determined, and attention has turned increasingly to the detailed properties of the electrons on the Fermi surface (FS), and in particular to their velocities and collision rates.

Traditionally, one assumes that collisions can be described by a relaxation time $\tau(\mathbf{k})$ associated with each state $\mathbf{k}$ on the FS. In transport problems, for example, we have to solve the Boltzmann equation

$$\frac{df_{\mathbf{k}}}{dt} = -\frac{\partial f_{\mathbf{k}}}{\partial \mathbf{r}} \cdot \mathbf{v} - \frac{\partial f_{\mathbf{k}}}{\partial \mathbf{v}} \cdot \dot{\mathbf{v}} + \left(\frac{\partial f_{\mathbf{k}}}{\partial t}\right)_{\text{coll}} \tag{1}$$

subject to the appropriate boundary conditions. Here $f_{\mathbf{k}}$ is the distribution function, which would be equal to the Fermi–Dirac function f_0 in the absence of applied fields. The first two terms on the right represent the rate of change of $f_{\mathbf{k}}$ due to diffusion and to applied fields, and offer no particular problems. The third term represents the effects of collisions, and can be written

$$(\partial f_{\mathbf{k}}/\partial t)_{\text{coll}} = -(1/kT) \int V(\mathbf{k}, \mathbf{k}')(\Phi_{\mathbf{k}} - \Phi_{\mathbf{k}'})\, d\mathbf{k}' \tag{2}$$

where

$$f_{\mathbf{k}} = f_{0,\mathbf{k}} - \Phi_{\mathbf{k}}\, \partial f_0/\partial \varepsilon \tag{3}$$

The term in $\Phi_{\mathbf{k}}$ represents the rate at which electrons are scattered out of state $\mathbf{k}$ into other states; the term in $\Phi_{\mathbf{k}'}$ represents the rate at which electrons are scattered *into* state $\mathbf{k}$ from other states, and the net value of $(\partial f/\partial t)_{\text{coll}}$ is the difference between the two. The term in $\Phi_{\mathbf{k}'}$ converts the Boltzmann equation into an integral equation, whose exact solution is virtually impossible except in the simplest cases. In the relaxation time approximation one replaces Eq. (2) by the very much simpler equation

$$(\partial f_{\mathbf{k}}/\partial t)_{\text{coll}} = -(f_{\mathbf{k}} - f_{0,\mathbf{k}})/\tau_{\mathbf{k}} \tag{4}$$

When, if at all, is this approximation valid? Depending on one's viewpoint, there are two possible answers: (a) always, and (b) never. In principle, at least, one can always solve the full Boltzmann equation [Eqs. (1) and (2)] for any problem, and once it has been solved one can insert the resultant values of $f_{\mathbf{k}}$ and $(\partial f/\partial t)_{\text{coll}}$

in Eq. (4) and simply use Eq. (4) to *define* a relaxation time $\tau_{\mathbf{k}}$ at each point on the FS. But this is not a very useful result because the $\tau_{\mathbf{k}}$ so defined will vary from one problem to the next, and even for a given problem it will vary, in general, with distance from the surface of the metal. To be of real value, $\tau_{\mathbf{k}}$ would need to be a function of $\mathbf{k}$ only, for a given sample at a given temperature, independent of the problem under consideration and independent of position in the sample. For such a "universal" relaxation time to exist, $(\partial f/\partial t)_{\mathrm{coll}}$ would need to be independent of the detailed form of $\Phi_{\mathbf{k}'}$; in other words, the second term in Eq. (2) would need to vanish identically on integration over $\mathbf{k}'$. Since the only condition on $\Phi_{\mathbf{k}'}$ is the particle conservation condition $\int \Phi_{\mathbf{k}'}(\partial f_{0,\mathbf{k}'}/\partial \varepsilon)\, d\mathbf{k}' = 0$, this will only occur if $V(\mathbf{k}, \mathbf{k}') = C(\partial f_{0,\mathbf{k}}/\partial \varepsilon) \cdot (\partial f_{0,\mathbf{k}'}/\partial \varepsilon)$. With this form of $V(\mathbf{k}, \mathbf{k}')$ every scattering process will be "catastrophic," in the sense that after scattering the electron is equally likely to reappear anywhere on the FS, and the relaxation time τ will be the same for all electrons, independent of $\mathbf{k}$. A universal relaxation time thus exists only under extremely restrictive conditions, and for practical purposes we can say that it never exists.

What does exist, quite generally, is a well-defined scattering time $\tau_s(\mathbf{k})$ which governs the rate at which electrons are scattered *out* of state $\mathbf{k}$. From Eqs. (2) and (4) we can write

$$
\begin{aligned}
1/\tau(\mathbf{k}) &= -(kT\,\partial f_{0,\mathbf{k}}/\partial \varepsilon)^{-1}\left[\int V(\mathbf{k}, \mathbf{k}')\, d\mathbf{k}' - \int V(\mathbf{k}, \mathbf{k}')\,(\Phi_{\mathbf{k}'}/\Phi_{\mathbf{k}})\, d\mathbf{k}'\right] \\
&= 1/\tau_s(\mathbf{k}) - 1/\tau_{\mathrm{in}}(\mathbf{k}) \qquad (5)
\end{aligned}
$$

and whereas the second term depends in a complicated way on the form of $\Phi_{\mathbf{k}'}$, the first term has a well-defined value independent of the form of the perturbation. Moreover, there are several experimental arrangements for which the inward scattering rate $1/\tau_{\mathrm{in}}$ becomes negligibly small, so that the measured relaxation time $\tau(\mathbf{k})$ is indeed equal to the scattering time $\tau_s(\mathbf{k})$. This will occur whenever the perturbation Φ produced by the applied fields is negligibly small except in one or two highly localized regions on the FS; within these regions inward scattering is negligible compared with outward scattering, and τ reduces to τ_s. This situation occurs, for example, in cyclotron resonance, magnetoacoustic oscillation studies, and the rf size effect (AKCR, MAO, and RFSE), and in all these cases the measured relaxation time should be τ_s to a good approximation.

We can reach the same conclusion by considering the actual trajectory of a given electron through the metal. If each scattering process is catastrophic in the sense that it completely randomizes the electron motion, the electron will completely forget its previous history at each collision, and the effective mean free path l_{eff} is simply the average distance between collisions. But if scattering is predominantly through small angles, the electron may retain some "memory" of its previous direction of motion over a substantially longer distance. Thus the dc electrical conductivity, which depends on the average distance traveled by an electron before forgetting its original direction of motion, will be governed by an effective mean free path l_{eff} substantially longer than the distance between individual collisions. But the effective free path for RFSE and similar experiments will not show a corresponding increase because even small-angle scattering suffices to remove an electron from the narrowly defined group contributing to the signal, and the effective free path in these experiments remains equal to the distance between individual collisions, corresponding to the scattering time τ_s.

In the de Haas–van Alphen (dHvA) effect and other phenomena that depend on orbit quantization in a magnetic field the collision broadening of the energy levels is again determined by τ_s, rather than by any longer relaxation time, for a similar reason: Even scattering through a small angle is enough to destroy the phase coherence of the wave function and so contribute to the collision broadening.

The dHvA effect has always been one of the most powerful tools for FS studies, and several groups have recently been using it to study the detailed anisotropy of $\tau_s(\mathbf{k})$ for scattering by both impurity atoms and dislocations. For this purpose it is much more powerful than the alternative techniques available, because, for example, RFSE signals can only be observed at very low defect concentrations—so low that it is difficult to characterize the defects precisely—whereas dHvA signals continue to be observable up to much higher concentrations. This point has been stressed by Springford[1] in an excellent review article on scattering rate measurements in the noble metals. Most of the work in this area has been done on the noble metals because at first sight one needs to know the velocity $v(\mathbf{k})$ at each point on the FS (as one does for the noble metals) in order to derive $\tau_s(\mathbf{k})$ from the orbit average $\overline{1/\tau_s(\mathbf{k})}$ which dHvA measurements yield. However, Arko and Mueller[2] have recently shown that one can extract $\tau_s(\mathbf{k})$ directly from measurements of this orbit average, in conjunction with measured cyclotron masses m^*, and have applied this technique to tungsten impurities in molybdenum.

I need not review the dHvA work in detail, because it is fully covered in the review papers of Coleridge[3] and Higgins[4] at this meeting, but I should just like to comment on the remarkable success that Coleridge has had in accounting for the detailed form of $\tau_s(\mathbf{k})$ in terms of the differential phase shifts produced by the impurity and by the host metal in the electron wave function. It really seems that impurity scattering can now be rather well understood in these terms.

One would naturally hope that the dHvA effect would be equally successful in studying the phonon scattering contribution to $\tau_s(\mathbf{k})$, but this turns out not to be the case, for a rather subtle reason. It was first reported by Palin and Shoenberg[5] at LT 12 that the apparent scattering time in mercury did not change appreciably between 1 and 17°K, though the phonon scattering rate changes enormously in this temperature range. This unexpected result has been explained by Engelsberg and Simpson[6] as an effect of the phonon enhancement of the effective mass. The strength of the dHvA signals depends on both the scattering rate and the effective mass: as the temperature is raised the phonon scattering tends to diminish the signal, but at the same time the phonon enhancement of the mass is also changing with temperature. These two effects are not independent, since they depend respectively on the imaginary part and the real part of the self-energy, and it turns out that they just compensate each other, so that the phonon scattering has little or no directly observable effect on the dHvA signals.

These subtle cancellation effects do not arise with the other methods of measuring scattering rates, and there is thus a pleasing complementarity between the dHvA effect and these other methods: The dHvA effect can be used to study defect scattering but not phonon scattering, and the other techniques can be used to study phonon scattering but not (because of the high sample purity needed) defect scattering. Of these other techniques, AKCR and MAO studies also yield average scattering rates around an orbit rather than scattering rates at a localized point on the FS, and I shall say little about them here, though there have been a number of AKCR studies of the temperature dependence of m^* and τ recently by Goy and others, some of which are reported at this meeting. I shall concentrate here on what seem to be the two most powerful techniques: RFSE and magnetic surface states.

RFSE signals arise when a thin, parallel-sided metal plate is placed in an rf field and a dc magnetic field H is applied parallel to the rf electric field but inclined at a small angle to the surface of the plate. Under these conditions electrons traveling almost parallel to H will pursue helical trajectories from one surface of the plate to the other. Just as in the Busch method for measuring e/m by immersing a cathode-ray tube in a longitudinal solenoid, these electrons will be periodically "refocused" as they travel along the field direction, and if the field strength is just right, this refocusing occurs just as they reach the far surface of the plate. A series of sharp spikes then appears in the surface impedance as a function of H whenever the focusing condition is satisfied.

In passing from one surface to the other the electrons have traveled a distance $d/\sin\phi$, where d is the plate thickness and ϕ is the angle at which H is tilted to the surface of the plate, and the signal strength is thus proportional to $\exp(-d/l_s\sin\phi)$, where $l_s = \tau_s v$ is the scattering length. By varying ϕ, one thus obtains a very direct measure of l_s for the localized group of electrons traveling almost parallel to H. The relevant group will indeed depend on the tilt angle ϕ, but since ϕ is typically varied between, say, 1° and 5°, the localization is still pretty good. In this way Gantmakher and Sharvin[7] first showed that the phonon contribution to $1/\tau_s$ varied approximately as T^3, as expected. In all experiments of this kind one expects a T^3 rather than a T^5 dependence because even phonons of low q scatter the electrons enough to remove them from the group contributing to the signal: the extra T^2 weighting factor which appears in the dc resistivity to allow for the ineffectiveness of small-angle scattering does not appear here. The observed dependence is indeed generally close to T^3. In their original measurements on Sn, Gantmakher and Sharvin found a $T^{3.3}$ dependence, and they attributed this slight departure from T^3 to the residual ineffectiveness of phonons of very small q, which fail to scatter the electrons out of the effective group. More recently Tsoi and Gantmakher[8] and Blaney and Parsons[9] have found a T^3 variation in K, and Trofimenkoff[10] has shown that the magnitude of the phonon scattering is in good agreement with theory.

On the other hand, Myers *et al.*[11] have found that for the lens surface in Cd the temperature dependence is faster than T^3 below 4°K, tending to T^5 at the lowest temperatures; and they have shown that the observed temperature dependence can be accounted for quite well by an approximate theory which takes account of the ineffectiveness of phonons of very small q. They have also shown that the observed anisotropy of phonon scattering over the lens surface can be accounted for remarkably successfully on a very simple model in which the anisotropy is attributed entirely to the elastic anisotropy of the material. This is the first detailed study of scattering anisotropy using the RFSE technique, and it shows great promise.

Still more promising, however, is the study of magnetic surface states, if one has the technical skill to prepare surfaces of the required perfection. It was observed long ago by Khaikin and by Koch that the surface impedance of a metal in an AKCR experiment sometimes shows curious oscillations at very low magnetic fields (typically below 50 G), and this remained unexplained until Nee and Prange[12] showed that they were due to rf-induced transitions between quantized surface states of the electrons. These states only exist if the surface finish is so perfect that an electron hitting the surface at a low incident angle is specularly reflected. Semiclassically, one can then think of the electron as skipping along the surface, alternately bouncing off the surface and being deflected back toward it by the weak field H. As Koch[13] has shown, one can obtain essentially the correct expression for the quantized energy levels by applying the Bohr–Sommerfeld quantization rule to the skipping orbits. It turns out that the positions of the lines depend on the quantity $v(\mathbf{k})/K(\mathbf{k})^{1/3}$, where $v(\mathbf{k})$ is the velocity of the skipping electrons and $K(\mathbf{k})$ is the radius of curvature of their **k**-space orbit at the relevant point. Here again the relevant group of electrons is well-localized in **k**-space: typically they cover a range of directions of **v** about $\frac{1}{2}^\circ$ by 5° in size. If $K(\mathbf{k})$ is known, one can thus infer $v(\mathbf{k})$ from the positions of the lines, and Doezema and Koch[14] have in this way determined $v(\mathbf{k})$ along the symmetry axes in Cu, and obtained excellent agreement with the earlier dHvA data of Halse and Lee.

More important, the linewidth is determined by the scattering rate, though the fitting process is a little laborious. In the words of Nee *et al.*,[15] "Values (of the fitting parameters) are chosen by trial and error to produce an aesthetically pleasing facsimile of the experimental tracing." However, the resultant fits are indeed aesthetically very satisfying, and again it is found that $1/\tau_s = \Gamma_0 + \gamma T^3$, as expected. Koch and Doezema [16] have determined the coefficient γ along the symmetry axes in Cu, and find that it exhibits considerable anisotropy, varying by a factor of 30 between the (110) direction, where it is low, and the necks at (111), where it is very high. Nowak and Lee[17] have recently carried out an APW pseudopotential calculation of γ in Cu, and their calculated values are generally in very good agreement with the experimental values.

It is worth recalling that the first conference devoted to electron mean free paths in metals only took place relatively recently, at Zürich in 1968. At that meeting[18] there was in fact rather little to report in the way of firm quantitative results on τ_s or l_s and their anisotropy, either theoretically or experimentally. Since then an impressive amount of progress has been made.

References

1. M. Springford, *Adv. Phys.* **20**, 493 (1971).
2. A.J. Arko and F.M. Mueller (to be published).
3. P.T. Coleridge, this volume.
4. R.J. Higgins, this volume.
5. C.J. Palin and D. Shoenberg, *Proc. 12th Intern. Conf. Low Temp. Phys. 1970*, Academic Press of Japan, Tokyo (1971), p. 513.
6. S. Engelsberg and G. Simpson, *Phys. Rev. B* **2**, 1657 (1970).
7. V.F. Gantmakher and Yu. V. Sharvin, *Soviet Phys.—JETP* **21**, 720 (1965).
8. V.S. Tsoi and V.F. Gantmakher, *Soviet Phys.—JETP* **29**, 663 (1969).
9. T.G. Blaney and D. Parsons, *J. Phys. C* **3**, 126 (1970).
10. P.N. Trofimenkoff, this volume.
11. A. Myers, S.G. Porter, and R.S. Thompson, *J. Phys. F* **2**, 24 (1972).
12. T.W. Nee and R.E. Prange, *Phys. Lett.* **25A**, 582 (1967).

13. J.F. Koch, *Phys. Kond. Materie* **9**, 146 (1969).
14. R.E. Doezema and J.F. Koch, *Phys. Rev. B* **5**, 3866 (1972).
15. T.W. Nee, J.F. Koch, and R.E. Prange, *Phys. Rev.* **174**, 758 (1968).
16. J.F. Koch and R.E. Doezema, *Phys. Rev. Lett.* **24**, 507 (1970); and to be published.
17. D. Nowak and M.J.G. Lee, *Phys. Rev. Lett.* **28**, 1201 (1972).
18. *Phys. Kond. Materie* **9**, J.L. Olsen, ed. (1969), p. 1.

De Haas–van Alphen Effect and Fermi Surface Changes in Dilute Alloys

P. T. Coleridge

Division of Physics, National Research Council of Canada
Ottawa, Canada

The series of low-temperature conferences have traditionally included papers on the de Haas–van Alphen (dHvA) effect in "pure" metals, Although such studies have usually concentrated on measuring the frequency of the oscillations to determine the size of the Fermi surface, it is also possible to look at the temperature dependence of the amplitude to measure the cyclotron mass, the field dependence of the amplitude to obtain estimates of the scattering, and the relative harmonic content to study Zeeman splitting of the Landau levels.

However, dHvA signals are also present in dilute alloys, although they vanish in more concentrated alloys. In fact, many of the specimens used in early dHvA studies would be called alloys by present standards, although the nature of such "alloys" would not be very well characterized. This paper will review work on dilute alloys where, in principle at least, the effects of defects are dominated by a single impurity of known concentration. It is obviously important in all such experiments that the specimens be characterized accurately, not just from the point of view of solute concentration and homogeneity, but also in terms of any possible clustering effects and presence of other defects such as dislocations.

In pure metals the frequencies of dHvA oscillations are directly related to extremal areas of the Fermi surface (FS) by the Onsager relationship. Accurate measurement of the various frequencies at a variety of orientations can be used to obtain Fermi surface dimensions with comparable accuracy, and, as expected, the volume of the Fermi surface found in this way accounts precisely for the total number of electrons per atom. In dilute alloys, where dHvA oscillations are still observed, the measured frequencies also define a Fermi surface which, in general, will differ from that of the host in both size and shape. The concept of an alloy Fermi surface and its relation to the average number of electrons per atom is not as obvious as for the pure metal, so we will just take it to be that quantity measured by the dHvA effect. It will be shown that it is now possible to determine alloy Fermi surfaces by using the same procedures that have proved so successful for pure metals and that the volume of this Fermi surface corresponds approximately to the average number of electrons per atom in the alloy. Moreover, it is possible to determine an effective potential that yields a band structure which, in the region of the Fermi level at least, fits the experimental data. In contrast to this pseudo-Bloch-wave character of the alloy, it will also be indicated how the dHvA effect gives detailed information on the way the impurity differs from the host atoms and how it therefore scatters the conduction electrons out of these pseudo-Bloch states.

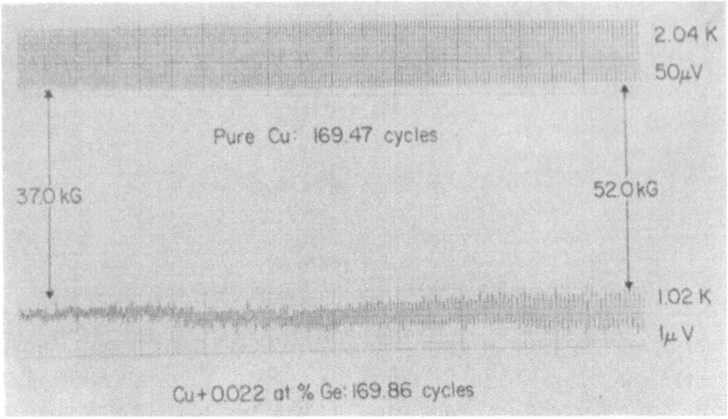

Fig. 1. Comparison of neck signals from a Cu(Ge) alloy with those from pure copper. The frequency has changed by about 0.2% and when the effect of temperature is taken into account the amplitude at 50 kG is down by a factor of about 400.

Experimentally, the most obvious effect of alloying is not the change in frequency but the reduction in amplitude due to scattering. This is demonstrated in Fig. 1, which compares the neck oscillations in a dilute Cu(Ge) alloy with those in pure Cu. The marked reduction in amplitude is typical and always provides an upper limit to the usable impurity concentration. In the present example this is about 0.02%. In more favorable systems and at higher magnetic fields one may introduce as much as 1–2% of impurity.

The difficulty of measuring the frequency changes produced in such dilute alloys is demonstrated by counting the number of oscillations between two fixed values; in our example the change is only 0.2%. To measure such a small frequency change with any reasonable accuracy requires the field to be measured with a precision of at least one in 10^4. Furthermore, it is difficult to align the specimen with the magnetic field to a precision much better than a few tenths of a degree, which produces in this case a further frequency error of about one in 10^4.

There are many alloy systems where the experimental restrictions are less stringent than the case just considered. In general, the larger the dHvA frequency, the smaller the fractional change in frequency and the more important orientation effects become. This is readily seen by considering the simplest model for predicting frequency changes, the rigid band (RB) model.

The definition used here for a rigid band model is essentially that of Heine.[1] Impurities of relative valence Z and concentration α, when introduced into a metal having density of states at the Fermi level of $n_0(\varepsilon_F)$, are assumed to have no effect except to alter the number of electrons in the conduction bands, leaving the shape of the bands unaltered. To first order, the increase in the Fermi energy is then given by

$$\Delta\varepsilon = Z\alpha/n_0(\varepsilon_F)$$

The definition of the cyclotron mass

$$m_c/m_0 = (\hbar^2/2\pi)\,\partial A/\partial\varepsilon$$

and the Onsager relation

$$F = (\hbar c/2\pi e)\,A$$

yields

$$\frac{\Delta F}{F} = \frac{c}{e\hbar}\,\frac{Z\alpha}{F}\,\frac{m_c/m_0}{n_0(\varepsilon_F)}$$

Alternatively, using the connection between density of states and thermal effective mass m_{th}, we may define

$$S = \frac{\Delta F/F}{\Delta n/n} = \left(\frac{4\pi}{3}\right)^{1/3}\frac{\hbar c}{e}\,\frac{n^{2/3}}{F}\,\frac{m_c}{m_{\text{th}}}$$

where n is the electron density in the pure metal. Typically, S is 6.2 for Cu necks and 0.69 for Cu bellies, compared with the free-electron value of 2/3. It is obvious from this expression why small sheets of Fermi surface (small F) are experimentally favored.

One important point concerns the electron–phonon enhancement of the effective masses; the derivation above assumes m_c (and m_{th}) to be band structure values, unenhanced by electron–phonon effects. If experimental values are used for these quantities, they will be enhanced over the band structure values by a factor of $1 + \lambda$, so the ratio m_c/m_{th} will be correct only if this enhancement is isotropic over the Fermi surface.

Early work on the dHvA effect in alloys as summarized by Heine,[1] although valuable in showing alloy experiments to be feasible and for distinguishing hole and electron sheets, was usually insufficiently accurate to allow any significant comparison with theory. Table I provides a summary of more recent work, and it is indeed apparent that although quite a variety of alloys has been studied, the measurements are restricted to small sheets of Fermi surface (major pieces of FS typically yield frequencies $> 10^8$G). Furthermore, we note that the valence difference between solute and solvent is usually less than or equal to two. In general, the results are in fair agreement with the RB model although there are significant deviations. The most dramatic effects occur in noncubic structures such as Be and Zn (hcp) and In (fc tetragonal), where a marked change in FS can be produced by a small change of c/a ratio [e.g., in Zn(Cu) this effect produces changes in the opposite sense to the prediction of the RB model[3]]. In some cases deviations from linearity were observed at higher concentrations.

The general (approximate) agreement with rigid band behavior, which of course implies that the band structure of the host will explain the Fermi surface, is almost completely confined to impurities with a valence difference of one or two. Results obtained on the Cu neck[13] emphasize this point because although Zn and Al appear approximately RB-like, Si and Ge are definitely not, and the transition metal impurities show no simple correlation with valence. Furthermore, in some other systems, e.g., Al(Mg),[4,5] there are deviations from RB behavior for some orbits, although others agree well.

A rather more obvious example of deviation from RB behavior is obtained by studying the neck size in noble metal alloys. For these alloys with zero valence

Table I

Ref.	Host	Solute	Approx. max freq, (10^6 G)	Estimated RB	Comments
2	Au	Ag, Pt	15	~0.6	Au(Ag) faster than linear interpolation
3	Zn	Cu, Al	5	—	c/a ratio very important
4	Al	Zn, Si, Ge Mg, Ag	3	1 ± 0.15	Zero change for Al(ZnGe)
5	Al	Mg	3	≤ 1	α oscillations ~60% RB
6	In	Cd	5	0.9 ± 0.1	Emphasized electron–phonon effects
7	In	Pb, Tl	5	1 – 2	c/a ratio important
8	Be	Cu	15	0.9 ± 0.1	Deviation at higher concentrations
9	Be	Cu	15	—	Detailed pseudopotential fit predominantly RB
10	Pb	Bi	50	1.0 ± 0.1	Some deviation at higher concentrations
11	Pb	In	18	1.0 ± 0.3	FS modified sufficiently to produce new frequency
12	Cu	Zn, Cd, Al, Pd, Ni	22	0.9 ± 0.3	Markedly less for Ni and Pd
13	Cu	Zn, Al, Ge, Si, Ni, Co, Fe, Mn, Cr	22	—	Significant deviations from RB for Ge, Si, and transition metals
14	Cr	Mn	23	—	New frequencies produced by q vector being commensurate with the lattice
15	Cr	V, Mn	23	~0	Frequency changes produced by change of q vector, not Fermi surface changes
36	Sb	Sn	—	—	—
37	Cd	Mg	—	—	—

difference the rigid band model implies zero change in frequency, so other effects, notably changes in lattice constant and "potential," become more obvious. Gross effects of lattice constant change can be allowed for by normalizing all frequencies to the appropriate free-electron cross section. A modified RB model would then imply zero change of neck size on alloying, but this is most unlikely to be correct because the neck is a different size in the three pure metals. We might therefore consider a simple model in which the effective potential in the alloy is the mean of the constituent potentials weighted according to concentrations, and in this case the neck areas interpolate linearly between the pure metal values. The experimental results of Springford and Templeton[16] are shown schematically in Fig. 2 and it is apparent that neither the rigid band model nor a simple mean potential model is adequate in

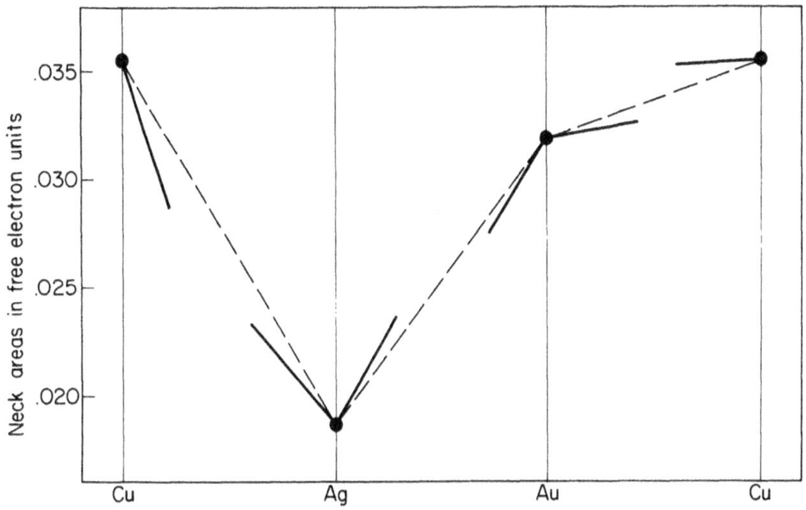

Fig. 2. Schematic representation of the rate of change of neck areas in the "noble" alloys using the data of Springford and Templeton,[16] which were obtained with alloy concentrations ~ 1%. The gross effects of lattice constant change have been allowed for by normalizing all areas to the cross section of the appropriate free-electron sphere. In this plot a rigid band model gives zero slope and a simple mean potential model corresponds to a linear interpolation between the pure metals.

explaining the actual behavior, which presumably reflects in some fairly complicated way the changes in band structure that occur on alloying. In the noble metals the position and width of the *d* band is probably the single most important factor. In a similar isoelectronic system, In(Tl), the results presented by Holtham and Parsons[7] suggest that reasonable agreement with experiment can be obtained by taking the mean potential of the alloy constituents but also taking account of the modified screening that results from the change in lattice constant.

In chromium[14,15] the frequency changes can be explained by taking the Fermi surface to be essentially unchanged and just considering changes in the *q*-vector (the vector describing the variations in the spin, which introduces multiple magnetic band gaps). The observed frequency changes are, in fact, four times larger and in the opposite sense to the prediction of the RB model.

Although many of these experiments involving small sheets of Fermi surface, particularly when valence differences of one or two are involved, can certainly be explained by a rigid band model modified to take account of lattice constant changes, and although the deviations from RB behavior that do occur can often be accounted for by constructing some new potential using fairly obvious prescriptions, it would nevertheless be desirable to measure accurately the changes in major pieces of Fermi surface and construct from such data the complete Fermi surface of the alloy. This would allow, among other things, a direct comparison of the FS volume with the average number of electrons per atom in the alloy and also a reliable comparison with band structure calculations.

This involves two separate problems: (i) measurement of the magnetic field sufficiently accurately (better than 1 ppm); and (ii) elimination of the errors associated with misorientation of the specimen. Measurement of the phase of the dHvA oscillations is relatively trivial. A number of groups have used NMR *in situ* to measure accurately the magnetic field during an experiment, and at Ottawa we have combined this technique with the facility to accurately align the dHvA specimen in a magnetic field.[17] Templeton discusses[18] the details of our technique and shows that with suitable care we can measure the frequencies corresponding to major pieces of Fermi surface with a precision relative to pure copper of ~1 ppm. For the two alloy systems we have studied, Cu(Al) and Cu(Ni), the frequency changes can be determined with an error of 1 or 2%, and there is a comparable error involved in measuring the impurity concentration of each sample. By analyzing the results with a Fourier expansion,[19] we can determine both the changes in k vector on the Fermi surface and also the volume of the alloy Fermi surface. Preliminary results show that Al increases the volume at the rate of 1.83 ± 0.05 electrons/atom, while Ni decreases it at the rate of 1.10 ± 0.05 electrons/atom.

The deviation of these values from the effective valence is probably associated with the effect of lattice strains. Apart from the effects of physical damage to the crystal, which can be avoided by suitable care in handling, the change of lattice constant associated with alloying is not homogeneous but is severely localized around the impurity atoms.[20] The effect of this on dHvA signals has been treated in an approximate fashion by Coleridge,[21]* and it certainly appears likely that a spurious effect on the frequency may be introduced by these strains scattering electrons in such a manner that the measured dHvA signals correspond to a preferentially sampled region of crystal away from impurities. In our experiments we have attempted to minimize such effects by choosing alloy systems in which strain effects (as characterized by the lattice constant change) are rather small.

We have also had some success in fitting the results to a band structure parameterized in terms of *s*-, *p*-, and *d*-phase shifts (analogous to pure-metal phase-shift fits; see, e.g., Ref. 22), so we tentatively conclude that, in principle, a potential can be determined from the experimental data such that a band structure calculated with this potential, using standard methods, explains the Fermi surface of the alloy as measured by the dHvA effect. Furthermore, at least in Cu(Al) and Cu(Ni), this is not just the rigid band structure of Cu. There are probably small discrepancies ($\lesssim 5\%$) between such a calculated Fermi surface and the measured "dHvA" Fermi surface, which arises from the special way the dHvA experiment samples the impurities.

Amplitude

In addition to the frequency of the dHvA signal, we can study the amplitude dependence. Under the most common experimental conditions the measured amplitude is given by

$$d^2M/dH^2 \propto TH^{-1/2} \exp\left[-K_0(m^*/m_0)(T+x)/H\right]$$

* Reconsideration of this calculation indicates that Ref. 21 probably overestimates this effect by a factor of order five.

where $K_0 = 146.9$ kG/°K and the Dingle or scattering temperature x is related to the scattering rate by $x = \hbar/2\pi k\tau$.

From the temperature dependence of the amplitude, the cyclotron mass m^* can be obtained, but because amplitude measurements are notoriously subject to experimental errors, it is difficult to achieve an accuracy of much better than 2% in determining m^*. The expected changes in m^* on alloying are much smaller than the frequency changes (for a free-electron rigid band model they are zero) so little of significance is to be expected from the very dilute alloys normally considered. Nevertheless, in Be(Cu 0.8%) Tripp and co-workers[9] found that for the neck frequency, which increased by a factor of three, the cyclotron mass did indeed show an increase of 50% as required by their "fixed band" model (i.e., a rigid-band model with higher-order terms included).

A very large effect from a different source has been observed in Al(Mn) by Paton.[23] With an alloy of 0.04% he observed a large increase in m^*, approximately 4%. In this case the increase is attributed to enhancement by localized spin fluctuations on the nearly magnetic impurity sites and the results agree quite well with parameters deduced from that theory.

In contrast to these measurements the field dependence of the amplitude, expressed as a Dingle temperature, is large and directly related to the average around the orbit of the scattering. In addition to the impurity scattering the measured Dingle temperature contains terms dependent on mosaic spread, dislocations, etc.; but with suitable precautions[24,25] it is possible to separate these terms and obtain only that part of the Dingle temperature appropriate to the impurity scattering. Springford and co-workers[26,27] analyze their results using a symmetrized Fourier expansion to obtain from the orbital averages a point-by-point plot of relaxation time. They are particularly careful, as is necessary in any such inversion, to obtain sufficient experimental data to ensure that their fit is meaningful compared to the precision of the data. As an example, in the Au(Ag) system, where the measured Dingle temperatures range from 9.8°K/at.% for the $\langle 100 \rangle$ belly to 2.8°K/at.% for the neck, the calculated fit for a 1% alloy using a four-term expansion is shown in Fig. 3.

In contrast to this empirical approach for fitting the scattering, there has recently been some success using analyses in terms of scattering phase shifts and the anisotropy over the FS of the wave functions of the metal.[28,29] Such a procedure provides fitting parameters (the scattering phase shifts) which are physically more significant than those derived from Fourier expansions, but it should be noted that the phase-shift approach will not allow for any small-angle scattering that may be associated with the impurity-induced strains mentioned above.

It should be emphasized that because scattering through extremely small angles is sufficient to destroy dHvA signals, the measured relaxation time is that appropriate to the total scattering, and should not be identified with the τ_{res} appropriate to resistivity. This is demonstrated rather vividly by studying the effect of dislocations on dHvA signals. Two groups of workers, Terwilliger and Higgins[30,31] and Coleridge and Watts,[32] have measured the Dingle temperatures corresponding to dislocations. The results, which show considerable anisotropy among various orbits, give a scattering rate $\sim 10^4$ times that deduced from the resistivity of dislocations. This implies that the average value of the angular weighting factor $(1 - \cos\theta)$

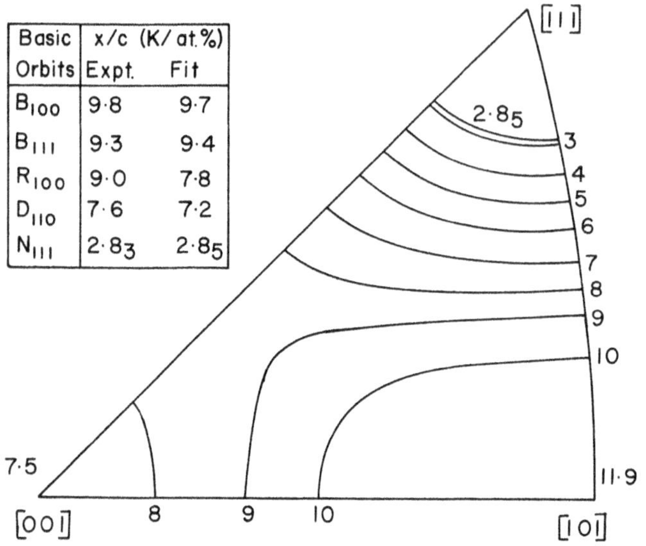

Basic Orbits	x/c (K/at.%) Expt.	Fit
B_{100}	9·8	9·7
B_{111}	9·3	9·4
R_{100}	9·0	7·8
D_{110}	7·6	7·2
N_{111}	2·83	2·85

Fig. 3. Inversion of dHvA scattering in Au(Ag) using the results of Lowndes *et al.*[26,27] The results of the four-term Fourier inversion have been replotted as an effective point-by-point Dingle temperature. The inversion used many more orbits than the five basic orbits shown.

in the resistivity is $\sim 10^{-4}$, corresponding to a root mean square scattering angle of $\sim 10^{-2}$ rad for dislocations.

One further technique using the amplitude is available. That is to extrapolate the amplitude dependence to infinite field (zero $1/H$) and compare the absolute amplitude so obtained with that in the pure host. Such measurements are useful for dealing with field-dependent scattering processes such as dephasing problems, or the case of magnetic impurities.

A more direct way of studying magnetic impurities is to look at the Zeeman splitting of the Landau levels. Normally the spacings between the Landau levels and the spin splitting are both proportional to the magnetic field, so the ratio of the two spacings is determined by the product of the g factor and the cyclotron mass, gm^*/m_0. When this is an odd integer, i.e., the spacing of all the spin-split Landau levels is equal, one has a so-called "spin-splitting zero" at which the amplitude of the fundamental dHvA signal vanishes, independent of field. If, however, the spins experience an additional constant exchange splitting, the spin-splitting zero will be field dependent and occur at a different value of m^*. An entirely equivalent description is to consider two Fermi surfaces, spin-up and spin-down, separated by the exchange splitting and producing beats. Exactly such a beat was observed by Hörnfeldt *et al.*[33] in the ferromagnetic Pd(Co) system and similar results can be seen in "Kondo" alloys, but with an antiferromagnetic exchange field. In this case it is easier to observe the spin-splitting zeros in pure Cu, which are removed by the addition of magnetic impurities. The zero is replaced by a minimum which is shifted in position (and dependent on magnetic field), the depth of which represents the differential scattering of the two spin states. Experiments of this nature have been performed on Cu(Cr), Cu(Fe), Cu(Mn), and Cu(Co) to study the magnitude of the exchange energy.[34,35]

These examples of the way the dHvA effect can be used to study the anisotropy of conduction electron scattering and the effects of localized moments show that it provides a powerful method* for probing an impurity with electrons of well-defined k values. By contrast, the ability to determine alloy Fermi surfaces means that the technique can also be used to investigate experimentally the effective potential that determines the "band structure" of alloys. One particular advantage of the method is that because these two complementary properties of the alloy can be determined simultaneously in the same alloy, comparison between them will be less sensitive to parameters such as exact solute concentration. The real limitation of the technique, of course, is the restriction to dilute alloys. For these experiments 1 % is a concentrated alloy and 5 % is unattainable.

References

1. V. Heine, *Proc. Phys. Soc.* **A69**, 505 (1956).
2. P.E. King-Smith, *Phil. Mag.* **12**, 1123, (1965).
3. R.J. Higgins and J.A. Marcus, *Phys. Rev.* **141**, 553 (1966).
4. J.P.G. Shepherd and W.L. Gordon, *Phys. Rev.* **169**, 541 (1968).
5. J.C. Abele and F.J. Blatt, *Phys. Rev. B* **1**, 1298 (1970).
6. R.J. Higgins, H.D. Kaehn and J.H. Condon, *Phys. Rev.* **181**, 1059 (1969).
7. P.M. Holtham and D. Parsons, this volume.
8. I.S. Goldstein, D.J. Sellmyer, and B.L. Averbach, *Phys. Rev. B* **2**, 1442 (1970).
9. J.H. Tripp, P.M. Everett, J.M. Fiske, and W.L. Gordon, *Phys. Rev. B* **2**, 1556 (1970).
10. J.R. Anderson and D.C. Hines, *Phys. Rev. B* **2**, 4752 (1970).
11. P.J. Tobin, D.J. Sellmyer, and B.L. Averbach, *J. Phys. Chem. Solids* **32**, 1721 (1971).
12. L.-F. Chollet and I.M. Templeton, *Phys. Rev.* **170**, 656 (1968).
13. P.T. Coleridge and I.M. Templeton, *Can. J. Phys.* **49**, 2449 (1971).
14. J.E. Graebner, in *Proc. 12th Intern. Conf. Low Temp. Phys., 1970,* Academic Press of Japan, Tokyo, (1971), p. 601.
15. E.J. Gutman and J.L. Stanford, *Phys. Rev. B* **4**, 4026 (1971).
16. M. Springford and I.M. Templeton, Private communication.
17. P.T. Coleridge, A.A.M. Croxon, G.B. Scott, and I.M. Templeton, *J. Phys. E: Sci. Instr.* **4**, 414 (1971); P.T. Coleridge and I.M. Templeton, *J. Phys. F: Metal Phys.* **2**, 643 (1972).
18. I.M. Templeton, this volume.
19. D. Roaf, *Phil. Trans. Roy. Soc.* **A255**, 135 (1962). M.R. Halse, *Phil. Trans. Roy. Soc.* **265**, 507 (1969).
20. J.D. Eshelby, *J. Appl. Phys.* **25**, 255 (1954).
21. P.T. Coleridge, *Phil. Mag.* **26**, 573 (1972).
22. M.J.G. Lee and V. Heine, *Phys. Rev. B* **5**, 3839 (1972).
23. B.E. Paton, *Can. J. Phys.* **49**, 1813 (1971).
24. M. Springford, *Adv. Phys.* **20**, 493 (1971).
25. D. Shoenberg, *Phys. Kond. Materie* **9**, 1 (1969).
26. D.H. Lowndes, K.M. Miller, R.G. Poulsen, and M. Springford, *Proc. Roy. Soc. (London)* (in press).
27. D.H. Lowndes, K.M. Miller, and M. Springford, *Phys. Rev. Lett.* **25**, 1111 (1970); K.M. Miller, R.G. Poulsen, and M. Springford, *J. Low. Temp. Phys.* **6**, 411 (1972); B. Bosacchi, J.B. Ketterson, and L.R. Windmiller, *Phys. Rev. B* **5**, 3816 (1972).
28. R. Harris, this volume.
29. P.T. Coleridge, *J. Phys. F: Metal Phys.* **2**, 1016 (1972).
30. D.W. Terwilliger and R.J. Higgins, *Phys. Lett.* **31A**, 316 (1970).

---

* A much more powerful technique for studying these alloys by looking at the line shape of the oscillations was described by Higgins[31] at this conference.

31. R.J. Higgins, this volume.
32. P.T. Coleridge and B.R. Watts, *Phil. Mag.* **24**, 1163 (1971).
33. S. Hörnfeldt, J.B. Ketterson, and L.R. Windmiller, *Phys. Rev. Lett.* **23**, 1292 (1969).
34. P.T. Coleridge, G.B. Scott, and I.M. Templeton, *Can. J. Phys.* **50**, 1999 (1972).
35. P.T. Coleridge and I.M. Templeton, *Phys. Rev, Lett.* **24**, 108 (1970).
36. A. Dunsworth and W.R. Datars, *Phys. Rev. B* **7**, 3435 (1973).
37. J. Boyd, R. Reschly, and W.L. Gordon, *Phys. Lett.* (to be published).

Electronic Structure of Impurities in Transition Metals

Junjiro Kanamori

Department of Physics, Faculty of Science
Osaka University, Toyonaka, Osaka, Japan

Introduction

The purpose of this paper is to discuss physical pictures of the electronic structure of transition and nontransition impurity atoms in transition metals on the basis of ab initio calculations which have been carried out so far for several cases with appropriately chosen potentials. The Anderson model is a very useful parametrization of the *d* virtual bound level of a transition impurity atom in normal metals. In the case of nontransition impurities, on the other hand, we do not have a simple physical picture of the electronic structure, particularly when they are in transition metals. There is much experimental evidence that the presence of the host *d* bands affects the electronic structure of nontransition impurities to a considerable extent. On the basis of a previously developed formalism[1] we present a physical picture that can explain experimental results consistently. The main subjects of the discussion are the NMR data of impurity nuclei and the magnetic properties of dilute alloys. The cases of transition impurities are also discussed briefly in connection with current theories of the appearance of the magnetic moments.

Effects of Nontransition Impurities

According to the measurements by Gerstenberg[2] and Lam and Myles,[3] the magnetic susceptibilities of Pd dilute alloys containing a few percent of both transition and nontransition impurities show a remarkable correlation to the valence of the impurities, falling on the same curve when plotted against the impurity concentration times the number of the valence electrons per atom. The impurities are (the assumed number of valence electrons is given in parentheses): Ag(1), Cd(2), Sb(5), Ti(4), V(5), Zr(4), Nb(5), Mo(6), Ta(5), W(6), etc.* A similar correlation is found in the changes of the saturation magnetization of ferromagnetic Ni,[4] as is shown in Table I. It is to be emphasized also that such a correlation with the valence of impurities is not found when the host metal is a transition metal other than Ni and Pd.† Another experimental fact which we are going to discuss is the slowing down of the nuclear relaxation of impurity nuclei of nontransition elements in transition metals pointed out by Kohzuki and Asayama.[5] In Table II we list some examples.[5-9] The measured $T_1 T$ of Al in Pd, for example, is about 13 times larger than

* According to Gerstenberg, Cr should be given the valence 3 instead of 6, and Re given 5 instead of 7.
† The correlation may be found in Pt and ferromagnetic Co, though no comprehensive data are available.

Table I. Decrease of Magnetization of Ni[4]

Solute	Valence	$-d\mu/dc$, μ_B
Cu	1	1.14
Zn	2	2.11
Al	3	2.80
Si	4	3.77
Ge	4	3.70
Sn	4	4.22
Sb	5	5.31

Table II. Data on T_1T for Impurity Nuclei

Solute	Host	T_1T, sec°K	Ref.
^{27}Al	Al	1.77 ± 0.1	6
	Ag	4.3 ± 0.8	9
	Pd	24.8 ± 3.0	5
	Ti	47.0 ± 4.0	9
	Zr	36.6 ± 7.5	9
^{63}Cu	Cu	$1.27 + 0.1$	6
	Pd	1.51 ± 0.05	9
^{109}Ag	Ag	9.6 ± 0.6	7
	Pd	22 ± 4	7
^{12}B	Au	$\geqq 60$	8
	Pd	$\geqq 185$	8
	Pt	$\geqq 270$	8

that of Al nuclei in pure Al or about six times larger than that of Al nuclei in Ag. The slowing down is observed also when Al is present as an impurity in other transition metals. It is found that the dependence of T_1T on the species of host metal (the measurement has been done so far for Ti, Zr, V, Cr, Fe, Co, Pd, and Ag) is quite similar between Al and Sn.[9] It is to be noted also that T_1T of Cu nuclei in Pd does not differ very much from the corresponding value in pure Cu.

Filling of the *d* Holes by the Impurity Valence Electrons

The magnetic susceptibility data of Pd alloys and the saturation magnetization data of Ni alloys have been interpreted by many authors as indicating that the valence electrons of impurity atoms are poured into the host *d* bands to fill the empty states. For example, in ferromagnetic Ni and also Co where the *d* bands of majority spin are completely full, the valence electrons will fill the minority-spin bands to produce a decrease of the saturation magnetization per impurity atom approximately equal to the valence of the impurity atom in units of the Bohr magneton. There are some experimental data which have been considered to support this picture. For example, the NMR data on Ni and Co alloys[10] show that there are several satellite lines on the low-frequency side of the main line of Co nuclei. The interpretation that these satellites arise from the immediate neighbors of impurity atoms is consistent with the picture, since the filling of the empty *d* orbitals will lead to smaller magnitudes of the internal field. It is hard theoretically, however, to imagine that the impurity atoms exist as bare ions without the valence electrons in a metal. According to soft X-ray spectroscopy[11] and isomer-shift measurement,[12] the nontransition impurity atoms seem to retain almost the same number of valence electrons as in the neutral state. The filling of the host *d* holes therefore must be "apparent." Previously Terakura and this author proposed an explanation of this "apparent" filling by the valence electrons, carrying out an ab initio calculation of the electronic structure of Al in ferromagnetic Ni.[13] In the following we point out that the same mechanism underlies the above-mentioned NMR and susceptibility data.

The original idea of the shielding mechanism of the nontransition impurity charge in transition metals can be traced back to Mott's review article,[14] where the possibility is mentioned that the valence orbitals of *s* and *p* symmetries of the impurity atom combine with the *d* orbitals of surrounding host atoms to form bonding and antibonding molecular orbitals. Suppose the Fermi level lies between the bonding and antibonding energy levels. If the weight of the impurity valence orbitals in the bonding orbital is appropriate, the impurity atom can retain enough electrons to be approximately electrically neutral. On the other hand, the *d* states, whose total number is approximately equal to the impurity valence, are pushed above the Fermi level, since in the empty antibonding state the relative weight between the host *d* orbitals and the impurity valence orbitals is just opposite to that in the bonding state. This may amount to the "apparent" filling of the *d* holes.

The above-mentioned idea, which, while attractive and justified somewhat in the cases of Ni and Pd, needs many refinements to be convincing. First of all we can question to what degree the impurity valence orbitals can be meaningful, since we know that the valence electrons have a broad energy distribution in normal metals. Also, the reason why the correlation to the impurity valence is not found in other transition metals, such as Fe, needs to be elucidated.

Ab Initio Calculations and Antiresonance in the Self-Energy

As is discussed in detail elsewhere,[1,11] we can actually introduce the effective valence orbitals of the impurity atom in the ab initio calculation of the electronic structure. In the following we assume the muffin-tin potential, in which only the potential at the (substitutional) impurity site is different from that of the host metal. The most convenient definition of the orbitals is given by the conditions that they satisfy the Schrödinger equation within the muffin-tin sphere at the impurity site and vanish on the surface of the sphere. Though these conditions define the core orbitals as well as those of much higher energies, we confine ourselves to those orbitals that correspond to the valence orbitals. It can be shown that the extra density of states, or, in other words, the change in state density brought in by an impurity atom, is given to a good approximation by

$$\Delta\rho_j = -(1/\pi)\,\mathrm{Im}\,(\partial/\partial E)\log(E - E_j - \Sigma_j) \qquad (1)$$

where j denotes the symmetry of the orbital; E_j is the eigenvalue determined by the abovementioned condition; Σ_j is the self-energy acquired through the mixing with the states of the host metal. Σ_j is expressed as

$$\Sigma_j = -\langle j|M \cdot \mathscr{G}^0 \cdot H'|j\rangle \tag{2}$$

where $\mathscr{G}^0$ is the Green function of the host energy bands; M and H', given by $M = E_j - H$ and $H' = V_{\text{imp}} - V_{\text{host}}$, correspond to the mixing elements, where H is the Hamiltonian and V is the atomic potential of either the impurity or the host.

In the case of nontransition impurities Σ_j turns out to be quite energy dependent. It is to be emphasized that the energy dependence of Σ_j is determined mainly by that of the host Green function, which is independent of the impurities. Figure 1 shows an example of the imaginary part of the calculated Σ_j. Generally $-\text{Im}\,\Sigma_j$ ($j = s$ or p) shows a hump below or in the lower half of the d bands; it is then followed by a dip which is broad for the s orbital, extending over the whole range of the d bands. For the p orbitals the bottom of the dip is situated around the top of the d bands. From Eq. (1) we can see that $-\text{Im}\,\Sigma_j$ gives a rough idea of the energy dependence of the extra number of states below a given energy in the energy range $|\Sigma_j| < |E - E_j|$. It can also give a rough measure of the energy distribution of an electron in the im-

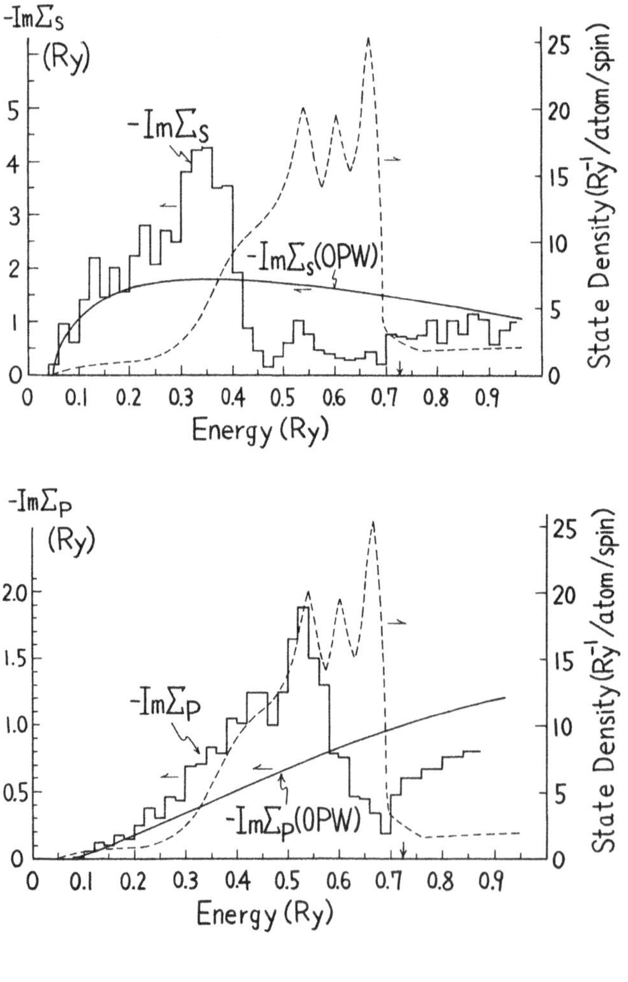

Fig. 1. Plot of $-\text{Im}\,\Sigma_j$ ($j = s$ and p) of Al as impurity in ferromagnetic Ni (for majority-spin states). The dashed curve represents the profile of the density of states of the host metal. Σ_j (OPW) is calculated by taking account of the mixing with the OPW states only. See Ref. 13 for details of the calculation.

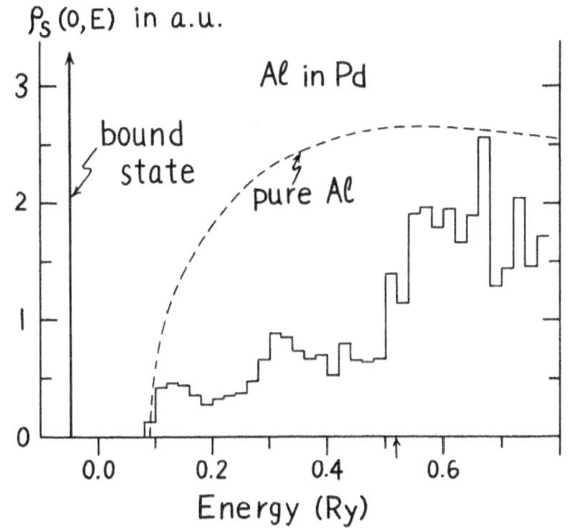

Fig. 2. The density of electrons per spin at an Al nucleus as impurity in Pd. The dashed curve represents a rough estimate of ρ_s (0, E) in pure Al.

purity orbital. Figure 2 shows the energy dependence of the electron density at the nucleus of Al in Pd. Because of the presence of the dip in $-\mathrm{Im}\,\Sigma_s$ the density is much smaller than the value which would be expected in the absence of the d bands, i.e., in an Al atom embedded in free-electron sea or in pure Al. As is shown in Fig. 3, the calculated T_1T of ^{27}Al explains the slowing down of T_1 quantitatively.[15] It can be shown also that Al is almost electrically neutral with the same potential used in the calculation of T_1. It may be worthwhile to comment on the apparent absence of the slowing down of the relaxation in the case of Cu in Pd. Figure 4 shows the energy distribution of electrons in the states of d symmetry within the muffin-tin sphere of Cu. Through the mixing with the d bands of Pd the distribution has a

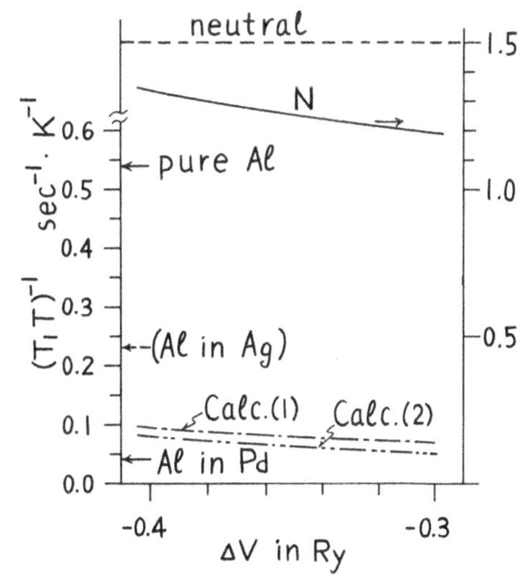

Fig. 3. $(T_1T)^{-1}$ of Al nuclei. Experimental data are shown by arrows on the left ordinate. Both calculations 1 and 2 are for Al as impurity in Pd. Since E_F is near a step in the histogram in Fig. 2, we estimate $\rho_s(0, E_F)$ by taking the average of two neighboring values in the calculation 1 and the lower value in the calculation 2. ΔV of the abscissa is an additive constant in the Al potential used in the calculation, which is defined by $V_{Al} = V_{Al}^0 + \Delta V$, with V_{Al}^0 representing the muffin-tin potential used in the band structure calculation of pure Al. The curve labeled N is the number of valence electrons per spin contained in the Wigner–Seitz sphere of Al in Pd and the dashed line indicates the value 1.5 which is necessary for electrical neutrality.

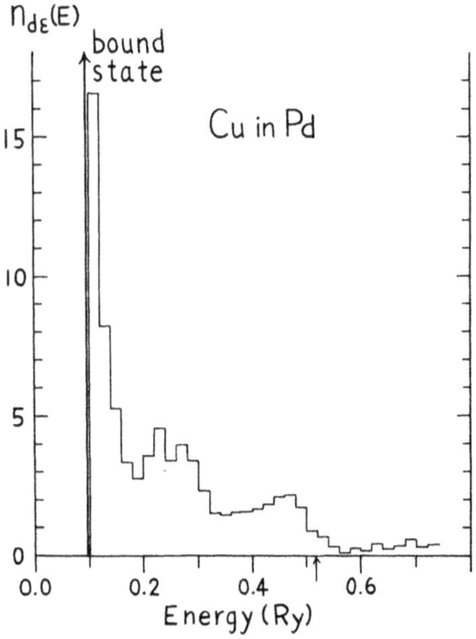

Fig. 4. The number of electrons per spin per unit energy in the states of *d* symmetry within the muffin-tin sphere of Cu as impurity in Pd.

finite amplitude at the Fermi level which makes the orbital relaxation mechanism rather important. The calculated $(T_1T)^{-1}_{\text{orb}}$ is large enough to compensate for the decrease of $(T_1T)^{-1}_s$. Since the dip extends over a wide energy range, we can expect the slowing down of the relaxation in other transition metals on the left of Pd in the periodic table, such as Ti and Zr (see Table II).

The dip in $-\text{Im}\,\Sigma_j$ and also in the density at the impurity nucleus corresponds to the antiresonance in the optical absorption. A typical model of the optical problem discussed by Fano[16] assumes that we have a virtual bound level as well as continuum states corresponding to the free states as excited states. The transition can occur from the ground state to either the virtual bound state or the continuum state, where in the former an electron can go either directly or indirectly via the continuum states to the virtual bound level. The interference among these processes gives rise to a dip accompanied by a hump in the absorption which is called the antiresonance. There is a one-to-one correspondence between the calculation of the self-energy and that of the absorption; the virtual level corresponds to the *d* bands and the continuum states to the free-electron-like branch (the OPW states) of the host energy bands. Adapting Fano's calculation to the present problem, we can conclude that the hump in $-\text{Im}\,\Sigma_j$ is higher and the dip is broader when the direct electron transfer between the impurity orbital and the host *d* bands is relatively strong compared to the width of the impurity level due to the mixing with the OPW states. Note that the direct transfer corresponds to the bonding between the *d* orbitals and the impurity orbitals.

The Shielding Mechanism of the Impurity Charge

Equation (1) tells us that the change of the number of states below the Fermi energy is related to $-\text{Im}\,\Sigma_j$ at that energy. If the Fermi energy is at the bottom of the antiresonance, the number of states which can accommodate electrons does not change appreciably even when the impurity potential for *s* and *p* electrons is deep. By calculating the number of electrons in the states of *s* and *p* symmetries within the impurity muffin-tin sphere, we can see, on the other hand, that electrons accumulate at the impurity site with such a potential as to keep approximately the electrical neutrality of the impurity atom. Since the number of states below the Fermi energy corresponds to the integration of the electron density over the whole space, the accumulation of electrons at the impurity site and the small change of the number of the occupied states can be consistent with each other only when the density of electrons decreases at surrounding atoms to produce an electron deficit there. The magnitude of the deficit is approximately equal to the excess of electrons in the *s* and *p* impurity orbitals compared to the host atom. This deficit, which produces an attractive potential, will be eventually shielded mainly by the electrons of the *d* bands. The apparent filling of the *d* holes in the minority-spin bands in ferromagnetic Ni is explained by this two-step mechanism. Of course if we could carry out a self-consistent calculation including not only the impurity potential but also the potentials at surrounding sites we would not have to divide the shielding mechanism into two steps. Since the perturbing potentials at surrounding sites do not seem to feed back appreciable effects on the electronic structure of the impurity atom, the present explanation is nevertheless quite meaningful, even quantitatively.

The antiresonance in the states of *p* symmetry has a relatively sharp dip whose bottom is close to the top of the *d* bands in the case of Al in Ni (and also in Pd). This fact may explain why the apparent filling of the *d* states by the impurity valence electrons is confined to those metals at the end of the transition series. In the case of Al in ferromagnetic Fe, for example, a preliminary calculation shows that the change of the number of states below the Fermi level due to the impurity potential can be almost sufficient to accommodate the valence electrons of the impurity atom, leading to no "apparent" filling of the *d* states.

The susceptibility of Pd dilute alloys is a more difficult subject, since it concerns the state density at the Fermi level instead of the number of occupied states. Assuming, however, that the perturbing potentials at surrounding sites are sufficiently shallow, we can explain it semiquantitatively by the same mechanism. In conclusion, it is highly desirable to check the calculated electronic structure by a direct measurement such as the photoemission of electrons. The bound state predicted in Figs. 2 and 4 will be one of the interesting subjects to consider.

Transition Impurities

The central theme in the problem of a transition impurity will be the magnitude of the Kondo temperature T_K. The ab initio calculation discussed in the preceding sections is at best the Hartree–Fock approximation, which we have to go beyond to discuss the Kondo temperature. Unfortunately, however, we have no quantitatively reliable theory which yields a smooth gradual transition from the nonmagnetic case (high T_K) to the magnetic case (low T_K) even for a simple electronic structure such as the Lorentzian density of states in the Anderson model. In ferromagnetic metals such as Fe and Ni, on the other hand, the ab initio calculation is very successful quantitatively if we assume an appropriate exchange potential. In the present author's opinion a similar calculation for the impurity problem will be meaningful in the sense that one will be able to establish a chemical rule between the magnitude of T_K and the electronic structure by use of the observed T_K and the calculated electronic structure of typical cases. So far no self-consistent calculation has been done for the impurity problem, though preliminary calculations were carried out for Ni in Cu by Harris[17] and Mn in Cu, Mn in Al, and Fe in Mo and Nb by Asano and Hamazaki.[18] In noble metals the virtual bound level of transition impurities deviates from the Lorentzian shape, since the extra density of states must vanish at the top of the *d* band, leading to a steeper rise on the lower energy side. The approximate half-width of Ni in Cu and of Mn in Cu seems to be about 0.4 eV, whereas that for Mn in Al is around 0.8 eV according to the calculations. The case of Mn in Al is known to be a typical moderately high T_K case ($T_K \simeq 530°K$). At present we cannot draw definite conclusions from these results. The electronic structure of transition impurities in transition metals is expected to be more complicated; we shall not go into discussions of this subject in this paper.

References

1. J. Kanamori, K. Terakura, and K. Yamada, *Prog. Theoret. Phys.* **41**, 1426 (1969); *ibid.* Suppl. No. 46, 221 (1970).
2. Von D. Gerstenberg. *Ann. Physik 7*, Folge 2, 236 (1958).
3. D.J. Lam and K.M. Myles, *J. Phys. Soc. Japan* **21**, 1503 (1966).
4. J. Crangle and M.J.C. Martin, *Phil. Mag.* **4**, 1006 (1955).
5. S. Kohzuki and K. Asayama, *J. Phys. Soc. Japan* **32**, 1237 (1972).
6. A. Abragam, *The Principles of Nuclear Magnetism*, Clarendon Press, Oxford (1961).
7. A. Narath, *J. Appl. Phys.* **39**, 553 (1968).
8. R.L. Williams, Jr., L. Pfeiffer, J.C. Wells, Jr., and L. Madansky, *Phys. Rev. C* **2**, 1219 (1970).
9. S. Kohzuki, N. Yokoyama, and K. Asayama, Private communication.
10. K. Asayama, *J. Phys. Soc. Japan* **18**, 1727 (1963); S. Kobayashi, K. Asayama, and J. Itoh, *J. Phys. Soc. Japan* **21**, 65 (1966).
11. A. Weger, G. Burri, and S. Steinemann, *Solid State Commun.* **9**, 1125 (1971).
12. L.D. Roberts, R.L. Becker, F.E. Obenshain, and J.O. Thomson, *Phys. Rev.* **137**, A895 (1965).
13. K. Terakura and J. Kanamori, *Prog. Theoret. Phys.* **46**, 1007 (1971).
14. N.F. Mott, *Adv. Phys.* **13**, 325 (1964).
15. K. Terakura and J. Kanamori, *J. Phys. Soc. (Japan)* **34**, 1520 (1973).
16. U. Fano, *Phys. Rev.* **124**, 1866 (1961).
17. R. Harris, *J. Phys. C, Solid State Phys.* **3**, 172 (1970).
18. S. Asano and M. Hamazaki (to be published).

2
Electron–Electron Interactions

Surface Paramagnons for Nearly Ferromagnetic Metals*

M. T. Béal-Monod,† P. Kumar, D. L. Mills,‡ H. Suhl, and R. A. Weiner§

Department of Physics, University of California—San Diego
La Jolla, California

The exchange enhancement of the susceptibility for a bounded interacting electron gas is examined. This problem is important in the study of the magnetic properties of thin metallic films and also in the statics and the dynamics of chemisorption and catalysis.

We begin by stating the integral equation satisfied by the susceptibility within the random phase approximation:

$$\chi(\mathbf{Q}, \omega, z, z') = \chi^0(\mathbf{Q}, \omega, z, z') + \sum_{z''} \chi^0(\mathbf{Q}, \omega, z, z'')\, U(z'')\, \chi(\mathbf{Q}, \omega, z'', z') \tag{1}$$

The meaning of the symbols is as follows: $\chi(\mathbf{Q}, \omega, z, z')$ is the Fourier transform of the space–time susceptibility with respect to the coordinates in the plane of the surface; $\mathbf{Q}$ is the two-component wave vector; and ω is the frequency). Due to the breaking of translational invariance by the presence of the surface normal to the z direction, the susceptibility is a function of z and z' separately rather than a function of $z - z'$, to which it, of course, reduces deep in the bulk. χ^0 refers to the susceptibility in the absence of interaction between the electrons and $U(z'')$ is the exchange interaction, henceforth taken to be local; its dependence on z'' allows for a possible variation of the exchange coupling with distance to the surface. The sum extends over the semiinfinite space occupied by the material. In this paper we consider only the static case $\omega = 0$.

The four basic processes depicted in Fig. 1 contribute to χ^0. Three of these involve reflections at the surface; the remaining one is the same as for the bulk. Note that in Fig. 1 in principle the one-electron propagator should be computed in the presence of the Hartree loops, including loops "reflected" at the surface; but this correction is neglected in the present paper.

Two models were used to describe χ^0. One is a free-electron model in which all wave functions vanish at a point near the surface whose location is determined by requirement of charge neutrality.[1] The other is based on a tight binding approximation.[2] Integrating the above equation over the variable z' and setting $Q = 0$ and $\omega = 0$ yields an integral equation for the magnetization per unit uniform applied magnetic field as a function of z. This equation was solved numerically. For both

* This research was supported in part by the Air Force Office of Scientific Research (AFSC) under Grant No. 71-2156 Mod. 01 and partially by Grant No. AFOSR 70-1936.

† At Université de Paris-Sud, Orsay, France.

‡ At University of California, Irvine, California.

§ At the Carnegie-Mellon University, Pittsburgh, Pennsylvania.

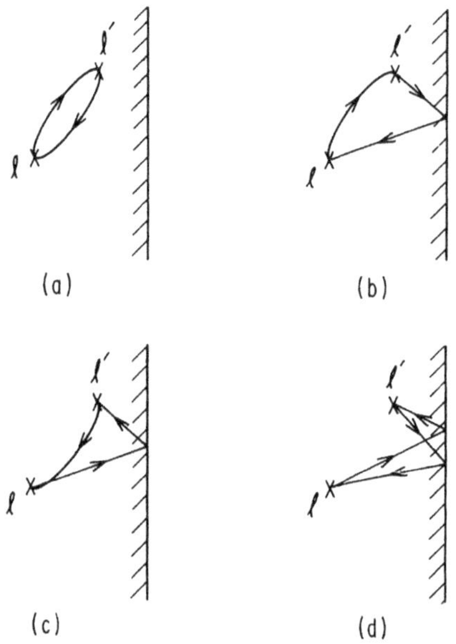

Fig. 1. The four processes involved in χ^0: (a) both electron and hole propagating from l to l' directly; (b) electron propagating directly while the hole reflects on the surface; (c) the converse; and (d) both electron and hole reflect from the surface.

models[3,4] we find that the magnetization rises to essentially its bulk value within a distance of $\sim 1/k_F$ from the surface. As in the case of corresponding magnetization for the noninteracting gas, there are, of course, oscillations rapidly damped toward the interior. Figure 2 shows $M(z)$ vs. z for various exchange parameters, in the case of model 1.

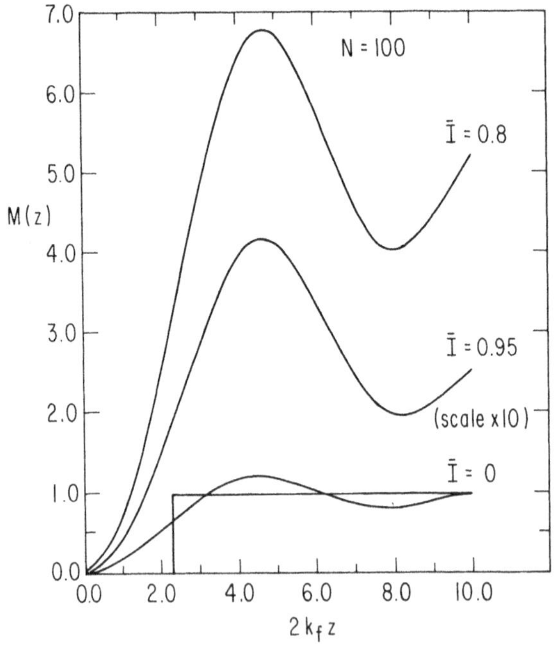

Fig. 2. $M(z)$ vs. $2k_Fz$ for various values of the exchange interaction. z is measured in units of $1/2k_F$.

Inclusion in χ^0 of only the loop in Fig. 1(a), though initially plausible (see Ref. 2), fails to yield the enhancement near the surface, although fully responsible for the enhancement in the interior. It would appear that high-momentum-transfer processes included in diagrams 1(b), 1(c), and 1(d) are essential features in describing the magnetization near the surface and its subsequent enhancement.

References

1. D.M. Newns, *Phys. Rev. B* **1**, 3304 (1970).
2. D.L. Mills, M.T. Béal-Monod, and R.A. Weiner, *Phys. Rev. B* **5**, 4637 (1972).
3. M.T. Béal-Monod, Pradeep Kumar, and H. Suhl, *Sol. St. Commun.* (to be published).
4. R.A. Weiner *et al.* (to be published).

The Dependence of the Electron–Electron Enhancement of the Effective Mass on Magnetic Field*

J. L. Smith and P. J. Stiles

Physics Department, Brown University
Providence, Rhode Island

Experiments were performed in the inverted surface of a Si MOSFET with a *p*-type substrate. This configuration was previously used in similar samples to study oscillatory magnetoconductance.[1] The results of these experiments showed that the electrons in the inverted surface of the Si behaved as a two-dimensional electron gas (2DEG). The dimensionality character comes about by way of the high electric field perpendicular to the surface. This results in such a steep triangular potential well that the motion perpendicular to the surface is quantized. The motion in the other two dimensions is not, and hence one has a 2DEG, with a two-dimensional density of states, independent of energy. Varying the electric field perpendicular to the surface then varies the surface electron density. The magnetoconductance measurements were performed with the magnetic field also perpendicular to the surface. The magnetic field quantized the motion parallel to the surface. As one changes the electron density, succeeding Landau levels are filled with an accompanying minimum in the conductivity. These are uniformly spaced in electron density, appropriate for a 2DEG. The temperature dependence of the amplitude of the oscillations is used to determine the effective mass, since it is a measure of the ratio of the thermal energy to the energy between Landau levels.

Later experiments were performed in which the magnetic field was not perpendicular to the surface.[2] Here it was demonstrated that the degeneracy of the Landau levels was proportional to the component of the magnetic field perpendicular to the surface. This is easily seen as proportional to the flux that threads the orbit and hence would determine the degeneracy as well as the cyclotron frequency. One would expect that the spin energy, however, would depend on the total field, since the spin, from the lack of spin–orbit coupling, can only sense the total field. When one tilts the field one changes the ratio of spin-splitting energy to Landau-splitting energy from a minimum of 0.2 to over one. It was from the case where this ratio was 0.5 that the effective g value was found to be larger than two and dependent on electron density. It was suggested by Janak[3] that this was due to electron–electron (*ee*) interaction.

Recently we demonstrated that the effective mass of electrons in a (2DEG)

* Work supported in part by the Advanced Research Projects Agency and the National Science Foundation, Grant No. GP-24130.

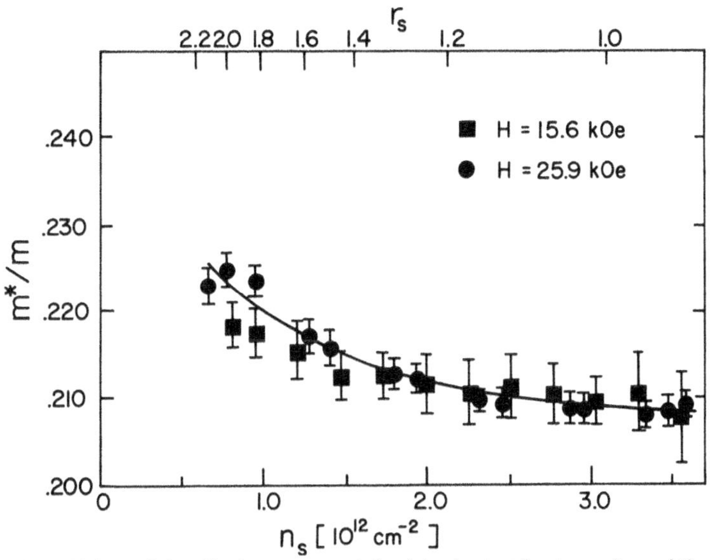

Fig. 1. Values of the effective mass as defined in the text for two values of the magnetic field perpendicular to the surface.

depended on the electron density.[4] This dependence was via the *ee* interaction rather than nonparabolicity of the energy bands involved. The results were obtained with the magnetic field normal to the 2DEG and field strengths of 15.6 and 25.9 kOe were used. Results are shown in Fig. 1. There is no statistically significant dependence on the value of the magnetic field. Previous measurements of the effective mass had shown that it was higher than the band mass by about 8%, but did not detect the small dependence on the electron density. The initial experiments in which the mass was measured with the magnetic field tilted relative to the normal gave results which differed significantly from the value obtained with the magnetic field normal to the 2DEG. Therefore we carried out experiments in which the component of the magnetic field normal to the surface was held constant. This ensured that the degeneracy of the Landau levels was constant and that the energy splitting was approximately so (exactly, if the mass shows no dependence on magnetic field). Values of the total magnetic field were then varied in some steps up to approximately 5.4 times the normal field. For a description of the experimental technique see Ref. 4. The results are shown in Fig. 2. Here we have plotted the points for each of five values of the total magnetic field, tilted at such an angle each time to keep the normal field constant. The statistical error bars have been left off in general for clarity. However, a few that are characteristic for each run are included. Examination will show that all, with the possible exception of the 80.4-kOe data, decrease with increasing n_s for $n_s > 10^{12}$ cm^{-2}. However, it is noticed that the average values for a given total magnetic field deviate more from each other than the dependence of the effective mass on electron density in this range. The question then arises: Is this an explicit dependence on the magnetic field or an implicit dependence via the ratio of spin splitting to Landau splitting?

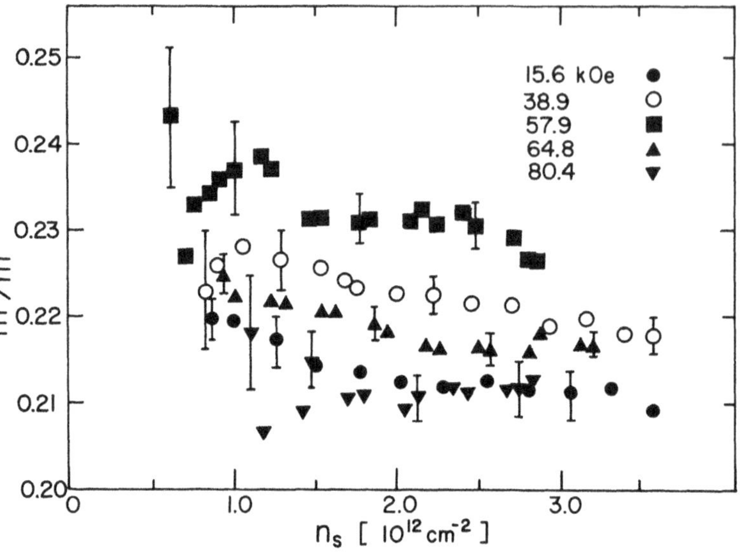

Fig. 2. The effective mass for different total magnetic fields but with the identical component normal to the 2DEG. The bars on the points are representative of the statistical uncertainty of the points. The apparent trend of decreasing uncertainty for higher electron density is real.

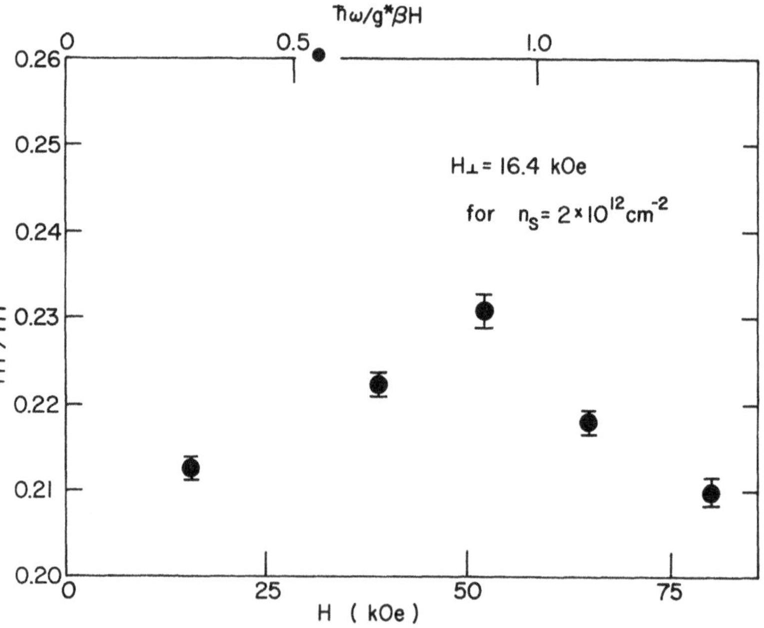

Fig. 3. The effective mass at a constant magnetic field normal to the 2DEG and a constant electron density, as a function of total magnetic field. The upper scale is approximate and relates the dependence of mass on the ratio of the spin energy to the Landau energy. The point at about 0.5 is explained in the text.

If we plot the value of the mass for a fixed density as a function of total magnetic field, we obtain Fig. 3. The upper scale is approximate and relates the dependence of mass to the ratio of the spin energy to the Landau energy. The point at $m^*/m = 0.26$ is from a lower electron density and is very near the ratio of splitting of 0.5. This is the position where the energy separation between spin levels for the same Landau level is the same as that for neighboring Landau levels. Hence the oscillations from Landau levels tend to zero amplitude at that point.

What would one expect for the effective mass measured in the above manner and for the g value as well? In a technical sense one asks what the excitation spectra is for quasiparticles under the conditions of a 2DEG with $E_F > \hbar\omega_c > kT > 0$. The energy of a quasiparticle is the normal kinetic energy plus the self-energy of the particle interactions with the Fermi sea and the interaction between quasiparticles. The case of a 2DEG in a magnetic field at $T = 0$ is significantly different than that in three dimensions. In three dimensions the ground state is composed of spin-split Landau levels which have a discrete energy separation as well as a continuous spectrum due to the unquantized third dimension. The quasiparticle excitation spectra is continuous also. Contrasting that, in two dimensions the ground state contains only a discrete spectrum, as does the excitation spectrum. The minimum separation in energy is of the order of $\hbar\omega_c$ for like spins and $g\beta H$ or $\hbar\omega_c - g\beta H$ for unlike spins.

In the two-dimensional case we are conveniently able to vary the electron density for a fixed magnetic field, and may completely fill a Landau level. In this case the next electron would have to be added to the next Landau level at an energy $\hbar\omega_c - g\beta H$ more than the previous one. If we half-filled a Landau level, then the next electron would be an energy $g\beta H$ higher than the previous one. In the absence of *ee* interaction and in all other cases the electrons would go in at the same energy as the previous electron.

In the case of partially filled or empty and half-filled Landau levels the interaction is expected to be different, since the ground state is different, with the maximum interaction occurring with half-filled levels, which brings up the final intriguing point. If we now increase the temperature above $T = 0$, then the occupation of the ground state and the number of quasiparticles change as kT approaches the quantizing energies. Therefore the mass would be temperature dependent as well.

For the case we have studied here $kT \leq 0.2\hbar\omega_c$, so its effect should be small but not necessarily negligible. However, kT may well be larger than $\hbar\omega_c - g\beta H$ in some of the cases studied here when the magnetic field is tipped.

The conclusions are the following: There was no statistically significant dependence observed in the work of Ref. 4, since it was done with a constant ratio of spin splitting to Landau splitting, as well as Landau splitting to kT. It is expected, however, that as the ratio of Landau splitting to kT is varied, the mass and the g value will vary. Second, if the ratio of Landau splitting to spin splitting varies for a relatively constant ratio of Landau splitting to kT, the effective mass and g value will vary.

Our future experiments will be carried out to much higher precision, by more precise measurement of the temperature and computer fitting of the experimental curves. We will also carry them out over a much larger temperature range, from 0.05 to 15°K.

References

1. A.B. Fowler, F.F. Fang, W.E. Howard, and P.J. Stiles, *Phys. Rev. Lett.* **20,** 901 (1966); and *J. Phys. Soc. Japan* (Suppl.) **21**, 331 (1966).
2. F.F. Fang and P.J. Stiles, *Phys. Rev.* **174**, 823 (1968).
3. J.F. Janak, *Phys. Rev.* **178**, 1416 (1969).
4. J.L. Smith and P.J. Stiles, *Phys. Rev. Lett.* **29**, 102 (1972).

Electron Spin Splitting in the Alkali Metals

B. Knecht, D. L. Randles, and D. Shoenberg

Royal Society Mond Laboratory
Cambridge, England

Introduction

Landau's Fermi liquid theory specifies the many-body interactions between electrons in a metal in terms of a number of phenomenological parameters which can in principle be determined by experiment. Such determinations are of interest both in providing checks of the internal self-consistency of the theory in its application to different experiments and in providing data to be interpreted in microscopic terms. One of these parameters, B_0, which relates to electron spin properties (it is the coefficient of the zeroth-order term in the Legendre expansion of the interaction), is involved in three different phenomena: the spin waves observed by Schultz and Dunifer,[1] the Pauli spin susceptibility, and the de Haas–van Alphen (dHvA) effect, whose amplitude is modified by the spin splitting of the Landau levels.

Both the last two phenomena essentially measure the spin-splitting factor g (defined so that the spin-splitting energy in field H is $ge\hbar H/2m_0c$, where m_0 is the free electron mass) and, as was pointed out by Kaplan and Glasser,[2] B_0 is related to g by

$$g = g_0/(1 + B_0) \qquad (1)$$

where g_0 is the spin-splitting factor as determined by CESR, which is *not* modified by many-body interactions. We have applied the dHvA method to the determination of g, and hence B_0, in the alkali metals, to which the theory should be particularly applicable because of the near sphericity of their Fermi surfaces.

Experimental Method

The modification of the dHvA amplitude by spin splitting can be thought of as due to interference between the oscillations coming from the spin-up and the spin-down electrons, which differ in phase by $\pi gm/m_0$ for the fundamental and by r times as much for the rth harmonic (here m is the cyclotron mass). Thus the amplitude of the rth harmonic is modified by the factor $\cos(r\pi S)$, where $S = \frac{1}{2}gm/m_0$, and we have in principle two methods of determining g from amplitude measurements: (1) by measuring the *ratio* of the amplitudes of two harmonics (in practice $r = 2$ and $r = 1$), and (2) by measuring the *absolute* amplitude of a single harmonic (in practice $r = 1$).

Both methods assume the reliability of the Lifshitz–Kosevich (LK) theory and require an accurate knowledge of the cyclotron mass m and the Dingle temperature

x, which can be obtained from measurements of the temperature and field dependences of the fundamental amplitude. The absolute amplitude approach, however, requires in addition knowledge of the factors common to all the harmonics, which, apart from universal constants, are the dHvA frequency F, the curvature factor $(\partial^2 a/\partial k_H^2)^{-1/2}$, and, of course, the calibration constant of the equipment and the sample volume. The first two are known to sufficient accuracy and it is only the last two which present any problem, as will be discussed below.

In order to obtain sufficient harmonic amplitude to make the ratio method practicable, it is essential to use as high H and low T as possible. However, in these conditions the absolute amplitude of $4\pi dM/dH$ becomes appreciable compared to unity, and this is just the condition for magnetic interaction (MI)[3] to become significant. Because the electrons see the field B rather than H, a simple harmonic oscillation of M as a function of $1/B$ contains harmonics when observed as a function of $1/H$. In other words, MI generates harmonics on its own account and this complicates the interpretation of the observed harmonic content of the oscillations.

Fortunately the *phases* of the harmonics generated by MI are different from those of the harmonics in the LK theory (which ignores MI), so, provided MI is not too dominant, it is still possible to separate the contributions from the two causes and make the ratio method work. A useful spin-off from this complication is that it provides an independent method of measuring the absolute amplitude of the fundamental oscillations, since the ratio of any pair of successive MI harmonic amplitudes involves only the absolute amplitude of dM/dH and the demagnetizing coefficient of the sample. The advantage of this method is that it eliminates the difficulties of calibrating the measuring equipment and measuring the sample volume. On the other hand, it involves amplitude and phase measurements of the rather feeble harmonic content and depends on the reliability of the assumption of strict homogeneity of M in the sample.

All three methods of determining g share the awkward ambiguity that at best it is only $\cos(\pi S)$ which is determined, and only $|\cos(\pi S)|$ unless the dHvA phase has been extremely precisely established. Thus any value of g given by

$$\tfrac{1}{2}gm/m_0 = \pm S \pm n \tag{2}$$

where n is any integer, will have the same magnitude of $\cos(\tfrac{1}{2}\pi gm/m_0)$. If, however, the sign of $\cos(\pi S)$ is known, the ambiguity is somewhat reduced since then only *even* values of n are admissible.

We have used the modulation method of measurement, and a number of practical points should be mentioned.

(a) To avoid skin-effect complications, the modulation frequency ω was made sufficiently low (typically of order 10^2 Hz for samples of diameter of order 1 mm) and detection at ω (effectively measuring dM/dH) was used. The necessary low frequency and small sample size give a signal-to-noise ratio a good deal below the optimum possible.

(b) Absolute calibration accuracy is usually limited by the precision with which the flux linkage between the sample and the pick-up coil can be determined. Accuracy of a few percent was achieved by reversing the roles of pick-up and modulation coils and then finding the ac current amplitude necessary to give the first Bessel zero in the dHvA signal.

(c) Determination of sample volume is complicated by the great reactivity of the alkali metals, which makes direct weighing impracticable. We relied at first on microscopic measurement of linear dimensions, but this can lead to quite appreciable errors, and we have now developed a rather better technique based on observing the magnetic effect of eddy currents induced in the sample at room temperature. This technique also provides a useful means of monitoring possible erosion of the mounted sample if it has been inadequately protected from the atmosphere.

(d) It is essential in both absolute and relative measurements to allow for the field dependence of amplitude; in practice this can be reliably done only if the logarithmic (Dingle) plot is accurately linear over a sufficient range of $1/H$ to allow reliable extrapolation to $1/H = 0$. This proved to be no problem in K, but in Rb the simple method of mounting the sample in grease produced nonlinear Dingle plots (presumably due to deformation when the grease froze). A mounting procedure in which the sample was constrained only by nylon fibers touching its opposite poles proved more successful, but in Cs we have not yet succeeded in avoiding strong beat effects and in Na there are special complications due to the martensitic transformation.

(e) In determining the absolute amplitude from the MI harmonic contribution, the shape of the sample is assumed to be accurately spherical; there may be appreciable error due to failure of this assumption.

Results

The results are summarized in Table I. We shall now discuss some of the details metal by metal.

K. These are at present the most reliable results, with good consistency between samples (the stated errors for each method are the rms errors of the means over five samples combined with the errors in m) and between methods (though the dHvA3 method may have appreciable systematic error, because the eddy current method of measuring volume was not yet available). As already explained, there is ambiguity in the g values, and *a priori* any value given by

$$g = \pm 0.486 \pm 1.653n$$

is as valid as the 2.82 shown in Table I. Thus ± 0.49, (± 1.17, ± 2.14), ± 2.82, ± 3.79, (± 4.47, ± 5.45), etc., would all be equally consistent with the data. However, the precise absolute dHvA phase determination by Templeton[4] indicates that $\cos(\pi S)$ is positive and this excludes the odd values of n, i.e., the bracketed values above. The value 2.82 gives $B_0 = -0.289$, in excellent agreement with the spin-wave determination, and has therefore been chosen as the most plausible one.

Rb. The results are for the sample that gave the best results (good Dingle plot, low x, and good consistency between methods). Results on two other samples suggested that the scatter between samples was comparable to that in K. No absolute dHvA phase determination is yet available, so it is not possible to reduce the ambiguity of the result $g = \pm 0.516 \pm 1653n$. The particular choice of 2.79, etc., shown in Table I has been somewhat arbitrarily made to agree best with the spin susceptibility value, and is also consistent with a preliminary comment by Schultz[5] that the spin wave value of B_0 in Rb is not very different from that in K.

Table I

	K		Rb		Cs		Na	
Method*	g	$-B_0$	g	$-B_0$	g	$-B_0$	g	$-B_0$
dHvA1	2.82 ± 0.03	0.289 ± 0.008	2.79	0.283	2.11, 2.97	0.05, 0.33	—	—
dHvA2	2.79 ± 0.04	0.283 ± 0.009	2.71	0.262	2.22, 2.86	0.10, 0.30	—	—
dHvA3	2.75 ± 0.05	0.273 ± 0.013	2.77	0.278	1.98, 3.10	−0.01, 0.35	2.73, 2.41	0.27, 0.1
χ/χ_0	2.61	0.23	2.58	0.22	2.17	0.08	2.81	0.29
SD	—	0.285 ± 0.020	—	—	—	—	—	0.215 ± 0.030
Theory	—	0.21, 0.24	—	—	—	—	—	0.14–0
m/m_0	1.21		1.21		1.57		1.24	
r_s (Bohr radii)	4.87		5.18		5.57		3.96	

* dHvA1: ratios of harmonics. dHvA2: absolute amplitude from MI harmonics. dHvA3: direct absolute amplitude. χ/χ_0: spin susceptibility (references in Ref. 5). SD: spin waves, Schultz and Dunifer.[1] Theory: estimates quoted by Rice.[5] Eq. (1) has been used to derive B_0 from g, and g_0 has been taken as exactly two for all the metals (the maximum deviation is of order $\frac{1}{2}\%$, for Cs). The values of m/m_0 are obtained from dHvA temperature dependence, except for Na, for which the cyclotron resonance value has been used.

Cs. Only two samples have so far been tried and both have shown strong beats (very irregular beats in one). Presumably these beats (about 100 oscillations per beat) are due to the presence of macroscopic regions of different dHvA phase (e.g., two crystals) and it is hoped that better sample preparation and handling will improve the situation. An attempt was made to interpret the results with regular beats on the assumption that all parts of the sample were in phase at the beat maxima. The rather poor consistency between the results

$$\pm 0.43 \pm 1.27n, \qquad \pm 0.32 \pm 1.27n, \qquad \text{and} \qquad \pm 0.56 \pm 1.27n$$

by the three methods reflects not only the uncertain reliability of interpretation when beats are present, but also the rather weak harmonic content of the oscillations. We have shown in Table I the two alternative values for each method corresponding to $n = 2$. One of these values (g fairly close to 2) agrees roughly with the spin-susceptibility value, while the other (g fairly close to 3) agrees with a preliminary comment by Schultz[5] that the spin-wave value of B_0 in Cs is also not very different from that in K.

Na. The amplitude of oscillations was very small and they could be observed over only a very limited range (from 100 to about 80 kG). This is probably basically because of the considerable phase smearing caused by the martensitic transformation, and indeed the estimated values of x (1.6 and 3.8°K for the only two samples studied so far) were much higher than for K (x typically 0.4°K). However, it is possible that a contributory factor to the low amplitude is a low value of $\cos(\pi S)$ and this is indeed suggested by the spin-wave result, which would give $\cos(\pi S) = 0.25 \pm 0.18$. Because of the small amplitude and high x it was not possible to study harmonic content, so we give in Table I only the direct absolute amplitude estimates for the two samples so far tried. These are not really consistent with each other (probably because the linearity of the Dingle plot is not maintained outside the range observed);

they correspond to

$$\pm 0.49 \pm 1.61n \qquad \text{and} \qquad \pm 0.81 \pm 1.61n$$

for the two samples, and the values shown in Table I (for $n = 2$) have been chosen to produce near agreement with the spin-wave and the spin-susceptibility estimates.

Discussion

It is hoped soon to improve the accuracy of the results, particularly as regards Cs and Na, by working at lower T, by improving sample preparation, and by improving the measuring method, and also to reduce the ambiguity of g values in Rb, Cs, and Na by absolute dHvA phase determination. Potentially the dHvA methods should be the most accurate for determining B_0.

In the meantime we may note that the spin-susceptibility estimates do not agree very well with the others (though their 10% possible error just about bridges the gaps, except perhaps for Cs), and that the theoretical estimates do not claim to be more than very rough. It is too early to attempt any serious correlation with fundamental parameters such as r_s, the mean interelectronic distance, though there is perhaps some suggestion that $|B_0|$ rises with r_s and passes through a maximum for $r_s \sim 5$. However, it should not be forgotten that B_0 is due to both electron–electron and to electron–phonon interaction, with

$$(1 + B_0) = (1 + B_{0\,\mathrm{ee}})(1 + B_{0\,\mathrm{eph}}) \qquad (3)$$

(see Rice[6]). The electron–phonon parameter $B_{0\,\mathrm{eph}}$ is positive and should be given by the mass enhancement factor, which is difficult to estimate. Thus any simple relation between $B_{0\,\mathrm{ee}}$ and r_s may well be confused by the variations of $B_{0\,\mathrm{eph}}$ which depend on lattice properties as well as electron density.

References

1. S. Schultz and G. Dunifer, *Phys. Rev. Lett.* **18**, 283 (1967) and private communication.
2. J.I. Kaplan and M.L. Glasser, *Phys. Rev.* **186**, 958 (1969).
3. D. Shoenberg, *Can. J. Phys.* **46**, 1915 (1968).
4. I.M. Templeton, *Phys. Rev. B* **5**, 3819 (1972).
5. S. Schultz, private communication.
6. T.M. Rice, *Phys. Rev.* **175**, 858 (1968).

Correlation Energy of Quantum Plasma in Low Magnetic Field: de Haas–van Alphen Behavior

M. L. Glasser and R. W. Danz

Battelle Memorial Institute
Columbus, Ohio

and

N. J. M. Horing

Department of Physics and Cryogenics Center
Stevens Institute of Technology, Hoboken, New Jersey

It is well known[1] that the random-phase approximation (RPA) calculation of correlation energy per particle ε_{cor} at zero temperature (degenerate limit) for an electron gas devolves upon the evaluation of the integral

$$\varepsilon_{\text{cor}} = (\hbar/4\pi\rho) \int [d\mathbf{p}/(2\pi)^3] \int_{-\infty}^{\infty} d\omega \{\ln [1 + \alpha(\mathbf{p}, \omega)] - \alpha(\mathbf{p}, \omega)\} \tag{1}$$

or, alternatively, integrating by parts,

$$\varepsilon_{\text{cor}} = \left(\frac{\hbar}{4\pi\rho} \right) \int \frac{d\mathbf{p}}{(2\pi)^3} \int_{-\infty}^{\infty} d\omega \frac{\omega\alpha(\mathbf{p}, \omega)\, \alpha'(\mathbf{p}, \omega)}{1 + \alpha(\mathbf{p}, \omega)} \tag{2}$$

The free-electron polarizability for imaginary frequency is represented here by

$$\alpha(\mathbf{p}, \omega) = 4\pi\alpha_0(\mathbf{p}, i\omega) = -(4\pi e^2/p^2)\, \text{Im}\, I(\mathbf{p}, i\omega + i0)$$

(in the notation of Ref. 2) and the above formal expression for ε_{cor} in terms of $\alpha(\mathbf{p}, \omega)$ is unchanged by the application of a magnetic field. All magnetic field effects are incorporated through the dependence of $\alpha(\mathbf{p}, \omega)$ on the magnetic field,[3,4] and this has been thoroughly investigated in connection with the evaluation of the RPA dielectric function in Ref. 2. The results obtained in this reference yield*

$$\alpha(\mathbf{p}, \omega) = -\frac{2\pi e^2}{p^2} \int_{-\infty}^{\infty} d\omega' \frac{f_0(\omega')}{\hbar^3} \int_{c - i\infty}^{c + i\infty} \frac{ds}{2\pi i} (\exp s\omega')\, \eta(s)\, \mathscr{A} \tag{3}$$

* Notation: $f_0(\omega)$ is the Fermi–Dirac distribution function; $\mathbf{H}$ is the magnetic field along the z axis; $\mathbf{p} = (p_z, \bar{p})$ is the wave vector; ζ = chemical potential = Fermi energy ε_F; $\beta = (KT)^{-1}$; ω_p is the classical plasma frequency $(4\pi e^2\rho/m)^{1/2}$; ω_c is the cyclotron frequency eH/mc; $p_F = (2m\varepsilon_F)^{1/2}/\hbar$, $k_{FT}^2 = 6\pi\rho e^2/\varepsilon_F$, $k_{FT}^2/p_F^2 = [(4/\pi)(4/9\pi)^{1/3}]\, r_s$.

where

$$\mathscr{A} = 2\frac{i}{\hbar}\int_0^\infty dt\, e^{-|\omega|t} \exp\left(-\frac{\hbar^2 p_z^2 s}{8m}\right) \exp\left(-\frac{\hbar \bar{p}^2}{2m\omega_c} \coth \frac{\hbar\omega_c s}{2}\right)$$
$$\times \left\{ \exp\left[-\frac{p_z^2}{8ms}(2t - i\hbar s)^2\right]\right.$$
$$\left.\times \exp\left[\frac{\hbar \bar{p}^2}{2m\omega_c} \frac{\cos[(\omega_c/2)(2t - i\hbar s)]}{\sinh(\hbar\omega_c s/2)}\right] - (i \to -i)\right\} \tag{4}$$

and $\eta(s)$ is essentially the Laplace transform of the free-electron density of states in a magnetic field

$$\eta(s) = (m/2\pi)^{3/2}\, \hbar\omega_c s^{-1/2} \coth(\hbar\omega_c s/2) \tag{5}$$

This RPA expression for the dynamic nonlocal free-electron polarizability is exact with respect to magnetic field dependence, and its structure embodies a host of longitudinal quantum magnetoplasma response phenomena. These include the two local principal plasmon modes in magnetic field, Bernstein-type plasmon resonances, "quantum" plasmon resonances, Debye–Thomas–Fermi static shielding contributions, anisotropic Friedel–Kohn "wiggle" static shielding contributions in magnetic field, etc. Moreover, in consequence of Landau quantization, one has de Haas–van Alphen (dHvA) oscillatory and monotonic quantum magnetic field effects which arise from the structure of this polarizability in the degenerate case, as well as classical magnetic field effects associated with the parameter ω_c/ω_p. Many of the phenomena cited above are relatively unimportant at low magnetic fields, particularly in the dominantly important low-wavenumber region of the correlation energy integral. In this connection one may reasonably argue that one may ignore quantum dynamic effects involved in the self-consistent internal plasma dynamics which determine the structure of $\mathscr{A}$ (for example, $1 \gg \hbar\omega_c/\zeta \to 0$), and thus replace $\mathscr{A}$ by its classical limit. This is to say, at low field and low wavenumber the plasma dynamics determining $\mathscr{A}$ is approximated well by a classical description. However, the statistical averaging involved is nevertheless *quantum* statistics, involving a sharply cut off Fermi function $f_0(\omega)$ in the ω' integral of $\alpha(\mathbf{p}, \omega)$, as well as a Landau-quantized density of states whose Laplace transform is $\eta(s)$. This description defines the "quasiclassical" model of the quantum plasma in magnetic field, whose physical content has been explored in depth.[5–7] To be specific, $\mathscr{A}_0$ of the quasiclassical model is obtained from $\mathscr{A}$ above by putting $\hbar \to 0$, with the result (appendix of Ref. 5)

$$\mathscr{A}_0 = 2s\left\{-1 + |\omega| \int_0^\infty dt\, e^{-|\omega|t} \exp(-p_z^2 t^2/2ms)\right.$$
$$\left.\times \exp[(\bar{p}^2/m\omega_c^2 s)(\cos\omega_c t - 1)]\right\} \tag{6}$$

The important low-wavenumber static and dynamic response features of the plasma correctly incorporated in $\mathscr{A}_0$ are as follows: The first term on the right of (6) yields the correct Thomas–Fermi static screening law.[6,7] The second term on the right of (6) has frequency dependence appropriate for describing[5] the two local principal plasmon modes in magnetic field, as well as Bernstein-type plasmon resonances for propagation perpendicular to the field. Higher-wavenumber phenomena ($p \sim p_F$) which are absent include "quantum" plasmon resonances, anisotropic Friedel–Kohn "wiggle" static shielding contributions, etc.; but since the dominantly important contributions to the correlation energy integral come from the low-wavenumber region, such higher-wavenumber phenomena can be safely neglected. Thus the quasiclassical model correctly incorporates all the low-wavenumber phenomena of importance in the correlation energy integral.

The structure of $\mathscr{A}_0$ may be further simplified in low magnetic field by noting that the excitation amplitudes of the Bernstein modes and the second principal plasmon mode (near ω_c in low field) are very small compared to the first principal plasmon mode (near ω_p in low field). This may be interpreted to mean that the zero-point plasmon energy contribution of the first principal plasmon (near ω_p) dominates the others for $\omega_c/\omega_p < 1$. Correspondingly, we may approximate $\omega_c t \sim \omega_c/|\omega| \approx \omega_c/\omega_p \ll 1$ in $\cos \omega_c t$ which appears in the second term of the right side of (6), and thus obtain

$$\mathscr{A}_0 = 2s\left[-1 + |\omega| \int_0^\infty dt \exp(-|\omega| t) \exp(-p^2t^2/2ms) \right] \tag{7}$$

or

$$\mathscr{A}_0 \cong -2s[1 - \pi^{1/2}x(\exp x^2)\operatorname{erfc} x] \equiv -2sF(x) \tag{8}$$

with

$$F(x) = 1 - \pi^{1/2}x(\exp x^2)\operatorname{erfc} x, \qquad x = (ms/2p^2)^{1/2}|\omega| \tag{9}$$

Within the framework of this low-field quasiclassical model the polarizability for imaginary frequency is given by

$$\alpha(\mathbf{p}, \omega) = \frac{4\pi e^2}{p^2}\left(\frac{m}{2\pi}\right)^{3/2} \hbar\omega_c \int d\omega' \frac{f_0(\omega')}{\hbar^3} \times \int_{c-i\infty}^{c+i\infty} \frac{ds}{2\pi i}(\exp s\omega')\, s^{1/2} \coth\left(\frac{\hbar\omega_c s}{2}\right) F(x) \tag{10}$$

where $x = |\omega|(ms/2p^2)^{1/2}$.

Now $\alpha(\mathbf{p}, \omega)$ may be divided into dHvA oscillatory and monotonic parts on the basis of contributions to the s integral from the isolated poles of $\coth(\hbar\omega_c s/2)$ at $s = s_n = 2\pi n i/\hbar\omega_c$ ($n = \pm 1, \pm 2, \ldots$) and from the branch cut along the negative real s axis associated with $s^{1/2}$, respectively. Thus $\alpha(\mathbf{p}, \omega) = \alpha_{\text{osc}}(\mathbf{p}, \omega) + \alpha_{\text{mon}}(\mathbf{p}, \omega)$. where, with k_{FT} the Thomas–Fermi wave vector and $x_n = |\omega|(ms_n/2p^2)^{1/2}$,

$$\alpha_{\text{osc}}(\mathbf{p}, \omega) = \frac{1}{4}\frac{k_{\text{FT}}^2}{p^2}\left(\frac{\hbar\omega_c}{2\varepsilon_{\text{F}}}\right)^{1/2} \operatorname{Re} \sum_{n=1}^{\infty} \left[\exp i\left(\frac{2\pi n\varepsilon_{\text{F}}}{\hbar\omega_c} - \frac{\pi}{4}\right)\right] \frac{F(x_n)}{n^{1/2}} \tag{11}$$

$\alpha_{\text{mon}}(\mathbf{p}, \omega)$ is given approximately at low-field by the same branch-cut integral which yields its zero-field semiclassical limit:

$$\alpha_{\text{mon}}(\mathbf{p}, \omega) = (k_{\text{FT}}^2/p^2)R(u), \qquad R(u) = 1 - u\tan^{-1}(1/u) \tag{12}$$

where $u = (|\omega|/p)(m/2\varepsilon_{\text{F}})^{1/2}$.

Recognizing that dHvA oscillatory corrections must be small, since

$$\alpha_{\rm osc}(\mathbf{p},\omega)/\alpha_{\rm mon}(\mathbf{p},\omega) \sim (\hbar\omega_c/\varepsilon_{\rm F})^{1/2} \ll 1$$

we expand $\varepsilon_{\rm cor}$ [Eq. (1)] in powers of $\alpha_{\rm osc}(\mathbf{p},\omega)$ to first order,* and thus identify the dHvA oscillatory part of the correlation energy as

$$\varepsilon_{\rm cor}^{\rm dHvA} = -\left(\frac{\hbar}{2\pi\rho}\right)\int\frac{d\mathbf{p}}{(2\pi)^3}\int_0^\infty d\omega\,\frac{\alpha_{\rm mon}(\mathbf{p},\omega)\,\alpha_{\rm osc}(\mathbf{p},\omega)}{1+\alpha_{\rm mon}(\mathbf{p},\omega)} \tag{13}$$

Upon rescaling the frequency according to $u = (|\omega|/p)(m/2\varepsilon_{\rm F})^{1/2}$ and introducing the quantity $a = (\varepsilon_{\rm F}s_n)^{1/2} = (i2\pi n\varepsilon_{\rm F}/\hbar\omega_c)^{1/2}$, we obtain

$$\varepsilon_{\rm cor}^{\rm dHvA} = \left(-\frac{\hbar}{2\pi\rho}\right)\frac{k_{\rm FT}^4}{8\pi^2}\left(\frac{\hbar\omega_c}{m}\right)^{1/2}\operatorname{Re}\sum_{i=1}^{\infty}\left[\exp i\left(\frac{2\pi n\varepsilon_{\rm F}}{\hbar\omega_c}-\frac{\pi}{4}\right)\right]\frac{1}{n^{1/2}}$$
$$\times\int_0^\infty du\,R(u)\,F(au)\int_0^{p_{\rm F}} dp\,\frac{p}{p^2+k_{\rm FT}^2R(u)} \tag{14}$$

The p integral has been cut off at $p_{\rm F}$ since the principal contribution comes from the lower-wavenumber region, and since the validity of the quasiclassical model is restricted to this region. The p integral is elementary, yielding

$$\int_0^{p_{\rm F}} dp\cdots = \frac{1}{2}\ln\{1+[p_{\rm F}^2/k_{\rm FT}^2R(u)]\},$$

whence we have the u integral as

$$\int_0^\infty du\cdots = \frac{1}{2}\int_0^\infty du\,R(u)\,F(au)\ln\{1+[p_{\rm F}^2/k_{\rm FT}^2R(u)]\} \tag{15a}$$

$$= (1/2a)\int_0^\infty dz\,F(z)\,R(z/a)\ln\{1+[p_{\rm F}^2/k_{\rm FT}^2R(z/a)\} \tag{15b}$$

Since $|a| \sim (\varepsilon_{\rm F}/\hbar\omega_c)^{1/2} \gg 1$ for low field, we may replace $R(z/a)$ with $R(0) = 1$, whence

$$\int_0^\infty du\cdots = (1/2a)\ln(1+p_{\rm F}^2/k_{\rm FT}^2)\int_0^\infty dz\,F(z) \tag{16}$$

The approximation $R(z/a)\to R(0) = 1$ in fact corresponds to purely static shielding contributions to the electron–electron interaction energy, in which dynamic shielding contributions through zero-point plasmon energies of the form $\frac{1}{2}\hbar\omega_p$ are negligibly small. It is quite clear that this calculation is committed to high densities, and accordingly we note that

$$p_{\rm F}^2/k_{\rm FT}^2 = [(\pi/4)(9\pi/4)^{1/3}](1/r_s) \gg 1$$

so that

$$\ln(1+p_{\rm F}^2/k_{\rm FT}^2)\to\ln(p_{\rm F}^2/k_{\rm FT}^2)\sim\ln r_s \tag{17}$$

* It is readily verified that the term of zero order in $\alpha_{\rm osc}(\mathbf{p},\omega)$ is just the Gell-Mann and Brueckner[8] result for correlation energy at zero field.

Finally, we have

$$\int_0^\infty dz\, F(z) = \frac{1}{2}\sqrt{\pi} \tag{18}$$

and we therefore obtain (per electron)

$$\varepsilon_{\text{cor}}^{\text{dHvA}} = (3\mathcal{R}/\pi^3)\ln(k_{\text{FT}}^2/p_{\text{F}}^2)\,\lambda\sum_1^\infty n^{-1}\sin(2\pi n/\lambda) \tag{19}$$

where $\mathcal{R}$ is the Rydberg unit and $\lambda = \hbar\omega_c/\varepsilon_{\text{F}}$. In conclusion, the result (19) yields the dHvA oscillatory correction (at low magnetic field) to the $\ln r_s$ term of the correlation energy at high density.

References

1. D. Pines, *Elementary Excitations in Solids*, Benjamin, New York, (1964).
2. N.J. Horing, *Ann. Phys.* **31**, 1 (1965).
3. N.J. Horing, in *Proc. Int. Conf. Stat. Mech.* (1968); *J. Phys. Soc. Japan* (Suppl.) **26**, 264 (1969).
4. N.J. Horing in *Electronic Structures in Solids*, E.D. Haidemenakis, ed., Plenum Press, New York (1969), p.209.
5. N.J. Horing and R.W. Danz, Bernstein Modes and the Quasiclassical Model of the Quantum Plasma in the Magnetic Field, *J. Phys. C* (to be published).
6. N.J. Horing in *Electronic Structures in Solids*, E.D. Haidemenakis, ed., Plenum Press, New York (1969), p. 223.
7. A.K. Das and E. de Alba, *J. Phys. C.: Solid State Phys.* **2**, 852 (1969).
8. M. Gell-Mann and K. Brueckner, *Phys. Rev.* **106**, 364 (1957).

Thermal and Magnetic Properties of Metals at Low Temperatures*

A. Isihara†

Theoretical Physics Institute
The University of Alberta, Alberta, Canada

Many-body theories of the low-temperature behavior of metals have been developed based more or less on the theory of the correlation energy of an electron gas at zero temperature. We have made improvements upon these theories as follows.

First, because the correlation energy has been evaluated in the past only for small r_s, we have tried to obtain large-r_s values from the ground-state energy, which is evaluated by the method of Padé approximants.[1] In this method the approximants are constructed in such a way that the small- and large-r_s expansions reproduce the known gas and bcc lattice results, respectively. We have found that the ground-state energy in the highest Padé approximation shows a minimum at around $r_s = 3.77$ and a singularity at around $r_s = 14.4$. At this singularity the energy is continuous and wiggles like a van der Waals curve, and the pressure shows a sharp cusp. If this singularity is associated with the crystallization of the electron gas, the density changes would be approximately 40%. The correlation energy obtained from the ground-state energy varies more strongly with r_s than those evaluated previously.

Second, we have succeeded in evaluating the specific heat by avoiding use of the zero-point correlation energy. Previous theories[2] of the specific heat are based on the approximate formula

$$c_v(T) = \frac{4k^2 T}{e^4}\left(\frac{d\varepsilon}{dp}\right)_{p_0}^{-1} \tag{1}$$

where the units are such that $\hbar = 1$ and $2m = 1$, and where the derivative comes from the density of the electronic levels at the Fermi surface of the electron gas in the absence of the Coulomb interactions.

We remark that Eq. (1) is not rigorous but is based on simplifying assumptions such as the smallness of the number of excited electrons. Correspondingly, the formula is characterized by p_0, the ideal-gas Fermi momentum. The specific heat is given by the momentum derivative of the ground-state energy instead of the usual temperature derivative of the internal energy. This is why the evaluation of the specific heat has been made along with that of the correlation energy.

* This work was supported by the National Research Council of Canada.
† Permanent address: Department of Physics, State University of New York at Buffalo.

We have approached the problem by evaluating the internal energy from the grand partition function for low but finite temperatures. This step corresponds to the generalization of the evaluation of the correlation energy to finite temperatures. One can show rigorously that the internal energy is given by

$$U - U_0 - kT^2(\partial \ln \Xi/\partial T)_{p_F} + (N - N_0)\,p_F^2 \tag{2}$$

where the subscript zero corresponds to an ideal gas, Ξ is the grand partition function which comes from the Coulomb interactions, and p_F is the true Fermi momentum in the presence of the interactions. According to the golden rule of the grand ensemble method, p_F is a function of the number of electrons N and the temperature. Therefore both terms in Eq. (2) contribute to the specific heat. If we neglect the difference $N - N_0$, which represents the number of excited electrons, and evaluate the energy in the limit $T \to 0$ by using the ring and exchange diagrams, the well-known expression for the correlation energy can be obtained.

The specific heat is evaluated if p_F is determined as a function of temperature and density, the latter being $p_0^3/3\pi^2$. The result is

$$c_v = c_0(1 + 0.166 r_s) \tag{3}$$

Hence the specific heat increases slightly with r_s. Such an increase has been observed for alkali metals.[3]

By extending the above calculation and applying the Sommerfeld method, one can evaluate the temperature dependence of the specific heat. A correction factor to the ideal gas specific heat is expected to appear in proportion to T^2.

Third, we have improved a similar theoretical situation in the case of the susceptibility.[4] The ground-state energy has been used in the past in such a way that it is reexpressed in terms of a polarization parameter in order to take into account the effects of a magnetic field. This parameter represents the difference in the numbers of electrons with spins parallel and antiparallel to the field. Since the final result for the susceptibility should not include the parameter, a variational method has been adopted for its elimination. Thus, as in the case of the specific heat, improvements in the evaluation of the susceptibility have been made parallel with those of the correlation energy.

We have made a very systematic approach to the magnetic behavior of an electron gas starting with weak fields.[5] In this approach no polarization parameter is introduced in an ad hoc manner but the total susceptibility is determined from the grand partition function expressed as a function of the field. The paramagnetic and diamagnetic parts are separated from each other by selecting spin-dependent terms.

In this evaluation the Fermi momentum is determined as a function of the density and field strength. That is, our theory explains the dependence of the Fermi momentum on the field strength. Such a field dependence has been observed in some alloys, such as Bi–Sb, and becomes significant as the field strength increases. For a weak field this decrease in p_F is mainly determined by $-a^4/12p_0^4$, where a^2 and p_0^2 represent, respectively, the energy due to the field and the Fermi energy. The susceptibilities are

$$\begin{aligned} \chi_p &= 2.5882 r_s^{-1} + 0.4289 - \cdots \\ \chi_d &= 0.8627 r_s^{-1} + 0.0525 + \cdots \end{aligned} \tag{4}$$

These expressions revise those previously reported.

For an intermediate field which corresponds to the de Haas–van Alphen (dHvA) effect p_F becomes an oscillating function of the magnetic field. Since p_F characterizes the dHvA oscillations, the final susceptibility expression consists of circular functions of circular functions. This gives an explanation of fine rippling oscillations superposed on main dHvA oscillations.

Finally, we have evaluated the exchange and correlation energies in the limit of intense magnetic field and low temperature that is known as the quantum strong magnetic field limit. In this limit the Fermi momentum decreases significantly with the field, and the system is characterized by a new parameter $2\pi^2 n/a^2$, n being the density.

References

1. A. Isihara and E.W.Montroll, *Proc. Nat. Acad. Sci.* **68**, 3111 (1971); A. Isihara (to be published).
2. M. Gell-Mann, *Phys. Rev.* **106**, 369 (1957); D. Pines, *Phys. Rev.* **92**, 626 (1953); D.F. DuBois, *Ann. Phys. (N.Y.)* **8**, 24 (1959); B. Ninham, *Ann. Phys. (N.Y.)* **28**, 220 (1964).
3. E.W. Collings, *J. Phys. Chem. Solids* **26**, 949 (1965).
4. D. Pines in *Solid State Physics*, F. Seitz and D. Turnbull, eds., Academic Press, New York (1955); M. Shimizu, *J. Phys. Soc. Japan* **15**, 2220 (1960); K.A. Brueckner and K. Sawada, *Phys. Rev.* **112**, 328 (1958); S.D. Silverstein, *Phys. Rev.* **130**, 1703 (1963).
5. A. Isihara, J.T. Tsai, and A. Isihara, *Phys. Rev. A* **3**, 990 (1971); J.T. Tsai, M. Wadati, and A. Isihara, *Phys. Rev. A* **4**, 1219 (1971); A. Isihara and J.T. Tsai, *Phys. Rev. B* **4**, 2280 (1971).

Excitonic Instability and Ultrasonic Attenuation under Strong Magnetic Fields in Bi*

H. Fukuyama†

Division of Engineering and Applied Physics
Harvard University, Cambridge, Massachusetts

and

T. Nagai

Department of Engineering
Kumamoto Kogyo University, Kumamoto, Japan

The attenuation coefficient of sound in semiconductors and semimetals under strong magnetic field has a series of peaks as a function of the magnetic field if the temperature is sufficiently low.[1] In Bi, as early as 1966, it was found by Sawada *et al.*[2] that the attenuation coefficient becomes anomalously large when two peaks, one due to electrons and the other to holes, are close. Although they did not investigate this problem any further, Mase and his collaborators[3] recently performed detailed experiments at lower temperatures focusing on these anomalies. They found that (a) only one of the peaks of the pair that lies on the lower side of the magnetic field has a strong temperature dependence around 1°K (asymmetry), and (b) the temperature dependence of this excess attenuation coefficient $\Delta\alpha$, above that expected by one-electron theory α_N, is given by

$$\Delta\alpha/\alpha_N = R\eta^{-\kappa}, \qquad \kappa = 0.49 \pm 0.06 \tag{1}$$

In Eq. (1) R is a number of the order of one and $\eta = (T - T_c)/T_c$. Experimental data fit with Eq. (1) as long as η is not too small, i.e., $\eta \gtrsim 0.1$ (strong temperature dependence).

We propose[4] a possible explanation of these phenomena based on excitonic instabilities.

The model Hamiltonian is

$$\mathscr{H} = \sum_{\alpha} \varepsilon_{\alpha}^{a} a_{\alpha}^{+} a_{\alpha} + \sum_{\alpha} \varepsilon_{\alpha}^{b} b_{\alpha}^{+} b_{\alpha} + V \int dr\, \psi_a^+ \psi_b^+ \psi_b \psi_a \tag{2}$$

where

$$\begin{aligned} \varepsilon_{\alpha}^{a} &= (N + \tfrac{1}{2})\,\omega_c + (p^2/2m^*) + E_a \\ \varepsilon_{\alpha}^{b} &= -[(N + \tfrac{1}{2})\,\omega_c + (p^2/2m^*) + E_b] \end{aligned} \tag{3}$$

* Supported in part by Grant No. GP-16504 of the National Science Foundation.
† Permanent Address: Tohoku University, Sendai, Japan.

$\psi_{a(b)}$ is the field operator of $a(b)$ band electrons. Coulomb interactions are approximated by the short-range force $V = 4\pi e^2/\varepsilon_0\kappa^2$, where ε_0 and κ are the static dielectric constant and inverse screening radius, respectively. As regards the Landau quantum number N, we consider $N = 1$ exclusively for electrons and holes, since this pair of states is of interest in the present experiments.

Treating excitonic instabilities in the mean field theory (ladder approximation) for the t-matrix between electrons and holes, we obtained the phase diagram, i.e., relationship between critical temperature T_c and

$$G \equiv \varepsilon^b_{N=1,p=0} - \varepsilon^a_{N=1,p=0} = -(3\omega_c + E_a + E_b)/2 \tag{4}$$

As was pointed out by Fenton,[5] T_c is around 1°K and is an order of magnitude larger than the estimate by Jerome *et al.*[6] in the absence of fields. We show the curve T_c vs. G in Fig. 1. Positive and negative G correspond to the peaks of lower and higher field sides, respectively. Note that T_c is always larger for $G \geqq 0$ than for $G \leqq 0$ with the same absolute magnitude. This explains the asymmetry of the experiments, point (a). For large G the present theory predicts a condensate with finite momentum as a pair in the field direction like the Fulde–Ferrell state[7] in superconductors.

By use of the t-matrix, or fluctuation propagator, determined above, we calculated the excess attenuation $\Delta\alpha$ due to fluctuations of the order parameter above T_c. This is shown to be written as Eq. (1) with $\kappa = 1/2$, which explains point (b).

Thus the present theory agrees with points (a) and (b) qualitatively. Quantitatively speaking, the scale of G in Fig. 1 is smaller by a factor of 2–4. In order to see whether there really exist excitonic phases or not in the present system, experiments at lower temperatures are desirable. Unfortunately, we do not know of any data in this region.

On the other hand, Fenton *et al.*[8] and Brandt and Chudinov[9] recently reported to have observed excitonic phase in Bi–Sb alloys under strong magnetic fields. Thus the existence of a correlated electron–hole system in Bi and Bi–Sb systems seems definite.

In analyzing these phenomena, the following two points seem important: the disorderedness of the lattice, and higher-order correlation between electrons and holes. As regards the first, it is known[10] that the scattering affects essentially the

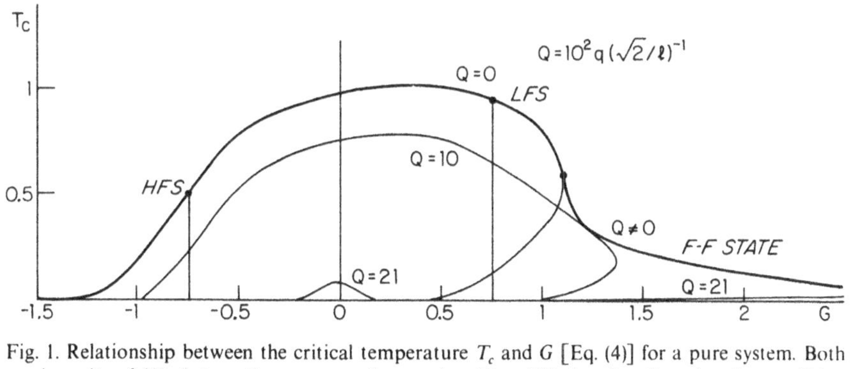

Fig. 1. Relationship between the critical temperature T_c and G [Eq. (4)] for a pure system. Both are in units of °K. Interactions are so chosen that $T_c = 1$°K for $G = 0$ under the condition $\omega_c = 10^2$ °K and $m^* = 10^{-2}m_0$ (m_0 is the free electron mass).

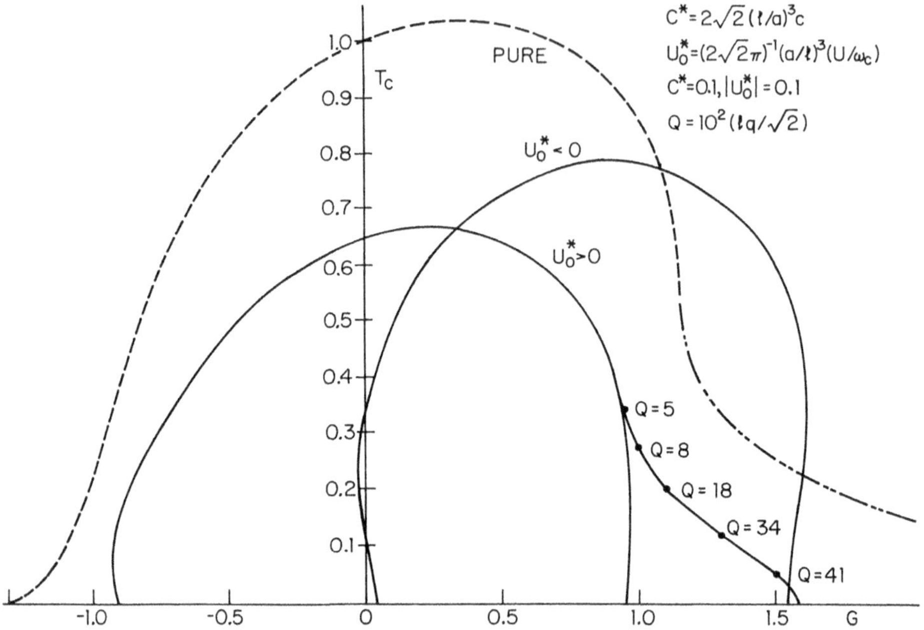

Fig. 2. T_c vs. G curve in the presence of a scattering potential, Eq. (5), treated in the CPA. Here c, l, and q are respectively, the atomic concentration of impurities, the radius of the Landau orbit, and the momentum in the direction of the magnetic field carried by an electron–hole pair; a is defined by $Na^3 = 1$, where N is the number of Bi atoms in a unit volume.

electronic state at the edge of Landau levels, where the attenuation mainly comes from. Figure 2 shows the phase diagram[11] in the presence of the following scattering potential:

$$\mathscr{H}' = U \sum_{l \in \text{ impurities}} \delta(r - R_l) \tag{5}$$

We treated Eq. (5) in the CPA,[12] which is an extension of the formalism by Kubo *et al.*[10] and is employed elsewhere.[13] In this figure c, l, and q are, respectively, the atomic concentration of impurities, the radius of the Landau orbit, and the momentum carried by an electron–hole pair in the direction of the field, and a is the lattice spacing, related to the number of atoms N in a unit volume by $Na^3 = 1$. Since we expect that electrons near the edge are localized to some extent in the sense of Anderson's notion of localization[14] with regard to motions in the direction of the field, it may be of interest to compare the present case with moderately doped semiconductors.[15]

As regards the second problem, that of higher-order correlation, it is noted that the values of the coherence length at $T = 0$ parallel and perpendicular to the fields, estimated by the present theory, are of the order of $l(\omega_c/T_c)^{1/2}$ and l, respectively. Then application of the Ginzburg criterion on the validity of the mean-field theory are shown to result in limited range of its applicability, $\eta \gtrsim 10^{-1\sim-2}$. This need to go beyond the mean field theory is followed by Abrikosov,[17] who, by use of the parquet approach, concluded the existence of a correlated electron–hole system at $T = 0$.

In reality, both of these may be connected. To reach definite conclusions, further investigations, from both theoretical and experimental sides are necessary.

References

1. V.L. Gurevich, *Soviet Phys.—JETP* **10**, 1190 (1960).
2. Y. Sawada, E. Burnstein, and L. Testardi, *J. Phys. Soc. Japan* **21**, 760 (1966).
3. Y. Matsumoto, T. Sakai, and S. Mase, *J. Phys. Soc. Japan* **28**, 1211 (1970); S. Mase, Y. Matsumoto, T. Sakai, and Y. Suido, in *Proc. 12th Intern. Conf. Low Temp. Phys.* 1970; Academic Press of Japan, Tokyo (1971); S. Mase and T. Sakai, *J. Phys. Soc. Japan* **31**, 730 (1971).
4. H. Fukuyama and T. Nagai, *Phys. Rev. B* **3**, 4413 (1971); *J. Phys. Soc. Japan* **31**, 812 (1971)..
5. E.W. Fenton, *Phys. Rev.* **170**, 816 (1968).
6. D. Jerome, T.M. Rice, and W. Kohn, *Phys. Rev.* **158**, 462 (1967).
7. P. Fulde and R.A. Ferrell, *Phys. Rev.* **135A**, 550 (1964).
8. E.W. Fenton, J.P. Jan, and A. Karlsson, *J. Low Temp. Phys.* **3**, 417 (1970).
9. N.B. Brandt and S.M. Chudinov, *Zh. Eksperim. i Teor. Fiz. Pisma* **13**, 146 (1971) [Soviet Phys.—*JETP Lett.* **13**, 102 (1971)]; *J. Low Temp. Phys.* **8**, 339 (1972).
10. R. Kubo, S. Miyake, and N. Hashitsume, *Solid State Physics*, F. Seitz and D. Turnbull, eds. Academic Press, New York (1965), Vol. 17, p. 269.
11. T. Nagai, Thesis, Kyushu Univ., 1972.
12. P. Soven, *Phys. Rev.* **156**, 809 (1967); B. Velicky, S. Kirkpatrick, and H. Ehrenreich, *Phys. Rev.* **175**, 749 (1969).
13. H. Hasegawa and M. Nakamura, *J. Phys. Soc. Japan* **26**, 1362 (1969); H. Shiba, K. Kanda, H. Hasegawa, and H. Fukuyama, *J. Phys. Soc. Japan* **30**, 972 (1971).
14. P.W. Anderson, *Phys. Rev.* **109**, 1492 (1958).
15. T. Ishiguro and N. Mikoshiba, in *Proc. Intern. Conf. on Physics of Semiconductors, Cambridge, 1970*, NBS, Springfield (1970), p. 569; M. Dutoit, *Phys. Rev. B* **3**, 453 (1971).
16. V.L. Ginzburg, *Fiz. Tverd. Tela* **2**, 1824 (1960) [*Soviet Phys.—Solid State* **2**, 1824 (1960)].
17. A.A. Abrikosov, *J. Low Temp. Phys.* **2**, 37, 175 (1970); **8**, 315 (1972).

Effects of the Excitonic Fluctuations above the Transition Temperature

M. T. Béal-Monod and K. Maki*

Laboratoire de Physique des Solides
Centre d'Orsay, Orsay, France

and

H. Fukuyama†

Division of Engineering and Applied Physics
Harvard University, Cambridge, Massachusetts

We report here on our study of the effects of the excitonic fluctuation on the various transport coefficients above the excitonic transition temperature T_N.[1] In our calculation we make use of the two-band model[2] to describe the excitonic correlation,

$$H = H_0 + H_I \tag{1}$$

$$H_0 = \sum_{\mathbf{p}} [\xi_a(\mathbf{p})\, a_{\mathbf{p}}^+ a_{\mathbf{p}} + \xi_b(\mathbf{p})\, b_{\mathbf{p}}^+ b_{\mathbf{p}}] + g \int \psi_a^+(\mathbf{r})\, \psi_a(\mathbf{r})\, \psi_b^+(\mathbf{r})\, \psi_b(\mathbf{r}) \tag{2}$$

$$H_I = \sum_i \int V(\mathbf{r} - \mathbf{r}_i)\, [\psi_a^+(\mathbf{r})\, \psi_a(\mathbf{r}) + \psi_b^+(\mathbf{r})\, \psi_b(\mathbf{r})] \tag{3}$$

where

$$\xi_a(\mathbf{p}) = (1/2m)\, p^2 - \mu_a, \qquad \xi_b(\mathbf{p}) = -(1/2m)\, p^2 + \mu_b \tag{4}$$

$$\psi_a^+(\mathbf{r}) = \sum \phi_{a\mathbf{p}}(\mathbf{r})\, a_{\mathbf{p}}^+, \qquad \psi_b^+(\mathbf{r}) = \sum \phi_{b\mathbf{p}+(\mathbf{k}/2)}(\mathbf{r})\, b_{\mathbf{p}}^+ \tag{5}$$

and $\phi_{a\mathbf{p}}(\mathbf{r})$ and $\phi_{b\mathbf{p}}(\mathbf{r})$ are the Bloch wave functions. We assumed here that the bands a and b are spherical but with different Fermi radii. g in the second term in H_0 is the effective Coulomb interaction between an electron in band a and a hole in band b. Furthermore, we also introduced here the random impurity-scattering potential through H_I, which simulates the lifetime effect due to the electron–phonon scattering.[3]

In order to study the effect of the excitonic fluctuation above the transition temperature (i.e., $T \geqq T_N$, where T_N is the transition temperature), we first have to determine the fluctuation propagator[1] or the correlation function of the excitonic

* Permanent Address, Department of Physics, Tohoku University, Sendai, Japan.
† On leave of absence from Department of Physics, Tohoku University, Sendai, Japan.

order parameter $\Psi(\mathbf{r})$:

$$\mathscr{D}(\mathbf{q}, \omega_\nu) = \langle [\Psi^+(\mathbf{r}), \Psi(\mathbf{r})] \rangle (\mathbf{q}, \omega_\nu) \tag{6}$$

where

$$\Psi^+(\mathbf{r}) = g\psi_a^+(\mathbf{r})\,\psi_b(\mathbf{r}) \tag{7}$$

Within the mean-field approximation the fluctuation propagator is given by[1]

$$\mathscr{D}^{-1}(\mathbf{q}, \omega_\nu) = N_0(\alpha\eta + \lambda q^2 + \lambda_0|\omega_\nu|) \tag{8}$$

where

$$\begin{aligned}
\eta &= \ln(T/T_N) \cong (T/T_N) - 1 \\
\alpha &= 1 - \mathrm{Re}\left[(x + iy)\,\psi^{(1)}(\tfrac{1}{2} + x + iy)\right] \\
\lambda &= \tfrac{1}{3}(\tau v)^2 \left\{\mathrm{Re}\left[\psi(\tfrac{1}{2} + x + iy) - \psi(\tfrac{1}{2} + \tfrac{1}{2}x + iy)\right] - \tfrac{1}{2}x\,\mathrm{Re}\,\psi^{(1)}(\tfrac{1}{2} + x + iy)\right\} \\
\lambda_0 &= (1/2\pi T)\,\mathrm{Re}\,\psi^{(1)}(\tfrac{1}{2} + x + iy) \\
x &= (1/2\pi\tau T), \qquad y = (1/4\pi T)(\mu_a - \mu_b)
\end{aligned} \tag{9}$$

and N_0 is the density of states of the electron in band a (or in band b, since the difference between them is negligible) at the Fermi level, τ is the electron lifetime due to the impurity scattering, v is the Fermi velocity, and $\psi(z)$ and $\psi^{(1)}(z)$ are the digamma and the trigamma functions, respectively. Here for simplicity we limit ourselves to the region $y \leqq y_0$, where the fluctuation propagator is largest at $\mathbf{q} = 0$,[4] indicating the transition into the commensurate spin-density wave state in the case of the itinerant antiferromagnetism. It is well known that the excitonic fluctuation gives rise to a large attenuation coefficient of ultrasound as first shown by Fukuyama and Nagai.[4,5] In particular, within the present model it is shown that the attenuation coefficients of both the longitudinal and the transverse sound wave diverge like $\eta^{-1/2}$ in the limit $ql \ll 1$ as the temperature approaches T_N, where q is the wave vector of the sound wave and l is the electronic mean free path.[6]

Here we consider the effect of the excitonic fluctuation on the electrical conductivity, thermal conductivity, and the nuclear spin–lattice relaxation rate. These contributions are calculated from the diagrams shown in Fig. 1, where the wavy lines are the fluctuation propagator and the solid lines are the electron propagator. We can discard the contribution from diagrams (d) and (e) (the so-called Aslamazov–Larkin term), since the contribution from these diagrams vanishes identically since the fluctuation does not carry electric current or entropy current. This is in sharp contrast to the case of the sound attenuation, where the AL term gives rise to the dominant contribution.[5,6]

The calculation of the diagrams (a)–(c) yields the contribution to the transport coefficient due to the fluctuation. The effects of the excitonic fluctuation are equivalent to the effects of the finite excitonic order parameter Δ,[7] if we expand the transport coefficients in powers of $|\Delta|^2$ and replace $|\Delta|^2$ by the average of the fluctuation

$$\langle|\Psi|^2\rangle = T\sum_q \mathscr{D}(\mathbf{q}, 0) = T\langle\mathscr{D}\rangle \tag{10}$$

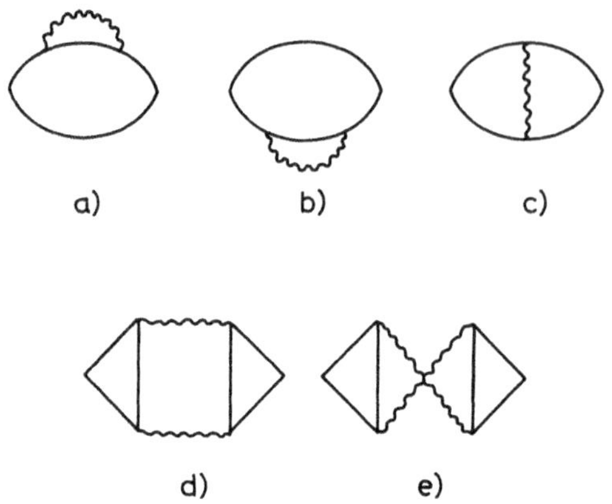

Fig. 1. Lowest-order contributions (a)–(c) of the fluctuation (wavy lines) to the transport coefficients, and the Aslamasov–Larkin diagrams (d, e). The full lines are the electron Green's functions. All the lines and vertices are renormalized by the impurity scattering.

We can evaluate $T\langle\mathscr{D}\rangle$ as

$$T\langle\mathscr{D}\rangle = \frac{T}{N_0}\int\frac{d^3q}{(2\pi)^3}\frac{1}{\alpha\eta+\lambda q^2} = \frac{1}{2\pi^2\lambda}\left[q_0 - \frac{\pi}{2}\left(\frac{\alpha\eta}{\lambda}\right)^{1/2}\right]\frac{T}{N_0} \tag{11}$$

where q_0 is a cutoff momentum.

Since $T\langle\mathscr{D}\rangle$ does not diverge at T_N (i.e., $\eta = 0$), but has a small cusp, the electrical conductivity and the thermal conductivity have small dips at T_N as given by

$$\sigma = \sigma_n\left\{1 - \frac{\tau}{2\pi T}\left[3\,\mathrm{Re}\,\psi^{(1)}(\tfrac{1}{2}+\rho) + x\,\mathrm{Re}\,\psi^{(2)}(\tfrac{1}{2}+\rho)\right]T\langle\mathscr{D}\rangle\right\} \tag{12}$$

and

$$K = K_n\big[1 + 3(\pi T)^{-2}\big(-2\tau + (1/2\pi T)\,\mathrm{Re}\{[\tfrac{3}{2}(1+\tau y/x)^2 + 1 + \tau y/x]\,\psi^{(1)}(\tfrac{1}{2}+\rho)$$
$$+ \tfrac{1}{2}x(1+\tau y/x)^2\,\psi^{(2)}(\tfrac{1}{2}+\rho)\}\big)\,T\langle\mathscr{D}\rangle\big] \tag{13}$$

where $\rho = x + iy$ and σ_n and K_n are the conductivities in the normal state. In the case of the nuclear spin–lattice relaxation rate we have to consider separately two cases[9]: the case where the excitonic order is spin singlet (nonmagnetic) and the case where the excitonic order is spin triplet (antiferromagnetic). In the former case we have

$$R/R_n = 1 + (\tau/2\pi T)\left[\mathrm{Re}\,\psi^{(1)}(\tfrac{1}{2}+\rho) + 3x\,\mathrm{Re}\,\psi^{(2)}(\tfrac{1}{2}+\rho)\right]T\langle\mathscr{D}\rangle \tag{14}$$

and in the latter case

$$R/R_n = 1 - (\tau/2\pi T)\left[\mathrm{Re}\,\psi^{(1)}(\tfrac{1}{2}+\rho) - 5x\,\mathrm{Re}\,\psi^{(2)}(\tfrac{1}{2}+\rho)\right]T\langle\mathscr{D}\rangle \tag{15}$$

where R_n is the Korringa relaxation rate in the normal state. In the singlet case the nuclear spin–lattice relaxation rate increases due to the fluctuation, while in the triplet case it decreases due to the fluctuation.

References

1. M.T. Béal-Monod, K. Maki, and H. Fukuyama, *J. Low Temp. Phys.* (to be published).
2. T.M. Rice, *Phys. Rev. B* **2**, 3619 (1970).
3. S.H. Liu, *Phys. Rev. B* **2**, 2664 (1970).
4. K. Maki and K. Nakanishi, *J. Low Temp. Phys.* **6**, 141 (1972).
5. H. Fukuyama and T. Nagai, *Phys. Rev. B* **3**, 4413 (1971); *J. Phys. Soc. Japan* **31**, 812 (1971).
6. K. Maki and H. Sato (to be published).
7. J. Zittartz, *Phys. Rev.* **165**, 605, 612 (1968).

3

Cyclotron Resonance

Far-Infrared Cyclotron Resonance in Potassium

S. J. Allen, Jr., L. W. Rupp, Jr., and P. H. Schmidt

Bell Laboratories
Murray Hill, New Jersey

One of the fundamental parameters describing the dynamics of quasiparticles in pure normal metals is the self-energy. The self-energy modifies the noninteracting electron energy–momentum relation to properly account for electron–electron and electron–phonon interactions.[1] Electromagnetic spectroscopy of the self-energy has been hindered by the fact that the frequency range of interest, close to the Debye frequency, is not conducive to the sensitive spectroscopy required to observe weak absorptive changes in metals. The most effective spectroscopic probe of the electron–phonon self-energy has been and is the tunneling measurement with a strong coupling superconductor.[2] Recently, however, progress has been made with electromagnetic spectroscopy. In particular, we note the observations of the Holstein volume absorption in superconducting Pb by Joyce and Richards.[3] Temperature-dependent cyclotron[4] and surface-state[5] resonances also probe the self-energy spectrum and have been used extensively to measure the temperature-dependent scattering rate.[5–10] Only recently have temperature-dependent mass shifts been observed by Sabo,[7] Goy,[8] Poulsen and Datars,[9] and Krasnopolin and Khaikin.[10] Frequency-dependent effects in cyclotron resonance[11] are more difficult to come by, but Goy *et al.*[8,12] have observed frequency-dependent masses and lifetimes in Hg, Pb, and In up to $\sim 15\ \text{cm}^{-1}$. In general the correct power law for the frequency effects is observed but the size of the shifts appears to be smaller than expected.

The experiments described here observe cyclotron resonance in K at 29.69, 32.12, 45.407, and $58.25\ \text{cm}^{-1}$. Although $15\ \text{cm}^{-1}$ short of the Debye peak, all four frequencies excite electron–hole pairs which sample an appreciable phonon density of states, and effects due to the frequency dependence of the self-energy should manifest themselves in the lifetime, if not the mass. Unlike the systems studied to date, K is a weakly coupled normal metal, but the case for studying these effects in K is most strongly carried by its being the archtypical normal metal whose simplicity makes a confrontation between theory and experiment hard to avoid. It has been extensively studied both theoretically[13] and experimentally (cyclotron resonance).[14] In particular, an extremely precise value of the microwave mass has recently been determined by Walsh,[15] $m^*/m = 1.217 \pm 0.002$, which is invaluable for this work.

The simple Fermi surface is important with regard to understanding cyclotron resonance at these frequencies, where the usual Azbel'–Kaner[16] or Chamber[17] theory does not appear to work. Further, far-infrared cyclotron resonance experiments in Cu and Be show that anomalous[18] results for the cyclotron mass can

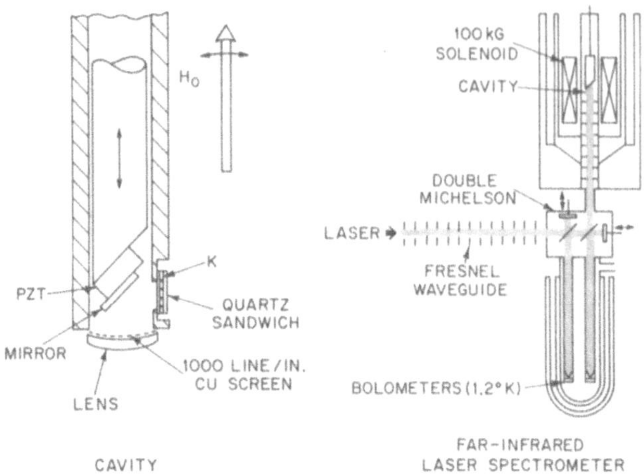

Fig. 1. Left: Nearly concentric Fabry-Perot reflection cavity. Right: Far-infrared laser spectrometer.

be obtained which are related to Fermi surface topology rather than frequency dependence of the self-energy. Potassium is free of such ambiguities.

The experiments were performed on a reflection cavity spectrometer, with balanced homodyne detection, developed to do sensitive reflection spectroscopy on real metals in the far infrared. The source of radiation is a CW laser oscillating at 29.69, 37.12, 45.407, 58.25, and 84.52 cm^{-1}. The heart of the apparatus is a nearly concentric Fabry-Perot reflection cavity (Fig. 1, left). The sample, forming one end of the cavity, lies just inside the radius of curvature of a 1000-line/in. spherical copper screen. The cavity is mechanically tuned to the laser frequency and then electrically locked on with a piezoelectric transducer. The cavity sits in one arm of a double Michelson interferometer which functions as a bridge. The bridge, cavity, and detector system can see changes as small as $1/10^6$ in reflectivity at the metal surface.

The sample was made from distilled potassium of resistivity ratio $\rho_{295}/\rho_{4.2} \approx$ 10,000 pressed between crystal quartz plates, 0.0105 in. thick, forming a sample $12 \times 6 \times 0.1$ mm. The sample was held at 4.2°K in a 110-kG magnet that can be tipped $\pm 0.8°$ with respect to the surface.

Typical experimental traces for E(infrared) $\perp H$ and $E \parallel H$ are shown in Figs. 2 and 3. The signals for $E \parallel H$ are approximately an order of magnitude weaker than $E \perp H$ and cannot be seen at 45 and 58 cm^{-1}. The $E \perp H$ signals seen at 45 and 58 cm^{-1} are similar to those seen at 29 and 32 cm^{-1} but are considerably broadened by the enhanced electron–phonon relaxation at these frequencies.

Theories of cyclotron resonance in the retardation regime, $1 \ll R_c/\delta < n^2$, have been given by Drew[19] and Meierovitch.[20] We attempt to use Drew's approach to extract a quantitative measure of the electron–phonon coupling. The change in surface admittance δY is given by

$$\delta Y = \int_0^\infty dx \int_0^\infty dx' E(x)\, \delta\sigma(x, x')\, E(x') \tag{1}$$

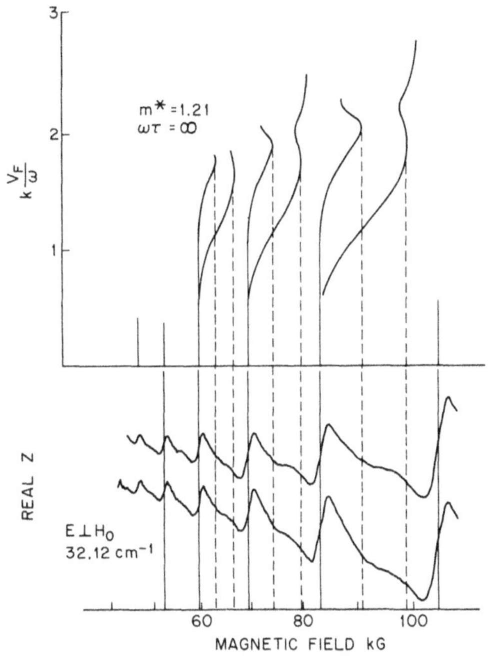

Fig. 2. Cyclotron resonance, $E \perp H$, 32.12 cm^{-1}. Infinite-medium dispersion relations for $\omega\tau = \infty$.

where $E(x)$ is the field in the sample off resonance and $\delta\sigma$ is a *small** change in conductivity on resonance.

Equation (1) can be integrated *exactly* for a diffuse boundary.

$$\delta Y^{zz} = \frac{ne^2}{m\omega_c} \sum_{n=-\infty}^{+\infty} A_n^{zz}(\alpha, R_c) \int \frac{dE}{\hbar\omega} \frac{f(E) - f(E + \hbar\omega)}{\gamma(E) + in}$$

where $f(E)$ is the Fermi function and

$$\gamma(E) = i\frac{\omega}{\omega_c}\left(1 + \frac{M(E) - M(E + \hbar\omega)}{\hbar\omega}\right) + \frac{\Gamma(E) + \Gamma(E + \hbar\omega)}{\hbar\omega_c}$$

$M(E)$ and $\Gamma(E)$ are the real and imaginary parts of the electron–phonon self-energy. $A_n^{zz}(\alpha, R_c)$ describes the retardation effects on cyclotron resonance and is given by

$$A_n^{zz}(\alpha, R_c) = (3/4\alpha)\int_0^{\pi} d\theta \sin\theta \cos^2\theta e^{-2\alpha R} J_n(-i\alpha R)\, J_n(i\alpha R)$$

where $R = R_c \sin\theta$ and J_n are Bessel functions, and α is the decay constant in $E(x) = e^{-\alpha x}$. The resonance is a sum of Lorentzians broadened and shifted by the electron–phonon interaction and weighted by a complex parameter $A_n^{zz}(\alpha, R_c)$. Assuming an electron–phonon λ of 0.1 and a $\omega = T = 0$ mass of 1.217,[15] we generate a line shape for $E \| H$ as shown in Fig. 3. Both the position and breadth of the leading

* Although all electrons in the bulk resonate at ω_c, they make only a small contribution near a diffuse surface, in the retardation regime.

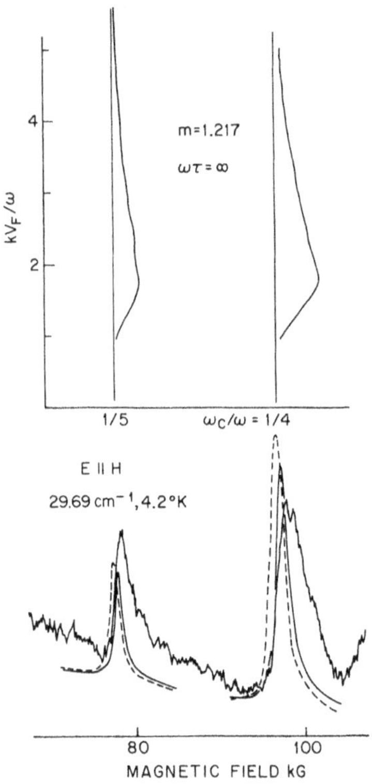

Fig. 3. Cyclotron resonance, $E \parallel H$, 29.69 cm^{-1}. Solid lines, Eq. (1), $\lambda = 0.1$. Dashed lines, Eq. (1), $\lambda = 0$, $1/\tau = 0.23\ cm^{-1}$.

edge agree very well with experiment. The attenuation of the subharmonics is reproduced, but there is considerable excess absorption on the high-field side of each resonance.

The calculation for $E \perp H$ gives essentially the same line shape as $E \parallel H$. In this case, however, there is comparatively more absorption and structure on the high-field side of resonance in the experimental traces. A calculation of the *infinite-medium* dispersion relations for cyclotron waves,[21] ignoring electron–phonon effects, shows that the excess absorption and structure correlate with the onset of real wave propagation in the bulk. The variational formulation of the problem, Eq. (1), ignores these effects and shows structure only near resonance.

The wave aspects notwithstanding, it appears that at least for $E \parallel H$, where there is no wave propagation for fields immediately below resonance, Eq. (1) may properly describe the onset of cyclotron resonance. If we fit the slope of the leading edge, normalized to the amplitude of the resonance, at the fourth and fifth subharmonics, at 29.69 and 32.12 cm^{-1}, to the predictions of Eq. (1), we can extract a λ (Table I). The values agree quite well with the accepted value determined from the low-temperature dc resistivity.[18] We have also computed (1) assuming $\lambda = 0$ but introducing a scattering rate comparable to that for $\lambda = 0.1$ (dashed lines in Fig. 3). The edge is clearly shifted to lower fields by approximately 0.5%, demonstrating that $\lambda = 0.1$ consistently gives both the position and breadth of the edge.

In summary, cyclotron resonance in potassium can be observed with a sensitive

Table I

λ, cm^{-1}	$n = 4$	$n = 5$
29.69	0.11	0.114
32.12	0.127	0.099

laser spectrometer at frequencies as high as 58.25 cm^{-1}. Although the high-field side of resonance is dominated by wave propagation in the bulk, it appears that the onset of resonance on the low-field side *may* be amenable to an analysis that ignores the wave effects. Proceeding with such an analysis, the electron phonon matrix element λ can be extracted from the position and breadth of the leading edge and is found to agree with determinations from resistivity.

Acknowledgments

We would like to acknowledge the useful suggestions and criticisms by W. M. Walsh, G. A. Baraff, P. M. Platzman, C. C. Grimes, T. M. Rice, and R. C. Dynes. The technical assistance of F. DeRosa is gratefully acknowledged.

References

1. J.R. Schrieffer, *Theory of Superconductivity,* Benjamin, New York (1964).
2. W.L. McMillan and J.M. Rowell, *Superconductivity,* Dekker, New York (1969), p. 561.
3. R.R. Joyce and P.L. Richards, *Phys. Rev. Lett.* **24**, 1007 (1970); G. Brandli, *Phys. Rev. Lett.* **28**, 159 (1972).
4. G. Grimvall, *Phys. Kond. Materie* **9**, 283 (1969); G. Grimvall, *J. Phys. Chem. Solids* **29**, 1221 (1968).
5. J.F. Koch and R.E. Doezema, *Phys. Rev. Lett.* **24**, 507 (1970).
6. P. Haussler and S.J. Welles, *Phys. Rev.* **152**, 675 (1966); T.W. Moore, *Phys. Rev. Lett.* **16**, 581 (1966).
7. J.J. Sabo, *Phys. Rev. B* **1**, 1325 (1970); P.B. Allen and M.L. Cohen, *Phys. Rev. B* **1**, 1329 (1970).
8. P. Goy, *Phys. Lett.* **31A**, 584 (1970); **32A**, 474 (1970); P. Goy and G. Weisbuch, *Phys. Rev. Lett.* **25**, 225 (1970).
9. R.G. Poulsen and W.R. Datars, *Solid State Commun.* **8**, 1969 (1970).
10. I. Krasnopolin and M.S. Khaikin, *Zh. Eksperim. i. Teor. Fiz. Pisma.* **12**, 76 (1970) [*Soviet Phys.—JETP Lett.* **12**, 54 (1970)].
11. H. Scher and T. Holstein, *Phys. Rev.* **148**, 598 (1966); P.B. Allen, in *Proc. 12th Intern. Conf. Low Temp. Phys.*, Academic Press of Japan, Tokyo (1971), p. 517.
12. P. Goy and B. Castaing, this volume.
13. T.M. Rice, *Phys. Rev.* **175**, 858 (1968) and references therein; P.B. Allen and M.J.G. Lee, *Phys. Rev. B* **5**, 3848 (1972).
14. C.C. Grimes and A.F. Kip, *Phys. Rev.* **132**, 1991 (1963); W.M. Walsh, Jr. and P.M. Platzman, *Phys. Rev. Lett.* **15**, 784 (1965); P.M. Platzman and W.M. Walsh, Jr., *Phys. Rev. Lett.* **19**, 514 (1967); **20**, 89 (E) (1968); P.M. Platzman, W.M. Walsh, Jr., and E-Ni Foo, *Phys. Rev.* **172**, 689 (1968).
15. W.M. Walsh, Jr., Private communication.
16. M.Ya. Azbel' and E.A. Kaner, *Zh. Eksperim. i Teor. Fiz.* **32**, 811 (1956) [*Soviet Phys.—JETP* **3**, 772 (1956)].
17. R.G. Chambers, *Proc. Phys. Soc.* **86**, 305 (1965).
18. S.J. Allen, Jr., *Bull. Am. Phys. Soc.* **17**, 41 (1972).
19. S.D. Drew, *Phys. Rev. B* **5**, 360 (1972).
20. B.E. Meierovich, *Zh. Eksperim. i Teor. Fiz.* **58**, 1412 (1970) [*Soviet Phys.—JETP* **31**, 756 (1970)].
21. E-Ni Foo, *Phys. Rev.* **182,** 674 (1969), and references therein; G.L. Dunifer, P.H. Schmidt, and W.M. Walsh, Jr., *Phys. Rev. Lett.* **26**, 1553 (1971).
22. R.C. Dynes and J.P. Carbotte, *Phys. Rev.* **175**, 913 (1968).

Cyclotron Resonances in Graphite by Using Circularly Polarized Radiation

Hiroyoshi Suematsu and Seiichi Tanuma

The Institute for Solid State Physics
The University of Tokyo, Roppongi, Minatoku, Tokyo, Japan

The present work is aimed at determining the signs of carriers, the cyclotron masses, and the band parameter γ_3 in graphite and also to study high-frequency transport phenomena by using well-defined circularly polarized radiation. A new assignment of the locations of electrons and holes in the Brillouin zone for graphite was made by Schroeder *et al.*[1] and has been confirmed by experimental investigations[2,3]: one elongated electron pocket centered at the K point and two equivalent hole pockets around the H points. These Fermi surfaces are known to have trigonally warped cross sections in the hexagonal plane, the degree of which depends on the parameter γ_3. When the cross section of the Fermi surface perpendicular to the magnetic field has a trigonal distortion, the sequence of harmonic resonances $(3N + 1)\,\omega_c = \omega$ is expected[4] under the diamagnetic resonance condition where $\omega_c = \pm eH/mc$ is the cyclotron angular frequency, including the sign of the carrier, and ω is the incident microwave angular frequency, which is taken as positive or negative for the left- or right-circular polarization, respectively. This sequence was observed by Galt *et al.*,[5] using circularly polarized radiation, and also by Williamson *et al.*[6] and O'Brien and Foner,[7] using linearly polarized radiation. The analysis of the intensity of these harmonic resonances should yield the γ_3 value.

Single-crystal graphite (SCG) was synthesized as segregations of flakes onto a carbon electrode from molten iron in the electric furnace of Toshiba Denko Company. Pyrolytic graphite (PG) was grown by Dr. R. J. Diefendorf in the G. E. Research Laboratory. All of these samples were given by Prof. Tsuzuku of Nihon University. Experiments were carried out with circularly polarized radiation at 23 and 69 GHz and with linearly polarized radiation at 120 GHz. The power ratios of one sense of the circularly polarized radiation to the opposite one were more than 30 and 10 at 23 and 69 GHz, respectively.[8]

The measurements were performed at a temperature of 1.0–1.6°K with the Faraday configuration and with the magnetic field parallel to the c axis. The field derivatives of microwave absorption dR/dH vs. magnetic field in SCG at 23.48, 69.08, and 118.35 GHz are shown in Fig. 1. A sequence of highly resolved resonances observed at 69.08 and 118.35 GHz is assigned to the majority electrons by means of plotting the integer $l(= 3N + 1)$ vs. the reciprocal magnetic field of the harmonic resonance, as shown in Fig. 2. The resonance fields in this figure are taken to be on the points of the maximum negative slopes of dR/dH. The linearity of the plot

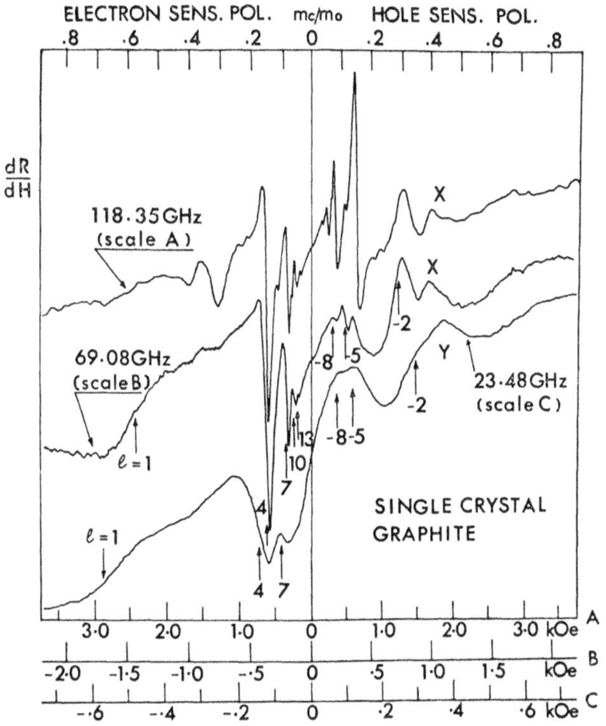

Fig. 1. The field derivatives of microwave absorption dR/dH vs. magnetic field in SCG. The left-hand (negative H) and right-hand (positive H) sides of the abscissa correspond to the circularly polarized radiation sensitive to electrons and holes, respectively. The measurement at 118.35 GHz was carried out with linearly polarized radiation. The field scales are normalized so that the resonance field for all the absorption lines is on the same vertical line. The arrows indicate the fundamental and harmonic resonance fields calculated from the slopes of the straight lines in Fig. 2.

is quite good for either of the higher frequencies, which indicates that the harmonic resonances arise from one Fermi surface with the trigonal warping. On the other hand, at 23.48 GHz the plot does not give a good linear relation, which is generally interpreted in terms of the line broadening due to the reduction of the $\omega\tau$ value, where τ is the relaxation time of the carrier. In Fig. 2 the fundamental resonance ($l = 1$) is on the electron-sensitive side of the abscissa. The fundamental resonances, which are expected to be of a nonresonant absorption due to the magnetoplasma effect, are in fact observed at -600, -1470, and -2870 Oe at 23.48, 69.08, and 118.35 GHz, respectively. From the slopes of the straight lines for 69.08 and 118.35 GHz in Fig. 2, the cyclotron mass is accurately determined to be $m_e/m_0 = 0.058 \pm 0.001$, while a slightly larger mass is obtained at 23 GHz due to the low $\omega\tau$ value. This value is in satisfactory agreement with the electron masses $m_e/m_0 = 0.058 \pm 0.02$ and 0.0567 determined from the SdH effect[9] and the band calculation (for $\gamma_3 = 0.25$ eV),[10] respectively. These facts strongly indicate that the observed carrier is assigned to the electron located at the K point in the Brillouin zone.

The band parameter γ_3 has been determined by Ushio *et al.*[11] as 0.21 ± 0.02 eV from fitting the calculated and experimental line shape of the harmonic resonances $l = 4$ and $l = 7$ at 23 GHz, since the ratio of the strength factors of these harmonics should be fairly sensitive to γ_3. The γ_3 value in the present studies is somewhat smaller than that obtained from the magnetooptic study, 0.29 ± 0.02 eV.[12] The disagreement is not understood at present.

In addition to the harmonic resonances described above, two remarkable structures are observed for the hole-sensitive polarization. At 23.48 GHz, unlike the case of the higher frequencies, a broad resonance (Y in Fig. 1) appears at a slightly higher field than that corresponding to the harmonic $l = -2$, which is the reason that the plot in Fig. 2 does not obey a linear relation. This structure has been interpreted by Ushio *et al.*[11] in terms of the enhancement of the hole resonance due to the nonlocal effect; the large spreading of resonance fields of the hole originating from the broad distribution of its cyclotron mass should be narrowed by the Doppler shift of resonance, because the resonance of the hole of smallest mass shifts to higher fields and that of the hole of largest mass to lower fields for a suitable propagation direction of a magnetoplasma wave. The hole resonance is calculated to become a rather sharp peak at 360 Oe at 23 GHz.

The other unidentified structures (X in Fig. 1) are observed at 980 and 1730 Oe at 69 and 118 GHz, respectively. This kind of structure is not explained by the

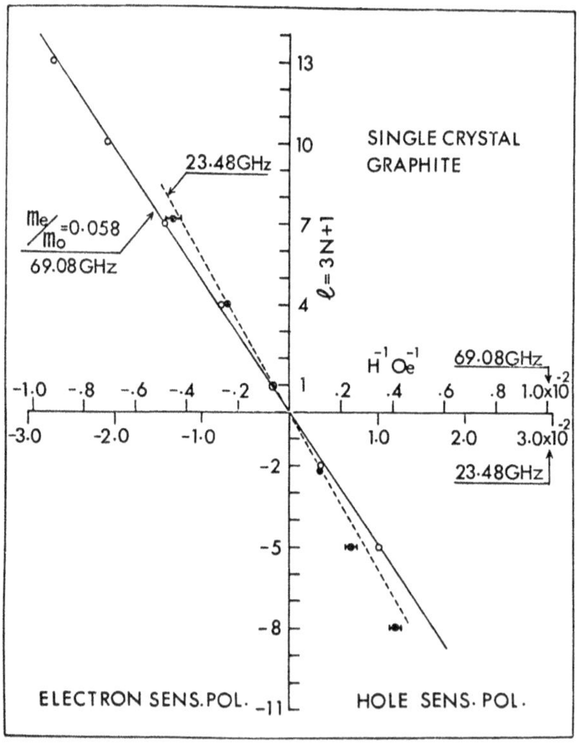

Fig. 2. Plots of $l = 3N + 1$ vs. the reciprocal field of the harmonic resonances in SCG. The field scale is normalized.

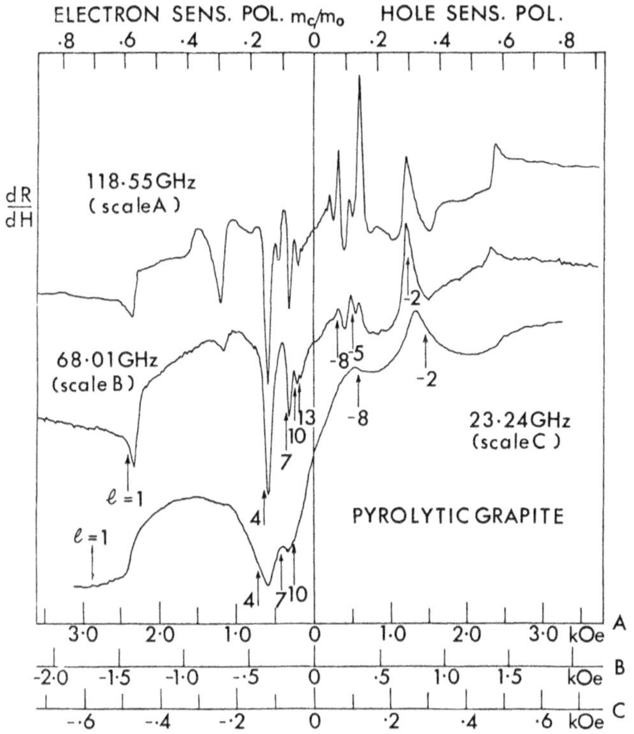

Fig. 3. The field derivatives of microwave absorption dR/dH vs. magnetic field in PG. Measurements were made with circularly polarized radiation at 23.24 and 68.01 GHz and with linearly polarized radiation at 118.55 GHz.

spatial dispersion effect of hole resonance because of the small Doppler shift at these higher frequencies, nor by the complicated, so-called "fin" parts of the Fermi surface because the orientation dependence of the field position of X is simply proportional to $1/\cos\theta$, as are those of the other harmonic resonances, where θ is the angle between the magnetic field and the c axis.[9] Recently Uda[13] has found that a structure very similar to X appears at the expected position when taking account of the electron–hole interaction, in particular, between the holelike electron having a discrete resonance spectrum at the $l = -2$ harmonic and the hole having a continuous resonance spectrum. One of the factors making this interpretation plausible is the fact that the resonance widths of X and the harmonic $l = -2$ at 118.35 GHz are almost the same as those at 69.08 GHz, while the other harmonic resonances are reduced by a factor of about 1.5 with increase of frequency when using the cyclotron mass scale instead of the real magnetic field scale. The line shape of X and the harmonic $l = -2$ calculated by Uda also give the same tendency because they are dependent on the mass distribution of holes rather than $\omega\tau$ if it is large enough. This structure in the resonance spectrum is remarkable experimental evidence of the many-body effect in an electron system, originating from the especially favorable band structure of the material.

On the other hand, in pyrolytic graphite (PG) the sequences of resonances of almost the same resolution as those of SCG are observed at 23, 68, and 118 GHz, as shown in Fig. 3. All the resonances at each frequency are interpreted in a satisfactory way by the trigonal warping of the electron Fermi surface. However, in the case of PG no additional structure is observed at each frequency. The difference in line shape between PG and SCG apparently arises from the difference of the structural imperfections; the misalignment of hexagonal planes in PG should make the movements of carriers along the *c* axis severely restricted.

Acknowledgments

We should like to thank Prof. U. Uemura, Dr. T. Uda, and Mr. H. Ushio of the University of Tokyo for their stimulating discussions and for kindly informing us of the results of their studies prior to publication.

References

1. P.R. Schroeder, M.S. Dresselhaus, and A. Javan, *Phys. Rev. Lett.* **20**, 1298 (1968).
2. J.D. Cooper, J.P. Smith, J. Woore, and D.A. Yaung, *J. Phys. C: Solid State Phys.* **4**, 442 (1970).
3. J.A. Woollam, *Phys. Rev. B* **3**, 1148 (1971).
4. Y. Inoue, *J. Phys. Soc. (Japan)* **17**, 808 (1962).
5. J.K. Galt, W.A. Yager, and H.W. Dail, *Phys. Rev.* **103**, 1586 (1956).
6. S.J. Williamson, M. Surma, H.C. Praddaude, R.A. Patten, and J.K. Furdyna, *Solid State Commun.* **4**, 37 (1966).
7. R.F. O'Brien and S. Foner, *Phys. Lett.* **25A**, 310 (1967).
8. H. Suematsu and S. Tanuma, *J. Phys. Soc. (Japan)* **33**, 1619 (1972).
9. D.E. Soule, J.W. McClure, and L.B. Smith, *Phys. Rev.* **134**, A453 (1964).
10. S.J. Williamson, S. Foner, and M.S. Dresselhaus, *Phys. Rev.* **140**, A1429 (1965).
11. H. Ushio, T. Uda, and Y. Uemura, *J. Phys. Soc. (Japan)* **33**, 1551 (1972).
12. P.R. Schroeder, M.S. Dresselhaus, and A. Javan, in *The Physics of Semimetals and Narrow-Gap Semiconductors*, D.L. Carter and R.T. Bate, eds., Pergamon Press (1971), p. 139.
13. T. Uda, Ph.D. Thesis, Univ. of Tokyo, 1972 (to be published).

Doppler-Shifted Cyclotron Resonance with Helicons and Acoustic Waves in Aluminum*

S. W. Hui and J. A. Rayne

Carnegie Mellon University
Pittsburgh, Pennsylvania

Doppler-shifted cyclotron resonance (DSCR) in a metal occurs whenever the cyclotron frequency of a group of electrons matches the apparent frequency of a penetrating wave. Both helicons and acoustic waves have been previously used for DSCR experiments in aluminum[1-3] but the data are incomplete. In this paper we report the results of more comprehensive measurements on super-purity aluminum and a detailed comparison with two current models[4,5] for its Fermi surface.

The single crystals used in this work were obtained from an ingot having a resistivity ratio of 14,800 supplied by the National Bureau of Standards, Boulder, Colorado. Samples with normals parallel to [001], [111], and [110] were spark-cut from the ingot, care being taken to remove any resulting surface damage by appropriate etching. Helicon DSCR were detected by surface impedance measurements between 0.45 and 3.5 MHz in fields up to 55 kG. Acoustic DSCR were measured by the pulse-echo method at frequencies between 30 and 210 MHz using both longitudinal and transverse waves. For the latter type of excitation, correction for rotation of the plane of polarization was made using the field-reversal method.[6]

In each case the DSCR condition is given by the equation

$$B_n = (q\hbar/2\pi ne)\,|\partial A/\partial k_B| = (q\hbar/2\pi ne)\,A' \tag{1}$$

Here q is the pertinent wave number and $|\partial A/\partial k_B|$ is the magnitude of the Fermi surface area derivative with respect to the component k_B of the electron wave vector along the magnetic field **B**. Zones on the Fermi surface having a stationary value of A' are dominant in DSCR, although their effectiveness depends on the nature of the electron–wave interaction. For the acoustic experiments n can assume integer values corresponding to a harmonic series of attenuation maxima or minima, whereas in the helicon experiments only the fundamental resonance ($n = 1$) is detectable.

Since the acoustic waves are not dispersive, the relevant q in Eq. (1) may be obtained from the known sound velocity and frequency. However, for helicons the value of q must be calculated from the dispersion relation (in MKS units)

$$q^2 = -i\omega\mu\sigma_{\pm} \tag{2}$$

where ω is the helicon frequency and $\sigma_{\pm} = \sigma_{xx} \pm i\sigma_{xy}$ are the elements of the conductivity tensor for circularly polarized excitation. Thus either $\sigma_{\pm}$ or equivalently

* Work supported by the National Science Foundation.

Table I. Comparison of $A'_m/2\pi$ for Effective Zones in Aluminum Derived from Empirical Fermi Surface Models and Various Experimental Techniques

		$A'_m/2\pi$, Å^{-1}					
		Calculated		Experimental			
Field direction	Effective zone	Ref. 4	Ref. 5	Longitudinal acoustic	Transverse acoustic	Helicon	Magneto-morphic[7]
[001]	A	0.93	1.04	1.05 ± 0.02 (0.82)[2]	1.16 ± 0.02* (1.08)[3]*	1.05 ± 0.05	0.97 ± 0.01
	B	1.8	1.8	1.83 ± 0.02	1.7 ± 0.1*	†	†
[111]	C	0.26	0.23	0.21*	0.22 ± 0.02	*	†
	D	0.75	0.75	†	†	*	—
	E	1.76	1.76	†	1.83 ± 0.01	1.75 ± 0.05	—
[110]	F	0.40	0.23	†	0.29*	0.4 ± 0.1	†
	G	1.10*	0.97*	0.84*	*	1.00 ± 0.05	0.82 ± 0.02

* Uncertain.
† Not observed.

the Fermi surface must be known, so that the measurement of A' from helicon data is a self-consistent process. The resulting values of A' computed from Eq. (1) for the helicon and acoustic data are given in Table I. Figure 1 illustrates schematically the effective zones on the Fermi surface of aluminum, to which the resonances have been assigned. The stationary values of A' corresponding to these zones, namely A'_m, have also been computed for two model Fermi surfaces of aluminum.[4,5] The results are presented in Table I, together with experimental data obtained from magnetomorphic experiments.[7]

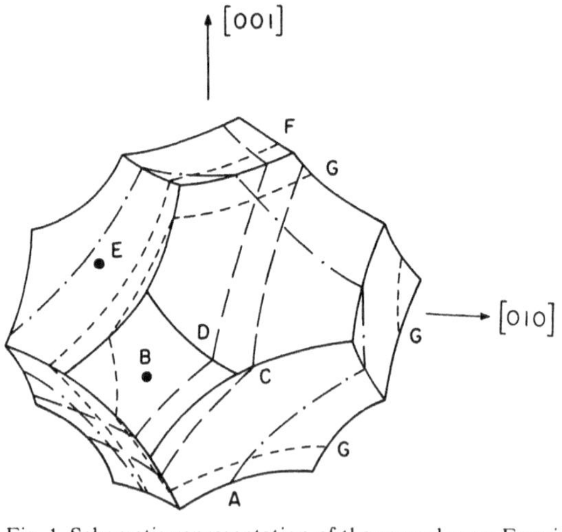

Fig. 1. Schematic representation of the second-zone Fermi surface for aluminum. Effective DSCR zones for the magnetic field along [110], [111], and [001] are shown as - - -, — —, and — · — lines, respectively.

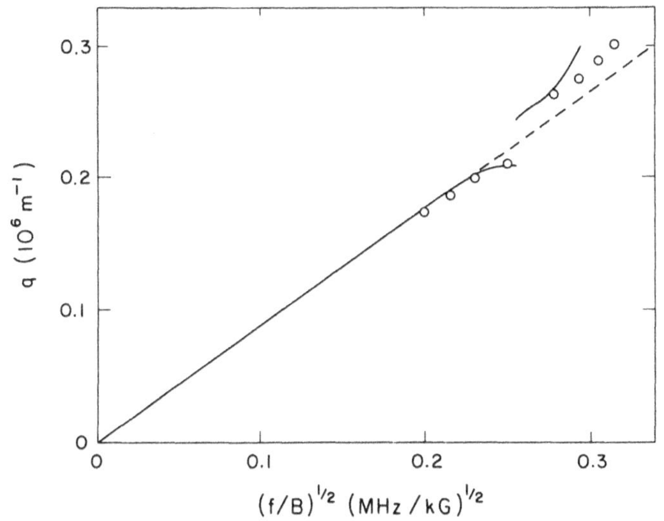

Fig. 2. Dispersion relation for helicons propagating along [111] at a frequency of 1.59 MHz. The solid and broken lines are calculated from nonlocal and local theory, respectively. Nonlocal effects above the window are caused by the zone D.

In the [001] field direction the major DSCR is due to zone A. The helicon dispersion and DSCR anomaly follow closely those computed from the Anderson model, which also gives good agreement with the longitudinal acoustic wave results. Since the harmonic numbers cannot be positively assigned, the transverse wave data for this field direction are somewhat uncertain. Zone B appears as a relatively small effect in both sets of data, particularly the transverse wave case, so that there is some uncertainty in the corresponding value of A_m'. Within experimental error it appears to agree with the predictions of both models.

For the [111] field direction zone E appears very strongly in both the helicon and transverse acoustic wave experiments. The negative nonlocal dispersion for the helicon data shown in Fig. 2 demonstrates clearly that the resonance zone is on a minority-carrier surface. Both the value of A_m' deduced from the observed helicon window (Fig. 2) and that from the acoustic data indicate that the zone is very free-electron-like. Figure 3 shows that zone C is also detected acoustically; the value of A_m' agrees well with the Anderson model. Due to the complicated dispersion beyond the helicon window region, the helicon data for zones C and D are uncertain.

The results for the [110] field direction are rather inconclusive. Despite the predicted large nonlocal effects due to zone G, the observed helicon dispersion relation remains linear. The field–frequency relation for both the F and G helicon anomalies appear to favor the Ashcroft model, but this result is not believed to be significant. Little information is available from the transverse acoustic data since the integer vs. $1/B$ plots do not pass through the origin, while the longitudinal wave data are almost featureless because of the extreme ineffectiveness of the relevant zones in this case.

As can be seen from Table I, the A_m' values derived from the present experiments

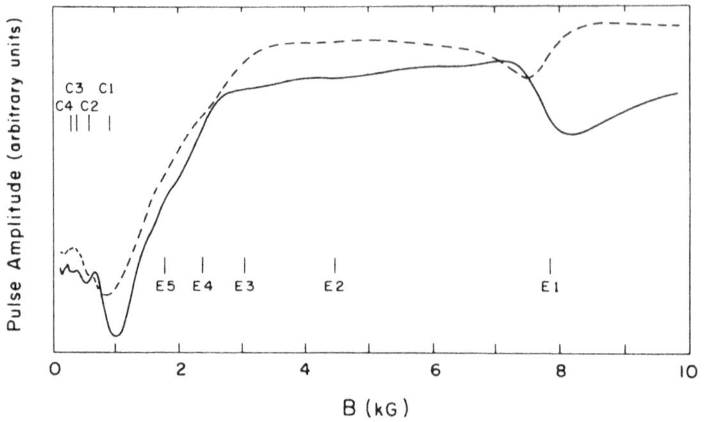

Fig. 3. Amplitude of the transmitted signal for transverse ultrasonic wave at 32 MHz with **q, B** along [111] and polarization along [1$\bar{1}$0]. Two DSCR series C and E are shown. The positions of the former were measured from an expanded trace not shown here. The full and dashed curves show the effects of reversing the applied magnetic field to correct for rotation of the plane of polarization.

are reasonably consistent. Due to the nondispersive nature of the acoustic waves, it is believed that the ultrasonic results are more accurate than those derived from helicon measurements. There is much better agreement with the Anderson model than with that due to Ashcroft, so that the former appears to be the preferred empirical representation of the Fermi surface for aluminum. No reason can be advanced for the discrepancy between the present results and those derived from magnetomorphic data.

Acknowledgments

We are grateful to Dr. J. R. Anderson, University of Maryland, for the computational results of his model Fermi surface. Thanks are also due to Dr. M. B. Kasen, Cryogenics Division, National Bureau of Standards, Boulder, for providing the high-purity aluminum used in this work.

References

1. J. L. Stanford and E.A. Stern, *Phys. Rev.* **144**, 534 (1966).
2. B.K. Jones, *Phil. Mag.* **9**, 217 (1964).
3. B.I. Miller, *Phys. Rev.* **151**, 519 (1966).
4. N.W. Ashcroft, *Phil. Mag.* **8**, 2055 (1963).
5. J.R. Anderson and S.S. Lane, *Phys. Rev. B* **2**, 298 (1970).
6. S.W. Hui and J.A. Rayne, *J. Phys. Chem. Solids* **33**, 611 (1972).
7. T. Alstadheim and R. Risnes, *Phil. Mag.* **18**, 885 (1968).

Dopplerons in Tungsten*

J. F. Carolan

Department of Physics, University of British Columbia
Vancouver, British Columbia, Canada

Some time ago we conducted a set of experiments[1] searching for effects of the Doppler-shifted cyclotron resonance (DSCR) mode,[2-5] the so-called Doppleron,[3] in compensated metals. Such a mode should exist in the region of strong DSCR. However, it is difficult to verify the existence of such a mode using ordinary linearly polarized experimental configurations because the standing waves excited in a metal slab have a field dependence similar to that found for the Gantmakher–Kaner (GK) size-effect oscillations.[6] Theoretically, there are other characteristic differences between Dopplerons and the GK oscillations. The Doppleron will generally be of larger amplitude and it will generally have a sharper onset with increasing applied magnetic field. These latter properties are difficult to fix quantitatively. Despite these difficulties, Fisher *et al.*[5] have identified the Doppleron in Cd through a careful analysis of the period, amplitude, and field dependence of the oscillations. They were assisted by the fact that the Doppleron was very strong in this case and the section of Fermi surface causing the oscillations was amenable to model calculations for comparison with experiment. However, such a convincing analysis is not feasible in other cases, and more evidence is required. To this end, there is one other aspect to consider, and that is the sense of circular polarization of the Doppleron versus the GK oscillation. The Doppleron, like the helicon, is a circularly polarized low-frequency wave. The GK oscillation is due to excitation of individual electrons across the sample with the characteristic cyclotron frequency. If one excites the sample at a low frequency with circularly polarized radiation, one would expect that the Doppleron would be found for only one sense of polarization, whereas the high-frequency GK oscillation would view the exciting field as quasistatic and would not show any polarization dependence.[3] Such behavior has been experimentally confirmed in In and Cd using circularly polarized rf transmission. The helicon for In and the Doppleron for Cd have one sense of polarization, while the GK oscillations occur for both senses.[1] Having established that one can differentiate between the Doppleron and the GK oscillation using circular polarization, we would like in this paper to examine the [110] direction in W. This provides an interesting test for the method because the DSCR is weaker than in the cases cited above. We begin with a discussion of the experimental procedure.

The circularly polarized transmission apparatus used in this experiment was similar to that used by others[7] for linearly polarized work except that instead of a single transmission coil, we used a pair of crossed coils driven 90° out of phase.

* This research was supported by the National Research Council of Canada.

The coils were 30 widely spaced turns of #40 wire wound on a $\frac{5}{8}$-in. square copper form $\frac{1}{8}$ in. thick. Since one coil was closer to the sample, the inputs were adjusted (approximate ratio 3:2) to produce equal pickup. The input level ranged up to 0.5 V. A pickup coil was situated in a copper cavity $\frac{1}{8}$ in. in diameter and $\frac{1}{4}$ in. in length with the number of turns chosen to tune the coil to the desired frequency range within the 50 kHz to 5 MHz range of the rest of the system. With our phase-sensitive detection scheme we have generally measured one component of the transmitted field, although we have recently added the capability to measure amplitude and phase directly. The degree of circular polarization achieved with this apparatus was about 95%. This was the actual polarization measured with helicons. Whether this residual mixing of polarization was caused by the coil and sample configuration or whether it was intrinsic to the sample has not been determined.

The tungsten samples used in this work were spark-cut and planed from a boule (RRR = 25,000) obtained from Westinghouse Research Laboratories. The 0.2-in.-diameter disk-shaped samples were then silver-soldered to a $\frac{1}{16}$-in.-thick piece of copper with a $\frac{1}{8}$-in.-diameter hole for the transmission. The copper was then in turn soft-soldered to the copper block containing the pickup coil. This drastic procedure was necessary to obtain effective isolation between the transmitting and receiving sides of the sample. For the circularly polarized experiment one must take care to eliminate any leakage which could result in a mixing of the senses of polarization. We have not made any detailed study of the effect of the silver-soldering except to note that the sample quality was reduced to the point that its resistance was residual at 4.2°K.

Figure 1 shows the circularly polarized transmission for the two senses of polarization which are obtained by reversing the magnetic field direction. The dc magnetic field is normal to the plane of the disk of thickness d. The overall pattern shown does not vary for tipping angles of the order of 5° from the normal. In the figure f is the applied frequency and b_{tr} is the transmitted rf magnetic field. In this case we are measuring only one component of b_{tr} at a constant reference phase chosen for maximum signal. One can see from the figure that near 8 kOe the amplitude of the lower curve is about three times that of the upper curve. We should remark that the isolation of the pickup coil in this case was not the 60–80 dB that we have observed with other metals. Furthermore, the amplitude ratio between the two polarizations has increased as we have improved the isolation, which implies that the actual amplitude ratio may be larger than observed here. As the field is increased above 8 kOe the signal in the lower curve gradually decreases until it has the same amplitude as the upper curve. In this high-field region we conclude that we have ordinary GK oscillations. On the lower-field side the oscillations have a relatively sharp onset at 5 kOe. In addition to the type of data shown in Fig. 1, we have also performed single-coil (linearly polarized) absorption experiments, which measure the surface resistance. These were performed prior to the silver-soldering for the transmission measurements. These absorption measurements confirm the sharp onset of the waves and that the frequency dependence of the onset is $f^{1/3}$ over the range 100 kHz to 3 MHz. These measurements also show smaller period oscillations in the low-field regime which are more than an order of magnitude smaller in amplitude than those shown in Fig. 1. These are presumably GK oscillations arising from other portions of the Fermi surface. The absorption data, with its better signal-

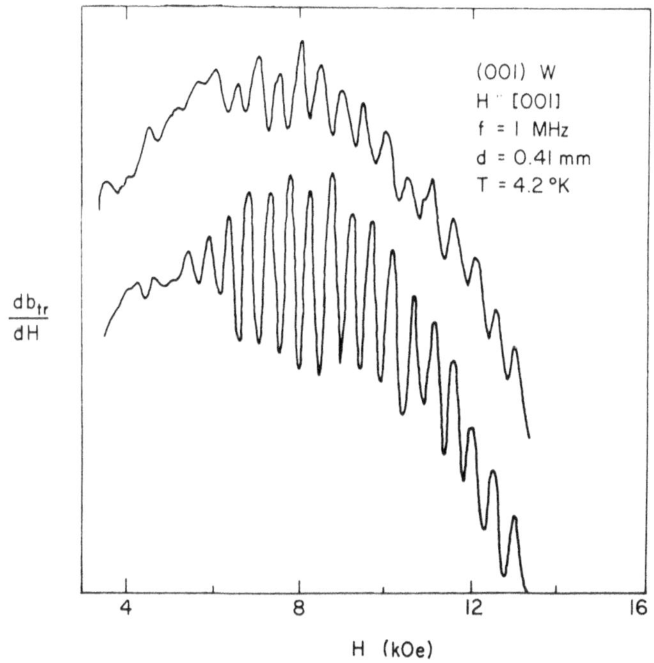

Fig. 1. The derivative of one component of the transmitted rf field for the electron sense of applied circular polarization (top curve) and for the hole sense (bottom curve).

to-noise ratio, allow us to make a better determination of period than is possible from the data in Fig. 1. We find that the period of the oscillations is constant, and within the experimental error of 1% the average value agrees with the results of other experiments and corresponds to the extremum of the derivative of the area for the hole octahedron.[8,9]

In discussing the data presented above, we are naturally interested in whether there is convincing evidence for the Doppleron. The polarization dependence, the sharp onset, the $f^{1/3}$ dependence of the onset, and the large amplitude of the oscillations in Fig. 1 relative to the smaller period oscillations are all positive evidence that the oscillations for the lower curve in Fig. 1 are Dopplerons. However, the fact that the period is constant to within experimental accuracy is not consistent with the ordinary Doppleron mode[3] and previous experiments.[5] At first sight this constant period would seem to imply that we are observing GK oscillations with the polarization dependence accounted for by an enhancement of the size-effect oscillations by the Dopplerons. This explanation is possible only if one assumes that the Doppleron propagates slightly but not all the way across the sample, and if one assumes diffuse reflection at the surfaces.[3] There is another explanation for the constant period, which is based on the details of the tungsten Fermi surface. The normal Doppleron[3] associated with a single DSCR results from a solution to the dispersion equation for which the imaginary part of the transverse conductivity, Im $\sigma_{\pm}$, does not change sign in the region of the Doppleron. There is a large dispersion in Im $\sigma_{\pm}$ which results in the characteristic period variation associated with the Doppleron. In tungsten the situation is slightly different. For values of the Fermi surface area derivative slightly larger than the hole octahedron extremum (see Fig. 3 of Ref. 8) there are two regions of electron orbits on the Fermi surface that undergo DSCR. This implies that for wave numbers slightly smaller than the value corresponding to the hole octahedron DSCR the Im σ_+ will change sign. Therefore the characteristic dispersion normally present where the Doppleron solution first appears will be sharply reduced and one would expect little period variation for the Doppleron. A detailed numerical computation is obviously required to check this qualitative argument.

In conclusion, we have observed a definite polarization dependence in the rf transmission which indicates the presence of Dopplerons. The question remaining is whether we are observing it directly or via an enhancement of the GK oscillations.

References

1. J.F. Carolan, *Bull. Am. Phys. Soc.* **15**, 802 (1970).
2. J.C. McGroddy, J.L. Stanford, and E.A. Stern, *Phys. Rev.* **141**, 437 (1966).
3. D.S. Falk, B. Gerson, and J.F. Carolan, *Phys. Rev. B* **1**, 406 (1970).
4. R.G. Chambers and V.G. Skobov, *J. Phys. F: Metal Physics* **1**, 202 (1970).
5. L.M. Fisher, V.V. Lavrova, V.A. Yudin, O.V. Konstantinov, and V.G. Skobov, *Zh. Eksperim. i Teor. Fiz.* **60**, 759 (1971) [*Soviet Phys.—JETP* **33**, 410 (1971)].
6. V.F. Gantmakher and E.A. Kaner, *Zh. Eksperim. i Teor. Fiz.* **48**, 1572 (1965) [*Soviet Phys.—JETP* **21**, 1053 (1965)].
7. C.C. Grimes and S.J. Buchsbaum, *Phys. Rev. Lett.* **12**, 357 (1964).
8. D.E. Soule and J.C. Abele, *Phys. Rev. Lett.* **23**, 1287 (1969).
9. S.W. Hui and J.A. Rayne, *J. Phys. Chem. Solids* **33**, 611 (1972).

Landau Damping in Degenerate Semiconducting Plasmas

P. Halevi

Instituto Nacional de Astrofísica, Optica y Electrónica
Tonantzintla, Puebla, Mexico

In a recent communication[1] (to be denoted I) we have pointed out that helicon waves, having a variable phase velocity, are particularly suitable for the study of Landau damping[2] in degenerate solid-state plasmas. For sufficiently high magnetic fields H their phase velocity ω/q is proportional to $H^{1/2}$. In metals, even for the highest magnetic fields attainable, the phase velocity is much smaller than the Fermi velocity v_{F}. In highly doped ($n \sim 10^{18}/\mathrm{cm}^3$) semiconductors, however, it is possible to achieve the condition $\omega/q \gtrsim v_{\mathrm{F}}$ for frequencies of the order of 10 GHz and magnetic fields of the order of several kG.

In the present work we shall generalize the results given in I for an isotropic carrier distribution (Fermi sphere) and, in addition, present numerical results for an anisotropic band structure.

Buchsbaum and Platzman[3] and, in a more compact form, Halevi and Rabinovitch,[4] have derived expressions for the nonlocal damping of helicon waves for a spherical Fermi surface, under the restriction $\omega\tau \ll 1$, where τ is the collision time. The generalization to arbitrary values of $\omega\tau$ is straightforward. We merely have to replace τ^{-1} by $\tau^{-1} - i\omega$ in Eq. (57) of Ref. 4. In addition, we take into account the existence of a small displacement current. The frequency being considered a real quantity, we obtain the following complex, first-order corrections to the wave vector:

$$\frac{\Delta q}{q} = \frac{1 + \cos^2\psi}{4|\cos\psi|}\frac{\omega\omega_c}{\omega_p^2} + \frac{i}{2|\cos\psi|}\frac{1 - i\omega\tau}{\omega_c\tau}\left[1 + \chi\left(\frac{q_z l}{1 - i\omega\tau}\right)\sin^2\psi\right] \tag{1}$$

$$\chi(z) = \frac{3}{8}\frac{z^2\tan^{-1}z}{z - \tan^{-1}z} - \frac{9}{8} \tag{2}$$

Here ψ is the angle between $\mathbf{H}$ and $\mathbf{q}$, ω_c is the cyclotron frequency eH/m^*c, ω_p is the plasma frequency $(4\pi ne^2/m^*\varepsilon_{\mathrm{L}})$, l is the mean free path $v_{\mathrm{F}}\tau$, and $q_z = q\cos\psi$.

The first term in Eq. (1) represents the correction due to the inclusion of the displacement current; it was derived under the condition $\omega\omega_c/\omega_p^2 \ll 1$. The imaginary part of the second term is the damping and its real part represents the high-frequency correction to the dispersion of the wave. Since the function χ goes to zero in the limit $q_z l/|1 - i\omega\tau| \to 0$, this function describes the nonlocal effects with which we are concerned in this paper:

$$\frac{\Delta q_{\mathrm{NL}}}{q} = \frac{i\sin^2\psi}{2\cos\psi}\frac{1 - i\omega\tau}{\omega_c\tau}\chi\left(\frac{q_z l}{1 - i\omega\tau}\right) \tag{3}$$

The last expression becomes independent of τ in the limit $\omega\tau \to \infty$:

$$\lim_{\omega\tau\to\infty} \frac{\Delta q_{\mathrm{NL}}}{q} = \frac{\omega \sin^2 \psi}{2\omega_c \cos\psi} \chi\left(i \frac{q_z v_{\mathrm{F}}}{\omega}\right) \tag{4}$$

We should point out, however, that $\omega\tau \gg 1$ does not, in general, ensure truly "collisionless" behavior. The reason is that local effects may be more important than nonlocal ones [first and second terms in the square brackets of Eq. (1), respectively]. On the other hand, it is clear that even the case $\omega\tau \ll 1$ may result in collisionless behavior (as is usually the case in metals), for the physical condition for collisionless behavior is that the distance l traversed by an electron between collisions, relative to the distance $(\omega/q)\tau$ traversed by the wave during the same period of time, is much larger than a wavelength. Thus the condition is

$$q_z l - \omega\tau \gg 1 \tag{5}$$

rather than $q_z l/|1 - i\omega\tau| \gg 1$ as usually expressed.* This means that for $\omega\tau \ll 1$ the behavior is collisionless for very slow waves ($\omega/q_z v_{\mathrm{F}} \ll 1$).

We also note that when the phase velocity exactly matches the Fermi velocity ($\omega/q_z v_{\mathrm{F}} = 1$) it is impossible, as a matter of principle, to obtain collisionless behavior, since we have $q_z l - \omega\tau = 0$ for however large (but finite) τ.

We may readily deduce the damping from Eq. (4). In the limit of very large collision time

$$\lim_{\omega\tau\to\infty} \mathrm{Im}\left(\frac{\Delta q_{\mathrm{NL}}}{q}\right) = \frac{3\pi}{32} F\left(\frac{\omega}{q_z v_{\mathrm{F}}}\right) qR \sin^2\psi \tag{6}$$

$$F(x) = \begin{cases} \left[\left(\frac{\pi}{2}x\right)^2 + \left(1 - \frac{1}{2}x \ln\frac{1+x}{1-x}\right)^2\right]^{-1}, & \frac{\omega}{q_z v_{\mathrm{F}}} \leqq 1 \\ 0, & \frac{\omega}{q_z v_{\mathrm{F}}} \geqq 1 \end{cases} \tag{7}$$

Here R is the cyclotron radius v_{F}/ω_c. The function F represents a generalization of Kaner and Skobov's[5] formula for the Landau damping of helicon waves. This formula was derived for $\omega/q_z v_{\mathrm{F}} \ll 1$, and, indeed,

$$\lim_{x\to 0} F(x) = 1 \tag{8}$$

On the other hand,

$$\lim_{x\to 1} F(x) = 0 \tag{9}$$

which means that the Landau damping gradually approaches zero as the phase velocity approaches the Fermi velocity. A more complete discussion of the isotropic case, including graphical analysis for finite $\omega\tau$, is given in I.

The experimental requirements for the realizability of suitable experiments

* These two conditions are *not* equivalent. Thus, for example, for $\omega\tau = 10$ and $\omega/q_z v_{\mathrm{F}} = 0.5$ we get $q_z l - \omega\tau = 10$ and $q_z l/|1 - i\omega\tau| \simeq 2$. The number 2 indicates *moderately* nonlocal behavior, while the actual behavior [given by Eq. (5)] is *extremely* nonlocal.

are quite demanding: high frequencies, high magnetic fields, high doping, and a high-mobility degenerate semiconductor. Lead telluride is a suitable candidate, and, in fact, Walpole and McWhorter[6] have done both theoretical and experimental work on this material. Their value of $\omega\tau$ was only 2; however, a mobility of 1×10^7 cm^2/V sec at 4.2°K and 7.8×10^{17} carriers/cm^3 has been reported.[7] With this sample and a frequency of about 20 GHz it is possible to have $\omega\tau \sim 25$.

The constant-energy surfaces of PbTe are four highly anisotropic ellipsoids. This anisotropy results[6] in two features which are absent in the isotropic case. One of them is that the Landau damping does not vanish for $\mathbf{H} \| \mathbf{q}$. The other one is that charge-density oscillations have to be taken into account in spite of a negligible displacement current. For $\mathbf{H} \| \mathbf{q} \| [001]$ our Eq. (3) has to be replaced by the following equation derived by Walpole and McWhorter[6]:

$$\frac{\Delta q_{\mathrm{NL}}}{q} = \frac{i(K-1)^2}{8(K+2)} \frac{q v_{\mathrm{F}}}{\omega_{ct}} G\left(\frac{ql}{1 - i\omega\tau}\right) \tag{10}$$

$$G(z) = \left(1 + \frac{1}{z^2}\right)^2 \tan^{-1} z - \left(\frac{5}{3z} + \frac{1}{z^3}\right) + \frac{1}{z^3} \frac{[(z + 1/z)\tan^{-1} z - 1]^2}{1 - i\omega\tau - (1/z)\tan^{-1} z} \tag{11}$$

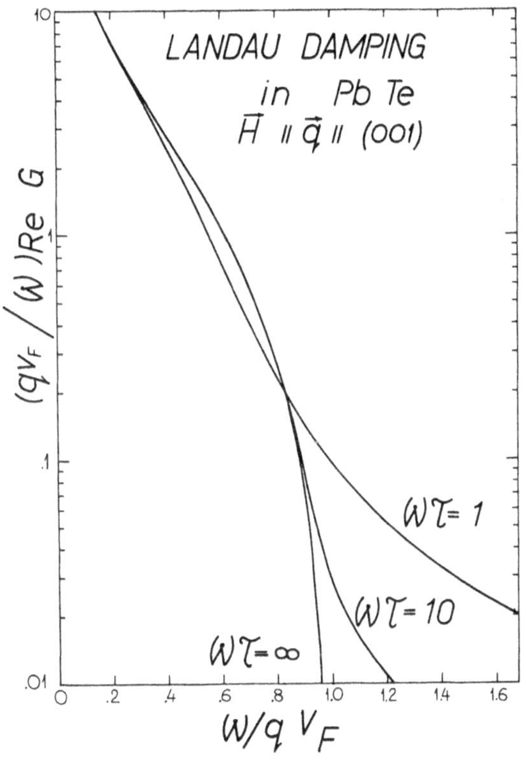

Fig. 1. The function $8(K+2)(K-1)^{-2}(\omega_{ct}/\omega)\,\mathrm{Im}\,(\Delta q_{\mathrm{NL}}/q)$, which determines the Landau damping of helicon waves in PbTe, as a function of the normalized phase velocity. Note logarithmic scale.

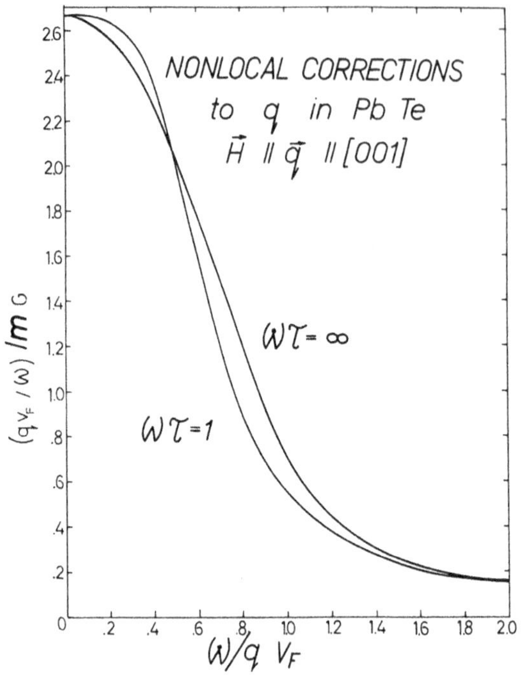

Fig. 2. The function $-8(K+2)(K-1)^{-2}(\omega_{ct}/\omega)\,\mathrm{Re}\,(\Delta q_{\mathrm{NL}}/q)$, which determines the nonlocal corrections to the wave vector in PbTe, as a function of the normalized phase velocity.

Here K is the anisotropy ratio of the ellipsoids of revolution ($K \simeq 10$ for PbTe), ω_{ct} is the cyclotron frequency corresponding to the transverse mass of an ellipsoid, and v_{F} is the Fermi velocity at the extremal point of any one of the ellipsoids.

The result for the damping is quite simple in the limit $\omega\tau \to \infty$:

$$\lim_{\omega\tau\to\infty} \mathrm{Im}\left(\frac{\Delta q_{\mathrm{NL}}}{q}\right) = \begin{cases} \dfrac{\pi(K-1)^2}{16(K+2)}\, qR_t\left(1-\dfrac{\omega^2}{q^2 v_{\mathrm{F}}^2}\right)^2, & \dfrac{\omega}{qv_{\mathrm{F}}} \leqq 1 \\ 0, & \dfrac{\omega}{qv_{\mathrm{F}}} \geqq 1 \end{cases} \tag{12}$$

where $R_t = v_{\mathrm{F}}/\omega_{ct}$ is the cyclotron radius corresponding to the transverse mass.

In Figs. 1 and 2, respectively, we plot the real and imaginary parts of the function $(qv_{\mathrm{F}}/\omega)\,G$, which are proportional to the Landau damping and associated nonlocal corrections to the wave vector, for various values of the parameter $\omega\tau$. The results are quite similar to those in the isotropic case I. Here, too, there is no first-order-like transition in the damping at $\omega/qv_{\mathrm{F}} = 1$ in the limit $\omega\tau \to \infty$. On the other hand, the peak in $\mathrm{Im}\,(\Delta q/q)$ at $\omega/qv_{\mathrm{F}} = 1$, found in the same limit for the isotropic case, is now absent. Generally speaking, the nonlocal corrections for the many-valley band structure are less sensitive to the value of $\omega\tau$ than for the spherical Fermi surface. We also note that for a finite $\omega\tau$ nonlocal damping exists in the region $\omega/qv_{\mathrm{F}} > 1$.

References

1. P. Halevi, *Sol. St. Commun.* **10**, 1189 (1972).
2. L.D. Landau, *J. Phys. USSR* **10**, 25 (1946); D. Pines, *Elementary Excitations in Solids*, Benjamin, New York (1964); E.A. Kaner and V.G. Skobov, *Advan. Phys.* **17**, 605 (1968).
3. S.J. Buchsbaum and P.M. Platzman, *Phys. Rev.* **154**, 395 (1967).
4. P. Halevi and K. Rabinovitch, *Phys. Rev. B* **1**, 1390 (1970).
5. E.A. Kaner and V.G. Skobov, *Zh. Eksperim. i Teor. Fiz.* **45**, 610 (1963) [*Soviet Phys.—JETP* **18**, 419 (1964)].
6. J.N. Walpole and A.L. McWhorter, *Phys. Rev.* **158**, 708 (1967).
7. J.F. Miller, J.W. Moody, and F.J. Reid, *J. Phys. Chem. Sol.* **28**, 1083 (1967).

Oscillatory Magnetooptical Effects in Group 5 Semimetal Alloys

M. J. Apps*

H. H. Wills Physics Laboratory
University of Bristol, England

Introduction

In this paper I will summarize the main results of a magnetoreflection (MR) study which has been carried out on single-crystal alloys of antimony with bismuth. Interest in the Sb-rich end of the group 5 alloy system was stimulated by previous investigations[1] of Sb itself, where peculiarities observed in the oscillatory interband MR spectra could not be fully explained in terms of existing band structure models for Sb and the crystallographically similar other group 5 semimetals, As and Bi. Tichovolsky and Mavroides[2] (TM) had carried out a MR study of Bi-rich BiSb alloys, providing significant experimental information in that complicated region of the alloy system (where a metal–insulator transition is observed). Relating this work to MR experiments in the pure elements As, Sb, and Bi, they renewed Mase's suggestion that a change in the spin–orbit coupling energy λ is the major factor responsible for band structure differences in group 5 and that alloying As (λ = 0.24 eV) with Sb (λ = 0.54 eV), or Sb with Bi (λ = 1.72 eV), causes an increase in λ in some way proportional to the alloy concentration. The TM results on concentrated bismuth alloys (corresponding to $\lambda \gtrsim 1.5$ eV) could not be expected to yield much information about Sb at λ = 0.54 eV. Indeed, the characteristics of the MR signals we had observed in Sb bore far less resemblance to those found[3] in Bi (or TM's alloys) than to those observed[4] in As. A study of the MR spectra in the Sb-rich alloys was thus expected to lead to a better understanding of the electronic structure of the pure elements as well as the alloys.

The work reported here was carried out on highly polished, damage-free single crystals of SbBi alloys of concentrations up to 12.5 at. % Bi. Samples oriented with the twofold crystallographic axis normal to the reflecting surface were mounted on a helium cold finger in a superconducting solenoid. Infrared reflectivity measurements were made using a two-beam spectrophotometer specifically designed for this work.[5]

Oscillatory Interband Magnetoreflection

The oscillations in the reflectivity characteristic of the interband MR effect arise from direct electronic transitions from an occupied valence band Landau

* Present address: Department of Physics, Simon Fraser University, Burnaby, British Columbia, Canada.

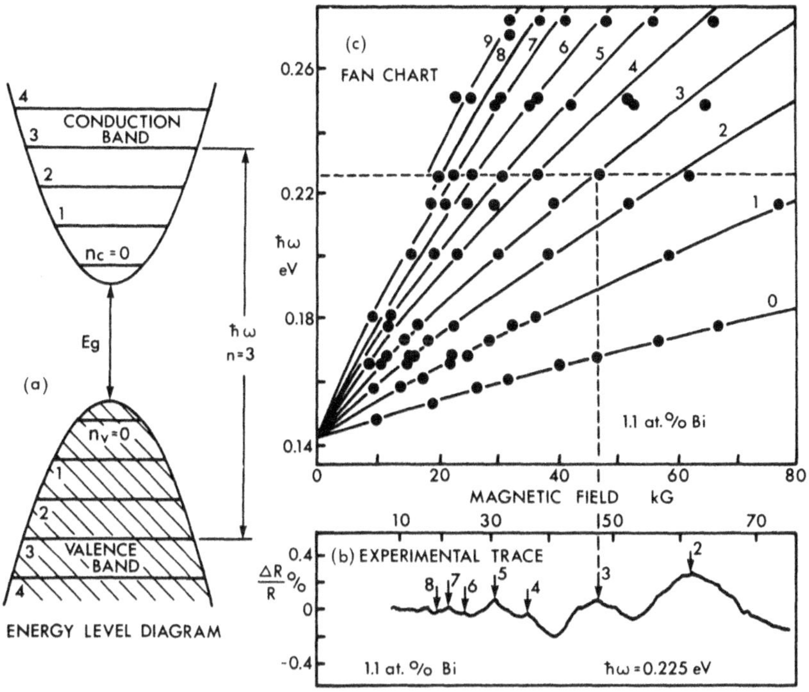

Fig 1. Interband magnetoreflection.

level to a corresponding unoccupied level in the conduction band. Such a transition, indicated in Fig. 1(a), causes a change in the optical conductivity $\sigma(\omega, H)$ and thereby in the optical reflectivity $R(\omega, H)$. Because the Landau level energy spacings are periodic in $1/H$, a series of oscillations are observed for a given photon energy. Energy conservation requires, for the simple band model shown,

$$\hbar\omega = E_g + (n_v + \tfrac{1}{2})(e\hbar/m_v c) H + (n_c + \tfrac{1}{2})(e\hbar/m_c c) H \tag{1}$$

where n_c, n_v, m_c, and m_v are, respectively, the Landau indices and cyclotron effective masses of the conduction and valence bands involved, $\hbar\omega$ is the incident photon energy, H is the magnetic field, and E_g is the energy gap separating the two bands.

It will be seen from Eq. (1) that a family of curves may be obtained by plotting the locus of the reflectivity oscillations as a function of $\hbar\omega$ and H. Figure 1(c) shows a typical fan chart obtained experimentally for a 1.1 at.% Bi alloy. By fitting the data represented by such fan charts to relations such as Eq. (1), certain basic band parameters may be obtained. In particular, the energy gap E_g is clearly the common intercept on the energy axis, while the slopes of the lines are related to the cyclotron masses of the bands involved.

Two distinct series of interband reflectivity oscillations were observed[1,6] in Sb. The most clearly observed of these is found at wavelengths less than ~ 8.3 μm in the near infrared, while the other series, considerably weaker, occurs at longer wavelengths (less than ~ 12.3 μm). Both these series have been observed in the SbBi alloys.

In presenting the alloy results, I first discuss the stronger of the two interband series. This series remains prominent throughout the entire alloy range studied, with typical maximum amplitudes of the order of $\Delta R/R \simeq 1\%$ at 60 kG. We have found that the characteristics observed for this series in pure Sb are preserved throughout the alloy range. In particular, the following observations are noted.

1. The MR oscillations exhibit the same anomalous line shapes as we observe in pure Sb. If our interpretation of the Sb signals[1] (in terms of a broken selection rule on n_v and n_c) is correct, this would indicate that the location of the interband gap remains at a point of low symmetry in the Brillouin zone.

2. The fine structure associated with the low-quantum-number reflectivity oscillations in Sb can be detected also in the alloys. However, the experimental resolution was somewhat lower for most of the alloys, so that what was commonly observed was a broadening of the low-quantum-number peaks characteristic of an unresolved triplet structure that is observed in Sb. If the explanation for the Sb results (spin splitting of the Landau levels) is adopted for the broadening of the lines in the alloys, we may conclude that the effective spin-splitting factor for these bands does not change by more than $\sim 25\%$ from the value 20 ± 2 found in Sb.

3. The curvature of the lines on the fan charts, indicating a departure from a quadratic dispersion relation, $E(k) \propto k^2$, for the bands, is well accounted for quantitatively by the Lax two-band model (a perturbation calculation in which the two bands are assumed to be interacting strongly enough with each other that the existence of other bands may be totally neglected). The solid lines shown in the fan chart of Fig. 1(c) represent a fit of the data points to the Lax model.

4. The reduced cyclotron mass of the two bands involved in the interband series appears to remain fairly constant but the energy gap between the bands shows a steady decrease with increasing Bi concentration. The decrease of the energy gap is, however, not as rapid as would be predicted by the band motion scheme forwarded by TM for the alloys if we assume that the spin–orbit coupling parameter increases linearly with Bi concentration.

The other interband series found in Sb is also observed in the alloys. Several factors make observation of this series difficult: First, the signal amplitudes are very small ($\Delta R/R < 0.2\%$) and occur at longer wavelengths, where source radiant energy is low. Second, a relatively large cyclotron mass restricts the observation of the low-quantum-number transitions to a narrow range of photon energies. Finally, the long-wavelength region where the transitions are found is also the domain of magnetoplasma phenomena; identification of the small interband oscillations is complicated by variations in background reflectivity due to magnetoplasma effects.

Nevertheless, numerically interpretable signals were observed in three of the alloys, giving band gap energies of 0.098 ± 0.002 eV for 1.1 at. % Bi, 0.095 ± 0.006 eV for 4.2 at. % Bi, and 0.070 ± 0.006 eV for 9.7 at. % Bi, indicating a steady decrease from 0.101 ± 0.001 eV obtained[1] for pure Sb. In the other alloys definite MR structure could be observed at energies below the prominent interband series gap energy but gave insufficient data to perform confident quantitative analysis.

For this interband series in Sb two sets of transitions of slightly different masses associated with the same band gap energy were distinguished. In the alloys the existence of additional structure indicates the presence of a second distinct, but unresolved, series across a common energy gap. This suggests that both the masses

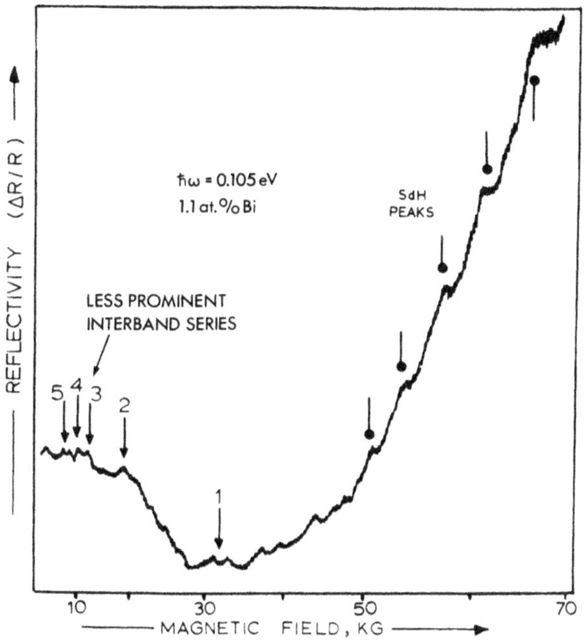

Fig. 2. Interband and intraband phenomena in an alloy specimen.

associated with this series in Sb increase with alloying and that we are unable to resolve the heavier-mass series in the alloys.

Intraband Effects

In addition to the interband effects just discussed, we have observed two magnetoplasma edges in the alloys. One edge, occurring at slightly lower photon energies than in pure Sb, was explicitly studied only in the 1.1 at. % Bi alloy but could be identified (as in Sb) with the (transport) observed electron carriers. The other edge, at rather higher photon energies, was observed in several of the alloys but quantitative interpretation proved difficult because of mixing of contributions by both interband and intraband mechanisms.

Optical Shubnikov–de Haas (SdH) oscillations,[7] arising from changes in the intraband contribution to the high-frequency conductivity as successive occupied Landau levels are swept across the Fermi surface with increasing magnetic field, were observed in Sb and, with equal clarity, in the 1.1 at.% Bi alloy. Figure 2 shows the SdH oscillations in this alloy riding on the magnetoplasma background at magnetic fields above the weaker interband series. Believed to be the first observation of the effect in any alloy, these oscillations give a dHvA period consistent with more accurate dHvA measurements by Freeman[8] on the same alloy.

The optical SdH effect has recently been studied in detail (in pure Sb) by Missel and Dresselhaus.[9] Although the mechanisms responsible for the phenomenon are more complex than was previously thought, very significant increases in signal amplitude ($\Delta R/R$ up to 100%) were obtained by cooling to lower temperatures.

If comparable S/N increases are to be found for alloy specimens, this could provide an experimental tool to give information when the related dHvA technique fails due to attenuation through high Dingle temperatures.

Acknowledgments

It is a pleasure to acknowledge the many contributions of Dr. D. A. Huntley to this work. Thanks are given also to Drs. P. Freeman, M. G. Priestley, and J. S. Shah, who helped in the growth of the alloy specimens and shared many stimulating discussions.

References

1. D.A. Huntley and M.J. Apps, *J. Phys. F: Metal Physics* **2**, L17–20 (1972).
2. E.J. Tichovolsky and J.G. Mavroides, *Sol. St. Commun.* **7**, 927 (1969).
3. M. Maltz and M.S. Dresselhaus, *Phys. Rev. B* **2**, 2877 (1970).
4. M. Maltz and M.S. Dresselhaus, *Phys. Rev.* **182**, 741 (1969).
5. M.J. Apps and D.A. Huntley, *Sci. Instr.* **5**, 1075 (1972).
6. M.S. Dresselhaus and J.G. Mavroides, *Optical Properties*, F. Abeles, ed. Amsterdam, North Holland (1966), p. 508.
7. M.S. Dresselhaus and J.G. Mavroides, *Sol. St. Commun.* **2**, 297 (1964).
8. P.V.N. Freeman, Ph.D. Thesis, University of Bristol, 1972.
9. F.P. Missel and M.S. Dresselhaus, *Phys. Rev. B* **5**, 1364 (1972).

4

Fermi Surfaces

The de Haas–van Alphen Spectrum and Fermi Surface of Iron*

David R. Baraff

Department of Physics, University of Toronto
Toronto, Ontario

Introduction

The research described in this paper represents the most extensive de Haas–van Alphen (dHvA) investigation in iron. It was undertaken to resolve some of the ambiguities of the present descriptions of iron in the ferromagnetic state. Although the itinerant nature of the electrons is well established, the band structure calculations are only in semiqualitative agreement. This is evident when a detailed comparison among several recent calculations is made.[1-5] Although the Fermi surface geometry determines the energy band structure only near the Fermi level, it imposes stringent requirements which a valid calculation must meet.[6]

This study has utilized a number of experimental techniques developed by Stark[7,8] and co-workers which have been proven effective in a number of metals, and in particular in nickel.[6,9] Those used in this study included (1) large-amplitude field modulation; (2) low-temperature cooling of the samples by either a ^3He refrigerator or a dilution refrigerator; (3) complete electromagnetic and acoustic shielding of the experimental apparatus by use of a quiet room; (4) use of a correlation analysis in conjunction with complex Fourier analysis of data recorded on magnetic tape and analyzed by computer.

The data obtained in this experiment were inverted to yield the k vectors of the Fermi surface.[10]† Guided by the size, shape, and location of the surface in the Brillouin zone, a coherent model for the Fermi surface was constructed. Where the data could not be inverted directly, a geometric model was constructed ad hoc which yields the areas to better than 1% in the (110) plane.

Model Band Structures and the Fermi Surface

One of the first major calculations was undertaken by Wood[1] using the augmented plane wave technique (APW) of Slater. He obtained a paramagnetic band structure; the calculation took no account of the exchange interaction band splitting.

* This work was supported in part by the Advanced Research Projects Agency and by the National Science Foundation.

† The computer program was written by S. Auluck and R. W. Stark and was based on the Mueller interpolation scheme and used the Kubic harmonic functions as derived by Falicov and Lee.[11]

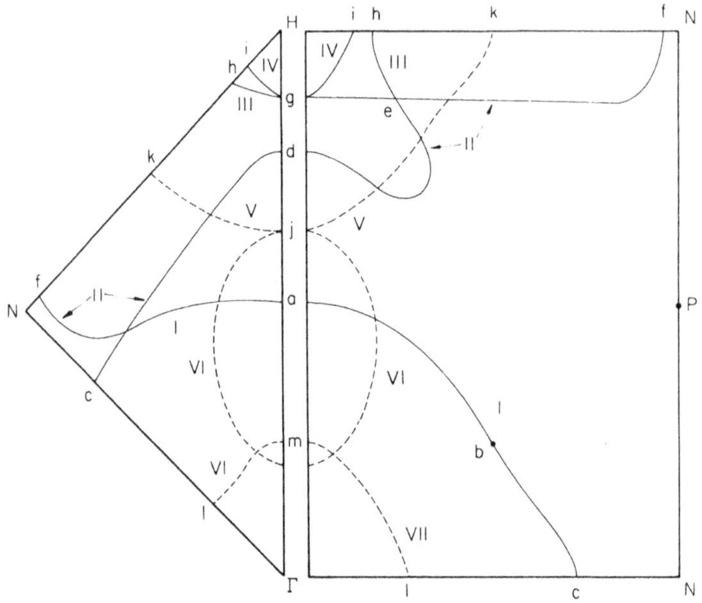

Fig. 1. Fermi surface cross sections in the (100) and (110) symmetry planes. Solid curves are majority-spin surface; dashed curves are minority-spin surface.

The ferromagnetic band structure was obtained by introducing a sufficient splitting to yield the correct magneton number.

The next major calculation was by Wakoh and Yamashita[2] (WY), who calculated a self-consistent ferromagnetic band structure which yielded a Fermi surface quite similar to that of Wood. The main difference was the slight k dependence of the bands.

Gold *et al.*[3] (GHPS) proposed a third semiempirical model for the band structure. The previous two calculations failed to consider spin–orbit effects. Gold *et al.* pointed out that the simultaneous presence of the spin–orbit effect and the exchange splitting account for some of the most interesting effects seen in the Fermi surface experiments. (See, for example, Hodges *et al.*'s[12] account of field-induced changes in the geometry of the X_5 hole pockets in nickel, or Stark's[9] account of mixed spin orbits in nickel.)

Maglic and Mueller[4] (MM) used a Wanier representation to interpolate the energy values calculated by WY. By incorporating the effects of spin–orbit splitting, they predicted a spin hybrid surface. Other pieces of the surface, where no overlap between sheets of opposite spin occur, are the same as found by WY.

The experimentally determined Fermi surface is shown in Fig. 1. The different sheets are accorded Roman numerals as follows.

I. The large spin-up (majority-spin) electron piece centered at Γ, and commonly referred to as the *s-d* piece, is found in all models. MM show that in some directions it will be coupled with the spin-down electron balls along Δ. GHPS imply that the effects of magnetic breakdown and tunneling will cause the unmixed surface to appear experimentally.

II. The large spin-up hole piece centered at H is called the hole arm piece. Both Wood and GHPS show these arms are pinched off before reaching N, whereas WY and MM show these arms continuous at N. GHSP also show these arms as degenerate with the *s-d* piece in the (100) plane, but WY and MM do not.

III. The intermediate spin-up hole pocket at H has some important degeneracies with the hole arm piece in the (111) direction; it may be degenerate in the (100) directions with the small spin-up hole pocket.

IV. The small spin-up hole pocket at H has dimensions which are quite sensitive to the exact location of the Fermi level. It is most important in determining the H_{25} energy.

V. The spin-down (minority-spin) hole octahedron is found in all models. MM predict a strong hybridization with the hole arm piece. We refer to this new topological form as the hybrid octahedron.

VI. The spin-down central electron sheet is found in all models with only minor variation. The main question lies in its relation to the electron balls located along the (100) symmetry line Δ. Only in Wood's model is there no overlap.

VII. The spin-down electron balls along Δ are predicted by all models. Other questions about these sheets are whether they contact the hole octahedron allowing open orbits in the (100) direction and whether or not a hybridization occurs between them and the opposite spin *s-d* piece.

VIII. The spin-down hole ellipsoids are predicted by all models except that of GHPS and Duff and Das,[5] who find them to be absent due to a reordering of the energy levels near N. These pieces are not shown in Fig. 1. In the other models they are centered at N and are similar to the hole ellipsoids observed in tungsten.

Results and Discussions

The body-centered cubic lattice of iron has high symmetry and data were taken with the applied field in the (100) and (110) planes only. Figure 2 shows the frequencies and areas observed in the (110) plane. (The extremal area in atomic units equals $2.673 \times 10^{-9}F$; F is the dHvA frequency in gauss.) The horizontal axis is the angular distance in degrees from the (100) direction. The crystal plane was oriented to better than $\frac{1}{2}°$. The branches are labeled in Greek letters, following GHPS where applicable.

The lowest two branches we observed have not been placed on the diagram since they were extremely well described by GHPS. Briefly, there are two sets of frequencies. The first set is attributed to an electron pocket along Δ; it is formed by the intersection of the electron balls with the central electron piece. The second frequency is the neck formed in the same way.

The branch labeled β is attributed to the central electron piece. GHPS were able to observe it only near (111); however, we found it continuous in the (110) plane. This does not contradict the assignment of the previous two branches. The amplitude of the β branch is quite diminished near the axes. It is probable that the orbit is completed by tunneling.

The γ branch is thought to be the orbit around the electron balls. This branch cuts off directly 9° from the axis. It is thought that the orbit intersects the *s-d* piece.

The ζ branch may also be attributed to the electron balls; however, this assignment is quite tentative.

The λ branch is quite easily identified as the small spin-up hole pocket at H, and the δ branch is the intermediate hole pocket at H. Both of these branches were followed over the entire plane. GHPS felt that the δ branch was the electron ball orbit in the (100) direction, but this assignment is incorrect. Also, the second harmonics of these two branches were followed. This was important for the λ branch, which overlaps the third harmonics of the lens in the (110) direction and the neck in the (100) direction.

The σ branch also is quite certainly identified; it is the orbit around the *s-d* piece. Although it is too small to be shown on the diagram, there is a splitting of about 1% near the (110) axis. This is due to a noncentral extremal which joins central orbit 6° from the axis in the (100) plane and causes the unusually large amplitude for this signal.

The α branch was observed by GHPS near (111). They concluded it was the central orbit on the hole octahedron. Actually, the topology observed is the hybrid piece whose geometry is similar to that of the unhybridized piece in the central region. Figure 3 shows a detailed fit of the observed frequencies with the frequencies calculated from a hybrid model. The agreement is excellent.

The dot labeled μ is a singular frequency. It is not observed $\frac{1}{2}°$ away from the (211) direction. It is attributed to a tunnel orbit around half of the intermediate spin-up hole pocket and half of the hybrid octahedron. It can only exist in the (211) direction because it is only in that direction that the tunnel orbit is also an extremal. The quantitative agreement is better than 1%.

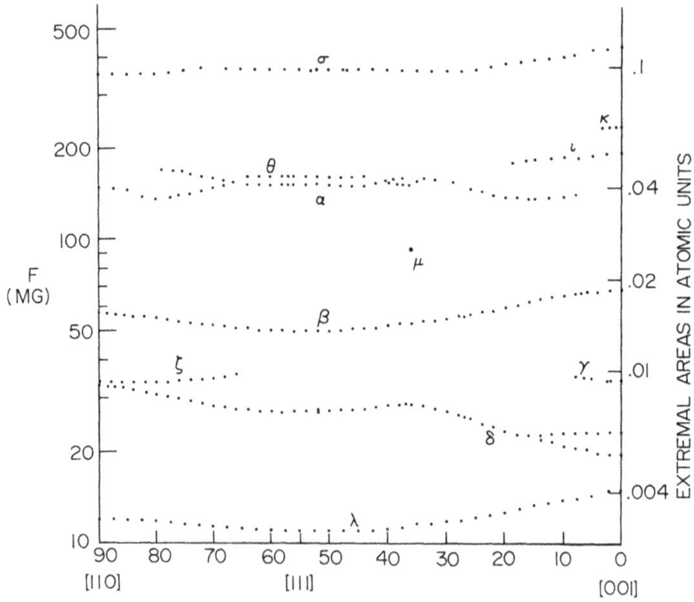

Fig. 2. The de Haas–van Alphen spectrum of extremal Fermi surface areas. Two ordinate scales are provided for convenience to both experimentalists and theoreticians.

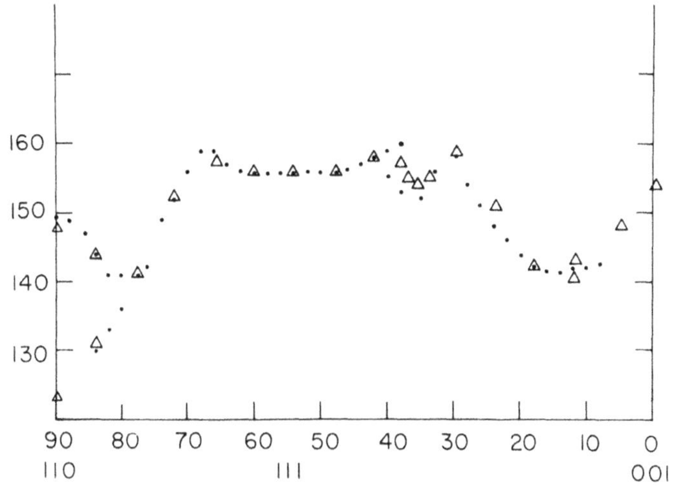

Fig. 3. The de Haas–van Alphen spectrum of the α-branch shown in enlarged detail. The dots are the experimental frequencies determined to an absolute accuracy of 1 %. The triangles are points calculated from the hybrid model.

The branches labeled θ and κ are thought to be extremals on the hybrid hole arm piece. The θ branch is a noncentral minimum in the (111) direction and the κ branch is a noncentral maximum in the (100) direction. This would indicate that the hole arms are pinched off as suggested by GHPS.

Conclusion

The experimental Fermi surface appears to be in semiquantitative agreement with that predicted by band structure calculations. The study has been complete enough to allow the unique conversion of area branches into k vectors for all of the nonhybridized pieces. This gives us confidence that the band structure calculations can be used as a basis for generating the hybrid pieces which account for the rest of the experimental data.

Acknowledgment

This work was carried out while the author was a graduate student at the University of Chicago working under the guidance of Prof. Royal W. Stark. Without the techniques he developed and without his leadership and guidance, this project could not have been done.

References

1. J.H. Wood, *Phys. Rev.* **126**, 517 (1962).
2. S. Wakoh and J. Yamashita, *J. Phys. Soc. Japan* **21**, 1712 (1966).
3. A.V. Gold, L. Hodges, P.T. Panousis, and D.R. Stone, *Intern. J. Magnetism* **2**, 357 (1971).
4. R. Maglic and F.M. Mueller, *Intern. J. Magnetism* **1**, 289 (1971).
5. K.J. Duff and T.P. Das, *Phys. Rev. B* **1**, 192 (1971).

6. D. Tsui, *Phys. Rev.* **164**, 669 (1967).
7. R.W. Stark, *Phys. Rev.* **162**, 589 (1967).
8. R.W. Stark and L.R. Windmiller, *Cryogenics* **8**, 272 (1968).
9. R.W. Stark, *Phys. Rev.* (to be published).
10. F.M. Mueller and M. Priestly, *Phys. Rev.* **148**, 638 (1966).
11. L. Falicov and M.J.G. Lee, *Proc. Roy. Soc.* **A304**, 319 (1968).
12. L. Hodges, D.R. Stone, and A.V. Gold, *Phys. Rev. Lett.* **19**, 655 (1967).

Some New Results on de Haas–van Alphen Effect in Cobalt

F. Batallan and I. Rosenman

Groupe de Physique des Solides de l'Ecole Normale Supérieure
Laboratoire associé au CNRS, Paris, France

In a recent paper[1] we reported low-frequency dHvA oscillations in hcp ferromagnetic cobalt in the range of $1-5 \times 10^6$ G. We observed two sets of dHvA frequencies: α, which was tentatively attributed to a pocket on the *LM* line in the spin-down FS, in agreement with Wakoh and Yamashita's (WY)[2] band structure calculation, and β, attributed to a neck at the Γ point in the spin-up FS, which could be found after shifting the spin-up level in the WY model. The existence of a Γ neck was confirmed by our high-field magnetoresistence studies,[3] which show the presence of a volcano-like structure in the angular variation.

Here we present some new results of our investigation of cobalt by the dHvA effect, mainly the set of frequencies ε in the $10-20 \times 10^6$ G range. An independent study of some of these frequencies has recently been reported by Anderson *et al.*[4]

Angular Variation

Figure 1 shows the angular variation of the ε frequencies. In the (0001) plane ε has three branches, with frequencies in the range of $9-17 \times 10^6$ G, which degenerate into two along the $\langle 10\bar{1}0\rangle$ and $\langle 11\bar{2}0\rangle$ high-symmetry directions (Table I). The lower, ε_3^1, is the dominant frequency, and the other two are obtained easily from the Fourier spectrum (Fig. 2). In the $(10\bar{1}0)$ plane we observe two branches. The lower, ε_1^1, has been seen from 90° to 39° to the *c* axis, at which angle it is superimposed on the β dHvA oscillation.[1] The higher, ε_1^2, is found in the Fourier spectrum from 90° to 78° to the *c* axis. Another set of frequencies, of about 14×10^6 G, lies between ε_1^1 and ε_1^2.

Interpretation

The ε set belongs to the *C* class in the Watts[5] classification. Since ε has three branches in the (0001) plane, it corresponds to *closed* surfaces in the Brillouin zone (BZ), implying that two branches should exist for any orientation in the $(10\bar{1}0)$ plane, except for the $\langle 0001\rangle$ direction, where they are degenerate. The least-squares fitting of the angular variation corresponds to six ellipsoids with a minimum cross section in the $\langle 1010\rangle$ direction; therefore the ε ellipsoids are located on the *LM* line in the BZ.

When referred to the FS model of WY, ε can only be found in the spin-down

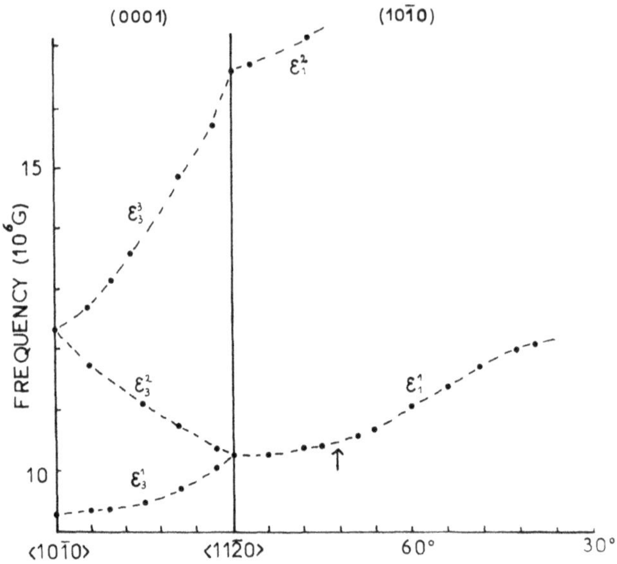

Fig. 1. Angular variation of the ε dHvA frequencies in the (0001) and (10$\bar{1}$0) planes. The arrow indicates the orientation corresponding to the anomalous disappearance of the dHvA effect.

FS, where the legs of the monster have a symmetry compatible with the ε frequency, but with a cross section two times too large. This is not surprising since in the WY calculation the position of the *s-p* bands relative to the *d* bands is too high.[1]

A clear interpretation of the dHvA results, especially of those related to the spin-down FS, necessitates a new band calculation. This computation is presently in progress.[6]

Anomaly in the Angular Amplitude Variation

When the magnetic field B is exactly in the (10$\bar{1}$0) plane, then the ε_1^1 dHvA amplitude presents an anomaly in the angular variation: Near 72° from the c axis the dHvA signal disappears and reappears within a range of 1°.

Tsui[7] has observed such an anomaly in the dHvA effect in nickel. Ruvalds and Falicov[8] have explained this as due to magnetic breakdown between parts of the FS with opposite spins. This occurs when two bands with opposite spins are

Table I. ε Frequency in Symmetry Directions

Nomenclature	Orientation	Frequency, 10^6 G
ε_3^1	$\langle 10\bar{1}0\rangle$	9.3
$\varepsilon_3^1, \varepsilon_3^2$	$\langle 11\bar{2}0\rangle$	10.3
$\varepsilon_3^2, \varepsilon_3^3$	$\langle 10\bar{1}0\rangle$	12.4
ε_3^3	$\langle 11\bar{2}0\rangle$	16.6

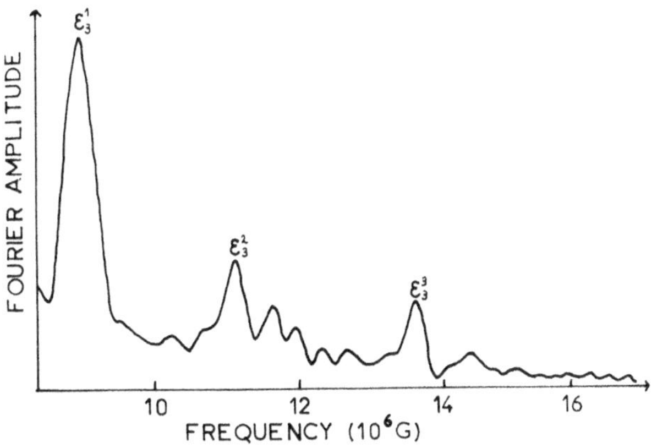

Fig. 2. Typical Fourier spectrum showing the presence of three dHvA frequencies in the (0001) plane.

mixed by a spin–orbit interaction, giving a band with an undefined spin. The dHvA frequency relative for such a point in the BZ disappears.

It is possible that a similar phenomenon occurs in cobalt, but this has to be confirmed by a band calculation.[6]

The study of this anomaly is in progress.

References

1. I. Rosenman and F. Batallan, *Phys. Rev. B* **5**, 1340 (1972).
2. S. Wakoh and J. Yamashita, *J. Phys. Soc. Japan* **28**, 1151 (1970).
3. F. Batallan and I. Rosenman, *Intern. Symp. Electronic Structure and Properties of Transition Metals, Kiev, May 1972.*
4. J.R. Anderson, J.J. Hudak, and R.D. Stone, *Proc. Magnetism and Magnetic Materials Conf.*, Am. Inst. of Phys., (1972), p. 477.
5. B.R. Watts, *Proc. Roy. Soc. (London)* **A282,** 521 (1963).
6. F. Batallan, I. Rosenman, and C. Sommers (to be published).
7. D.C. Tsui, *Phys. Rev.* **164**, 669 (1967).
8. J. Ruvalds and L.M. Falicov, *Phys. Rev.* **172**, 508 (1968).

The de Haas–van Alphen Effect in YZn, an Alloy of the Beta-Brass Type

J.-P. Jan

National Research Council of Canada
Ottawa, Canada

The Fermi surface of various ordered alloys of the beta-brass type at the equiatomic composition has been the subject of several recent studies, both theoretical and experimental (for bibliography see Ref. 1). So far the alloys studied have been of two types, containing either a monovalent and a divalent atom (e.g., β'-CuZn), or a transition metal and a trivalent element (e.g., β'-PdIn). In both cases the Fermi surface appears to be a reasonable distortion of the nearly-free-electron model based on three electrons per unit cell, although several difficulties still remain concerning the interpretation of experimental results in terms of theoretical band structures.

We report here on the observation of the de Haas–van Alphen (dHvA) effect in β'-YZn, an ordered alloy with five conduction electrons per unit cell. A single crystal with a residual resistance ratio of 60, the preparation of which has been described by Schiltz *et al.*,[2] was kindly supplied by Dr. J. F. Smith of the Ames Laboratory, USAEC, Iowa State University. The dHvA effect was observed by the modulation method, in fields up to 100 kG provided by a superconducting solenoid. An X-ray study in liquid helium did not reveal any phase transformation on cooling, and allowed the determination of the lattice parameter: $a = 3.558$ Å at 4.2°K.

The dHvA signals were weak and difficult to observe, in spite of the residual resistance ratio of 60, which is reasonably large for an alloy. This is not surprising, since the absolute resistivity at room temperature is relatively large, approximately 30 $\mu\Omega$-cm. Measurements were made in the $\{100\}$ and $\{110\}$ planes. A low-frequency $\alpha = (1.746 \pm 0.002) \times 10^6$ G was observed in all directions in the $\{110\}$ plane. This frequency is constant within experimental uncertainty, and most likely corresponds to a small spherical sheet of Fermi surface containing 5.88×10^{-4} charge carriers per unit cell. The corresponding cyclotron effective mass (measured in the $\langle 100 \rangle$ direction) is 0.237 ± 0.001 in units of the free-electron mass. A very weak branch β of $7\text{–}8 \times 10^6$ G is seen in most directions, in the form of a few oscillations at the highest field, but it has not been observed at lower fields, where a larger number of oscillations would have allowed a more accurate determination of the frequency.

Several branches labeled γ are also observed, and are shown in Fig. 1. The two branches near $\langle 110 \rangle$ in the $\{110\}$ plane probably arise from the splitting of a doubly degenerate branch, caused by a small misorientation of the sample. This is consistent with the observations in the $\{100\}$ plane: Near $\langle 110 \rangle$ the signals are exceedingly weak, and beats are observed, which are quite sensitive to a small tilt in orientation

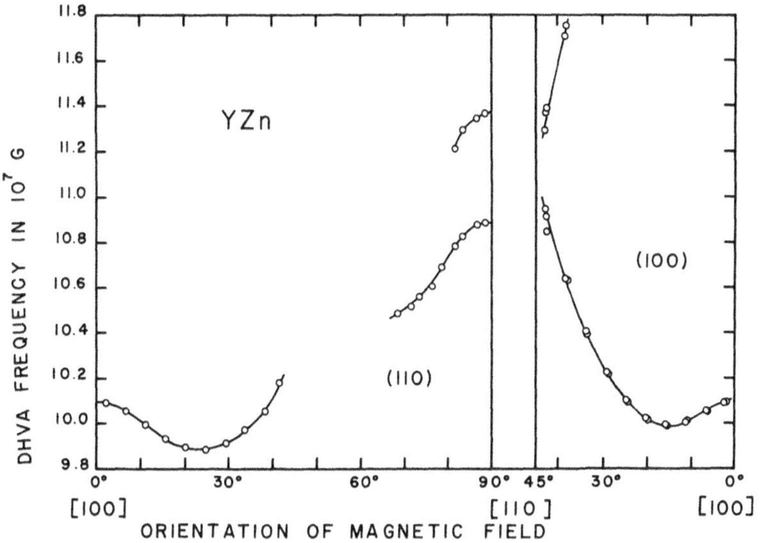

Fig. 1. The γ oscillations in YZn.

away from the {100} plane. The branches near ⟨100⟩ are of a larger amplitude, remarkably reproducible from sample to sample, and do not show any evidence of beats.

The symmetry of the α branch would assign the spherical sheet to points Γ or R in the Brillouin zone (center and corner of the cube, respectively), and the behavior of the γ branches is consistent with an open surface whose closed orbits have the symmetry of M or X (middle of the cube edge and middle of the face, respectively), whereas nothing definite can be said about the β branch. We shall examine these results in terms of two very different theoretical models of the Fermi surface, the nearly-free-electron (NFE) model and the result of an APW band structure calculation.

The NFE model corresponding to five electrons per unit cell predicts the following: In the first band there are very small hole pockets at the corners R of the Brillouin zone; in the second band there are slightly larger hole pockets at R and a hole closed sheet with the general shape of a cube at Γ, the center of the zone; the third band is multiply connected, made of intersecting electron "butterflies" centered at M, the middle of the cube edge (cf. Fig. 2, upper part); the fourth band is made of independent electron "cigars" running along the cube edges and centered at M.

There is another model, based on the APW band structure calculation of Belakhovsky *et al.*[3] Their APW bands, shown in Fig. 3, are very different from free-electron bands. This is not surprising, since yttrium has the neutral atomic configuration $4d^1 5s^2$. In this model the first band is full, and the second band is almost full, with small hole pockets at M. The third band is an open sheet, sketched in Fig. 2, and completely different from the corresponding NFE band, and the fourth band is empty.

The α branch could be explained by the hole pockets at R in the first or second band of the NFE model; their size has the right order of magnitude, although small

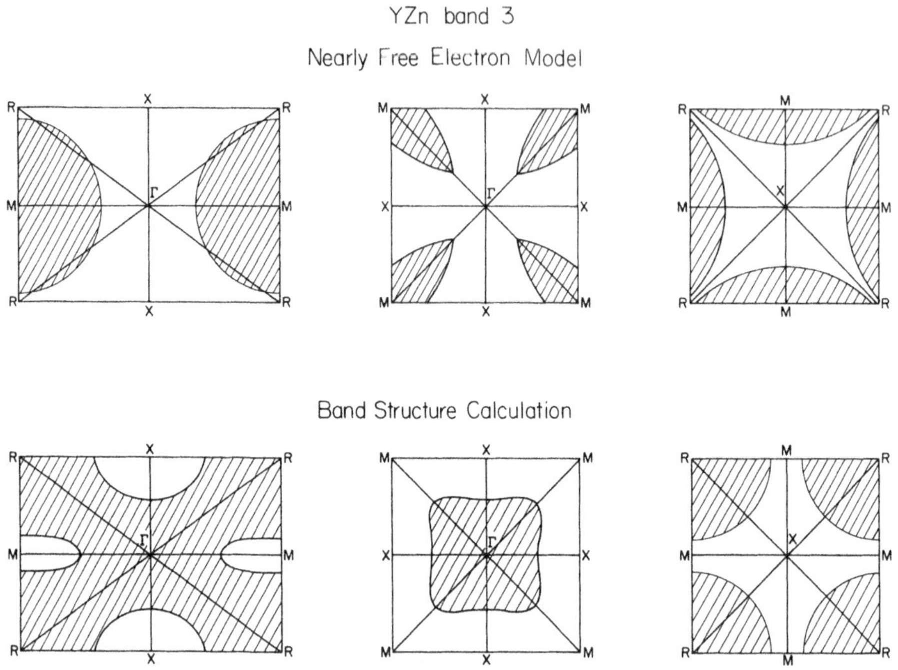

Fig. 2. Theoretical third band of YZn; comparison between the NFE model and a model based on the band structure calculation of Ref. 3.

NFE pockets generally tend to be wiped out by the lattice potential. The APW model does not predict any spherical pocket at R or Γ; it is very unlikely that the small hole pocket at M in the second band could be exactly spherical, because M does not exhibit cubic symmetry. However, a small lowering of point $R_{1(2')}$ in the APW bands, a minor modification, could give rise to a small pocket of cubic symmetry, which would be spherical in the first approximation.

The γ branches come quite close to the predictions of the NFE model in the third band. They have approximately the correct angular variation, order of magnitude, and degeneracy. Their interpretation in terms of the APW model appears to be much more difficult. The only point of agreement is in the $\langle 100 \rangle$ direction, where the APW model predicts a frequency of about 11×10^7 G, corresponding to an orbit centered at Γ (Fig. 2). The APW model, however, predicts a rapid increase of this frequency away from $\langle 110 \rangle$, whereas the observed branch decreases slightly at first and does not vary much. In addition, the APW model does not seem to predict any doubly degenerate orbit in the $\langle 110 \rangle$ direction. The elusive β branch might correspond to holes at Γ in the second band of the NFE model, and does not correspond to anything in the APW model, but it is really not sufficiently known to allow any definite interpretation at this stage.

In conclusion, the NFE model can account for the experimental results better than the APW model. Indeed, if no APW model had been available, a reasonable distortion of the NFE model would have seemed adequate. This result is quite surprising indeed, since the presence of yttrium would certainly justify the existence

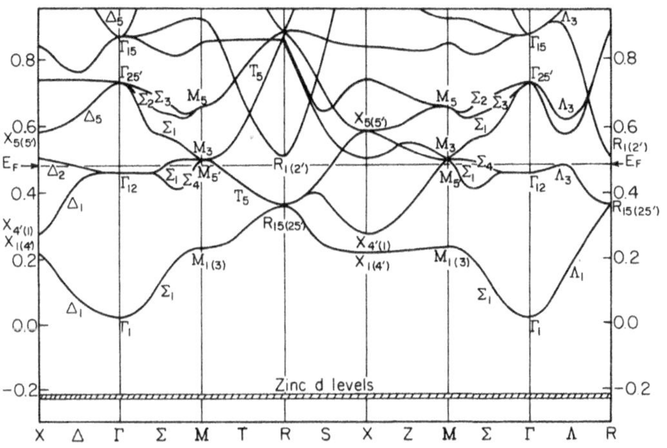

Fig. 3. Energy bands of β'-YZn, after Belakhovsky *et al.*[3]

of a Fermi surface quite different from the NFE surface. One could speculate that charge transfer might play a significant role to alter the APW bands (the calculation of Ref. 3 was not made self-consistent in this respect) or even that the $4d^1 5s^2$ configuration for Y is not a good starting point for the APW calculation.

Acknowledgments

The author is grateful to Dr. J. F. Smith of the Ames Laboratory, USAEC, Iowa State University, for supplying the samples, to Dr. J. Pierre for having provided the manuscript of Ref. 3 before its publication, and to Miss N. L. Martin for her assistance with the determination of the lattice parameter and the calculation and construction of Fermi surface models.

References

1. J.-P. Jan and C.M. Perrott, *J. Low Temp. Phys.* **8**, (3/4), 195 (1972).
2. R.J. Schiltz, Jr., T.S. Prevender, and J.F. Smith, *J. Appl. Phys.* **42**, 4680 (1971).
3. M. Belakhovsky, J. Pierre, and D.K. Ray, *Phys. Rev. B* **6**, 939 (1972).

Note Added in Proof: For more complete accounts, see J.-P. Jan, *Phys. Rev. B* **8**, 3590 (1973) and J.-P. Jan, N.L. Martin, and A. Wenger, *Phys. Rev. B* (15 Feb. 1974).

De Haas–van Alphen Effect in Technetium*

A. J. Arko, G. W. Crabtree, S. P. Hörnfeldt, J. B. Ketterson, G. Kostorz, and L. R. Windmiller

Argonne National Laboratory
Argonne, Illinois

We report the first observation of de Haas–van Alphen (dHvA) oscillations in single crystals of technetium, a radioactive metal whose longest-lived isotope (^{99}Tc, obtained from nuclear fuel reprocessing) decays by soft β^- emission with a half-life of 2.1×10^5 yr. Gram quantities of ^{99}Tc were obtained from ORNL and purified by electron-transport and electron-beam zone-refining.[1] Carbon was the major impurity. Single crystals in the form of 1/8-in. rods were obtained in the zone-refining process with a residual resistance ratio of ~ 100. The dHvA specimens were cut from these rods by spark erosion (with final dimensions $\sim 0.5 \times 0.5 \times 2$ mm) and electropolished until all damage was removed. For handling purposes and to avoid contamination of equipment, the specimens were coated with varnish. Measurements were made at temperatures down to 0.3°K and fields up to 60 kG, utilizing the usual field-modulation technique. Data were obtained from both field sweeps and field rotations. Because of the low resistance ratio of the specimens, the data are probably not complete even in the two planes reported.

Figure 1 shows the Brillouin zone for the hcp structure. A complete band structure calculation for Tc is not available at this writing; however, calculations by D. Koelling of this laboratory are underway. The overall band structure is similar to that of the element Re, which is just below Tc in the periodic table. The dHvA data for Re are reasonably complete.[2,3] The Fermi surface of Re as drawn by Mattheiss,[4] is shown in Fig. 2. Note that Re has three hole surfaces located at the L point (h_5, h_6, and h_7). The Tc calculations indicate that these same three surfaces should also be present in Tc. The topology of e_8 is such that it probably does not extend to the ΓMK plane. The structure of the levels near Γ and K is somewhat different between Re and Tc and thus the nature of the surface in the vicinity of these points is quite uncertain.

The frequencies of oscillations (in atomic units) vs. the field direction in the $(\bar{1}100)$ and (0001) planes are shown in Fig. 3. The different families of curves are labeled by letters, and the subscripts denote different branches of a family. Effective masses for a number of these frequencies and several directions of H are given in Table I. The accuracy of the effective masses is estimated to be $\pm 2\%$.

A plot of P^2 (where P is the reciprocal of the dHvA frequency) vs. $\cos^2\theta$ of the a_i branch yields a straight line in both the $(\bar{1}100)$ and the basal planes (i.e., ellipsoidal in shape). In spite of the fact that only a single branch is observed (for only a limited

* Work performed under the auspices of the U.S. Atomic Energy Commission.

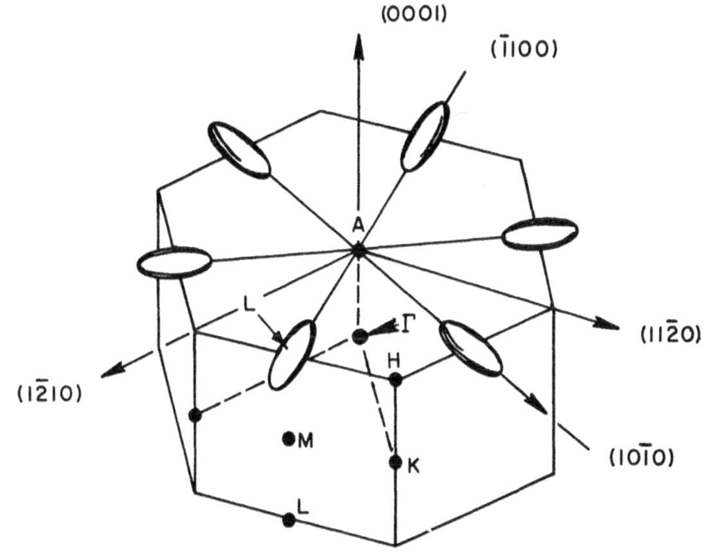

Fig. 1. The hcp Brillouin zone showing the most probable locations of the ellipsoids.

range of angles) in the basal plane, it seems likely that these branches arise from a set of closed surfaces. From the symmetry of the basal plane data these surfaces are in class C or D in the notation of Watts.[5] The possibilities for class C include the point L. Since the h_5 surface of Re has roughly the same size and anisotropy as that observed for the a_i branch, we tentatively make this identification. (These ellipsoids are shown in Fig. 1.) The ratio of the semimajor to semiminor axes of these ellipsoids is 5:1; the corresponding ratio in Re is 3:1. Along [10$\bar{1}$0], for example, the minimum

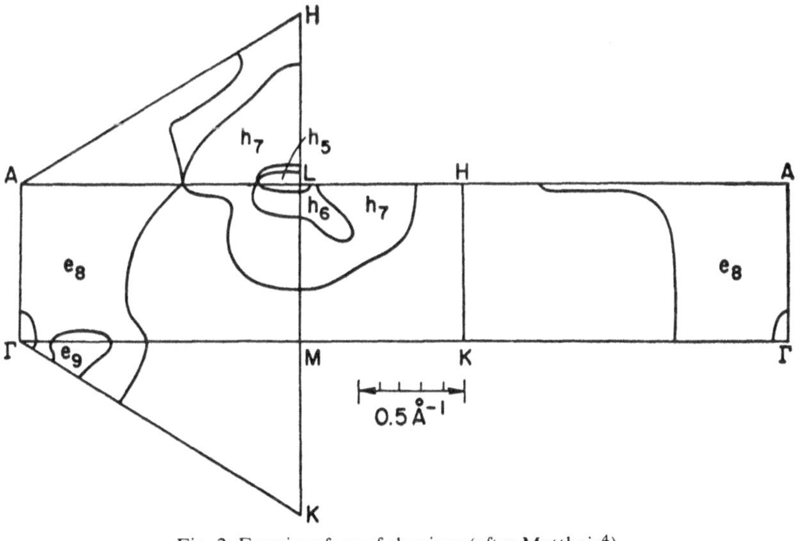

Fig. 2. Fermi surface of rhenium (after Mattheis[4]).

Table I. Effective Masses at Selected Points*

Frequency	Plane	Angle	m^*/m_0
a_1	(0001)	0	0.23
		15	0.21
		30	0.20
a_2	(0001)	15	0.26
b	(0001)	0	0.63
		7	0.61
		27	0.65
c	(0001)	0	0.80
		20	0.71
		30	0.68
	$(\bar{1}100)$	87	0.71
		90	0.78
d_1	(0001)	0	0.81
		30	0.78
d_2	(0001)	15	0.81
		30	0.83
d_1	$(\bar{1}100)$	21	1.07
e	(0001)	0	0.77
f_1	(0001)	0	0.90
		15	0.97
		30	1.11
f_2	(0001)	0	1.00
g_1	$(\bar{1}100)$	58	1.03
		64	1.00
		67	0.93
		90	0.96
g_2	$(\bar{1}100)$	58	1.09
		73	1.01
l	$(\bar{1}100)$	36	1.65
m_1	$(\bar{1}100)$	18	1.71
m_1	(0001)	0	1.40
		12	1.40
		18	1.35
		30	1.39
m_2	(0001)	12	1.64
		18	1.67
k	$(\bar{1}100)$	56	2.40

* All angles are measured from the $[11\bar{2}0]$ axis. Accuracy is $\pm 2\%$. Here m^*/m_0 is the ratio between the measured effective mass and the mass of a free electron.

frequency values are 0.00234 a.u. for Tc and 0.00206 a.u. for Re; the maximum values along $[11\bar{2}0]$ are 0.0120 a.u. (extrapolated) for Tc and 0.0067 a.u. for Re.

Branch a_3 is not observed near $[11\bar{2}0]$. A similar loss of signal for the corresponding frequency in Re was interpreted as resulting from magnetic breakdown between the ellipsoids h_5 and a set of dumbbell-shaped surfaces h_6 centered at L which by group theory are degenerate[5] along the AL line. It is also possible that the loss of signal is merely a consequence of the magnetization vector for the ellipsoids being unfavorable with respect to the pickup coil. Nevertheless, the possibility of breakdown will be investigated further.

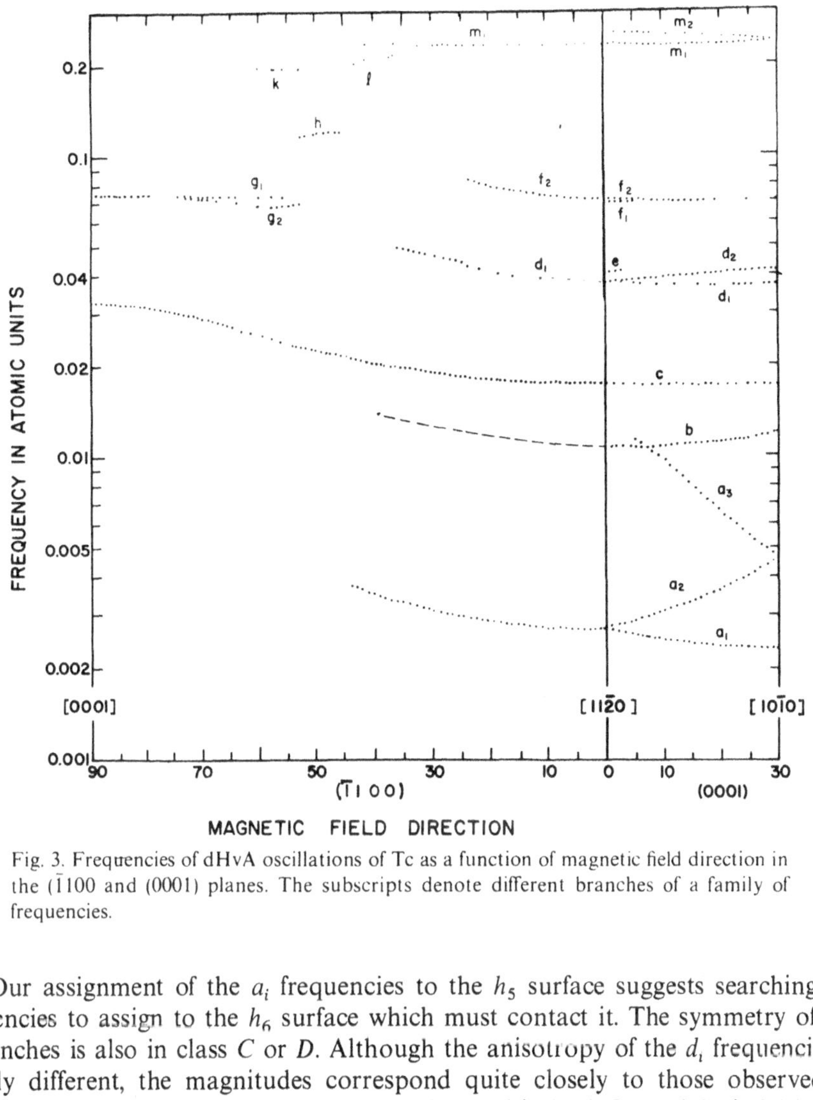

Fig. 3. Frequencies of dHvA oscillations of Tc as a function of magnetic field direction in the ($\bar{1}$100 and (0001) planes. The subscripts denote different branches of a family of frequencies.

Our assignment of the a_i frequencies to the h_5 surface suggests searching for frequencies to assign to the h_6 surface which must contact it. The symmetry of the d_i branches is also in class C or D. Although the anisotropy of the d_i frequencies is slightly different, the magnitudes correspond quite closely to those observed in Re for the h_6 surface. The mean frequency observed in both Re and Tc is 0.041 a.u. The e branch could correspond to the off-central dumbbell orbit observed in Ref. 2.

The m frequencies are also consistent with a location at the L point; their magnitude suggests that they may be associated with the remaining L-centered h_7 surface. The mean frequency is 0.22 a.u. vs. 0.205 for the corresponding Re surface.

The c frequency is the only branch which was observed for all field angles. It has the shape of a lens which is nearly isotropic in the basal plane. The anisotropy of this surface falls in class A; it must be located on the $A\Gamma$ line, or at H or K. The preliminary Tc band structure calculations do not indicate the existence of such a surface near these symmetry locations. However, as stated, the level structure near Γ and K differs considerably from Re and may thus be quite sensitive to small changes in the assumed crystal potential. The b frequency appears also to be lens-shaped, however, with more anisotropy in the basal plane. Similar arguments apply to its location and sensitivity to crystal potential.

The remaining frequencies are observed over too limited a range of angles to facilitate identification at this time. Many of the remaining questions should be answered when more complete band structure calculations (with a density-of-states histogram) become available. We also hope to extend these measurements to higher fields in the near future.

Acknowledgment

We would like to thank C. H. Sowers for technical assistance.

References

1. G. Kostorz and S. Mihailovich, in *Proc. 12th Intern. Conf. Low Temp. Phys., 1970,* Academic Press of Japan, Tokyo (1971), p. 341.
2. A.S. Joseph and A.C. Thorsen, *Phys. Rev.* **133**, A1546 (1964).
3. A.C. Thorsen, A.S. Joseph, and L.E. Valby, *Phys. Rev.* **150**, 523 (1966).
4. L.F. Mattheiss, *Phys. Rev.* **151**, 450 (1966).
5. B.R. Watts, *Proc. Roy. Soc.* **A282,** 547 (1964).

From Fermi Surface Data to One-Electron Potentials

J. F. Janak

IBM Thomas J. Watson Research Center
Yorktown Heights, New York

Introduction

To find the theoretical Fermi surface of a solid, one starts with a one-electron crystal potential which is used to find the energy bands. (In the APW and KKR methods of energyband calculation the potential can be represented by energy- and angular-momentum-dependent scattering phase shifts.) In self-consistent band calculations the charge density is also found from the wave functions and used as part of the input for constructing the crystal potential. Once the energy bands are known, integrations over k space are required to obtain values for de Haas–van Alphen (dHvA) areas and cyclotron masses for orbits on the Fermi surface.

In this paper we will first compare various methods for performing the k-space integrations, an important link in the computational chain going from the potential to the Fermi surface properties. Our real interest, however, is the much larger issue of what can be inferred from experimental Fermi surface information about the crystal potential. The part of this potential which is least understood is that due to exchange and correlation effects. Although no general and exact prescription is known for these effects, there are at least two practicable prescriptions within the local approximation that appear to provide reasonable estimates of these effects and a good starting place for more exact and elaborate prescriptions. One prescription is due to Slater,[1] and the other follows from theorems due to Kohn and Sham,[2] for finding the exchange-correlation potential from the electronic charge density with no recourse to experimental data. Rather than finding a set of scattering phase shifts which lead to agreement with experimental dHvA areas, as has been done to remarkably high accuracy for copper,[3,4] we want to test the ab initio prescriptions for the exchange-correlation potential by comparing them to experimental Fermi surface data, and see if we can infer how the approximations used in obtaining these prescriptions have to be improved.

Fermi-Surface Calculations

Existing methods for computing Fermi surface properties include the following.

1. Methods conveniently called call ray methods,[4-6] which construct de Haas–van Alphen areas by numerically integrating

$$A = \frac{1}{2}\int_0^{2\pi} [k(E, \psi)]^2 \, d\psi \tag{1}$$

and then construct cyclotron masses $(\partial A/\partial E)$ by numerical differentiation. The values of $k(E, \psi)$ required to perform the integral for a particular Fermi surface cross section are obtained by solving the energy band problem for a fixed ψ (that is, along some ray in reciprocal space) for a few energies near the Fermi energy.

2. Methods based on the algebraic expressions given by Gilat and Raubenheimer for the area of that portion of any arbitrary plane within a cube or other polyhedron. These methods[7] for constructing de Haas–van Alphen areas proceed as follows: One divides reciprocal space into a large number of polyhedra and adds up the areas of the portions of the field plane in the appropriate polyhedra inside the Fermi surface. The masses can then be obtained by numerical differentiation.

3. A third method based on an algorithm for performing line integrals.[8] If after dividing reciprocal space into a large number of cubes one assumes that the energy surface in each cube is a plane, the portion of an orbit in any cube becomes a straight line, and integrals over the orbit become sums of integrals over straight-line segments, each of which can be performed algebraically.

These three methods are to some extent complementary: The ray methods can be carried to very high accuracy (for example, calculations of de Haas–van Alphen areas to five significant figures have been made[3,4] for copper), but the positions of the extremal orbits have to be known ahead of time. (For simple Fermi surfaces, these positions are determined by symmetry.) The line integral method will work on an arbitrary Fermi surface, and automatically finds all the extremal cross sections by finding the zeros of $\partial A/\partial k$, which is also a line integral. However, the line integral method requires knowing the gradients of the energy bands, and present attainable accuracy is of the order of 1%. The Gilat–Raubenheimer method can also be automated, but suffers from difficulties in those polyhedra which are cut by both the Fermi surface and the field plane, where one wants the area of only part of the field plane.

Automated Fermi surface calculations are highly desirable in investigating new materials with complicated Fermi surfaces, but the high accuracy inherent in the ray methods is also a desirable property. No doubt, the ray method can be made more general, and the other methods can be made more accurate, but the simplest expedient for obtaining very high-accuracy results on a new material is probably to use the automated methods to locate the extremal cross sections, and then use the ray method to compute the areas and masses.

One-Electron Potentials

Kohn and Sham[2] have shown that the electron–electron interaction in an electron system can be handled exactly by solving an effective one-particle problem of the form

$$[-\tfrac{1}{2}\nabla^2 + V(r) + \mu_{xc}(r)]\,\psi_i(r) = \varepsilon_i\psi_i(r) \tag{2}$$

In this equation $V(r)$ is the Coulomb (Hartree) potential [the solution of Poisson's equation given the nuclear positions and the electron charge density $n(r)$], and $\mu_{xc}(r)$ is the exchange-correlation potential, which is a function of $n(r)$. The charge density $n(r)$ is obtained from

$$n(\mathbf{r}) = \sum_{\varepsilon_i \leqq \mu} |\psi_i(r)|^2 \tag{3}$$

where μ is the Fermi energy, and the entire problem is solved self-consistently. The difficulty in applying these formulas is that the exact functional dependence of μ_{xc} on $n(\mathbf{r})$ is not known, so that the whole statement is an existence theorem. Furthermore, there is no guarantee that the surface $\varepsilon_i = \mu$ appearing in Eq. (3) has anything to do with the real Fermi surface.

Sham and Kohn[9] have investigated the form of μ_{xc} when the density $n(\mathbf{r})$ is slowly varying. In the lowest order, μ_{xc} becomes a local function of the density [i.e., μ_{xc} at the point $\mathbf{r}$ depends on $n(\mathbf{r})$ at the same point $\mathbf{r}$], and the form of this function can be obtained from the theory of the homogeneous electron gas. Sham and Kohn were also able to show that in this limit the Fermi surface defined by Eq. (3) is the same as the experimental Fermi surface. Thus a comparison of theoretical and experimental dHvA areas forms a valid test of local approximations to the exchange-correlation potential.

The exchange-correlation potential is assumed to be of the form

$$\mu_{xc}(r) = 3(3/\pi)^{1/3}\,\alpha[n(r)]^{1/3} \tag{4}$$

where μ_{xc} is in rydbergs if $n(r)$ is in atomic units. In Slater's $X\alpha$ method,[10] α is taken to be a constant, and it has been suggested that values of α obtained by fitting to atomic calculations should also be used for solids. Hedin and Lundqvist[11] have followed the Kohn–Sham approach, and have obtained the exchange-correlation potential from the results of Singwi *et al.*[12] for the homogeneous electron gas. (We shall call this the KSSHL potential.) In this case μ_{xc} still has the form given above, except that α is a function of the density n, increasing from two-thirds for high densities to slightly more than unity for low densities.

Copper and Palladium

Self-consistent calculations of the Fermi surface of copper have been carried out using the KSSHL exchange-correlation potential and using the $X\alpha$ potential for various values of α. The results for the errors in the (111) belly and neck areas are shown in Fig. 1. It turns out that many of the energies obtained using the KSSHL potential are quite close to those which would have been obtained for $X\alpha$ (with $\alpha = 0.7$), and thus the KSSHL results are plotted in Fig. 1 at this value of α. It is seen from the figure that both ab initio potentials, the KSSHL potential and the $X\alpha$ potential with $\alpha = 0.706$ (obtained from the virial theorem for the atom) or $\alpha = 0.721$ (obtained from minimizing the Hartree–Fock energy), lead to neck areas about 20% too large; the value of α required to fit the Fermi surface neck with the $X\alpha$ potential is $\alpha = 0.77$. (The belly orbit is much less sensitive to changes in the exchange-correlation potential because of the constraint on the volume of the Fermi surface.) The energy-band calculations were carried out nonrelativistically in the muffin-tin approximation. Estimates of these corrections indicate that they are too small by at least a factor of two to explain the 20% errors in the ab initio calculations, and this has led us to question the validity of the local approximations to the exchange-correlation functional.[13]

We have also performed self-consistent calculations of the Fermi surface of palladium using the KSSHL potential. The results are similar: errors in extremal

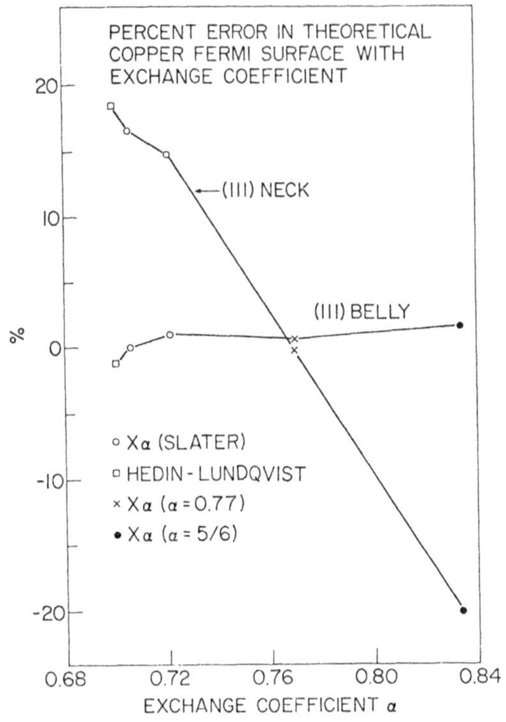

Fig. 1. Self-consistent calculation of the errors in the belly and neck areas.

areas for the fourth- and fifth-band (*d*-band) hole surfaces are less than about 20%, while the errors for the sixth-band (electron) surface are considerably smaller. These results are encouraging since the flatness of the *d* bands implies that relatively small errors in eigenvalues will lead to large errors in the Fermi surface shape. The *d* bands, in other words, form quite a stringent test of the crystal potential.

Fundamental theory thus needs further refinement before completely accurate ab initio prescriptions for obtaining the crystal potential are known. It is certainly encouraging, however, that the presently available theories yield from first principles results which are in agreement with experiment to within 20%.

Acknowledgment

The results described here were obtained in collaboration with A. R. Williams and V. L. Moruzzi.

References

1. J.C. Slater, *Phys. Rev.* **81**, 385 (1951).
2. W. Kohn and L.J. Sham, *Phys. Rev.* **140**, A1133 (1965).
3. D. Nowak and M.J.G. Lee, *Phys. Rev. B* **5**, 2851 (1972).
4. J.C. Shaw, J.B. Ketterson, and L.R. Windmiller, *Phys. Rev. B* **5**, 3894 (1972).

5. J.B. Ketterson, F.M. Mueller, and L.R. Windmiller, *Phys. Rev.* **186**, 656 (1969).
6. M.J.G. Lee, *Phys. Rev.* **178**, 956 (1969).
7. O. Jepson and O.K. Anderson, *Sol. St. Commun.* **9**, 1763 (1971).
8. J.F. Janak, *Sol. St. Commun.* **10**, 833 (1972).
9. L.J. Sham and W. Kohn, *Phys. Rev.* **145**, 561 (1966).
10. J.C. Slater, *Int. J. Quantum Chem.* **55,** 403 (1971).
11. L. Hedin and S. Lundqvist, *J. Phys. C* **4**, 2064 (1971).
12. K.S. Singwi, A. Sjolander, M.P. Tosi, and R.H. Land, *Phys. Rev. B* **1**, 1044 (1970).
13. J.F. Janak, A.R. Williams, and V.L. Moruzzi, *Phys. Rev. B* **6,** 4367 (1972).

De Haas–van Alphen Oscillations in Titanium

G. N. Kamm

Naval Research Laboratory, Washington, D.C.

and

J. R. Anderson

University of Maryland, College Park, Maryland

Introduction

We have observed de Haas–van Alphen (dHvA) oscillations in hcp titanium. To our knowledge this is the first time the dHvA effect has been observed in this metal.

Titanium is an interesting metal for a number of reasons.

1. It is the transition metal with the lowest atomic number Z (with the exception of scandium, which might be classed as a transition metal). The next hcp metal in the periodic table is cobalt, which has complications due to its normally ferromagnetic state. Because titanium has a low Z, it should be relatively simple to carry out first-principles band structure calculations ignoring the effects of spin–orbit coupling.

2. A number of measurements and calculations have been made on zirconium,[1–5] in the same column of the periodic table, and we expect titanium and zirconium to have similar electronic structures. Further, as we will point out, there are similarities in the pattern of dHvA oscillations. Thus it may well be possible to compare theory with experiment for both metals without the complication of a large spin–orbit coupling term. (Because the hcp structure has a twofold degeneracy on the hexagonal faces of the Brillouin zone, it may be necessary to include this coupling in some of the calculations.)

3. Only recently have adequate samples been prepared which allow dHvA studies. Since our samples have a resistivity ratio of only about 175 and exhibit a substantial mosaic structure, we may have missed a number of dHvA oscillations, but the data are sufficiently complete that tests of various band structure calculations are possible. In fact, we find that our results cannot be adequately described by any of the existing electronic structure computations.[5-8]

4. The Fermi surface sheets of titanium are characterized by high effective masses, and the existing band calculations show that in the vicinity of the Fermi level the energy bands are relatively flat. As a theoretical consequence, the size and shape of the calculated Fermi surfaces will be sensitive to small changes in the bands and Fermi level. For example, Altman and Bradley[5] found drastic changes in their calculated Fermi surfaces resulting from small modifications of the potential.[5] Ex-

perimentally, the high effective masses mean that the amplitudes of dHvA oscillations will be greatly reduced compared to other typical metals.

In the next section a brief description of the experimental details is given; preliminary results are then given, and a comparison of the data with existing band calculations is given in the final section.

Experiment

The dHvA data are measured by means of the usual field modulation technique of Schoenberg and Stiles[9] as adapted for low frequency by Stark and Windmiller.[10] A superconducting magnet of 93 kG was used in these experiments and the signals were detected at the second harmonic of the modulation frequency of about 100 Hz. The data were plotted and recorded digitally for computer analysis. Figure 1 shows a typical x–y recorder plot of the oscillations with the resulting interpolated Fourier transform spectrum. An optimum filtering window was chosen to minimize sideband response.[11] Data were taken in the three principal symmetry planes (0001), (10$\bar{1}$0), and (11$\bar{2}$0) (Miller–Bravais crystallographic notation).[12]

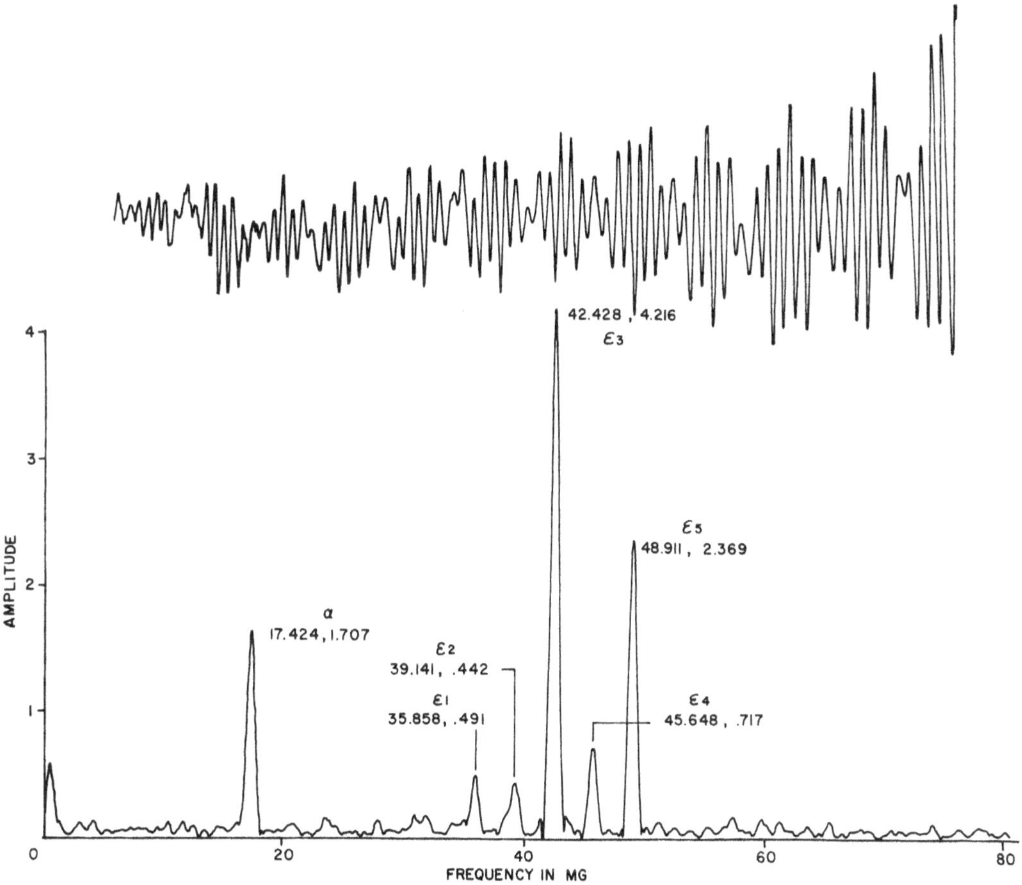

Fig. 1. dHvA oscillation pattern and resulting spectrum for H parallel to the [11$\bar{2}$0] direction. Noted are the oscillations, frequencies, and amplitudes.

Specimens were prepared by spark-cutting from bulk single crystals obtained from Alfa Crystals (Ventron Corporation) which had been grown by electron–beam zone-refining of iodide crystal bar titanium. They were etched to remove material damaged by the spark-cutting process. The mosaic structure of these crystals probably limited the amplitude of signals where the frequency changed rapidly with angle. The effects are also shown in the relatively high Dingle temperature of about 5°K.

Results

As illustrated in Fig. 2, the dHvA oscillations for titanium fall into five sets of oscillations, which are labeled α, β, γ, δ, and the group $\varepsilon 1$, $\varepsilon 2$, $\varepsilon 3$, $\varepsilon 4$, and $\varepsilon 5$. The nomenclature is consistent as far as possible with that used for zirconium by Thorsen and Joseph,[1] who observed oscillations similar to our α and β frequencies.

The α group is most readily explained. These oscillations occur at all angles, are the strongest ones observed, and are of essentially uniform amplitude. They clearly represent a closed, nearly spherical surface independent of other portions of the Fermi surface.

The group β is particularly strong for $H \parallel [0001]$ but becomes rapidly weaker away from that direction, the oscillations are not seen farther than 22° from [0001], where the frequency is increasing more rapidly with angle. This branch fits a cylinder with axis along [0001], as shown by the dashed curves in Fig. 2.

The segments γ and δ have been observed over such a limited range of angles that we have not attempted to relate them to a Fermi surface. Like the other branches, they are observed both in the $(11\bar{2}0)$ and the $(10\bar{1}0)$ planes.

Observed throughout the basal plane and in the $(11\bar{2}0)$ and $(10\bar{1}0)$ planes to within about 30° of the c axis is the group of similar oscillations labeled $\varepsilon 1$, $\varepsilon 2$, $\varepsilon 3$, $\varepsilon 4$, and $\varepsilon 5$. Their frequencies are nearly equally spaced and nearly independent of angle. In more detail their behavior can be described as follows: In the basal plane

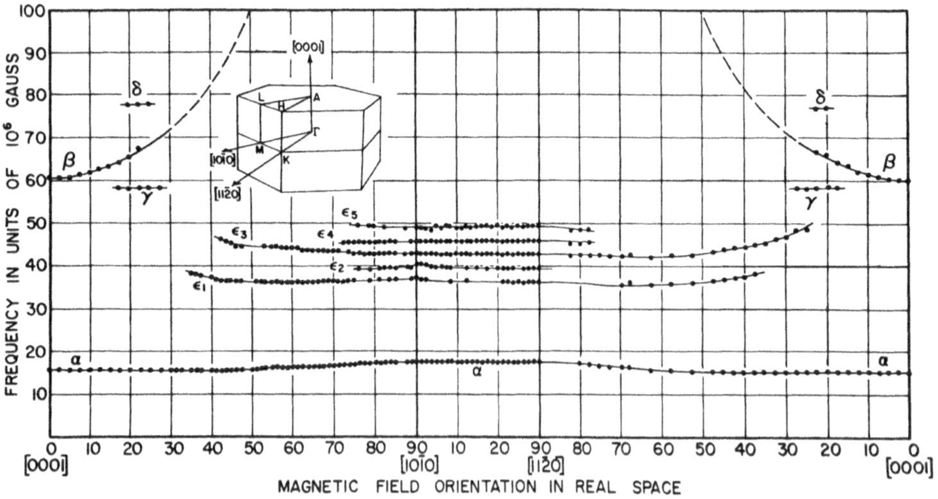

Fig. 2. Angular variation of the observed dHvA oscillation frequencies for the principal symmetry planes.

the $\varepsilon 1$ and $\varepsilon 2$ frequencies rise about 3% between $[11\bar{2}0]$ and $[10\bar{1}0]$; most of the increase occurs within 5° of $[101\bar{0}]$. Over this same angular range $\varepsilon 3$ rises about 1% while $\varepsilon 4$ and $\varepsilon 5$ each fall about $\frac{1}{2}$%. As the magnetic field is rotated out of the basal plane the frequencies also change very little. At angles greater than 15° or 20° we have been able to follow only the $\varepsilon 1$ and $\varepsilon 3$ oscillations. These frequencies begin to increase more rapidly at about 40° from the basal plane. In contrast to the α oscillations, the amplitude behavior of the ε group is complex even in the basal plane where the frequencies are most constant. All of the components have a maximum amplitude at $[10\bar{1}0]$; $\varepsilon 3$ and $\varepsilon 4$ are the strongest. In the $[11\bar{2}0]$ direction $\varepsilon 1$, $\varepsilon 3$, and $\varepsilon 5$ show maxima while $\varepsilon 2$ and $\varepsilon 4$ have minima and here $\varepsilon 3$ and $\varepsilon 5$ are the strongest oscillations. All components show a minimum at about 10° from $[10\bar{1}0]$.

A summary of our frequencies and of effective masses is given in Table I for the principal symmetry directions and for an intermediate direction in the $(10\bar{1}0)$ plane.

Interpretation and Comparison with Band Calculations

Of the four band structure calculations for titanium[5-8] which can be compared with our experimental results, those by Mattheiss,[6] while suggestive, are not given

Table I. De Haas-van Alphen Frequencies and Effective Masses*

Orientation	Oscillation	dHvA Frequency, 10^6 G	Effective mass ratio μ
[0001]	α	15.7	1.0
	β	60.5	2.0
$[10\bar{1}0]$	α	17.5	1.7
	$\varepsilon 1$	37.0	2.7
	$\varepsilon 2$	40.3	—†
	$\varepsilon 3$	43.0 } ‡	2.7
	$\varepsilon 4$	45.7 }	
	$\varepsilon 5$	49.0	—†
$[11\bar{2}0]$	α	17.5	1.55
	$\varepsilon 1$	36.0	2.8
	$\varepsilon 2$	39.3	—†
	$\varepsilon 3$	42.6	2.6
	$\varepsilon 4$	45.8	—†
	$\varepsilon 5$	49.1	2.7
$(10\bar{1}0)$§	α	15.8	1.25
	$\varepsilon 1$	37.2	3.4
	$\varepsilon 2$	44.1	3.0

* Uncertainty in frequency ±1%. Uncertainty in effective mass rato 0.1.

† For these oscillations the signal is too weak to reliably measure the spectral peaks.

‡ Effective masses unresolved.

§ In the $(10\bar{1}0)$ plane $47\frac{1}{2}$° from the [0001] direction.

in sufficient detail to allow a direct comparison with experiment. The two APW calculations by Hygh[7,8] and Welch show hole surfaces at Γ in the third and fourth bands. After suitable modification (by pinching off the cylindrical structure between Γ and A), the portion in their third band could reasonably correspond with the α oscillations and the larger cylindrical portion in their fourth band might account for the β oscillations. The other portions of their surfaces, however, do not correlate with our experiments. Altman and Bradley,[5] using the cellular method, have calculated the band structure and Fermi surface of titanium for three slightly different potentials. One of these gives a hole surface at Γ in the fourth band which correlates well with our α oscillations. The same model also has two electron-type surfaces at K, one in the fifth band and a somewhat smaller one in the sixth band. If such surfaces were to approximate ellipsoids of revolution overlapped in the manner suggested in Fig. 3, they might provide an explanation of the oscillations $\varepsilon 1$ to $\varepsilon 5$. In this case accidental degeneracies or very small gaps between the two surfaces would be required and the resulting breakdown effects might help to explain the complex amplitude variation of these oscillations. The disappearance of these oscillations at angles near [0001] could be a consequence of an increasing effective mass as noted in Table I as well as possibly of geometric and magnetic breakdown effects.

We have somewhat arbitrarily located these intersecting "ellipsoids" at K in Fig. 3 to correspond with the results of Altman and Bradley. If we assume, as appears to be the case, that there is no branching at the symmetry directions of the field, then, according to Watts,[13] such a surface might be centered at Γ, A, Δ, K, or H. This model has no other surfaces and therefore would not account for the β, γ, and δ oscillations observed experimentally.

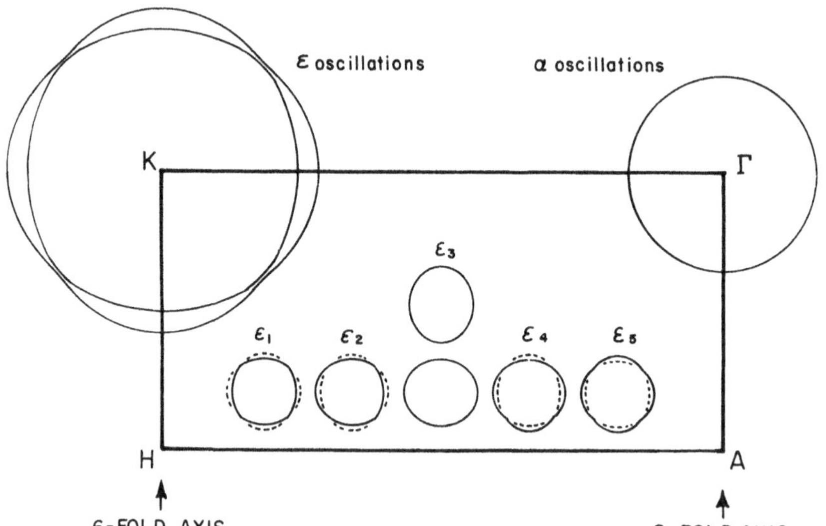

Fig. 3. Appearance and proposed location of portions of the titanium Fermi surface observed experimentally and suggested by a model of Altman and Bradley. The inset shows how breakdown orbits can give rise to five components such as $\varepsilon 1$ to $\varepsilon 5$. Other possibly larger portions of the Fermi surface are not shown.

Conclusions

Preliminary dHvA measurements of frequencies, effective masses, and amplitudes have given an explanation of several portions of the Fermi surface of titanium. There exist additional and possibly larger portions of the surface which are seen only in part. This work should provide guidelines for theoretical band calculations of titanium, since none of the existing theories appears adequate to provide an interpretation of the data.

References

1. A.C. Thorsen and A.S. Joseph, *Phys. Rev.* **131**, 2078 (1963).
2. J.E. Schirber, *Phys. Lett.* **33A**, 172 (1970).
3. S.L. Altman and C.J. Bradley, *Phys. Rev.* **135**, A1253 (1964).
4. T.L. Loucks, *Phys. Rev.* **159,** 544 (1967).
5. S.L. Altman and C.J. Bradley, *Proc. Phys. Soc. (London)* **92**, 764 (1967).
6. L.F. Mattheiss, *Phys. Rev.* **134**, A970 (1964).
7. E.H. Hygh and R.M. Welch, *Phys. Rev. B* **1**, 2424 (1970).
8. R.M.Welch and E.H. Hygh, *Phys. Rev. B* **4**, 4261 (1971).
9. D. Schoenberg and P. Stiles, *Proc. Roy. Soc. (London)* **A281**, 62 (1964).
10. R.W. Stark and L.R. Windmiller, *Cryogenics* **8**, 272 (1968).
11. W.B. Mosley, NRL Report 7179 (1971).
12. *Acta Met.* **15**, 387 (1967).
13. B.R. Watts, *Proc. Roy. Soc. (London)* **A282**, 521 (1963).

The de Haas–van Alphen Effect and the Electronic Structure of the Transition Metals*

L. R. Windmiller, S. Hörnfeldt,† J. Shaw,‡
G. Crabtree, and J. B. Ketterson

Argonne National Laboratory
Argonne, Illinois

In recent years many techniques have been developed for the study of the Fermi surface of metals. Of these techniques, the de Haas–van Alphen (dHvA) effect has emerged as the most powerful for the accurate determination of Fermi momenta and Fermi velocities. With recent improvements dHvA measurements are now also yielding additional information on the scattering time, g factor, and pressure or strain dependence of electrons at the Fermi energy.§

Many of the recent improvements in the detection of the dHvA effect are instrumental in nature. The availability of stable, high-field superconducting magnets has allowed the observation of dHvA orbits associated with quite high effective masses. Of equal importance has been the development of the field modulation technique.[2] This technique achieves very high sensitivity in addition to naturally producing very effective spectrometer action enabling separation of a wealth of individual dHvA frequency components. Very high absolute precision in the measurement of dHvA frequencies is routinely achieved using *in situ* NMR calibration; very precise differential measurements of dHvA frequency anisotropy are obtained using the field rotation method. A further increase in sensitivity has recently been realized by measurements of the effect at reduced temperatures produced by ^{3}He evaporative cooling,[3] and the employment of ^{3}He–^{4}He dilution refrigerators should increase sensitivity even further.

Coupled with the above improvements has been a steady progress in materials purification and crystal growth. Suitable samples of most of the simple and transition metals can now be prepared. The rare earth and actinide metals, however, are an area where considerable improvement in samples is still required. Finally, a very important feature of the dHvA effect is that these high-precision measurements are actually performed, because of the high phase of the oscillations, on very accurately characterized groups of electrons on the Fermi surface.

In the past the major disadvantage of dHvA measurements was that they produced orbital averages of Fermi surface parameters rather than point properties.

* Work performed under the auspices of the U.S. Atomic Energy Commission. Invited paper.

† Visiting Scientist, Uppsala University, Uppsala, Sweden.
‡ Present Address: Thomas J. Watson Research Laboratory, IBM. Yorktown Heights, New York.

§ For a comprehensive review see Gold.[1]

For example, extremal cross-sectional areas, cyclotron effective masses, and orbital relaxation times are measured rather than Fermi radii, Fermi velocities, and point relaxation times. This objection has now been largely overcome by the many inversion procedures that have recently been developed. Some of these inversion procedures depend upon the expansion of Fermi surface properties in some basis set and the adjustment of the parameters of the model to obtain close agreement with the dHvA data. Inversion procedures have now been developed using as parameters the coefficients in either a spherical harmonic or Fourier series expansion.[4,5] More recently the phase shifts which occur in the Korringa–Kohn–Rostoker (KKR) or the augmented plane wave (APW) methods of calculating electronic band structures have been used as disposable parameters.[6] This technique compares dHvA experimental areas with the theoretical areas calculated from initial phase shifts which describe only roughly the geometry of the Fermi surface. The differential properties of the calculation are then exploited in order to derive corrections to the phase shifts in an iterative procedure which converges rapidly to a set of phase shifts which best fit the data. It developed, unexpectedly, that quality of fit to the dHvA data is nearly independent of the choice of the difference between the Fermi energy and the muffin-tin zero energy. The Fermi surface of a transition metal is then tabulated by giving the energy dependence of a few phase shifts. The degree to which an *ab initio* band structure calculation agrees with the dHvA effect data may be determined by comparing the band structure phase shifts (at the band structure Fermi energy) with those tabulated from the parameterization scheme (at the same energy). The KKR or APW parameterization schemes are not only capable of inverting dHvA areas into phase shifts, but they can also fit effective cyclotron masses or dHvA pressure data, yielding the derivatives of the phase shifts with energy or lattice constant, respectively. From the phase shift derivatives we can deduce the Fermi velocity or the change of the Fermi radius with lattice constant, respectively.

Since a major reason for acquiring complete and precise Fermi surface information is to enable detailed comparison with theory (in the sense of both band structure calculations and the effect of the anisotropy of the Fermi surface on other physical properties of the metal), these inversion procedures cannot be viewed as just an unpleasant necessity in the interpretation of dHvA data, but are the logical output of any detailed investigation of the Fermi surface of a metal. From preliminary investigations in those transition metals where detailed dHvA data are presently available it appears possible that the entire body of experimental information on Fermi surface properties can be effectively tabulated in terms of a rather small number of parameters.[7] Efficient computer algorithms and numerical procedures are now at the disposal of the investigator to effectively utilize such a tabulation to reproduce, for his own needs, those anisotropic quantities needed for a variety of other calculations.

In the simple metals accurate first-principles energy band structure calculations may be obtained using the orthogonalized plane wave technique; the resulting bands are largely free-electron-like. The presence of d electrons causes the transition metal band structures to have a mixed tight-binding, nearly-free-electron-like character; little progress was made in these metals until the more powerful APW and KKR techniques were applied to the problem. The results have on the whole been quite successful. Several self-consistent calculations have now been performed, thus elim-

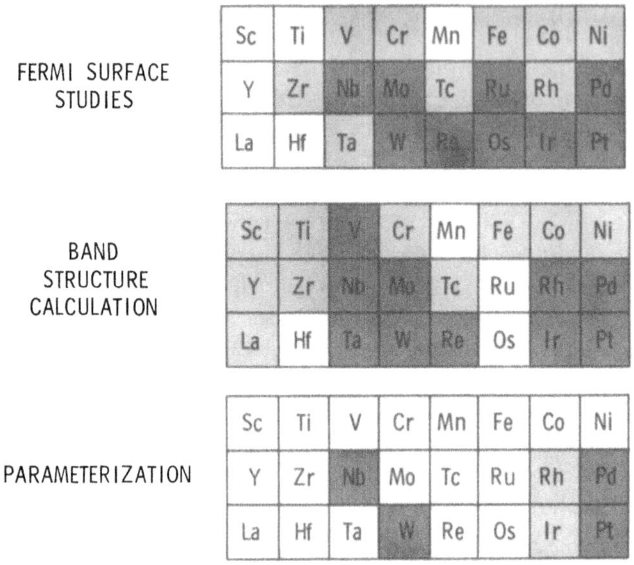

Fig. 1. Charts of the transition metals illustrating the state of completion of experimental and theoretical study. Explanation of shading is given in the text.

inating the uncertainty associated with the appropriate atomic configuration to be used in a non-self-consistent calculation.[8] An important problem remains, however. It is not known at this time to what extent exchange and correlation effects can be represented by a local potential. Several approximations are currently popular. Since the dHvA effect represent the most precise information available concerning the electronic structure of a metal, it is expected that the agreement which can be obtained between accurate self-consistent band structure calculations and precise dHvA measurements will be the most stringent test of the local exchange–correlation approximation. The generalization of the APW and KKR techniques to treat non-muffin-tin potentials is rapidly developing and should pose no problem in the future. Clearly, detailed Fermi surface studies are indispensible to the advancement of our theoretical understanding of the electronic properties of the transition metals.

It is interesting to note how far along we are toward being able to produce a complete tabulation of the Fermi surface properties of the transition metals.* What is needed in each metal is a detailed experimental dHvA study, a band structure calculation to help interpret the complex experimental results and fix starting phase shifts for the parameterization scheme, and finally a complete, accurate parameterization of the Fermi surface properties. Figure 1 shows three charts of the transition metal series of the periodic table. The top chart reviews the dHvA experimental state of each metal. Heavy shading indicates complete enough understanding to perform an effective parameterization of the Fermi surface, while light shading indi-

* For a review of this field up to about 1970 see Cracknell.[9]

cates enough experimental work is available to only understand the general features of the Fermi surface. The middle table reviews the state of band structure calculations in each metal. Heavy shading indicates those metals where substantial agreement between the calculations and experiments have been achieved, while light shading indicates those metals where band structure calculations are available but agreement has not yet been reached with experiment. The bottom table reviews the state of parameterization in each metal. Heavy shading indicates effective fits are available (the fits may not in some cases be as accurate as the data input, but in all cases reproduce rather well the overall features of the Fermi surface properties), while light shading indicates metals currently being parameterized but not yet converged to effective fits.

These charts show quite clearly that there is much work left to do. Hopefully, some of the metals in the table can be shaded in on the basis of results presented at this conference. Very difficult theoretical and experimental problems exist in the ferromagnetic metals iron, cobalt, and nickel, so that despite a great deal of work having been done, our understanding of them at this time is in a relatively primitive state. Perhaps adequate understanding must await the advances gained from the study of the other transition metals.

The fcc platinum group transition metals, Pt, Pd, Ir, and Rh have been studied extensively in recent years.[10-13] Detailed relativistic APW calculations of the Fermi surface have been performed for this group by Andersen.[14,15] An APW-combined interpolation scheme calculation of the Pd bands is also available.[16] For the first three metals very detailed experimental Fermi surface studies have been performed. The agreement of the experimentally determined Fermi surface properties with the theoretical predictions is striking. Even for the case of Rh, which has not been as extensively studied as the other three metals, available experimental data support completely the existing theoretical model.

Iridium, the most recently studied metal of this group, is a good example of the striking agreement between experiment and theory. The energy bands for Ir (as drawn by Andersen[15]) are shown in Fig. 2; a $5d^7 6s^2$ potential with full Slater exchange was employed. In order to investigate the Fermi surface properties, the k vectors on three closely spaced constant-energy surfaces in the vicinity of the Fermi surface were calculated in detail. At each energy the number of carriers and the areas of the extremal dHvA orbits for symmetry directions were obtained by numerical integration. The Fermi energy was then determined by interpolating to that value of energy such that the difference between the volumes of the electron and hole surfaces corresponded to one state per atom. The extremal areas, effective masses, and the density of states at the Fermi energy were then determined by interpolation. The Ir band structure of Fig. 2 predicts four distinct sheets (i.e., unrelated by symmetry) to the Fermi surface. Two hole sheets are centered at the points X of the Brillouin zone. These correspond to two sets of three pockets, each set related by the cubic symmetry of the lattice. The smaller pocket ($X3$), which approximates quite closely to a prolate ellipsoid of revolution about the ΓX line, is situated entirely within the larger pocket ($X4$), whose shape is distinctly different from an ellipsoid. Two large electron sheets centered at Γ are predicted. The smaller of these sheets ($\Gamma 6$), which is situated wholly within the larger ($\Gamma 5$), approximates a sphere in shape whose surface is slightly indented along the ΓK or [110] directions. A similar surface, differing

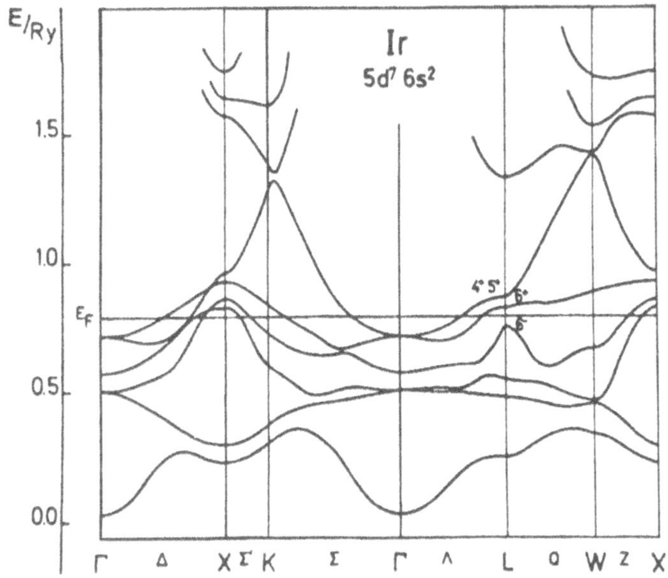

Fig. 2. The iridium band structure as drawn by Andersen.[15]

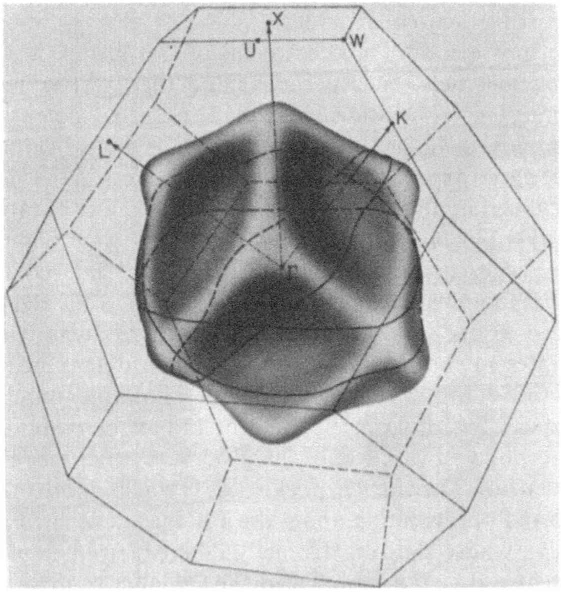

Fig. 3. The electron surface derived from the Γ6 band as drawn for Pd.

only in scale, is also present in the other Pt group transition metals. This surface, drawn for the case of Pd, is shown in Fig. 3, with the central extremal orbits for fields along the main symmetry axes illustrated. The larger Γ-centered sheet also approximates a sphere in shape. However, for this sheet the anisotropy is more pronounced, there being large bumps in the [110] directions.

In order to achieve a detailed comparison between experimental results and the band structure calculation, we obtained the calculated k vector nets from Andersen. Computer programs were written to use these nets in order to compute the area and effective mass of orbits on the Fermi surface as a function of magnetic field direction and position of orbit cross section with respect to surface inversion center. Interpolation between net values was employed to find the appropriate radius and energy derivative at each point integrated.

The anisotropy of the extremal areas of the smaller electron sheet (Fig. 3) for magnetic fields in the (110) plane is shown in Fig. 4. The solid line gives the computed extremal areas. Near [111] this surface supports two extremal areas; one, the smaller area, corresponds to the central cross section through the inversion center, and a second, the larger one, corresponds to a non-central cross section. The deviation of the theoretical curve from the experimental points varies between 0% and about 1%, the experimental areas being consistently the larger and indicat-

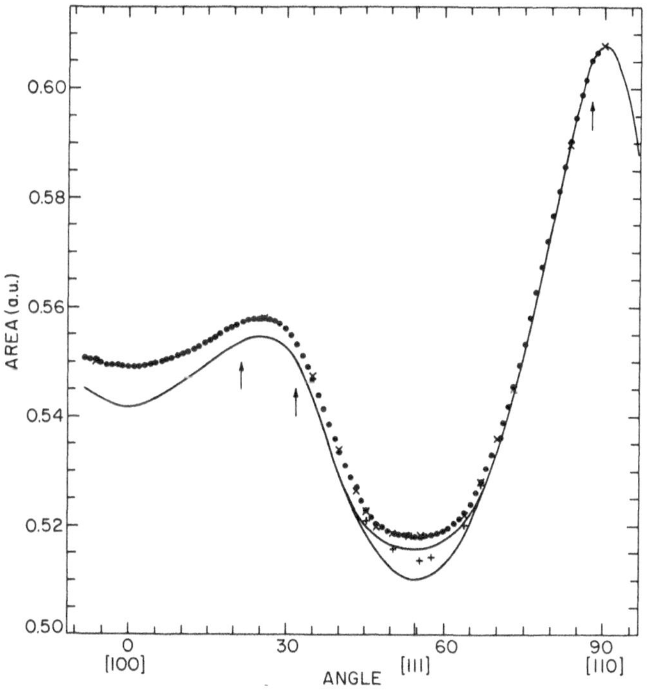

Fig. 4. The anisotropy of the extremal areas of the Γ6 electron sheet in Ir for magnetic fields in the (110) plane. The solid lines give the computed theoretical fit.

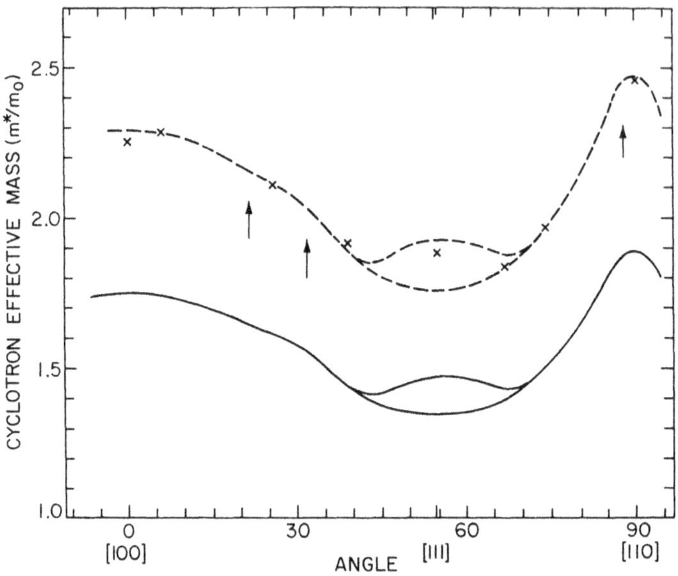

Fig. 5. The anisotropy of the effective cyclotron mass of the Γ6 electron sheet in Ir magnetic fields in the (110) plane. The solid line gives the computed theoretical anisotropy, while the dashed line is derived from this by multiplication by the factor 1.31.

ing a surface slightly more isotropic than expected from the calculation. Such detailed agreement with an *a priori* band structure calculation is exceptional. Excellent agreement with effective cyclotron mass calculations is also achieved if we assume the mass enhancement factor is isotropic. This is shown in Fig. 5, where the solid line gives the anisotropy of the calculated effective cyclotron mass and the dashed line is derived from the theoretical curve by multiplying by the factor 1.31.

Table I gives an overall view of the comparison of the experimental results with the band structure calculations. The only substantial disagreement occurs in the results for the smallest ellipsoidal pocket. However, this is expected since very small changes in the potential can change the size of this sheet considerably. The larger pocket centered at X is fairly accurately predicted, especially with regard to its detailed anisotropy. For example, this surface is slightly dumbbell-shaped in the (110) plane, so that it will have a noncentral extremal orbit for field directions near [110]. A detail of this is shown in Fig. 6.

From the above example it is clear that band structure calculations are capable of producing bands which accurately represent the Fermi surface properties as determined experimentally from the dHvA effect. The application of the phase shift parameterization technique to Ir will be discussed in the following paper. As an example of the results of a phase shift parameterization, Fig. 7 shows the anisotropy of the radius vector of the Γ-centered electron sheet of the Fermi surface of Pd. This is the sheet similar to the one in Ir that was discussed above and depicted in Fig. 3. The experimental anisotropy is derived from a spherical harmonic inversion of the

Table I. Comparison of Experimental Results for Ir with RAPW Band Calculations by Andersen and Mackintosh[14]

		Extremal areas, a.u.				Cyclotron effective masses, m^*/m_0				
		Central orbits		Noncentral orbits		Central orbits		Noncentral orbits		
		Theory	Experiment	Theory	Experiment	Theory	Experiment	Theory	Experiment	Mass enhancement factor
Γ6	[100]	0.541	0.5492 ± 0.0005	—	—	1.75	2.25 ± 0.02	—	—	1.29
	[110]	0.607	0.6062 ± 0.0010	—	—	1.86	2.46 ± 0.01	—	—	1.30
	[111]	0.510	0.5138 ± 0.0005	0.516	0.5180 ± 0.0005	1.34	—	1.47	1.88 ± 0.02	1.28
Γ5	[100]	1.507	1.425 ± 0.005	0.872 0.880	0.873 ± 0.001 0.877 ± 0.001	2.55	—	1.86 2.04	2.47 ± 0.01	1.33
	[110]	1.162	1.185 ± 0.002	1.190	1.193 ± 0.002	2.26	—	2.49	2.99 ± 0.04	1.20
	[111]	1.448	—			3.02	—	—	—	
$X3$	[100]	0.0261	0.01832 ± 0.0004	—	—	0.21	0.228 ± 0.003	—	—	1.1
		0.0388	0.02884 ± 0.0005	—	—	0.24	0.28 ± 0.01	—	—	1.2
	[110]	0.0293	0.02135 ± 0.0005	—	—	0.22	0.26 ± 0.01	—	—	1.2
		0.0405	0.02973 ± 0.0006	—	—	0.26	0.31 ± 0.01	—	—	1.2
	[111]	0.0318	0.02294 ± 0.0004	—	—	0.22	0.25 ± 0.01	—	—	1.1
$X4$	[100]	0.0915	0.1013 ± 0.0002	—	—	0.65	0.91 ± 0.02	—	—	1.4
		0.1467	0.1483 ± 0.0005	—	—	1.05	—	—	—	—
	[110]	0.1065	0.1121 ± 0.0003	—	—	0.76	—	—	—	—
		0.1098	0.1159 ± 0.0002	0.1150	0.1193 ± 0.0002	0.72	0.93 ± 0.01	0.78	0.98 ± 0.01	1.3
	[111]	0.1195	0.1242 ± 0.0002	—	—	0.83	1.28 ± 0.06	—	—	1.5

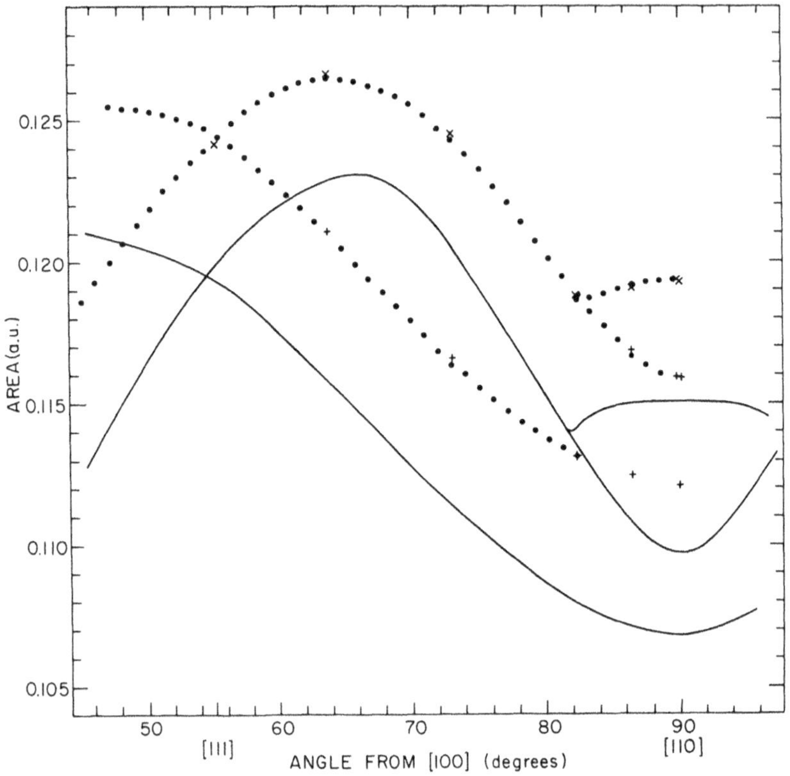

Fig. 6. The anisotropy of the extremal areas of the $X4$ hole pocket in Ir for magnetic fields in the (110) plane near [110]. The full lines are the computed theoretical fits.

dHvA data in seven different asymmetry planes. This fit is believed to be accurate to about 0.1 %, except, perhaps, near [111]. The squares are the APW calculation of Andersen,[15] while the circles are the APW interpolation scheme calculation of Mueller *et al.*[16] The fit derived from the KKR parameterization technique using $E = 0.7735(2\pi/a)^2$, $n_{s1/2} = -0.3468$, $n_{p1/2} = n_{p3/2} = -0.07150$, $n_{d3/2} = -0.3131$, and $n_{d5/2} = -0.3737$ is shown by the crosses. This fit gave a rms error of 0.7 % when fitting all observed symmetry direction areas (on all three sheets). The KKR parameterization technique with only five adjustable parameters is capable of fitting the experimental dHvA data within the limits of experimental error.

It is clear from the above examples that through the proper combination of theoretical and experimental techniques our knowledge of the Fermi surface can be quite complete. With the availability of accurate parameterization techniques (such as the KKR-APW phase shift representation) it is no longer necessary to resort to grossly oversimplified models of the Fermi surface when attempting to interpret experimental data (e.g., transport data). In the next few years we may hope to see the majority of the transition metal group reduced to a collection of muffin-tin phase shifts. With a proper understanding of one-electron phenomena one may then hope to go on to treat many-electron phenomena (e.g., superconductivity and magnetic susceptibility).

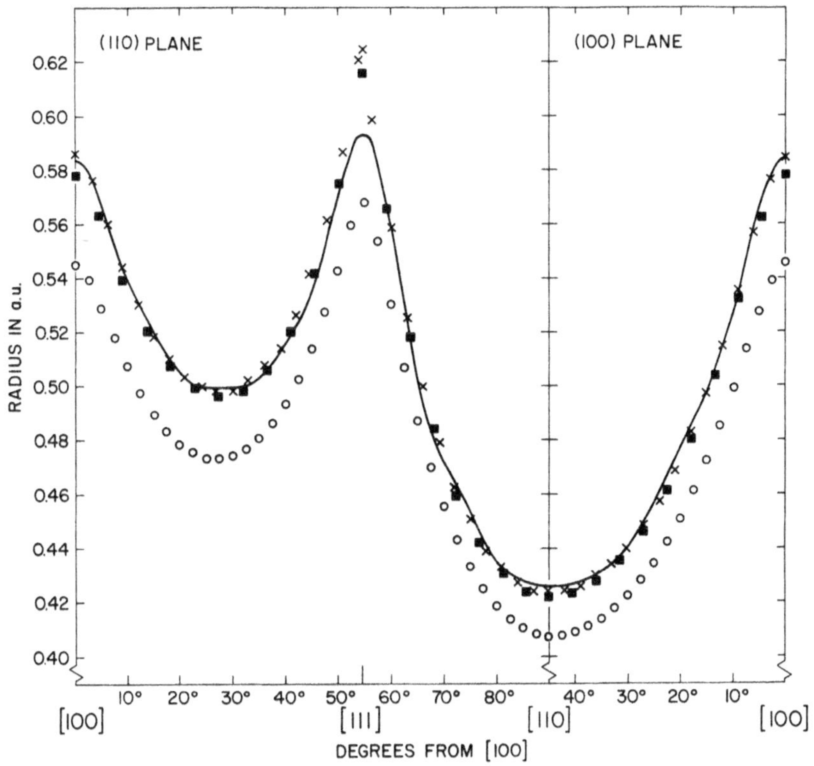

Fig. 7. The anisotropy of the Fermi radius of the Γ-centered electron sheet of Pd in the (100) and (110) planes. The solid line is a spherical harmonic inversion,[11] the circles are the Andersen APW calculation,[15] the squares are the Mueller *et al.* APW-interpolation scheme calculations,[16] and the crosses are the KKR inversion scheme results.

Acknowledgment

We are grateful to O. K. Andersen for supplying us with the unpublished results of his calculations.

References

1. A.V. Gold, in *Solid State Physics, Vol. 1: Electrons in Metals,* J.F. Cochran and R.R. Haering, eds., Gordon and Breach, New York (1968), pp. 39–126.
2. R.W. Stark and L.R. Windmiller, *Cryogenics* **8**, 272 (1968).
3. L.R. Windmiller and J.B. Ketterson, *Rev. Sci. Instr.* **39**, 1672 (1968).
4. F.M. Mueller, *Phys. Rev.* **148**, 636 (1966); J.B. Ketterson, L.R. Windmiller, S. Hornfeldt, and F.M. Mueller, *Sol. St. Commun.* **6**, 851 (1968); A. Aurbach, J.B. Ketterson, F.M. Mueller, and L.R. Windmiller, ANL Report No. 7659.
5. M.R. Halse, *Phil. Trans. Roy. Soc. (London)* **265**, 507 (1969); B. Bosacchi, J.B. Ketterson, and L.R. Windmiller, *Phys. Rev. B* **4**, 1197 (1971).
6. M.J.G. Lee, *Phys. Rev.* **178**, 953 (1969); **187**, 901 (1969); J.C. Shaw, J.B. Ketterson, and L.R. Windmiller *Phys. Rev. B* **5**, 3894 (1972).
7. J.C. Shaw, J.B. Ketterson, and L.R. Windmiller, Proc. Intern. Symp. on Atomic, Molecular, and Solid-State Theory, Sanibel Island, Florida, *Int. J. Quantum Chem.* **6** (to be published).

8. R.M. Welch and E.H. Hygh, *Phys. Rev. B* **4**, 4261 (1971); D.A. Papaconstantopoulos, J.R. Anderson, and J.W. McCaffrey, *Phys. Rev. B* **5**, 1214 (1972); M. Yasue, E. Hayashi, and M. Shimizu, *J. Phys. Soc. Japan* **29**, 1446 (1970); J.W.D. Connolly, *Int. J. Quant. Chem.* **IIS**, 257 (1968); S. Wakoh and J. Yamashito, *J. Phys. Soc. Japan* **21**, 1712 (1966); S. Wakoh, *J. Phys. Soc. Japan* **20**, 1894 (1965); J.R. Anderson, private communication.
9. A.P. Cracknell, *Adv. Phys.* **20**, 1 (1971).
10. J.B. Ketterson and L.R. Windmiller, *Phys. Rev. B* **2**, 4813 (1970).
11. L.R. Windmiller, J.B. Ketterson, and S. Hornfeldt, *Phys. Rev. B* **3**, 4213 (1971).
12. J.J. Grodski and A.F. Dixon, *Phys. Rev.* (to be published); S. Hornfeldt, J.B. Ketterson, and L.R. Windmiller (to be published).
13. P.T. Coleridge, *Proc. Roy. Soc.* **A295**, 458 (1966); J.B. Ketterson, L.R. Windmiller, and S. Hornfeldt, *Phys. Lett.* **26A**, 115 (1968).
14. O.K. Andersen and A.R. Mackintosh, *Sol. St. Commun.* **6**, 285 (1968).
15. O.K. Andersen, *Phys. Rev. B* **2**, 883 (1970).
16. F.M. Mueller, A.J. Freeman, J.O. Dimmock, and A.M. Furdyna, *Phys. Rev. B* **1**, 4617 (1970).

Phase Shift Analysis of Iridium*

J. C. Shaw,† J. B. Ketterson, and L. R. Windmiller

Argonne National Laboratory
Argonne, Illinois

Introduction

There has been considerable recent interest in the fitting of phase shifts to de Haas–van Alphen (dHvA) data.[1–3] We will present here results on attempts to fit recent dHvA data for iridium,[4] where we have included the effects of spin–orbit coupling to take into account the removal of degeneracies in the four sheets of the iridium Fermi surface (FS).

Attempts to fit more complicated multisheeted surfaces, where spin–orbit effects are important, was prompted by the considerable success in the accurate representation of the noble metal dHvA data.[3] Root-mean-square (rms) deviations between calculated and experimental FS areas were found to be within the experimental error of one part in 10^4 for the copper data of Coleridge *et al.*[5] As has been emphasized elsewhere, the basic value of this representation lies in its use as a physically based inversion method, whereby the anisotropy of dHvA data is converted into Fermi radii. The advantages are several fold, but most important is that few parameters are needed for representing the several sheets that may exist for complicated FS topologies. Recent results for palladium and platinum[6] have resulted in fits with rms errors within experimental uncertainties; however, data for all orbits do not exist, because of the large cyclotron effective masses for orbits on the open surface of these metals. This is not the case for Ir, where complete data exist for all of the four known sheets. Thus Ir provides a rigorous test of the muffin-tin potential approximation implicit in the parameterization of FS data using partial wave phase shifts.

In addition, we reported in an earlier paper[3] that the cyclotron mass anistropy in the noble metals could be characterized by using the energy derivatives of the phase shifts to adjust the slopes of the dispersion relations at the Fermi level. Fits to the cyclotron effective mass anisotropy gave rms deviations within experimental uncertainties. The resulting velocities (again, we are using the fitting as an inversion scheme) were in good agreement with the results of other workers.

The attempts to fit both areas and cyclotron effective masses for iridium have been less encouraging. Area fits have given rms deviations of approximately 5%, well outside experimental uncertainties. In addition, the fitting of cyclotron mass data has not been very fruitful. In both cases, however, the quality of fit is independent

* Work performed under the auspices of the United States Atomic Energy Commission.

† Present address: IBM, Thomas J. Watson Research Center, Yorktown Heights, New York.

of the height of the Fermi level above the muffin-tin zero, defined by the parameter $E = E_F - V_{MTZ}$, where E_F is the Fermi energy and V_{MTZ} is the muffin-tin zero.

Method

The Korringa–Kohn–Rostoker[7] (KKR) method has been utilized in calculating dispersion relations. Kohn and Rostoker have shown that the secular equation

$$\det|\mathsf{M}^{SEC}| = \det|A_{LL'}(\mathbf{k}, E) + E^{1/2}\delta_{LL'} ctn\eta_l(E)| = 0 \quad (1)$$

results from the application of a variational procedure to the integral equation $\Lambda(\mathbf{k}, E, \psi) = V - VGV$. Here the η_l are the phase shifts characterizing the scattering of electrons from a spherical muffin-tin potential. The $A_{LL'}(\mathbf{k}, E)$ are the usual "structure" constants containing all information about the crystal structure. In addition, the angular momentum indices are denoted by $L \equiv (l, m)$. The value of Λ is zero at a root given by $E(\mathbf{k})$. One can show that Λ may be expanded as

$$\Lambda(\mathbf{k}, E, \boldsymbol{\eta}(E)) = -\sum_{\alpha} \lambda^{(\alpha)}(\mathbf{k}, E, \boldsymbol{\eta}(E))\, B_\alpha^+ B_\alpha \quad (2)$$

where the $\lambda^{(\alpha)}$ are eigenvalues of M^{SEC} and the B_α are related to the eigenvectors. The variational problem now reduces to

$$\frac{\partial \Lambda}{\partial B_\beta} = -\lambda^{(\beta)} B_\beta^+ = 0 \quad (3)$$

Thus, either $\lambda^{(\beta)} = 0$ or $B_\beta^+ = 0$ at a root and $\Lambda \to 0$ as a single $\lambda^{(\alpha)} \to 0$. The particular α corresponding to a sheet of the FS is invariant with $\mathbf{k}$ and one can concentrate on the behavior of one particular eigenvalue during the course of FS calculations involving a given sheet.

Derivatives of $\lambda^{(\beta)}$ with respect to E, $\mathbf{k}$, and $\boldsymbol{\eta}$ can be obtained by using the Hellman–Feynmann theorem.[8] These derivatives allow the direct computation of the velocity and various derivatives of areas and volumes.[3] These quantities can, in turn, be used to compute the phase shift derivatives of the least-squares-fit functional for fitting area data, for example. These techniques have been described in detail in Ref. 3.

For the spin orbit case, the secular matrix is given by

$$\det|B_{\kappa\mu;\kappa'\mu'}(\mathbf{k}, E) + E^{1/2}\,\delta_{\kappa\kappa'}\,\delta_{\mu\mu'} ctn\eta_j(E)| = 0 \quad (4)$$

where the usual rules apply for relating κ and μ to the total angular momentum, $j = l \pm \frac{1}{2}$.[9] Thus, while three phase shifts η_0, η_1, and η_3 suffice to describe the FS of copper, five phase shifts $\eta_{s1/2}$, $\eta_{p1/2}$, $\eta_{p3/2}$, $\eta_{d3/2}$, and $\eta_{d5/2}$ suffice to describe the FS of Pd and Pt.

The velocity can be derived by straightforward chain-rule differentiation and is given by

$$\nabla E(\mathbf{k}, \boldsymbol{\eta}, E)|_{\lambda^{(\alpha)}=0} = -\Delta_k \lambda^{(\alpha)}/(\partial\lambda^{(\alpha)}/\partial E) \quad (5)$$

where

$$\left.\frac{\partial\lambda^{(\alpha)}(\mathbf{k}, E, \boldsymbol{\eta}(E))}{\partial E}\right|_{\lambda^{(\alpha)}=0} = \left(\frac{\partial\lambda^{(\alpha)}}{\partial E}\right)_{\eta} + \frac{\partial\lambda^{(\alpha)}}{\partial\boldsymbol{\eta}}\cdot\boldsymbol{\eta}'_E \quad (6)$$

and where $\boldsymbol{\eta}'_E = \partial\boldsymbol{\eta}/\partial E$.

The $\lambda^{(\alpha)}$ is that eigenvalue pertaining to the FS sheet under consideration and η is a collection of phase shifts in either l representation (no spin–orbit) or j representation (with spin–orbit). The η'_E can be used as adjustable parameters to fit cyclotron mass data. This has been done successfully for copper and gold.

Results

In Table I we give ten experimental areas and the corresponding calculated quantities. These fits were completed at several different values of $E = E_F - V_{MTZ}$; however, fits at only two values of E are given. In general, the quality of fit is poorer than the experimental uncertainty, particularly so for the larger Γ-centered electron surface. The calculated values obtained from an *ab initio* calculation by Andersen and Mackintosh[10] are also given. It is seen that in general the fitted quantities are in better agreement, but the improvement is disappointing. These quantities have been obtained using $l_{max} = 3$, but ignoring the spin–orbit splitting of the f levels, i.e., six parameters with $\eta_{f5/2} = \eta_{f7/2}$. The η's were varied at a fixed value of E until the phase shift derivative of the rms functional approached 10^{-3}.

Computer generated orthographic projections of the radii are shown in Fig. 1 for the X-centered hole ellipse and distorted ellipse for bands 4 and 3, and the smaller

Table I. A Comparison of Experimental dHvA Areas and Masses for Iridium with the Phase-Shift-Fitted Values and the Theoretical Values Calculated by Andersen and Mackintosh[10]

		Areas, a.u.				Mass, m^*/m_0		Number of electrons	
		Present work		Exp.	Ref. 10	Present work	Exp	Present work	Ref. 10
		E_1	E_2						
Band 1, Γ_1	⟨100⟩	0.556	0.554	0.549	0.541	2.26	2.26		
	⟨110⟩	0.602	0.604	0.606	0.607	2.50	2.46	0.241	0.240
	⟨111⟩	0.511	0.510	0.514	0.510	1.84	1.88		
Band 2, Γ_2	⟨100⟩	1.457	1.456	1.425	1.507	—	—	0.806	0.837
	⟨110⟩	1.136	1.136	1.185	1.162	—	—		
Band 3, X_1	⟨100⟩	0.101	0.100	0.101	0.0915	−0.91	−0.91	−0.0726	−0.0677
	⟨110⟩	0.113	0.113	0.112	0.1098	—	—		
Band 4, X_2	⟨100⟩	0.0188	0.0183	0.0183	0.0261	−0.227	−0.228	−0.0065	−0.0093
		0.0283	0.0279	0.0288	0.0388	−0.295	−0.280		
	⟨110⟩	0.0215	0.0211	0.0214	0.0293	−0.246	−0.260		
							Total	0.968	1.000

$E_1 = 0.4000(2\pi/a)^2$; $E_2 = 0.845(2\pi/a)^2$.

E_1: $\eta_{s1/2} = 0.9895$, $\eta_{p1/2} = 0.5667$, $\eta_{p3/2} = 0.0923$, $\eta_{d3/2} = -0.1051$, $\eta_{d5/2} = -0.1577$, $\eta_{f5/2} = \eta_{f7/2} = -0.00022$.

E_2: $\eta_{s1/2} = 0.0514$, $\eta_{p1/2} = 0.3209$, $\eta_{p3/2} = -0.1616$, $\eta_{d3/2} = -0.3831$, $\eta_{d5/2} = -0.5202$, $\eta_{f5/2} = \eta_{f7/2} = -0.0044$.

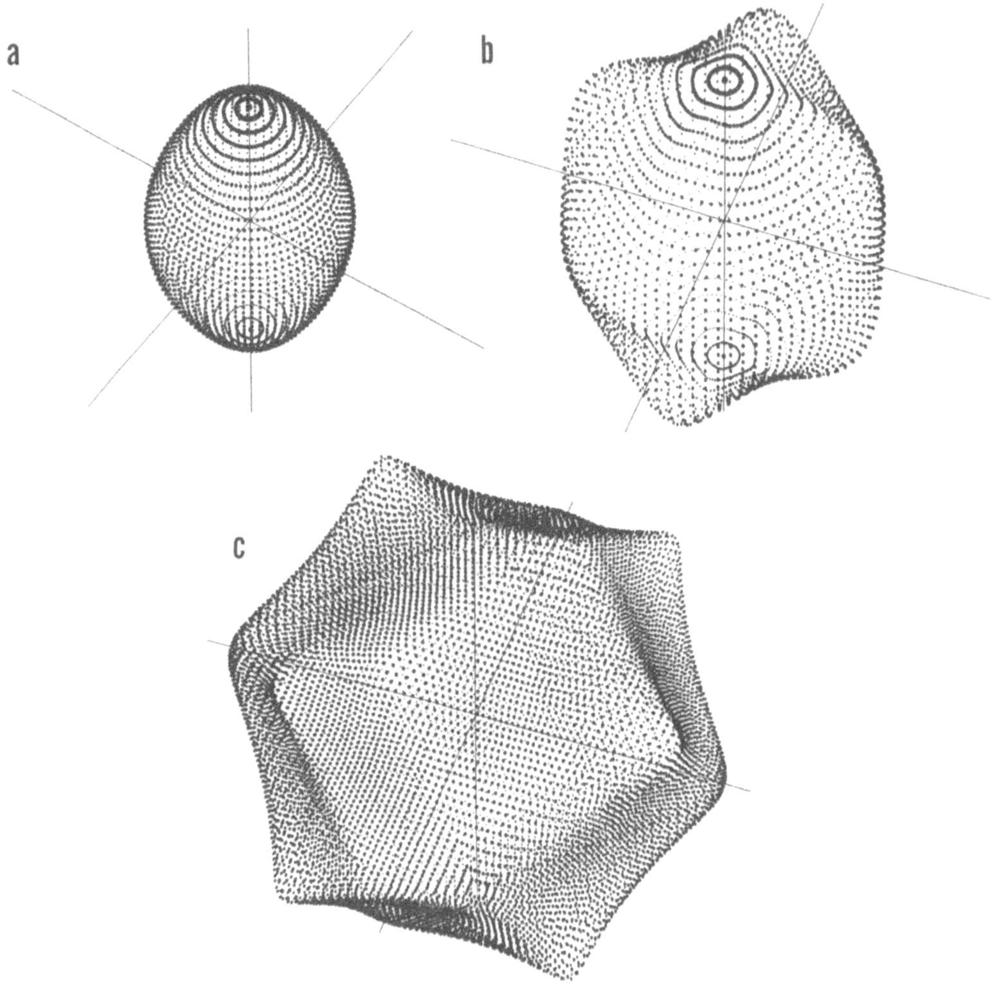

Fig. 1. (a) X-centered hole ellipsoid in iridium, (b) X-centered hole sheet in iridium, (c) smaller Γ-centered electron sheet in iridium.

Γ-centered electron sheet of band 1. The Γ-centered sheet was generated by computing 274 points in the 1/48 irreducible wedge taking 88 sec on an IBM 360/91. The other points were generated by applying the 48 operations of the O_h symmetry group. Similarly the radii for the point X were generated in a 1/16 irreducible wedge and the operations of D_{4h} applied to generate the other points. A Stromberg-Carlson SC-4020 was used for the final graphic display.

Finally, Table I also gives the values for the cyclotron effective masses. We have found that to get physical values for the velocities, fits using $\eta_{p1/2} = \eta_{p3/2}$ and $\eta_{f5/2} = \eta_{f7/2}$ were required. We have found, however, that the inverted velocities were not invariant with $E = E_F - V_{MTZ}$. The quality-of-fit to the areas using five phase shifts does not differ significantly from that shown in the table.

Conclusion

It is clear that, while the quality of fit is less than desired, the interpretation of the experimental data seems to be reasonably self-consistent. Whether the results warrant the inclusion of non-muffin-tin corrections is to be the subject of further investigation. As with other investigations, we have found that the quality-of-fit and the resulting Fermi radii are quite insensitive to the choice of $E_F - V_{MTZ}$. Because of the rigid band behavior of the transition series, this has made possible the selection of starting phase shifts from the *ab initio* band calculation of another metal of the same crystal structure in the transition series.

References

1. B. Segall and F.J. Ham, *Meth. Comp. Phys.* **8**, 251 (1968); M.J.G. Lee, *Phys. Rev. Lett.* **24**, 501 (1971); J.F. Cooke, H.S. Davis, and R.F. Wood, *Phys. Rev. Lett.* **25**, 28 (1970).
2. J.C. Shaw, J.B. Ketterson, and L.R. Windmiller, *Bull. Am. Phys. Soc.* **16**, 314 (1971).
3. J.C. Shaw, J.B. Ketterson, and L.R. Windmiller, *Phys. Rev. B* **5**, 3894 (1972).
4. L.R. Windmiller, S. Hornfeldt, J. Shaw, G. Crabtree, and J.B. Ketterson, this volume.
5. P.T. Coleridge, A.A.M. Croxon, G.B. Scott, and I.M. Templeton, *J. Physics E* **4**, 414 (1971).
6. J.C. Shaw, J.B. Ketterson, and L.R. Windmiller, Proc. Intern. Symp. on Atomic Molecular, and Solid State Theory, Sanibel, Island, Florida, *Int. J. Quant. Chem.*, **6**, (to be published).
7. J. Korringa, *Physica* **13**, 392 (1947); W. Kohn and N. Rostoker, *Phys. Rev.* **94**, 1111 (1954).
8. P.O. Löwdin, *J. Mol. Spectry.* **13**, 326 (1964).
9. M.E. Rose, *Relativistic Electron Theory*, Wiley, New York (1961).
10. O.K. Andersen and A.R. Mackintosh, *Sol. St. Commun.* **6**, 285 (1968).

Hole Octahedron Fermi Surface Curvature in Tungsten

D. E. Soule* and J. C. Abele†

Douglas Advanced Research Laboratory
Huntington Beach, California

Introduction

The Fermi surface (FS) curvature has been investigated using the spatial resonance of the magnetomorphic size effect.[1] The curvature depends on the orbital momentum of electrons orbiting about the magnetic field **H**. The magnetomorphic period ΔH is related to the average component of the orbital momentum $\langle m^* v_H \rangle$ in the direction of **H** by

$$\Delta H = (2\pi c/ed \sec \theta) \langle m^* v_H \rangle \tag{1}$$

where d is the sample plate thickness and θ is the angle between **H** and the plate normal. This orbital momentum in turn is equivalent to the orbital area derivative

$$\langle m^* v_H \rangle = (1/2\pi) |\partial A/\partial k_H| \equiv R \tag{2}$$

where A is the cross-sectional area of the FS perpendicular to **H** and k_H is the component of the wave vector parallel to **H**. In the limit of a spherical FS, R approaches the radius of curvature and the curvature is $1/R^2$. In certain noncentral regions on the FS, R_{coh} is defined, where $\partial R/\partial k_H = 0$ and the resulting coherent orbital bands lead to observable signals in the transport properties. The types of bands satisfying this condition include (I) limiting point orbits, (II) inflection zone orbital bands, (III) constant-derivative orbital bands, and (I-III) extended limiting point orbital bands. Comparisons can then be made between the observed $R_{\mathrm{obs}} = [ed(\sec \theta)/2\pi c]$ ΔH from Eq. (1) and the $R_{\mathrm{coh}} = (1/2\pi) |\partial A/\partial k_H|$, the orbital area derivative calculated from representative FS models.

Experimental Results and Analysis

In this investigation magnetomorphic measurements have been extended on (100) and (110) tungsten single crystals ($RRR = 42{,}000$) with thicknesses ranging from 90.5 to 230 μ. Oscillations in the Hall voltage were digitally recorded and computer-analyzed using the Fourier transform. The increased resolution of this technique has resulted in the observation of several additional orbital bands. The orientation dependence of R_{obs} for three crystals is shown in Fig. 1 for the (100) and (110) magnetic field rotation planes.

* Present address: Western Illinois, University Macomb, Illinois.
† Present address: Lewis and Clark College, Portland, Oregon.

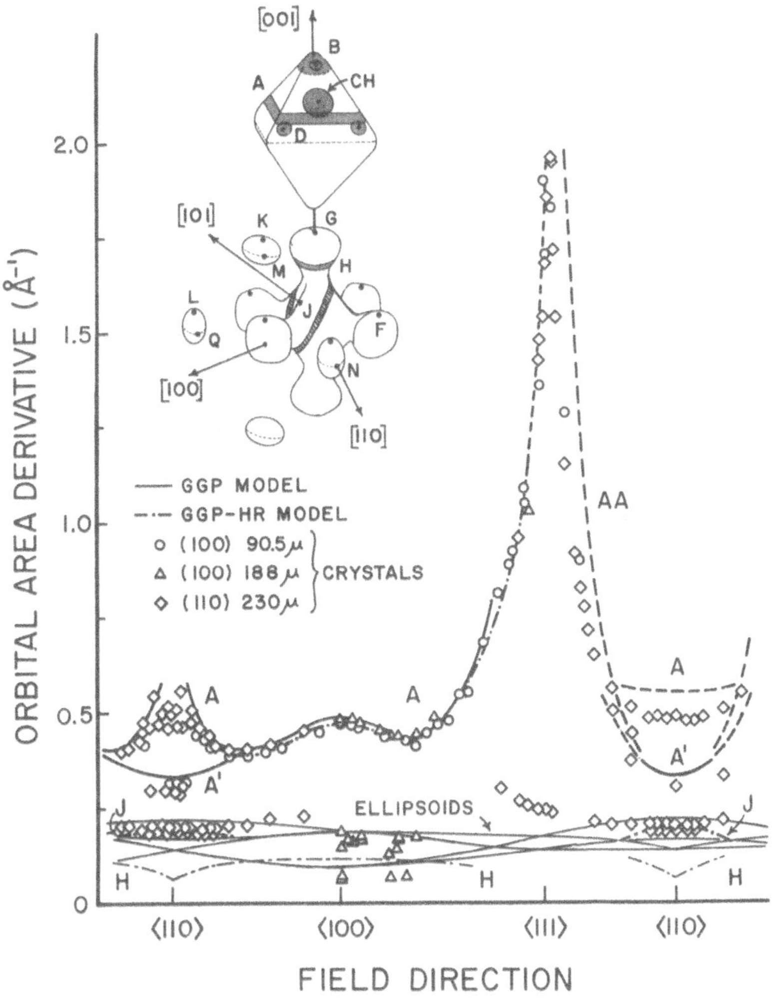

Fig. 1. Observed and calculated orbital area derivatives as a function of the magnetic field direction in the (110) and (100) principal planes. Data are for three single crystals oriented parallel to the (100) and (110) planes. The labeled branches refer to the orbital bands identified on the Fermi surface and are of types described in Fig. 2.

The dominant term in these measurements was identified[1] as that due to a type II inflection zone band *A* on the hole octahedron, as located on the FS in the figure. An analysis of the expected R_{coh} behavior was performed by a numerical computer differentiation using iteration procedures on the Girvan, Gold, and Phillips[2] (GGP) parametrized FS model, originally determined by a least-squares fit to their de Haas–van Alphen (dHvA) data. Calculations were made for the hole octahedron, hole ellipsoids, and selected regions on the electron jack. Primary interest, however, was centered on the behavior of orbital bands on the hole octahedron. For this case the GGP model is

$$|k_x|^m + |k_y|^m + |k_z|^m = P_4^m \tag{3}$$

where

$$m = P_5 \sum_{i=x,y,z} \left[1 - \left(\frac{k_i}{k} \right)^2 \right]^{P_6}$$

and the parameters P_4, P_5, and P_6 were set by a fit to the central extremal area dHvA values for the (100) and (110) principal field rotation planes. As shown in Fig. 1, there is good agreement between the observed R_{obs} data and the calculated R_{coh} for coherent orbital bands for the GGP model over most of the (100) and (110) field rotation planes.

A second FS model is due to Hui and Rayne[3] who adjusted the parameters of the GGP model, maintaining the central cross-sections relative to the dHvA effect constant, to conform more closely to their Doppler-shifted acoustic cyclotron resonance (DSCR) measurements. Reference to Fig. 1 shows that the agreement between our magnetomorphic results and this model is further improved over that of the GGP model. The hole octahedron parameters required only slight modification, whereas those of the electron jack had to be modified appreciably. This was found mainly for the J band on the jack body, which was reduced in magnitude and angular dependence (this band was originally identified as I in Ref. 1).

A third, more generalized FS model using the Mueller inversion scheme[4] was calculated[1] for the hole octahedron based on the dHvA data.[2,5] This method involved least-squares fitting the dHvA data to the cross-sectional area cubic harmonic expansion. The resulting FS had flatter (111) facets on the hole octahedron and slightly sharper ⟨100⟩ tips, as was also found for the GGP-HR model. Several results from this model are shown in Table I.

Virtually all of the above types of orbital bands were observed in the present measurements, showing the degree of applicability of the magnetomorphic size effect. This is demonstrated for the hole octahedron in Fig. 2, where the calculated R is plotted as a function of k_H. At ⟨100⟩ the most dominant term is the type II inflection zone band A. In the (110) plane shown in Fig. 2(a), good agreement was found between R_{obs} and R_{coh} as this term rises to a dominant peak at ⟨111⟩. A type III band at B on the octahedron may exist, though the data do not clearly confirm it. In addition, the orbital area derivative from the Mueller inversion model did not show a well-defined constant-derivative band in this region on the octahedron. In Fig. 2(b) in the (100) plane, a type III band A' branches off near ⟨110⟩, as is confirmed by the data. Figure 2(c) continues in the (110) plane from the ⟨110⟩ direction to show a progression from a type III A' band, along with the A band, to a new type of orbital band AA (called type IV), that branches into maximum *and* minimum peaks in R. Physically it appears reasonable that both peaks should support observable coherent orbital bands, though the data show only a median branch that cannot be assigned specifically to either one. The type II limiting band in Fig. 2(c) represents the upper limit of the calculated R vs. k_H curve and strongly indicates a maximum near the limiting point in progressing continuously toward ⟨111⟩, as does the ⟨111⟩ curve in Fig. 2(a), which continues to rise, approaching an almost infinite slope very close to the limiting point. Further calculations are necessary to establish this unequivocally, since the derivative is particularly sensitive to slight surface perturbations near a limiting point.

As shown in Fig. 1, there is a good agreement between these data and the GGP

Table I. Calculated and Observed Orbital Area Derivatives for Coherent Orbital Bands in Tungsten

			Calculated $R_{coh} = (1/2\pi)\lvert\partial A/\partial k_H\rvert$, $Å^{-1}$			Observed $R_{obs} = (ed/2\pi c)(\sec\theta)\,\Delta H$, $Å^{-1}$
Orientation	Fermi surface		GGP model	GGP-HR model	Mueller inversion model	
H ∥ ⟨100⟩	Hole octahedron	(band *A*)	0.482	0.470[3]	0.491 ± 0.049	0.472 ± 0.004
		(band *B*)	0.181	—	—	?
	Electron jack					
	Body	(band *J*)	0.357	—	—	—
	Neck edges	(band *H*)	0.10	0.120[3]	—	—
	Knobs	(*LP-F*)	>0.89	—	—	—
		(*LP-G*)	>2.41	—	—	—
	Hole ellipsoids	(*LP-K*)	0.192	—	—	?
		(*LP-L*)	0.092	—	—	
H ∥ ⟨111⟩	Hole octahedron					
	Concave facet (*ELP-CH*)		0.826	—	2.24 ± 0.31	1.96 ± 0.04 (Concave facet and Trigonal bulges)
	Trigonal bulges (*LP-D*)		0.727	—	0.826	
	Limiting band *A*		≥1.98	—	—	
H ∥ ⟨110⟩	Hole octahedron	(band *A*)	≥0.609, 0.55[3]	—	—	0.48 ± 0.03
		(band *A'*)	0.334	—	—	0.32 ± 0.03
	Electron jack body					
		(band *J*)	0.224[3]	0.200[3]	—	0.187 ± 0.003, 0.188[7]
	Hole ellipsoids	(*LP-Q*)	0.223			0.203 ± 0.003, 0.200[6]
		(*LP-N*)	0.169	—	—	
		(*LP-M*)	0.142			

and GGP-HR models over most of the angular field direction range for the *A* band on the hole octahedron. In the vicinity of ⟨110⟩, however, the data show a complex branching not predicted by these FS models. Also, near ⟨111⟩ further refinement is required to resolve the precise behavior of the orbital bands near and at limiting points such as *D* and around the ⟨100⟩ vertices.

In addition we have compared these results with the DSCR measurements[3] and with the Gantmakher–Kaner measurements of Girvan and Canepa,[6] which also depend on the orbital area derivative. For the DSCR data the agreement is satisfactory, as observed primarily for the *A* band, except that the DSCR measurements do not show type III bands such as *A'* or type IV bands at *AA*. The identification of a *J* band about ⟨110⟩ is due to Hui and Rayne,[3] where the DSCR measurement could distinguish the sign of the carrier to be negative. We observe two signals at 0.203 and 0.187 $Å^{-1}$ at ⟨110⟩, as shown in Table I. The former agrees with the value of 0.200 $Å^{-1}$ from the Gantmakher–Kaner measurements[6] and could be related to the *LP-Q* hole ellipsoid limiting point orbit with a value of 0.223 $Å^{-1}$ at ⟨110⟩. The latter agrees with the 0.188 $Å^{-1}$ DSCR measured value[7] identified as the *J* band. Also, the DSCR data reveal a second holelike branch *C* that lies close to the first signal at ⟨110⟩. A further means of discrimination between these branches is the angular extent away from ⟨110⟩ over which a given branch exists. For example,

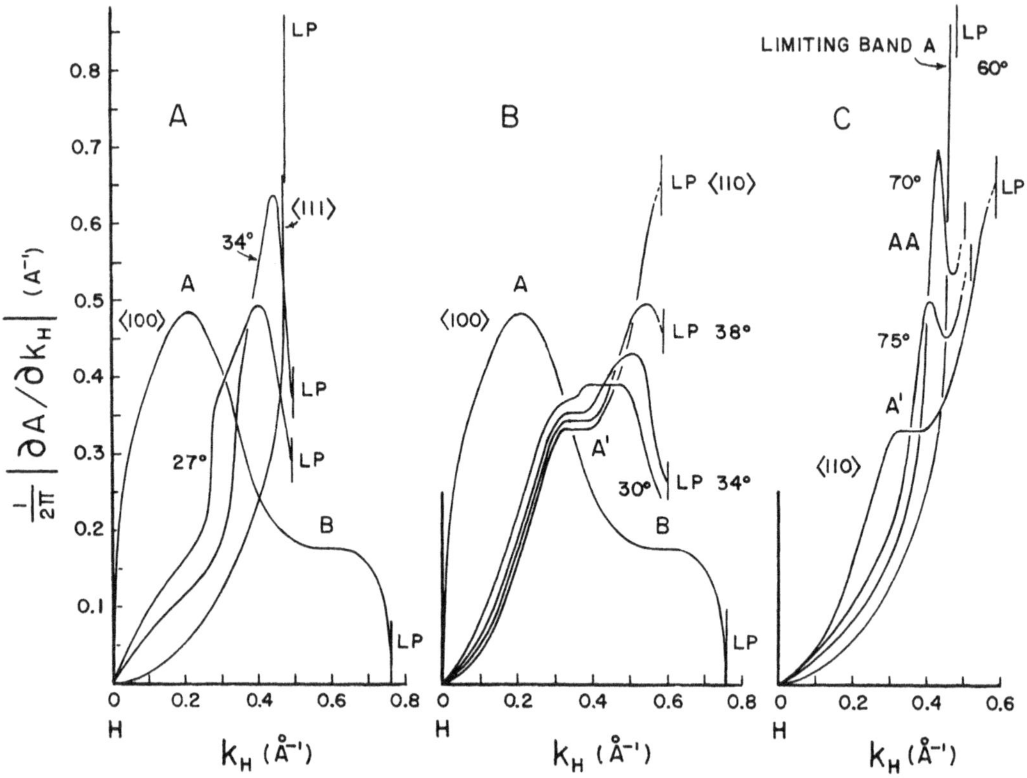

Fig. 2. Calculated orbital area derivative behavior on the hole octahedron Fermi surface from the GGP model. The plots are versus the wave vector parallel to the magnetic field, as measured from the center of the Fermi surface at H, where A and C are for the (110) plane and B is for the (100) plane. The angles are marked from ⟨100⟩.

the J band must be limited by the knobs on the electron jack. Whether the curvature $d^2R_{\rm coh}/d\theta^2$ is positive or negative is also relevant. A definite identification of these branches therefore requires further resolution. The lower points in the ⟨100⟩ region in Fig. 1 could possibly be associated with hole ellipsoids, since there is a fourfold limiting point degeneracy at 0.192 Å^{-1}. At ⟨111⟩, however, the lower observed branch does not fit any existing orbital zone. The predicted H band on the neck of the electron jack knob[1] was not observed here, but was seen in the DSCR measurements.

Acknowledgments

We would like to thank Dr. Hui for communicating his data and calculated results, Mr. Eicher for his technical assistance, Dr. Stamires for allowing use of his magnet, and Mr. H. Meriwether and Drs. Schwartz, Stogren, and Lawrence for helpful discussions.

References

1. D.E. Soule and J.C. Abele, *Phys. Rev. Lett.* **23**, 1287 (1969); D.E. Soule, *Bull. Am. Phys. Soc.* **16**, 336 (1971).
2. R.F. Girvan, A.V. Gold, and R.A. Phillips, *J. Phys. Chem. Solids* **29**, 1485 (1968).
3. S.W. Hui and J.A. Rayne, *Phys. Lett.* **33A**, 459 (1970); *J. Phys. Chem. Solids* **33**, 611 (1972).
4. F.M. Mueller, *Phys. Rev.* **148**, 636 (1966).
5. D.M. Sparlin and J.A. Marcus, *Phys. Rev.* **144**, 484 (1966).
6. R.F. Girvan and P.C. Canepa, *Bull. Am. Phys. Soc.* **14**, 29 (1969).
7. S.W. Hui, private communication.

Magnetic Breakdown and Quantum Oscillations in the Magnetoresistance of Tin*

J. K. Hulbert and R. C. Young

Department of Physics, University of Birmingham, Birmingham, England

Our recent work on quantum oscillations in the magnetoresistance of tin has concentrated on two areas: (1) the fast oscillations for **H** parallel to the c axis, and (2) the oscillations from nonextremal cross sections of the δ orbit in complex junctions, for **H** between 10° and 23° from the c axis. The most interesting feature of area 1, the observation of Pippard's zone oscillation, has already been published,[1] so this paper will concentrate on area 2.

Transverse magnetoresistive oscillations arising from complex junction mechanisms have been discussed by Young.[2,3] We have investigated the oscillations from the third zone δ orbit for **H** between 10° and 23° from [001] and show that a full explanation requires not only the complex junction, but also a family of discrete networks to be termed the "climbing orbits."

The angle between the [110] and the plane perpendicular to **H** was 3.5°. The orientation of our specimen was

$$\theta = 87.2° \pm 0.5°, \qquad \phi = 39.8° \pm 0.5°$$

where θ is the polar angle of the specimen axis with respect to the [001] and ϕ is the azimuthal angle from [001] to the plane containing the specimen axis and the [001]. With this orientation the aperiodic open orbits on the fourth zone are closed by junctions along the [110], the current direction being along the specimen axis, itself near [110]. Magnetic fields were typically 70 kG and temperatures approximately 1.2°K.

Fourier analysis of the oscillations generally revealed two frequencies: the "fundamental" at the frequency of the extremal δ^1, and a "second harmonic" at 2–2.5 times δ^1; the angular dependences of these amplitudes and frequencies are displayed in Figs. 1 and 2, respectively. Figure 1 shows a large peak in amplitude at 20°. At this point, which coincided with the type II anomaly of Anderson and Young,[4] the waveform was sinusoidal and its Fourier transform showed a single peak with only 3% of an exact second harmonic. The amplitude of the waveforms dropped suddenly within 19.7° and the waveforms became more distorted by second harmonic. We also observed at an angle less than 19.7° the onset of high-frequency ζ oscilla-

* The support of the Science Research Council of Great Britain is gratefully acknowledged.

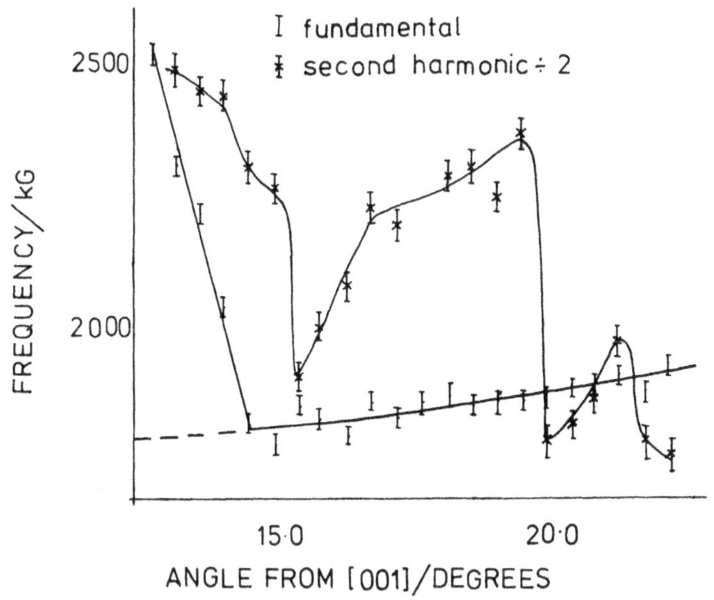

Fig. 1. Angular dependence of the amplitudes of "fundamental" and "second harmonic" components.

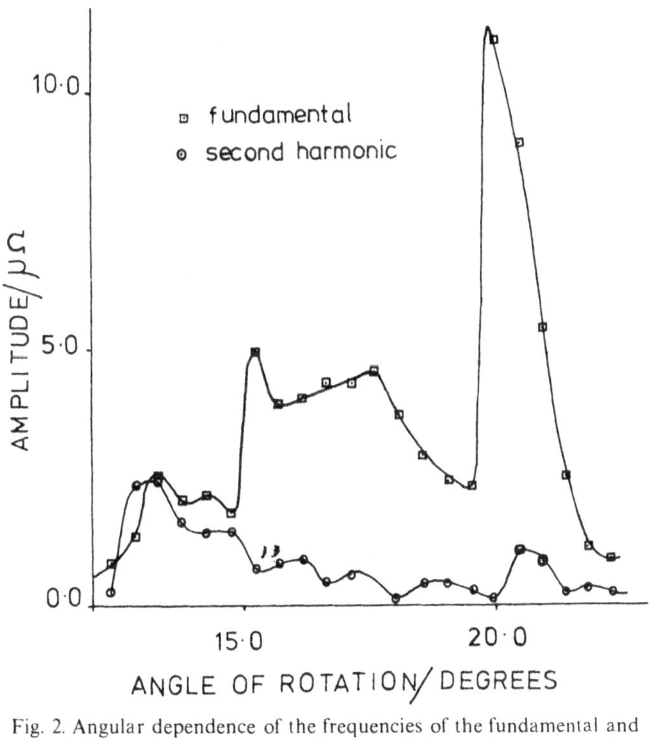

Fig. 2. Angular dependence of the frequencies of the fundamental and second harmonic components (the second harmonic frequency has been divided by two).

tions. These remained in the waveforms until 11.9°, at which point all the oscillations disappeared, leaving an almost monotonic increase of resistance with field.

If open orbits on the fourth zone are closed by the simple junction of Fig. 3(a), then, for a given field direction and an extremal δ, the closest angle to the [001] at which this mechanism can exist is determined by the existence of a ζ orbit between the δ orbit and the top of the sheet. For a direction along [110] this cutoff will occur at 20.3° (using Craven's pseudopotential surface[5]), in good agreement with the observed drop at 19.7° in the amplitude of the fundamental component. When this simple junction cannot exist, the complex junction of Fig. 3(b), with two δ orbits, has been proposed.[2] The predicted properties—a dominant second harmonic of a nonextremal area, and fast oscillations from the large orbit in the junction—agree well with what we see below 14–15°. Our cutoff angle for all the oscillations of 12.6° is in good agreement with the predicted 13.1°. The striking variation of the fundamental in this region cannot possibly be due to the angular dependence of the extremal area; it must arise from nonextremal areas in the complex junction. As the cutoff is approached the available range of k_H for the Fresnel integral becomes narrow, so that a nonextremal area no longer has its fundamental washed out. Just at the cutoff the nonextremal cross section is precisely selected, so that our experimental value for the nonextremal frequency of the third zone Fermi surface, for **H** at 12.6° from the c axis and for $k_H = 0.28 \times 10^8$ cm^{-1}, is 2490 kG. From Craven's model we estimate 2420 kG, while the extremal δ^1 frequency is 1810 kG.

The presence of oscillations between 14° and 19.7° defies explanation in terms of the junctions so far discussed, and a still more complex scheme of closure must be invoked. The fact that the second harmonic frequency is still greater than twice that expected from an extremal δ^1 orbit suggests the mode shown in Fig. 3(c). Bearing in mind that there is an angle of 3.5° between the [110] and the plane perpendicular to **H**, we can see that an electron moving from the top of the fourth zone sheet to

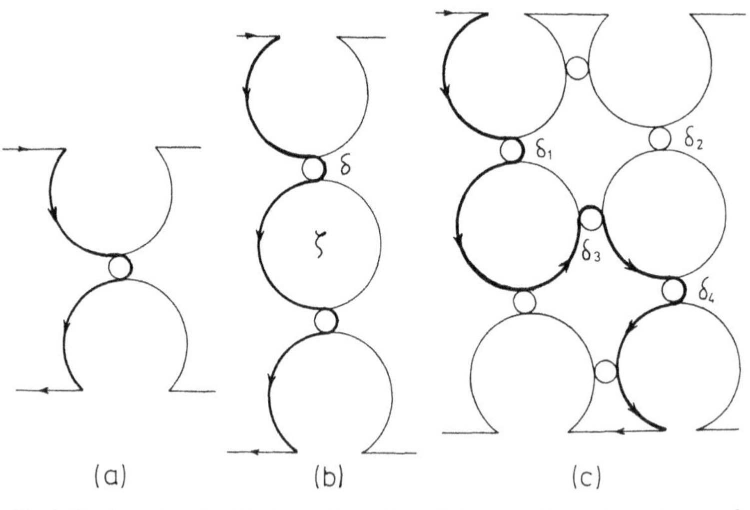

Fig. 3. Closing networks: (a) the simple junction, (b) the complex junction of Young,[2] (c) the climbing orbit, in which the electron is able to keep to regions of lower energy gap, at δ_1 and δ_4, by making a sideways step via δ_3.

the bottom may regain height by moving in the [110], hence the term "climbing orbit." At the same time a mechanism using the δ_1 orbit to tunnel instead of δ_2 is favored because it keeps the electron near the median plane of the Fermi surface and hence raises the breakdown probability. The climbing is made at δ_3 and since this is in the median plane, tunneling is quite probable. δ_3 is also extremal and explains the presence of the normal fundamental frequency throughout this region. It is the δ_1 and δ_4 areas that give rise to the second harmonic in this region in a similar way to the complex junction. This climbing mode will have an onset of oscillation at a greater angle than that for the mode using δ_2 and δ_4. It can be shown to exist between 14.3° and 16.8° on Craven's model and this is in reasonable agreement with the experimental region 15.2–18.1° where a large fundamental component is observed. Since tunneling in the median plane is highly probable, networks using more than one δ_3 orbit become possible, in particular a two-stage network would theoretically exist over the range 15.5–19.6°, hence explaining the remaining region of Figs. 1 and 2. The Fourier peaks observed in this region are poorly resolved, indicative of the large number of components to be expected from such a complicated network. Clearly the complex junction of Fig. 3(b) can still exist over this region, but it requires breakdown at a larger energy gap as the angle increases and so fails to compete with the climbing orbits, while they in turn are superseded by the simple junction as soon as it is allowed at 19.7°.

The climbing orbits have an obvious converse in that the electron may lose height by moving in the [110] direction, giving rise to the "descending orbits," and these would exist nearer to the c axis than the other networks described so far. This form of network was not observed on this particular specimen since it may be shown that for this orientation the networks are cut off by ζ orbits before they are allowed by breakdown.

Conclusion

We have shown that the detailed behavior of the angular dependences of both amplitude and frequency for the low-frequency δ oscillations may be explained by using specific networks to close aperiodic open orbits on the fourth zone. We have also shown that the topology of the situation gives rise to oscillations from extremal combinations of nonextremal areas and in one case from a completely nonextremal area, giving a δ frequency of 2490 kG, which compares favorably with a theoretical estimate of 2420 kG.

References

1. J.K. Hulbert and R.C. Young, *Phys. Rev. Lett.* **27**, 1048 (1971).
2. R.C. Young, *Phys. Lett.* **30A**, 510 (1969).
3. R.C. Young, *J. Phys. C* **4**, 458 (1971).
4. J.G. Anderson and R.C. Young, *Phys. Rev.* **168**, 696 (1968).
5. J.E. Craven, *Phys. Rev.* **182**, 693 (1969).

Quantum Oscillations in the Thermal Transport Properties of Tin*

R. C. Young

Department of Physics, University of Birmingham
Birmingham, England

This paper describes a low-field (up to 25 kG) investigation of quantum oscillations in the thermal magnetoresistance and thermoelectric emf of a single crystal of tin. Oscillations were seen from orbits δ^1, δ^2, and τ (identified from Craven and Stark[1]). The oscillations in δ^1, which have already been reported in the thermoelectric emf,[2] correspond to the oscillations introduced in the electrical resistance by magnetic breakdown.[3] They are here analyzed in terms of the conductivity tensors used to explain the electrical resistance[4] and the general treatment of transport properties given by Ziman.[5] The much smaller amplitude oscillations from δ^2 and τ, which have not been previously reported, are not due to magnetic breakdown but represent the usual Shubnikov–de Haas effect.

The large-amplitude resistance oscillations fall into two groups: sinusoidal oscillations from the closing of aperiodic open orbits by breakdown, and resonant oscillations when open orbits were formed by breakdown. This work will be confined to closing networks, where the resistance oscillations are explained by introducing into σ_{xx} a term proportional to $-(b/H^2)\cos\phi$, where $b = \exp(-2\pi/\omega\tau)$ and $\phi = \hbar A/eH$ for an orbit of area A. The open orbit runs along **x** and **H** is along **z**.

The field and temperature gradient are related to the electrical and thermal currents by:[5]

$$\mathbf{J} = \boldsymbol{\sigma}\mathbf{E} + \gamma(-\operatorname{grad} T/T) \tag{1}$$

$$\mathbf{U} = \gamma\mathbf{E} - \kappa \operatorname{grad} T \tag{2}$$

where

$$\boldsymbol{\sigma} = \int_0^\infty \boldsymbol{\sigma}(\varepsilon)\,(-\partial f/\partial\varepsilon)\,d\varepsilon \tag{3}$$

$$\gamma = (1/e)\int_0^\infty \boldsymbol{\sigma}(\varepsilon)\,(\varepsilon - \varepsilon_F)\,(-\partial f/\partial\varepsilon)\,d\varepsilon \tag{4}$$

$$\kappa = (1/e^2 T)\int_0^\infty \boldsymbol{\sigma}(\varepsilon)\,(\varepsilon - \varepsilon_F)^2\,(-\partial f/\partial\varepsilon)\,d\varepsilon \tag{5}$$

and $f(\varepsilon)$ is the Fermi–Dirac distribution.

* This work was supported by a grant from the Science Research Council.

Substituting the sinusoidal form of σ_{xx} and writing $A(\varepsilon) = A(\varepsilon_F) + (\varepsilon - \varepsilon_F)/h\omega$ leads to oscillating terms with field and temperature dependences given by

$$\sigma_{xx} \sim -(b/H^2)\cos(f_0/H)\,D(a) \tag{6}$$

$$\gamma_{xx} \sim -(b/H^2)(kT/e)\sin(f_0/H)\,dD/da \tag{7}$$

$$K_{xx} \sim -(b/H^2)(k^2T/e^2)\cos(f_0/H)\,d^2D/da^2 \tag{8}$$

where f_0 is the de Haas–van Alphen frequency, $D = \pi a/\sinh(\pi a)$, and $a = 2\pi kT/h\omega$. It will be assumed that all nonoscillating terms in γ can be neglected and that other terms in κ are related to the corresponding terms in σ by the Wiedemann–Franz law. Equations (1) and (2) can then be solved, using the same approximations as in Ref. 4, to give for $J = 0$ a thermal gradient and emf (defining $\mathbf{E} = P\mathbf{U}$) with field and temperature dependences

$$\rho_{yy}^{\text{th}} \sim -(b/T)\cos(f_0/H)\,d^2D/da^2 \tag{9}$$

$$P_{yy} \sim -(b/T)\sin(f_0/H)\,dD/da \tag{10}$$

Similar integrals apply for opening networks, but because of the resonant form of σ_{xx} they do not reduce to such simple expressions.

The function D, dD/da, and d^2D/da^2 are displayed in Fig. 1, and perhaps the most interesting prediction is that the amplitude of the oscillations in the thermal resistance should pass through zero (with a phase change of π) at a field and temperature given by $\pi a = 1.62$. Experimental results to illustrate this have already been reported.[6] The phase change of π was indeed found experimentally (actually from $-4°$ for $\pi a \ll 1.62$ to $187°$ for $\pi a \gg 1.62$, both relative to the electrical resistance). The relative phase of the thermoelectric emf was 80°, in good accord with the predicted 90°. The line shape of the thermal oscillations was sinusoidal, as predicted.

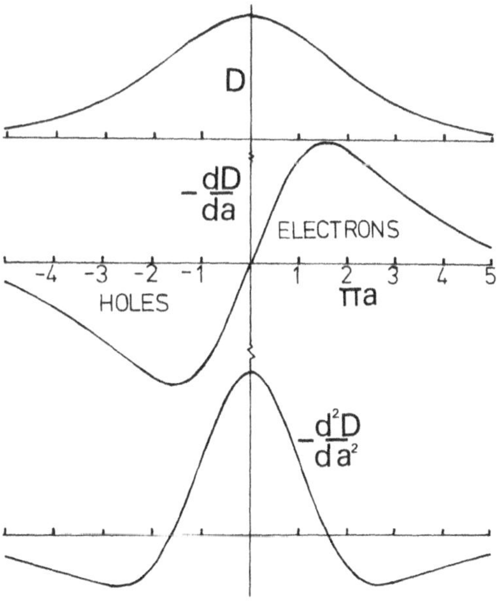

Fig. 1. The functions D, $-dD/da$, and $-d^2D/Da^2$ plotted against πa, where $D = \pi a/\sinh(\pi a)$ and $a = 2\pi^2 kT/h\omega$.

The prominent features of the oscillations in thermoelectric emf are (1) large amplitude (up to 20–30 $\mu V/^{\circ}K$) with good sinusoidal shape, and (2) the observation that they are one or two orders of magnitude greater in amplitude than the field-dependent background emf. Figure 2 illustrates the differential character of thermoelectric effects—a striking contrast to resistance oscillations, which form only 10–20% of a strongly field-dependent background.

For large πa, where sinh approximates to exponential, the field dependence of amplitude should give a straight line for $\log(AH)$ vs. $1/H$. This shows in both examples I and II of Fig. 2, for large n, i.e., low fields. This will fall off at high fields, where only the factor b of Eq. (10) will still contribute to this law. The predicted maximum is only expected at $\pi a = 1.62$ if there is no scattering. Clearly the effect of b is to push this peak to higher fields, presumably above 25 kG.

Effective masses were estimated from all three properties. The thermal properties gave values of $0.11–0.12 m_0$, well in line with other values, but the electrical resistance gave 0.14, while previous resistance work at higher fields[4] gave 0.16. Clearly this discrepancy needs further investigation.

The theory so far developed is valid when $H > H_0$ and magnetic breakdown is almost complete. From Barklie,[7] H_0 is estimated as 11.3 kG, while breakdown effects in the nonoscillatory resistance were not complete till 20 kG. Thus it was hoped, by following the oscillating thermoelectric emf to low fields, to see the onset of breakdown. However, example II of Fig. 2 displays the "high-field" behavior even at as low a field as 6 kG, where the oscillations start to be swamped by noise and drift. This is not entirely unexpected because a detailed analysis shows the criterion for the onset of oscillations to be $P\omega\tau = 1$, and even for the extended orbits involved this can probably be satisfied at quite low fields. It does not, however, seem consistent with the nonoscillating resistance.

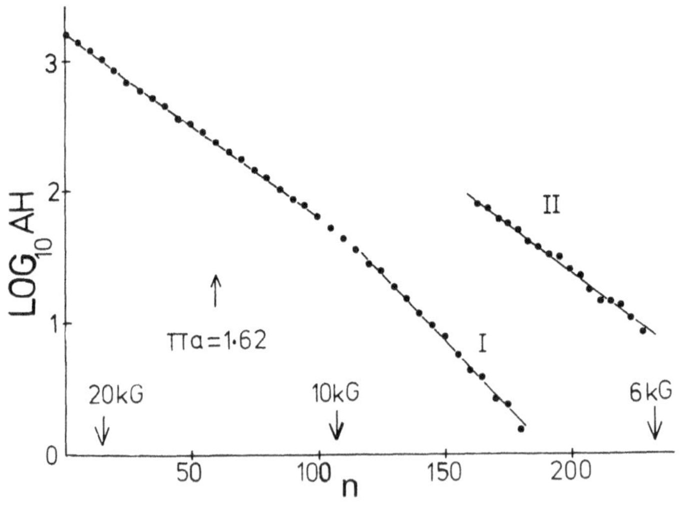

Fig. 2. Plots of $\log_{10}(AH)$ vs. n for thermoelectric oscillations of amplitude A and harmonic number n. Example I was taken over the full field range at $T = 1.34^{\circ}K$, while II was restricted to low fields and was from a different specimen.

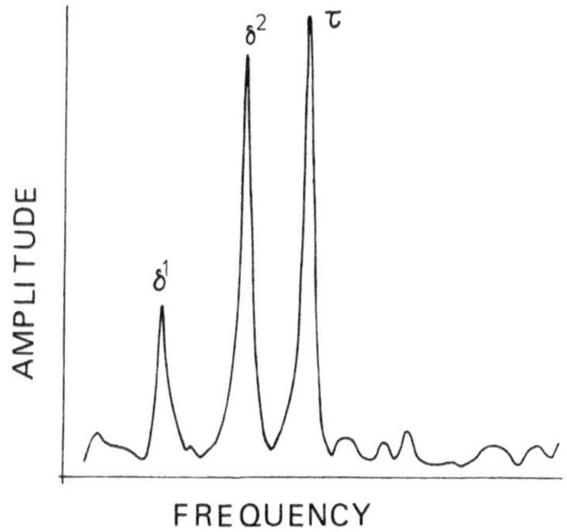

Fig. 3. A Fourier transform of oscillatory thermoelectric emf in a region of **H** where breakdown was unimportant.

The large-amplitude sinusoidal oscillations caused by magnetic breakdown are only seen when **H** is about 22° from the *c* axis. Well inside this region, and away from symmetry planes, the oscillations of thermoelectric emf were of amplitude only about 1 % of the largest breakdown oscillations, although still well resolved. A Fourier analysis is shown in Fig. 3, where it can be seen that δ^1 has a smaller amplitude than the contributions at frequencies corresponding to δ^2 and τ. Since neither of these orbits is in a region of magnetic breakdown, they must represent the regular Shubnikov–de Haas effect, brought out by the high sensitivity of thermoelectric properties.

There are two mechanisms which could contribute to the Shubnikov–de Haas effect: (1) modulation of the density of states of actual carriers, and (2) modulation of the relaxation time of carriers, which is usually regarded as the principal contributor.[8] In the region where we see δ^2 and τ, σ_{xx} is dominated by open orbits, so that contribution (1) from closed orbits must be negligible and the oscillations must be due to the modulation of the relaxation time of electrons on open orbits.

The possibility that mechanism (1) contributes via σ_{xy} is ruled out by the absence of harmonics in Fig. 3. Thus we are able to demonstrate that in the absence of magnetic breakdown the principal mechanism of Shubnikov–de Haas oscillations in tin is indeed the modulation of relaxation time.

Acknowledgment

Dr. J. K. Hulbert kindly carried out the Fourier analysis.

References

1. J.E. Craven and R.W. Stark, *Phys. Rev.* **168**, 849 (1968).
2. J.A. Woollam and P.W. Schroeder, *Phys. Rev. Lett.* **2**, 81 (1968).
3. R.C. Young, *Phys. Rev.* **152**, 659 (1966); *Phys. Lett.* **27A**, 539 (1968).
4. R.C. Young, *J. Phys. C: Solid State Phys.* **4**, 458, (1971).
5. J.M. Ziman, *Principles of the Theory of Solids,* Cambridge Univ. Press (1964).
6. R.C. Young, *Phys. Lett.* **36A**, 355 (1971).
7. R.C. Barklie, *Phil. Mag.* **23**, 175 (1971).
8. H.J. Trodahl and F.H. Blatt, *Phys. Rev.* **180**, 706 (1969).

5

Fermi Surfaces—Stress Effects

The Stress Dependence of the Magnetic Susceptibility in Zinc between 4.2 and 300°K*

H. R. Ott and R. Griessen

Laboratorium für Festkörperphysik
Eidgenössische Technische Hochschule, Zürich, Switzerland

The steady magnetic susceptibility in zinc is known for its anisotropic temperature dependence.[1,2] In the direction perpendicular to the hexagonal axis the susceptibility is almost constant, but a pronounced diamagnetic maximum is found at about 110°K parallel to the hexagonal axis. It has been suggested[3] that this behavior is due to changes of the topology of the Fermi surface, namely the disappearance of the third-band needles. A manifestation of such an effect should also be observable in the magnetostriction along the hexagonal axis, since such length changes are a measure of the stress dependence of the susceptibility.

This is expressed by the relation

$$\frac{l_i(H) - l_i(0)}{l_i(0)} = \frac{H^2}{2}\left[\chi \frac{1}{V}\left(\frac{\partial V}{\partial \sigma_i}\right) + \frac{\partial \chi}{\partial \sigma_i}\right] \tag{1}$$

where l_i is the length of the sample and σ_i is a tension in the i direction. The first term in the bracket is very small in the case of zinc and can be neglected.

We have measured the steady magnetostriction of a high-purity (6N) single crystal of zinc along the hexagonal axis. The sample had the form of a cylinder of 17 mm length and 6 mm diameter and the cylinder axis was parallel to the hexagonal axis of the crystal lattice. The length changes were measured with the three-terminal capacitance method due to White[4] in the temperature range from 4.2°K to room temperature. The magnetic field direction was parallel to the cylinder axis of the sample and magnetostrictive length changes were induced by varying an external magnetic field from 0 to 22 kOe. In Fig. 1 the measured strains are shown as a function of temperature. The relatively large scattering of the experimental data is due to thermal instabilities and to the unusual large thermal expansion coefficient of zinc along the hexagonal axis. At low temperatures, where de Haas–van Alphen oscillations occur, we have taken the mean magnetostriction.

With formula (1) we can calculate the stress dependence of $\chi_{\parallel}$ as a function of temperature. We have used these data to evaluate $\chi_{\parallel}(T)$ for the case of a crystal with an applied uniaxial tension of 100 bar along the hexagonal axis. This result is shown in Fig. 2; it may be seen that the main effect of such a tension is a shift of

* This work was in part financially supported by the Schweizerischer Nationalfond zur Förderung der Wissenschaft.

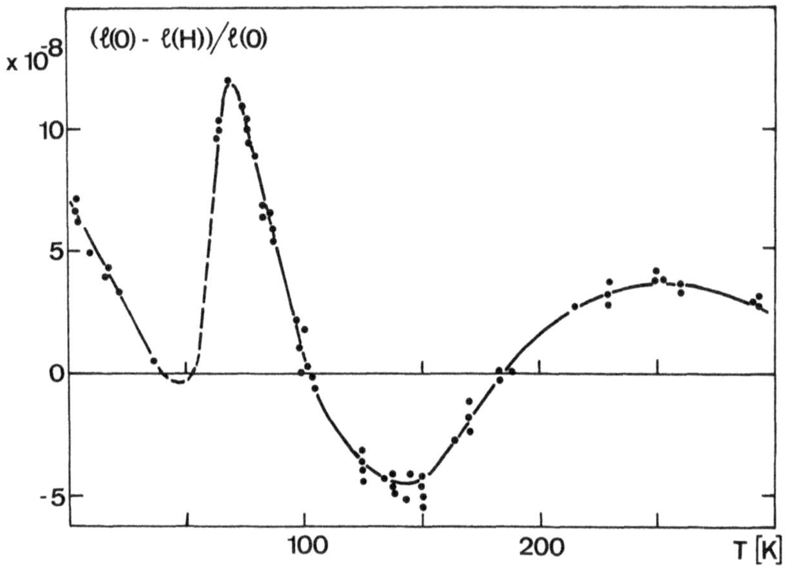

Fig. 1. Steady magnetostriction of a zinc single crystal along the hexagonal axis. The magnetic field is parallel to the hexagonal axis.

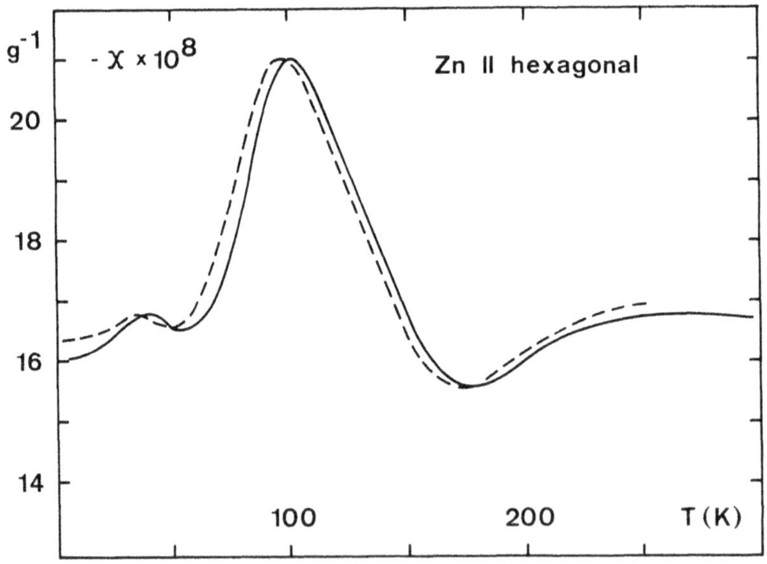

Fig. 2. (— —) Shift of $\chi_{\parallel}(T)$ induced by a tension of 100 bar parallel to the hexagonal axis as calculated from the stress dependence $\partial\chi_{\parallel}/\partial\sigma$. (——) Result of Berlincourt and Steele.[2]

the diamagnetic peak toward lower temperatures. This effect can be explained by assuming that the susceptibility depends mainly on the relative position of the Fermi energy E_F and the lower edge of the third band. This means that the minimum extremal cross section of the third-band needles is a measure of the anomalous part of the susceptibility. A uniaxial tension parallel to the hexagonal axis increases the volume. Consequently, the Fermi energy becomes smaller but the dimensions of the Brillouin zone increase in the directions perpendicular to the hexagonal axis. Hence at low temperatures the Fermi energy lies closer to the bottom of the third band for the stressed case than for the undeformed case. The Fermi energy will therefore reach the lower edge of the third band at a lower temperature, and also the disappearance of the third-band needles is shifted toward lower temperatures.

A possible check of this simple explanation can be made by a numerical evaluation of the temperature where the needles disappear. We have done this using a three-OPW calculation and experimental data on the temperature dependence of the linear thermal expansion coefficients[5] and of the compliance factors.[6] The result is shown in Fig. 3 for tensions of 0, 500, and 1000 bar. From this calculation we may conclude that a tension of 100 bar leads to a shift of the diamagnetic peak toward lower

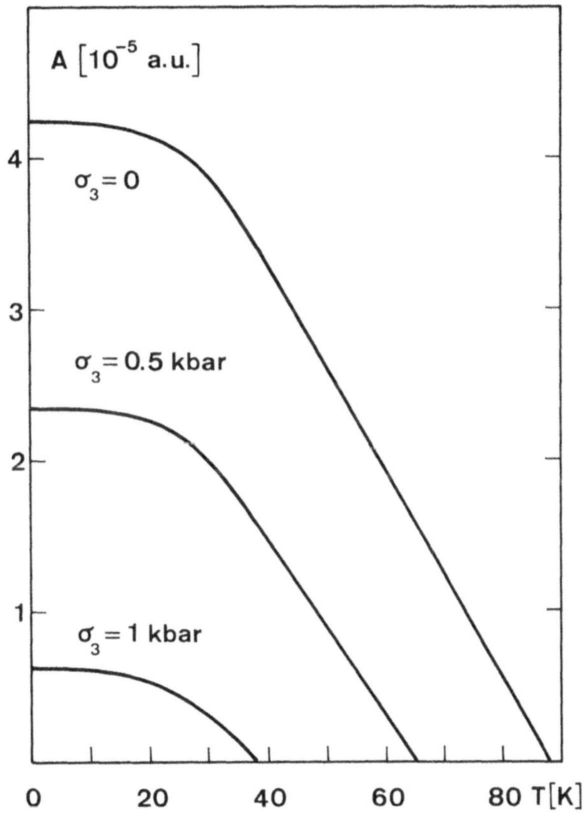

Fig. 3. Minimum extremal cross section of the third-band needles for various tensions as a function of temperature.

temperatures by 4.7°K. This is in good agreement with the shift obtained from the measured stress dependence of $\chi_{\parallel}$ of approximately 5°K, which is indicated in Fig. 2. From the three-OPW calculation it may also be seen that for a tension of nearly 1200 bar parallel to the hexagonal axis the third-band needles would disappear even for temperatures below 4.2°K. This is in good agreement with data on the stress dependence of the minimum extremal area of the third zone of the Fermi surface.[7] Such uniaxial stresses lie, of course, far beyond the flow limit of a zinc single crystal and experiments of this kind can hardly be carried out. The same effect can, however, be observed by alloying pure metals and measurements of this kind have been reported by Verkin and co-workers for indium alloys[8] and for Cd–Mg alloys[3] and also by Brandt for bismuth alloys,[9] where the susceptibilities show a peak as a function of the effective valence number.

In conclusion, we may say that our experiment and the relevant three-OPW calculation support the argument that the occurrence of a diamagnetic maximum of the susceptibility along the hexagonal axis is due to a change in the topology of the Fermi surface of zinc.

Acknowledgments

We are grateful to Prof. J. L. Olsen for his constant interest and support during this work. Thanks are also due to Dr. R. S. Sorbello for many helpful and stimulating discussions.

References

1. J.A. Marcus, *Phys. Rev.* **76**, 621 (1949).
2. T.G. Berlincourt and M.C. Steele, *Phys. Rev.* **95**, 1421 (1954).
3. B.I. Verkin, L.B. Kuznicheva, and I.V. Svechkarev, *Zh. Eksperim. i Teor. Fiz.* **54**, 74 (1968) [*Soviet Phys.–JETP* **27**, 41 (1968)].
4. G.K. White, *Cryogenics* **1**, 151 (1961).
5. R.D. McCammon and G.K. White, *Phil. Mag.* **11**, 1125 (1965).
6. G.A. Alers and J.R. Neighbours, *J. Phys. Chem. Solids* **7**, 58 (1958).
7. R. Griessen and A. Kündig, *Sol. St. Commun.* **11**, 295 (1972).
8. B.I. Verkin and I.V. Svechkarev, *Zh. Eksperim. i Teor. Fiz.* **47**, 404 (1964) [*Soviet Phys.—JETP* **20**, 267 (1965)].
9. N.B. Brandt, Doctoral dissertation, Moscow State University (1963).

Effect of Pressure on the Knight Shift of $AuGa_2$

H. T. Weaver, J. E. Schirber, and A. Narath

Sandia Laboratories
Albuquerque, New Mexico

The anomalous temperature dependence[1] of the conduction electron hyperfine interaction with ^{71}Ga in $AuGa_2$ has prompted considerable experimental work investigating various electronic and magnetic properties of this material. Briefly, it was found that the Knight shift K of ^{71}Ga changed from a small negative value at 4°K to a rather large positive value at room temperature. This large change in K was explained by a band model proposed by Switendick and Narath[2] in which a flat band of strong Ga 4s character which is completely filled at absolute zero is depopulated as the temperature increases. These authors speculated that either thermal depopulation or raising this band relative to the Fermi energy via the thermal expansion could be the driving mechanism for depopulation. Subsequent measurement of the pressure dependence of Fermi surface cross sections by Schirber and Switendick[3] indicated that the depopulation was not a result of thermal expansion. On the contrary, band calculations as a function of lattice spacing,[3] together with the experimental de Haas–van Alphen (dHvA) work,[3,4] indicated that under sufficient lattice compression the band in question passes through the Fermi level from below, creating a second band hole surface. The occurrence of this electron transition was verified by Schirber,[5] who measured the superconducting transition temperature of $AuGa_2$ as a function of hydrostatic pressure and found a large, almost discontinuous increase near 6 kbar. The ^{71}Ga Knight shift was expected [4,5] to display similar pressure effects.

In the present note we report measurements of the Knight shift and spin–lattice relaxation time of ^{71}Ga in $AuGa_2$ as a function of hydrostatic pressure at several temperatures. We will attempt to interpret these results in terms of the band model and electron transition picture.

Samples were prepared by grinding scraps from single-crystal boules used in earlier dHvA and superconducting transition temperature work.[3-5] The metal powder was mixed with KCl and placed into a $\frac{1}{8}$-in.-diameter coil which served as transmitter and receiver. This coil was inserted into a Be–Cu cylinder in which pressures were generated by careful isobaric freezing of He.[6] Resonance frequencies were determined at low pressures from beat frequency measurements, but the resonance became so broad near the transition that a spin-echo profile was used to determine the center of the resonance. Spin–lattice relaxation times were measured by monitoring the magnetization recovery following a comb of saturating pulses. Rotating fields up to 100 G were employed.

Figure 1 shows the Knight shift as a function of pressure at 4°K. Near the knee

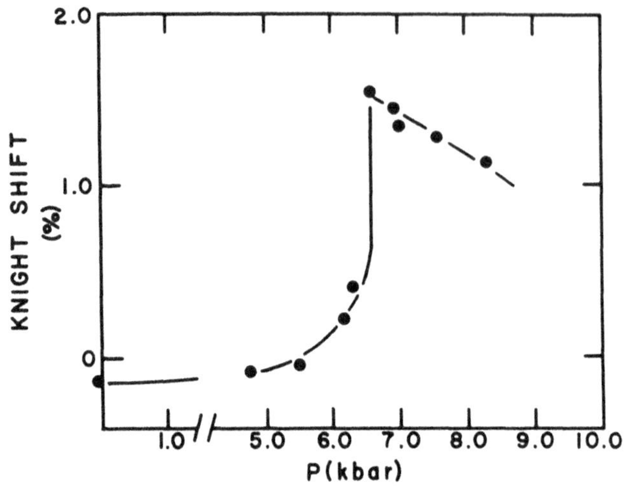

Fig. 1. Pressure dependence of the Knight shift of ^{71}Ga in $AuGa_2$ at 4°K.

of the curve the resonance lines broadened to about 200 G and were asymmetric, with a tail extending toward shift values observed in the transition region for pressures on either side of the transition. Above 7 kbar the lines became symmetric. Comparison of the ^{69}Ga and ^{71}Ga spectra indicated that the broadening was mainly due to Knight-shift inhomogeneity. Some quadrupole effects were evident in the magnetization recovery, but the main broadening was magnetic in origin. The line skewing or asymmetry probably reflects local strains in the material.

The spin–lattice relaxation rate behavior is qualitatively similar to that of the Knight shift (Table I). However, the rapid decrease in the rate above the transition is surprising.

Figures 2 and 3 show the temperature dependence of K and $(T_1T)^{-1}$ at ~6.9 kbar. For comparison, the data of Jaccarino *et al.*[1] are also shown. Note that the low-temperature, high-pressure Knight shift and relaxation rate are much larger than the room-temperature values.

Table I. Summary of NMR Data as a Function of Pressure at 4°K

P, kbar	K, %	$(T_1T)^{-1}$, (sec °K)$^{-1}$	κ
0.0	−0.13(2)	1.5(1)	—
4.6	−0.10(2)	2.5(2)	—
5.5	−0.04(2)	—	—
6.2	0.41(5)	6.2(6)	—
6.2	0.23(5)	10. (2)	—
6.6	1.53(10)	17. (2)	0.20
6.9	1.32(10)	22. (3)	0.36
6.9	1.44(10)	24. (3)	0.33
7.6	1.26(10)	10. (2)	0.18
8.3	1.13(10)	4. (1)	0.09

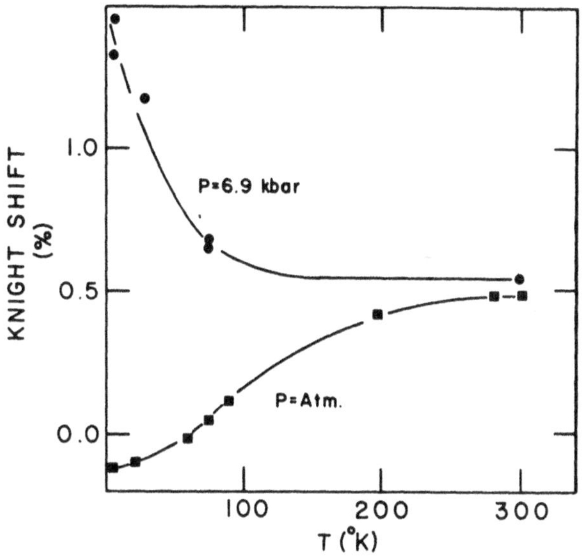

Fig. 2. Temperature dependence of the Knight shift of ^{71}Ga in $AuGa_2$ at pressures indicated. The atmospheric pressure data are from Ref. 1.

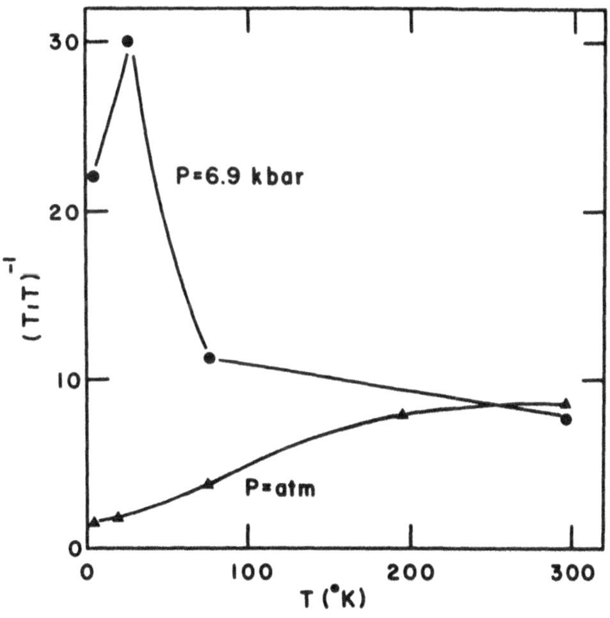

Fig. 3. Temperature dependence of the spin–lattice relaxation rate of ^{71}Ga in $AuGa_2$ at pressures indicated. The atmospheric pressure data are from Ref. 1.

Qualitatively, the nuclear resonance data are satisfactorily explained, as were the dHvA and superconducting transition temperature results, by the existence of a sharp increase in the density of states located a few millivolts below the Fermi level at low temperature and zero pressure. Rigid band motion of this band edge with increasing pressure produces a rapid increase for the NMR parameters as the Fermi energy E_F is intersected. On the other hand, increasing temperature expands the lattice and tends to push the band edge further below the Fermi energy so that a thermal smearing of E_F is needed to allow hyperfine interactions with electrons in this band. The reduced values of K and $(T_1 T)^{-1}$ at room temperature compared with their maximum values at low temperatures support this interpretation. More detailed consideration of the data requires a somewhat more speculative characterization of the band structure and its volume dependence.

For s-contact hyperfine interactions with an independent-electron gas at low temperatures the Knight shift is proportional to the density of states at the Fermi energy, $N(E_F)$, and the relaxation rate is proportional to $N^2(E_F)$. Consequently, the Korringa product $K^2 T_1 T$ is independent of band effects. If we make the reasonable assumption that the p character of the conduction electrons in $AuGa_2$ is measured by the atmospheric pressure, low-temperature Knight shift, it is clear that above the transition the dominant hyperfine mechanism is the direct contact interaction. However, by defining

$$\kappa = (K^2 T_1 T)_{\text{free electron}} / (K^2 T_1 T)_{\text{measured}}$$

and tabulating this quantity for pressures above the transition, we see (Table I) that κ deviates from unity. A similar observation was made for simple metals[8] where $\kappa \approx 0.5$. Such reductions in κ can be attributed to electron–electron interactions, which usually enhance K more than $(T_1 T)^{-1/2}$. Our values for κ are much smaller than for simple metals, leading to the conclusion that we are dealing with a strongly exchange-enhanced system.

With this interpretation there remains the puzzling decrease in κ as the pressure increases above the transition. This simply reflects the fact that $(T_1 T)^{-1/2}$ is decreasing much faster than K. We recall that K measures the infinite-wavelength ($\mathbf{q} = 0$) response of the electron gas, whereas $(T_1 T)^{-1}$ depends on all $\mathbf{q}$ involved in electron–nuclear scattering processes.[8] It follows that any change in shape of the Fermi surface which alters the q-dependence of the response function can lead to changes in κ.

With regard to $AuGa_2$, band calculations indicate[9] formation at the transition of a Fermi surface sheet consisting of Γ-X oriented needles intersecting at Γ. This system would appear more two-dimensional than three-dimensional, leading to an enhancement of the response function (see, e.g., Ref. 10) for nonzero $\mathbf{q}$ and consequently to an increased κ. As the pressure increases, the sheet "rounds out" to a shape more closely resembling the free-electron construction, producing a decrease in κ as experimentally observed. This suggests that the electron–electron enhancement at the transition is even larger than the κ values of Table I indicate.

The decrease in K at the highest pressures (Fig. 1) can be caused by pushing a sharp peak in the density of states through the Fermi level or, alternatively, by a slight deviation from rigid band behavior. Since the Fermi level would tend to be pinned in a high density-of-states region, the second alternative appears more attractive.

Although the Knight shift at 6.9 kbar monotonically decreases with T, the relaxation rate goes through a maximum. Figure 3 shows only one data point to support such a contention, but we have made rate measurements at 27°K and 7.5 kbar which confirm this result. Under the assumption of a nearly rigid band, the only way both T_1T and K can decrease with increasing temperature is for $N(E)$ to be larger than $N(E_F)$ at values of E within kT of E_F. Proper placement of E_F relative to the band edge allows $(T_1T)^{-1}$, which varies as $N^2(E)$, to weight the increase in $N(E)$ initially more than is possible for K, which varies directly as $N(E)$. A variety of shapes can meet these requirements.

Acknowledgments

We acknowledge useful conversations with A. C. Switendick and J. P. Van Dyke and the technical assistance of B. R. Hansen and R. L. White.

References

1. V. Jaccarino, M. Weger, J.H. Wernick, and A. Menth, *Phys. Rev. Lett.* **21**, 1811 (1968).
2. A.C. Switendick and A. Narath, *Phys. Rev. Lett.* **22**, 1423 (1969).
3. J.E. Schirber and A.C. Switendick, *Sol. St. Commun.* **8**, 1383 (1970).
4. J.E. Schirber, *Phys. Rev. B* **6**, 333 (1972).
5. J.E. Schirber, *Phys. Rev. Lett.* **28**, 1127 (1972).
6. J.E. Schirber, *Cryogenics* **10**, 418 (1970).
7. J.-P. Jan, W.B. Pearson, V. Saito, M. Springford, and I.M. Templeton, *Phil. Mag.* **12**, 1271 (1965).
8. A. Narath and H.T. Weaver, *Phys. Rev.* **175**, 373 (1968).
9. A.C. Switendick, private communication.
10. E. Ehrenfreund, E.F. Rybaczewski, A.F. Garito, and A.J. Heeger, *Phys. Rev. Lett.* **28**, 873 (1972).

Magnetic-Field-Induced Pseudo Phase Transition in Cadmium

D. E. Hagen, Wayne E. Tefft, and D. M. Sparlin

Department of Physics, University of Missouri—Rolla
Rolla, Missouri

The essential feature of the electron "transition" proposed by Lifshitz[1] is the nonanalytic behavior of the electron density of states function at the singular energy E_S corresponding to a topological change of the band structure. In the absence of level broadening due to scattering, this nonanalyticity results in singularities in the thermodynamic properties as the singular point E_S is moved past the Fermi energy.

The energy shift of each of the spin-split sheets of a band is $\pm g\mu H$. In order to generate a Lifshitz "transition," we must achieve magnetic fields such that

$$H_K = |E_S - E_F|/g\mu \tag{1}$$

where H_K is the "kink" field.

In order to experimentally achieve such fields, either g is extraordinarily large or the separation $|E_S - E_F|$ is very small, or both. Band structure calculations for cadmium[2] indicate that the states K_7, K_8, and K_9 are closely spaced (enhanced g factor) and bracket the Fermi energy (small $|E_S - E_F|$). Suppressing subscripts to emphasize the functional form, the magnetic-field-induced strain at zero stress and constant temperature is determined by

$$\varepsilon(H, T) = -\left(\frac{\partial F(H, \sigma, T)}{\partial \sigma}\right)_{H,T}\Bigg|_{\sigma = 0} \tag{2}$$

The magnetostriction therefore depends upon the stress (and magnetic field) dependence of the free energy, $F(H, \sigma, T)$.

We assume that the free energy may be separated into the sum of two terms. The first is the contribution due to those states which are near an E_S and have a large g factor and small $E_S - E_F$ (those in the immediate neighborhood of K in the Brillouin zone of cadmium, for instance). The second is due to all other states in the Brillouin zone not included in the first. The Fermi energy is assumed constant under the applied field. We obtain for the strain contribution of a single band of the above description the equation

$$\varepsilon = -\sum_i \frac{\partial E_i/\partial \sigma}{1 + \exp[(E_i - E_F)/kT]} \tag{3}$$

Assuming a rigid band approximation for the stress dependence of the energy levels near K, we write

$$\partial E_i/\partial\sigma = E_d(\partial\varepsilon/\partial\sigma) = E_d s \tag{4}$$

where E_d is an average deformation potential constant for states in the neighborhood of K_7, K_8, and K_9 and s is the appropriate elastic compliance coefficient for cadmium.[6]

We also assume that g is independent of $\mathbf{k}$ in the neighborhood of K. The sum over Landau levels in the neighborhood of K is replaced by an integral over the zero-field density of states due to the high quantum number involved.[3] The singular portion of the zero-field density-of-states function is assumed to be of the form $(E - E_S)^{1/2}$ for $E > E_S$ and zero otherwise.[1] The resulting single-band strain contribution due to such states near S is

$$\varepsilon_S = A\left\{\int_{E_S - g\mu H}^{\infty} \frac{(E - E_S + g\mu H)^{1/2}\, dE}{1 + \exp\left[(E - E_F)\, kT\right]} + \int_{E_S + g\mu H}^{\infty} \frac{(E - E_S - g\mu H)^{1/2}\, dE}{1 + \exp\left[(E - E_F)/kT\right]}\right\} \tag{5}$$

where

$$A = \left[-2^{1/2}(m^*)^{3/2}/2\pi^2\hbar^3\right] E_d s \tag{6}$$

and m^* is the density-of-states effective mass. The singular point in question seems to be level K_9.[2] This level is associated with the thickness of the web at the center of the trifoliate orbit.[3]

The complete function used to fit the experimental data is

$$\varepsilon(H) = C_1 + C_2H^2 + C_3H^4 + C_4H^6 + C_5\left\{\int_0^{\infty} \frac{x^{1/2}\, dx}{1 + \exp(x + C_6 - C_7H)} + \int_0^{\infty} \frac{x^{1/2}\, dx}{1 + \exp(x + C_6 + C_7H)}\right\} \tag{7}$$

The constant term is included to determine the recorded voltage corresponding to zero strain. The next three terms constitute the even-power series expansion of the analytic part of the magnetostriction. The final term is ε_S from Eq. (5) with a change of variable $x = (E - E_S \pm g\mu H)/kT$ (the upper sign for the first integral and the lower sign for the second), and $C_5 = (kT)^{3/2} A$, $C_6 = (E_S - E_F)/kT$, and $C_7 = g\mu/kT$.

The available magnetostriction data for cadmium[2] were fit to Eq. (7) using a standard nonlinear least-squares routine. A plot of ε/H^2 vs. H^2 as shown in Fig. 1 is a severe test of the quality of the fit. Oscillation of the data about the fitted curve, caused by thermal drift, was exaggerated at low fields by the division of ε by H^2. The anamolous contribution [Eq. (5)] is the difference between the dashed and solid curves in Fig. 1.

The parameters C_5, C_6, and C_7 are all predicted to have a definite temperature dependence. Although our temperature-dependent studies are not yet complete, one sample was recently observed at 4.0 and 1.2°K with H along the c axis. Both C_6 and C_7 increased by about 25% in going to the lower temperature, thereby de-

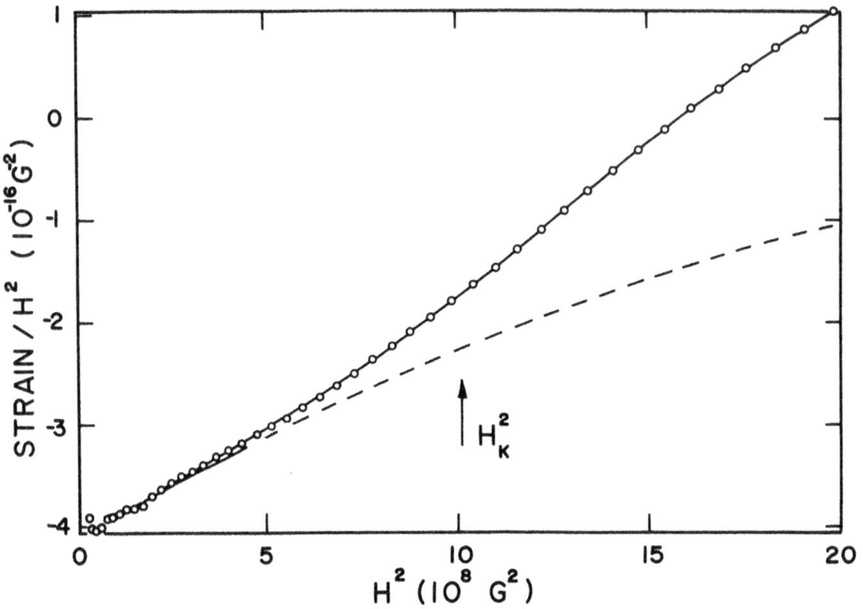

Fig. 1. A typical plot of ε/H^2 vs. H^2. The circles indicate the experimental data and the solid line indicates the best fit of the function shown in Eq. (8) to these points. The dashed line indicates the classical magnetostriction without the anomalous contribution [Eq. (6)]. The location of the "kink" field is shown by the vertical arrow labeled H_K^2.

creasing the transition width, as expected, while C_5 decreased by about 40%. These results are consistent with a Dingle temperature[7] of 10° from the behavior of C_6 and C_7. The variation of C_5 gives $T_0 = 5°K$; however, C_5 is the least reliable of all the coefficients and is discounted strongly.

The c-axis value of $C_6 = 9.0$ leads to the relation $(E_S - E_F)(\text{meV}) = \frac{3}{4}T'(°\text{K})$. For an effective temperature of $T' = 14°\text{K}$, $(E_S - E_F) = 10$ meV. Band structure calculations for cadmium having a c/a ratio near 1.865 give a value for $(E_S - E_F)$ of 120 meV for state K_9.[2] The magnetostriction samples are observed to creep for several thermal cycles before stabilizing to give the best measurements, thus increasing c/a and reducing $K_9 - E_F$.

Since $C_7 = g\mu/kT'$ contains both the g factor and the effective temperature, g is best determined from the "kink" field $H_K = C_6/C_7 = (E_S - E_F)/g\mu$. The "kink" field H_K rises from 30 ± 1 kG near the c axis to 45 ± 5 kG about 60° away. Solving for g in convenient units gives

$$g = 172(E_S - E_F)(\text{meV})/H_K(\text{kG}) \tag{8}$$

Using the c-axis value of $H_K = 30$ kG, one obtains $g = 57$ for $E_S - E_F = 10$ meV. This is a plausible value for states near K when compared with the g factors for the needles in zinc.[5]

Since the strain was measured in the basal plane, we use[9] $s = s_{11} = 1.26 \times 10^{-12}$ cm^2/dyne to obtain

$$C_5 = 0.546 \times 10^{-8}(m^*/m_0)^{3/2} E_d(T')^{3/2} \tag{9}$$

where E_d is in eV and T' in °K. Using the c-axis value of 7.5×10^{-8} for C_5, $E_d \approx 10$ eV, and $T' = 14$°K gives $m^*/m_0 = 0.09$. This value is within the range of effective masses one observes in other materials of comparable band gaps.

The angular variation of C_5 should be the same as $(T')^{3/2}$ or $(C_6)^{-3/2}$. A calculation of the ratio of C_5 to $(C_6)^{-3/2}$ showed a great deal of scatter but no very definite systematic trend. This result was taken to be consistent with a direct proportionality relation.

According to Bennett and Falicov,[5] the g factor for states near K_9 is of the form

$$g(\theta) = [B^2 + (A^2 - B^2)\cos^2\theta]^{1/2} \tag{10}$$

A plot of H_K^{-2} vs. $\cos^2\theta$ is shown in Fig. 2. This is consistent with Eq. (10), with $A = 4.3(E_S - E_F)$ and $B = 2.2(E_S - E_F)$, where $E_S - E_F$ is in meV.

As a final comment, we note that the present model predicts an anomalous contribution to the susceptibility. Since the magnetization, strain, stress, and magnetic field are related by

$$(\partial M/\partial \sigma)_H = (\partial \varepsilon/\partial H)_\sigma \tag{11}$$

we obtain for the anomalous magnetization

$$M_S = C_8\left[\int_0^\infty \frac{x^{1/2}\,dx}{1 + \exp(x + C_6 - C_7 H)} - \int_0^\infty \frac{x^{1/2}\,dx}{1 + \exp(x + C_6 + C_7 H)}\right] \tag{12}$$

where $C_8 = (2^{1/2}/2\pi^2)\,g\mu(kTm^*/\hbar^2)^{3/2}$.

Although only an order-of-magnitude estimate for C_8 is possible from the present analysis, the other parameters in Eq. (12) should be about the same as those obtained from the magnetostriction. A measurement of the monotonic magnetization along the c axis at magnetic fields in the range 10–100 kG would certainly be of great value in testing the validity of the crude model presented here.

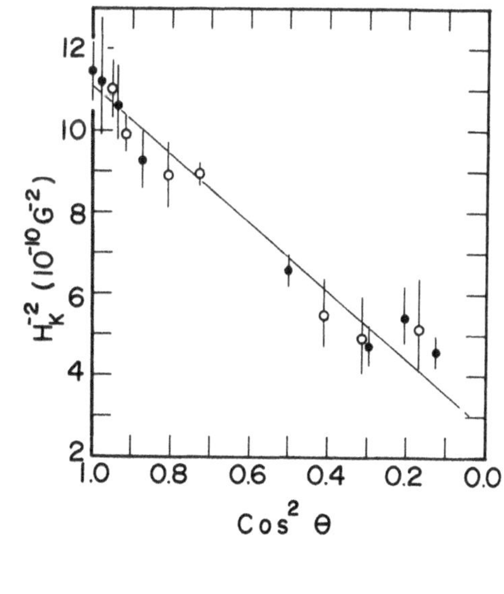

Fig. 2. The solid line is a least-squares straight-line fit to the open circle data and yields $H_K^{-2} = (2.85 + 8.06\cos^2\theta)(10^{-10}\,G^{-2})$. The error bars are the standard deviations of the values of H_K^{-2} as obtained from curve fits of the magnetostriction data to Eq. (7). The open circles are for strain along $[10\bar{1}0]$ with transverse magnetic field. The solid dots are for strain along $[11\bar{2}0]$ with transverse magnetic fields in a separate sample. They yield $H_K^{-2} = (2.69 + 8.41\cos^2\theta)$ $(10^{-10}G^{-2})$.

References

1. I.M. Lifshitz, *Zh. Eksperim. i Teor. Fiz.* **38**, 1569 (1960) [*Soviet Phys.—JETP* **11**, 1130 (1960)].
2. R.W. Stark and L.M. Falicov, *Phys. Rev. Lett.* **19**, 795 (1967); S. Auluck, private communication.
3. D.C. Tsui and R.W. Stark, *Phys. Rev. Lett.* **16**, 19 (1966).
4. J.M. Carter and D.M. Sparlin, *Phys. Rev. Lett.* **26**, 700 (1971); Thesis, Univ. of Missouri—Rolla (unpublished).
5. Alan J. Bennett and L.M. Falicov, *Phys. Rev.* **136**, A998 (1964).
6. C.W. Garland and J. Silverman, *Phys. Rev.* **119**, 1218 (1960).
7. R.B. Dingle, *Proc. Roy. Soc.* **A211**, 517 (1952).

6

Lifetimes

Audiofrequency Size Effect and Determination of the Electron Mean Free Path

H. H. A. Awater and J. S. Lass

Fysisch Laboratorium, Katholieke Universiteit Nijmegen, The Netherlands

The audiofrequency size effect occurs when the classical skin depth* $\delta_{CL} = (\frac{1}{2}\mu_0\omega/\rho_b)^{-1/2}$ of an oscillatory magnetic field penetrating a conducting slab is comparable with the thickness of the slab d. The induced electric field and currents can be thought of as a sum of all antisymmetric standing wave modes of the metal slab, which are excited with various intensities depending upon the frequency. The damping of each mode is proportional to $(\lambda_n/\pi\delta_{CL})^2$, where λ_n is the wavelength of the standing wave, $\lambda_n = 2d/(2n - 1)$, so that for sufficiently low frequencies only the first mode or the first two modes are damped appreciably. We have chosen to speak of the audiofrequency size effect because this occurs at angular frequencies $\omega \lesssim 1/\tau_1$ (or $\omega \lesssim 1/\tau_2$), where

$$\tau_n = \frac{1}{(2n-1)^2}\frac{\mu_0 d^2}{\pi^2 \rho_b} \tag{1}$$

which for typical samples of high-purity metal is around 10–1000 Hz.

The physical interest of this effect is that the expression for the surface impedance at these low frequencies is sufficiently simple to allow the bulk resistivity ρ_b and the mean free path l to be extracted from measurements of the surface impedance. The solution of the problem in the case of nonlocal conductivity ($l \gtrsim \delta_{CL}$) can be found essentially along the lines of the theory of the anomalous skin effect.[1] The surface impedance is

$$Z = -i\mu_0\omega d 4\pi^{-2} \sum_{n=1}^{\infty} (2n-1)^{-2}(1 - i\omega\tau_n)^{-1} \tag{2}$$

where the summation for $n \geqq 2$ can be approximated by a constant $K = 0.1226$ for $\omega \leqq 1/\tau_2$. The time constants τ_n are given by an expression similar to (1) where $1/\rho_b$ must be replaced by suitable Fourier transformation coefficients of the conductivity.[2] These time constants also correspond to the decay times discussed by Brandli and Cotti[3] in a similar size effect. We can make use of their tabulated calculations for τ_1 and τ_2 in the case when not only is the conductivity nonlocal, but

* Units: $\mu_0 = 4\pi \times 10^{-9}$ V sec A^{-1} cm^{-1}, ρ_b in Ω-cm.

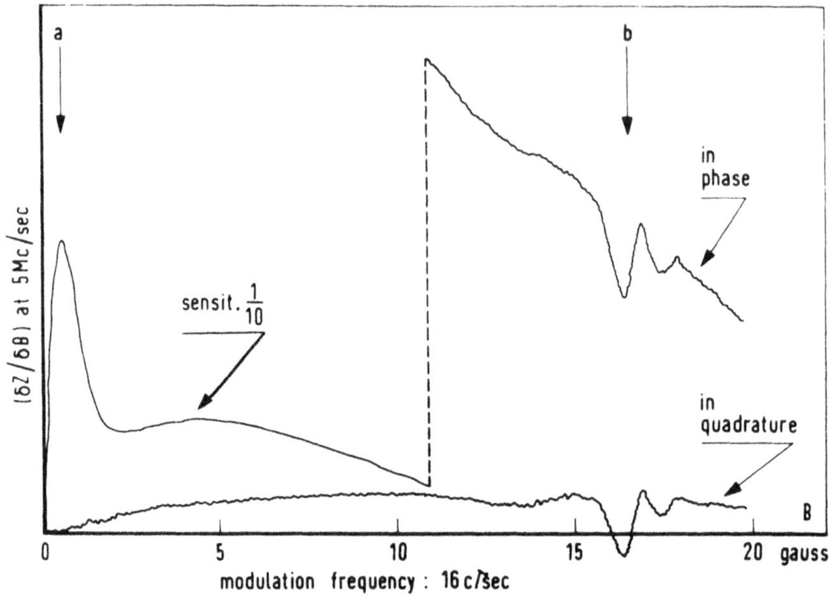

Fig. 1. Two Gantmakher experiment field sweeps with distinctive peaks of the radiofrequency surface impedance derivative. In phase with the modulation we observe a large sharp peak (a) and a gradual bump, while the signal in quadrature (note ten times higher sensitivity) shows no signs of this peak. However, the Gantmakher resonance (b) has sizable components on both curves, indicating a signal lagging by $\phi = 37°$. The resonance corresponds to a minimum orbit on the third-zone dumbbell of the FS of white tin.

also the diffuse scattering of electrons at the surface must be added to the resistance of the bulk.*

To determine τ_1 and τ_2 experimentally, it is sufficient to know the phase ϕ of the complex surface impedance, $\tan \phi = \mathrm{Im}(Z)/\mathrm{Re}(Z)$, as a function of frequency, necessitating only relative measurements. Even the simplest version of a surface impedance measurement, via the mutual impedance of two coils surrounding the sample,[4] requires an accurate determination of the dead space between the sample and the secondary coil. This can be difficult, especially if the sample cannot be made superconducting.

Therefore we investigated a new technique of measuring $\tan \phi$, which requires considerably more instrumentation, but gives us the possibility of avoiding many of the physical shortcomings of the more direct measurements.

In the following example of the audiofrequency size effect, the incomplete penetration of the low-frequency field is observed by studying the influence it has on Gantmakher resonance curves.[5] In the Gantmakher experiment (the parallel-field radiofrequency size effect) a modulated radiofrequency signal is picked up by a coil wound tightly around a thin metal slab, the modulation being caused by a low-frequency (ω) magnetic field applied parallel to the surface of the slab. Two kinds of peaks of the signal as function of a dc field can be distinguished (see Fig. 1). The one

* For diffuse scattering Eq. (2) represents an approximate solution which is exact when we assume, as in Ref. 2, that the sinusoidal standing-wave modes are uncoupled. The error involved has been shown to be at most 1% for τ_1.

at the lowest fields is due to field-induced surface states[6] which we can think of as electrons skipping along trajectories lying within 1 μm of the surface. The contribution of these states to the signal will be modulated by the oscillating field at the surface H_{surf}. On the contrary, the electrons responsible for the Gantmakher resonance peak at higher fields move on trajectories which reach exactly from one surface to the other and whose dimensions are modulated by the oscillating field averaged across the sample $\bar{H}$. Since the complex surface impedance is defined as

$$Z = |Z|\, e^{i\phi} = -i\mu_0\omega(d/2)\,\bar{H}/H_{\text{surf}} \tag{3}$$

we see that the phase lag between the two signal peaks is equal to ϕ.

Measured values of $\tan\phi$ from two field sweeps (see Fig. 1) at various modulation frequencies are given in Fig. 2. The errors were caused by poor reproducibility rather than low sensitivity. The sample was a single-crystal slab of tin with the surfaces normal to [010] and $H \parallel [001]$.

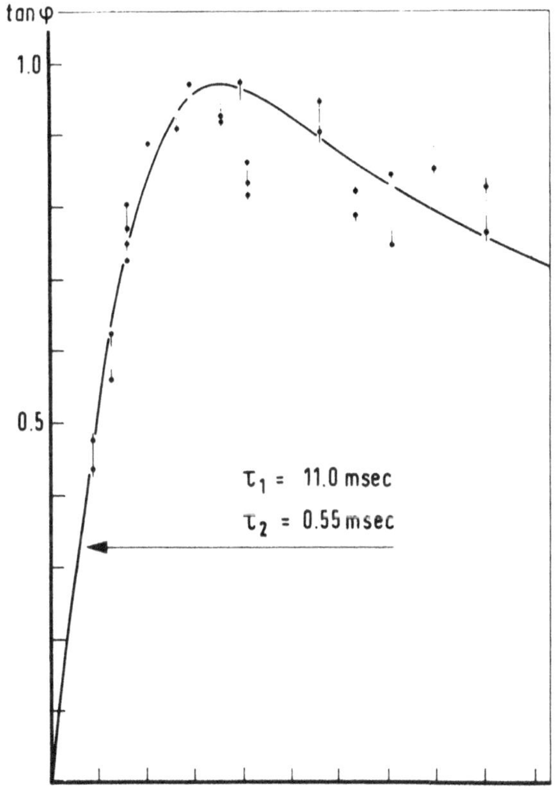

Fig. 2. Values of $\tan\phi$ for various modulation frequencies determined by evaluating curves like the one shown in Fig. 1. A curve of Eq. (4) has been fitted using two adjustable parameters τ_1 and τ_2.

The data have been fitted by a curve derived from Eq. (2):

$$\tan\phi = \omega(y\tau_1 + \tau_2)(yz + 1)^{-1} \tag{4}$$

where $y = 9(1 + \omega^2\tau_2^2)(1 + \omega^2\tau_1^2)^{-1}$ and $z = 1 + K(1 + \omega^2\tau_1^2)$. The values of the time constants τ_1 and τ_2 found in this way were used to determine the bulk resistivity $\rho_b = 7 \times 10^{-11}$ Ω-cm and the mean free path $l = 1 \times 10^{-1}$ cm (both values $\pm 50\%$) with the help of tables in Ref. 3 ($p = 0$). The present accuracy does not represent an improvement compared with the eddy-current decay method.

The uncertainties of these values canceled out somewhat in the product $\rho_b l = (7 \pm 0.7) \times 10^{-12}$ Ω-cm^2.* This reflects the fact, not mentioned by Brandli and Cotti,[3] that their numerical data lead to the following very good approximation, for $d/l \lesssim 10$,

$$\rho_b l = (5 \times 10^{-11})[(1/\tau_2) - (8.1/\tau_1)]\, d^3 \tag{5}$$

independent of the surface-scattering parameter p. This insensitivity to the surface scattering of a very low-frequency size effect is reminiscent of the anomalous skin effect.

The present method demonstrates that the audiofrequency size effect can serve as a useful corollary to the radiofrequency size effect in studying the electron mean free path.

Acknowledgments

The authors wish to thank Prof. A. R. de Vroomen and Dr. A. P. van Gelder for their useful advice. One of us (J.S.L.) is grateful for the hospitality of the University of Nijmegen.

References

1. G.E.H. Reuter and E.H. Sondheimer, *Proc. Roy. Soc. (London)* **A195**, 336 (1948).
2. P. Cotti, *Phys. Lett.* **4**, 114 (1963); *Phys. Kond. Materie* **3**, 40 (1964).
3. G. Brandli and P. Cotti, *Helv. Phys. Acta* **38**, 801 (1965); *J. Low Temp. Phys.* **1**, 153 (1969).
4. K.R. Lyall and J.F. Cochran, *Phys. Rev.* **159**, 517 (1967).
5. H.H.A. Awater and J.S. Lass (to be published).
6. K.E. Sibbald, A.L. Mears, and J.F. Koch, *Phys. Rev. Lett.* **27**, 14 (1971).
7. B.N. Alexandrov, *Zh. Eksperim. i Teor. Fiz.* **43**, 399 (1962) [*Soviet Phys.—JETP* **16**, 286 (1963)].

* Compare with 10×10^{-12} Ω-cm^2 from dc size-effect measurements[7] and the anomalous skin effect value[4] 15×10^{-12} Ω-cm^2 for the inverse of the free area of the Fermi surface $\rho_b l$.

De Haas–van Alphen Effect Studies of Electron Scattering in *Cu*H

B. Lengeler and W. R. Wampler

Kernforschungsanlage Jülich, Institut für Festkörperforschung Jülich, Germany

In the last few years several investigations of electron scattering in dilute alloys of the noble metals have been done by means of the de Haas–van Alphen effect, especially for Cu.[1–7] In this paper we report on the measurement of the Dingle temperatures for five different orbits in the system copper containing dissolved hydrogen. This system differs from the above-investigated alloys in the following points. First, hydrogen as an impurity has the simplest electronic structure of all elements. Second, hydrogen is the only element that is dissolved into interstitial lattice sites in copper (to a significant amount). This fact strongly influences the scattering of the conduction electrons by the H impurities and therefore the Dingle temperatures. The solubility and the diffusion coefficient of hydrogen in Cu have been studied extensively.[8,9] We have loaded our crystals at 870 and 1050°C at pressure up to 60 atm. To freeze in the hydrogen the samples were quenched to 77°K at a rate of 1000°C/sec. This cooling rate guaranteed that the concentration of frozen-in vacancies is always negligible compared to the hydrogen concentration. The maximum hydrogen concentration reported here is 320 ppm H (i.e., 320 atoms of H for 10^6 atoms of Cu). The amount of hydrogen dissolved was determined by heat extraction[8,9] with an accuracy of $\pm 5\%$. High-quality single crystals grown by the Czochralski technique in our crystal laboratory[10] as rods 2 mm in diameter were used throughout this investigation. No further treatment of the surface of the cylindrical samples was needed. The residual resistivity ratio of the samples as grown is 2000. Annealing in an atmosphere of 1×10^{-4} Torr O_2 increases this ratio to 20,000. We have measured the de Haas–van Alphen effect by a field modulation technique.[11] To ensure uniform penetration of the modulation field throughout the sample, a modulation frequency of 30 Hz was used with detection at the second harmonic. The maximum field strength available is 68 kG. The field homogenity is 1×10^{-5} in a sphere 1 cm in diameter. A rotation mechanism allowed the crystal to be turned through 360° around an axis perpendicular to the magnetic field. In this plane an angle can be set to within 3 min.

At given temperature T, field strength H, and sample orientation the oscillatory part of the magnetization is given by[11]

$$M = A(T, H) \sin [2\pi(F/H) - \phi]$$

when only one dHvA frequency F is effective for that orientation and when the harmonic content in M can be neglected. The frequency F is related to the extremal cross section A_{ex} of the Fermi surface by $F = (\hbar c/2\pi e) A_{ex}$. The dHvA amplitude $A(T, H)$ is given by

$$A(T, H) = B \frac{T}{H^{1/2}} \frac{\exp(-\lambda m^* X/H)}{\sinh(\lambda m^* T/H)}, \qquad \lambda = 146.9 \text{ kG/°K}$$

The constant B is only dependent on the crystal orientation. Here $m^* = m_c/m_0$ is the ratio of the cyclotron mass for the considered orbit to the free-electron mass m_0. Measuring the amplitude $A(T, H)$ at fixed temperature T as a function of the magnetic field, a plot of $\ln [AH^{1/2} \sinh(\lambda m^* T/H)]$ vs. $1/H$ gives the Dingle temperature X when the cyclotron mass m_c is known.[12] Figure 1 shows Dingle plots for four different extremal orbits in a Cu crystal ⟨110⟩ containing 72 ppm of hydrogen. Measurements of this kind have been performed for eight different concentrations C_H of hydrogen and for the following five orbits: neck, belly (111), belly (100), dog's bone, turning point 16° from (100) in a plane ⟨110⟩. Figure 2 shows the Dingle temperature for these orbits as a function of C_H. Every point in this graph is the average of either four or two values since the different orbits appear for symmetry reasons four or two times when the crystal is rotated through 360°. The values at $C_H = 0$ are measured in a Cu crystal quenched in an atmosphere of He in the same way as for a quench in hydrogen, to see the influence of rapid cooling. It is known that He is practically insoluble in Cu.[13] We believe that the Dingle temperatures are measured with an accuracy of ± 0.05°K. Table I shows the Dingle temperatures per at.% H for these five orbits in Cu. The first point to notice is the large absolute value for X, i.e., the hydrogen is a strong scattering center in Cu. Comparison with divalent (non-transition metal) impurities like Zn and Cd[3] gives a value of only 20°K/at.%

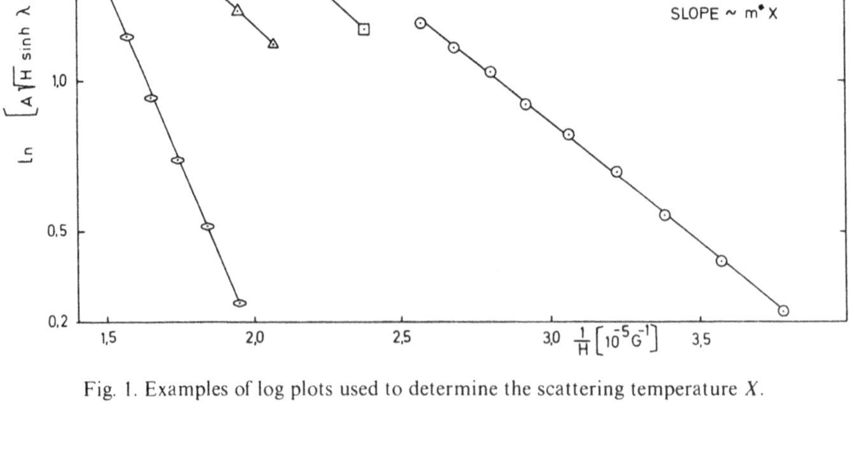

Fig. 1. Examples of log plots used to determine the scattering temperature X.

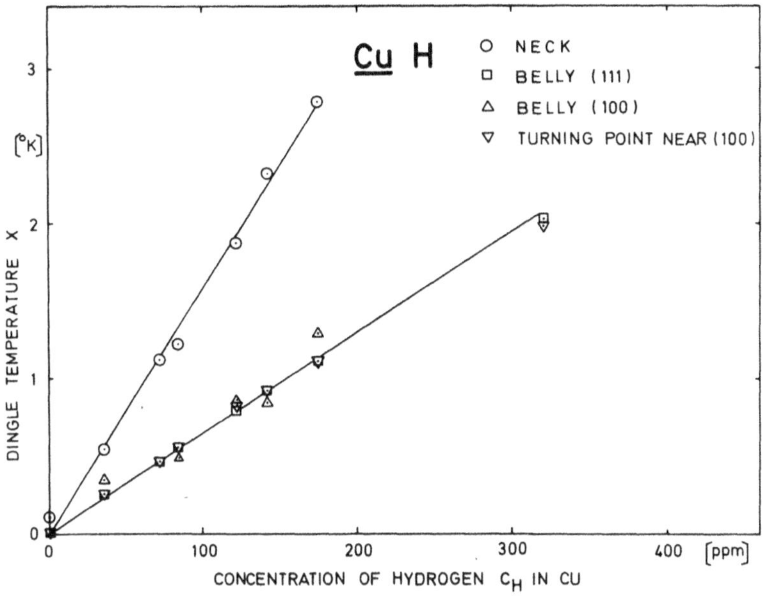

Fig. 2. Dingle temperature X vs. hydrogen concentration in *Cu*H for four different orbits.

for the neck and belly orbits. Another point to notice is that all three belly orbits have the same Dingle temperature $X = 65°\text{K/at.}\%$. The value of X_{neck} is larger by a factor of 2.5. The value of X for the dog's bone is much larger, too, although the error here is rather large. The anisotropy of the scattering can be qualitatively understood in terms of the character of the electron wave functions on different parts of the Fermi surface. It is known from the band structure of Cu[14] that the electrons on the neck regions are *p*-like, that they are *s*-like with strong admixture of *d* for the (110) regions, and *s*-like with some *p* and *d* admixture for the (100) regions. The neck electrons should therefore be scattered strongly by an interstitial impurity like the hydrogen in Cu. The rather large value of the belly Dingle temperatures indicates that the screening length of the proton is large so that the *s*-like belly electrons are

Table I. Dingle Temperatures per at. % H in Cu for Five Different Orbits

Orbit	Dingle Temperature X, °K, per at. % H in Cu
Neck	159 ± 8
Belly (111)	65 ± 4
Belly (100)	66 ± 6
Turning point near (100) in ⟨110⟩	65 ± 4
Dog's bone	140 ± 25

scattered by the hydrogen, too. If it is possible to deduce from the Dingle temperature data the distribution of the screening charge around the proton, then it is possible to better understand the solubility of hydrogen in Cu.[15]

Acknowledgements

The authors would like to thank Prof. W. Schilling for encouraging this work, Dr. W. Uelhoff and co-workers for supplying the single crystals, and Mr. J. Keppels and K. H. Klatt for careful experimental assistance.

References

1. L.F. Chollet and I.M. Templeton, *Phys. Rev.* **170**, 656 (1968).
2. M. Springford, J.R. Stockton, and I.M. Templeton, *Phys. Kond. Materie* **9**, 15 (1969).
3. I.M. Templeton, P.T. Coleridge, and L.F. Chollet, *Phys. Kond. Materie* **9**, 21 (1969).
4. D.H. Lowndes, K. Miller, and M. Springford, *Phys. Rev. Lett.* **25**, 1111 (1970).
5. M. Springford, *Adv. Phys.* **20**, 493 (1971).
6. E. Mann, *Phys. Kond. Materie* **12**, 210 (1971).
7. W. Hehl and E. Mann, *Phys. Stat. Sol.* **48**, 271 (1971).
8. W. Eichenauer, W. Löser, and H. Witte, *Z. Metallk.* **56**, 287 (1965).
9. C.L. Thomas, *Trans. Metall. Soc. AIME* **239**, 485 (1967).
10. H. Fehmer and W. Uehlhoff, *J. Crystal Growth* **13/14**, 257 (1972).
11. D. Shoenberg and P.J. Stiles, *Proc. Roy. Soc.* **A281**, 62 (1964).
12. M.R. Halse, *Phil. Trans. Royal Soc.* **265**, 507 (1969).
13. R.S. Barnes and D.T. Mazey, *Proc. Roy. Soc.* **A276**, 1360 (1963).
14. J.S. Dugdale and M. Bailyn, *Phys. Rev.* **157**, 485 (1967).
15. Y. Ebisuzaki and M. O'Keeffe, *Progr. Solid State Chem.* **4**, 187 (1967).

Quantum Interference Study of Magnetic Breakdown and Scattering Lifetimes in Magnesium*

C. B. Friedberg† and R. W. Stark

Department of Physics and The James Franck Institute
The University of Chicago, Chicago, Illinois
and Department of Physics, University of Arizona
Tucson, Arizona

We have previously reported[1] the observation of large-amplitude oscillations in the transverse magnetoresistance of Mg which result from the direct interference of coherent electron states. We remarked that this interference effect has certain unique properties which render it a useful tool for studying both electron quasiparticle lifetimes and magnetic breakdown in Mg and in dilute Mg alloys. In the following paragraphs we present the preliminary results of such studies.

Let us first illustrate the origin of the interference oscillations by considering a simple (but hypothetical) example. Figure 1 shows an open trajectory which has been constructed so that a section of the trajectory has two branches labeled A and B, which join at magnetic breakdown junctions J_1 and J_2. In an applied magnetic field $\mathbf{H}$, electron quasiparticles move from point 1 to point 2 on the trajectory via the solid-line paths A and B, with finite probability amplitudes for each. The total transmission probability $|\langle 2|1\rangle|^2$ can be calculated from the following rules in the absence of any scattering.

1. The transmitted and reflected wave functions $|t\rangle$ and $|r\rangle$ at a magnetic breakdown junction are given by

$$|t\rangle = p, \qquad |r\rangle = (1 - p^2)^{1/2} \exp(i\pi/2) \tag{1}$$

where

$$p^2 = \exp(-H_0/H)$$

and H_0 is the magnetic breakdown parameter as usually defined. This ensures both the conservation of probability and the orthogonality of $|r\rangle$ and $|t\rangle$ at the junctions. Although other choices of the absolute phases are possible, we select this convention in analogy with the phase changes which accompany the transmitted and reflected portions of a light wave at a partially silvered mirror.

2. The phase change along a path segment L due to the vector potential $\mathbf{A}$ is

* Supported by the National Science Foundation. Invited paper.

† Fannie and John Hertz Foundation Fellow. Present address: Department of Physics, University of Virginia, Charlottesville, Virginia.

given by

$$\phi_L = (e/\hbar c) \int_L \mathbf{A} \cdot dl \tag{2}$$

We may now calculate the transmission probability $|\langle 2|1\rangle|^2$ for Fig. 1 and we find that

$$|\langle 2|1\rangle|^2 = |pe^{i\phi_A}p + (1 - p^2)^{1/2} e^{i\pi/2} e^{i\phi_B}(1 - p^2)^{1/2} e^{i\pi/2}|^2$$

where we have taken J_1 and J_2 to be identical magnetic breakdown junctions with transmission probabilities p^2 onto path A. This expression easily simplifies to

$$|\langle 2|1\rangle|^2 = p^4 + (1 - p^2)^2 - 2p^2(1 - p^2)\cos(\phi_A - \phi_B)$$

which is of the form

$$|\langle 2|1\rangle|^2 = I_0 + I_1 \cos\theta \tag{3}$$

Here we have defined the phase difference $\theta \equiv \phi_A - \phi_B = (e/\hbar c)H\mathscr{A}$, where $\mathscr{A}$ is the area enclosed by paths A and B at field strength H. Thus the current along the open trajectory contains a noninterference portion I_0 and an interference portion $I_1 \cos\theta$. Note that phase coherence is a requisite for the presence of the interference term, since this depends solely on the phase difference $\phi_A - \phi_B$. Furthermore, the periodicity of the interference term is proportional to H^{-1} since the area $\mathscr{A}$ decreases as H^{-2}. By choosing the axis of our magnetoresistance sample such that it is parallel to the open orbit in question, and by applying current in this direction as well, the open orbit will largely dominate the magnetoconductivity and the interference oscillations will be readily observable as a function of H.

Before leaving this simple model let us investigate the effects of finite temperature on the amplitude of these oscillations. For this discussion we will continue to assume the ideal case of no scattering—that is, infinite quasiparticle lifetime τ. Then each quasiparticle has a well-defined energy, and hence a well-defined trajectory, as it traverses the interference region. At finite temperatures there is a distribution of such states about the Fermi energy, but each quasiparticle still has a definite energy.

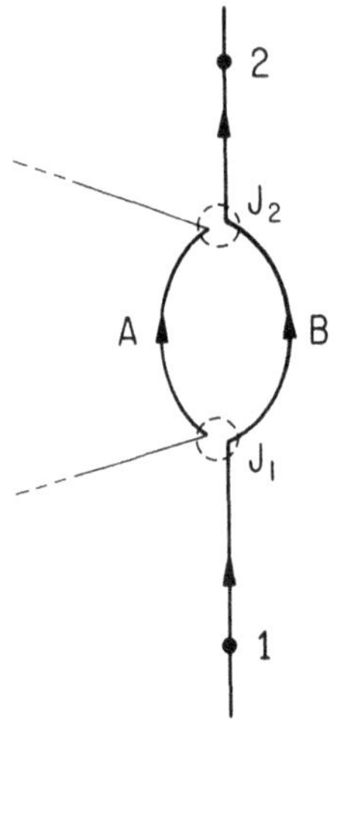

Fig. 1. Hypothetical interference paths A and B in configuration space which make up a portion of an open trajectory. J_1 and J_2 are magnetic breakdown junctions.

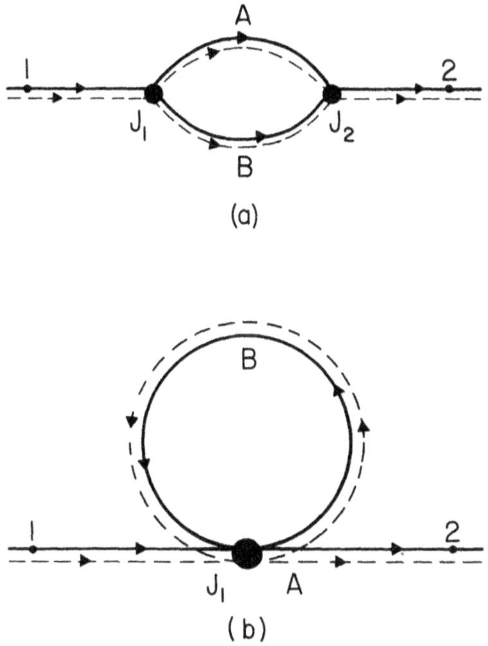

Fig. 2. Two hypothetical open trajectories for electron quasiparticles in momentum space with energy $E = E_F$ (solid lines) and $E = E_F + \Delta E$ (dashed lines).

For example, the solid-line trajectories in Fig. 2(a) could represent the trajectory in **k** space for a quasiparticle with energy $E = E_F$, while the dashed-line trajectory would be appropriate for a slightly different energy $E = E_F + \Delta E$.

In calculating the contribution of this open trajectory to the conductivity,[2] one invokes the linearized Boltzmann equation and then integrates over energy with the weighting factor $\partial f_0/\partial E$, where f_0 is the unperturbed Fermi–Dirac distribution function. The energy dependence $I(E)$ of the interference term is factorable, giving

$$I(E) = \int (-\partial f_0/\partial E) \cos[\phi_A(E) - \phi_B(E)]\, dE \tag{4}$$

where $\phi(E) = \phi(E_F) + \phi(\Delta E)$ is the phase change appropriate for a trajectory of energy E. In the case where $\phi_A(\Delta E) = \phi_B(\Delta E)$, the energy-dependent portion of the phase change cancels, and the integral reduces to

$$I(E) = \cos[\phi_A(E_F) - \phi_B(E_F)] \tag{5}$$

with no reduction in amplitude. The kT broadening of the Fermi surface has no effect on these "pure" interference oscillations.

For the more usual case let us consider the other extreme. Figure 2(b) shows another hypothetical open trajectory in **k** space which has a junction J_1 with a closed orbit. Let A represent the direct path from point 1 to point 2, with path B adding one traverse of the closed orbit. The relevant oscillatory term will then involve, as above, the phase difference $\phi_A(E) - \phi_B(E)$. In this case, however, the energy-dependent phase difference is clearly not zero, but rather

$$\phi_A(\Delta E) - \phi_B(\Delta E) = (\hbar c/eH)\,\Delta A_k = (2\pi c/e\hbar H)\,m^*\Delta E \tag{6}$$

ΔA_k being the difference in area of the closed **k**-space orbits of energy E_F and $E_F + \Delta E$. Putting this into the Boltzmann equation and integrating over energy, one obtains a term $I(E)$ in the conductivity,

$$I(E) = (X/\sinh X)\cos[\phi_A(E_F) - \phi_B(E_F)] \tag{7}$$

with

$$X = 2\pi^2 kT/\hbar\omega_c = (2\pi^2 ckT/eH\hbar)m^*$$

This is just the usual amplitude reduction factor due to finite-temperature broadening of the Fermi surface, which is generally characterized by the effective mass. This factor is appropriate for those phenomena which may be described in terms of standing wave (quantized) states, such as in the de Haas–van Alphen[3] and de Haas–Shubnikov effects.

Figure 3 shows the trajectories which give rise to interference in Mg single crystals. The darker lines represent the basic interferometer trajectories, which are similar to those in Fig. 1. There is, however, an additional junction at the exact center of the interferometer which arises from crystal symmetry and has a magnetic breakdown parameter H_2 unequal to those (H_1) at the ends of the interferometer. The transmission probability for this double interferometer can be calculated exactly as illustrated for Fig. 1, except there are now four possible paths which contribute

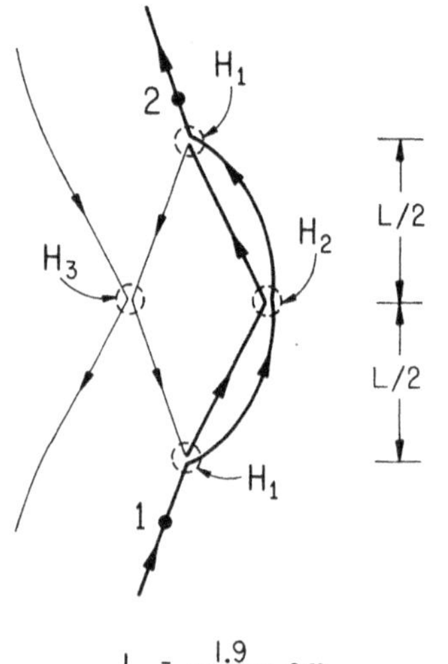

Fig. 3. Actual interference trajectories for magnesium when **H** is parallel to a $[10\bar{1}0]$-type crystal axis (not to scale).

to $\langle 2|1\rangle$. The result is

$$|\langle 2|1\rangle|^2 = I_0 + I_1 \cos\theta + I_2 \cos 2\theta \tag{8}$$

where

$$I_0 = (1 - p_1^2)^2(1 - p_2^2) + 4p_1^2p_2^2(1 - p_1^2) + p_1^4(1 - p_2^2)$$
$$I_1 = 4p_1p_2(1 - p_1^2)^{1/2}(2p_1^2 - 1)(1 - p_2^2)^{1/2}$$
$$I_2 = -2(1 - p_1^2)(1 - p_2^2)p_1^2$$

with $p_i^2 = \exp(-H_i/H)$, $i = 1, 2$, and θ corresponding to the area between the arms of half of the interferometer (i.e., the single loop). Here again we have a noninterference portion I_0 as well as two interference contributions. I_1 is the portion of the interference current which results from wave transfer across the central junction, while the second harmonic I_2 is that portion for which Bragg reflection takes place at the central junction. It is important to note that $I_1 = 0$ when $p_1^2 = 1/2$, that is, when there is equal amplitude on all four arms of the interferometer. This is due to the symmetry of the interferometer and specifically results from a cancellation between waves on trajectory segments differing by two Bragg reflections and hence by relative phase π. This vanishing of the fundamental interference frequency allows us to experimentally determine H_1 *directly*.

Another feature of the trajectories in Fig. 3 is the presence of a "feedback loop" which passes through the magnetic breakdown junction H_3 and enables certain wave components to interfere on multiple passes through the interference region. It is easily seen that this feedback loop introduces higher harmonic content to the interference current (i.e., $n\theta$ terms) as well as standing wave components similar to those discussed in connection with Fig. 2(b). These standing wave components are, distinguished, of course, from the interference portion by their strong temperature dependence. We should also point out that there is another path leading to the interferometer through tunneling at junction H_3. Amplitudes incident from this source must also be included in calculating interference oscillation amplitudes.

Figure 4 shows the transverse magnetoresistivity of a Mg single crystal as a function of H where the direction of $\mathbf{H}$ is perpendicular to the plane of the trajectories in Fig. 3 and the current direction $\mathbf{J}$ is along the open trajectory. The lower curve was taken at a temperature of 4.2°K and the upper curve at 1.5°K. The low-frequency oscillations in both curves correspond to the θ-component of the interference current. There is also visible a smaller amount of the 2θ (second harmonic) component, especially at the lower field strengths. The standing wave components are evident in the 1.5°K data and appear as the high-frequency oscillations. The amplitude of these oscillations is strongly temperature dependent due to the damping factor in Eq. (7) appropriate for an effective mass $m^* \sim 0.20m_0$. On the other hand, the interference oscillations which are visible on these curves suffer virtually no measurable change in amplitude in this temperature range. They are insensitive to the intrinsic broadening of the Fermi distribution.

The stringent requirement of phase coherence for observation of the interference oscillations means that the quasiparticle lifetime τ must be longer than the time t required for the quasiparticle to traverse the interference region, since *any* scattering event is catastrophic in the sense of destroying phase coherence. Mathematically, we can write the probability for a quasiparticle to retain phase coherence as $\exp(-t/\tau)$, so that the wave amplitude should be reduced by $\exp(-t_L/2\tau)$ for each path segment L. For example, in our simple model of Fig. 1, the interference term will contain a factor

$$\exp[-(t_A + t_B)/2\tau]$$

t_A and t_B being the traversal times along paths A and B, respectively. These time intervals are proportional to H^{-1}.

For the Mg interferometer the traversal times along the various path segments are known from the detailed band structure of the metal. Hence a measure of the amplitude of the interference oscillations gives us information about the quantum lifetime τ. In particular we have both measured experimentally and calculated numerically the ratio of the oscillation amplitudes for the fundamental (θ) and second harmonic (2θ) interference components. The calculation takes into account all wave components which pass through the interference region up to four times via the feedback loop. Contributions from coherent and incoherent wave amplitudes are tabulated in the proper proportion using the exponential factor given above. The calculation thus involves a total of four parameters—those characteristic of the magnetic breakdown junctions, H_1, H_2, and H_3, and the quantum lifetime τ.

Experimentally we have observed both the θ and 2θ interference components down to fields as low as 1.25 kG at 1.5°K using field modulation techniques. We have made accurate measurements of the *ratio* of the amplitudes of these two frequencies down to 2.0 kG in pure Mg. The amplitude ratio is chosen for comparison to the

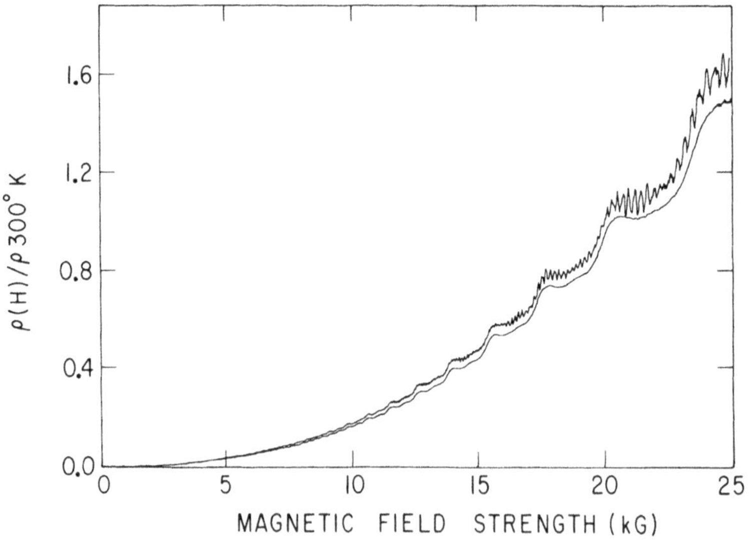

Fig. 4. Transverse magnetoresistivity of a magnesium single crystal for **H** along a [10$\bar{1}$0]-type axis and the current **J** along a [11$\bar{2}$0]-type axis. The temperature T is 4.2 K for the lower curve and 1.5°K for the upper curve. The low frequency oscillations which appear with equal amplitude in both curves are due to interference. The high-frequency oscillations, which exhibit strong temperature dependence, result from standing wave components.

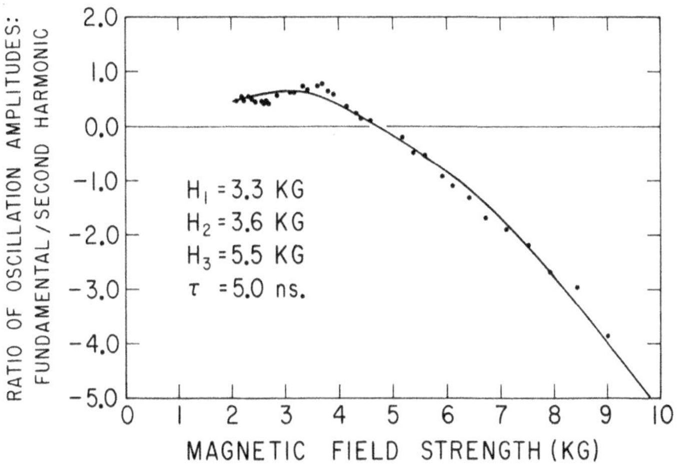

Fig. 5. Interference oscillation amplitude ratio (fundamental/second harmonic) for high-purity Mg at 1.5 K. The smooth curve is the result of our calculation as described in the text.

calculation rather than the absolute amplitudes since the former is independent of the background resistivity.

The resulting amplitude ratio for high-purity Mg is shown in Fig. 5 for the field range 2–10 kG. The points are experimental while the smooth curve is the result of our calculation. The fit was obtained by variation of the four parameters H_1, H_2, H_3, and τ. The most notable feature of these data is the vanishing and subsequent reversal of sign of the fundamental (θ) amplitude, and hence the ratio, at a value of $H = 4.8$ kG. As remarked above, this cancellation occurs when $p_1^2 = \exp(-H_1/H) = 1/2$. This fixes H_1 at 3.3 kG independent of models for the full conductivity and independent of Fermi statistics. The remaining parameters are determined by fitting the data points for the entire field range in Fig. 5. The results for the curve given are $H_2 = 3.6$ kG, $H_3 = 5.5$ kG, and $\tau = 5.0$ nsec.

At $H = 1.25$ kG the length L of the interferometer is ~ 0.02 mm and the presence of the interference oscillations at this field strength indicates truly macroscopic phase coherence. However, even for these large distances the traversal time t is only on the order of 0.04 nsec, a small fraction of the lifetime quoted above. In fact, even at 2 kG the amplitude ratio is highly insensitive to lifetime effects for $\tau \geqq 0.1$ nsec. One would have to go to $H \sim 10$ G in order to increase t sufficiently for scattering to play a dominant role. Since the background resistivity and absolute amplitudes decrease as approximately H^2, measurements in this range of field are beyond our sensitivity. The loss of interference signal below 1.25 kG is not a lifetime effect. If it were, the relative amplitude of the fundamental to second harmonic would be a rapidly increasing function with decreasing H. This is clearly not the case in Fig. 5. The data are consistent with the quoted value of $\tau = 5.0$ nsec, although this value must be considered an "order-of-magnitude" number in determining τ from this experiment alone.*

* Recent magnetoresistance measurements for Mg with **H** parallel to [0001] indicate the presence of Pippard's quasiparticle or magnetic band states for the phase-coherent coupled orbit network. These data imply that a lifetime $\tau \geqq 5.0$ nsec is quite realistic.[4]

By introducing coherence limiting perturbations into the pure Mg, one can drastically reduce the quantum lifetime. We have grown single crystals of dilute Mg host alloys—specifically, Mg–Cd and Mg–Zn. Similar measurements for these samples yield values of τ in the range where the lifetime may be more accurately determined. Typical data give values of $\tau \sim 0.02$ nsec with $H_1 = 3.4$ kG, $H_2 = 4.8$ kG, and $H_3 = 5.5$ kG for a Mg–Cd alloy containing on the order of 1 ppm (atomic) Cd, and $\tau \sim 0.015$ nsec, $H_1 = 3.6$ kG, $H_2 = 5.0$ kG, and $H_3 = 5.5$ kG for a Mg–Zn alloy with roughly the same concentration. Thus the effect of impurity content on the order of 1 ppm is to drastically reduce the quantum lifetime (by two orders of magnitude) and also to cause increases in the magnetic breakdown parameters.

Since the lifetime τ can be expressed as

$$1/\tau = (1/\tau_0) + (1/\tau_{e-e}) + (1/\tau_{e-ph})$$

or

$$1/\tau = (1/\tau_0) + AT^2 + BT^3 \tag{9}$$

where τ_0 is the impurity and defect-induced lifetime, one could expect to find variations in the measured amplitude ratio as a function of temperature. Hopefully the electron–electron (T^2) contribution and the electron–phonon (T^3) contribution can be separated. Our preliminary measurements in the temperature range 1.5–4.2°K show only small variations in the absolute amplitudes even at the lowest practical field strengths. These variations are only slightly above the noise level. However, the amplitude ratio can be fit as $\exp(-aT^x)$, where $2 \leqq x \leqq 3$.

In summary, we have made the first direct measurement of magnetic breakdown parameters of a metal by exploiting the symmetry of the interferometer trajectories. We have also determined the quantum lifetime τ of the electron states in high-purity Mg and in very dilute Mg alloys from measurements of the ratio of amplitudes of the fundamental and second harmonic interference oscillations. Preliminary measurements of the temperature dependence of the lifetime τ have been made in the limited range of 1.5–4.2°K. The relatively small changes observed in τ should become more easily measurable at higher temperatures (say 20°K), unobscured by the kT broadening factor which dominates amplitude measurements to a large degree in other experiments. We have, in fact, observed the interference oscillations with sizable amplitude in the dc magnetoresistance at temperatures $T \sim 15$°K. Since the intrinsic lifetime in our pure Mg crystals is very long (5×10^{-9} sec), there is therefore good reason to believe that by extending our measurements to these higher temperatures we will be able to separate the T^3 component to the amplitude variation and thus unmask the electron–electron scattering lifetime.

References

1. R.W. Stark and C.B. Friedberg, *Phys. Rev. Lett.* **26**, 556 (1971).
2. A.B. Pippard, *Proc. Roy. Soc. (London)* **A287**, 165 (1965).
3. A.V. Gold, in *Electrons in Metals, Simon Fraser Summer School Lectures, Solid State Physics, 1963*, J.F. Cochran and R.R. Haering, eds., Gordon and Breach, New York (1968). Vol. 1.
4. A.B. Pippard, *Phil. Trans. Roy. Soc.*, **A256**, 317 (1964); C.B. Friedberg and R.W. Stark (to be published).

Electron Scattering in Metals As Seen in the Landau Quantum Oscillations*

R. J. Higgins, H. Alles, Y. K. Chang, and D. W. Terwilliger

Department of Physics
University of Oregon, Eugene, Oregon

A variety of Fermi surface mapping techniques are now well established.[1,2] The strength of these techniques is that only particular groups of electrons show up, selected by a phase cancellation or similar extremal condition. Since the measurement itself acts as a spectrometer, comparison with theory is direct and fairly unambiguous, especially by comparison with macroscopic transport properties which average over the entire Fermi surface. The principal technique to be discussed is the de Haas–van Alphen effect, an oscillation in magnetic susceptibility which is a periodic function of the inverse of the applied magnetic field, and related effects which have their origin in the quantization of electron orbits in a magnetic field (Landau levels). This class of measurements will be referred to here as the Landau quantum oscillations (LQO).

Interest in fermiology has recently turned from the geometric toward the dynamical properties of electrons, since it is the electron's response to its environment (becoming a quasiparticle in the process) which determines transport, magnetic, and superconducting behavior. The quasiparticle effective mass m^* (and the related Fermi velocity v_F) is measurable by cyclotron resonance or by the temperature dependence of the LQO amplitudes. The quasiparticle lifetime is also directly measurable, for example, by the dependence of LQO amplitudes on magnetic field. This technique is only recently emerging as a useful one, because of the problems in making reliable amplitude measurements. However, recent results are encouraging, giving direct evidence on problems of scattering of electrons from phonons, lattice imperfections, and impurities.[3] The results of these measurements should be of importance to more general problems such as the alloy problem and the origins of magnetism, because of certain special features of LQO information: (1) The LQO act as a spectrometer, singling out well-defined groups of electrons. (2) The LQO measure the actual lifetime of a state, unweighted by the scattering angle as in resistivity. As a result, the LQO relaxation time "sees" small-angle scattering events. (3) Orbits on different locations of the FS may have wave functions of different symmetry character. Comparisons of scattering from different orbits may therefore reveal much about the nature of the scattering.

* Research supported by the National Science Foundation under Grants GP-25532, GH-33775, and GJ-32786. Invited paper.

Scattering Temperatures in "Real" Crystals

The results of scattering temperature (X) measurements in the past have often had a bad reputation. Scattering temperatures in nominally pure crystals ranged about 1 deg, when purity alone ought to have given a factor of ten lower, and did not decrease with increasing purity. The X value varied from sample to sample, and alloying results from different labs did not agree either in size or anisotropy. These fluctuations, generally ascribed to crystal mosaic structure, obscured the results of such scattering measurements. This situation is now rather well understood, and a well-defined list of procedures exists for ensuring reliable X results.[4] The results of scattering temperature measurements in crystals with a carefully controlled array of lattice dislocations give surprisingly large values (orders of magnitude larger than resistivity would indicate) and suggest that the phase-coherent Landau level is a sensitive indicator of small-angle scattering.[5]

Dingle Scattering Temperature Errors Due to Crystal Substructure. It is necessary to be precise in defining what one means by mosaic structure. A nominally single crystal generally consists of many subgrains, regions with a relative rotation of a fraction of a degree. Adjacent regions may contribute out of phase, due to the angular dependence of FS cross section, and this phase-smearing error is field dependent, giving an error in X.[6] *Within* a subgrain, nonuniform arrays of lattice dislocations may occur in growth or handling. Within such an imperfect region the dHvA amplitude is wiped out, so that relatively perfect regions dominate. But since the "active" volume is again field dependent, there is an X error.

The X error due to subgrain phase smearing gives in general a spurious *anisotropy* to the results. This follows since the X error goes as $F''\phi^2$ at a symmetry direction and as $F'\phi$ away from symmetry (where ϕ is a measure of the subgrain tilt), and the second term is generally a good deal larger than the first. An example calculated for copper is shown in Fig. 1. As a result, one should be suspicious of scattering anisotropy results which show a sharp dip at symmetry directions, until the subgrain structure has been well determined for the samples. The most reliable X maps use only data from symmetry directions (and turning points) where this error is least, but even

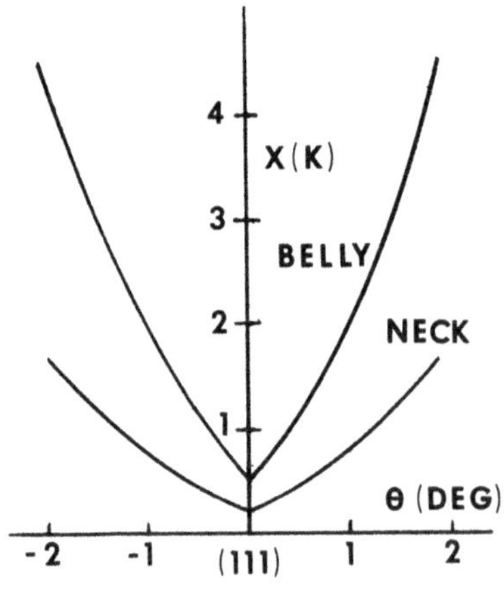

Fig. 1. X error due to crystal subgrain structure, shown as a function of magnetic field direction. In this example experimental values of F' and F'' for copper were used and a root mean square subgrain tilt of 0.5° was assumed.

then one requires accurate crystal orientation ($<0.1°$) for even the best crystals (subgrain tilts $<0.1°$). Miller *et al.*[7] have suggested that the X error can be used to *characterize* the mosaic structure, once the X map has been fitted using data at symmetry directions.

Similarly, in a study of impurity scattering it is essential to characterize both pure and alloy specimens, since the crystal substructure may be a rapid function of impurity concentration. In normal crystal growth constitutional supercooling can lead to a network of tilted subgrains absent in a pure crystal grown under identical conditions. If there is a substantial size mismatch, the impurity can also introduce extra strain dephasing, either in the form of extra lattice dislocations or in an elastic strain field surrounding each impurity. Coleridge[8] has shown that this last situation can lead to a contribution to X which is itself field dependent, and may be an appreciable fraction of what is measured. In *Cu*–Au, which should in principle be a simple system to understand, this contribution is said to be nearly 50%, and may well account for the discrepancy of a factor close to two in interpreting the measured magnitudes via potential scattering.

The measured residual resistance ratio is not at all a good monitor of the scattering seen by a Landau level when long-range (small-angle) scattering disturbances are present. For example, oxygen-annealing a noble metal crystal with transition metal impurities can dramatically increase the measured residual resistance ratio, presumably tying up the impurity into a more weakly scattering oxide complex. Yet the measured X values shown in the second column of Table I remain higher by several tenths of a degree than what one would estimate due to other impurities and dislocations. Presumably, accompanying the impurity complex is a long-range elastic strain field which readily dephases the Landau level. By contrast, in a crystal which is pure and relatively perfect it is indeed possible to obtain scattering temperatures of a few hundredths of a °K, well within the estimate of impurity background, as shown in the third column of Table I.

A random array of lattice dislocations also produces an increase in X far larger than resistivity[5] would indicate, as shown in the fourth column of Table I. From the point of view of scattering studies in general, these measurements confirm the supposition that the 1 deg or so uncertainty often prevalent in X measurements has an origin in "mosaic structure."

Dislocations and the dHvA Effect in Copper. A systematic study of the effect of lattice dislocations on X in copper by Terwilliger and Higgins[4] (TH) reveals X values which are both large and anisotropic, varying by a factor of four from neck to belly. The interpretation of these results leads to a better understanding of what is meant by a scattering event as seen by a Landau level, and sheds some new light on the old problem of electron scattering from a dislocation strain field.

The direction of this anisotropy follows the pattern observed by Koch and Doezema[9] in surface-state scattering by *phonons*, and we note that the dislocation may be viewed as a frozen phonon field. However, the anisotropy also follows that observed for the response of the FS *cross section* to a *homogeneous* elastic strain.[10,11] Watts[12] has interpreted these results by phase-smearing arguments on a microscopic scale combining the spatially varying dislocation strain field with the sensitivity of the FS to strain. The theory could be adjusted to fit the TH results quite

Table I. Effect of Oxygen Annealing on Initially "Dirty" and "Clean" Crystals*

	Crystal			
	Cu 859 "dirty"	Cu Blind baffle "clean"	Cu 812 "dislocated"	
Estimated impurities, ppm	10 other + (20 Fe)	8 other + (1 Fe)	14 other + (< 1 Fe)	
Residual resistance ratio (before) → after oxygen anneal	(70) → 2500	(1400) → 4300	(2000) → 2500	
Dislocation density, 10^7 cm^{-2}	0.3	0.01	2.5 ± 0.5	
X_N^{est}, °K				
Other impurities	0.04	0.03	0.04	0.04
Dislocations	0.10†	0.00	[0.00]‡	0.80 ± 0.15†
SUM	0.14	0.03	[0.04]‡	0.85 ± 0.15†
X_N(meas), °K	0.28 ± 0.05	0.03 ± 0.04	1.0 ± 0.1	
X_B^{est}, °K				
Impurities	0.06	0.02	0.03	0.03
Dislocations	0.03	0.00	[0.00]‡	0.20 ± 0.04†
SUM	0.09	0.02	[0.03]‡	0.23 ± 0.04†
X_B(meas), °K	0.28 ± 0.05	0.02 ± 0.04	0.4 ± 0.1	

* Estimates of X (impurity) are made on the assumption that the resistivity after oxygen annealing is a measure of the concentration of non-transition metal impurities, labeled "other," and that the change in resistivity on oxygen annealing is a measure of the Fe content.

† X from direct measurement.

‡ X estimated from resistivity.

well, fitting the results of Coleridge and Watts[13] (with rather fewer data and some inhomogeneity of D) rather less well. The parameters coming from the fit were in plausible agreement with estimates from the existing data on FS response to homogeneous elastic strain.

It would seem at first that these are two conflicting explanations for this experiment, one as scattering, the other as dephasing. On closer examination, it appears that there is no clear distinction between small-angle scattering and dephasing within an orbit. The orbits are fairly large compared with dislocation spacing, and since the strain field is long range ($1/r$), an electron samples a wide variation of strain *within* a given orbit. Rather than a summation of nonequivalent orbits (the usual phase-smearing idea), we are in a regime where dephasing occurs within individual orbits. Thus the distinction between inhomogeneous broadening and homogeneous broadening of the Landau levels (borrowed from the language of resonance measurements) is not appropriate here. The scattering angle required to destroy phase memory is quite small, of order $1/n$, where n is the phase F/H, which may typically be 1000

* We were reminded of Pippard's arguments by P. T. Coleridge.

or more. This result follows by applying Pippard's notions[14]* of the phase change on switching orbit centers (developed for the magnetic breakdown situation) to the present case, as shown in Fig. 2. Here the switch in orbit centers occurs continuously due to the elastic shift of the lattice in the region near the dislocation. This gives the scattering angle criterion just mentioned, and a general prescription for the reduction in dHvA amplitude identical in form with that assumed by Watts.[12]

Watts assumed that the electrons travel within a strain distribution, due to the dislocations, which has Gaussian form. Although plausible, this leads to a prediction of nonlinear dependence of X on D. The TH data[4] show no clear evidence for the curvature predicted by the Watts theory, fitting a simple straight line with about a factor of two less error over the range of D we believe reliable. But the difficulty in determining D and its fluctuations makes this a hard test to pin down. The same theory predicts a field-dependent X, manifested in a curvature of the usual log plot of amplitude. The data display a factor of two less error in fitting to a conventional (field-independent) X than to the Watts theory, but since the accuracy in amplitude measurements is also of the same order (1%), this test is just barely significant.

What one is trying to do is distinguish between two models for the Landau level line shape, but with experimental information existing only over a narrow range of magnetic field. The problem is analogous to making a Kramers–Kronig transform in optical measurements when data exist only over a finite wavelength range. Although deviations from linearity over a narrow field range may be small enough to escape detection, the differences in what one defines as X (the slope) may be large, since the relationship between X and scattering strength is sensitive to choice of line shape.[5] In view of the present limitations in interpreting experiment, we are led to look for ways of using more of the information available, in an attempt to characterize Landau level line shape. A promising approach seems to be the use of the harmonic content of the LQO oscillations, i.e., wave shape analysis.

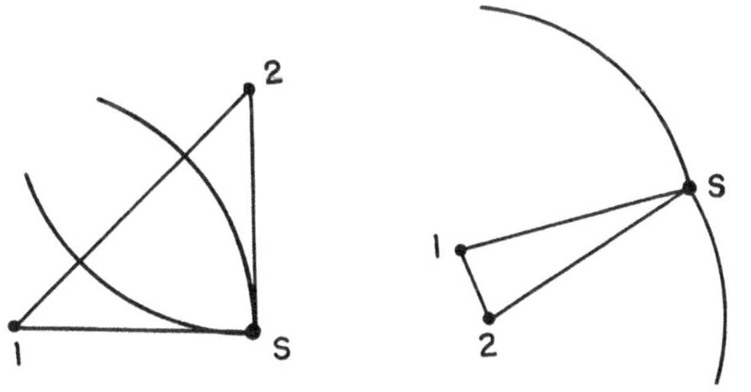

Fig. 2. Orbit center switching model for small-angle scattering. Left: Orbit center switching upon Bragg reflection at S. The phase change on switching is the area 1-S-2, as applied by Pippard[14] to the magnetic breakdown problem. Right: The analogous geometry when the orbit center switch is due to small-angle scattering at S.

Scattering Information from Landau Quantum Oscillation Wave Shape and an Application to the Magnetic Impurity Problem

The central question of the magnetic impurity problem is to understand the conditions under which a transition metal impurity will display a magnetic moment, and how this is connected with the host band structure and impurity–host interaction.[15,16] Part of this larger problem is the resistance minimum phenomenon, or Kondo effect, whose origin is still the subject of some debate between adherents of spin-compensated state and spin-fluctuation models. LQO techniques have special assets for such studies which are only beginning to be exploited.

1. The extremum condition singles out a well-defined group of electrons. Observation of scattering *anisotropy* (inaccessible by ordinary transport) may arise because of the differing symmetry character of wave functions on different locations on the FS. Explanation of such anisotropy may therefore reveal the basic origin of the interaction, in a way resembling an experimental partial wave analysis.

2. Since the LQO measure the actual lifetime of the state, unweighted by scattering angle, small-angle scattering events indicative of long-range interactions are accessible. This ought to be of special interest in probing the T and H dependences of the scattering below the resistance minimum, giving a handle on the question of the long-range spin polarization cloud.

3. In favorable cases the LQO can resolve spin-split Landau levels. If one could resolve the scattering for both spins separately (analogous to a spin-polarized beam in nuclear physics), a great deal of information on the location of the impurity level and the impurity–electron interaction would come out which is normally lost in averaging in ordinary measurement such as resistivity.

An elegant example of how useful LQO techniques may be is the work of Lowndes *et al.*,[17,18] who measured the τ map for Au–Fe (where the orbital average X was unfolded to give a pointwise τ). There is a large (5 : 1) anisotropy, and the scattering is dramatically large at [101]. This is the near-neighbor direction in real space where core overlap is most important, and the Au wave function is most d-like. The most d-like electrons are therefore most strongly coupled to the (d-like) Fe impurity, an important clue to the origin of a variety of magnetic impurity phenomena. The connection with wave function symmetry has been explicitly verified in a calculation for Cu alloys by Blaker and Harris[19] and also by Coleridge.[20]

Coleridge and Templeton[21] have observed a spin-splitting beat in Cu–Cr appearing as a function of magnetic field. Spin-split zeros are observed in LQO whenever

$$(m^* g/m_0) = 1, 3, 5, \ldots$$

Ordinarily, a spin zero is observed by tipping the field until the m^*g combination comes out right, and the observation of a beat at fixed angle as a function of H requires a field-dependent g factor. Coleridge and Templeton have interpreted this observation to give an estimate for the exchange energy coupling electron to impurity, and have extended these observations to follow spin zeros with other impurities.[22] With Fe, in contrast to Cr, the ordinary spin zeros rapidly disappear, which is evidence for spin anisotropy of scattering (and indirect evidence about the location of the Fe impurity level about E_F).

We have begun a series of measurements of scattering in magnetic impurity systems, using waveform analysis to extract rather more information than comes from the usual Dingle plot.[23] For example, when X is field dependent, an X obtained as the slope of the usual Dingle plot has little meaning. However, it is possible to obtain a better measure of X and its field dependence through the relative LQO harmonic amplitudes:

$$X = -\frac{mH}{m^*K}\ln\left[\frac{A_2\sqrt{2}\cos(\phi/2)\sinh 2x}{A_1\cos\phi\sinh x}\right]$$

where $K = 2\pi^2/ck_B/e\hbar = 146.9$ kG/°K, $\phi = \pi g m^*/m$, and $x = Km^*T/mH$.

In cases where spin-split beats are observed, second harmonic spin-split beats occur twice as often in H, and interlace the first harmonic beats, offering a way to follow the field dependence of the exchange energy. Finally, a measurement of the relative phase of the LQO harmonics offers a rather general way to measure spin anisotropy of scattering, even when dramatic beat minima are absent.

In this experiment a laboratory computer is used not only to digitize and analyze data, but also to control many phases of the experiment. Using the computer, it is also possible to do phase-sensitive detection of quadrature components of many harmonics of the modulation frequency *simultaneously*. This is of special importance in recovering LQO harmonic information, which rapidly disappears due to impurity scattering, since each successive LQO harmonic is reduced by an *additional* factor

$$\exp(-2\pi^2 k_B X/\mu_B^* H)$$

Using the spectrometer possibilities of the modulation technique, however, it is possible to enhance relative harmonic content by a factor of about 10^5 by detecting at higher harmonics. We are therefore able to measure LQO second-harmonic amplitudes, to an accuracy of a few percent, which are only about one part in 1000 of the Lifshitz–Kosevich magnetization. An example is shown in Fig. 3. Simultaneous detection of multiple modulation harmonics is essential in preserving LK harmonic phase information, and is also an asset in unraveling several other effects which can alter LQO harmonic content, such as magnetic interaction and eddy currents.

Using these techniques, we have directly measured a strong-spin anisotropy of scattering in Cu–Fe, using the relative LQO phases, as shown in Fig. 4. The quantity plotted is the relative harmonic phase $2\theta_1 - \theta_2$, where θ_1 is the first harmonic phase and θ_2 is the second harmonic phase determined by a Fourier analysis. This quantity is independent of the location in $1/H$ of the data window, i.e., does not require knowledge of the absolute phase or zero-point phase of the Landau levels. Shown for comparison is a measured relative phase for pure copper, which has the appropriate value of $\pi/4$ as predicted by the Lifshitz–Kosevich theory.

The origin of the $A_1 : A_2$ relative phase shift may be seen in a phasor diagram (Fig. 5) for the magnetization amplitudes of both spins for the first and second dHvA harmonics. The anisotropy parameter[24] δ provides a simple relation between the four amplitudes:

$$A_1^{\uparrow}/A_1^{\downarrow} = e^{2\delta}, \qquad A_2^{\uparrow}/A_2^{\downarrow} = e^{4\delta}$$

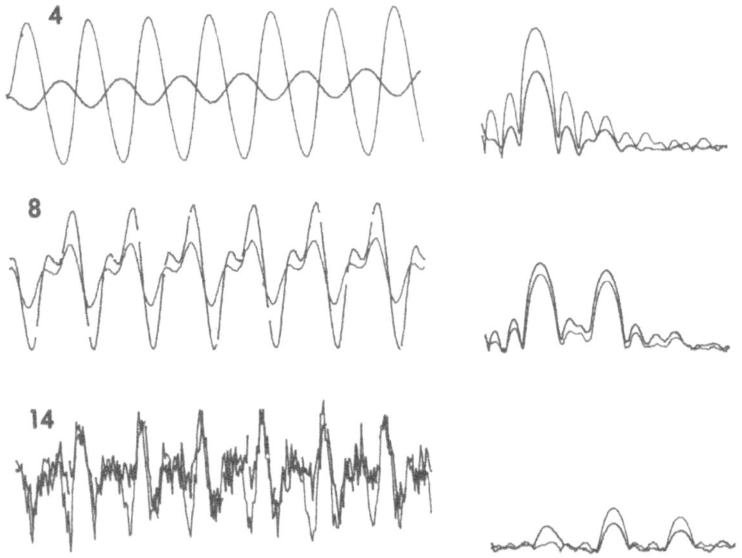

Fig. 3. Example of simultaneous phase-sensitive detection of multiple harmonics, for a Cu–100 ppm Fe crystal. Left: Neck oscillations as seen at quadrature phases in the 4ω, 8ω, and 14ω channels, demonstrating increasing dHvA harmonic content in the higher channels. Right: Fourier transform (amplitude on a log scale) of the same data.

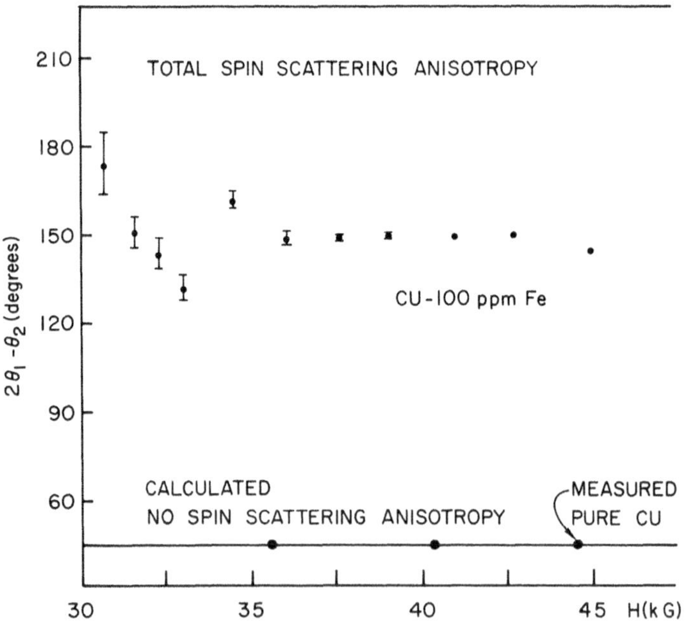

Fig. 4. Measured relative phase $2\theta_1 - \theta_2$ for neck orbit in Cu–100 ppm Fe, demonstrating spin-scattering anisotropy. Shown for comparison are measured relative phases for pure Cu.

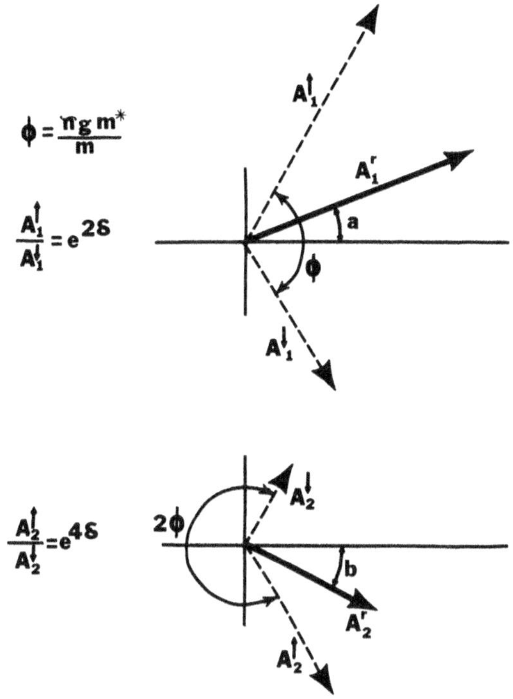

Fig. 5. Vector plot illustrating phase relationships for first and second harmonic contributions to the Landau quantum oscillations in the presence of spin-scattering anisotropy.

where δ is related to the Dingle temperature by

$$X^{\uparrow\downarrow} = X \pm (\hbar\omega_c/2\pi^2 k_B)\,\delta$$

Elementary trigometric manipulations yield the following relations among the quantities in Fig. 5:

$$a = \tan^{-1}\left[\left(\tan\frac{\Phi}{2}\right)\frac{e^{2\delta}-1}{e^{2\delta}+1}\right], \qquad b = \tan^{-1}\left[\left(\tan\Phi\right)\frac{e^{4\delta}-1}{e^{4\delta}+1}\right]$$

$$A_1^r = A_1^{\uparrow}\cos(\tfrac{1}{2}\Phi - a) + A_1^{\downarrow}\cos(\tfrac{1}{2}\Phi + a)$$

$$A_2^r = A_2^{\uparrow}\cos(\Phi - b) + A_2^{\downarrow}\cos(\Phi + b)$$

The quantities measured are

$$A_1^r/A_2^r \qquad \text{and} \qquad 2\theta_1 - \theta_2 = 2a - b \mp \tfrac{1}{4}\pi$$

where the minus (plus) sign is for a maximum (minimum) orbit on the Fermi surface.

In CuFe the ln A_1 vs. $1/H$ and A_2 vs. $1/H$ plots at 1.2°K are straight lines in the range of 25–45 kG, and give a Dingle temperature of about 1°K. From this an A_1/A_2 ratio may be calculated using Lifshitz–Kosevich theory. The measured

A_1/A_2 ratio and measured phase $2\theta_1 - \theta_2$ may then be used with the LK calculated A_1/A_2 ratio to deduce δ and Φ as a function of magnetic field.

These preliminary results demonstrate the potential of waveform analysis for the study of magnetic impurities in metals.

References

1. J.M. Cochran and R.R. Haering (eds.), *Solid State Physics, Vol. I, Electrons in Metals,* Gordon and Breach, New York (1968).
2. J.M. Ziman, (ed.), *The Physics of Metals, Vol. I, Electrons,* Cambridge Univ. Press, Cambridge (1969).
3. M. Springford, *Adv. Phys.* **20**, 493 (1971).
4. D.W. Terwilliger and R.J. Higgins, *J. Appl. Phys.* **43,** 3346 (1972).
5. D.W. Terwilliger and R.J. Higgins, *Phys. Rev. B* **7,** 667 (1973).
6. D. Shoenberg, *Phil. Trans. Roy. Soc.* **A255**, 85 (1962).
7. K.M. Miller, R.G. Poulsen, and M. Springford, *J. Low Temp. Phys.* **6**, 411 (1972).
8. P. Coleridge, *Phil. Mag.* **26,** 573 (1972); see also Corrigendum.
9. J.F. Koch and R.E. Doezema, *Phys. Rev. Lett.* **24**, 507 (1970).
10. D. Shoenberg and B.R. Watts, *Phil. Mag.* **15**, 1275 (1967).
11. I.M. Templeton, *Proc. Roy. Soc.* **A292**, 413 (1966).
12. B.R. Watts, *Phil. Mag.* **24**, 1151 (1971).
13. P.T. Coleridge and B.R. Watts, *Phil. Mag.* **24**, 1163 (1971).
14. A.B. Pippard, in Ref. 1.
15. A.J. Heeger, in *Solid State Physics,* Vol. 23, F. Seitz, D. Turnbull, and H. Ehrenreich, eds., Academic Press, New York (1969).
16. J. Kondo, in *Solid State Physics,* Vol. 23, F. Seitz, D. Turnbull, and H. Ehrenreich, eds., Academic Press, New York (1969).
17. D.H. Lowndes, K.M. Miller, R.G. Poulsen, and M. Springford, *Proc. Roy. Soc.* **A331,** 497 (1973).
18. D.H. Lowndes, K.M. Miller, R.G. Poulsen, and M. Springford, this volume; *Phys. Rev. Lett.* **25,** 1111 (1970).
19. J.W. Blaker and R. Harris, *J. Phys. C* **4,** 569 (1971); R. Harris, *J. Phys. F* **3,** 89 (1973).
20. P.T. Coleridge, this volume.
21. P.T. Coleridge and I.M. Templeton, *Phys. Rev. Lett.* **27**, 507 (1971).
22. P.T. Coleridge, G.B. Scott, and I.M. Templeton, *Can. J. Phys.* **50,** 1999 (1972).
23. H. Alles, R.J. Higgins, and D.H. Lowndes, *Phys. Rev. Lett.* **30,** 705 (1973).
24. G. Simpson and B.E. Paton, *Phys. Lett.* **34A**, 180 (1971).

Anisotropy of Electronic Lifetime Due to Phonon Scattering in Copper*

David Nowak

Lawrence Livermore Laboratory
University of California, Livermore, California

and

Martin J. G. Lee

The James Franck Institute and the Department of Physics
The University of Chicago, Chicago, Illinois

As an approach to the interpretation and correlation of experimental Fermi surface data, the augmented-plane-wave (APW) pseudopotential has proven very successful.[1-3] The parameters, the phase shifts $\eta_l(E_F)$, are few in number, and a relatively small amount of experimental information is needed to obtain an accurate model of the Fermi surface. The numerical values of the phase shifts depend on the Fermi energy parameter E_F, which measures the Fermi energy with respect to the mean interstitial potential. Recent studies have shown that for a wide range of values of E_F, constant-energy surfaces for copper can be constructed whose extremal areas agree with the experimental Fermi surface data[4] to better than one part in 10^4.[5] A further advantage of the phase shift approach is that the trend of Fermi surface distortions is readily understood in terms of the systematic variation of the phase shifts within the periodic table.[6]

The experimental Fermi surface distortions give information only about the Fourier coefficients of the lattice potential associated with the first few reciprocal lattice vectors. The analytic expression for the APW pseudopotential can nevertheless be extrapolated to yield the Fourier coefficients for arbitrary **q**. In the low-q limit the behavior of the pseudopotential is determined by screening considerations. If the pseudopotential is to describe correctly the screening of long-wavelength density fluctuations, it must satisfy

$$\left\langle \lim_{\mathbf{q}\to 0} V_{\mathbf{k},\mathbf{k}+\mathbf{q}} \right\rangle = 1/D(E) \tag{1}$$

where the angular brackets denote a Fermi surface average of the low-q limit of the pseudopotential and $D(E)$ is the density of states at the Fermi surface. It is found that the behavior of the APW pseudopotential in the low-q region depends on the choice of the otherwise arbitrary Fermi energy parameter E_F. In particular, it is

* Work supported in part by facilities provided by The Advanced Research Projects Agency for materials research at the University of Chicago.

possible to construct a pseudopotential which satisfies Eq. (1). Such a pseudopotential is said to be of pseudoatom form.[7]

In the APW formulation the matrix element of the electron–phonon interaction between states on a surface of constant energy can be reduced to an integral over the surface of the muffin-tin sphere.[8] It has been shown that the observed Fermi surface distortions yield all the information necessary to evaluate these surface integrals and to deduce the matrix elements of the electron–phonon interaction between states on the Fermi surface.[9] If, as in the alkali metals, the Fermi surface is almost spherical, coupling by transverse phonons contributes to a negligible extent to those matrix elements which describe normal electron–phonon processes. However, on strongly aspherical regions of a nonspherical Fermi surface, such as that of copper, coupling by transverse phonons contributes significantly. Because the Fermi surface of copper is strongly influenced by nonlocal effects associated with the d bands which lie about 2 eV below the Fermi energy, the low-q limit of the APW pseudopotential in copper is markedly anisotropic. By appropriate choice of the Fermi energy parameter, an APW pseudopotential has been constructed which satisfies the long-wavelength screening condition of Eq. (1). This pseudoatom pseudopotential is an especially promising starting point for studies of the electron–phonon interaction in copper because the treatment of screening is purely empirical. In the low-q region the form of the pseudopotential is determined by the known response to a long-wavelength density fluctuation, and in the structure region it is determined by the known distortions of the Fermi surface. Calculations by Chui suggest that the errors in extending this treatment of the screening to the intermediate region should be rather small.[10]

At low temperature the inverse lifetime of an electron in a quantum state $\mathbf{k}$ on the Fermi surface takes the form

$$1/\tau(\mathbf{k}) = \Gamma_0 + \gamma(\mathbf{k})\, T^3$$

where the coefficient $\gamma(\mathbf{k})$ is proportional to the rate of scattering by thermally excited phonons. In order to test the APW pseudopotential description of the electron–phonon interaction, we have calculated the anisotropy of γ over the Fermi surface of copper. In the low-temperature limit γ is given by[11]

$$\gamma(\mathbf{k}) = \frac{1.31\Omega k_B^3}{M V_\mathbf{k} \hbar^4} \sum_\sigma \left\langle \frac{|V_\sigma(0, \mathbf{k})|^2}{C_\sigma^4} \right\rangle \tag{2}$$

where Ω is the volume of the primitive unit cell, M is the ion mass, $V_\mathbf{k}$ is the Fermi velocity of the state $\mathbf{k}$, $V_\sigma(0, \mathbf{k})$ is the small-angle scattering limit of the pseudopotential, and C_σ is the velocity associated with the phonon mode of polarization σ. Of the anisotropic quantities in Eq. (2), the Fermi velocity and the velocity of sound vary by no more than factors of two over the Fermi surface of copper, and the anisotropy of γ is dominated by the anisotropy of the small-angle scattering limit of the pseudopotential. The value of γ was calculated at various points on the Fermi surface using 30-APW expansions of the wave functions.[11] On the belly of the Fermi surface, where the matrix element involves only terms coupled by longitudinal phonons, the low-q limit of the pseudopotential is very nearly independent of the orientation of $\mathbf{q}$. Close to the necks, where transverse phonons become important, the matrix element must be averaged over all possible orientations of $\mathbf{q}$. The anisotropy of the inverse lifetime in the (100) and (110) symmetry zones is shown in Fig. 1, and a contour diagram for general directions is shown in Fig. 2. The anisotropy of

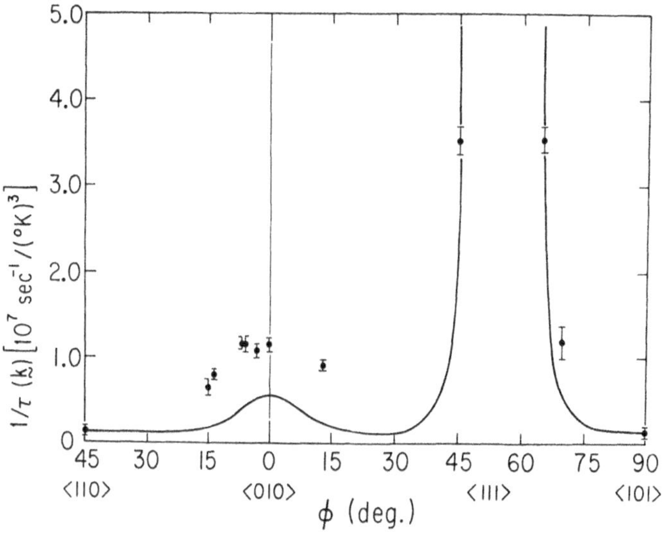

Fig. 1. Anisotropy of the inverse lifetime in symmetry zones of the Fermi surface of copper. The points represent experimental data of Koch and Doezema.[13] The experimental points represent averages over an angular range of about $\pm 6°$.

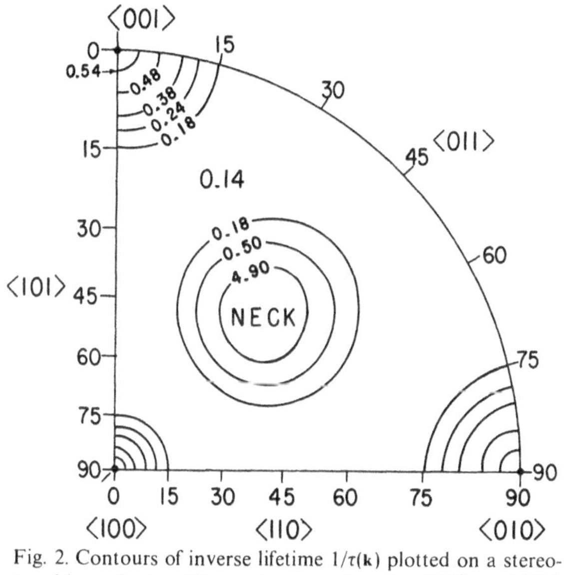

Fig. 2. Contours of inverse lifetime $1/\tau(\mathbf{k})$ plotted on a stereographic projection. The contours are in units of $10^7 \mathrm{sec}^{-1}/°\mathrm{K}^3$.

γ on the belly is a consequence of variations in the long-wavelength limit of the pseudopotential. The large increase in scattering on the necks is caused by coupling to transverse phonons.

Recent experiments[12–14] have yielded measurements of the relaxation time averaged over small regions of the Fermi surface. In Fig. 1 experimental lifetimes derived from studies of surface-state resonances in copper by Koch and Doezema[13] are compared with our results. The overall agreement is satisfactory, although our

Table I. Comparison of Orbital Averages of the Inverse Lifetime with Radiofrequency Size-Effect Results of Gantmakher and Gasparov

	10^7 sec^{-1} °K^{-3}	
	Ref. 14	Present work
$1/\tau$ (belly $\perp \langle 100\rangle$)	0.30	0.23 ± 0.05
$1/\tau$ (dog's bone $\perp \langle 100\rangle$)	0.67	0.67 ± 0.17

calculation underestimates the absolute scattering rate close to $\langle 100\rangle$. Gantmakher and Gasparov[14] have recently measured the scattering rate averaged around certain orbits on the Fermi surface of copper by means of the radiofrequency size effect. It is straightforward to derive the corresponding orbital averages from our calculations. A comparison with some of their preliminary data is set out in Table I. Apparently the discrepancy for the $\langle 100\rangle$ belly orbit arises because our calculations underestimate the scattering rate close to $\langle 100\rangle$, as is suggested by the comparison with the results of Koch and Doezema in Fig. 1.

The overall agreement with the experimental data shows that the anisotropy of the lifetime of electrons in states on the Fermi surface of copper due to scattering by thermal phonons can be explained by a phase-shift pseudopotential description of the matrix elements of the electron–phonon interaction. The anisotropy on the belly is correlated with variations in the sensitivity to scattering by longitudinal phonons (i.e., with the deformation potential). The very short lifetime on the necks is a consequence of the large increase in scattering by transverse phonons on this region of the Fermi surface. This agreement suggests that the APW pseudopotential may provide a useful empirical description of the electron–phonon interaction in other metals of the transition series.

References

1. M.J.G. Lee, *Phys. Rev.* **187,** 901 (1969).
2. N.J. Coenen and A.R. de Vroomen, *J. Phys. F* **2**, 487 (1972).
3. J.C. Shaw, J.B. Ketterson, and L.R. Windmiller, *Phys. Rev.* **5**, 3894 (1972).
4. P.T. Coleridge, A.A.M. Croxon, G.B. Scott, and I.M. Templeton, *J. Phys.* **E4**, 414 (1971); **4**, 928 (1971).
5. D. Nowak and M.J.G. Lee. *Phys. Rev.* **5**, 2854 (1972).
6. J.C. Shaw, J.B. Ketterson, and L.R. Windmiller, *Int. J. Quantum Chem.* **6**(6), (1972).
7. M.J.G. Lee and V. Heine, *Phys. Rev. B* **5**, 3839 (1972).
8. D.C. Golibersuch, *Phys. Rev.* **157**, 532 (1967).
9. P.B. Allen and M.J.G. Lee, *Phys. Rev. B* **5**, 3848 (1972).
10. S.K. Sinha, private communication.
11. D. Nowak, *Phys. Rev. B* **6**, 3691 (1972).
12. P. Haussler and S.J. Welles, *Phys. Rev.* **152**, 675 (1966).
13. J.F. Koch and R.E. Doezema, *Phys. Rev. Lett.* **24**, 507 (1970); and to be published.
14. V.F. Gantmakher, private communication.

7

Alloys

Maps of Electronic Lifetime Anisotropy for Dilute Alloys of Ag, Cu, Zn, and Fe in Au from dHvA Dingle Temperature Measurements*

D. H. Lowndes

Department of Physics
University of Oregon, Eugene, Oregon

K. M. Miller

Materials Science Division
Berkeley Nuclear Labs, Berkeley, Gloucestershire, England

R. G. Poulsen

Magnetic Domain Devices Department
Bell-Northern Research, Ottawa, Canada

and

M. Springford

School of Mathematical and Physical Sciences
University of Sussex, Falmer, Brighton, England

Introduction

The scattering of conduction electrons in metals and alloys has been recognized as an important problem for some time. It is generally appreciated theoretically that the details of the anisotropy of the electron scattering, described, for example, by an anisotropic lifetime or an anisotropic mean free path, do play an important role in determining the magnitudes of transport coefficients[1–3]; for example, the electrical resistivity, the magnetoresistance, and the thermopower all depend on the anisotropic electronic lifetime. However, direct measurements of the various transport coefficients themselves have so far not provided us with a coherent and detailed picture of the underlying scattering anisotropy in the case of scattering by impurity atoms, precisely because the ordinary transport coefficients embrace all of the anisotropic features of the conduction process, and therefore the scattering also, inside an integral expression. In order to extract information on electron scattering anisotropy from transport measurements, it has been necessary to use very simple models for the scattering anisotropy. Present conclusions concerning electron scattering anisotropy based on measurements of the ordinary transport properties have recently been reviewed by Springford.[3]

* The support of this project by the Science Research Council (U.K.) is gratefully acknowledged, as is support received by one of us (D.H.L.) from the Research Corporation and The National Science Foundation (U.S.). Invited paper.

In contrast to measurements of transport coefficients, there are *direct* experiments in which the measured property refers to a well-defined and readily identified group of electrons at the Fermi surface: the Landau quantum oscillations arising from electrons at the extremal cross sections of a Fermi surface, for example. The dHvA effect is one such experiment, and it has several advantages for the detailed mapping out of local electronic lifetimes: First, the phase $\phi = 2\pi F/H$ of the dHvA oscillations in metals remains large (even in magnetic fields of order 10 T). This has the consequence that orbitally averaged lifetimes determined experimentally are only very limited averages over narrow orbits on the Fermi surface, making the determination of local lifetimes, using the inversion technique described below, relatively straightforward. Second, the dHvA effect is believed to measure a true electronic lifetime (in the sense of a reciprocal Landau level width), rather than a transport relaxation time, which includes additional weighting factors. Thus, comparison of the measured lifetime anisotropy with theoretical calculations is relatively straightforward, since it is precisely this level width (or *total* transition rate) which arises naturally in calculations of impurity scattering based on band structure wave functions for dilute alloys. Finally, the Dingle temperature measured by the dHvA effect is primarily a measure of scattering by impurity atoms and other imperfections, *but not by phonons.*[4] Thus it is not necessary to face the problem of separately determining the effects of phonon and impurity scattering.

It is also important to mention one potential drawback of the dHvA method for electron scattering measurements: Since the measurements are carried out in high fields and over a finite field range, it may be difficult *in the case of field-dependent scattering processes* (for example, by magnetic impurities) to make a direct comparison of dHvA lifetimes and those determined in zero-field transport measurements.

We describe in this paper the application of the dHvA effect to explore the anisotropy of the lifetime of conduction electrons over the Fermi surface and we illustrate the method by applying it in some detail to gold. The method permits *local* values of the electronic lifetimes $\tau(\mathbf{k})$ to be deduced directly from the Dingle temperatures measured by the dHvA effect, and we have obtained the variation of $\tau(\mathbf{k})$ over the Fermi surface for scattering due to the presence of Ag, Cu, Zn, or Fe as impurity separately.

In the next section we describe the assumptions of the method and present the inversion equation which is used to "unfold" the reciprocal-lifetime orbital averages measured by the dHvA effect, to give local lifetime values. Then we give a brief synopsis of a technique developed to correct measured Dingle temperatures for the effect of phase-smearing due to mosaic microcrystalline structure—the principal source of error in these measurements. In this section we also briefly review the principal experimental features and techniques used. The principal experimental results and their discussion are then presented. The results take the form of topographical maps showing the variation of $\tau(\mathbf{k})$ over the Fermi surface in Au for dilute concentrations of Ag, Cu, Zn, or Fe as impurity.

A letter on a preliminary version of the present method has been published,[6] and several purely calculational aspects of the present method are described in more detail elsewhere.[5,7]

Method and Inversion Equation

The dHvA effect is the oscillatory component of the magnetic susceptibility of a metal which is observed under conditions of adequate crystalline purity and perfection, low temperature, and high magnetic field. The dHvA oscillations are periodic in reciprocal field, and the temperature and field dependence of the amplitude of the oscillations yield information about a well-defined group of electrons: namely those whose **k** vector moves in a narrow band of cyclotron orbits centered on extremal cross sections of the Fermi surface in a plane normal to the applied magnetic field. The oscillatory magnetization due to a single extremal cross section of the Fermi surface may be expressed as follows:[8]

$$\mathbf{M} = D(\mathbf{H}) \sum_{r=1}^{\infty} I_r K_r S_r \sin\left(\frac{2\pi r F}{H} + \beta\right) \tag{1}$$

where

$$D(\mathbf{H}) = -\frac{e\hbar A}{4\pi^4 m^*}\left(\frac{2\pi e H}{\hbar}\right)^{1/2}\left(\frac{\partial^2 A}{\partial k_H^2}\right)^{-1/2}$$

$$I_r = \frac{(r\lambda m^* T/H)}{\sinh(r\lambda m^* T/H)}$$

$$K_r = \exp(-r\lambda m x/H); \qquad S_r = \cos(r\pi g^* m^*/2); \qquad F = (\hbar/2\pi e)A$$

In these expressions r is an integer harmonic index in the Fourier sum for M, β is a phase constant which is independent of H at least for the conditions of high phase (i.e., $2\pi F/H \gg 1$) which normally prevail, F is the dHvA frequency, A is the extremal cross-sectional area of the Fermi surface normal to H, and m^* and g^* are the fully enhanced cyclotron mass and orbital g factor, respectively, though it should be noted that the separate enhancements of g^* and m^* due to electron–phonon coupling *cancel* in the *product* g^*m^*, which is then independent of electron–phonon coupling.[4]

The effect of electron scattering appears through the terms K_r, in which λ is a constant ($= 14.69\ T/°K$), x is the Dingle temperature, and m, it should be noted, is the band mass including electron–electron interactions but *excluding* the electron–phonon renormalization.[4] Under our usual experimental conditions the fundamental ($r = 1$) term in Eq. (1) dominates, and the effect of electron scattering is given by the term K_1. From this it follows that a graph (or "Dingle plot") of ln (dHvA amplitude) vs. $1/H$ is sufficient to determine the Dingle temperature x for the particular extremal orbit from which each oscillation arises, provided that one knows the electron *band* mass m (as specified above) for the orbit.

The Dingle temperature is a measure of the total electron scattering rate due to collisions with lattice defects and impurity atoms.[9] Equivalently, it measures the broadening of the electronic Landau energy levels due to these collisions. Brailsford[10] has related x to an average around the extremal orbit on the Fermi surface of the true mean collision time $\tau(\mathbf{k})$ such that

$$x = (\hbar/2\pi k_B)\langle 1/\tau(\mathbf{k})\rangle \tag{2}$$

where $\langle\ \rangle$ denotes a suitable orbital average (see below). This result is intuitively as expected if all scattering events cause the states $|\mathbf{k}\rangle$ and $|\mathbf{k}'\rangle$ to be phase-incoherent. It can be shown that the limiting angle of scatter for such a condition to be true will be of order N^{-1}, where $N = F/H$, and thus for metals at normal magnetic fields (5–10 T) the limiting angle of scatter will be very small ($\sim 10^{-2}$–10^{-4} rad). Thus nearly all scattering events are equally weighted in contributing to x. For the particular case of scattering of electrons by phonons, however, this statement needs some qualification. Palin[11] and Simpson and Englesberg[4] have demonstrated that, with respect to the term K_r in Eq. (1), the electron–phonon renormalization of x and m have nearly completely cancelling effects on the dHvA amplitude. Consequently, in the sense that x does not measure $\langle 1/\tau_{\text{e-ph}}(\mathbf{k})\rangle$, it does not reflect the *total* scattering rate. In the present experiments, however, we are concerned only with situations where the scattering of electrons is dominated by the presence of substitutional impurity atoms of known types, i.e., well-characterized dilute alloys. Thus by a "true reciprocal lifetime" we mean a lifetime defined as the reciprocal total transition rate due to collisions with impurity atoms, and any other residual lattice defects which may be present. An important feature of the present results on the magnitudes of the *local* electronic lifetimes $\tau(\mathbf{k})$ is that, in contrast to results deduced from the ordinary transport properties, they do permit a comparison between theory and experiment *prior* to the stage when the Boltzmann equation or its equivalent needs to be solved.

If we now suppose that Dingle temperatures can be measured for each of a set of different, but very well-defined extremal orbits, then it should be possible to deduce a map showing the local electronic scattering rate $\tau(\mathbf{k})^{-1}$ from the measured set of orbital averages. To do this, (1) we must specify the way in which averaging of the scattering rate takes place around an orbit, and (2) we require an inversion equation to unfold the *set* of orbital averages and yield values of the local scattering rate.

The inversion procedure is based on the premise that each local lifetime $\tau(\mathbf{k})$ contributes in Eq. (2) to the orbital average $\langle 1/\tau(\mathbf{k})\rangle$ for the cyclotron orbit in proportion to the amount of time dt spent by the wave vector $\mathbf{k}$ in the corresponding region of the Fermi surface. Then

$$\left\langle \frac{1}{\tau(\mathbf{k})} \right\rangle = \frac{1}{T_c} \oint \frac{dt}{\tau(t)} \tag{3}$$

where T_c is the cyclotron period in a magnetic field, and where the variation of $\tau(\mathbf{k})$ around the orbit is represented through the implicit time dependence $\mathbf{k}(t)$, t being time.

From the Lorentz force equation we may write that

$$dt = \frac{\hbar}{eH} \frac{d\mathbf{k}}{V_\perp(\mathbf{k})}$$

where $V_\perp(\mathbf{k})$ is the projection of the Fermi velocity onto the orbital plane and $d\mathbf{k}$ is an element of arc on the cyclotron orbit. If $k_\perp$ and the angle η specify the position of a representative point on the cyclotron orbit relative to some arbitrary origin in the orbital plane, it is easy to see that

$$dk = dk_\perp = k_\perp\, d\eta / \{[\mathbf{V}(\mathbf{k}) \cdot \mathbf{k}_\perp] / V_\perp(\mathbf{k})\, k_\perp\}$$

So we obtain, finally,

$$x = \frac{\hbar}{2\pi k_B} \oint \frac{W(\eta)\, d\eta}{\tau(\eta)} \Bigg/ \oint W(\eta)\, d\eta \tag{4}$$

where $W(\eta)$ is an anisotropic weight function,

$$W(\eta) = k_{\perp}^{2}/\mathbf{V}(\mathbf{k})\cdot \mathbf{k}_{\perp} \tag{5}$$

and

$$2\pi m^* = \oint W(\eta)\, d\eta \tag{6}$$

with all quantities in free-electron units.

The $W(\eta)$ may readily be computed from a knowledge of the Fermi surface geometry and the Fermi velocity of the noble metals, for example, from the results of Halse,[12] Bosacchi *et al.*,[13] Lee,[14] and Christiansen.[15] We have computed the angular variations of the $W(\eta)$ for a number of orbits in gold, using the Au5 model of the Fermi surface of Halse, and the results for various orbits are shown in Fig. 1. The trajectories of the representative points of electrons in these orbits over the Fermi surface are mapped onto the stereographic projection in Fig. 2. It should be noted

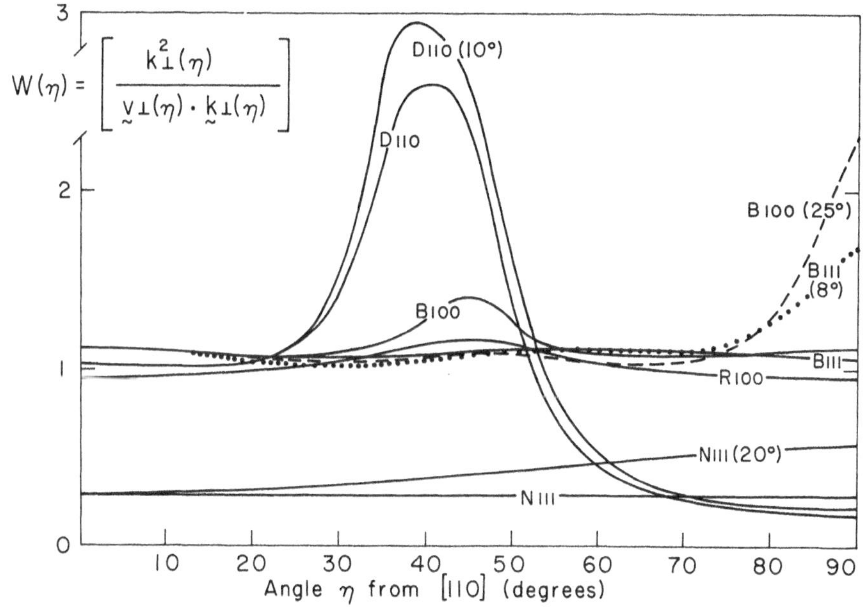

Fig. 1. Variation of the weight function W, as defined with reference to Eq. (5), as a function of the angle η which specifies the position of the representative point around an extremal orbit. The origin of η is the [110] direction. The W are given for orbits in the principal symmetry directions, and also for some "tipped" orbits, the corresponding trajectories over the Fermi surface being shown in Fig. 2. The "tipped" orbits are as follows: $B_{100}(25°)$ = [100] belly tipped 25° toward [100] in a (110) plane; $B_{111}(8°)$ = [111] belly tipped 8° toward [111] in a (110) plane; $D_{110}(10°)$ = [110] dog's bone tipped 10° toward [111] in a (110) plane; $N_{111}(20°)$ = [111] neck tipped 20° toward [110] in a (110) plane.

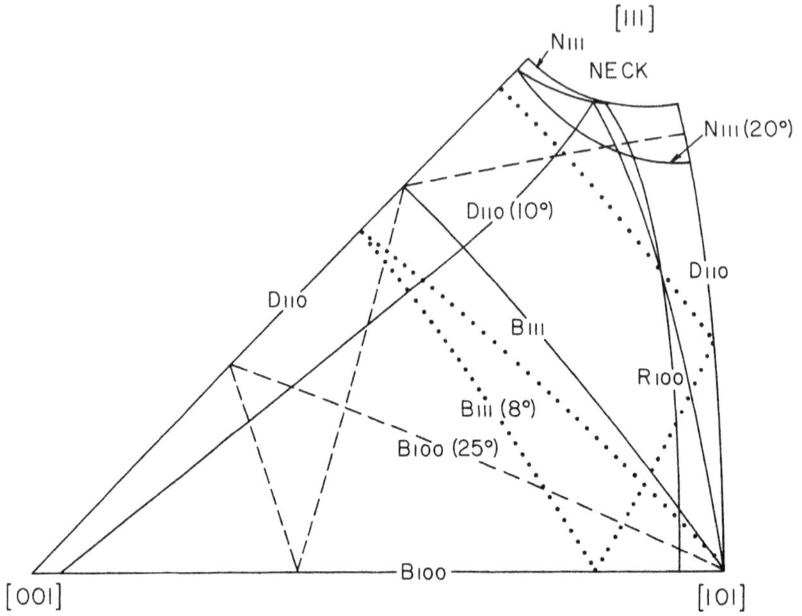

Fig. 2. The trajectories over the Fermi surface of some extremal symmetry and "tipped" orbits in Au. The key is as for Fig. 1, which shows the weight functions $W(\eta)$ of the orbits. The figure shows only the basic 1/48th unit of the stereographic projection.

that in computing the $W(\eta)$ we have used the true quasiparticle velocities as deduced by Halse as is appropriate when an orbitally averaged quantity is desired. The $W(\eta)$ are quite anisotropic and the orbits selected are seen to give good angular coverage over the Fermi surface. Using Eq. (6), m^* may be evaluated and used either as a consistency check by comparison against measured values of m^*, or as a substitute for measured values of m^* when determining the Dingle temperatures for orbits where m^* has not been determined experimentally.

The determination of $\tau(\mathbf{k})$ now reduces to the problem of best fitting a set of x_m, measured for a variety of extremal orbits, by self-consistent inversion of the corresponding integrals defined in (4), and this was accomplished in the following way.

A set of x_m are measured, the orbits being selected such that the trajectories give as comprehensive a coverage over the Fermi surface as possible (see Fig. 2). A general representation $f(C_i, \mathbf{k})$, for $\tau(\mathbf{k})$ is chosen and the parameters C_i are adjusted so as to give the best fit of the measured x_m to corresponding values computed from Eq. (4) using an iterative least-squares fitting procedure. The representation must preserve crystal symmetry and should be chosen to yield as good a fit to the experimental data with as few adjustable parameters C_i as possible. This is because meaningful fits can be achieved only when the number of adjustable parameters is less than the number of orbits for which x_m is measured, and for typical specimens having appreciable mosaic spread accurate measurements of x_m can generally only be made for field directions for which the dHvA frequency has a turning point, as discussed below and in Refs. 5 and 7.

For noble metal alloys we have found that a cubic harmonic representation for $\tau(\mathbf{k})$ is unsuccessful in that a large number (>12) of terms C_i are required to obtain a reasonable fit to the experimental measurements. On the other hand, substantially better fits have been obtained, with fewer coefficients, using a symmetrized Fourier series representation, similar to that used by Halse in connection with the noble metal Fermi surfaces. Because of this we have chosen the Fourier series representation

$$\begin{aligned}\tau[(k_x, k_y, k_z)]^{-1} &= \sum_{lmn} C_{lmn} \cos(\tfrac{1}{2} lak_x) \cos(\tfrac{1}{2} mak_y) \cos(\tfrac{1}{2} nak_z) \\ &= C_{000} + C_{110}(\textstyle\sum \cos \tfrac{1}{2} ak_x \cos \tfrac{1}{2} ak_y) + C_{200}(\textstyle\sum \cos ak_x) \\ &\quad + C_{211}(\textstyle\sum \cos ak_x \cos \tfrac{1}{2} ak_y \cos \tfrac{1}{2} ak_z) + C_{220}(\textstyle\sum \cos ak_x \cos ak_y) \\ &\quad + C_{310}(\textstyle\sum \cos \tfrac{3}{2} ak_x \cos \tfrac{1}{2} ak_y + \textstyle\sum \cos \tfrac{3}{2} ak_y \cos \tfrac{1}{2} ak_x) \\ &\quad + C_{222}(\cos ak_x \cos ak_y \cos ak_z) + \cdots\end{aligned} \tag{7}$$

where $l + m + n$ is even for the fcc lattice, $(\sum)$ denotes a sum over permutations of x, y, and z, a is the lattice constant, and k_x, k_y, and k_z are the k-space coordinates of a point on the Fermi surface.

The best fit to the experimental data was decided as follows. For a given number M of coefficients C_i in Eq. (7), the best fit of the computed Dingle temperatures x_f to the measured values x_m was ascertained by the method of least squares. The significance of the best fit for this number of coefficients C_i could then be determined from tables of the χ^2 distribution for the particular number of degrees of freedom $N - M$, with χ^2 given by

$$\chi^2 = \sum_1^N [(x_f - x_m)/\sigma_m]^2 \tag{8}$$

σ_m being the standard deviations of the measured values x_m and the sum is over the N orbits included in the fit. This procedure was then repeated for different numbers of coefficients C_i and the best fit was selected as that which had the highest significance.

We may rewrite Eq. (7) in a compact form,

$$[\tau(k_x k_y k_z)]^{-1} = \sum_{l,m,n} \tau_{lmn}^{-1} \tag{9}$$

and then, using Eq. (4), calculate the contribution to x per unit coefficient in (9) for any orbit. To simplify future computational effort, we have computed and published elsewhere[7] the contributions to X from each of the first seven terms in Eq. (9). Results are given for the seven "basic orbits," i.e., those for which $\partial F/\partial \theta = 0$, and may therefore be readily used to yield preliminary maps of $\tau(\mathbf{k})$ directly from measured Dingle temperatures, using Eq. (7) and the results of Halse.[12]

We emphasize that the Fourier series representation (7) was chosen empirically, its merits being that it conforms with the crystal symmetry requirements and generally enables a fit to be made to the experimental results, to within experimental error, using only a few terms in the expansion for $\tau(\mathbf{k})^{-1}$ (typically 3–5). However, the *method* described here is quite general. For example, Eq. (7) could perhaps more meaningfully be replaced by a sum of "partial lifetimes" based on a knowledge of the s, p, d, etc., charge distributions over the Fermi surface of the host metal, when such calculations, using realistic wave functions, are available. Thus our method is quite general and may in principle be employed in metals of any crystal symmetry.

Experimental Techniques

The many possible sources of error in the measurement of Dingle temperatures have been reviewed elsewhere.[3,7,16] Before presenting our results it should be mentioned that the most important spurious contribution to the Dingle temperature comes from the variation of the dHvA phase over the sample volume, which has its origin, in our experiments, primarily in the mosaic spread in crystalline orientation, which is found quite generally, even in nominally single crystals. It must be emphasized that this mosaic spread typically amounts to only a few minutes of arc (i.e., $\sim 10^{-3}$–10^{-4} rad) but it has drastic consequences in producing a spurious enhancement of the Dingle temperature for all extremal orbits for which $\partial F/\partial\theta \neq 0$. Dingle temperature measurements with the magnetic field parallel to a crystal symmetry axis are then free of this spurious enhancement, but measurements for arbitrary orientations are in general spuriously enhanced.

In phase smearing, the spurious contribution to x arises from variations $\delta\phi$ of the dHvA phase over the sample volume. Typically, $\phi = 2\pi F/H$ is $\sim 10^3$–10^5 and is therefore $\gg 2\pi$. The effect of phase smearing on X is negligible if $\delta\phi \ll 1$. Shoenberg[16] has shown that phase smearing may be incorporated in Eq. (1) by replacing $\sin(\phi + \beta)$ by a summation over the sample volume

$$\sum_{\text{volume}} \sin(\phi + \delta\phi + \beta) = f(H)\sin(\phi + \beta) \tag{10}$$

and has derived the form of the amplitude correction factor $f(H)$ corresponding to phase distributions of Lorentzian, Gaussian, or linear forms. (This theory is valid only if $\delta\phi \ll 1$ over dimensions comparable to the cyclotron orbit.) For each of these phase-smearing models we have calculated the contribution Δx to x, resulting from the effects of phase smearing, from the slope $-\lambda m^* \Delta x$ of the best least-squares computer fit of a straight line to a plot of $\ln f(H)$ vs. H^{-1}, in analogy with the method for the determination of x as described in the previous section. The results of these calculations are summarized in Table I and Fig. 3, and are given in more detail elsewhere.[5]

In the present work we have used these results to determine quantitatively the degree of mosaic spread in our samples. Briefly, what is done, for a single specimen,

Table I. Dingle Temperature Enhancement Due to Mosaic Spread*

dHvA phase distribution function	dHvA amplitude correction factor $f(H)$	Dingle temperature enhancement $\Delta X(m^*, H)$
Lorentzian $[1 + (\delta\varphi H/H_0)^2]^{-1}$	$\exp(-H_0/H)$	$H_0/146.9m^*$
Gaussian $\exp[-(\delta\varphi H/H_0)^2]$	$\exp(-H_0/2H)^2$	$(H_0/2)^2/146.9m^*\langle H\rangle$
Linear $\delta\varphi = 2H_0/H$	$(H/H_0)\sin(H_0/H)$	(See text)

* $H_0 = 2\pi F' \Delta\theta$ (for mosaic spread): $\langle H\rangle$ is the characteristic magnetic field used in determining X.

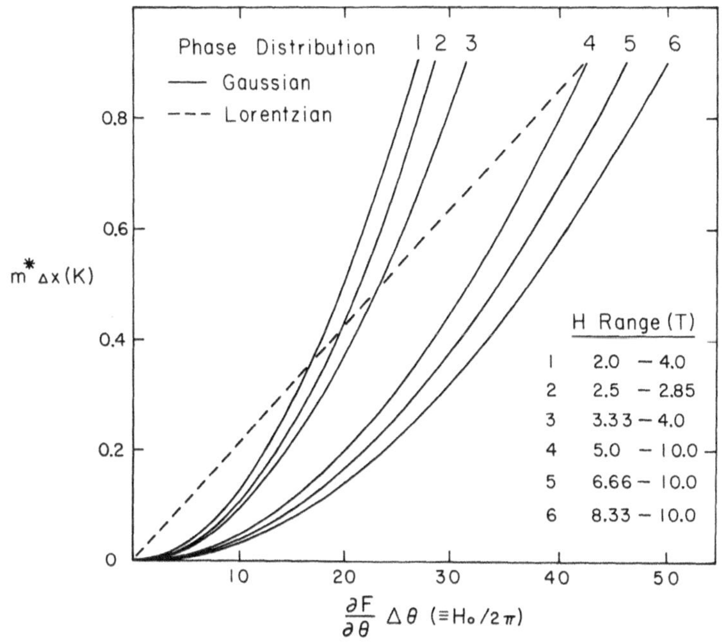

Fig. 3. The product $m^* \Delta x$ (m^* being the effective mass and Δx the error in the Dingle temperature) as a function of $H_0/2\pi$, where H_0 relates to the phase spread $\delta\phi$ and is defined by $H_0 = 2\pi F' \Delta\theta$, where $\Delta\theta$ is the characteristic mosaic spread angle. Note that for a Gaussian distribution of phase the correction Δx is field dependent, while for a Lorentzian distribution it is not.

is to measure a set of approximately 20 Dingle temperatures for different orientations in the (110) plane, including the six orientations for which $\partial F/\partial\theta = 0$.[5,7] The data for these six "basic orbits" are used then to obtain a preliminary map for $\tau(\mathbf{k})$. As described above, the enhancements Δx of the other measured Dingle temperatures are then calculated, these being the differences between the measured values and those predicted using the preliminary map of $\tau(\mathbf{k})$. If this set of Δx can be understood consistently, within experimental error, in terms of a single mosaic spread parameter ζ, then the additional Dingle temperatures, corrected for mosaic spread enhancement, are combined with the six "basic orbits" to produce a final set of x's for that sample. For the cases presented here this procedure was possible, although the parameter characterizing the mosaic spread varied from crystal to crystal. (This procedure was not possible for crystals which had been strained by thermal cycling or by mishandling. In such cases no single value of characteristic mosaic spread angle ζ could be found to account for the values of Δx.)

This procedure was repeated for other samples and at different solute concentrations. The Dingle temperature per at.% of solute for a given orbit could then be determined from a best straight-line fit to a graph of x vs. c. In this way errors arising from phase smearing having its origin in field inhomogeneity or crystal strain were minimized, although, as shown below, these contributed negligibly in the present work. A final map of $\tau(\mathbf{k})$ was produced by inversion of the set of 20 values of x/c. Thus these final maps are normalized to give Dingle temperatures for one given at.% of solute, and are ultimately derived from several specimens containing different solute atom concentrations, and from 15 to 20 orbits for each concentration, corrected for the effect of mosaic spread.

For studies of electron–impurity scattering the Δx due to inhomogeneity of field magnitude, solute concentration, or crystal strain can be estimated from the zero-concentration intercept of plots of x versus solute concentration. The Δx due to strain may be reduced by spark-cutting the samples slowly (at low spark energy), carefully annealing them, and avoiding strain or thermal cycling in the dHvA apparatus. With these precautions for gold we find zero-concentration intercepts of $\pm 0.2°$K are readily obtained, and these intercepts are often randomly distributed about the origin and are zero within experimental error. Two representative examples may be seen in Fig. 4, which shows Dingle temperature measurements as a function of concentration in the system Au(Ag).

Our measurements were carried out using a field modulation dHvA spectrometer which has been described in detail elsewhere.[7] Measurements were made at sufficiently low modulation frequency (generally 120 Hz) so that skin effects were avoided, with detection at the second harmonic. A 10-T Oxford Instruments Nb_3Sn superconducting magnet was used with field inhomogeneity about two parts in 10^5 over the specimen volume. Provision was made for accurately moving the dHvA probe along, and rotating the probe about, the field direction so that one could readily locate the postition of maximum magnetic field homogeneity by observing changes in phase of the dHvA oscillations with such motion at constant field. Care was taken to adjust the sample temperature appropriate to the impurity concentration so that the harmonic content in the dHvA oscillations and magnetic interaction effects introduced negligible errors. The measurements were made with the magnetic field in the (110) plane to within $\pm\frac{1}{2}°$ at both the symmetry directions and a number of intermediate orientations.

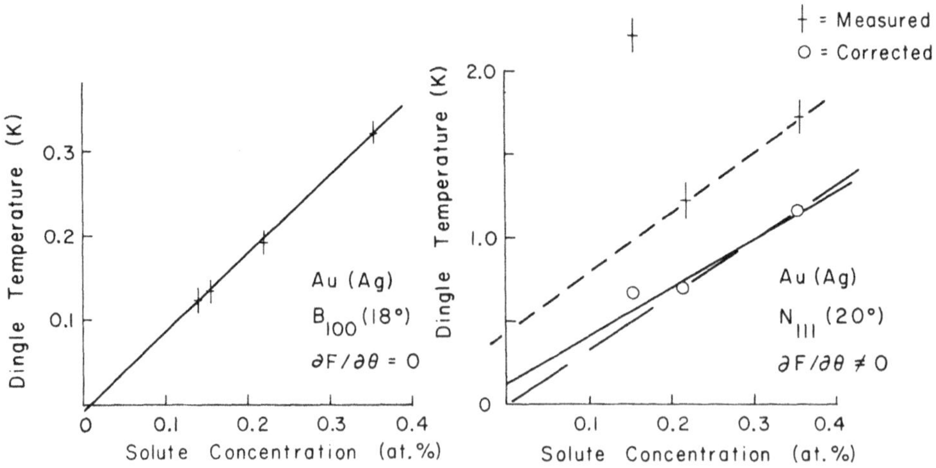

Fig. 4. Concentration dependence of Dingle temperatures in Au(Ag), (a) for the [100] belly tipped 18° in a (110) plane, where $\partial F/\partial\theta = 0$, and (b) for the neck orbit at 20° from [111] toward [110] in a (110) plane. For the latter the measured values are enhanced by mosaic spread of crystalline orientation by amounts which differ for each concentration. The corrected values, shown as circles, display the approximately linear dependence upon concentration.

The gold alloys studied were grown in ultrapure graphite moulds by Metals Research Ltd., Melbourn, Cambridge, cut oversize by very slow spark erosion, carefully etched to size in warm aqua regia, and finally annealed in high vacuum at 600°C for several days to minimize straining of the specimen. The specimen dimensions were such that even at helium temperatures the specimen was not squeezed and hence strained by contraction of the ivory specimen holder. Straining of the specimen in the ivory holder was further minimized by fixing it with a little grease at one end only, slowly cooling the specimen, and never thermally cycling the specimen beyond a few °K. Several portions of each alloy were spectrographically analyzed for trace impurities by the Johnson Matthey Co., Ltd., and the ingots were rejected unless the estimated total contribution to the Dingle temperature for the assembled trace impurities was negligible in comparison to that from the added solute. Actual solute concentrations in the particular samples used for measurement were determined after the dHvA experiments by dc measurement of the residual resistance ratio. For Au(Ag), Au(Cu), and Au(Zn) the resistivity per at.% of solute was taken as 0.36×10^{-8}, 0.45×10^{-8}, and 0.95×10^{-8} Ω-m, respectively,[17] and for Au(Fe), 10.3×10^{-8} Ω-m.[18]

Results

The principal experimental results are the final lifetime maps presented in Figs. 5–8. Tables comparing in detail the measured lifetimes with those calculated from these maps have been published elsewhere.[7]

It should be noted that while there is some knowledge of the electron–phonon mass renormalization factors $\lambda_0(k)$ in gold[15] they are not yet sufficiently complete that a reliable interpolation may be made to yield appropriate orbital averages $\langle \lambda_0 \rangle$. Throughout this work, therefore, we have used the fully enhanced cyclotron effective mass values (or measured cyclotron masses) m^*, given by

$$m^* = m(1 + \langle \lambda_0 \rangle)$$

where m refers to the mass enhanced only by electron–electron effects. In consequence, the Dingle temperatures deduced from the Dingle plots using the masses m^* will also be "diminished" values x^* given by

$$x^* = x/(1 + \langle \lambda_0 \rangle)$$

and the conduction electron lifetimes determined from these x^* should also properly be referred to as "diminished lifetimes." We note also that while the average electron–phonon renormalization is small in Au, namely $\langle \lambda_0 \rangle \approx 10\%$,[15] it probably exhibits, like copper, quite large variations over the Fermi surface.

Au(Ag). This system was extensively investigated; three dilute alloys were used, of solute concentrations close to 0.16, 0.22, and 0.36 at.%, the precise values being determined for each crystal by RRR measurement. Some important conclusions follow from the measured concentration dependence of x, some examples of which are shown in Fig. 4. For the six "basic orbits" x is directly proportional to concentration and extrapolates to zero concentration of solute to within experimental error. The only exception to this is for the D_{110} orbit, for which a least-squares fit to the experimental points yields a positive value of $x = 0.2°$K at $c = 0$, which lies just outside the experimental error, estimated to be at worst 0.15°K. With this exception these results indicate that the "background scattering," which might be attributed predominantly to inhomogeneities in the magnetic field or to strain in the crystals, is small and less than the random error in each measurement. In contrast to this behavior for the "basic orbits," however, Fig. 4(b) depicts the measured experimental points at a field orientation for which $\partial F/\partial\theta \neq 0$ and, as can clearly be seen, it is only after correction for the mosaic spread enhancement of x that the concentration dependence of x is sensible. The largest value of the intercept at $c = 0$ was $+0.2$°K for the D_{110}. However, the assembled values of this intercept for all orbits averaged to zero and in view of this the origin in Fig. 4 was consequently treated as an experimental point in arriving at the measured values of x/c.

For crystals cut from each of the three ingots the mosaic spread enhancement of x could be accounted for in terms of a Gaussian spread of phase. Since, however, this is an arbitrary assumption, the numerical values of the mosaic spread parameter may be used only as a guide. It was not possible to distinguish between different models for phase smearing by mosaic spread, e.g., Gaussian and Lorentzian, but evidently the assumption of uniform spread about all directions was reasonable in these cases. For the ingots of nominal solute concentration 0.16, 0.22, and 0.36 at.% the mosaic spread parameter ζ was found to be $(3.6 \pm 5) \times 10^{-4}$, $(1.8 \pm 0.5) \times 10^{-4}$, and $(1.8 \pm 0.5) \times 10^{-4}$ rad respectively. The most significant reciprocal lifetime inversion was accomplished with four coefficients in the Fourier expansion, the final lifetime map for Au(Ag) being given in Fig. 5. Using a four-coefficient Fourier representation, as for Fig. 5, the experimental measurements are fitted with an rms

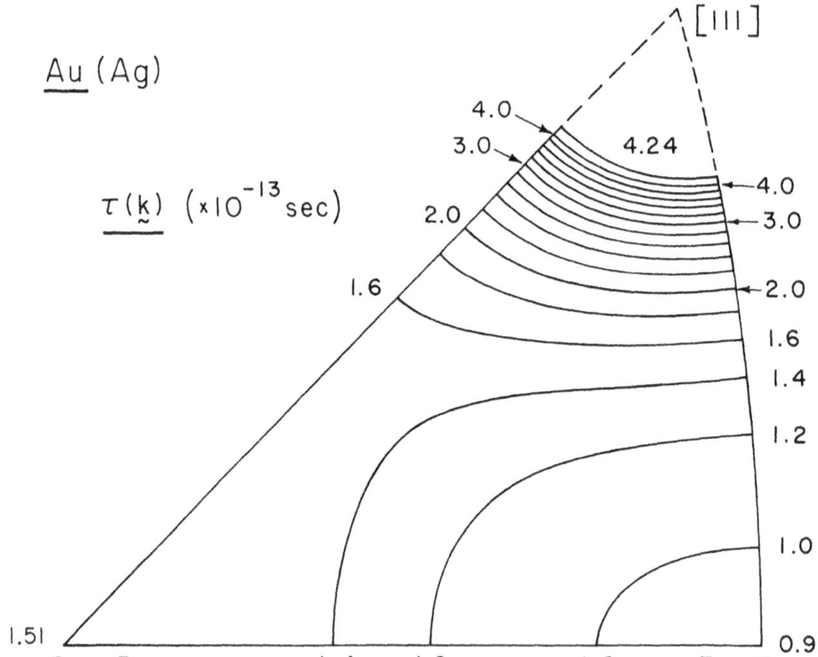

Fig. 5. Final map showing the variation of $\tau(\mathbf{k})$ over the Fermi surface in Au(Ag) per at.% of solute. Units are 10^{-13} sec, and the map shows the basic 1/48th part of the stereographic projection. The map is based upon a four-term Fourier series representation for $\tau(\mathbf{k})^{-1}$ for which the coefficients C_i are listed in Table II.

error of 5.6%, although for one orbit, the "four-rosette" R_{100}, fitted and measured values are rather more discrepant. Increasing the number of Fourier coefficients in the inversion procedure much above four left the map in Fig. 5 essentially unchanged except around [001], where the Fermi surface is least well sampled by the network of orbits used in the inversion.

Au(Cu). Experiments were conducted in alloys of concentrations close to 0.24, 0.47, and 0.79 at.%. For samples of the highest concentration, however, measurements could only usefully be made on the "neck" orbits in view of the very small dHvA signals derived from the larger sections. The analysis of results proceeded as for Au(Ag), values of ζ for alloys of concentration 0.24 and 0.47 at.% being $(6 \pm 1) \times 10^{-4}$ and $(5 \pm 1) \times 10^{-4}$ rad, respectively. The final map of $\tau(\mathbf{k})$ using a four-term Fourier representation is presented in Fig. 6. In this case the rms error of fitted to experimental Dingle temperatures was 4.8%.

Au(Zn). Considerable difficulties were encountered in preparing samples of adequate crystalline perfection. At solute concentrations greater than ~0.1 at.% crystalline substructure on a relatively large scale was encountered, which gave rise to anomalous amplitudes and spurious zeros in crystal-rotation plots of the dHvA effect. The results for this system are therefore based on measurements at a single concentration of 0.093 at.% solute, in which reliable results were consistently obtained.

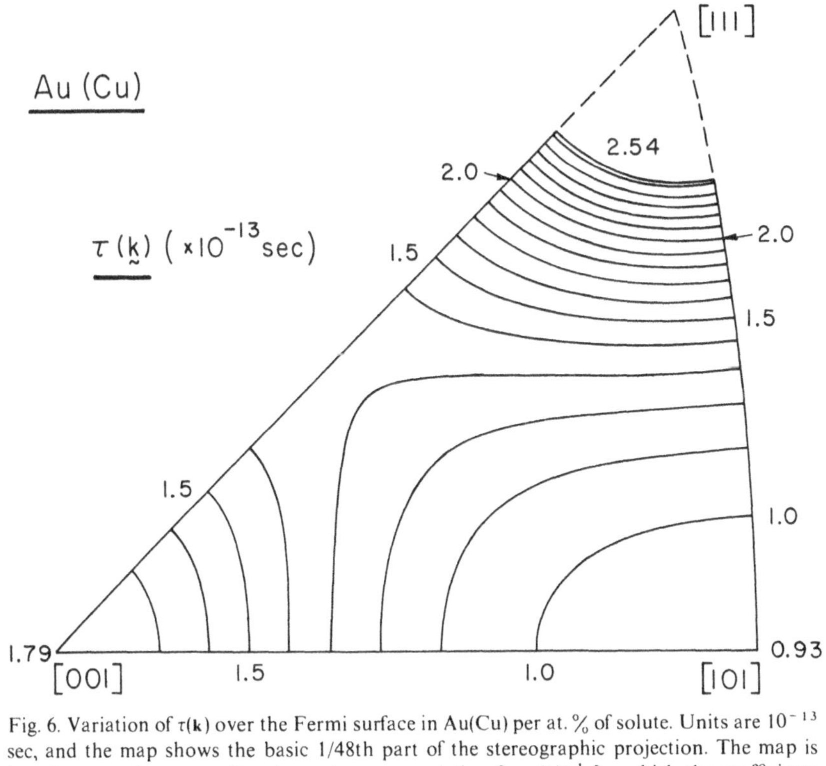

Fig. 6. Variation of $\tau(\mathbf{k})$ over the Fermi surface in Au(Cu) per at.% of solute. Units are 10^{-13} sec, and the map shows the basic 1/48th part of the stereographic projection. The map is based upon a four-term Fourier series representation for $\tau(\mathbf{k})^{-1}$ for which the coefficients C_i are listed in Table II.

It is assumed in this that for the six "basic orbits" the measured values correspond wholly to the presence of zinc and that measurements for all other orbits will have been enhanced only by the effects of mosaic spread (i.e., other phase-smearing contributions, such as field inhomogeneity, are insignificant). Based on the previous results for Au(Ag) and Au(Cu), this is expected to be a good approximation. A good fit to experimental results (5.1% rms error) was achieved using four Fourier coefficients, the final map of $\tau(\mathbf{k})$ being shown in Fig. 7. Crystals cut from this ingot were characterized by a value of the mosaic spread parameter of $(1.5 \pm 0.5) \times 10^{-4}$ rad.

Au(Fe) (Figs. 8 and 9). For electron scattering by magnetic impurities, departures from the basic theory [Eq. (1)] are expected to occur and anomalies in the magnetic field dependence of the dHvA amplitude have been observed, for example, by Coleridge and Templeton in Cu(Cr) and by Paton, Hedgecock, and Muir in Zn(Mn). Our lifetime mapping work was not conceived as a detailed investigation of such scattering anomalies over the Fermi surface, or as a function of magnetic field and temperature, but rather as an examination of the features of the scattering anisotropy in a variety of dilute alloys under one set of experimental conditions. The results then refer to an alloy concentration of 170 ppm, the residual resistivity being

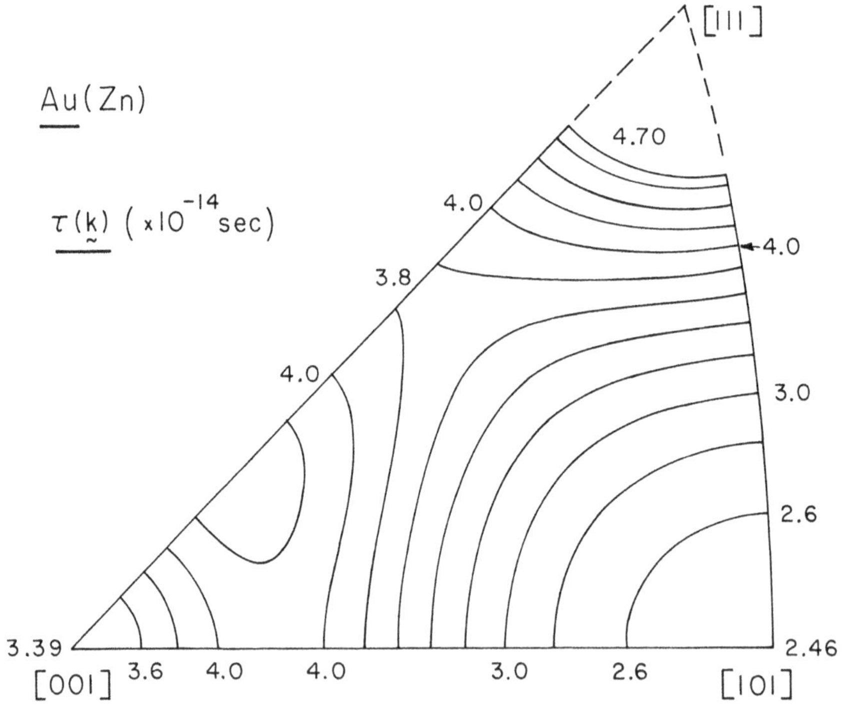

Fig. 7. Variation of $\tau(\mathbf{k})$ over the Fermi surface in Au(Zn) per at. % of solute. Units are 10^{-14} sec, and the map shows the basic 1/48th part of the stereographic projection. The map is based upon a four-term Fourier series representation for $\tau(\mathbf{k})^{-1}$ for which the coefficients C_i are listed in Table II.

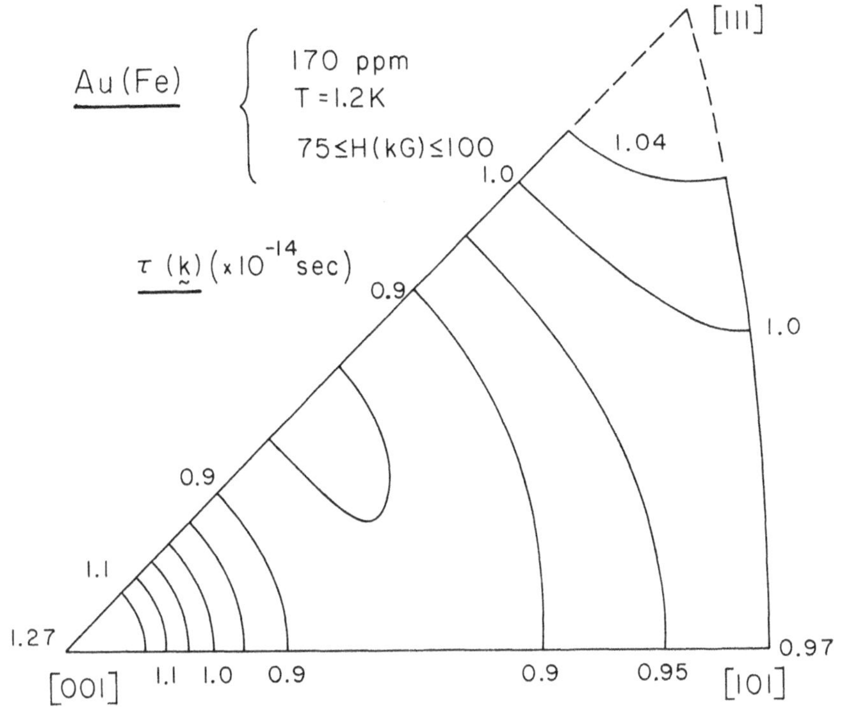

Fig. 8. Variation of $\tau(\mathbf{k})$ over the Fermi surface in Au(Fe) per at. % of solute. Units are 10^{-14} sec and the map shows the basic 1/48th part of the stereographic projection. The map is based upon measurements in an alloy of solute concentration = 170 ppm and a four-term Fourier series representation for $\tau(\mathbf{k})^{-1}$ for which the coefficients C_i are listed in Table II.

1.76×10^{-5} Ω-m. For crystals cut from this ingot the characteristic mosaic spread parameter was $(1.8 \pm 0.5) \times 10^{-4}$ rad, a value typical of dilute gold-based alloys. Experiments were performed between 1.2 and 2.0°K, the Kondo temperature T_K for dilute Au(Fe) being expected to be in the neighborhood of 0.2°K.[18]

The observed Dingle plots are typified by Fig. 9, in which it is seen that a change of slope occurs over a small range of field at ~8.0 T. Such anomalous behavior was observed for all orbits with the exception of the [111] "neck," although the change in slope and field at which it occurred were both temperature and orientation dependent. It is evidently of interest to investigate this system further, in particular with regard to the temperature and magnetic field dependences of the anomaly; we are carrying out these experiments. However, for the purpose of analyzing the present experiments in terms of possible scattering anisotropy and treating Fe impurities in a manner analogous to the other dilute alloys already described, we have taken for the Dingle temperature the value obtained from the *high-field* slope, with the result shown in Fig. 8. (The *increase* in slope obtained by using the slope at *low* fields corresponds to an apparent increase in x of ~5–20%, depending on the particular orbit.) A four-term Fourier representation for $\tau(\mathbf{k})^{-1}$ was used.

Collected in Table II are the final fitted Fourier coefficients which represent the variation of $\tau(\mathbf{k})$ over the Fermi surface as displayed in the final maps of Figs. 5–8.

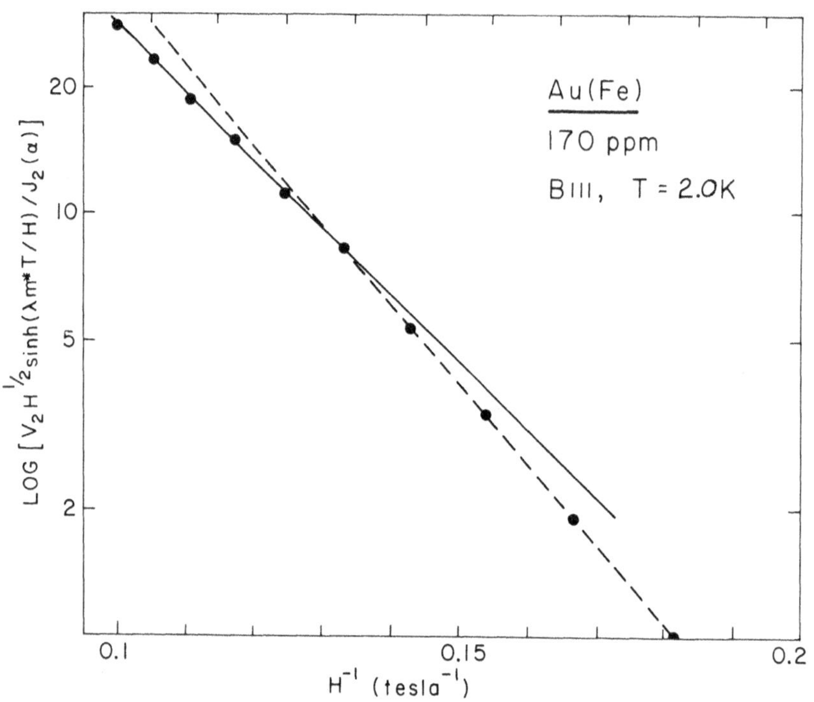

Fig. 9. Anomalous field dependence of dHvA amplitude, as revealed by a Dingle plot in an Au–170 ppm Fe alloy. Measurements refer to the [111] belly investigated at 2.0 K.

To determine the significance of the final fitted results from tables of the χ^2 distribution, we need to know the standard deviation σ_m of the measured Dingle temperatures. Considering both the random error of measurement and the several possible sources of systematic error,[7] we estimate that measured Dingle temperatures are at most in error by 0.15°K. Taking as an estimate for σ_m the value 0.1°K, the significance of the final fits (Figs. 5–8) is in each case $\gtrsim 90\%$.

Table II. Summary of Coefficients in the Fourier Series Representation for $\tau(\mathbf{k})^{-1}$, Corresponding to the Best Fit to the Experimental Results for Au(Ag), Au(Cu), Au(Zn), and Au(Fe)*

Fourier coefficients	Au(Ag)	Au(Cu)	Au(Zn)	Au(Fe)
C_{000}	14.6272	15.3560	25.794	215.43
C_{110}	1.0288	1.2803	− 132.67	418.56
C_{200}	5.0821	4.6109	8.7548	17.935
C_{211}	17.9638	17.8937	114.99	− 170.31

* In each case the coefficients refer to 1 at. % of solute, and higher-order coefficients are zero.

Discussion

Au(Ag) and Au(Cu). Qualitatively, the results for Au(Ag) and Au(Cu) are similar, although the extent of the anisotropy is greatest for Au(Ag), being 4.5 : 1 as against 2.7 : 1 for Au(Cu). Since potential scattering is expected to be relatively small in both of these homovalent alloys, the observed differences then reflect both the differences in scattering contributions of relativistic origin and the contribution from the lattice distortion which accompanies the inclusion of Cu, but not Ag, into Au. Detailed calculations by Brown and Morgan[21] indicate that relativistic contributions are important in Ag-based alloys, and this should be even more the case for Au-based alloys.

Silver and copper differ dramatically in the manner in which they alloy with Au: Ag is incorporated with essentially no change in lattice constant ($\Delta a/ac = 0.0063$), while rather large lattice distortion accompanies the inclusion of Cu ($\Delta a/ac = 0.091$).[22] As our lifetime maps show, the electronic lifetime variations are nearly identical for Ag and Cu impurities over much of the Fermi surface, but lifetimes differ by nearly a factor of two in the neck region near [111], the lifetime being shorter here for scattering by Cu impurities. This is consistent with what is expected for the effect of lattice distortion, since the effect of lattice distortion has been observed to be greatest in the noble metals for electron states near the Brillouin zone boundary.[23,24]

A more detailed comparison of our maps for Au(Ag) and Au(Cu) is relevant at this point because of the information such a comparison provides regarding the physical origin of the effective Dingle temperatures which accompany alloying, and the relative importance of lattice strain. Two components to the lattice strain field may be distinguished: One arises from any mismatch in size between impurity and host atoms, the other from the presence of defects such as grain boundaries and dislocations in the sample. The fact that there must be these two distinct types of strain effects has been realized by workers in the dilute alloy field for several years. In fact, it is customary to attribute the residual Dingle temperatures found in pure metals, or the zero-concentration Dingle temperature intercept in dilute alloys, to phase smearing arising from dislocations or mosaic structure. Phillips and Gold[25] and Terwilliger and Higgins[24] have shown that the residual Dingle temperature in pure metals can be reduced to very nearly zero by careful crystal preparation, annealing, and handling, as we have similarly shown for the zero-concentration Dingle temperature intercept for dilute alloys. As described earlier in connection with our corrections for mosaic spread enhancement, the very high phase of dHvA oscillations has the consequence that very small relative changes in phase (for example, due to strain fields) over the sample volume can cause destructive interference of signals from different parts of the sample. This produces a field-dependent decrease of dHvA amplitude which appears experimentally as an apparent enhancement of electron scattering. For the same reason true small-angle scattering events, through angles so small (typically $\ll 1°$) as to have negligible effect on the electrical resistivity, are also "catastrophic" for the phase coherence of dHvA signals. The problem in describing strain effects is then to distinguish between true small-angle scattering events and phase smearing, if possible.

The effect of impurity-induced strain has been calculated recently by Coleridge.[26]

He considers that this form of scattering is small-angle scattering but that it can be *calculated* by a phase-smearing argument. His phase-smearing calculation then involves summing contributions to the dHvA signal from all regions of the crystal [as in Eq. (10)], using as a model the strain field surrounding a substitutional atom in an isotropic elastic continuum. He finds that there is a "linear region" (at high magnetic fields and low concentration) in which the amplitude reduction due to strain dephasing behaves like a genuine Dingle temperature.

In Table III we compare the predictions of Coleridge's theory for the "linear region" with our observed differences in neck and belly Dingle temperatures in Au(Ag) and Au(Cu). (We have used Coleridge's sign and scaling conventions for the various parameters, and obtained numerical values for Au from Ref. 23.) If we attribute all of the difference in the Dingle temperatures for corresponding orbits in Au(Ag) and Au(Cu) to the effect of impurity-induced strain, then it appears that the actual effect of impurity-induced strain is nearly an order of magnitude *less* than that predicted by Coleridge's calculation. This conclusion is tentative, but does appear to be well beyond a possible error of a factor ~ 2 in estimating the parameters used in our calculation, and the likely differences in the potential scattering produced by Cu and Ag impurities. This result then suggests that the use of a phase-smearing approach to calculate the effect of impurity-induced strain may be incorrect. A general criterion for the phase-smearing approach to be valid is that the change in phase $\delta\phi$ over dimensions comparable to a cyclotron orbit be much less than one.[16] For values of magnetic field and solute concentration normally encountered in dHvA measurements (and within Coleridge's "linear region") it seems unlikely that this condition is met. For example, at $H = 5$ T the dimension of a belly orbit in Au is $\sim 10^4$ lattice parameters, and so for solute atom concentrations in the range 0.01–1.0% the orbit encloses an area containing a number of impurity cells of order 500 and samples a number n of impurity cells (of order 10) in completing one orbit. In this regime the calculation of the effect of impurity-induced strain is not straightforward, but a crude model suggests that Coleridge's result could be reduced in an extreme case by a factor of as much as $1/\sqrt{n}$, and that the effect upon x will become *nonlinear* in concentration *well before* leaving Coleridge's "linear regime." These arguments suggest, then, that the effect of solute-induced lattice strain on the Dingle

Table III. Comparison of Measured Dingle Temperatures (per at.% of solute) with the Effect Predicted for Impurity-Induced Strain in Ref. 26, and Estimating Parameters from the Data of Ref. 23*

Orbit	Strain calculation			Experiment		
	x_S, at.%		ΔX_S, at.%	x_S, at.%		ΔX_S, at.%
	Au(Ag)	Au(Cu)		Au(Ag)	Au(Cu)	
B_{100}	~ 0.36	~ 5.2	~ 4.8	9.7	10.0	0.3
B_{111}	~ 0.36	~ 5.2	~ 4.8	9.4	10.7	1.3
N_{111}	~ 1.3	~ 18.8	~ 17.5	2.85	4.79	1.94

* $x_S \sim 250k_a\langle\mu\rangle/m^*$ (°K/at.%). k_a is 0.0063 for Au(Ag), and 0.091 for Au(Cu). $\langle\mu\rangle \simeq 0.24$.

temperature may be negligible in Au(Ag), Au(Zn), and Au(Fe). However, solute-induced strain in Au(Cu) may contribute up to 10% for the belly, and 40% for the neck, of the measured Dingle temperatures, since $\Delta a/a$ is large and potential scattering is rather weak.

In comparing our experimentally determined lifetime maps with theoretical calculations it is necessary to compare with calculations based on a rigorous approach to the electronic band structure and wave functions for the host metal and for the dilute alloy system. It would be very useful to have available theoretical maps expressing the variation of the scattering rate over the Fermi surface as a sum of partial scattering rates, based on a knowledge of the *s*, *p*, and *d* charge distributions over the Fermi surface of the host metal and depending on differences between host and solute atom scattering phase shifts. Such calculations, using rigorous band structure wave functions, are not yet available for Au, but have been carried out by Harris[27] and Coleridge[28] for Cu. Their results for pure *s*-wave scattering in Cu are very similar to our maps for Au(Ag) and Au(Cu) (except on the neck near [111]). The point of this comparison is to make clear the detail which both experiment and theory are now capable of achieving in describing fundamental scattering processes in dilute alloys, and to point up the need for extending rigorous calculations to heavy elements such as Au. This is important if we are to fully understand the relative importance of potential scattering, lattice distortion, and relativistic effects.

Au(Zn). For Zn (heterovalent impurity) in Au we expect considerable potential scattering, in addition to relativistic effects and small-angle scattering due to impurity-induced strain. As discussed above, we expect the latter to be relatively small for Au(Zn). The lifetimes of Fig. 7 are, on average, 20 times shorter than for the homovalent alloys, and though many qualitative features of the variation over the Fermi surface are similar, the lifetime variation is considerably more complicated between [111] and [001].

Au(Fe). For the case of Au(Fe), as depicted in Fig. 8, we find the scattering to be of order 100 times stronger than for the homovalent alloys, and to be essentially constant over the Fermi surface. This is to us a very surprising result, since by an extension of Ziman's[1] arguments pertaining to scattering in the noble metals we might expect that for scattering dominated by a 3*d* transition metal impurity such as Fe the resulting scattering anisotropy would mirror the distribution of *d* charge over the Fermi surface. The recent calculations of Coleridge and Harris have indicated that in Cu the *d*-charge distribution is appreciably anisotropic, being more concentrated around the ⟨110⟩ directions, although with some contribution at the necks. In view of the similarities in the band structures of the noble metals, this is also expected to be qualitatively the case in Au, and was indeed the interpretation given to a preliminary map of the lifetime anisotropy in Au(Fe) at 4.2°K[6] (shown in Fig. 10), which *does* show strong scattering in the [101] region of the Fermi surface. A simple interpretation suggests that what we are observing, then, by means of the lifetime mapping technique are local changes in the scattering rate $\tau(\mathbf{k})^{-1}$ and that these are largest near [101]. The temperature and field dependences of the scattering presumably have their origin in spin-scattering processes associated with the presence of a temperature- and field-dependent ($\sim \mu H/kT$) effective magnetic moment on

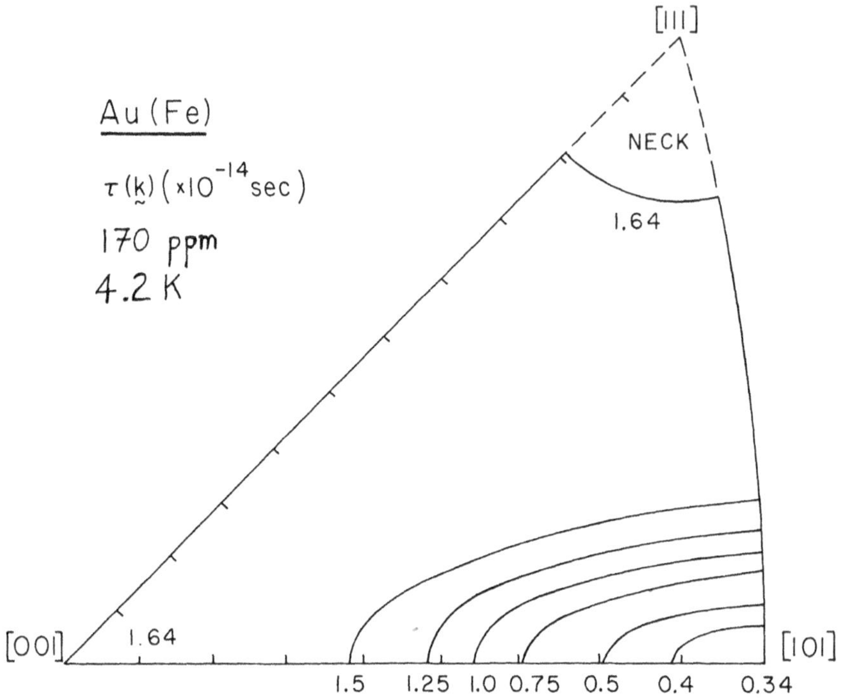

Fig. 10. Preliminary lifetime map for Au(Fe) at 4.2 K (from Ref. 6), based on Dingle temperatures for only five orbits and using an early version of the lifetime inversion method.

the impurity site. However, the anomalous magnetic field dependence of the dHvA amplitude could conceivably also be associated with a "beat," of the type seen in *Cu*Cr by Coleridge and Templeton. We are presently carrying out experiments to resolve these interpretations as well as some other complicating factors.[7]

Acknowledgments

We are indebted to Dr. D. Shoenberg for advice and critical comments, to Dr. D. Randles for helpful discussions and assistance with computing, and to Dr. R. J. Higgins for helpful discussions. For permitting us to see accounts of their work prior to publication, we should like to thank P. T. Coleridge, R. Harris, J. B. Ketterson, and I. M. Templeton.

References

1. J.M. Ziman, *Adv. Phys.* **10**, 1 (1961) ; *Phys. Rev.* **121**, 1320 (1961).
2. B.R. Coles, *Phys. Rev.* **101**, 1254 (1956).
3. M. Springford, *Adv. Phys.* **20**, 493 (1971).
4. G. Simpson and S. Engelsberg, *Phys. Rev. B* **2**, 1657 (1970).
5. K.M. Miller, R.G. Poulsen, and M. Springford, *J. Low Temp. Phys.* **6**, 411 (1971).
6. D.H. Lowndes, K.M. Miller, and M. Springford, *Phys. Rev. Lett.* **25**, 1111 (1970).
7. D.H. Lowndes, K.M. Miller, R.G. Poulsen, and M. Springford, *Proc. Roy. Soc. (London)* **A331**, 497 (1973).
8. A.V. Gold, in *Solid State Physics, The Simon Fraser University Lectures, Volume I. Electrons in Metals,* J.F. Cochran and R.R. Haering, eds., Gordon and Breach, New York.

9. R.B. Dingle, *Proc. Roy. Soc. A* **211**, 517 (1952).
10. A.D. Brailsford, *Phys. Rev.* **149**, 456 (1966).
11. C. Palin, Ph.D. Dissertation, Cambridge Univ., England, 1971.
12. M.R. Halse, *Phil. Trans. Roy. Soc.* **A265**, 507 (1969).
13. B. Bosacchi, J.B. Ketterson, and L.R. Windmiller, *Phys. Rev. B* **4**, 1197 (1971).
14. M.J.G. Lee, *Phys. Rev. B* **2**, 250 (1970).
15. N.E. Christensen, *Sol. St. Commun.* **9**, 749 (1971).
16. D. Shoenberg, *Phil. Trans. Roy. Soc.* **A255**, 85 (1962); *Phys. Kond. Materie* **9**, 1 (1969).
17. F.J. Blatt, *Solid State Phys.* **4**, 318 (1957).
18. P.J. Ford, T.E. Whall, and J.W. Loram, *Phys. Rev. B* **2**, 1547 (1970); P. Ford, Ph.D. Dissertation, Univ. of Sussex, England, 1970.
19. P.T. Coleridge and I.M. Templeton, *Phys. Rev. Lett.* **24**, 108 (1970).
20. B.E. Paton, F.T. Hedgecock, and W.B. Muir, *Phys. Rev. B* **2**, 4549 (1970).
21. H.R. Brown and G.J. Morgan, *J. Phys. F* **1**, 132 (1971).
22. W.B. Pearson, *Lattice Spacings and Structures of Metals and Alloys* Pergamon, New York (1964).
23. I.M. Templeton, *Proc. Roy. Soc.* **292A**, 413 (1966); D. Shoenberg and B.R. Watts, *Phil. Mag.* **15**, 1275 (1967).
24. D. Terwilliger and R.J. Higgins, *Phys. Lett.* **31A**, 316 (1970); and to be published.
25. R.A. Phillips and A.V. Gold, *Phys. Rev.* **178**, 932 (1969).
26. P.T. Coleridge, *Phil. Mag.* (to be published).
27. R. Harris (to be published).
28. P.T. Coleridge (to be published).

The Fermi Surface of Dilute Indium Alloys

P. M. Holtham

Department of Physics, University of British Columbia
Vancouver, British Columbia, Canada

and

D. Parsons

Department of Physics, University of Bristol
Bristol, England

Indium is a convenient metal in which to study the effects of alloying. From the experimentalist's point of view, being in group IIIb of the periodic table, it is possible to introduce impurities which either add conduction electrons or remove them. The Fermi surface of indium is reasonably simple compared with many metals and there are several small cross-sectional areas that are extremely sensitive to small changes in Fermi level ε_F and hence are most suitable for studying the effect of alloying. From the theoretical standpoint, indium has a simple crystal structure (face-centered tetragonal), the Fermi surface can be adequately described using a local pseudopotential model,[1] and with an atomic number of 49 the effects of spin–orbit interaction are not too large.

In this work we report de Haas–van Alphen measurements of the changes that occur in the third-zone electron surface when indium is alloyed with lead or thallium. Measurements were made using the torque balance technique in magnetic fields of up to 27 kG produced by a conventional electromagnet.

The third-zone indium Fermi surface[1] is sketched in Fig. 1 and consists of four "β-arms" elongated along the directions [110] and joined at the points C'. Two sets of de Haas–van Alphen frequencies have been studied in the alloys, the first set arising from the bellies or maximum sections of the arms (medium frequencies $\sim 4.6 \times 10^6$ G) and the second arising from the orbits at the junction (low frequencies $\sim 1 \times 10^5$ G). Data have been taken on both sets of frequencies in alloys of up to $1\frac{1}{2}\%$ concentration. Data were taken with the magnetic field both in the plane perpendicular to [100] and perpendicular to [001].

Indium–Lead

The effect of adding lead is to increase the electron concentration and therefore increase ε_F. Although the lattice parameters change such that the c/a ratio increases, and the volume of the unit cell increases, thus decreasing ε_F, the overall change is such that we expect the size of the Fermi surface to increase. In Fig. 2 we show part

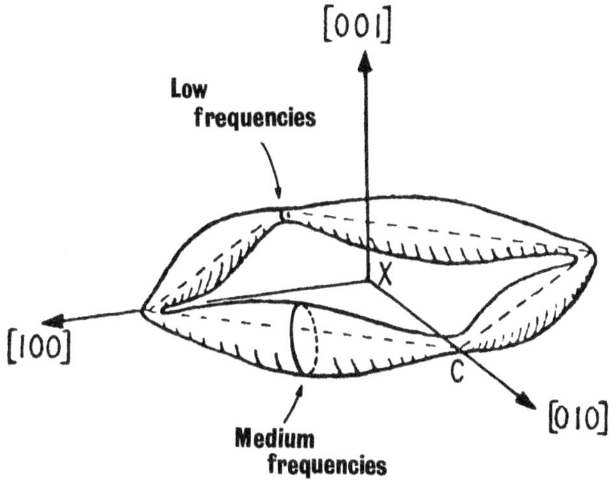

Fig. 1. The third-zone electron Fermi surface for indium (after Hughes and Shepherd[1]).

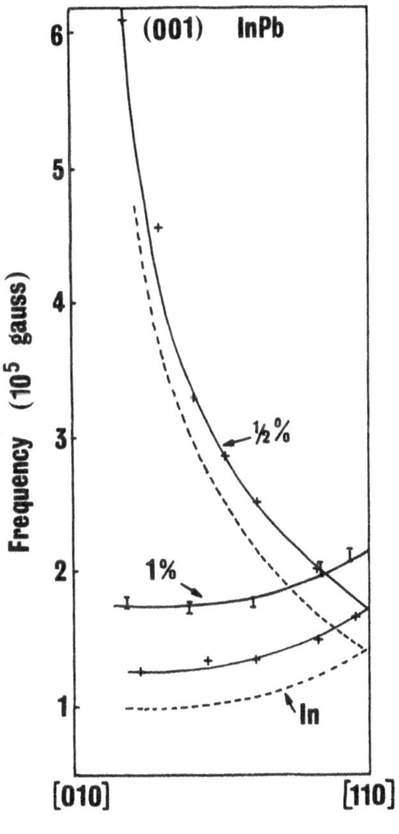

Fig. 2. Low-frequency branches in (001) for indium–lead alloys (concentrations are by weight).

of the low-frequency data taken in one of the planes studied. We see that the changes are quite large; for example in a 1% alloy at [100] the frequency has been increased by about 80%.

We find that a rigid band model (which takes account solely of changes in Fermi level) gives fair agreement with the changes in these low frequencies.

Indium–Thallium

Although this is an isoelectronic alloy, the change in atomic volume causes ε_F to decrease. The part of the data shown in Fig. 3 indicate that we again have large changes in frequency. The slight increase in frequency near [100] together with data taken in the (001) plane may suggest a slight thickening of the junction region, giving rise to two extremal cross sections, although we have not as yet observed the continuation of the lower-frequency branch.

Medium Frequencies

We have also studied the changes in the frequencies that arise from the β-arm maximum. These changes ($\sim 2\%$ for a $\frac{1}{2}\%$ InPb alloy and for a 1% InTl alloy) are

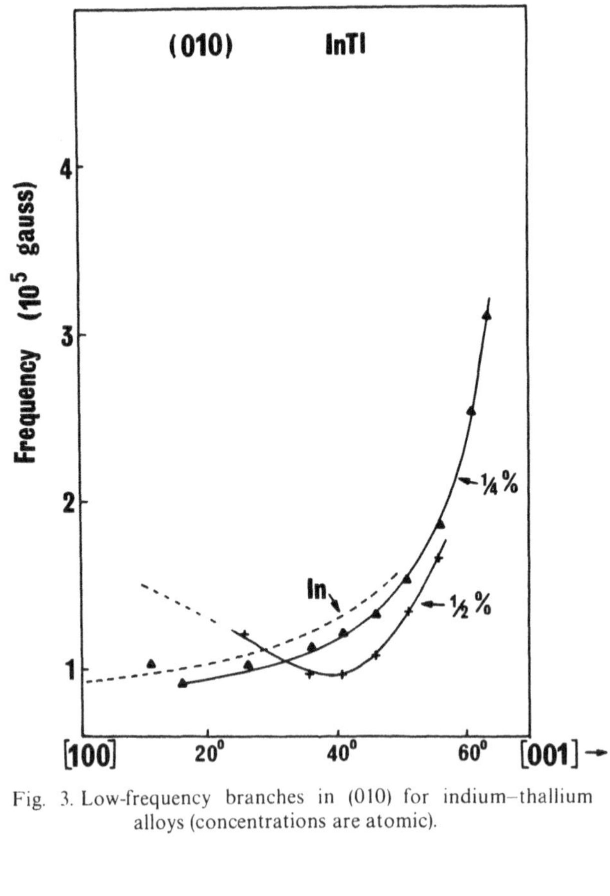

Fig. 3. Low-frequency branches in (010) for indium–thallium alloys (concentrations are atomic).

not so great as for the junction orbits, but we are able to detect a difference from pure indium. Here we find that the rigid band model (for InPb) underestimates the change by about a factor of two.

Theoretical Models

These results show that the rigid band model is unsatisfactory in explaining the overall changes in the Fermi surface. Clearly in alloying we are changing the potential in which the electrons move and we should take this into account. To do this, we develop a pseudopotential model for indium which includes the effect of spin–orbit coupling. The model, which differs from that used by Hughes and Shepherd[1] for indium, is similar to the calculations of Holtham and Priestley.[2]

Matrix elements take the form

$$\langle \mathbf{k}'s'|H|\mathbf{k}s\rangle = k^2\,\delta_{\mathbf{k}'\,\mathbf{k}}\,\delta_{s',s} + U(g)\,S(g)\,\delta_{\mathbf{k}'-\,\mathbf{k},\mathbf{g}}\,\delta_{s',s} - i\lambda_{4p}D_{4p}(k')\,D_{4p}(k)\,S(g)\,\delta_{\mathbf{k}'-\,\mathbf{k},\mathbf{g}}(\hat{\mathbf{k}}' \wedge \hat{\mathbf{k}})\cdot\langle s'|\boldsymbol{\sigma}|\,s\rangle$$

We have tried a prescription for determining $U(g')$ for the alloy of the form

$$U(g') = (1-x)\,(\Omega_a^{\mathrm{In}}/\Omega_a^{\mathrm{alloy}})\,U^{\mathrm{In}}(g') + x\,(\Omega_a^{\mathrm{Pb}}/\Omega_a^{\mathrm{alloy}})\,U^{\mathrm{Pb}}(g')$$

where x is the concentration of the impurity; the volume factors are necessary because of the definition

$$U(g) = (1/\Omega_a)\int[\exp(-i\mathbf{g}\cdot\mathbf{r})]\,U(r)\,d^3r$$

In addition, the $U^{\mathrm{In}}(g')$ will not be the same pseudopotential as in the pure metal since the lattice parameters, and hence the g values, have changed. We therefore have to move along our form-factor curve to determine our new $U^{\mathrm{In}}(g')$. Similarly, the $U^{\mathrm{Pb}}(g')$ can be determined and a band structure calculation made, using the above prescription.

The indium coefficients are taken from our own calculation of the indium Fermi surface, the lead coefficients are taken from the work of Anderson *et al.*[3] and those for thallium are taken from the work of Holtham and Priestley.[2]

We find that this is still not sufficient to give complete agreement with experiment, and we are at present undertaking calculations which allow for screening. Since the electron concentration is changing, the screening changes, and this may affect the potential considerably. To allow for screening, we need to go back and "unscreen" the pure metal potential, then estimate $U^{\mathrm{bare}}(g')$ for the constituents, and finally rescreen, using the dielectric function appropriate to the alloy environment. Preliminary calculations using this approach suggest the agreement with experiment may be far better.

References

1. A.J. Hughes and J.P.G. Shepherd, *J. Phys. C: Solid State Phys.* **2**, 661 (1969).
2. P.M. Holtham and M.G. Priestley, *J. Phys. F: Metal Phys.* **1**, 621 (1971).
3. J.R. Anderson, W.J. O'Sullivan, and J.E. Schirber, *Phys. Rev. B* **5**, 4683 (1972).

High-Precision dHvA Studies in Dilute Copper Alloys*

I. M. Templeton

National Research Council of Canada
Ottawa, Canada

Introduction

Until recently the state-of-the art limits of precision for Fermi surface (FS) cross-section measurement by the de Haas–van Alphen (dHvA) effect has been somewhere between 0.1 and 0.01% (100–1000 ppm). Let us examine the effect of such a limit on the possibility of studying alloying changes in the noble metals. Table I has been set up for the operating conditions $H = 50$ kG and $T = 1°$K, where we can say by experience that an upper limit for Dingle or scattering temperature consistent with a tolerable signal-to-noise ratio is about 2°K for a belly orbit and 4°K for a neck orbit. From our published figures[1] for scattering per atomic percent of solute we can calculate the maximum concentrations of, for example, Al and Ni for which there are usable signals from the $\langle 111 \rangle$ neck and belly orbits. If we then make a reasonable guess at the changes of electron concentration per atomic percent of solute and use a rigid band calculation, which for copper predicts that the fractional changes in neck and belly orbits are respectively about six times and two-thirds the fractional change of electron concentration, then it is clear that while the relatively large changes in the small, sensitive neck cross section can be evaluated reasonably well,[1] no useful measurements of changes in *major* cross sections of the FS are possible if the precision is limited to 0.01%. Clearly a detailed study of alloying changes requires in fact a precision of a few ppm, an improvement of about two orders of magnitude.

New Technique

Coleridge and Templeton have recently developed a technique[2,3] whereby such precision can be reached, and have started to use it in studying the effects of alloying on the FS of copper. Let us examine briefly some of the requirements and restrictions in such a measurement, and the means used in dealing with them. The requirements of high homogeneity of magnetic field are obvious and will not be discussed here. Sample orientation is clearly of great importance: it is of little use comparing any other than symmetry direction orbits or other orbits which may be precisely defined in situ. It is impossible to mount a sample to better than a few tenths of a degree, and for some orbits such a misorientation may well correspond to an error of as much as 100 ppm in cross section. We therefore orient the samples in situ by tilting to an appropriate turning point of dHvA phase with angle. Since we can usually define such a turning point to a few thousandths of a cycle, the orienta-

* Invited paper.

Table I. Dilute Cu-Based Alloys: Maximum Concentrations for Acceptable Signal Levels and Corresponding Cross-Section Changes*

Solute	Orbit	x/c, °K/at.%	Maximum c, at.%	Probable ΔV, el/at	Expected $\Delta A/A$, %
Al	⟨111⟩ Neck	80	0.05	+ 2	0.6
Al	⟨111⟩ Belly	64	0.03		0.04
Ni	⟨111⟩ Neck	25	0.16	− 1	− 0.96
Ni	⟨111⟩ Belly	50	0.04		− 0.027

* $H \approx 50$ kG, $T \approx 1$°K, $x_N \leq 4$°K, $x_B \leq 2$°K. The values of x/c are taken from Ref. 1.

tion for a major cross section (total phase $\sim 10^4$) is good to at least one part in 10^6. (See Ref. 3 for diagram of probe and of orientation technique.)

Having achieved the required precision of orientation, the next step is to determine the relationship between dHvA phase and magnetic field with equal precision. The field is measured with a small NMR sample of aluminum powder in epoxy resin which surrounds the dHvA specimen (shown diagrammatically in Fig. 1). We lock the rf oscillator frequency to the center of the first derivative of the NMR

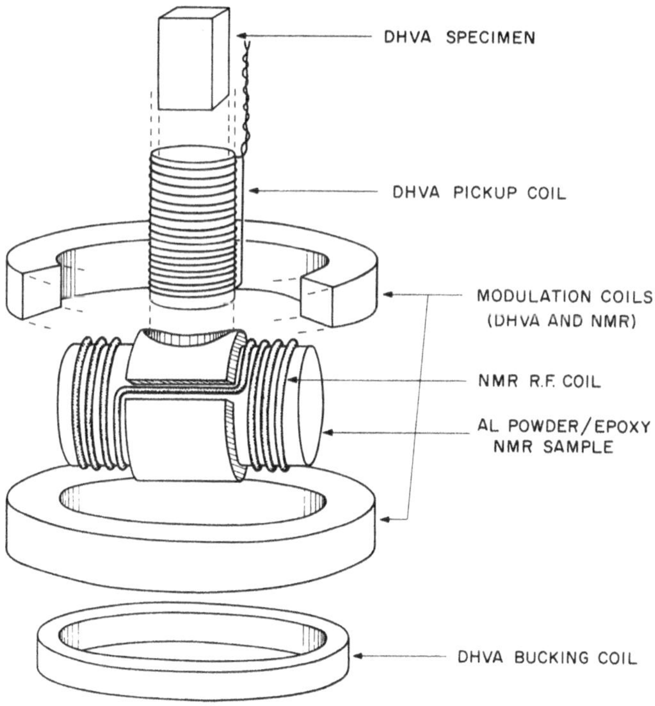

Fig. 1. Exploded view of the dHvA and NMR modulation and pickup system.

resonance, adjust the error signal to zero, and measure the frequency: in a persistent field the stability is about one part in 5×10^6. If we heat the persistent switch on the superconducting solenoid to just above its critical value with the power supply slightly offset, we can drift the field very slowly downward. We record the dHvA oscillations as a function of time, stop the drift by switching off the heater at, or close to, appropriate zero-crossing points (fractional phase = 0.25 or 0.75 cycle) and measure the field at these points in terms of the NMR frequency. This is illustrated and discussed further in Ref. 3, though the illustration is rather idealized in that only one component is present. We repeat this measurement at several convenient values of magnetic field.

So far, then, we have a set of values for fractional phase vs. field, with both phase and field defined to better than 1 ppm. Assuming that we already know the frequency F_0 to somewhat better than 0.1 %, we can calculate the corresponding values of total phase (as F_0/H) knowing that the integer parts of these numbers are correct to a few cycles in 10^4. We then determine the differences between the measured and calculated *fractional* parts of the phase and plot these differences as a function of reciprocal field. This yields a straight line whose slope gives a correction to the estimated frequency and whose intercept at $1/H = 0$ should (and in practice does[3]) give the infinite-field phase factor ϕ_∞. That is, if the oscillatory part of the dHvA effect is given by $\cos\{2\pi[(F/H) + \phi_\infty]\}$, the plot is essentially of $(F - F_0)/H$ vs. $1/H$. Figure 2 illustrates this process (actually for the dog's bone orbit in silver) and Fig. 3 shows the results of an early set of measurements on samples cut from a Cu(Al) 100 ppm ingot, giving the actual difference of dHvA phase in cycles between alloy and pure host on our six standard orbits as a function of reciprocal field. Note that, as expected, electron orbits increase and hole orbits decrease in area (dHvA frequency) as the electron concentration is increased.

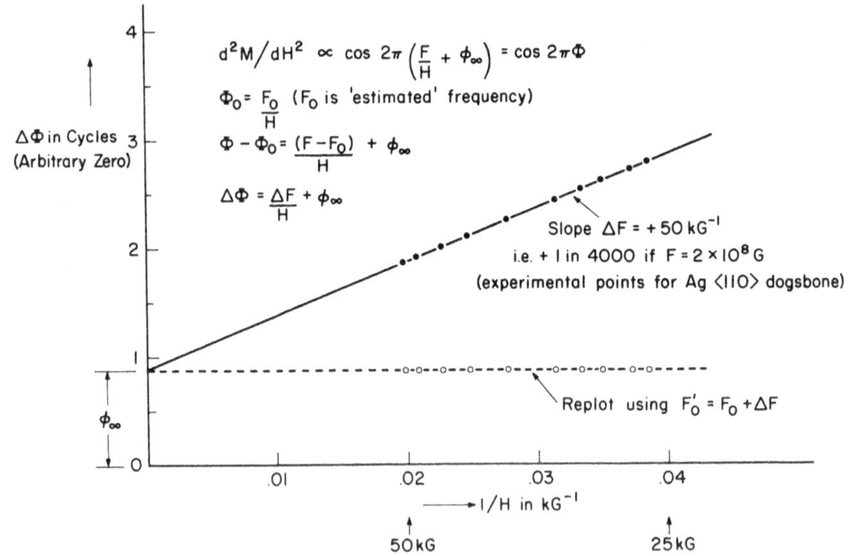

Fig. 2. Use of differences between observed and estimated fractional phases to obtain a correction to a trial frequency value.

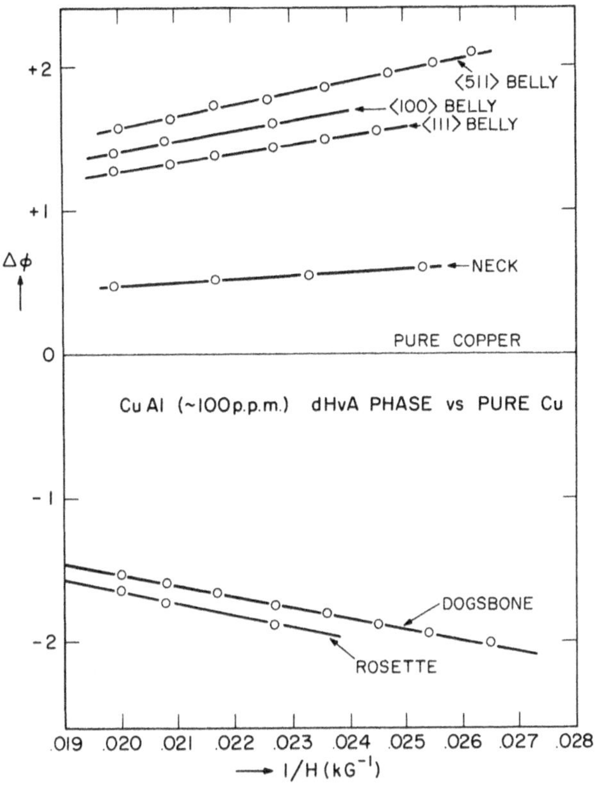

Fig. 3. Phase differences in cycles between pure Cu and Cu + ~ 100 ppm Al for a standard set of orbits. The orbit labeled < 511 > is the belly turning point on the (110) plane near < 100 >.

As can be seen from the plots, the precision is indeed high, and in fact we find that the determination of solute concentration in individual samples is one of our larger sources of error.

Interpretation

We must now decide what use to make of the results, which we express, say, as change of cross section per 100 ppm of solute having made appropriate (generally rather small) corrections for the effect of lattice constant change. Halse[4] has fitted the FS of copper using both five-term and seven-term Fourier expansions. (In practice one finds that these involve four and five independent coefficients, respectively.) If on the basis of such a model we determine the derivatives of our set of six chosen cross sections with respect to these four or five coefficients, we can make a least-squares fit to our observed cross-section changes and determine the corresponding volume and radius changes in the model FS. Figure 4 shows the result of such fits to the same six pieces of data for Cu(Ni) 100 ppm: We plot contour lines of changes normal to the FS as parts per million of a free-electron sphere radius. The volume

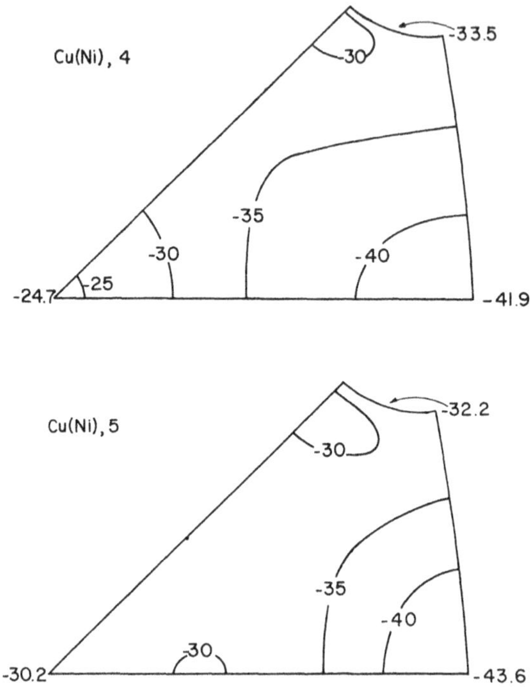

Fig. 4. Changes normal to the Fermi surface in ppm of a free-electron sphere radius for 100 ppm Ni in Cu. Comparison between four-term and five-term Fourier expansion fits to observed cross-section changes. Note: The data shown here and in Fig. 5 are preliminary and may be modified somewhat when the full set of measurements is completed.

changes indicated by these two fits are very similar, being -110 and -107 ppm for the four- and five-coefficient fits, respectively, but there are some differences in the radial variations. However, examination of the details of the two fits shows that the inclusion of an extra term not only fails to improve the rms error, but there is also a marked reduction in the statistical significance of the fit. It thus seems probable that the extra radial variations apparent with five coefficients are in fact nonphysical. Concentrating, then, on the presumably more realistic four-coefficient fits, we find that on fitting Cu(Al) 100 ppm the volume change is $+183$ ppm compared with the $+200$ ppm one might expect for a trivalent solute. In Fig. 5 we compare Cu(Ni) and Cu(Al), plotting this time the changes as fractions of the rigid-band change* for the appropriate change of volume. We find that while both Ni and Al appear to change the radius at $\langle 100 \rangle$ by less than the rigid-band figure, there is a fairly strong contrast between the amounts of change at $\langle 110 \rangle$ and the neck, nickel causing markedly less change at the neck and markedly more at $\langle 110 \rangle$ than aluminum. The changes of FS *shape* can be summarized by saying that the addition of Ni, while

* We take as our rigid-band model one in which the cross-section changes are proportional to m^*.

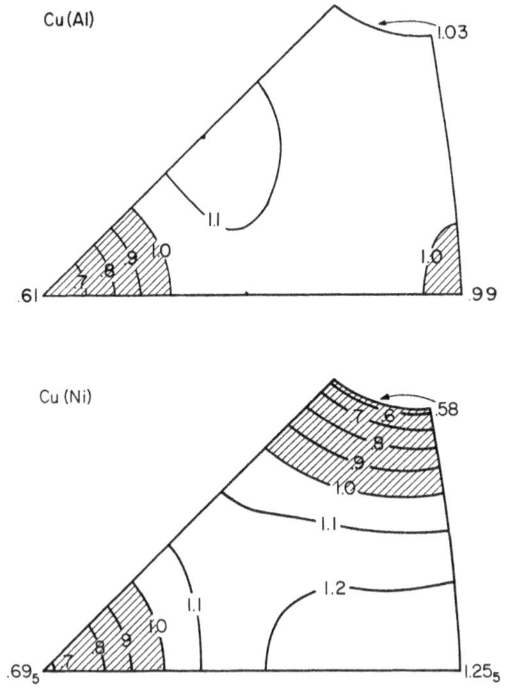

Fig. 5. Changes normal to the Fermi surface in Cu(Al) and Cu(Ni) as a fraction of the rigid band change giving the same change of volume. A four-term Fourier expansion fit has been used in each case.

it decreases the FS volume, increases the distortion fairly markedly, particularly at $\langle 110 \rangle$ and at the neck, while Al causes a mild reduction in distortion. In the case of Cu(Ni) there is optical evidence[5] for a well-defined virtual bound state lying some 0.8 eV below the Fermi level, and it seems reasonable to suppose that its presence is responsible for the increased FS distortion. The ~ 0.1 and ~ 0.2 electron/atom discrepancies can presumably be accounted for by saying that in an alloy the Fermi surface volume as measured by a given technique does not necessarily have to account for all the conduction electrons[6]: in fact it is very reasonable to suppose that our measurement samples only the coherent part of the band structure.

It should perhaps be mentioned here that the Fourier expansion fits referred to above are only empirical models without physical basis other than crystal structure and symmetry. Coleridge has had some success very recently in fitting our results to a phase-shift model based on an APW band structure calculation: it appears that only the first three phase shifts ($l = 0, 1, 2$) are needed to give a good fit to our observations.

Conclusions

The results outlined here are only a beginning: other systems of interest would be Ag–Au, which has change of potential and d-band position without change of lattice constant or electron concentration; Cu(Zn) and Cu(Si), for comparison with Cu(Al) at different valences; and Cu(Pd), which has a very sharply defined virtual bound state[5] lying appreciably lower than that of Ni, for comparison with Cu(Ni).

To summarize, while the precision of standard dHvA techniques is inadequate for any detailed study of the effects of alloying, our new high-precision technique can provide the required two orders of magnitude improvement in precision and should prove to be a most useful tool in future alloying studies.

References

1. P.T. Coleridge and I.M. Templeton, *Can. J. Phys.* **49**, 2449 (1971).
2. P.T. Coleridge, A.A.M. Croxon, G.B. Scott, and I.M. Templeton, *J. Phys. E: Sci. Instr.* **4**, 414 (1971).
3. P.T. Coleridge and I.M. Templeton, *J. Phys. F: Metal Phys.* **2**, 643 (1972).
4. M.R. Halse, *Phil. Trans. Roy. Soc.* **265**, 507 (1969).
5. N.R. Comins, private communication, 1972.
6. E.A. Stern, *Phys. Rev.* **168**, 730 (1968).

The Electrical Resistivity of Dilute Nickel Alloys

D. Greig and J. A. Rowlands*

Physics Department, The University of Leeds
Leeds, England

There have recently been a number of advances in our understanding of conduction in ferromagnetic metals. The model used is a two-current model, in which the conduction electrons are believed to consist of two groups of s-like electrons of opposite spin—the spin-up and spin-down electrons.[1-5] For nickel it is assumed that the spin-down d holes do not carry any current, but that the presence of high-density vacant d states tends to enhance the scattering probability of the spin-down (minority) electrons compared to that of the spin-up (majority) electrons. We therefore expect the conductivity ratio $\sigma(\uparrow)/\sigma(\downarrow)$ to be substantially greater than unity. Evidence from a range of transport properties has shown that this is indeed the case for the impurity conductivity ratio $\alpha = \sigma_0(\uparrow)/\sigma_0(\downarrow)$ in *Ni*–Co, *Ni*–Fe, *Ni*–Rh, *Ni*–Ru, and *Ni*–Mn.[1,2,4,6]

There are, however, other types of alloy in which this ratio is completely different: (i) *Ni*–Cr, *Ni*–V, *Ni*–Zr, and *Ni*–Ti.[2,6,7] For these alloys, a virtual bound state near the Fermi level in the up direction greatly reduces $\sigma_0(\uparrow)$, and $\alpha < 1$. (ii) *Ni*–Pd. In this system the evidence suggests[2] that $\alpha \simeq 1$. In a homovalent alloy the scattering potential is very localized, and the spin-down transitions into d states—which lie at some distance from the cores—simply do not take place. (iii) *Ni*–Cu. We should now like to present evidence showing that this system is quite different from *all* the others, in what the conductivity ratio depends, not only on the *type* of impurity, but also on the *concentration*.

Large values of α are always associated with large deviations from Matthiessen's rule (DMR) which are roughly proportional to T^2. The contrast between the behavior of the resistivity in *Ni*–Cu alloys on the one hand and more typical nickel alloys on the other is shown in Fig. 1, in which we have plotted both new and previously published data on the temperature-dependent resistivity ρ_i at 20°K as a function of impurity resistivity ρ_0. The values of ρ_i are obtained for each alloy by subtracting ρ_0 from the measured resistivity ρ_m. Full details of the measurements at 20°K and several other selected temperatures are given in Table I.

For "pure" nickel at 20°K, $\rho_i = 0.009$ $\mu\Omega$-cm,[8] so that all values of ρ_i greater than this in Fig. 1 represent DMR. These deviations clearly fall into three categories: (i) Very large values are seen in *Ni*–Co and *Ni*–Fe. (ii) Rather small values are observed in *Ni*–Pd. In both cases (a) ρ_i does not depend very much on ρ_0, and (b) the concentration-dependent part can be shown to vary as T^3. Both these features are

* Present address: Physics Department, University of Alberta, Edmonton, Alberta, Canada.

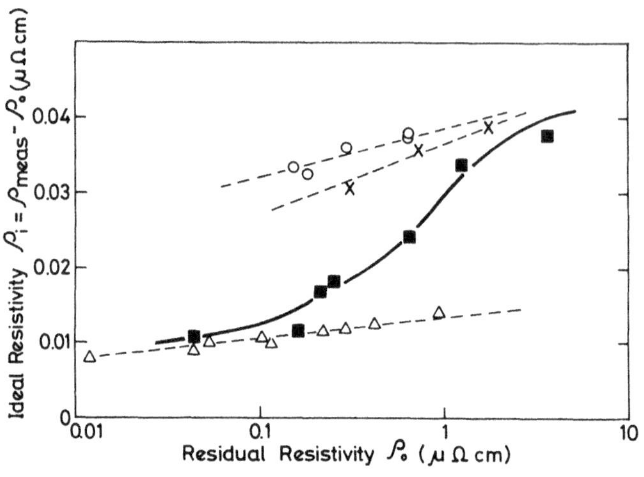

Fig. 1. Variation of ρ_i with ρ_0 at 20°K for *Ni*–Pd (Δ), *Ni*–Co (○), *Ni*–Fe (X), and *Ni*–Cu (■).

Table I. Measurements of ρ_0 and $\rho_i = \rho_m - \rho_0$ in Dilute Nickel Alloys

Alloy (at.%)	ρ_0, $\mu\Omega$-cm	ρ_i, $\mu\Omega$-cm						
		6.0°K	10.0°K	20.0°K	30.0°K	30.5°K	40.0°K	40.2°K
+ 0.03 Pd*	0.012		0.0020	0.0084	0.0258	—	0.0682	—
+ 0.25 Pd	0.0537	0.0006	0.0022	0.0104	—	0.0311	—	0.0754
+ 0.3 Pd*	0.045	—	0.0020	0.0090	0.028	—	0.0726	—
+ 0.5 Pd	0.1010	0.0006	0.0022	0.0109	—	0.0332	—	0.0796
+ 1.0 Pd*	0.119	—	0.0020	0.0100	0.0318	—	—	—
+ 1.0 Pd	0.2227	0.0006	0.0025	0.0118	—	0.0363	—	0.0866
+ 1.5 Pd	0.2884	0.0006	0.0025	0.0123	—	0.0378	—	0.0897
+ 2.0 Pd	0.4099	0.0006	0.0025	0.0129	—	0.0394	—	0.0937
+ 5.0 Pd	0.9426	0.0008	0.0027	0.0142	—	0.0459	—	0.1091
+ 0.01 Cu*	0.168‡	—	0.0021	0.0112	0.0323	—	0.0805	—
+ 0.04 Cu*	0.044	—	0.0023	0.0115	0.0320	—	0.0738	—
+ 0.16 Cu†	0.209	—	0.004	0.017	0.048	—	0.101	—
+ 0.3 Cu*	0.247	—	0.0038	0.0182	0.0504	—	0.111	—
+ 0.8 Cu†	0.627	—	0.005	0.024	0.062	—	0.125	—
+ 1.6 Cu†	1.261	—	0.009	0.034	0.078	—	0.150	—
+ 4.9 Cu†	3.701	—	0.011	0.038	0.087	—	0.167	—
+ 1.0 Co*	0.153	—	0.0066	0.0336	0.0869	—	0.179	—
+ 1.0 Co	0.185	0.00184	0.0066	0.0327	0.0879	—	—	—
+ 2.0 Co*	0.297	—	0.0069	0.0363	0.0935	—	0.198	—
+ 5.0 Co*	0.647	—	0.0081	0.0372	0.1039	—	0.214	—
+ 5.0 Co	0.660	0.00224	0.0074	0.0381	0.1045	—	—	—
+ 0.5 Fe*	0.307	—	0.0083	0.0308	0.0773	—	0.155	—
+ 2.0 Fe*	0.713	—	0.0084	0.0359	0.0978	—	0.205	—
+ 5.0 Fe*	1.80	—	0.006	0.039	0.105	—	0.208	—

* Measured by T. Farrell.
† Measured by J. P. Harrison.
‡ This sample clearly contains an additional impurity.

also found experimentally in dilute Al alloys,[9] although the absolute magnitude of the deviations in such alloys is an order of magnitude less than those in *Ni*–Co and *Ni*–Fe. This is the special feature of these latter alloys that we attribute to the large values of α. (iii) Variable values are found in *Ni*–Cu which approach two different limits for low and high solute concentrations. Furthermore, the concentration-dependent part of DMR in this instance varies as T^2.

We interpret the data in *Ni*–Cu as evidence of a *change* in α from $\alpha \sim 1$ when the copper content c is less than 0.1% to $\alpha \gtrsim 10$ when $c > 1\%$, brought about by a relative reduction in $\sigma_0(\downarrow)$ over this concentration range. We suggest that the reason for this is precisely the same as that first given by Mott[10] to explain the unusual concentration dependence of ρ_0 in *Pd*–Ag. He suggested that *s-d* scattering only occurs at Ag atoms when their concentration has increased sufficiently for the periodic potential to be perturbed throughout the whole lattice. These ideas have been applied again comparatively recently to interpret resistivity data in *Pd*–Ag.[11,12] If the same argument is applied to the spin-down band in *Ni*–Cu, we see that scattering will change with increasing copper content from *s-s*—as in the spin-up band—to *s-d*, giving rise to a substantial increase in α. This change will be observed as a concentration-dependent "T^2" DMR.

Such an effect will also be reflected in other transport properties.

Residual resistivity. Kondorskii *et al.*[13] found the unexpected result that up to 25% Cu in nickel, ρ_0 was *linear* in c, and showed no evidence of the 25% reduction at that concentration predicted by the Nordheim $c(1 - c)$ relationship. This anomaly can be readily explained if we assume, again following Mott, that the component of resistivity arising from *s-d* scattering is proportional to $c^2(1 - c)$. In this case $\rho_0(\uparrow) = Ac(1 - c)$ and $\rho_0(\downarrow) = Ac(1 - c) + Bc^2(1 - c)$, so that for large values of Bc the value of ρ_0 at a given concentration can be as much as twice the value predicted by the Nordheim formula. In fact, if we assume $B = 2A$, the resulting ρ_0 vs. c curve is remarkably linear.

Thermopower. Thermoelectric effects are difficult to interpret, but we have some earlier evidence[14] of a phonon-drag component that increases with c for very dilute *Ni*–Cu, and this can also be explained by an increase in the value of α.

References

1. A. Fert and I.A. Campbell, *Phys. Rev. Lett.* **21**, 1190 (1968).
2. T. Farrell and D. Greig, *J. Phys. C* **1**, 1359 (1968).
3. I.A. Campbell, *Phys. Rev. Lett.* **24**, 269 (1970).
4. F.C. Schwerer and J.W. Conroy, *J. Phys. F* **1**, 877 (1971).
5. M.C. Cadeville and J. Roussel, *J. Phys. F* **1**, 686 (1971).
6. J. Durand and F. Gautier, *J. Phys. Chem. Solids* **31**, 2773 (1970).
7. I.A. Campbell, A. Fert, and O. Jaoul, *J. Phys. C: Metal Phys.* (Suppl.) S95 (1970).
8. G.K. White and S.B. Woods, *Phil. Trans. Roy. Soc. A* **251**, 273 (1959).
9. A.D. Caplin and C. Rizzuto, *J. Phys. C* **3**, L117 (1970).
10. N.F. Mott, *Proc. Roy. Soc. A* **156**, 368 (1936).
11. B.R. Coles and J.C. Taylor, *Proc. Roy. Soc. A* **267**, 139 (1962).
12. J.S. Dugdale and A.M. Guénault, *Phil. Mag.* **13**, 503 (1966).
13. E.I. Kondorskii, O.S. Galkina, and L.A. Chernikova, *Zh. Eksperim. i Teor. Fiz.* **7**, 741 (1958).
14. T. Farrell and D. Greig, in *Proc. 10th Intern. Conf. Low Temp. Phys., 1966*, VINITI, Moscow (1967), Vol. 4, p. 96.

The Hall Resistivity in Dilute Crystals of *Cd*–In and *Cd*–Ag: An Impurity Problem for Intersheet Scattering*

D. A. Lilly and A. N. Gerritsen

Physics Department, Purdue University
Lafayette, Indiana

Several years ago the dependence of the Hall resistivity ρ_{21} on temperature, magnetic field, and crystallographic direction was reported.[1] The results were generally confirmed later,[2] in particular for the observed sign reversal of ρ_{21} below 10°K for pure cadmium and dilute alloys of *Cd*–Zn. This reversal was explained by Young *et al.*[3] with a model based on small-angle scattering of carriers between parts of the second-zone hole sheet in the ΓKM (basal) plane. This scattering results in a net increase of electron orbits, thereby disturbing the carrier compensation of the material. Semiquantitative agreement between theory and experiment was achieved by comparison between $\omega\bar{\tau}$ ($\bar{\tau}$ is the average scattering time when $H = 0$) and $\omega\mathscr{T}$, where $\mathscr{T}$ is the intersheet scattering time, adjusted to fit the data. Here ω is the relevant cyclotron frequency for the orbit in question. The explicit dependence of $\mathscr{T}$ on temperature and impurity concentration is unknown, but $\mathscr{T}$ increases with the former and decreases with the latter.[3]

Since the compensation can be disturbed by changing the e/a ratio as well, we investigated single crystals of *Cd*–Ag and *Cd*–In grown from the melt. In contrast to the *Cd*–Zn alloys, in which the sign reversal of ρ_{21} was observed at higher temperatures than in pure cadmium, these samples displayed no sign reversal with solute concentrations of 0.04 at.%. The investigations were continued for lower impurity concentrations, thereby abandoning the search for e/a effects.

The results for $\rho_{21}(T)$ ($\mathbf{H} \parallel \langle 0001 \rangle$, $\mathbf{J}$ in the basal plane) together with the residual resistance data for several *Cd*–Zn alloys and the present alloys are given in Figs. 1 and 2, respectively. Hall resistance data for pure cadmium are included in Fig. 1 for comparison. Note that the residual ρ_{21} becomes more negative as sample purity increases. In Fig. 2 it can be seen that ρ_{21} remains positive for the alloy samples, with the exception of the most dilute alloy (solute concentration < 5 ppm). In Fig. 2 $\omega\bar{\tau} \geqq 4$ for $T = 10$°K and $H = 23$ kG. The resistivity $\rho_{11}(H = 0, T)$, of all samples varied as $a + bT^n$ ($n = 5.0 \pm 0.2$) for $1.4°\text{K} < T < 10°\text{K}$.

All samples exhibit an increase in ρ_{21} for $T < 30$°K. At $T \approx 15$°K, ρ_{21} reaches a maximum, the origin of which has not been established. Saeger and Lück[4] have suggested that localized Umklapp processes could be responsible for the maximum in ρ_{21} that has been observed in polycrystalline cadmium and in other metals (e.g.,

* Research supported by the National Science Foundation.

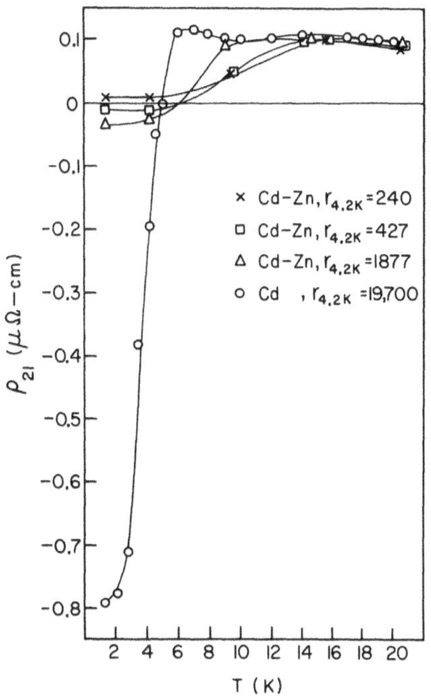

Fig. 1. Hall resistivity $\rho_{21}(H = 20\text{kG})$ vs. temperature for pure cadmium and *Cd*–Zn alloys. *Cd*–Zn data taken from Ref. 1.

Cu, Zn). Hurd and Alderson[5] have associated a similar maximum they observed in Hall resistivity data of Cu and *Cu*–Zn alloys with transitions for certain electrons from the low- to high-field region. Both suggestions are inadequate, however, to account for the appearance of a second maximum around 7°K in the present results. This second maximum is followed by a sharp decrease, ending in a positive, residual value $\rho_{21}(0)$ which increases with increasing sample purity.

Considering $\omega\bar{\tau} > 1$, one expects that the general formal conditions for large effects in ρ_{21} as encountered for the *Cd*–Zn alloys are met and that similar $\rho_{21}(T)$ behavior should be observed. Although a detailed consideration of the magnitude of the intersheet scattering time on type of impurity is not possible, the theory predicts a continuous decrease in $\rho_{21}(0)$ when the purity increases. The results shown in Fig. 2, namely the increase of $\rho_{21}(0)$ with increasing purity, followed by a sharp decrease for a relatively small change in purity, can not be deduced from the formalism of the intersheet scattering theory. We therefore feel that in the present alloys the intersheet scattering is inhibited for a range of low impurity concentrations and becomes manifest only in samples of purity comparable to that of pure cadmium. It will be shown that the observed behavior is in qualitative agreement with observations on the carrier mean free path (mfp) in cadmium.

Recent measurements,[6,7] on the carrier mfp for several closed cyclotron orbits in single crystals of pure cadmium ($r_{4.2} \approx 35{,}000$) show that the mfp varies according to the relation

$$l^{-1}(T) - l^{-1}(0) = CT^n \tag{1}$$

which can be written

$$\tau^{-1}(T) - \tau^{-1}(0) = BT^n \tag{2}$$

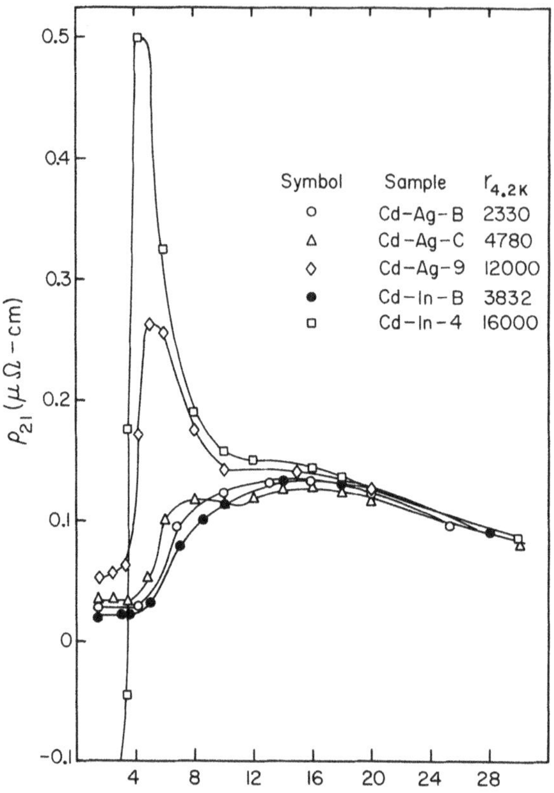

Fig. 2. Hall resistivity $\rho_{21}(H = 23\text{kG})$ vs. temperature for *Cd*–In and *Cd*–Ag alloys.

Here $n = 3$ for holes on the second-zone hole sheet and $n = 5$ for electrons on the third-zone "lens." This temperature dependence was observed for $\mathbf{H} \parallel \langle 0001 \rangle$ involving the same closed orbits that control ρ_{21}.

In the high-field region ($\omega\bar{\tau} \gg 1$) the Hall resistivity for a compensated metal like cadmium can be written[8] in the two-band approximation as

$$\rho_{21} = (\omega_h \tau_h - \omega_e \tau_e)\, \rho_{11}(H = 0, T) \tag{3}$$

$$= \text{const} \times \left(\frac{\omega_h \tau_h}{\omega_e \tau_e} - 1 \right) \tag{3a}$$

for $\rho_{11}(H = 0, T) \propto \tau_e^{-1}(T)$. In the present case $\rho_{21} > 0$ even when $\tau_e = \tau_h$ because the inequality of the electron and hole masses is such that $\omega_h > \omega_e$. The temperature deviation in ρ_{21} from $\rho_{11}(T)$ then arises from the differences between $\tau_e(T)$ and $\tau_h(T)$. The observation that ρ_{21} increases with decreasing temperature indicates that even though $\tau_e(T)$ increases faster, the increase in the magnitude of $\tau_h(T)$ dominates in the temperature range under consideration.

Figure 3 is a qualitative plot of $\omega\tau$ on double logarithmic scale that will illustrate the arguments used to account for the observed behavior of $\rho_{21}(T)$. The lines repre-

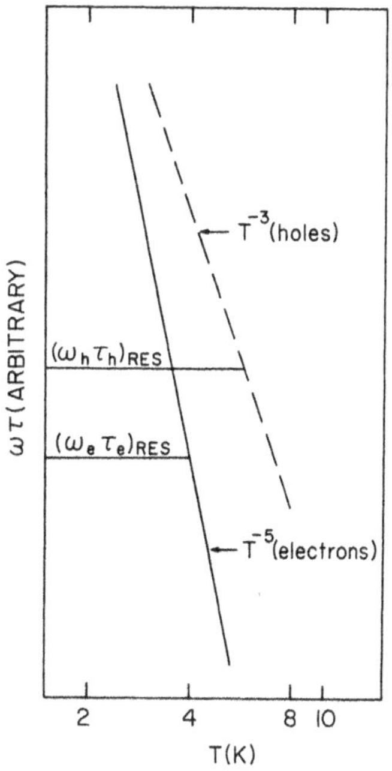

Fig. 3. Log $\omega\tau$ vs. log T for electrons and holes using the mfp temperature dependence as given in Ref. 6.

sent Eq. (2), so that $\omega_h \tau_h(T) > \omega_e \tau_e(T)$. According to Eq. (3), when the residual values for τ_h and τ_e have been reached ρ_{21} will be positive when $(m_e^*/m_h^*)\, \tau_h(0) > \tau_e(0)$ is satisfied. Since[3] $(m_e^*/m_h^*) \simeq 1.8$, $\tau_h(0)$ could be less than $\tau_e(0)$ for $\rho_{21}(0)$ still to assume a positive value. If, indeed, $\tau_h(0) < \tau_e(0)$, then it is reasonable to assume that $\tau_h(0)$ is reached at a higher temperature than $\tau_e(0)$. This assumption leads consequently to a temperature range in which $\omega_e \tau_e$ increases while $\omega_h \tau_h$ = const, leading to a decrease in ρ_{21}.

If one is willing to accept the qualitative argument that indeed a different temperature dependence for hole and electron scattering times governs the behavior of $\rho_{21}(T)$, then the sign reversal of ρ_{21} observed in sample Cd–In–4 (<5 ppm indium) fits in the suggested explanation if the lines in Fig. 3 cross at about 3.5°K. Such a crossover would not account, however, for the large negative values of ρ_{21} that are observed. For this effect the explanation in terms of intersheet scattering remains the most satisfactory.

A quantitative analysis of $\rho_{21}(T)$ meets with severe difficulties. For example, Eq. (3) leads to the condition $\rho_{21} \propto H$, for fixed temperature. The observations, however, show that for T between 4.2 and 10°K, ρ_{21} for H = 20 kG can be twice as great as the value extrapolated linearly from the low-field data. Moreover, this effect is larger for more pure samples.

One must conclude that the present results with *Cd*–Ag and *Cd*–In alloys introduce the problem of the inhibition of intersheet scattering in Cd by these impurities in amounts of a few ppm. This inhibition is in marked contrast to the case with zinc as solute, which enhances such scattering.

Several possibilities present themselves as a reason for this inhibition. A particular configuration of the solute atoms could produce a very small change in the radius of the Fermi sphere, which could have a large effect on the intersheet scattering. Or, without producing any traceable effect on the Fermi surface, impurities such as Ag or In that are not readily soluble in Cd may assume lattice positions such that carriers are scattered away from the areas favorable for intersheet scattering. In general it is interesting to note that although these small impurity concentrations do not affect the bulk conductivity properties, they do influence in a striking manner a property governed by specific areas of the Fermi surface.

References

1. O.P. Katyal and A.N. Gerritsen, *Phys. Rev.* **178**, 1037 (1969).
2. H. Schwarz, *Phys. Stat. Sol.* **39**, 515 (1970).
3. R.A. Young, J. Ruvalds, and L.M. Falicov, *Phys. Rev.* **178**, 1043 (1969).
4. K.E. Saeger and R. Lück, *Phys. Kond. Materie* **9**, 91–97 (1969).
5. C.M. Hurd and J.E.A. Alderson, *J. Phys. Chem. Solids* **32**, 175 (1971).
6. P.D. Hambourger and S. Huang, "Electron Mean Free Paths in Cadmium," presented at the San Francisco APS Meeting, January 1972 (unpublished).
7. P.D. Hambourger, this volume.
8. L.M. Falicov, A.B. Pippard, and Paul R. Sievert, *Phys. Rev.* **151**, 498 (1966).

Calculations of de Haas–van Alphen Lifetimes in Copper-Based Alloys

R. Harris

McGill University
Montreal, Quebec, Canada

The most recent de Haas–van Alphen data in dilute alloys based on Cu, Ag, and Au have been analyzed in terms of an anisotropic lifetime $\tau_0(\mathbf{k})$ for an electron at a point $\mathbf{k}$ on the Fermi surface.[1] This lifetime is proportional to the inverse of the Dingle temperature, which in turn is related to the amplitude of the dHvA signal. In similar fashion, recent magnetoresistance measurements in dilute alloys based on Cu and Ag have been analyzed via the Boltzmann equation in terms of an anisotropic relaxation time $\tau(\mathbf{k})$ for an electron at a point $\mathbf{k}$ on the Fermi surface.[2]

The present paper is concerned with the theoretical treatment of the scattering. In particular, it examines the relationship between $\tau_0(\mathbf{k})$ and $\tau(\mathbf{k})$, and identifies situations where these two quantities should be identical. A discussion of the existing data for dilute alloys based on Cu uses realistic band-structure data for Cu: It shows that the theory gives a consistent picture of these alloys, although the data are not complete enough to justify a detailed analysis. For one alloy, *Cu*–Ni, the data are sufficiently good that a detailed comparision is possible: Agreement is very good.

In a previous paper[3] the KKR model of electrons in a periodic lattice[4] is extended to describe the scattering of Bloch electrons by a spherical perturbing potential. The theory depends upon the muffin-tin model for a dilute alloy, where not only the solvent atoms but also the solute atoms are represented by nonoverlapping potentials, and the region between the spheres corresponds to a constant potential. In the dilute limit the scattering is caused by just one solute atom, and the t-matrix elements can then be written in terms of the Bloch functions $\psi(\mathbf{k})$. The t-matrix $t(\mathbf{k}, \mathbf{k}')$ thus contains band-structure information of the most detailed kind.

The transition probability $P(\mathbf{k}, \mathbf{k}')$ from Bloch state $\mathbf{k}$ to Bloch state $\mathbf{k}'$ is related closely to the t-matrix. Furthermore, since the dHvA lifetime $\tau_0(\mathbf{k})$ is a measurement of the total scattering rate from $\mathbf{k}$ to all other points on the Fermi surface, it may be written as

$$[\tau_0(\mathbf{k})]^{-1} = \sum_{\mathbf{k}'} P(\mathbf{k}, \mathbf{k}') = -2\,\mathrm{Im}\,[t(\mathbf{k}, \mathbf{k}')]$$

The second equality of this equation follows from the optical theorem of conventional scattering theory. Thus $\tau_0(\mathbf{k})$ is related strongly to the detailed band structure of the solvent metal. Evaluation of $\tau_0(\mathbf{k})$ is straightforward when the scattering is dominated only by one-phase shift, since in this case a contour map of $\tau_0(\mathbf{k})$ represents just a mapping of the s-, p-, or d-like character of the wave function $\psi(\mathbf{k})$ of the solvent metal.[3]

Most of the available data for copper-based alloys are expressed only as the ratio of average values of $\tau_0(\mathbf{k})$ around particular orbits. The $\langle 111 \rangle$ belly orbit and the neck orbit are usually used, and the data thus yield a single ratio τ_B/τ_N. The theory predicts that the largest value of this ratio would be in an alloy dominated by p-wave scattering, and would have the value ~ 2.4. This value is approached, but not exceeded, by the heterovalent solutes Zn, Al, Ga, Ge, and Si.[1] In a similar way a minimum value (approaching zero) for τ_B/τ_N corresponds to an alloy dominated by s-wave scattering, and this is more nearly appropriate to the transition metal solutes Co, Fe, Mn, and Cr.[1]

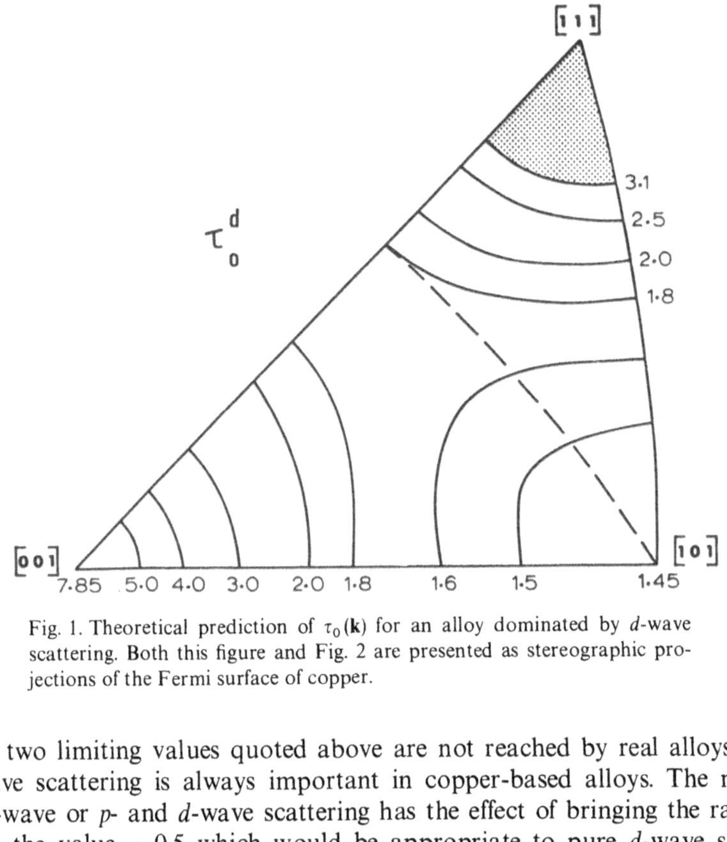

Fig. 1. Theoretical prediction of $\tau_0(\mathbf{k})$ for an alloy dominated by d-wave scattering. Both this figure and Fig. 2 are presented as stereographic projections of the Fermi surface of copper.

The two limiting values quoted above are not reached by real alloys because the d-wave scattering is always important in copper-based alloys. The mixing of s- and d-wave or p- and d-wave scattering has the effect of bringing the ratio τ_B/τ_N closer to the value ~ 0.5 which would be appropriate to pure d-wave scattering. Thus values of τ_B/τ_N lying between 0 and 0.5 correspond to the mixing of s- and d-wave scattering, and values between 0.5 and 2.4 to the mixing of p and d.

The alloy *Cu*–Ni appears to be dominated only by d-wave scattering, since the value τ_B/τ_N is ~ 0.5.[1] Further support for this statement comes from the magnetoresistance data for this alloy, which are analyzed in terms of an anisotropic relaxation time $\tau(\mathbf{k})$.[2] It can be shown that $\tau(\mathbf{k})$, as a solution of the Boltzmann equation containing $P(\mathbf{k}, \mathbf{k}')$ as the scattering term, becomes identical with $\tau_0(\mathbf{k})$ when the scattering is dominated by one phase shift.[3] Since this is the case in *Cu*–Ni, the experimental map of $\tau(\mathbf{k})$ should correspond well with the theoretical calculation of $\tau_0(\mathbf{k})$ for this alloy. The excellent correspondence is shown by a comparison of

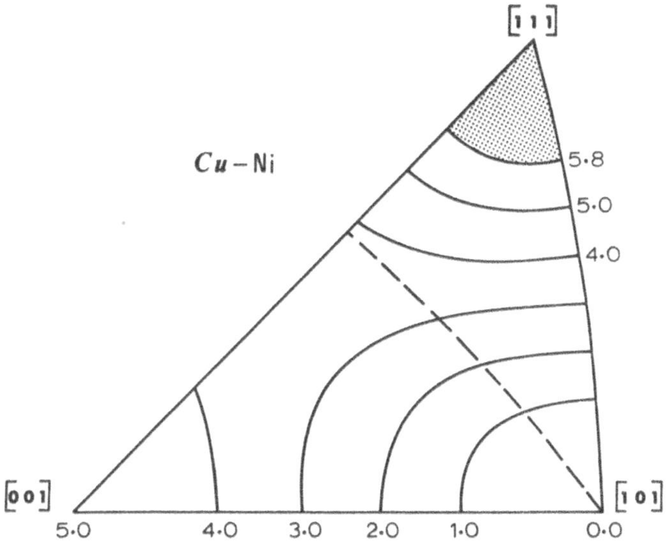

Fig. 2. Experimental map of $\tau(\mathbf{k})$ for *Cu*–Ni from Ref. 2.

Figs. 1 and 2, which differ only by a trivial scaling factor. A more complete account of this work will be published elsewhere.[5]

References

1. M. Springford, *Adv. Phys.* **20**, 493 (1971); P.T. Coleridge and I.M. Templeton, *Can. J. Phys.* **49**, 2449 (1971); D.H. Lowndes, K. Miller, and M. Springford, *Phys. Rev. Lett.* **25**, 1111 (1970).
2. A. Atkinson, *J. Phys. F* **1**, 863 (1971).
3. J.W. Blaker and R. Harris, *J. Phys. C* **4**, 569 (1971).
4. J.L. Beeby, *Proc. Roy. Soc. A* **302**, 113 (1967); R. Harris, *J. Phys. C* **3**, 172 (1970).
5. R. Harris, *J. Phys. F* (to be published).

8

Electron–Phonon Interaction

The Phonon-Exchange Vertex in Transition Metals—A Localized Approach

H. Gutfreund and A. Birnboim

The Racah Institute of Physics
The Hebrew University of Jerusalem, Israel

Appel and Kohn have suggested in a recent paper[1] (hereafter referred to as AK) that the natural way of formulating the Cooper-pairing mechanism in transition metals with narrow d bands is by employing the Wannier or the atomic site representation rather than the Bloch representation. The former emphasizes the atomic nature of the tightly bound d electrons, which are generally believed to be responsible for superconductivity in the transition metals.

The present report describes an application of this approach to niobium and vanadium crystals. We discuss in particular the irreducible phonon-exchange vertex, which plays the main role in the theory. In view of the various approximations made, and due to the fact that there is not a single parameter in the theory which is adjusted to get a desirable answer, one can hardly hope to obtain correct transition temperatures. What is hoped is that such a calculation will give reasonable estimates of T_c and, what is more important, will display the trends in the variation of T_c from one material to another.

The superconducting transition is connected with the appearance of a solution of the homogeneous Dyson equation for the vertex function. The AK formulation begins with the transformation of this equation to the site representation. For a pair of electrons with center-of-mass momentum $\mathbf{q} = 0$ (Cooper pair) it has the form (a nondegenerate band is assumed)

$$\Gamma(\mathbf{n}, \omega) = -(1/\beta) \sum_{\omega' \mathbf{n}'} \Gamma(\mathbf{n}', \omega') \left[\sum_{\mathbf{m}} F(\mathbf{m}, \omega') I(\mathbf{n}' - \mathbf{m}, \omega'; \mathbf{n}, \omega) \right] \tag{1}$$

where $\beta = 1/kT$ and the frequency sum runs over the values $\omega' = (\pi/\beta)(2n + 1)$. The function F is the convolution of two single-particle Green's functions in the site representation. Following AK, we neglect $F(\mathbf{m}, \omega)$ for $\mathbf{m} \neq 0$, while

$$F(0, \omega) = \int_{\varepsilon_{\min}}^{\varepsilon_{\max}} \frac{N(\varepsilon)\, d\varepsilon}{\varepsilon^2 + \omega^2} \tag{2}$$

where $N(\varepsilon)$ is the electronic density of states and $\varepsilon_{\min}$ and $\varepsilon_{\max}$ are the lowest and highest band energies, respectively. The last factor I is a contracted irreducible vertex defined by

$$I(\mathbf{n}', \omega'; \mathbf{n}, \omega) = (1/N) \sum_{\mathbf{n}_1 \mathbf{n}_2} I(\mathbf{n}_2 + \mathbf{n}', \mathbf{n}_2, \omega'; \mathbf{n}_1 + \mathbf{n}, \mathbf{n}_1, \omega) \tag{3}$$

Each term in the sum on the rhs is the irreducible vertex describing the transition of two electrons initially on sites $\mathbf{n}_2, \mathbf{n}_2 + \mathbf{n}'$ to sites $\mathbf{n}_1, \mathbf{n}_1 + \mathbf{n}$ while an ionic vibration propagates between two sites, the indices of which are summed over. The basic approximation of AK is that of the "contact model," which neglects all terms with $\mathbf{n}, \mathbf{n}' \neq 0$. This means that only transitions in which the two electrons are always on the same site are important. Such an assumption is plausible in view of the short range of the electron–phonon interaction and the tightly bound electrons in question. Its validity will be demonstrated by an explicit calculation.

The irreducible vertex consists of a Coulomb part and a phonon-exchange part. In the contact model the first is given by

$$U = \sum_{\mathbf{n}} \int \Phi^*(\mathbf{r} - \mathbf{n}) \Phi^*(\mathbf{r}' - \mathbf{n}) V(|\mathbf{r} - \mathbf{r}'|) \Phi(\mathbf{r}) \Phi(\mathbf{r}') \, d^3\mathbf{r} \, d^3\mathbf{r}' \tag{4}$$

and the latter by

$$I_{\text{ph}}(\omega) = L(\omega) \sum_{\mathbf{n}_1 \mathbf{n}_2, \alpha} |\langle \mathbf{n}_1 | \nabla_\alpha V | \mathbf{n}_2 \rangle|^2 \tag{5}$$

where

$$L(\omega) = \frac{1}{M} \int \frac{g(\omega') \, d\omega'}{\omega^2 - (\omega')^2} \tag{6}$$

$g(\omega')$ is the normalized phonon density of states and M is the ionic mass. The matrix elements are

$$\langle \mathbf{n}_1 | \nabla_\alpha V | \mathbf{n}_2 \rangle = \int \Phi^*(\mathbf{r} - \mathbf{n}_1) \nabla_{\mathbf{r}_\alpha} V(r) \Phi(\mathbf{r} - \mathbf{n}_2) \, d^3\mathbf{r} \tag{7}$$

and $\Phi(\mathbf{r} - \mathbf{n})$ is an atomic wave function around site $\mathbf{n}$. Equation (5) was derived by AK in the usual tight-binding formulation of the electron–phonon interaction in which the Bloch wave functions are linear combinations of atomic orbitals centered around the lattice points. It was recently suggested[2,3] that this should be modified by taking combinations of atomic orbitals around the ions. Doing that, one allows the electrons to follow adiabatically the motion of the ions and thereby one takes into account in zeroth order all the interaction between an electron and the neighboring ion. The main difference between the usual formulation and the modified one is that the latter excludes one-center integrals. This does not concern us here because we consider only intraband transitions and for these the one-center integrals in Eq. (5) vanish due to parity arguments. Another difference between the two formulations is in the way the two-center integrals [$\mathbf{n}_1 = 0$ or $\mathbf{n}_2 = 0$ in Eq. (5)] appear. In the contact model the modified formulation leads to the same expression as in Eq. (5), except that the elements $\langle \mathbf{n} | \nabla_{\mathbf{r}} V(r) | \mathbf{0} \rangle$ are replaced by $\nabla_{\mathbf{n}} \langle \mathbf{n} | V(r) | \mathbf{0} \rangle$. It will turn out that the latter are, as a rule, smaller than the former. The other type of integrals which we shall consider, namely the degenerate three-center integrals [$\mathbf{n}_1 = \mathbf{n}_2 \neq 0$ in Eq. (5)] appear identically in both formulations. A detailed derivation and discussion of these points will appear in a more extended publication.

In the contact model the set of equations in (1) reduces to a single equation

$$\Gamma_1(\omega) = -\frac{1}{\beta} \sum_{\omega'} F(0, \omega') I_{\text{ph}}(\omega - \omega') \times \left[\Gamma_1(\omega') - \frac{(U/\beta) \sum_{\omega''} \Gamma_1(\omega'') F(0, \omega'')}{1 + N(0) U \log(\alpha \beta \varepsilon_b)} \right] \tag{8}$$

where $\Gamma_1(\omega)$ is that part of Γ which vanishes as $\omega \to \infty$, $N(0)$ is the density of states on the Fermi surface, ε_b is an "effective" bandwidth, and α is a number of order one. The part which is constant as $\omega \to \infty$ (that which comes from the Coulomb vertex) has been eliminated from this equation. To get an estimate of T_c from Eq. (8), AK make a crude approximation, replacing $I_{\text{ph}}(\omega - \omega')$ by an average negative constant $-I$ for $\omega, \omega' < \omega_0$ and by zero otherwise. This leads to a BCS-type formula

$$kT_c = 1.13\omega_0 \exp[-1/N(0)(I - \tilde{U})] \tag{9}$$

where

$$\tilde{U} = \frac{U}{1 + N(0)\, U \log(\alpha\varepsilon_b/\alpha'\omega_0)} \tag{10}$$

and α' is a number of order one.

Let us now see how the result of AK is related to McMillan's theory.[4] It will turn out that $I_{\text{ph}}(\omega)$ is almost constant up to a frequency somewhat below ω_D. Thus, whether we solve Eq. (8) or make the approximation leading to Eq. (9), the result will be determined mainly by $I_{\text{ph}}(0)$. In view of Eqs. (5) and (6), we can write

$$-N(0)\, I_{\text{ph}}(0) = \langle 1/\omega^2 \rangle\, N(0) \langle J^2 \rangle / M \tag{11}$$

where $\langle \omega^{-2} \rangle$ is an average over the phonon density distribution and $\langle J^2 \rangle$ denotes the sum of the matrix elements in Eq. (5). This form is reminiscent of the parameter λ in McMillan's theory. There, $\langle \omega^{-2} \rangle$ is replaced by $\langle \omega \rangle^{-2}$ and $\langle J^2 \rangle$ is a Fermi-surface average of terms like $|\langle \mathbf{k} | \nabla_\alpha V | \mathbf{k}' \rangle|^2$. Transforming these terms to the site representation and keeping only those in which the two right-hand lattice sites are the same and so are the left-hand sites ("contact approximation") one gets essentially the same expression for $\langle J^2 \rangle$ as in AK. McMillan derives a semiempirical equation for T_c, which he uses to evaluate λ from known values of T_c. In discussing the results of our calculations, we shall compare our $-N(0)\, I_{\text{ph}}(0)$ with McMillan's empirical λ. Equation (9), which was obtained by a rough approximation, is essentially the weak coupling limit of McMillan's equation for T_c.

We have calculated the phonon-exchange vertex for V and Nb assuming that the d electrons are described by the threefold degenerate Γ'_{25} band built from the atomic orbitals $\phi_1 \propto xy$, $\phi_2 \propto xz$, $\phi_3 \propto yz$. The only effect of the band degeneracy is that in a cubic crystal $I_{\text{ph}}(\omega)$ should be replaced by

$$I_{\text{ph}}(\omega) = \tfrac{1}{3}[I_{\text{ph}}^{11}(\omega) + 2 I_{\text{ph}}^{12}(\omega)] \tag{12}$$

where

$$I_{\text{ph}}^{1i}(\omega) = L(\omega) \sum_{\mathbf{n}_1, \mathbf{n}_2, \alpha} |\langle \mathbf{n}_1^1 | \nabla_\alpha V | \mathbf{n}_2^i \rangle|^2 \tag{13}$$

The upper index denotes an atomic orbital.

The function $L(\omega)$ was calculated from experimental phonon density distribution curves for vanadium[5] and niobium,[6] and is plotted in Figs. 1 and 2. The only dependence of $I_{\text{ph}}(\omega)$ on ω is through $L(\omega)$. We find that $[ML(0)]^{-1/2}$ is equal to 220°K for V and to 170°K for Nb. These numbers should be compared with McMillan's values of $\langle \omega^2 \rangle^{1/2}$, which are 290°K for V and 230°K for Nb.

The calculation of J^2 is a much harder problem and its details will be described

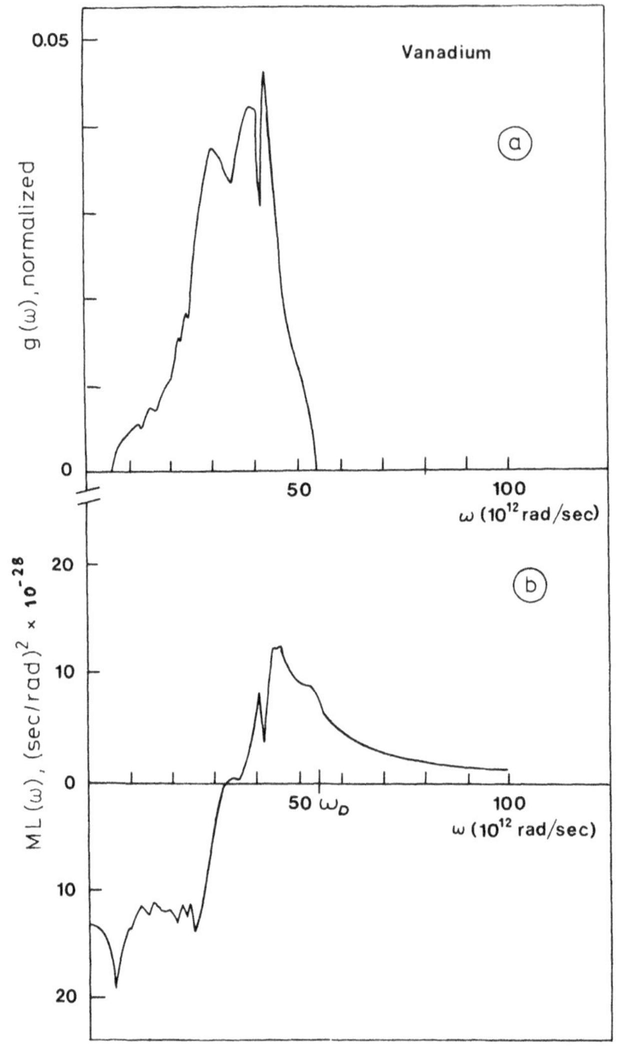

Fig. 1. (a) The experimental phonon frequency distribution and (b) the function $ML(\omega)$ for V.

elsewhere. Here we refer only to the results. The atomic functions and potentials were taken from the Herman–Skillman tables.[7] We have checked the validity of the contact model by calculating explicitly the largest contracted vertex [Eq. (3)] which is of a noncontact type, namely $I(\mathbf{n}, \omega; 0, \omega)$ with $\mathbf{n} = (1,0,0)$. It is smaller by a factor of about 10 than $I(0, \omega; 0, \omega)$. Thus the contact approximation is a reasonable one provided that one can approximate the potential in the crystal by a sum of atomic potentials and the Wannier functions by atomic orbitals.

The calculated values of $-N(0)\,I_{\text{ph}}(0)$ are listed in Table I. The contributions of the two-center and the degenerate three-center integrals are given separately. As mentioned before, $-N(0)\,I_{\text{ph}}(0)$ is expected to be of the same order as McMillan's

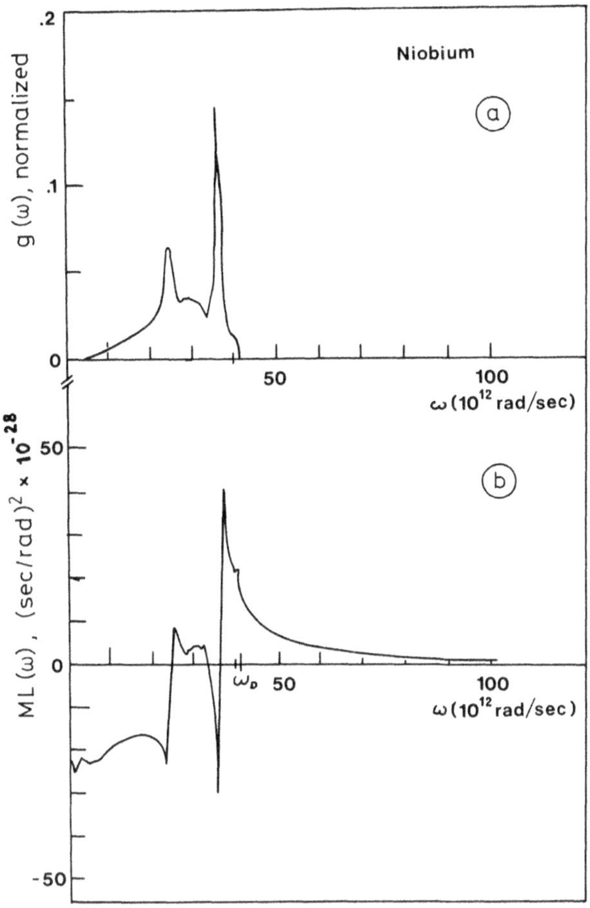

Fig. 2. Same as Fig. 1. for Nb.

parameter λ, although they are not identical. The first row shows the results in the usual tight binding formulation. The numbers are much too high to give a reasonable T_c. The modified tight binding formulation brings us closer to the desired order of magnitude due to the considerable reduction of the two-center integrals (second

Table I

	Vanadium, $\lambda = 0.60$			Niobium, $\lambda = 0.82$		
Method*	Two-center integrals	Degenerate three-center integrals	Total	Two-center integrals	Degenerate three-center integrals	Total
TB	14.0	4.7	18.7	9.7	3.9	13.6
MTB	2.3	4.7	7.0	4.3	3.7	8.0
SMTB	0.6	0.2	0.8	1.9	1.1	3.0

* TB: tight binding; MTB: modified tight binding; SMTB: including screening (see text).

row). A further reduction which leads to the expected order of magnitude is obtained when the screening of the atomic potential by the s electrons in the crystal is taken into account. We have screened the Herman–Skillman potential in the Thomas–Fermi approximation at distances beyond the last maximum of the d orbitals. The results are given in the third row. The screening affects particularly the three-center integrals, because in these the main contribution comes from the neighborhood of the ion with the two electrons, where the potential is screened. In the two-center integrals the major contribution comes from the vicinity of the vibrating ion and there the potential is not screened. This result agrees with the conjecture of Barisic[3] that the two-center integrals are dominant in determining the coupling of tightly bound electrons with phonons.

The calculation described here should be repeated for other transition metals. As one goes to elements with more d electrons it may become necessary to include all five d functions. The transition temperature itself should be found from Eq. (8). To this end, one needs the function $F(0, \omega)$ [Eq. (2)], which may be derived from some reasonable band structure calculation. An effort in these directions is in progress.

Acknowledgments

This work has benefited from many discussions with Prof. M. Weger. One of us (H.G.) is grateful to Prof. W. Kohn for several discussions during the preliminary stages of this work.

References

1. J. Appel and W. Kohn, *Phys. Rev. B* **4**, 2162 (1972).
2. T.K. Mitra, *J. Phys. C* **2**, 52 (1969).
3. S. Barisic, *Phys. Rev. B* **5**, 932 (1972).
4. W.L. McMillan, *Phys. Rev.* **167**, 331 (1968).
5. B. Mozer, K. Otnes, and H. Palevsky, in *Proc. Intern. Conf. Lattice Dynamics*, R. Wallis, ed., Pergamon Press (1965), p. 63.
6. Y. Nakagawa and A.D.B. Woods, *Phys. Rev. Lett.* **11**, 271 (1963).
7. F. Herman and S. Skillman, *Atomic Structure Calculations*, Prentice Hall, New Jersey (1963).

T_c in Dilute Nb–Ta Alloys Estimated from Resistivity Measurements

Ö. Rapp and M. Pokorny

*Department of Physics, Royal Institute of Technology
Stockholm, Sweden*

Background

It has been found[1] that the electronic specific heat coefficient γ behaves anomalously in the Nb–Ta alloy system, increasing at first up to about 4 at.% Ta and then decreasing monotonically with further addition of Ta. This maximum in γ is small but would clearly be reflected in a maximum in the superconducting transition temperature T_c of about 10°K in a conventional calculation based on constant electron–phonon interaction. No corresponding maximum in T_c has been observed, however, which has been taken as one piece of experimental evidence[2] that the density of states is irrelevant in determining T_c.

In a simple model[3] the electron–phonon coupling V_{ph} is estimated from γ and resistivity data. With this approach it was possible to explain qualitatively why there is a minimum in T_c in the Ta–V alloy system.[4]

Here this model is again used to investigate whether the observed anomaly in γ of Nb–Ta alloys could be included in a calculation of T_c together with the estimated V_{ph}. A more detailed report of this work is given elsewhere.[5]

Experimental

A set of alloys in the concentration range up to 20 at.% Ta was prepared in a conventional arc furnace. Following an initial cold-work, the alloys were homogenized, cold-drawn to about 0.4 mm diameter, and annealed. Wires are preferred in this experiment to enable us to measure conveniently the temperature derivative of resistivity $d\rho/dT$ in the temperature region where the resistivity is linear, which is the resistive property required in the model. High-temperature resistance measurements were made in an oven under a vacuum below 10^{-6} Torr in the region 220–280°C, and 0°C resistivities were determined between fixed edges in an ice bath. The result for $d\rho/dT$ at 250°C is shown in Fig. 1. The points represent values calculated from these measurements at compositions sometimes intermediate to the alloy concentrations used. With this scheme a more arbitrary interpolation in the specific heat data[1] was avoided. Finally, T_c was measured by a resistive technique.

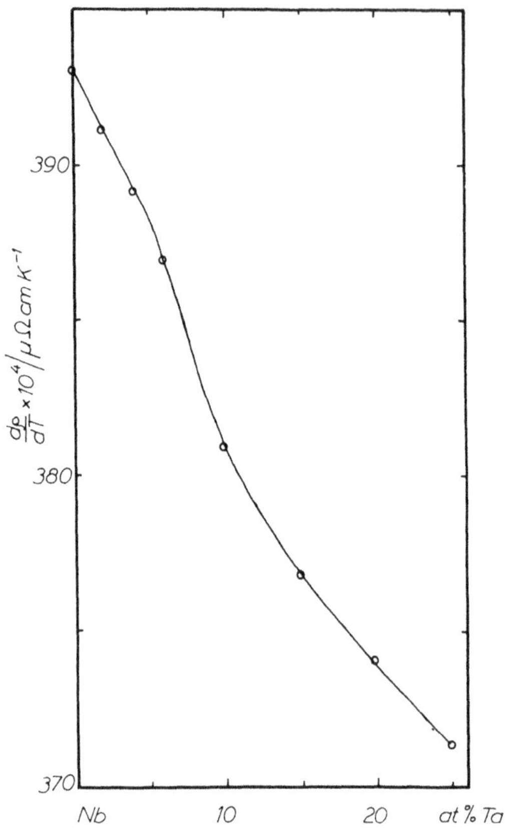

Fig. 1. $d\rho/dT$ at 250°C vs. Ta concentration.

Analysis

V_{ph} is estimated from the following expression[4]:

$$V_{ph} = \text{const} \times (d\rho/dT)\left[1 + 5N_{bs}(0)\right]^{-1} \tag{1}$$

$N_{bs}(0)$ is the band structure density of states. This estimate is arrived at[3] by considering a simple Mott picture of a transition metal, neglecting *d-d* scattering and *U* processes and making some simplifying assumptions about average matrix elements for the different scattering processes. A constant low density of states in the *s* band is also assumed.

$N_{bs}(0)$ is obtained from the "specific heat" density of states $N_\gamma(0)$ by the renormalization

$$N_{bs}(0) = \left[1/(1 + \lambda)\right] N_\gamma(0) \tag{2}$$

where

$$\lambda = N_{bs}(0)\, V_{ph} \tag{3}$$

is the coupling constant.

$N_\gamma(0)$ is taken from Ref. 1 and $d\rho/dT$ from Fig. 1. Values of λ, $N_{bs}(0)$, and V_{ph} were calculated from Eqs. (1)–(3) by iteration. Starting with $\lambda = \text{const} = \lambda(\text{Nb})$ in Eq. (2), this process converged to the third decimal place within three steps. The constant in Eq. (1), taken as a constant over the alloy system, was adjusted to give the observed T_c for Nb from McMillan's formula.[6]

The results for $N_{bs}(0)$, V_{ph}, and λ are given in Fig. 2. A maximum in $N_{bs}(0)$ is seen of much the same nature as that in γ in Ref. 1. This maximum corresponds to a minimum in V_{ph}, so that the product λ becomes a monotonic decreasing function of the Ta concentration, as shown by the points in the lower part of the figure. The solid curve given for comparison shows what we may call experimental λ's. These are values calculated from the observed T_c and Debye temperature in Ref. 1 using McMillan's formula[6]

$$T_c = \frac{\theta}{1.45} \exp - \frac{1.04(1+\lambda)}{\lambda - \mu^*(1+0.62\lambda)} \tag{4}$$

where μ^* is taken to be 0.13, as usual.

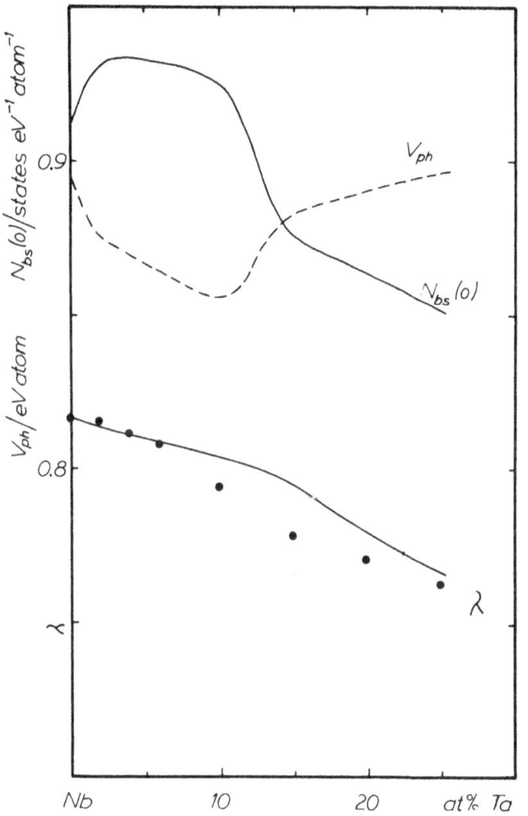

Fig. 2. Upper: $N_{bs}(0)$ and V_{ph} vs. Ta concentration as calculated from Eqs. (1)–(3). Lower: λ vs. Ta concentration; points, calculated as above; solid curve, experimental from observed T_c and θ and Eq. (4).

The agreement between calculated and observed λ is rather good, certainly better than what might be expected from the rather crude model. It can further be observed from Figs. 1 and 2 that the steeper part in the λ vs. concentration curve at around 10 at.% Ta corresponds well to a similar feature in $d\rho/dT$. Such a correlation may be inferred from Eqs. (1) and (3) when $N_{bs}(0)$ is fairly large, and indeed $\lambda \approx d\rho/dT$ is a fairly good estimate of λ in this case.[5]

With the calculated λ from Fig. 2 and smoothed θ values from Ref. 1, T_c can be calculated from Eq. (4). Again $\mu^* = 0.13$ is used. The result is shown by the solid curve in Fig. 3. Points are experimental results from different work.

The quantitative agreement is about 5%, but this should not be stressed too much in view of the sensitivity to the different approximations used. For example, the replacement of a suitably averaged phonon frequency by θ in Eq. (4) or the assumption of a constant μ^* are both approximations which will modify the results. Also, the experimental θ points are already somewhat scattered,[1] which gives some arbitrariness to the smoothing procedure. As a further example, although Eq. (1)

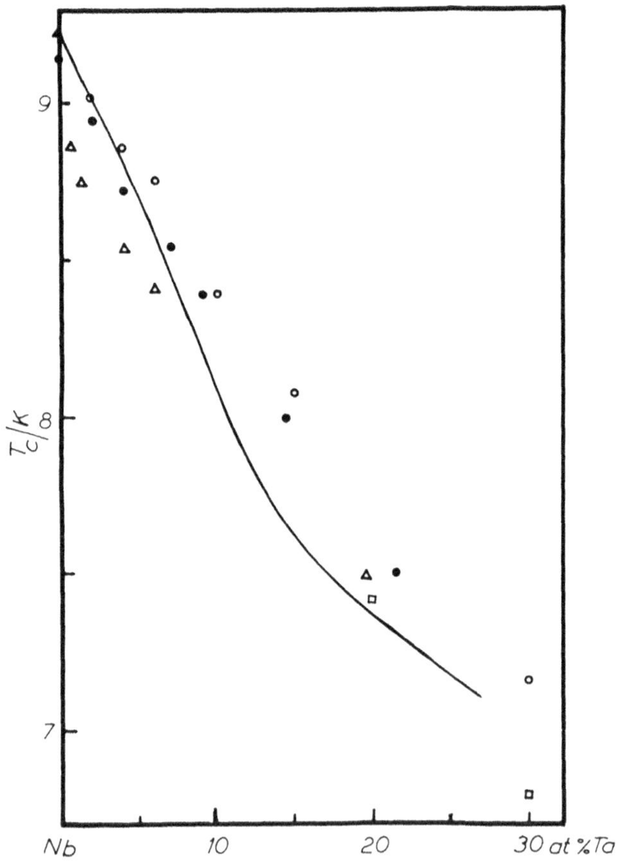

Fig. 3. T_c for Nb–Ta alloys. Points are experimental: Open circles, Ref. 1; triangles, Ref. 7; squares, Ref. 8; solid circles, present work. Full curve is present calculation.

may well be of the correct form,[5] the number 5 in the denominator is uncertain and might be too small.

In conclusion, therefore, it is found that in the concentration range studied the observed monotonic decrease of T_c with increasing Ta concentration in Nb is qualitatively obtained with a calculation from McMillan's formula with observed values of θ, λ, and $d\rho/dT$ and one adjustable parameter.

References

1. Y. Kimura, T. Ohtsuka, T. Matsui, and T. Mizusaki, *Phys. Lett.* **29A**, 284 (1969).
2. J.J. Hopfield, in *Proc. Int. Conf. Science of Superconductivity, Stanford, 1969,* F. Chilton, ed., North-Holland, Amsterdam (1971), pp. 41 ff.
3. G. Gladstone, M.A. Jensen, and J.R. Schrieffer, in *Superconductivity,* R. Parks, ed., Marcel Dekker, New York (1969), Vol. 2, pp. 697 ff.
4. Ö. Rapp, *Sol. St. Commun.* **9**, 1 (1971).
5. Ö. Rapp and M. Pokorny, *Phys. Scripta* **6**, 200 (1972).
6. W.L. McMillan, *Phys. Rev.* **167**, 331 (1968).
7. A. Ikushima and T. Mizusaki, *J. Phys. Chem. Solids* **30**, 873 (1969).
8. J.M. Corsan and A.J. Cook, *Phys. Lett.* **28A**, 500 (1969).

Electron–Phonon Interaction in Lead and Indium: Study by High-Frequency Cyclotron Resonance*

P. Goy and B. Castaing

Groupe de Physique des Solides, Ecole Normale Supérieure
Paris, France

Introduction

The essential information concerning both the electron–phonon interaction and the phonon spectrum is contained in the function $\alpha^2(\omega) F(\omega)$, where $\alpha(\omega)$ represents the energy-dependent electron–phonon coupling constant and $F(\omega)$ the phonon density of states for phonons of frequency ω. The quantity $\alpha^2(\omega) F(\omega)$ has been experimentally determined for lead and indium by the tunneling technique.[1] There is a peak in energy due to the transverse phonons near $E_t \approx 4$ meV for lead and 6 meV for indium (Fig. 1).

The changes of the electron–phonon interaction strength are expected to be observable on m^* and τ when the distance of the quasiparticle from the Fermi level ($\hbar\Omega$ or kT) is of the order of the typical phonon energy E_t. Observation of the frequency effects clearly requires very high microwave frequencies Ω, which until recently were not available. In addition, such high frequencies are needed in the temperature experiment to observe well-defined cyclotron resonances, with the short relaxation time at finite temperature T.

The purpose of this paper is to present an Azbel'–Kaner[2] cyclotron resonance experiment which measured the frequency and temperature dependences of the cyclotron effective mass m^* and of the electronic relaxation time τ in lead and indium. These experiments, which provide new experimental evidence for the mass enhancement due to the electron–phonon interaction, were performed at frequencies from 70 to 460 GHz and at temperatures between 1.5 and 6°K.

Brief Theory

At the Fermi level, i.e., $T = 0$ and $\omega = 0$, where ω is the energy distance from the Fermi level, the observed cyclotron mass m^* is generally larger than the band calculated value m_b by approximately a factor of two for lead and indium. This discrepancy is mainly attributed to the electron–phonon interaction.[3]

The electron–phonon interaction strength can be measured by means of the renormalization coefficient λ defined via the observed mass m^*:

* Invited paper.

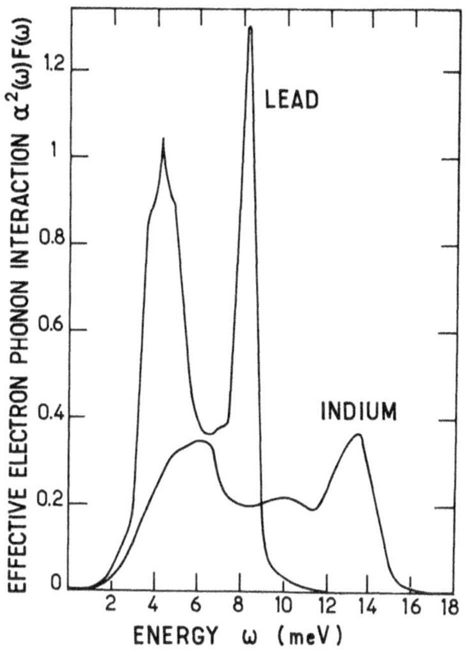

Fig. 1. The electron-phonon density of states $\alpha^2 F(\omega)$ for lead and indium deduced from superconducting tunneling experiments by McMillan and Rowell.[1]

$$m^* = (1 + \lambda)\, m_b \tag{1}$$

λ is approximately 1.5 for lead and 0.9 for indium.[1]

The presence of the electron–phonon interaction modifies the dispersion relation of the free carriers. The eigenvalues of the Hamiltonian do not correspond purely to electrons but, because of the interaction, to mixed electron–phonon states. For this quasiparticle the electron is called dressed or surrounded by a cloud of virtual phonons.

In terms of the band structure let us call $\varepsilon_{\mathbf{k}}$ the single-particle energy, where $\mathbf{k}$ is the wave vector. Taking into account the interaction, the new energy $E_{\mathbf{k}}$ is the single-particle energy $\varepsilon_{\mathbf{k}}$ plus a complex term $\Sigma_{\text{e-ph}}$ called the self-energy:

$$E_{\mathbf{k}} = \varepsilon_{\mathbf{k}} + \Sigma_{\text{e-ph}}(\mathbf{k}, E_{\mathbf{k}}) = \varepsilon_{\mathbf{k}} + M_{\text{e-ph}}(\mathbf{k}, E_{\mathbf{k}}) - i\Gamma_{\text{e-ph}}(\mathbf{k}, E_{\mathbf{k}}) \tag{2}$$

Assuming that the $\mathbf{k}$ dependence of the real part M of Σ is negligible, λ can be obtained at the Fermi level by the following equation:

$$\lambda(T = 0, \omega = 0) = -\,\partial M_{\text{e-ph}}(\omega)/\partial\omega\big|_{E_F} \tag{3}$$

The electron–phonon interaction distorts the band structure $E_{\mathbf{k}}$, relative to the single-particle band structure $\varepsilon_{\mathbf{k}}$, in a very small region of typical energy $\hbar\omega \approx 2E_t$ about the Fermi energy. The cyclotron resonance mass is proportional to the derivative of the Fermi surface section A with respect to energy, taken at the Fermi energy E_F:

$$m^* = \frac{\hbar^2}{2\pi} \frac{\partial A}{\partial \varepsilon}\bigg|_{E_F} \tag{4}$$

Therefore this technique samples the distorted region and gives a mass which deviates from the single-particle mass m_b.

Such deviations are expected to be both temperature and microwave frequency dependent since as Ω increases a wider range of the distorted region is sampled, while as T increases the distortion of $E_{\mathbf{k}}$ changes. Thus the enhancement factor λ can vary with frequency[4,5] and temperature.[6,7] With increasing T or Ω, $\lambda(T, \Omega)$ first increases, reaches a maximum, then decreases down to zero. For temperature variations[6,7] the maximum is reached at lower energy ($kT \approx E_t/4$) than for frequency variations[4,5] ($\hbar\Omega \approx E_t$). In the cyclotron resonance technique, because of the correlated relaxation time decay, the only observable shifts will correspond to increasing λ.

The predicted maximum relative shifts $\Delta\lambda/\lambda$ are nearly the same in many metals.[7] The observable quantity is the effective mass m^*. According to Eq. (1), the relative change in m^* will be given by the following equation:

$$\Delta m^*/m^* = [\lambda/(1 + \lambda)]\, \Delta\lambda/\lambda \tag{5}$$

Any shift in λ can be seen in m^*, but λ has to be large to give an easily observable shift in m^* because of the factor $\lambda/(1 + \lambda)$.

Several significant experiments on the T variation have been performed in the metals Zn,[8] Pb,[9,10] and Hg,[11] not including semimetals.

In this paper we present the first experiments which have been performed as a function of T and Ω together. Lead and indium have been chosen because of their large renormalization coefficients λ and their low typical phonon energies E_t.

The previous calculations concerning the changes of the electron–phonon interaction have been made either versus frequency[4,5] or versus temperature.[6,7] In the following calculation T and Ω effects are predicted together in the ranges $\hbar\Omega \leq E_t/2$ and $kT \leq E_t/8$, where cyclotron resonance is observable.

The classical cyclotron resonance conductivity is[2]

$$\sigma \propto \coth \{\pi[(i\Omega/\omega_c) + (1/\omega_c\tau)]\} \tag{6}$$

The electron–phonon gas conductivity is[4]

$$\sigma' \propto \int d\varepsilon \frac{f(\varepsilon) - f(\varepsilon + \hbar\Omega)}{\hbar\Omega} \times \coth \pi \left[i\frac{\hbar\Omega + M(\varepsilon) - M(\varepsilon + \hbar\Omega)}{\hbar\omega_c} + \frac{\Gamma(\varepsilon) + (\Gamma\varepsilon + \hbar\Omega)}{\hbar\omega_c} \right] \tag{7}$$

The factor $[f(\varepsilon) - f(\varepsilon + \hbar\Omega)]/\hbar\Omega$ is not zero only in a small energy range of the order of $kT + \hbar\Omega$. In this range variations of $M(\varepsilon) - M(\varepsilon + \hbar\Omega)$ are small compared to $\Gamma(\varepsilon) + \Gamma(\varepsilon + \hbar\Omega)$, which represents the width of resonance. Thus it is a good approximation to write

$$\sigma' \propto \coth \pi \left[i\Omega \frac{1 + \lambda(T, \Omega)}{\omega_c} + \frac{1 + \lambda(T, \Omega)}{\omega_c \tau(T, \Omega)} \right] \tag{8}$$

Equation (8) corresponds to the classical case [Eq. (6)] with a renormalized mass [Eq. (1)] and with

$$\lambda(T,\Omega) = \int d\varepsilon \frac{f(\varepsilon) - f(\varepsilon + \hbar\Omega)}{\hbar\Omega} \frac{M(\varepsilon) - M(\varepsilon + \hbar\Omega)}{\hbar\Omega} \tag{9}$$

$$\frac{\hbar}{\tau(T,\Omega)} = \frac{1}{1 + \lambda(T,\Omega)} \int d\varepsilon \frac{f(\varepsilon) - f(\varepsilon + \hbar\Omega)}{\hbar\Omega} [\Gamma(\varepsilon) + \Gamma(\varepsilon + \hbar\Omega)] \tag{10}$$

At $\Omega = 0$ and $T = 0$ Eq. (9) corresponds to Eq. (3). In some work[6,7] Eq. (3) is used to obtain $\lambda(T, \Omega = 0)$. According to Eq. (9), our calculation gives a $\lambda(T, \Omega = 0)$ variation which is much more rapid (by approximately a factor of two as compared to Ref. 6) because of the distribution term contribution.

In Eq. (10) the factor $[1 + \lambda(T,\Omega)]^{-1}$ has been omitted in Refs. 6 and 12. For this reason the predicted $\tau(T, \Omega = 0)$ is too small by a factor of 2.5 in lead.

The first-order formula for the self-energy is

$$\Sigma_{\text{e-ph}}(\varepsilon) = \hbar \int d\varepsilon' \int d\omega\, \alpha^2 F(\omega) \frac{1 - f(\varepsilon') + N(\omega)}{\varepsilon - \varepsilon' - \hbar\omega + i\delta} + \frac{f(\varepsilon') + N(\omega)}{\varepsilon - \varepsilon' + \hbar\omega + i\delta} \tag{11}$$

where f is the electron Fermi distribution, N is the phonon Bose distribution, and δ is a positive infinitesimal.

Theoretical Relaxation Time Dependences. Equations (10) and (11) give

$$\begin{aligned} \frac{\hbar}{\tau(T,\Omega)} &= \frac{2\pi\hbar}{1 + \lambda(T,\Omega)} \int d\varepsilon \frac{f(\varepsilon) - f(\varepsilon + \hbar\Omega)}{\hbar\Omega} \\ &\quad \times \int d\omega\, \alpha^2 F(\omega) [2N + 1 + f(\varepsilon + \hbar\omega) - f(\varepsilon - \hbar\omega)] \\ &= \frac{2\pi\hbar}{1 + \lambda(T,\Omega)} \int d\omega \frac{\alpha^2 F(\omega)}{\hbar\Omega} \\ &\quad \times \int d\varepsilon [f(\varepsilon) - f(\varepsilon + \hbar\Omega)] [f(\varepsilon + \hbar\omega) + f(\hbar\omega - e)] \\ &\quad + \frac{4\pi\hbar}{1 + \lambda(T,\Omega)} \int d\omega \frac{\alpha^2 F(\omega)}{\exp(\hbar\omega/kT) - 1} \end{aligned} \tag{12}$$

$$\begin{aligned} \frac{\hbar/\tau(T,\Omega)}{2\pi/[1 + \lambda(T,\Omega)]} &\simeq \int_0^{\Omega} d\omega \frac{\Omega - \omega}{\Omega} \alpha^2 F(\omega) + 2 \int d\omega \frac{\alpha^2 F(\omega)}{\exp(\hbar\omega/kT) - 1} \\ &\quad + \int \frac{dw}{h\Omega} \frac{(\omega - \Omega)\, \alpha^2 F(\omega)}{\sinh[h(\omega - \Omega)/2kT]} \exp - \left| \frac{\hbar(\omega - \Omega)}{2kT} \right| \end{aligned} \tag{13}$$

The relaxation time dependence is mainly due to the low-energy part of $\alpha^2 F(\omega)$. This part is not precisely determined by tunneling experiments but we may suppose a Debye variation $\alpha^2 F(\omega) = a(\hbar\omega)^2$. Since the variations of $\lambda(T,\Omega)$ are small, Eq.

(13) gives

$$\frac{\hbar}{\tau(T,\Omega)} = \frac{30a}{1+\lambda(0,0)}\left[(kT)^3 + \left(\frac{\hbar\Omega}{3.9}\right)^3 + 0.68\hbar\Omega(kT)^2\right] \tag{14}$$

We find here the classical T^3 dependence,[2] a similar Ω^3 dependence, and a crossed term ΩT^2 corresponding to phonons emitted at the frequency $\omega' = \Omega + (k/\hbar)\, T$ by electrons excited by the microwave.

Theoretical Cyclotron Mass Dependences. Since we are interested only in low temperatures, we use the classical approximation

$$f(\varepsilon) = \Theta(\varepsilon) - (\pi^2/6)\, k^2 T^2 \delta'(\varepsilon) \tag{15}$$

where Θ is the Heaviside function and δ' is the derivative of the Dirac distribution. Equations (9), (11), and (15) give

$$\lambda(T,\Omega) = A(\Omega) + (\pi^2/6)\, k^2 T^2 B(\Omega) \tag{16}$$

with

$$A(\Omega) = 2\int d\omega \frac{\alpha^2 F(\omega)}{\omega}\left(\frac{\omega^2}{\Omega^2}\log\left|1-\frac{\Omega^2}{\omega^2}\right| + \frac{\omega}{\Omega}\log\left|\frac{\omega+\Omega}{\omega-\Omega}\right|\right)$$

and

$$B(\Omega) = \frac{8}{\hbar^2}\int d\omega \frac{\alpha^2 F(\omega)}{\omega}\,\frac{1}{\omega^2-\Omega^2}$$

If we expand $A(\Omega)$ and $B(\Omega)$ in powers of Ω, we obtain

$$A(\Omega) = 2\int d\omega \frac{\alpha^2 F(\omega)}{\omega} + \frac{\Omega^2}{3}\int d\omega \frac{\alpha^2 F(\omega)}{\omega^3} \tag{17}$$

Since $\hbar\Omega$ is small compared to the first phonon peak energy E_t,

$$B(\Omega) \simeq \frac{8}{\hbar^2}\int d\omega \frac{\alpha^2 F(\omega)}{\omega^3} \tag{18}$$

If we take into account the low-energy part of the phonon spectrum $\alpha^2 F(\omega) = a(\hbar\omega)^2$, Eqs. (16)–(18) give $T^2 \log T$ and $\Omega^2 \log \Omega$ dependences for $\lambda(T,\Omega)$. The $\log T$ and $\log \Omega$ terms are noticeable only at very low T and Ω. In fact, $\alpha^2 F(\omega)$ deviates very soon from the Debye spectrum (Fig. 1), and the parabolic low-energy contribution can be neglected in our frequency and temperature ranges. Then Eqs. (16)–(18) give

$$\lambda(T,\Omega) = 2\int d\omega \frac{\alpha^2 F(\omega)}{\omega} + \frac{4\pi^2}{3\hbar^2}\int d\omega \frac{\alpha^2 F(\omega)}{\omega^3}\left[(kT)^2 + \left(\frac{\hbar\Omega}{2\pi}\right)^2\right] \tag{19}$$

In Eq. (19) the first term $2\int d\omega\,[\alpha^2 F(\omega)/\omega]$ is the well-known renormalization coefficient at the Fermi level $\lambda(0,0)$ according to Eq. (3). The second term gives a parabolic dependence of the mass versus T and Ω. At the same energy $kT = \hbar\Omega$ the temperature is much more effective than the frequency. The T and Ω dependences are quite independent [no crossed term $T\Omega$ in Eq. (19)].

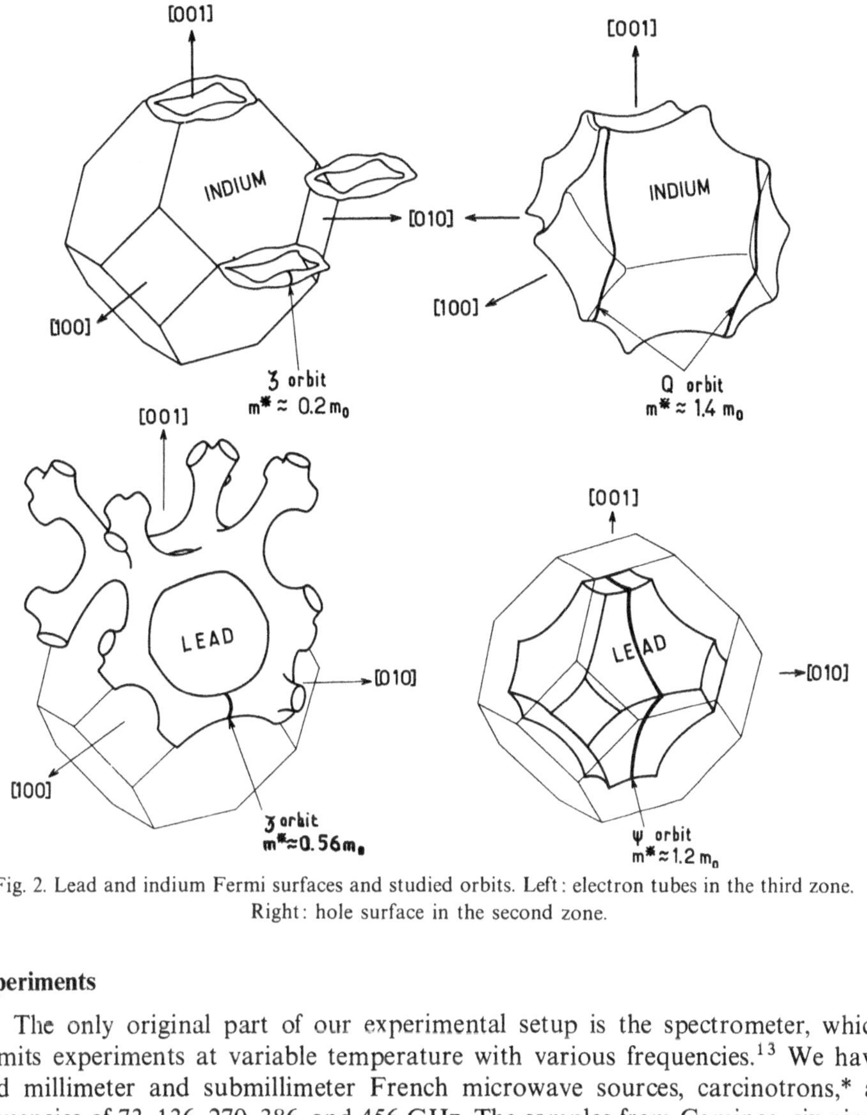

Fig. 2. Lead and indium Fermi surfaces and studied orbits. Left: electron tubes in the third zone. Right: hole surface in the second zone.

Experiments

The only original part of our experimental setup is the spectrometer, which permits experiments at variable temperature with various frequencies.[13] We have used millimeter and submillimeter French microwave sources, carcinotrons,* at frequencies of 73, 136, 270, 386, and 456 GHz. The samples from Cominco, six nines purity, have been grown against optically-polished quartz plates.[14] Their resistivity ratio is of the order of 10^5 between room temperature and 1.5°K.

A previous theory[12] predicts a magnetic field dependence. No change in either m^* or τ has been observed on successive subharmonics in the magnetic field range 8–27 kG.

In each metal the two orbits studied [ζ and ψ in lead,[15] ζ and Q in indium[16] (Fig. 2)] give the same $m^*(T, \Omega)$ dependences: The mass enhancement depends on the high-energy part of $\alpha^2 F(\omega)$, which is not orbit dependent.

* Thomson-C.S.F. Co.

On the contrary, relaxation time variations are orbit dependent: In lead τ_ψ decreases more rapidly than τ_ζ by a factor of 1.5. In indium τ_Q decreases more rapidly than τ_ζ by about one order of magnitude. This fact can be explained by the difference between transverse and longitudinal phonons coupling to the electrons. For a spherical Fermi surface there is no coupling between transverse phonons and electrons at low energy: Umklapp processes occur only at higher energy. However, in metals such as lead and indium the Fermi surface cuts the zone boundaries and here transverse phonons are well coupled to the electrons. Thus a given orbit has a strong (or soft) variation of τ^{-1} vs. T^3 if a large (or small) part of this orbit is close to the zone boundaries.

In the following we consider only the results concerning ζ orbits traced around electron tubes in the third zone with the magnetic field H along the tube axis, $H \| [110]$.

Experimental Cyclotron Mass Dependences. For a fixed frequency the positions of the subharmonics are shifted toward higher field when the temperature T increases, showing the correlated mass increase (Fig. 3). The mass increase with T is parabolic

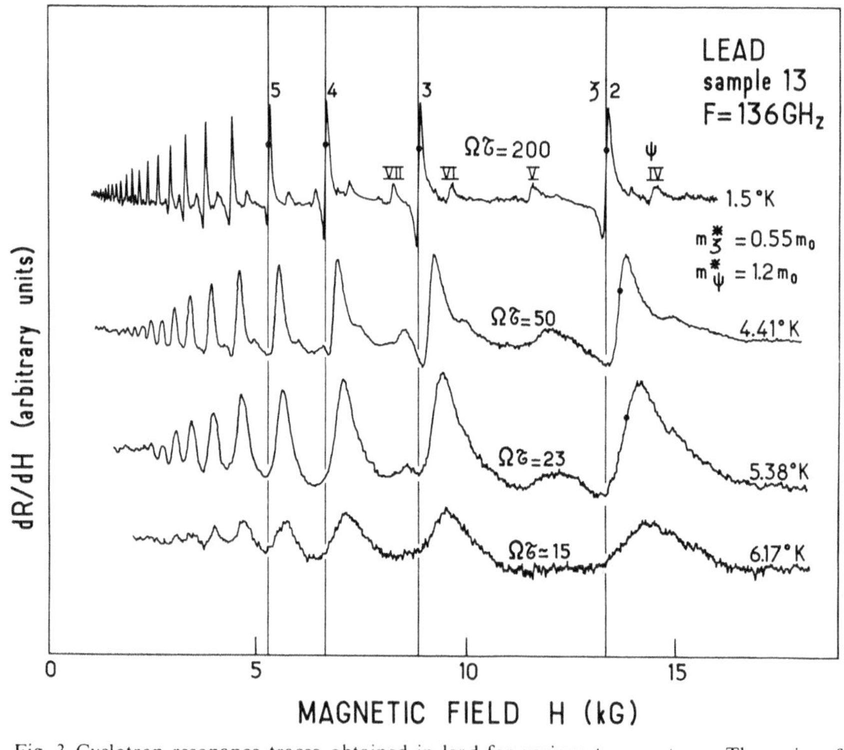

Fig. 3. Cyclotron resonance traces obtained in lead for various temperatures. The series of large peaks is due to the ζ orbit. Arabic numbers give their Azbel'–Kaner subharmonic order. The series of small peaks is due to the ψ orbit (Roman numbers). The vertical lines are the low-temperature positions of the subharmonics due to ζ orbit. The black dots indicate the resonance field positions. Above 6°K (bottom) the resonance is broadened too much to permit any precise measurement.

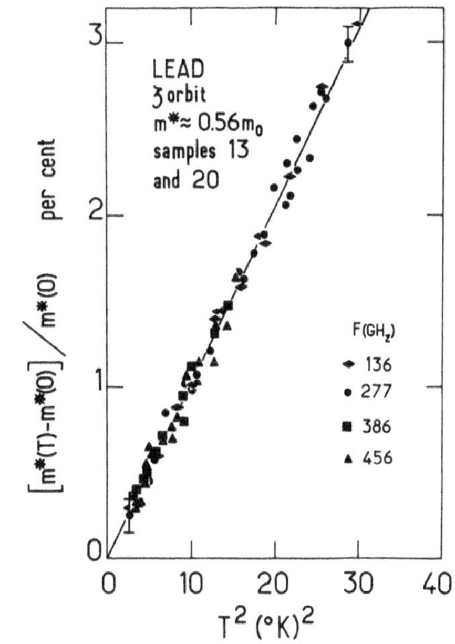

Fig. 4. Effective mass enhancement as a function of T^2 obtained at various frequencies in lead.

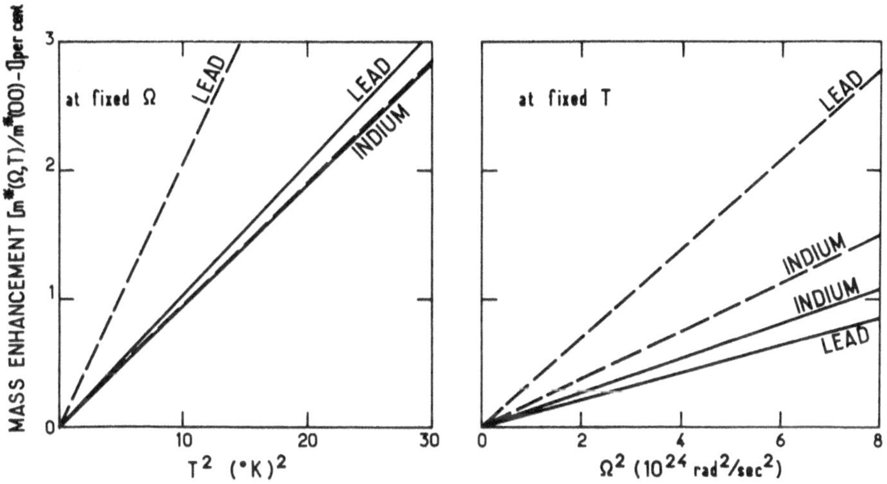

Fig. 5. Summary of the effective mass dependences in lead and indium. Dashed lines correspond to Eq. (19) and solid lines correspond to experimental results.

and quite independent of Ω (Fig. 4), as predicted in Eq. (19). For a fixed temperature the mass increase with Ω is also parabolic. Figure 5 summarizes lead and indium results.

Experimental Relaxation Time Dependences. For a fixed frequency the resonance width increases with temperature, showing the relaxation time decay (Fig. 3). To obtain the pure electron–phonon relaxation frequency, we have to subtract the impurity relaxation frequency (obtained at $T = 0$ and $\Omega = 0$), from the experimental relaxation frequency given by the linewidth. As predicted in Eq. (14), the measured relaxation frequency is cubic in T and Ω.

In lead the slope $\tau^{-1}(T)/T^3$ increases with frequency. Therefore a crossed term appears, but this term is $\Omega^3 T^3$ rather than ΩT^2. In addition, the frequency effect is smaller than predicted. The experimental variations in lead for $\Omega/2\pi \leq$ 460 GHz and $T \leq 4.2°\text{K}$ are

$$\frac{\hbar}{\tau(T,\Omega)} = \frac{3.25}{(k51°\text{K})^2}\left[(kT)^3 + \left(\frac{\hbar\Omega}{7}\right)^3 + (\hbar\Omega)^3\left(\frac{T}{25°\text{K}}\right)^3\right] \tag{20}$$

In indium $\tau^{-1}(T)$ can vary faster than T^3 at high frequency (Fig. 6, left), because of the peculiar position of the considered ζ orbit in phase space, which is very close to the hole Fermi surface (Fig. 7a). An electron can diffuse to the hole surface, so a new scattering process appears.[17] This scattering process needs a phonon of wave vector larger than the minimum distance between the electron tube and the hole surface, i.e., a phonon energy $\hbar\omega_1 \geq 1$ meV (Fig. 7b). In our low temperature range this energy ω_1 is available only with phonons emitted at the frequency $\omega' = \omega + (k/\hbar)T$ by electrons excited by the high-frequency microwave [crossed term in Eq. (14)]. For the considered ζ orbit the $\alpha^2 F(\omega)$ function can be approximated

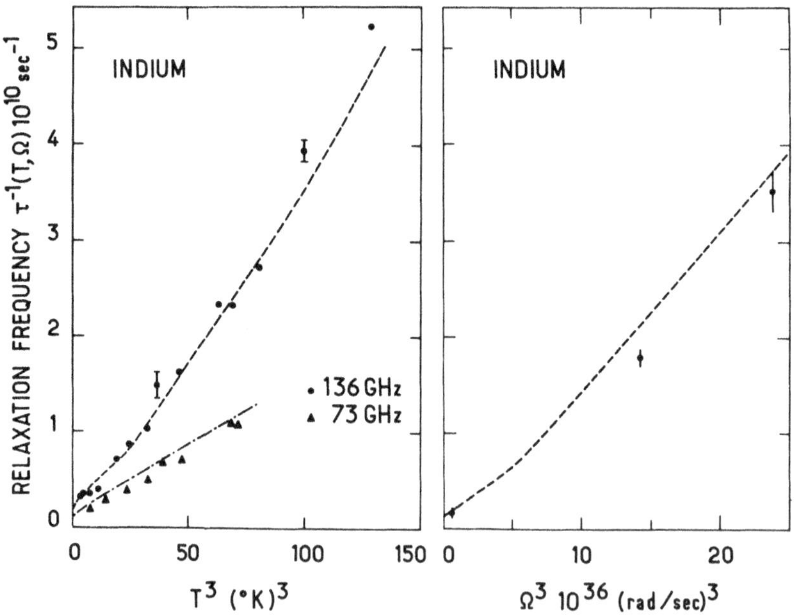

Fig. 6. Relaxation frequency variations for ζ orbit in indium. Dashed lines correspond to Eqs. (13) and (21)–(24).

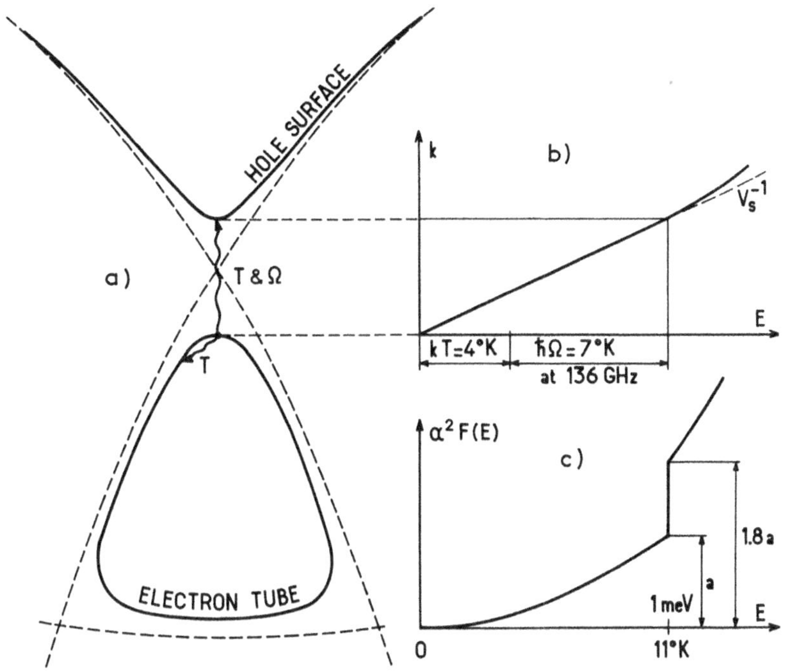

Fig. 7. A new diffusion mechanism at finite frequency for ζ orbit in indium. (a) Cross section of the Fermi surface. (b) Dispersion relation for phonons. V_S is the velocity of sound. (c) The $\alpha^2F(\omega)$ function, taking into account the new diffusion mechanism.

as follows (Fig. 7c):

$$\alpha^2F(\omega) = a(\hbar\omega)^2, \qquad \omega < \omega_1 \tag{21}$$

$$\alpha^2F(\omega) = b(\hbar\omega)^2, \qquad \omega \geq \omega_1 \tag{22}$$

The value of a can be deduced from the tunneling determination[1] of $\alpha^2F(\omega)$ (Fig. 1). We find

$$a = 5.3 \times 10^{-3} \quad (\text{meV})^{-2} \tag{23}$$

To obtain the good fit of experimental $\tau(T, \Omega)$ points with Eqs. (13), (21), and (22) (Fig. 6, dashed lines), we take

$$b = 1.8a \tag{24}$$

Conclusion

Temperature and frequency variations of the electron–phonon interaction have been studied in lead and indium by cyclotron resonance at high frequency. All predicted effects are observed but the frequency effects are generally smaller than expected. Because of the larger λ and the lower E_t lead should give larger effects than indium, but experimental results do not show any significant difference.

Acknowledgments

Thanks are due to Drs. T. Yéou, A. Schaer, and G. Guérin from the Thomson-C.S.F. Co.

References

1. W.L. McMillan and J.M. Rowell, *Superconductivity*, R.D. Parks, ed., Marcel Dekker, New York (1969).
2. M. Ya. Azbel' and E.A. Kaner, *J. Phys. Chem. Solids* **6**, 113 (1958).
3. S. Nakajima and M. Watabe, *Progr. Theoret. Phys. (Kyoto)* **29**, 341 (1963); **30**, 772 (1963).
4. H. Scher and T. Holstein, *Phys. Rev.* **148**, 598 (1966).
5. P.B. Allen, *Phys. Rev. B* **3**, 305 (1971).
6. G. Grimvall, *Phys. Kond. Materie* **9**, 283 (1969).
7. P.B. Allen and M.L. Cohen, *Phys. Rev. B* **1**, 1329 (1970).
8. J.J. Sabo, Jr., *Phys. Rev. B* **1**, 1325 (1970).
9. I. Krasnopolin and M.S. Khaikin, *Zh. Eksperim. i Teor. Fiz. Pisma* **12**, 76 (1970) [*Soviet Phys.—JETP Lett.* **12**, 54 (1970)].
10. P. Goy, *Phys. Lett.* **32A**, 474 (1970).
11. R.G. Poulsen and W.R. Datars, *Sol. State Commun.* **8**, 1969 (1970).
12. P.B. Allen, in *Proc. 12th Intern. Conf. Low Temp. Phys., 1970*, Academic Press of Japan, Tokyo (1971), p. 517.
13. P. Goy and B. Castaing, *Phys. Rev. B* **7**, 6609 (1973).
14. Yu. V. Sharvin and V. F. Gantmakher, *Cryogenics* **5**, 278 (1965).
15. R.T. Mina and M.S. Khaikin, *Zh. Eksperim. i Teor. Fiz.* **45**, 1304 (1963) [*Soviet Phys.—JETP* **18**, 896 (1964)].
16. R.T. Mina and M.S. Khaikin, *Zh. Eksperim. i Teor. Fiz.* **51**, 62 (1966) [*Soviet Phys.—JETP* **24**, 42 (1966)].
17. B. Castaing and P. Goy, *J. Phys. C* **6**, 2040 (1973).

Electron–Phonon Interaction and Electronic Properties of Zinc*

Philip G. Tomlinson and James C. Swihart

Department of Physics, Indiana University
Bloomington, Indiana

Introduction

The electron–phonon interaction has a profound effect on the electronic properties of metals, providing the interaction which produces superconductivity, producing a scattering mechanism in electronic transport phenomena, and enhancing the electronic mass which is observed in the electronic specific heat and cyclotron resonance. Certain of these properties, such as the mass enhancement averaged over the Fermi surface, can be calculated rather well for simple metals using rather simple models, as was first shown by Swihart *et al.*[1] for the case of the mass enhancement of Pb, Hg, and Al, and independently by Ashcroft and Wilkins[2] for Pb, Al, and Na. The superconducting transition temperature is another isotropic property that can be dealt with by a simple model in the case of simple metals.[1,3-5] However, any attempt to predict theoretically anisotropic properties, such as the mass enhancement or superconducting energy gap as a function of position on the Fermi surface, or anisotropic transport properties, must consider more sophisticated models.

The elements that enter a calculation of electron–phonon effects are (1) the phonon spectrum, including not only the frequencies but also polarization vectors of the normal modes, (2) the electron–phonon interaction, which is a function of the initial and final electron states as well as of the energy and polarization of the phonon, and (3) the properties of the Fermi surface from which and to which the electrons are scattered. For the isotropic properties the following simplifications have been previously made[1-5]: (1) an isotropic or even Debye model is taken for the phonons, (2) the electron–phonon interaction is treated as a constant, (3) the Fermi surface is treated as a free-electron sphere, and (4) the electron states are considered to be single orthogonalized plane waves (OPW's). In certain cases corrections are made for the nonspherical Fermi surface by multiplying by the ratio of the density of states for the real Fermi surface to that for a free-electron sphere.

A few treatments have gone beyond the simplifications discussed above.[6-11] Balsley and Swihart considered tin,[6] Leavens and Carbotte considered aluminum,[7] Rice and Sham studied potassium,[8] and Truant and Carbotte studied zinc.[9] In each of these cases realistic phonon spectra and electron–phonon interactions were taken, but the electrons were treated as consisting of a single OPW on a spherical

* Research supported by the National Science Foundation.

Fermi surface. Bennett[10] considered the anisotropic superconducting energy gap in lead, taking into account the anisotropy due to the phonon spectrum and considering in an approximate way the effect of the band structure. However, he considered the electron–phonon interaction to be a constant. He argued that the band structure should have a small effect on the anisotropic gap. Klemens *et al.*,[11] in treating the anisotropic electrical conductivity of tin, considered that the most important contribution to the anisotropy is the anisotropic Fermi surface. We shall see that this effect is also important in zinc.

In this paper we explore the electron–phonon interaction in zinc. We believe this is the first calculation for a simple metal which deals with the phonons, the electron–phonon interaction, and the electrons realistically. In particular, we shall compare the results for the real Fermi surface and several OPW's with those for a single OPW on a spherical Fermi surface.

The Phonon Model

The phonon frequencies and polarization vectors used in these calculations were determined by the nonlocal Stark–Falicov (SF) pseudopotential.[12] Having been fit to Fermi surface data, the SF pseudopotential is given only at values of momentum transfer **Q** equal to reciprocal lattice vectors. We use an interpolation supplied to us by P. B. Allen which agrees with the fitted points of Stark and Falicov and with the Heine–Animalu form factor[13] at low **Q** values. Using a single scaling factor, we bring the phonons into relatively good agreement with the data, except for the lower-frequency branches. These branches are very important in determining electron–phonon properties, so this discrepancy will have a serious effect upon our results. However, the conclusions we draw concerning effects of band structure should remain completely valid.

Since pseudopotential phonon calculations require much computing time, we utilize methods from QUAD[14] to determine the phonon frequencies and eigenvectors at a finite number of points in the Brillouin zone and interpolate using a quadratic least-squares technique.

Band Structure

Properties related to the electronic band structure include the shape of the Fermi surface, anisotropic band mass, and electron–phonon matrix elements. The SF pseudopotential is particularly well-suited for these calculations since it has been fit extremely accurately to Fermi surface data.[12] In these calculations the number of OPW's used is determined by the proximity of the respective Bragg planes to the point in question. At some points two OPW's are sufficient, while at others we use as many as seven. We feel that inclusion of more OPW's would have a negligible effect on the final results.

In Fig. 1 we display the zinc Fermi surface in the extended and reduced zone schemes. As can be seen, it is nearly spherical except near Bragg planes. However, many of the pieces one would expect from a one-OPW Fermi surface are found not to exist.[12] In fact, the ratio of the area of our Fermi surface to a free-electron one is about 0.5.

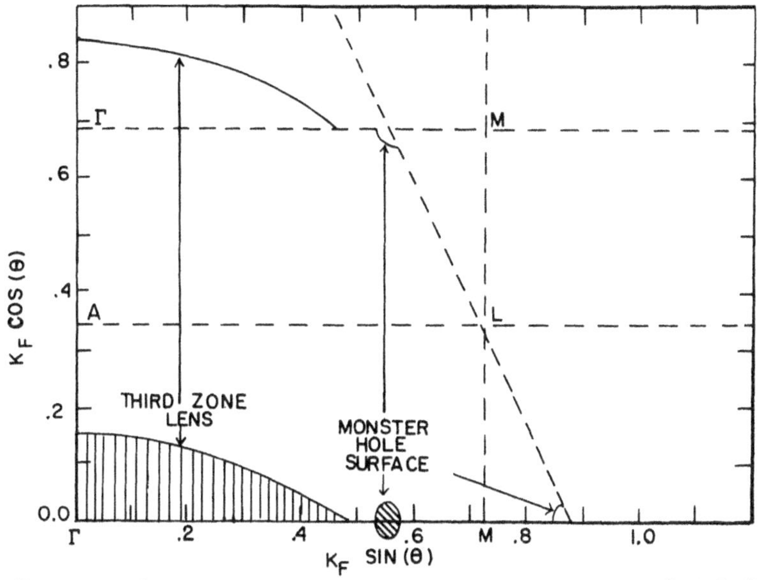

Fig. 1. Intersection of the Fermi surface of zinc with the plane containing the [0, 0, 0, 1] and [1, 0, $\bar{1}$, 0,] directions. Surfaces are shown in both the extended and reduced zone schemes. Dashed lines represent Bragg planes.

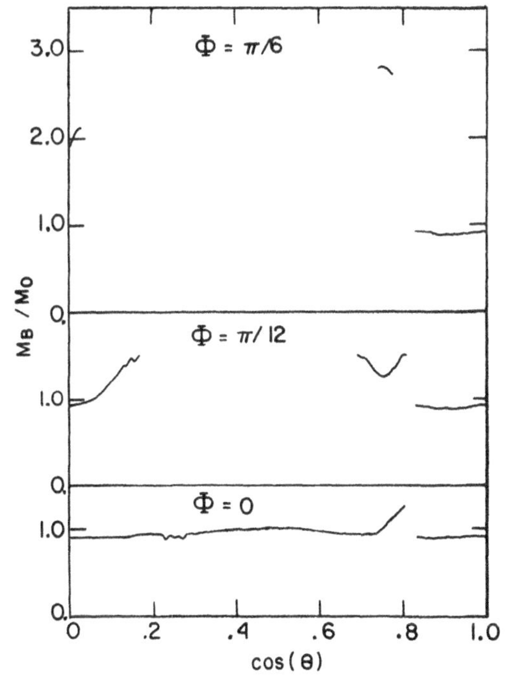

Fig. 2. Anisotropic band mass as a function of position on the Fermi surface. Here θ and ϕ are polar coordinates. The c axis is the polar axis and ϕ is measured from the [1, 1, $\bar{2}$, 0] direction.

For the anisotropic band mass we utilize the formulas

$$\hbar\mathbf{k} = m_B\mathbf{v}, \qquad \mathbf{v} = \hbar^{-1}\,\nabla_{\mathbf{k}}\,E_{\mathbf{k}}$$

We calculate v_F by taking the derivative of the band energy along a normal to the Fermi surface.

In Fig. 2 we present our results for the band mass as a function of position on the Fermi surface.

Effective Mass Renormalization

The phonon contribution to the effective electron mass at the Fermi surface is usually expressed in terms of the parameter λ, where

$$m^*(\mathbf{k}) = m_B(\mathbf{k})\left[1 + \lambda(\mathbf{k})\right]$$

and

$$\lambda(\theta, \phi) = 2\int_0^\infty (d\omega/\omega)\,\alpha^2F(\omega, \theta, \phi)$$

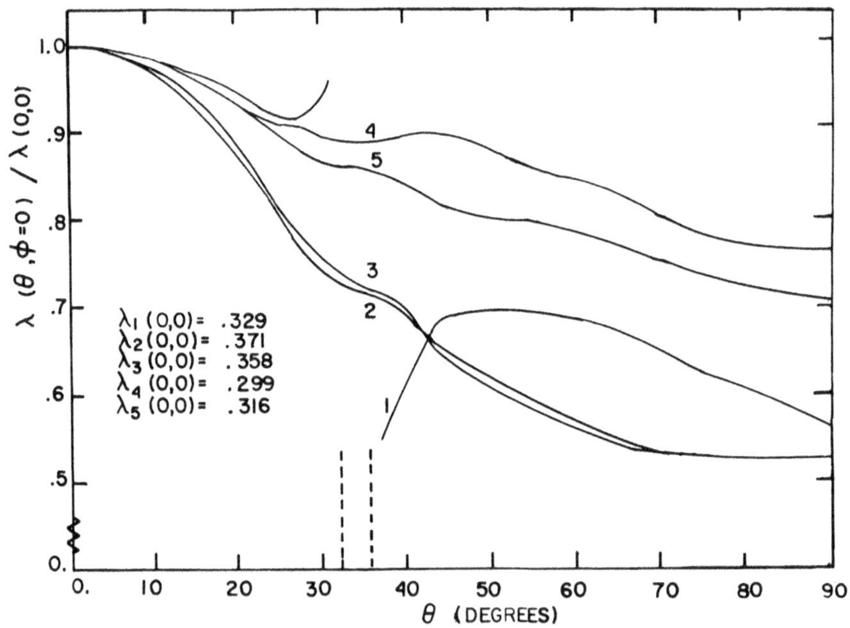

Fig. 3. Anisotropic electron-phonon renormalization constant λ as a function of polar coordinate θ with $\phi = 0$. Curve 1 uses the many-OPW Fermi surface and electron–phonon matrix element. Curve 2 uses the many-OPW Fermi surface but only one OPW for the matrix element. Curve 3 uses a one-OPW matrix element and the many OPW Fermi surface projected onto a sphere. Curves 4 and 5 represent scattering on a spherical Fermi surface. All but curve 5 use the anisotropic band mass. For values of θ between the dashed vertical lines there are no electron states.

with

$$\alpha^2 F(\omega, \theta, \phi) = \sum_{\lambda} \int_{\mathrm{FS}} \frac{d^2k'}{(2\pi)^3 v_{\mathrm{F}}(\theta', \phi')} |g_{\mathbf{k},\mathbf{k}'\lambda}|^2 \,\delta(\omega - \omega_{\mathbf{p}\lambda})$$

where θ and ϕ are the polar coordinates of the vector $\mathbf{k}$ at the Fermi surface; $\mathbf{q} + \mathbf{G} = \mathbf{k}' - \mathbf{k}$, with $\mathbf{G}$ a reciprocal lattice vector; $g_{\mathbf{k},\mathbf{k}'\lambda}$ is the electron–phonon coupling constant between states $\mathbf{k}$ and $\mathbf{k}'$ and phonon mode λ; v_{F} is the Fermi velocity; and $\omega_{q\lambda}$ is the frequency of phonon mode λ at point $\mathbf{q}$ in the first Brillouin zone.

We have calculated λ at 11 points where the Fermi surface intersects the plane containing the $[0, 0, 0, 1]$ and $[1, 1, \bar{2}, 0]$ directions. Results for the various assumptions regarding band structure effects are shown in Fig. 3. As can be seen, the five curves seem to fall into three categories, indicating which factors are most important for zinc. As far as the Fermi surface goes, the shape of the surfaces that exist are not important, as shown by the similarity of curves 2 and 3. However, both differ qualitatively from 4 and 5, both of which arise from a spherical Fermi surface. Curves 4 and 5 differ from one another only by the effect of the anisotropic band mass employed by 4 and not by 5. Curve 1, which includes all the refinements we have developed—a many-OPW Fermi surface, many-OPW electron–phonon coupling, and anisotropic band mass—differs qualitatively from all others.

Since zinc is one of the polyvalent metals whose Fermi surface is most nearly free-electron-like, we conclude that our results show that in order to understand anisotropic properties, the effects of band structure should not be ignored.

References

1. J.C. Swihart, D.J. Scalapino, and Y. Wada, in *Proc. 9th Intern. Conf. Low Temp. Phys., 1964,* Plenum Press, New York (1965), p. 607; *Phys. Rev. Lett.* **14**, 106 (1965).
2. N.W. Ashcroft and J.W. Wilkins, *Phys. Lett.* **14**, 285 (1965).
3. P. Morel and P.W. Anderson, *Phys. Rev.* **125**, 1263 (1962).
4. J.P. Carbotte and R.C. Dynes, *Phys. Rev.* **172**, 476 (1968).
5. P.B. Allen and M.L. Cohen, *Phys. Rev.* **187**, 525 (1969); P.B. Allen, M.L. Cohen, L.M. Falicov, and R.V. Kasowski, *Phys. Rev. Lett.* **21**, 1794 (1968).
6. J.F. Balsley and J.C. Swihart, in *Proc. 12th Intern. Conf. Low Temp. Phys., 1970,* Academic Press of Japan, Tokyo (1971), p. 303; J.F. Balsley, Ph.D. Thesis, 1969, unpublished.
7. C.R. Leavens and J.P. Carbotte, *Ann. Phys.* **70**, 338 (1972).
8. T.M. Rice and L.J. Sham, *Phys. Rev. B* **1**, 4546 (1970).
9. P.T. Truant, Ph.D. Thesis, 1972; P.T. Truant and J.P. Carbotte (to be published).
10. A.J. Bennett, *Phys. Rev.* **140**, A1902 (1965).
11. P.G. Klemens, C. Van Baarle, and F.W. Gorter, *Physica* **30**, 1470 (1964).
12. R.W. Stark and L.M. Falicov, *Phys. Rev. Lett.* **19**, 795 (1967).
13. A.O.E. Animalu and V. Heine, *Phil. Mag.* **12**, 1249 (1965).
14. F.M. Mueller, J.W. Garland, M.H. Cohen, and K.H. Bennemann, Argonne National Lab. Report ANL-7556 (1969).

The Temperature Dependence of the Electron Mean Free Path in the Radiofrequency Size Effect of Potassium

P. N. Trofimenkoff

Department of Physics and Theoretical Physics Institute
University of Alberta, Edmonton, Alberta, Canada

In addition to the measurement of caliper dimensions of the Fermi surfaces of many metals, the radiofrequency size effect (rfse) has been used to obtain values of the electron mean free path (mfp) in several metals. In particular, the temperature (T) dependence of the electron mfp, $l_T(T)$, in potassium has been extracted from results of measurements of the variation of the rfse line amplitudes with T.[1,2] Here the relationship between $l_T(T)$ as obtained from the rfse experiments and the calculated phonon-limited mfp, $l_{\text{ep}}(T)$, will be discussed.[3]

The various types of rfse experiments and their interpretations have been extensively discussed elsewhere.[4-6] Briefly, and referring to Fig. 1, the sample is a thin plate of rather pure metal with thickness d typically a fraction of a millimeter. A coil provides rf excitation; detection is through the oscillator circuit that includes the coil. The plate is placed at liquid helium temperatures in a constant magnetic field. For the parallel field rfse the constant magnetic field $\mathbf{H}_P$ is parallel to the surface of the plate; for the tilted field experiments the constant magnetic field $\mathbf{H}_T$ is inclined at an angle ϕ relative to the surface of the plate.

The rfse lines are due to spatial resonances that occur in the plate and that alter the surface impedance of the plate. These resonances can be understood in a

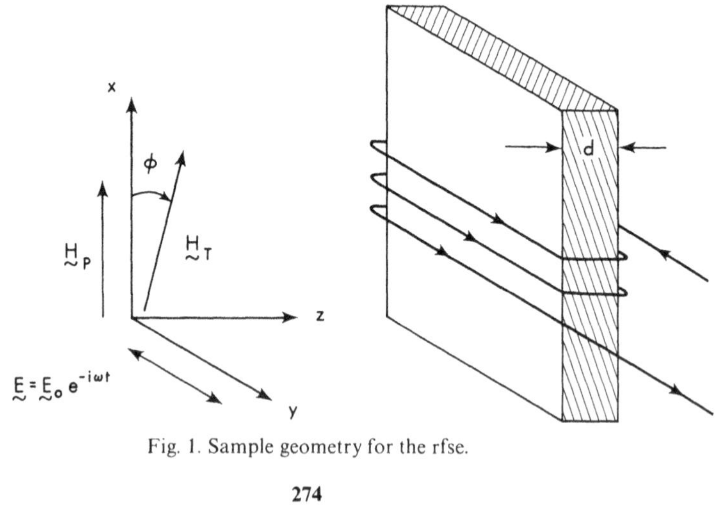

Fig. 1. Sample geometry for the rfse.

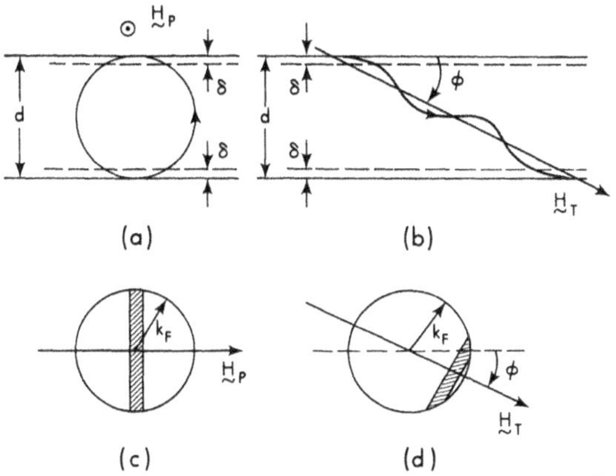

Fig. 2. Spatial trajectories for (a) parallel field and (b) tilted field rfse and phase space position of resonant electrons for (c) parallel field and (d) tilted field rfse.

semiclassical picture of the electron quasiparticle excitations within the plate. With reference to Fig. 2(a, b) the rf electromagnetic field is damped in the skin depth region $\delta \cong d/100$. In the parallel field rfse (Fig. 2a) a spatial resonance occurs for that value of H_P such that an electron with momentum on the central section of the Fermi surface (Fig. 2c) and excited in the skin depth on one side of the plate is directed to the skin depth on the second side and thereby contributes to a current splash at the second surface. In the tilted field rfse (Fig. 2b) an electron with momentum near the elliptic limiting point of the Fermi surface (Fig. 2d) and excited in the skin depth on one side of the plate follows a helical trajectory and contributes to a current splash in the skin depth at the second side for appropriate values of H_T and ϕ. The electrons in the two bands on the Fermi surface depicted in Fig. 2(c, d) dominate in the resonances due to the large number of electrons in these regions that can contribute to the current splashes and their selection in the skin depth by virtue of their velocities being parallel to the surface ("effective" electrons).

Since an electron excited in the skin depth on one side of the plate travels to the second side over a distance λ in the absence of any significant electromagnetic field, it is expected[7] that the amplitude A of the rfse line has the dependence

$$A = A_0 e^{-\lambda/l} \tag{1}$$

where

$$1/l = (1/l_0) + [1/l_T(T)] \tag{2}$$

with l_0 the mfp extrapolated to $T = 0°$K.

If the T-dependent part of the mfp is due to scattering of the electrons out of the resonant bands by phonons and only one-phonon emission and absorption processes are accounted for, then

$$l_{ep}(T) = v_F \tau_{ep}(E = 0, T) \tag{3}$$

where v_F is the Fermi velocity and $\tau_{ep}(E = 0, T)$ is the relaxation time for an electron at the Fermi level ($E = 0$). For general energies E a golden rule calculation gives[8]

$$\frac{1}{\tau_{\text{ep}}(E, T)} = \frac{2\pi}{\hbar} \int_0^{\omega_0} d\omega \, \alpha^2(\omega) F(\omega) [2N_0(\omega, T) + 1 - f_0(E - \omega, T) + f_0(E + \omega, T)] \tag{4}$$

where f_0 and N_0 are, respectively, the equilibrium electron and phonon distribution functions, $\hbar$ is Planck's constant divided by 2π, and $\alpha^2(\omega) F(\omega)$ is the Eliashberg function, including the electron–phonon coupling function and the phonon density of states $F(\omega)$, which is zero above ω_0.

A one-orthogonalized-plane-wave calculation of $\alpha^2(\omega) F(\omega)$ using the Heine–Abarenkov pseudopotential for the electron–ion cross section and a Born–von Kármán force constant model, fitted to the phonon dispersion curves obtained from inelastic neutron scattering measurements and sound velocity data, has been made for the lattice dynamics[9,10] according to

$$\alpha^2(\omega) F(\omega) = \frac{1}{N} \sum_{\lambda} \int_{<2k_F} \frac{d^3q}{(2\pi)^3} L_\lambda(\mathbf{q}) \, \delta[\omega - \omega(\mathbf{q}; \lambda)] \tag{5}$$

where

$$L_\lambda(\mathbf{q}) = \frac{1}{4} \frac{m}{M} \frac{|\mathbf{q} \cdot \xi(\mathbf{q}; \lambda)|^2}{k_F q \omega(\mathbf{q}; \lambda)} |w(\mathbf{q})|^2 \tag{6}$$

Here N is the ion number density, m and M are the electron and ion masses, respectively, k_F is the Fermi wave vector, $w(\mathbf{q})$ is the screened ion pseudopotential, and

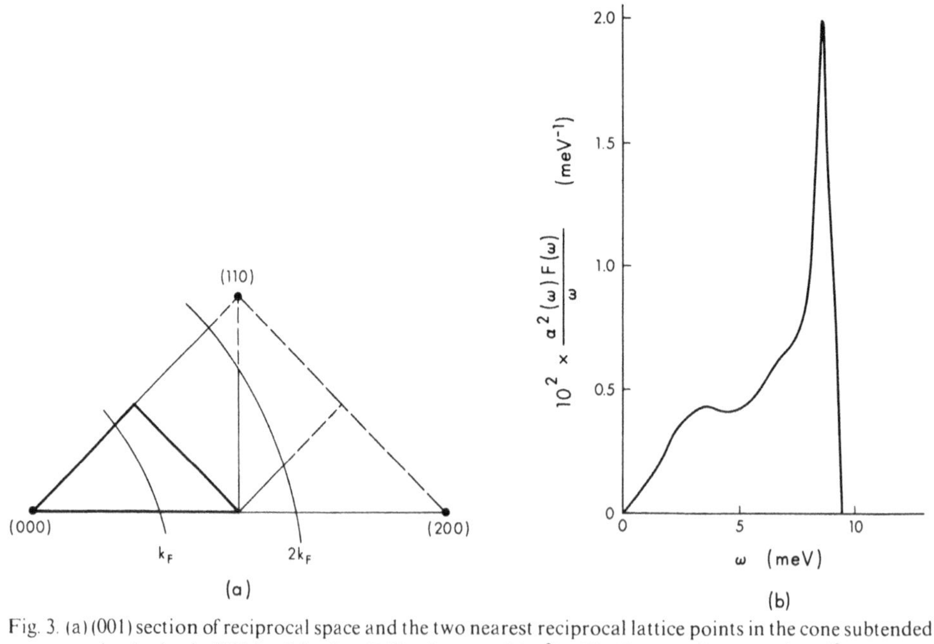

Fig. 3. (a) (001) section of reciprocal space and the two nearest reciprocal lattice points in the cone subtended by the basic 1/48th of the first Brillouin zone. (b) $\alpha^2(\omega)F(\omega)/\omega$ vs. ω for potassium.[10]

$\omega(\mathbf{q};\lambda)$ and $\xi(\mathbf{q};\lambda)$ are the phonon energy and polarization vector for wave vector $\mathbf{q}$ and polarization branch λ. The integral in Eq. (5) extends out to $2k_F$, as depicted in Fig. 3(a). From Fig. 3(a) it is evident that Umklapp processes are excluded at low energy since the Fermi surface lies entirely within the first Brillouin zone. The results of the calculation are reproduced in Fig. 3, where $\alpha^2(\omega)F(\omega)/\omega$ is plotted vs. ω. It is evident that at low energies $\alpha^2(\omega)F(\omega) \cong c\omega^2$, where c is found to be 1.2×10^{-3} meV^{-2}; for low temperatures only the low-energy part of $\alpha^2(\omega)F(\omega)$ is sampled in Eq. (4).

With the above quadratic choice for $\alpha^2(\omega)F(\omega)$, one obtains[8]

$$1/\tau_{\text{ep}}(E = 0, T) = (14\pi/\hbar)\,\zeta(3)\,c(kT)^3 \tag{7}$$

where $\zeta(3) = 1.202\ldots$ and k is Boltzmann's constant. This gives $\tau_{\text{ep}} \cong 1.62 \times 10^{-8}T^{-3}$ sec °K^3, and with $v_F = 0.86 \times 10^8$ cm/sec one obtains

$$l_{\text{ep}}(T) \cong 1.40\,T^{-3}\ \text{cm}\ {}^\circ\text{K}^3 \tag{8}$$

The result obtained from the parallel field rfse by Tsoĭ and Gantmakher[1] is $l_T(T) = bT^{-3}$ where b ranged from 2.7 to 3.0 cm °K^3; from the tilted field rfse Blaney and Parsons[2] obtained $l_T(T) = bT^{-3}$ with $b = 2.7$ cm °K^3.

In conclusion, the agreement between $l_{\text{ep}}(T)$ and $l_T(T)$ obtained from the rfse experiments is within a factor of two. That the experimental mfp is longer is most likely due to the circumstance that for some of the electron–phonon processes, particularly the very small-angle scattering processes, one-phonon absorption or emission is insufficient to remove an electron from the resonant band.

References

1. V.S. Tsoĭ and V.F. Gantmakher, *Soviet Phys.—JETP* **29**, 663 (1969).
2. T.G. Blaney and D. Parsons, *J. Phys. C* **3**, 126 (1970).
3. P.N. Trofimenkoff, *Sol. St. Commun.* **11**, 743 (1972).
4. V.F. Gantmakher, in *Progress in Low Temperature Physics*, C.J. Gorter, ed., North-Holland, Amsterdam (1966), Vol. 5.
5. W.M. Walsh, Jr., in *Solid State Physics*, J.F. Cochran and R.R. Haering, eds., Gordon and Breach, New York (1968), Vol. 1.
6. R.G. Chambers, *Phys. Kond. Materie* **9**, 171 (1969).
7. V.F. Gantmakher and Yu.V. Sharvin, *Soviet Phys.—JETP* **21**, 720 (1965).
8. J.W. Wilkins, Lecture notes on observable many-body effects in metals, Nordita, 1968.
9. J.P. Carbotte and R.C. Dynes, *Phys. Rev.* **172**, 476 (1968).
10. J.P. Carbotte, R.C. Dynes, and P.N. Trofimenkoff, *Can. J. Phys.* **47**, 1107 (1969).

Electronic Mean Free Paths in Cadmium

P. D. Hambourger

Department of Physics, Cleveland State University
Cleveland, Ohio

Introduction

We have measured the temperature dependence of the mean free path for three orbits on the Fermi surface of cadmium using the size-dependent oscillatory magnetoresistance effect.[1,2] This effect, observed when the magnetic field **H** is normal to or at an oblique angle to the sample faces, arises from electrons that travel across the sample on distorted helical trajectories without scattering in the interior. If many electrons at the Fermi level have helical trajectories of the same pitch, the conductivity becomes an oscillatory function of magnetic field, going through one oscillation when the field is changed by the increment which adds one more turn to the spiral.[3] This requires that there be an extremum or discontinuity in the function $\partial S/\partial k_z$, the derivative of Fermi surface orbit area with respect to the component of electron wave vector along **H**.

Since scattering from impurity atoms or defects occurs at random, the average contribution of scattered electrons to the oscillatory current is zero, so the amplitude of the oscillations in the sample conductivity is proportional to the number of electrons that travel across the sample without scattering. By measuring the amplitudes of the size-effect oscillations, we can therefore obtain information about the mean free paths of electrons located in discrete bands on the Fermi surface.

In the presence of the size effect the conductivity tensor consists of the usual monotonic part plus a small oscillatory term. Munarin *et al.*[1] have shown that the transverse oscillatory magnetoconductivity is of the form (neglecting harmonics)

$$\sigma_{yy} = A_0 \{\exp[-(d \sec \alpha)/l(T)]\} \sin(2\pi F H + \phi) \qquad (1)$$

where $l(T)$ is the mean distance the electron travels in the magnetic field direction between scatterings, d is the sample thickness, α is the angle between **H** and the normal to the sample face, and A_0 is an amplitude factor. The coordinates are chosen so the $\hat{z}$ direction is parallel to **H**. Since one actually measures the sample *resistivity* rather than the conductivity, it is necessary in principle to determine all the *monotonic* magnetoresistivity tensor elements, since these also change with temperature and affect the amplitude of the observed resistivity oscillations. It is not possible to measure most of these tensor elements on the same sample, but certain approximations can be made, particularly since the variation of the monotonic resistivity with temperature is small compared to the variation of the amplitude of the oscillations. Since cadmium is a compensated metal and the field directions used in this experiment are those for which there are no open orbits, we assume that all off-diagonal elements

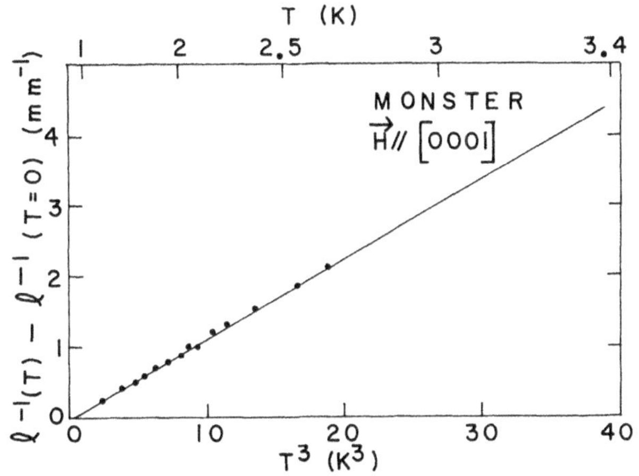

Fig. 1. Temperature dependence of $l^{-1}(T)$ for an orbit on the second-band hole surface with **H** ∥ [0001].

of the monotonic tensor are small compared to the very large quadratic transverse magnetoresistivity. We further assume that the monotonic transverse resistivity elements ρ_{xx}^0 and ρ_{yy}^0 are proportional to each other at all temperatures. Inverting the conductivity tensor and making use of Eq. (1), we find that

$$l^{-1}(T) - l^{-1}(T=0) = K + (d \sec \alpha)^{-1} \ln\left[A(T)/(\rho_{xx}^0)^2\right] \quad (2)$$

where $A(T)$ is the amplitude of the oscillations in the transverse magnetoresistivity ρ_{xx} and K is a constant determined by extrapolation to $T = 0$. Although the last assumption may not be correct in the case of anisotropic mean free paths, the error introduced is small because in this experiment the oscillation amplitude varies by a factor of 10–100 over the range of temperatures covered, while ρ_{xx}^0 varies by only a factor of about 1.5.

Experimental Procedure

The samples had resistance ratios of 30,000–40,000 at 4.2°K and were approximately 1 mm thick.* At each temperature and magnetic field orientation the oscillation amplitude and total sample resistivity were measured separately. The oscillatory part of the magnetoresistance was measured by the low-frequency field modulation technique.[1] Appropriate choice of modulation amplitude and field orientation allowed the recording of one oscillatory period at a time without interference from others. The monotonic resistivity was measured by the usual dc method. Most data were taken with a magnetic field of 7 kOe, although some measurements were made at 4 and 12 kOe; the resulting values of $l^{-1}(T) - l^{-1}(T=0)$ were identical to within experimental error.

Results

Two highly complex orbits on the second-band "monster" and one simple orbit on the third-band "lens" were studied. One of the monster orbits, observed

* Samples were those used in the work referred to in Ref. 2.

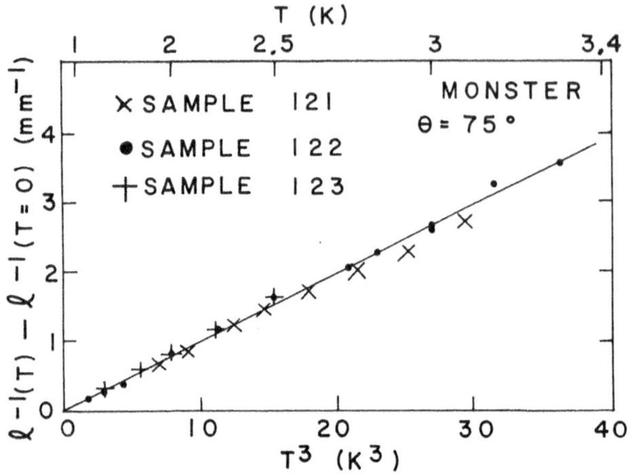

Fig. 2. Temperature dependence of $l^{-1}(T)$ for an orbit on the second-band hole surface with **H** 75° from [0001] in the (10$\bar{1}$0) plane; data from three samples are shown.

with **H** ∥ [0001], is located just above the ΓKM plane.[2] As shown in Fig. 1, $l^{-1}(T)$ was found to vary accurately as T^3. Data taken on two samples are in good agreement. Results for the other monster orbit, centered near the ΓKM plane[2] and observed with **H** tipped 75° from [0001] in the (10$\bar{1}$0) plane, are shown in Fig. 2. Again $l^{-1}(T)$ varies as T^3, and the results from the three samples measured are in good agreement.

The lens orbit, located rather near the edge of the lens,[2] was studied with **H** 77.5° from [0001] in the (10$\bar{1}$0) plane (Fig. 3). Here we found that $l^{-1}(T)$ closely follows a T^5 dependence. This is consistent with the results of Naberezhnykh and Tsymbal,[4] who measured the mean free path for the central orbit on the lens with **H** in the

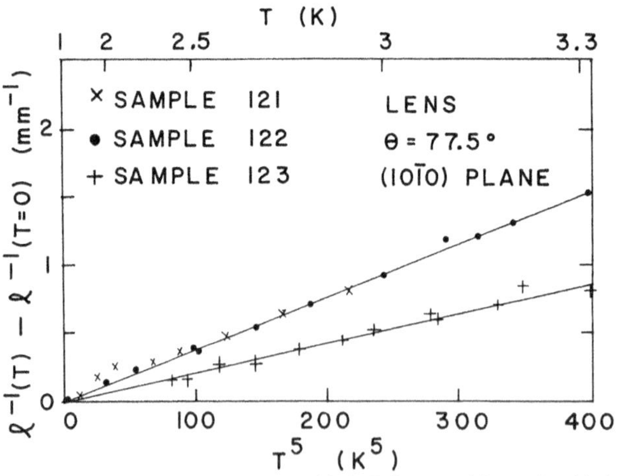

Fig. 3. Temperature dependence of $l^{-1}(T)$ for an orbit on the third-band electron surface ("lens") with **H** 77.5° from [0001] in the (10$\bar{1}$0) plane; data from three samples are shown.

basal plane, using the Gantmakher rf size effect, and also observed a T^5 dependence. An unexplained feature of our data is the variation of the slope of the T^5 line from sample to sample, in spite of the fact that data taken on the monster orbit (Fig. 2) for nearly the same field orientation were almost identical in all three samples.

The T^3 dependence observed for the monster orbits implies that a collision with a single phonon is sufficient to eliminate an electron's contribution to the oscillatory current, while the T^5 dependence for the lens orbit suggests that multiple scatterings may be required to deflect a lens electron sufficiently to eliminate its effect.[4] This seems consistent with the highly convoluted shape of the monster electron trajectories as opposed to the rather simple shape of the trajectories of the lens electrons.

The sample dependence observed in the lens oscillations (Fig. 3) suggests that the variation of mean free path with temperature is affected by impurities or defects, although it is not clear why the monster electrons are not likewise affected. Measurements on dilute cadmium alloys are planned in order to study the effect of controlled amounts of impurities.

References

1. J.A. Munarin, J.A. Marcus, and Philip E. Bloomfield, *Phys. Rev.* **172**, 718 (1968).
2. P.D. Hambourger and J.A. Marcus, *Bull. Am. Phys. Soc.* **14**, 400 (1969); *Phys. Rev. B* (to be published).
3. R.G. Chambers, in *The Physics of Metals, 1. Electrons*, J.M. Ziman, ed., University Press, Cambridge (1969), p. 191.
4. V.P. Naberezhnykh and D.T. Tsymbal, *Zh. Eksperim. i Teor. Fiz. Pisma* **5** (9), 319 (1967) [*Soviet Phys.—JETP Lett.* **5**, 263 (1967)].

Electromagnetic Generation of Acoustic Waves Using Metal Films at Low Temperatures*

Y. Goldstein and A. Zemel

The Racah Institute of Physics
The Hebrew University of Jerusalem, Israel

The problem of electromagnetic generation of transverse acoustic waves in a metal has received considerable theoretical attention over the last few years. To calculate the conversion efficiency of the process, most of the authors[1-5] considered mainly the direct electromagnetic force on the lattice ions. Abeles[6] and later Southgate[7] included, in addition to the direct electromagnetic force, the force on the ions arising from the scattering of electrons at the metal surface. Abeles considered a simplified model of a thin film, neglected the electronic scattering within the film, and assumed diffuse surface scattering. Southgate considered a semiinfinite metal but included bulk scattering and allowed for partially specular surface scattering. In the limit of long mean free path and diffuse surface scattering the two calculations are in good agreement and yield conversion efficiencies comparable to our experimental values. The theory is extremely sensitive to the value of the electromagnetic penetration depth and to the fraction n/n_0 of the conduction electrons that can be scattered by the surface. In the normal state $n/n_0 = 1$. In the superconducting state, however, if pair scattering at the surface conserves momentum, n/n_0 is expected to decrease with temperature. In addition, in the superconducting state the electromagnetic penetration depth also decreases with decreasing temperature. Thus one expects a sharp decrease in the conversion efficiency of electromagnetic to acoustic energy upon the transition of the metal film from the normal into the superconducting state.

Using indium films, Abeles[6] was the first to observe experimentally electromagnetic generation of acoustic waves in a metal. His experiments were performed in the vicinity of the superconducting transition temperature T_c of indium. Above T_c the conversion efficiency α of electromagnetic to acoustic power was found to be very small (3×10^{-7}). Below T_c the conversion efficiency decreased even further and the signal disappeared in noise. We succeeded in obtaining with indium a conversion efficiency of 5×10^{-5}, two orders of magnitude higher than attained by Abeles. Because of the high conversion efficiency we were able to study the phenomena down to $0.4T_c$ and preliminary results have already been published.[8] In addition we succeeded in generating electromagnetically transverse acoustic waves using Sn films; preliminary measurements show a behavior similar to that of indium.

* This work was supported in part by The Central Research Fund of the Hebrew University of Jerusalem.

However, because of the lower conversion efficiencies, we were able to study quantitatively the phenomena only down to $0.95T_c$.

The experimental arrangement was similar to that in Ref. 8. A thin metal film ($d \approx 1300$ Å in the case of In and $d \approx 1600$ Å in the case of Sn) was vacuum-deposited on an optically polished high-purity silicon rod. The axis of the rods was along either the [110] or [100] direction and their end faces were parallel within 6 sec of arc. The silicon face with the metal film on it was pressed against a 0.6-cm-diameter hole in the bottom wall of a high-Q resonant rectangular microwave cavity. The cavity was excited with 9.02-GHz microwave pulses of 1 μsec duration and 150 W peak power. The power reflected and/or radiated from the cavity was detected by a sensitive microwave detector and displayed on an oscilloscope. Following the initial excitation pulse several echoes were observed delayed by times corresponding to the flight of transverse acoustic waves in silicon. Because the echo amplitude decreased exponentially, we define the insertion loss (IL) as the ratio of the echo power, extrapolated to the time $t = 0$, to the microwave input power.

Above T_c, IL was independent of microwave power. Below T_c, on the other hand, IL was strongly power dependent. This is illustrated in Fig. 1, where we plotted IL for In and Sn as a function of the microwave power (both expressed in dB) at temperatures close to T_c. At the highest microwave power (0 dB) IL has the same values (86 dB for In and 100 dB for Sn) as above T_c. At somewhat lower powers IL decreases until it becomes constant at intermediate values (the shoulder region). At still lower powers IL decreases again until it attains its final value at low powers. The power dependence of IL is interpreted as follows. At the arrival of the high-power microwave pulse the superconducting film is "quenched" and becomes normal. At temperatures close to T_c at the highest powers the film acts as normal even at the generation of the first echo pulse (and sometimes also the second one). The

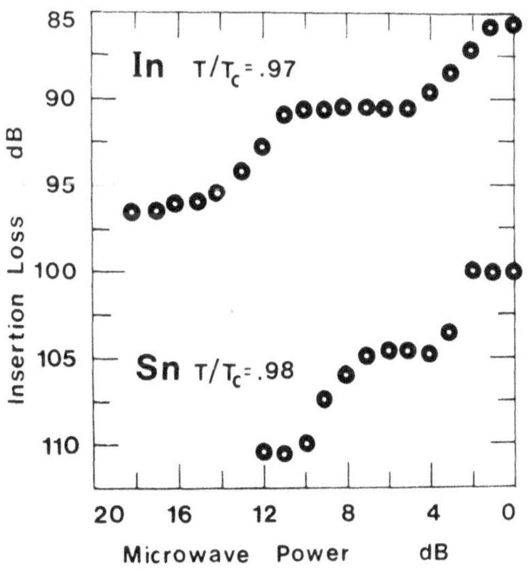

Fig. 1. The dependence of IL on microwave input power: 0 dB on the abscissa corresponds to the maximum microwave power (150 W).

measured IL is the product of the conversion efficiency for generating the sound wave by the microwave pulse and the conversion efficiency for generating the echo pulse (at a later time) by the sound wave. At high powers IL $= \alpha_n^2$, where α_n is the conversion efficiency of indium in the normal state. As the power is decreased the metal film becomes gradually superconducting at the time of the echo pulse. Hence IL decreases (increases in dB) until it eventually becomes equal to $\alpha_n\alpha_s$, where α_s is the conversion efficiency of the superconductor. Decreasing the power further does not affect IL (the "shoulder" region) until a power level is reached at which the film is not quenched completely at the arrival of the excitation pulse. Now IL again decreases with decreasing power until it attains its final value α_s^2. Thus to determine α_s, one can use the value of IL at either low powers or in the "shoulder" region.

Above T_c (up to 4.2°K) the conversion efficiencies for both metals were independent of the temperature T. The value for In of α_n was 5×10^{-5} (43 dB) and for Sn was $\sim 10^{-5}$ (50 dB). Figure 2 shows the temperature dependence of $(\alpha_s)^2$ for indium. The points were obtained from the values of IL at low powers and the triangles from the values of IL in the shoulder region. The theoretical curves in Fig. 2

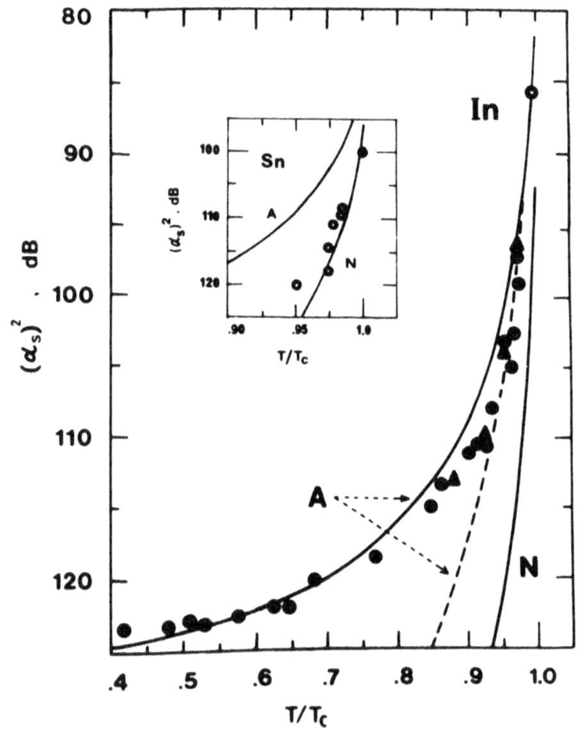

Fig. 2. The measured conversion efficiency (squared) for indium vs. the reduced temperature. The theoretical curves were calculated on different assumptions as explained in the text. The insert shows the temperature dependence of $(\alpha_s)^2$ for Sn close to T_c.

were calculated from the theory for α^2 derived by Abeles.[6] We expressed[8] the penetration depth $|1/\kappa|$ in terms of the conductivity using the theory of the skin effect. The curve labeled N corresponds to the normal limit while the curves labeled A correspond to the extreme anomalous limit. The temperature dependence of the conductivity in the superconducting state was calculated from the theory of Mattis and Bardeen.[9] In calculating the two solid curves we assumed that the fraction of electrons that can be scattered at the surface $n/n_0 = 1$. In calculating the dashed curve, we assumed, according to the two-fluid model,[10] $n/n_0 = (T/T_c)^4$. In all these calculations we neglected the contribution to the acoustic wave due to the direct interaction of the electromagnetic field with the lattice ions. For indium at our frequencies this interaction yields an insertion loss more than three orders of magnitude below the experimental values.

Inspecting Fig. 2, we see that the temperature dependence of the experimental data agrees well with the solid curve labeled A, obtained for the anomalous limit and with $n/n_0 = 1$. Thus the entire temperature dependence of this curve arises from the temperature dependence of the electromagnetic penetration depth. Because the anomalous limit becomes a better approximation the lower the temperature, we fitted (the solid) curve A to the experimental results at low temperatures. The fit yields the values of 600–650 Å for $|1/\kappa|$ at $T = 0.4T_c$. This value is in reasonable agreement with the values for the superconducting penetration depth of indium published in the literature.[6,11,12] At higher temperatures the experimental data deviate somewhat from curve A and are bracketed by the two solid theoretical curves. This is expected because with increasing temperatures the approximation of the extreme anomalous limit becomes progressively worse.

The insert in Fig. 2 shows the temperature dependence of the conversion efficiency for Sn films. The general behavior of Sn was similar to that of indium. However, in our experiments with Sn we found a high attenuation between successive echoes and only three echoes could be observed. Thus the value of $\alpha_n = 10^{-5}$ which depends on the extrapolation to $t = 0$ is somewhat uncertain. In addition, because of the lower value of α_n and the high attenuation we were unable to measure α_s below $0.95T_c$. The experimental data follow curve N (normal limit, $n/n_0 = 1$), which seems to be reasonable at high reduced temperatures. However, we feel that the results on Sn are preliminary, and possibly the conversion efficiency can be raised by improving the experiment.

In conclusion, the good agreement between our results for indium and the theory of Abeles[6] shows that at our frequencies the generation of the acoustic waves is due primarily to diffuse surface scattering of the conduction electrons. Furthermore, to fit the theory to the experiment, we had to assume that $n/n_0 = 1$, and thus we conclude that superconducting pair scattering at the surface does not conserve momentum. In the case of Sn the experimental data are as yet inconclusive as to whether diffuse surface scattering persists below T_c. We would also like to point out that because of the relatively high conversion efficiencies and the momentum transfer by pair scattering at the surface, it may very well be that transverse phonon generation is one of the important loss mechanisms in very high-Q superconducting cavities. In the case of indium, for example, the residual microwave surface resistance at 10 GHz due to this mechanism is calculated to be $1–2 \times 10^{-7}\ \Omega$.

References

1. M.I. Kaganov and V.B. Fiks, *Fiz. Metal. Metalloved.* **19**, 489 (1965) [*Phys. Metals Metallogr.* **19**, 8 (1965)].
2. J.J. Quinn, *Phys. Lett.* **25A**, 522 (1967).
3. R.C. Alig, *Phys. Rev.* **178**, 1050 (1969).
4. J. Halbritter, *J. Appl. Phys.* **42**, 82 (1971).
5. C. Passow, *Phys. Rev. Lett.* **28**, 427 (1972).
6. B. Abeles, *Phys. Rev. Lett.* **19**, 1181 (1967).
7. P.D. Southgate, *J. Appl. Phys.* **40**, 22 (1969).
8. Y. Goldstein and A. Zemel, *Phys. Rev. Lett.* **28**, 147 (1972).
9. D.C. Mattis and J. Bardeen, *Phys. Rev.* **111**, 412 (1958).
10. M. Tinkham, *Phys. Rev.* **129**, 2413 (1963).
11. A.M. Toxen, *Phys. Rev.* **127**, 382 (1962).
12. J.I. Gittleman, S. Bozowski, and B. Rosenblum, *Phys. Rev.* **161**, 398 (1967).

Magnetoacoustic Effect in Niobium*

J. R. Leibowitz

Catholic University of America, Washington, D.C.

E. M. Alexander†

University of Maryland, College Park, Maryland

and

G. V. Blessing,‡ T. L. Francavilla,† and J. R. Peverley

Catholic University of America, Washington, D.C.

We recently reported[1,6] the detection of magnetoacoustic§ (MA) oscillations in niobium. Present results suggest that under appropriate circumstances the transverse wave MA interaction is more selective among the available extremal calipers than the longitudinal wave interaction. A consequence in these instances is that oscillation interference is reduced or removed and the ql constraint that must be met to ensure sufficiently large oscillation number n to discriminate interfering periods is thereby relaxed; here q is the magnitude of the acoustic wave propagation vector and l is the electron mean free path. Accordingly, the results in niobium suggest that the generally neglected transverse wave MA interaction deserves reexamination. For most orientations of acoustic wave and field **H** the oscillation number n was 6–7 for 345–435 MHz transverse waves (T waves) and 3–4 for 510 MHz longitudinal waves (L waves). The doubling of n in the more recent T-wave work makes possible a more critical examination of the identification of extremal calipers with Fermi sheets in Nb and the characteristics of the T-wave magnetoacoustic interaction in the case of a multisheet Fermi surface.

As is well known, when the MA oscillations are associated with extremal orbits the periodicity in reciprocal magnetic field $\Delta(1/H)$ measures the extremal **k**-space caliper dimension Δk in accordance with the relation $\Delta k = e\lambda/[ch\,\Delta(H^{-1})]$, where λ is the acoustic wavelength; the direction of Δk is that of $\mathbf{q} \times \mathbf{H}$. Of course, in most actual Fermi surfaces a given **H** orientation will in general be associated with not one but several calipers, each corresponding to a given oscillation amplitude, period, and relative phase. The resulting interference pattern is not readily analyzable, in general, into component periods and corresponding calipers when n is not large. At the same time the need for large n entails the requirement $ql \gg 1$: for the appear-

* Work supported in part by AFOSR 70-1931.

† Present address: U.S. Naval Research Laboratory, Washington, D.C.

‡ Present address: U.S. Naval Weapons Laboratory, Dahlgren, Va.

§ In this paper we refer to the MA effect for which **H** is normal to **q**.

ance of n oscillations it is in practice required[2] that $ql \approx 4n$. The ql constraint has served to limit the usefulness of the MA technique for fermiology (as compared, say, to the de Haas–van Alphen method) since it demands specimens of very high purity.

The free-electron theory[3] for the MA effect shows that in the case of T waves the orientation of wave polarization $\boldsymbol{\varepsilon}$ relative to $\mathbf{H}$ affects the amplitude of the oscillations. There are predicted well-defined oscillations for $\mathbf{H} \perp \boldsymbol{\varepsilon}$, but only oscillations of relatively small amplitude showing no extrema for the case $\mathbf{H} \parallel \boldsymbol{\varepsilon}$. The fact that, unlike the L-wave case, rotation of $\mathbf{H}$ in the plane normal to $\mathbf{q}$ involves at the same time a change of relative orientation of $\mathbf{H}$ and $\boldsymbol{\varepsilon}$ in the case of T waves—and hence introduces a "geometric anisotropy" in addition to that of the Fermi surface—undoubtedly was a contributing factor historically in the general neglect by experimentalists of T waves for Fermi surface studies. However, we suggest that in light of the present experimental results in niobium, together with current understanding of the factors influencing oscillation amplitude in real metals, a reexamination is in order.

It has been shown (see the review in Ref. 2) that, aside from its dependence on carrier lifetimes, the longitudinal MA oscillation amplitude for given (large) n depends on the Fermi surface curvatures at caliper end points evaluated (i) along the cyclotron orbit and (ii) along the direction of $\mathbf{H}$. While the analysis referred explicitly to the case of L waves, the dependence of amplitude on Fermi surface curvatures is preserved for the T-wave interaction. However, an additional factor influencing amplitude only for the T-wave case is associated with wave polarization $\boldsymbol{\varepsilon}$. It was noted earlier that in free-electron theory T-wave MA amplitude depends on the relative orientation of $\mathbf{H}$ and $\boldsymbol{\varepsilon}$. Pippard[4] has shown that in the "real-metals" case the problem is more complicated in general, but that the dependence of the T-wave interaction on $\boldsymbol{\varepsilon}$ is greatly simplified when $ql \gg 1$ and $\mathbf{q}$ and $\boldsymbol{\varepsilon}$ lie in directions of high lattice symmetry. The data suggest a strong caliper selectivity determined by $\mathbf{v}_F \cdot \boldsymbol{\varepsilon}/v_F$ evaluated at caliper end points; here $\mathbf{v}_F$ is the Fermi velocity.

Figures 1 and 2 show, respectively, 510-MHz L-wave data for $\mathbf{q}$ along [100], and 210-MHz T-wave MA oscillations for $\mathbf{q} \parallel [100]$ and $\boldsymbol{\varepsilon} \parallel [010]$. The L-wave data are compared with the relatively low-frequency T-wave data (435 MHz was the highest T-wave frequency examined) so as to permit comparison of L-wave and T-wave oscillations having roughly comparable ql values, and thereby remove the possibility that "ql effects" are responsible for the qualitative differences between the MA oscillations of Figs. 1 and 2. Note the evidence in the L-wave case for oscillation interference, the character of which is very sensitive to rotation of $\mathbf{H}$ in the (001) plane; in contrast the T-wave oscillations display a single dominant period in this range of n values. Space limitations do not permit showing the 345-MHz T-wave results: 6–7 oscillations for all orientations of $\boldsymbol{\varepsilon}$ were obtained ($\mathbf{H}$ rotated through 360° in approximately 5° intervals). In all cases there was obtained a rather "unmixed" period, the value of which changed "smoothly" with rotation of $\mathbf{H}$ through the entire range of field angles.

Figure 3 represents schematically the projections onto the (100) plane of the so-called "effective zones" corresponding to $\mathbf{q} \parallel [100]$; the effective zones are the loci on the Fermi surface for which $\mathbf{v}_F$ is normal to $\mathbf{q}$. The projections were computed from the band structure calculations by Mattheiss[5] for Nb. In Fig. 3 the calipers

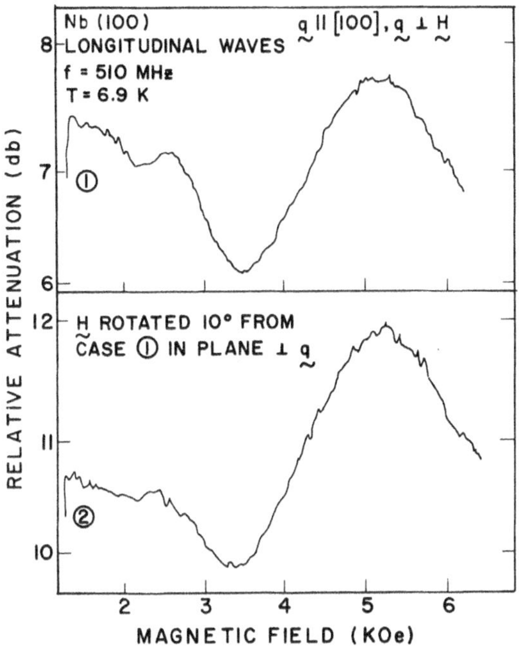

Fig. 1. Longitudinal wave magnetoacoustic effect (**q** ⊥ **H**) for frequency of 510 MHz showing strong interference from competing calipers. Note sensitivity of the "interference pattern" to small change in orientation of **H**.

Δk are the extremal diameters perpendicular to a given **H** direction. The four-armed closed and open figures centered on Γ are associated with the so-called "octahedral" and "jungle gym" sheets,[5] respectively. The remaining (closed) figures correspond to the so-called "distorted ellipsoids" (ELL). We shall find it useful to consider these in two groups: those labeled I and II correspond to ELL sheets bisected by the (100) plane through Γ, while those labeled III–VI belong to the ELL which are canted relative to and do not intersect the central (100) plane.

Analysis of the T-wave data for $\mathbf{q} \parallel [100]$ and $\varepsilon \parallel [010]$ yields a most intriguing result. The experimentally determined extremal calipers Δk corresponding to all field angles, rotated through 360° in 5° intervals, agree everywhere within about 2% with the calculated Δk values corresponding to ELL III—or, equivalently ELL V (see the closed figures labeled III and V in Fig. 3). The results suggest that calipers associated with all of the other sheets are relatively unimportant. The unimportance of the octahedral sheet may perhaps be attributed to the fact that the **k**-space curvatures about the effective zones are relatively large (see Fig. 3). It might be argued, too, that the absence of the open and closed extremal calipers of the jungle gym sheet is attributable, in part, to the close approach of the extremal orbits to the octahedral surface and the attendant likelihood of destruction of phasing between the acoustic wave and the electron orbits by interband scattering. The apparent relative unimportance of ELL I and II calipers is more difficult to understand. But our interest is directed to those sheets which do contribute, and we turn our attention to the remaining surfaces ELL III–IV. They are entirely equivalent from the point of view of **q**, which demands a fourfold rotational symmetry for the oscillations as **H** is rotated through 360°. The apparent preference for ELL III (or, equivalently, ELL V) in the aforementioned T-wave results can only be attributed to the introduction of **ε** in the (100) plane.

It is useful in this connection to examine the selectivity criterion cited earlier. We then find for the case **ε** ∥ [010] that $|\mathbf{v}_F \cdot \boldsymbol{\varepsilon}/v_F|$ in the region of caliper end points is everywhere greater for ELL III (or V) than for ELL IV (or VI).

The case **q** ∥ [100] and **ε** ∥ [011] was examined for consistency with the above inferences pertaining to the **ε** ∥ [010] case. It is clear that for **ε** ∥ [011] both ELL III and IV contribute and that, consequently, oscillation interference should be observed for all **H** orientations except those corresponding to the ⟨011⟩ directions in the (100) plane. For these special directions the extremal calipers of III and IV are

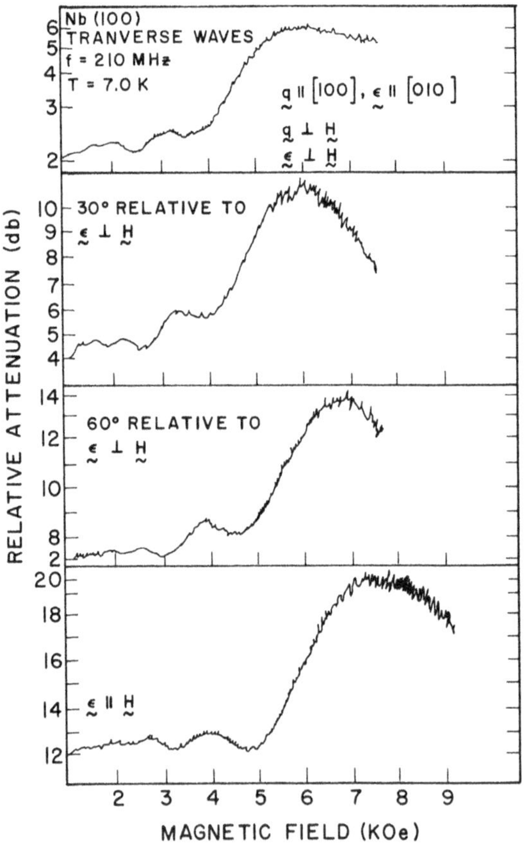

Fig. 2. Transverse wave magnetoacoustic oscillations at 210 MHz. which corresponds roughly to the same ql value as in the case of longitudinal waves at 510 MHz given in Fig. 1. Note the striking difference from the longitudinal wave oscillations. There is now seen a relatively pure harmonic sequence over the whole range of **H** orientations. from **H** ⊥ **ε** to **H** ∥ **ε**.

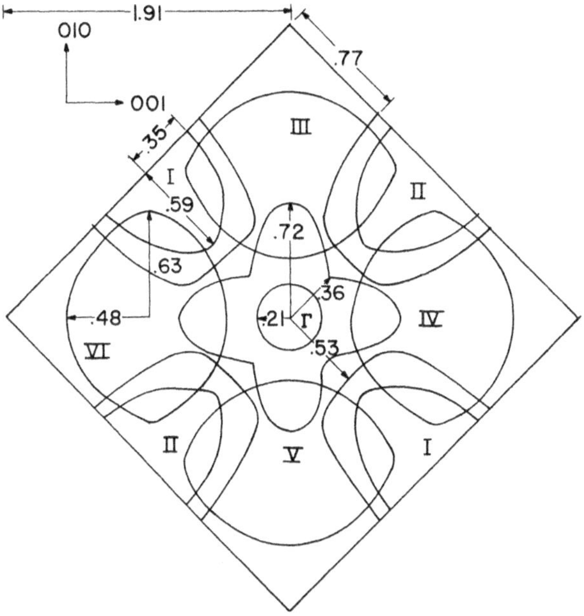

Fig. 3. Effective zones (in reciprocal angstroms) of the Nb Fermi surface for the case **q** ∥ [100]. Extremal calipers for each sheet are the calipers normal to **H**, which lies in the plane of the figure.

degenerate, and a pure oscillation corresponding to $\Delta k = 1.04$ Å^{-1} is predicted. The data satisfy expectations: The value determined for the degenerate Δk is 1.05 ± 0.02 Å^{-1}; the oscillation number n was too small to permit analysis of the component periods corresponding to the "mixed" orientations.

In summary we find that under appropriate conditions the T-wave MA effect may provide greater caliper selectivity than the L-wave interaction and, thereby, in relaxing the requirement for large n (and large ql), enhance the usefulness of the MA effect for fermiology.

Acknowledgments

We are grateful to Dr. R. E. Reed of Oak Ridge for the Nb samples, and to the Aspen Center for Physics, where some of this work was completed.

References

1. J.R. Leibowitz, E. Alexander, and J.R. Peverley, *Phys. Lett.* **44A**, 298 (1973).
2. J. Mertsching, *Phys. Stat. Sol.* **16**, 267 (1966).
3. T. Kjeldaas, Jr. and T. Holstein, *Phys. Rev. Lett.* **2**, 340 (1959); M.H. Cohen, M.J. Harrison, and W.A. Harrison, *Phys. Rev.* **117**, 937 (1960).
4. A.B. Pippard, *Proc. Roy. Soc.* **A257**, 165 (1960).
5. L.F. Mattheiss, *Phys. Rev. B* **1**, 373 (1970).
6. J.R. Leibowitz, G.V. Blessing, and E.M. Alexander, *Phys. Rev. B* (to be published).*

* **Note Added in Proof:** Reference 6 is a more detailed study of the magnetoacoustic problem in Nb.

9

Electronic Properties

Electrical and Structural Properties of Amorphous Metal—Metal Oxide Systems

J. J. Hauser

Bell Laboratories
Murray Hill, New Jersey

The metal–metal oxide films studied in this work were deposited using the getter-sputtering technique.[1] The target preparation was similar to that used for the $Al-Al_2O_3$ system.[1] The resistivity of the films was measured as a function of temperature between 1 and 300°K in three different ways: a four-point contact method measuring the voltage drop created by a constant current (as low as 10 nA) in the plane of the film for resistances up to 10^{10} Ω; a two-point contact method measuring the current flowing as a result of a constant applied voltage (90 V) for resistances up to 10^{15} Ω; and finally, a capacitance configuration where the current flowing across the film thickness as a result of a fixed applied voltage between two gold electrodes is measured.

All the films studied in the present experiment can be considered as amorphous or possibly as quasiamorphous, as we shall see when the susceptibility experiments are examined. Indeed, the X-ray diffraction patterns showed no resolvable peaks and the electron diffraction patterns consisted at best of very broad diffuse rings which suggest submicroscopic grains with a diameter smaller than 20 Å (see Fig. 4 of Ref. 1).

$Al-Al_2O_3$

The temperature dependence of the resistivity of $Al-Al_2O_3$ films is well approximated by a relationship of the form

$$R = R_0 \exp[(T_0/T)^{1/n}] \quad (1)$$

where the exponent n can take values between 1 and 4. The fit to relation (1) is convincingly established in Fig. 1. Since the data were obtained down to 1°K, Fig. 1 establishes conclusively that no fractional power other than 1/4 can be used in relation (1). We shall now turn our attention to $Al-Al_2O_3$ films with increasing resistivities. In Fig. 2 one observes that films with room-temperature resistivities as high as 5.5 Ω-cm are well represented by relation (1) with $n = 4$ and a maximum value for T_0 of 10^5 °K. The fit is excellent between 77 and 1°K, but it is obvious from Fig. 4 of Ref. 1 that the room-temperature point cannot be included. When $\rho(RT) \simeq 10$ Ω-cm the data can no longer be fitted with $n = 4$; indeed an excellent fit can be achieved with $n = 3$, as shown in Fig. 4 of Ref. 1. For higher resistivities (a few tens of Ω-cm) the data can be fitted by relation (1) with $n = 2$. It is also clear

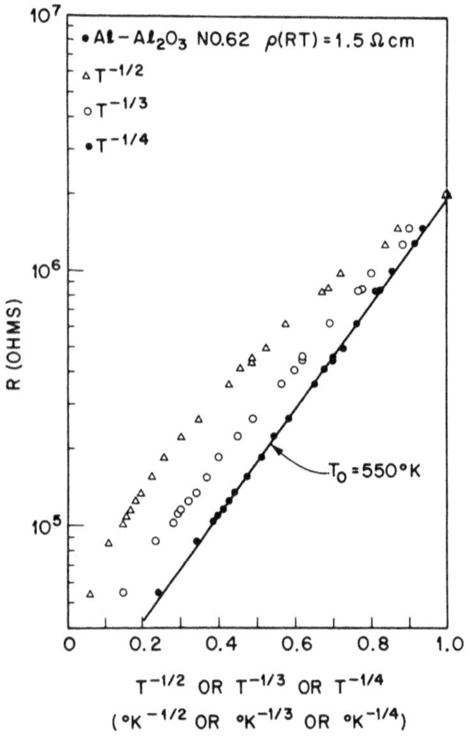

Fig. 1. Temperature dependence of the resistance fitted to $R = R_0 \exp[(T_0/T)^{1/n}$ with $n = 2, 3, 4$ for an Al_2O_3 film with $\rho(RT) = 1.5\ \Omega$-cm.

from Fig. 2 that for a given n value the slope is steeper (i.e., T_0 is larger), the higher the room-temperature resistivity of the sample. If one now considers films with the highest resistivities (a few hundreds of Ω-cm) it can be seen from Fig. 2 that the temperature dependence is somewhat slower than a pure exponential ($n = 1$) but somewhat faster than that given by relation (1) with $n = 2$. Consequently, it seems that with increasing $\rho(RT)$ (i.e., with increasing Al_2O_3 content) the temperature dependence of the resistance approaches a pure exponential.

Films with room-temperature resistivity larger than $\sim 1\ \Omega$-cm are no longer superconducting by resistive measurements. Nevertheless, such films with resistivities as high as 320 Ω-cm (Al–Al_2O_3 No. 71) become superconducting at 1.2°K in a susceptibility measurement. Since the strength of superconducting signal at 1.2°K is quite weak, one is led to assume that the superconductivity arises from submicroscopic Al grains (size smaller than 20 Å), as inferred from the fact that the electron diffraction patterns are those of a totally amorphous material.

The ac resistance of an Al–Al_2O_3 film was measured between two gold electrodes. The data show that the ac resistance decreases with frequency as $\nu^{-0.7}$, which reflects a conduction by hopping[2]; the frequency exponent should be 1 for single hops and 0.5 for multiple hops.

The experimental results in the Al–Al_2O_3 system can be summarized in the following manner: The temperature dependence of the resistance of all Al–Al_2O_3 films studied can be represented by relation (1) with $n = 4$ over a wide range of

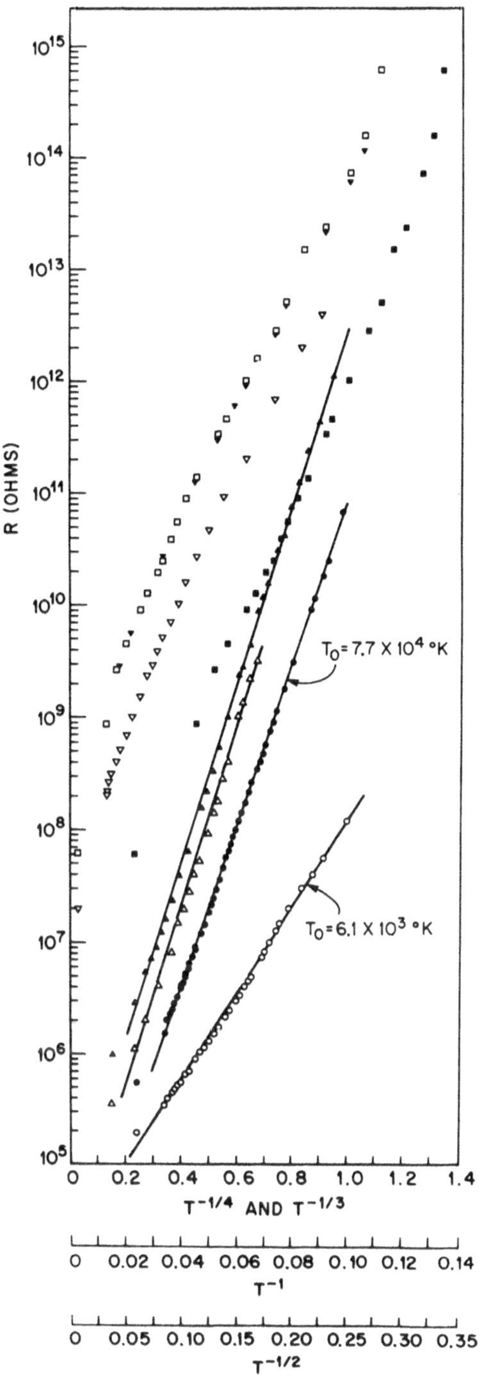

Fig. 2. Various temperature dependences of the resistance for $Al–Al_2O_3$ films ranging in room-temperature resistivities between 5 and 640 Ω-cm.

compositions. As the Al_2O_3 content is increased (for $\rho > 5$ Ω-cm) the exponent n decreases progressively toward one, which is the expected exponential behavior for a doped oxide. This behavior, along with the ac results, suggests that the conduction in such films is similar to the thermally activated hopping mechanism described for other amorphous materials.[3-7] The susceptibility experiments reveal the presence of submicroscopic Al grains in which the energy levels could be quantized as a result of size quantization.* These energy levels could be the localized states between which the electrons hop. If one considers the quantized levels in a 20-Å-diameter Al grain, such levels would be smeared by kT at about 80°K, thus explaining why the room-temperature data often seemed to deviate from the fitted curves. Although a hopping mechanism qualitatively explains the data, it seems very hard to reconcile the low observed values of T_0 with the percolation theory of Ambegaokar *et al.*[4] The minimum theoretical value of T_0 which one can obtain with reasonable values of the parameters is 2×10^4 °K. Consequently, although this theory could explain in detail films with $\rho(RT) > 5$ Ω-cm, the parameters could not be adjusted to fit the films with the lower resistivities.

Sn–SnO_2

Sn–SnO_2 films with resistivities smaller than 1 Ω-cm can be considered as quasiamorphous since the electron diffraction patterns show broad, diffuse rings. This is in contrast with Sn–SnO_2 films with $\rho > 1$ Ω-cm, which, just like all Al–Al_2O_3 films, yield a totally diffuse electron diffraction pattern. The temperature dependence of the resistance for films with $\rho < 1$ Ω-cm can be fitted by relation (1) with $n = 1$. All Sn–SnO_2 films with $\rho < 1$ Ω-cm display, when measured by susceptibility, a superconducting transition around 3.8°K. The relatively large strength of the superconducting signal indicates, in agreement with electron diffraction, that such films are granular with a grain size of approximately 20 Å.

While Al–Al_2O_3 films displayed a superconducting transition for ρ as high as 320 Ω-cm, Sn–SnO_2 films with $\rho > 1$ Ω-cm show no sign of superconductivity. Consequently, such films are totally amorphous and are even devoid of submicroscopic grains. This fact was confirmed by NMR measurements† on a scraped 26-μm film ($\rho = 44$ Ω-cm) which indicated a zero Knight shift. This does not exclude the possibility of a small concentration of Sn clusters which could still behave as hopping centers. The temperature dependence of the resistance for Sn–SnO_2 films with resistivities >1 Ω-cm is shown in Fig. 3. It is clear from Fig. 3, where $n = 1, 2, 4$ has been attempted for Sn–SnO_2 No. 34, that $n = 2$ is the best fit over eight orders of magnitude change in resistance. As in the Al–Al_2O_3 system, where one found that a pure exponential behavior ($n = 1$) is approached for high resistivities (high Al_2O_3 content), one observes an exponential dependence of R on T for Sn–SnO_2 films with $\rho \simeq 10^3$ Ω-cm. The activation energy of 25 meV is quite reasonable for conduction in a heavily doped oxide.

* A 20-Å-diameter grain of Al contains 750 atoms which results in quantized levels approximately 7 meV apart.

† I am indebted to W. W. Warren for these NMR measurements.

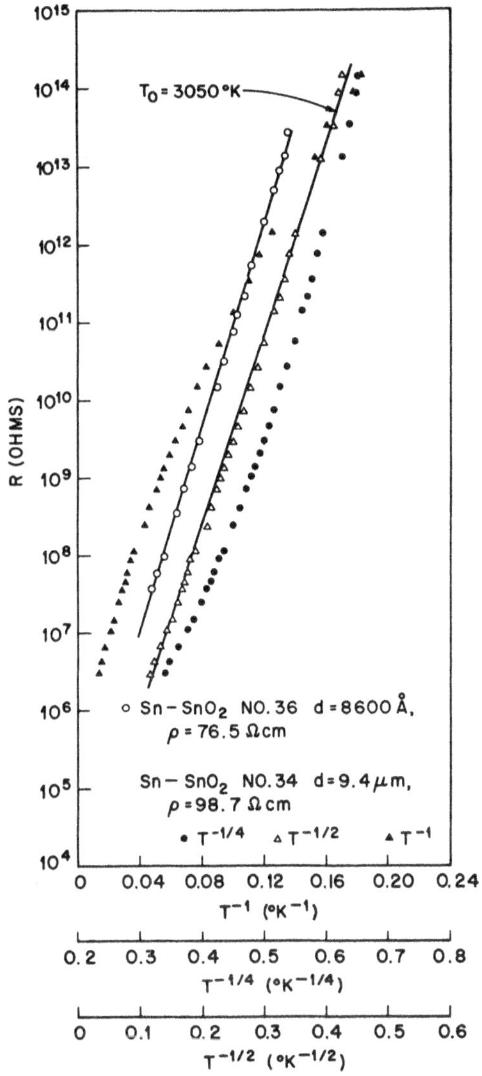

Fig. 3. Various temperature dependences of the resistance for Sn–SnO_2 films deposited at 300°K.

One may summarize these results by stating that the temperature dependence of the resistance of Sn–SnO_2 films can be well described by relation (1) with $n = 2$ for most of the films studied; $n = 1$ for both extremes, the metal-rich side, or granular films, and the oxide-rich side, or doped dielectric. The major difference from the Al–Al_2O_3 system is the absence of a region with $n = 4$.

We shall now consider the measurement of a Sn–SnO_2 film in a capacitance configuration between two gold electrodes. The temperature dependence of the resistance for such a measurement ($n = 2$ and $T_0 = 3080$°K) is in very good agreement with the measurements in a planar geometry (see, for example, Fig. 3). The ac resistance decreases with frequency as $\nu^{-0.8}$, which is similar to the dependence found in many glasses[2] and corresponds to a hopping mechanism of conductivity. At sufficiently high electric fields $\log I$ is a linear function of $V^{1/2}$.

Similar results were found for the $Ta-Ta_2O_5$ and the Ni–NiO systems.

Acknowledgments

I would like to thank B. I. Halperin for helpful discussions and J. H. Wellendorf for his technical assistance.

References

1. J.J. Hauser, *Phys. Rev. B* **3**, 1611 (1970), and references therein.
2. K.L. Chopra and S.K. Bahl, *Phys. Rev. B* **1**, 2545 (1970).
3. N.F. Mott, *J. Non-Cryst. Solids* **1**, 1 (1968); *Phil. Mag.* **19**, 835 (1969).
4. V. Ambegaokar, B.I. Halperin, and J.S. Langer, *Phys. Rev. B* **4**, 2612 (1971).
5. P.A. Walley, *Thin Solid Films* **2**, 327 (1968).
6. M. Morgan and P.A. Walley, *Phil. Mag.* **23**, 661 (1971).
7. P.A. Walley and A.K. Jonscher, *Thin Solid Films* **1**, 367 (1968).

Lead Phthalocyanine: A One-Dimensional Conductor

K. Ukei, K. Takamoto, and E. Kanda

The Research Institute for Iron, Steel and Other Metals
Tohoku University, Sendai, Japan

One-dimensional metallic compounds are important from a theoretical and possibly practical point of view. It has been rigorously proved[1] that long-range order does not exist in a Fermi system for one dimension at finite temperatures. In real one-dimensional substance the dimensionality is not as mathematically strict as in this theoretical proof. It is interesting to examine the electrical and magnetic properties of a "one-dimensional superconductor," if it exists. Little[2] inferred that in one-dimensional superconductive samples the resistance would drop smoothly and rapidly toward zero as the temperature tends to 0°K. Moreover, he proposed that superconductivity by a nonphonon mechanism may be possible in a linear macromolecule.[3] By any mechanism, whether a phonon mechanism or not, a one-dimensional metal may be considered a prerequisite to a one-dimensional superconductor. Besides, a one-dimensional conductor is important theoretically in itself.

Recently a few kinds of one-dimensional metallic compounds, such as TCNQ salts[4] and Krogmann salts (i.e., square planar complexes of platinum),[5-7] were reported. Especially conclusive evidence was given for the metallic behavior of Krogmann salts.[6,7] But the existence of a one-dimensional superconductor has not yet been reported. In the present work, as a step toward finding a one-dimensional superconductor, the temperature dependence of the electrical resistivity and magnetic susceptibility of lead phthalocyanine (PbPc) has been examined down to the liquid helium temperatures.

PbPc was prepared by the reaction of PbO and phthalonitrile in refluxing α-chloronaphthalene.[8] The test cell for measuring the electrical resistivity parallel to the *c* axis is shown in the insert of Fig. 1. The PbPc crystals were grown by sublimation techniques upon the bottom silver electrode on a glass substrate maintained at about 250°C in a stream of nitrogen at 5–7 Torr, and then the top silver electrode was vacuum-deposited on the crystals. These crystals are composed of needle crystals of various lengths and look like dark green velvet. The *c* axes of these crystals are thought to be almost perpendicular to the glass substrate, based on consideration of the crystal habit. The temperature dependence of the electrical resistivity parallel to the *c* axis is metallic and nearly constant at liquid helium temperatures, as shown in Fig. 1. The specific resistivity of PbPc parallel to the *c* axis cannot be exactly determined because only the order of magnitude of the film thickness can be estimated, due to the roughness of its surface; moreover, the effective area for conduction is uncertain in a crystal of this kind of compound. Its value was estimated to be

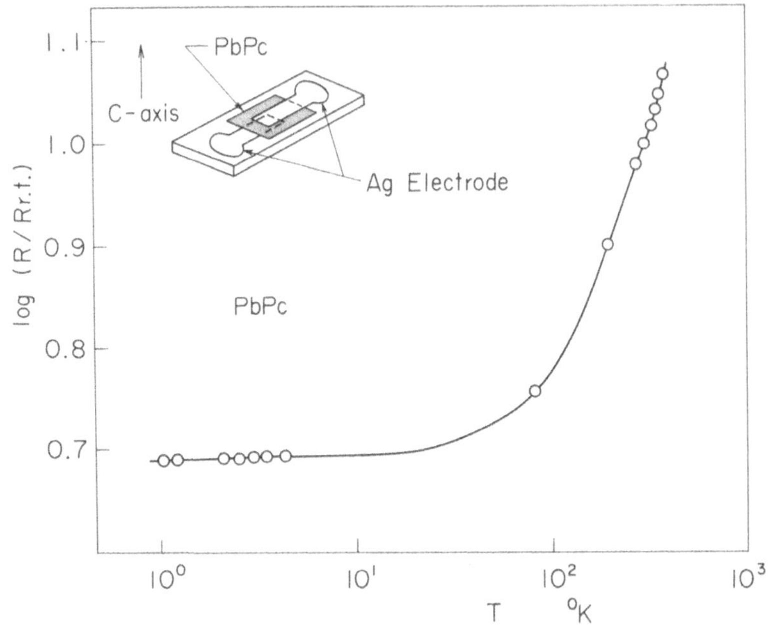

Fig. 1. The temperature dependence of the electrical resistivity of PbPc parallel to the *c* axis.

less than 10^4 Ω-cm. The temperature dependence of the electrical resistivity parallel to the *c* axis was also measured using a single crystal, and found to be semiconductive. Similarly to the case of Krogmann salts,[5,6] this result is thought to be due to cracks or lattice defects of the crystal, which inevitably occur in this kind of compound and have a decisive effect on the electrical conductivity. The temperature dependence of the electrical resistivity perpendicular to the *c* axis was measured between 20 and 90°C. In this case test cells were made by growing PbPc crystals on the vacuum-deposited silver electrodes split into two. The results follow the well-known relationship

$$R = R_0 \exp(E/2kT)$$

where $E = 0.66$ eV. It is possible that this semiconductive character of the temperature dependence of the electrical resistivity perpendicular to the *c* axis results from the crystalline boundaries as well as having an intrinsic origin.

The crystal structure determined is shown in Fig. 2. In this crystal PbPc molecules are stacked linearly in the direction of the *c* axis normal to the molecular plane, and this stacking structure bears a strong resemblance to those of Krogmann salts.[5] In the case of other metal phthalocyanines—for instance, CuPc—the molecular planes are not perpendicular to the stacking direction, and the distance between the molecular planes is about 3.4 Å, which is shorter than that of PbPc.[9] In the case of PbPc the molecular plane is normal to the stacking direction and the intermolecular distance parallel to this direction is not the distance for closest packing. This suggests that a path of the electron overlap of lead atoms is created parallel to the

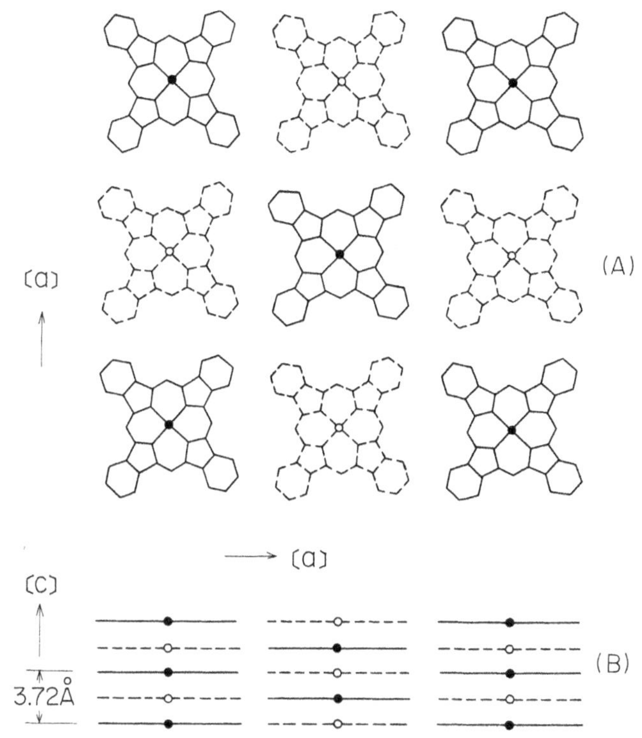

Fig. 2. The crystal structure of PbPc. (A) The plane figure: (B) the front view; in (A) the dashed lines show the molecules $c/2$ below the plane of the paper(c plane) and in (B) the open circles represent the Pb atoms which lie in the plane $a/2$ below the plane of the paper (a plane).

c axis. The interatomic distance of lead atoms in PbPc (3.72 Å) is slightly larger than that in lead metal (3.48 Å). From a consideration of this intermolecular structure and the metallic character of the temperature dependence of the electrical resistivity parallel to the c axis, the conduction band may be thought to arise from the overlap of the lead orbitals which point above and below the molecular plane. The intermolecular structure perpendicular to the c axis suggests a semiconductive conduction normal to the c axis. In this way this substance may be considered to be a one-dimensional conductor.

The temperature dependence of the electrical resistivity of PbPc parallel to the c axis was further measured down to 0.3°K by immersing the test cell in liquid ^{3}He. The measured value of the electrical resistivity decreases feebly and gradually as the temperature decreases below 1°K, but the rate of decrease in the electrical resistivity differs from sample to sample. Below 4.2°K the magnetic susceptibility of powdered PbPc was also measured by an ac bridge method down to 0.3°K. The magnetic susceptibility of the annealed PbPc is temperature independent between 4.2 and 1°K, but it decreases slightly and gradually as the temperature is lowered below 1°K (Fig 3). The decrement between 1 and 0.3°K is of the order of

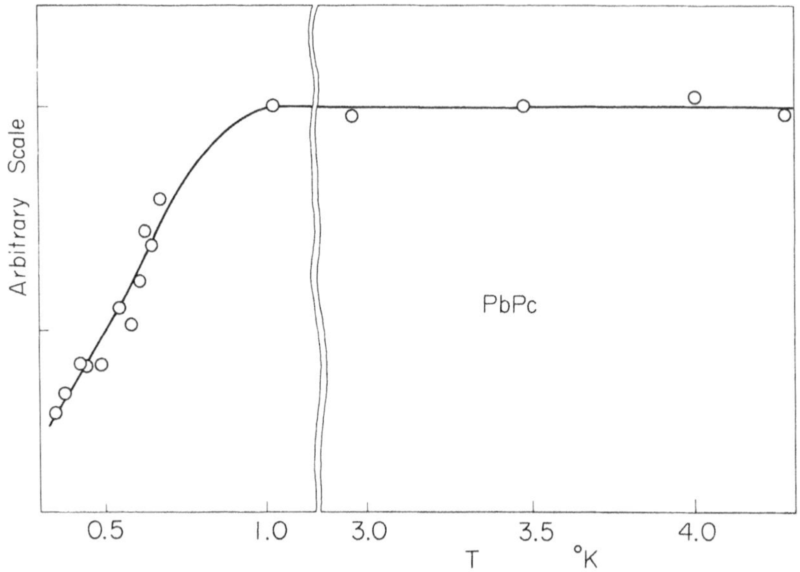

Fig. 3. The temperature dependence of the magnetic susceptibility of annealed PbPc powder.

magnitude of 10^{-4} of the perfect diamagnetism. We consider that this decrement of the magnetic susceptibility might be due to the occurrence of diamagnetism superposed upon the temperature-independent paramagnetism. In some cases the unannealed sample did not show the above-mentioned occurrence of the diamagnetism below 1°K, but this sample also showed such diamagnetism after it had been annealed at 360°C for about 20 hr. The X-ray powder patterns before and after this heat treatment show that the (220) and (111) reflections are intensified by the heat treatment. This means that an ordering process of the crystalline lattice proceeds during the heat treatment. The ordering along the *c* axis favors metallic conduction. The measured value of the magnetic susceptibility between 1 and 4.2°K shows a paramagnetism of the order of magnitude of 10^{-5} emu/g. This paramagnetism might be due to Pauli paramagnetism of the conduction electrons. In such a case this would imply a relatively narrow band.

As mentioned above, the electrical resistivity parallel to the *c* axis and the magnetic susceptibility decrease gradually below 1°K. With regard to this superconductor-like anomaly (though feeble) below 1°K, two possible origins are considered, an extrinsic and an intrinsic one. As an example of the former, one might consider the presence of lead metal which could result from the decomposition of PbPc. It has been reported that the superconductive transition temperature of ultrathin lead films of 10 Å thickness becomes nearly equal to 1°K.[10] Therefore the above-mentioned anomaly below 1°K might be possible if there were traces of metallic lead particles of such a size in the sample. If the anomaly results intrinsically, it may be ascribed to the fluctuations of the superconductive order parameter.

References

1. P.C. Hohenberg, *Phys. Rev.* **158**, 383 (1967).
2. W.A. Little, *Phys. Rev.* **156**, 396 (1967).
3. W.A. Little, *Phys. Rev.* **134**, A1416 (1964).
4. W.J. Siemons, P.E. Bierstedt, and R.G. Kepler, *J. Chem. Phys.* **39**, 3523 (1963); R.G. Kepler, *J. Chem. Phys.* **39**, 3528 (1963); R.W. Tsien, C.M. Huggins, and O.H. Le Blanc, Jr., *J. Chem. Phys.* **45**, 4370 (1966); R.M. Vlasova and A.I. Gutman, *Fiz. Tverd. Tela* **13**, 2808 (1971) [*Soviet Phys.—Solid State* **13**, 2356 (1972)].
5. K. Krogmann and H.D. Hausen, *Z. Anorg. Allg. Chem.* **358**, 67 (1968).
6. D. Kuse and H.R. Zeller, *Phys. Rev. Lett.* **27**, 1060 (1971).
7. H.P. Geserich. H.D. Hausen, K. Krogmann, and P. Stampfl, *Phys. Stat. Sol.* (a) **9**, 187 (1972).
8. W.J. Kroenke and M.E. Kenney, *Inorg. Chem.* **3**, 251 (1964).
9. J.M. Robertson and I. Woodward, *J. Chem. Soc.* **1937**, 219; J.M. Robertson and I. Woodward, *J. Chem. Soc.* **1940**, 36; C.J. Brown, *J. Chem. Soc.* (A) **1968**, 2488.
10. M. Strongin, R.S. Thompson, O.F. Kammerer, and J.E. Crow, *Phys. Rev. B* **1**, 1078 (1970).

Nonequilibrium Effects in Normal Metal–Insulator–Metal Junctions at Low Temperature

J. G. Adler, H. J. Kreuzer, P. N. Trofimenkoff and W. J. Wattamaniuk

Department of Physics
University of Alberta, Edmonton, Alberta, Canada

A great deal of effort has been devoted to the theoretical and experimental study of low-bias structure in the dynamic conductance $\sigma = dI/dV$ and its derivative $d\sigma/dV$ of normal metal–insulator–metal tunneling junctions.[1] This paper deals with the very small conductance minima in the vicinity of zero bias and below the phonon peaks[2] that have been observed for a variety of combinations of metal electrodes.[3] We consider the nonequilibrium aspects of tunneling and present a transport model that explains the very small structure that occurs near zero bias.[4]

In the usual semiclassical single-particle-transfer picture tunneling is restricted to electrons with momenta in the forward direction (perpendicular to the barrier) and is assumed to occur between equilibrium electron distributions in the metal electrodes on either side of the barrier. In the presence of a current a nonequilibrium situation exists and the electron relaxation rates in the metal electrodes have to be taken into account. Scattering of electrons out of the forward direction in the electrodes can occur by means of elastic impurity scattering with a constant relaxation time τ_i and inelastic phonon scattering with an energy- and temperature-dependent relaxation time τ_{ep} (E, T), where E is the energy measured relative to the chemical potential and T is the temperature (see Fig. 1). In addition, boundary scattering at the ends of the electrode allow us to write the following two rate equations in the steady state in the right (R) and left (L) electrodes:

$$\frac{\partial f_R(E,T)}{\partial t} = 0 = \frac{D}{\tau_0}[f_L(E,T) - f_R(E,T)] - \delta f_R(E,T)\left(\frac{1}{\tau_{ep}^R(E,T)} + \frac{1}{\tau_i^R}\right) \tag{1a}$$

$$\frac{\partial f_L(E,T)}{\partial t} = 0 = -\frac{D}{\tau_0}[f_L(E,T) - f_R(E,T)] - \delta f_L(E,T)\left(\frac{1}{\tau_{ep}^L(eV-E)} + \frac{1}{\tau_i^L}\right) \tag{1b}$$

where $f_L(E,T) = f_0(E - eV, T) + \delta f_L(E,T)$ and $f_R(E,T) = f_0(E,T) + \delta f_R(E,T)$ are the steady-state occupations in the left and right electrodes as opposed to the equilibrium distributions $f_0(E - eV, T)$ and $f_0(E,T)$; D is the intrinsic barrier penetrability and $1/\tau_0$ is the decay constant, so that D/τ_0 is a transition rate for electrons crossing the barrier.

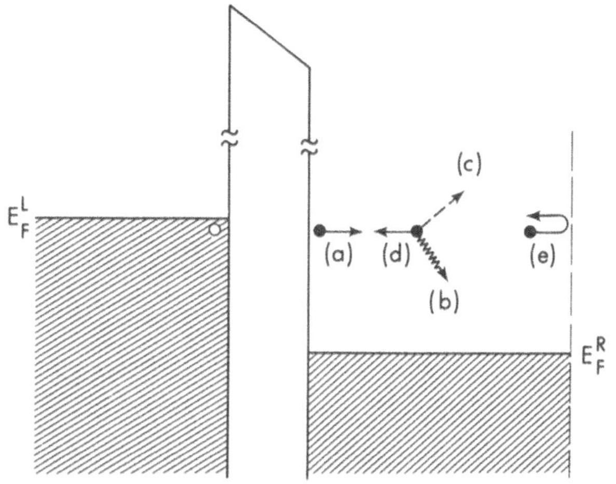

Fig. 1. An electron that tunnels through the barrier will leave a hole in the left electrode and occupy a state (a) in the right electrode. This electron will travel away from the barrier in the right electrode and may be scattered by various processes: (1) Inelastic scattering with emission or absorption of a phonon (b) will alter both the electron's energy and momentum; this type of process will be described by an energy- and temperature-dependent relaxation time $\tau_{ep}(E,T)$, where E is the energy measured relative to the chemical potential of the right electrode and T is the temperature. (2) Elastic scattering by impurities, (c) and (d), will alter the electron's momentum only and will be accounted for by a constant relaxation time τ_i. (3) Boundary scattering, if specular (e), will reverse the component of the electron's momentum perpendicular to the barrier.

Solution of Eqs. (1a) and (1b) provides

$$f_L(E, T) - f_R(E, T) = [f_0(E - eV, T) - f_0(E, T)][1 - g(E)] \tag{2a}$$

where the blocking factor $g(E)$ is given by

$$g(E) = \frac{D/\tau_0}{(D/\tau_0) + \{1/[\tau_L(eV - E, T) + \tau_R(E, T)]\}} \tag{2b}$$

where $1/\tau = (1/\tau_{ep}) + (1/\tau_i)$. In the calculation of the current I, σ, and $d\sigma/dV$ the above expression for $f_L(E, T) - f_R(E, T)$ is used rather than $f_0(E - eV, T) - f_0(E, T)$. The results of such calculations show that a minimum in σ appears at zero bias and a peak at $\sim$ 1–2 mV appears in $d\sigma/dV$.

Experimentally, the zero-bias structure is observed in a region up to 1 or 2 mV and the maximum change in σ is $\simeq 0.1\%$ (Fig. 2). At $E = 1$ meV in Al, τ_{ep}(1 meV, $T = 0$°K) $\simeq 10^{-10}$ sec. The time τ_0 is equated to $2l/v_F$, where l is the electrode thickness and v_F is the Fermi velocity to obtain $\tau_0 \simeq 2 \times 10^{-13}$ sec for $l = 2000$ Å. D is obtained from $g(E) \simeq 10^{-3}$ to be $\sim 2 \times 10^{-6}$ if the impurity time τ_i is equated to the electron–phonon time τ_{ep} at the peak position. This value of D corresponds to a current density of $\simeq 30$ mA/mm^2 at 1 mV, which, for a 50-Ω junction and a tunneling current of 0.02 mA at 1 mV, leads to an effective tunneling area of $\simeq 10^{-3}$ mm^2.

The above theory shows that the peak height and position of $d\sigma/dV$ are a function of temperature. Measurements of the temperature dependence of the peak heights

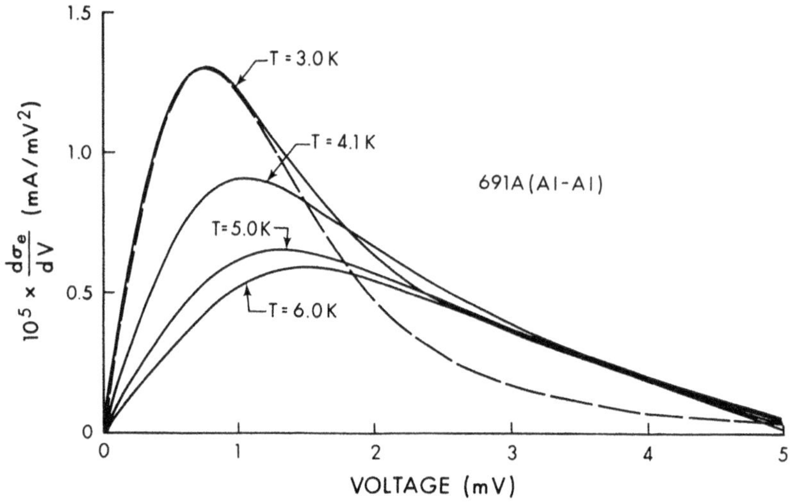

Fig. 2. Even part of $d\sigma/dV$ as a function of V for different temperatures. The solid lines represent the experimental data and the dashed line is the theoretical result for $T = 3°K$.

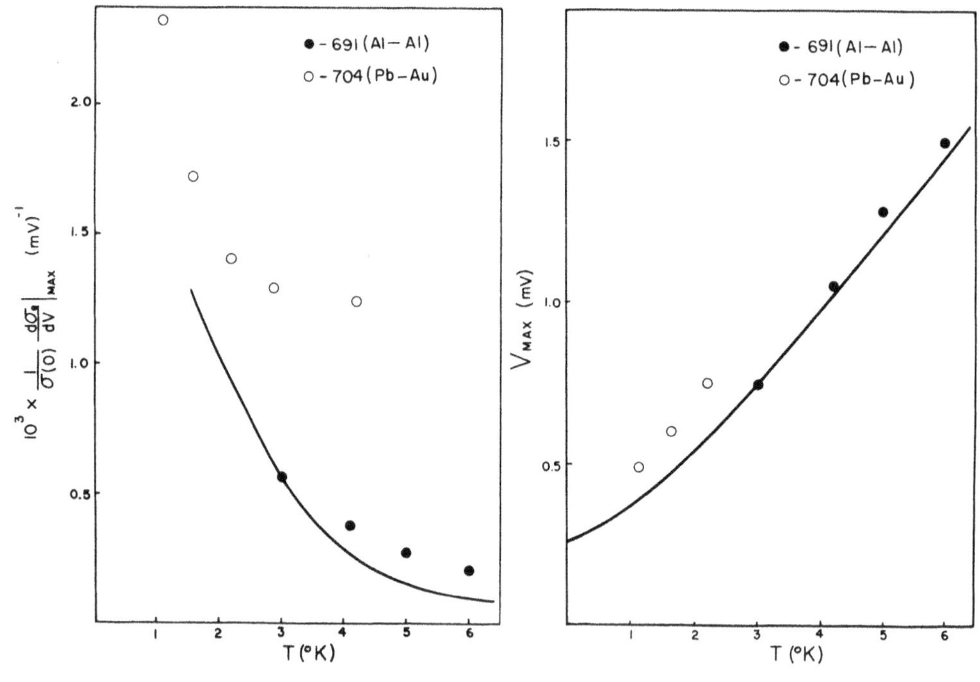

Fig. 3. Plot of maximum height $(d\sigma_e/dV|_{max}$ and position of peak V_{max} as a function of temperature. The solid curves represent theoretical results obtained using the values of D/τ_0 and τ_i given in the text for the 3°K curve for the Al–Al junction (solid circles). Open circles are experimental data for a Pb–Au junction for which time was insufficient to allow a theoretical calculation to be made.

and positions of $d\sigma/dV$ in Al–Al_xO_y–Al junctions have been made (Fig. 3) and the theoretical and experimental results are in good agreement, provided D/τ_0 and τ_i are suitably chosen. Using the 3.0°K data we obtain $D/\tau_0 = 0.72 \times 10^7$ sec^{-1} and $\tau_i = 3.5 \times 10^{-10}$ sec. The obtained values of D/τ_0 and τ_i are reasonable and realistic. Thus the theory and experiments together yield information on D/τ_0 and τ_i, or D/τ_0 and τ_{ep} if an independent measurement of τ_i is made. The results of similar measurements and calculations for superconducting junctions will be presented in the near future.

References

1. C.B. Duke, in *Solid State Physics*, Suppl. 10, *Tunneling in Solids*, F. Seitz, D. Turnbull, and H. Ehrenreich, eds., Academic Press, New York (1969).
2. R.C. Jaklevic and J. Lambe, *Phys. Rev.* **165**, 821 (1968); J.M. Rowell, W.L. McMillan, and W.F. Feldman, *Phys. Rev.* **180**, 658 (1969).
3. T.T. Chen and J.G. Adler, *Sol. St. Commun.* **8**, 1965 (1970).
4. P.N. Trofimenkoff, H.J. Kreuzer, W.J. Wattamaniuk, and J.G. Adler, *Phys. Rev. Lett.* **29**, 597 (1972).

Low-Temperature Light Scattering Studies of Spin Diffusion and Motional Narrowing in Wide-Gap Semiconductors

J. F. Scott*

Department of Physics and Astrophysics
University of Colorado, Boulder, Colorado

P. A. Fleury

Bell Telephone Laboratories
Murray Hill, New Jersey

and

T. C. Damen

Bell Telephone Laboratories
Holmdel, New Jersey

Introduction

Following in the original studies of spontaneous spin-flip scattering in the narrow-bandgap semiconductors InSb, InAs, and PbTe,[1-3] the successful production of a tunable spin-flip laser in InSb operating in the 6- and 11-μm regions of the infrared[4-5] has led to renewed interest in the basic spin dynamics in such semiconductors. The most complete study of linewidths in InSb is that of Brueck,[6] whose analysis shows explicitly the role of orbital collisions which preserve the spin state; Brueck's interpretation is that of motional narrowing.[7] A complementary analysis has been given by Auyang,[8] using Kubo's formalism[9] for collision-modified densities of states.

More recently we have studied both spontaneous and stimulated spin-flip scattering in the wide-bandgap semiconductors CdS and ZnSe.[10]

As a spin-flip Raman laser, CdS has a much smaller tuning range than InSb (about 3 cm^{-1}). However, the coherent emission at the idler wavelength (1.2–2.0 mm) may prove to be technologically useful as a tunable source for the far IR.

Of more direct physical interest are the differences between spin behavior in InSb and wide-gap materials like ZnSe and CdS. Spin–orbit coupling in the latter is several orders of magnitude weaker than in InSb. In addition, CdS and ZnSe are very strongly polar (polaron coupling constant $\alpha = 0.4$–0.7), while InSb is not ($\alpha = 0.02$). These properties should affect spin lifetimes. For example, in InSb the strong spin–orbit coupling makes the spin–lattice relaxation time T_1 very short. Brueck estimates it as 10^{-10} sec and as comparable to the collision time τ.

* Work supported in part by National Science Foundation contract No. GH-34681.

By comparison, the weak spin–orbit coupling in CdS produces a conduction electron g value near 2 ($g = 1.8$) and a very long spin–lattice relaxation time, about five orders of magnitude longer than the collision time, even when the stronger electron–phonon interactions are considered.

In two recent papers we have analyzed the linewidth and line shape of the CdS and ZnSe spin-flip spectra as functions of temperature, field, and momentum transfer.[10] In the present paper we wish to summarize data for additional samples having different mobilities, and to compare our experimental results with the theories of Brueck and Auyang.

Experimental

The data reported here are all for a $\mathbf{q} \cdot \mathbf{H} = 0$ geometry. Scattering angles were varied within the xy plane where z is the field direction $\hat{H}$. Linewidth measurements were made at 5145 Å with an Ar laser emitting 100–400 mW. Magnetic fields between 50 and 100 kG were employed, with no change in linewidth or shape noted as a function of field. The measurements reported in Table I were at temperatures above 70°K; at lower temperatures freeze-out occurred for the lightly doped samples, i.e., the electrons became bound to shallow donors. The linewidths and shapes were independent of sample geometry; i.e., $\hat{H} \parallel \hat{c}$ was the same as $\hat{H} \perp \hat{c}$, where $\hat{c}$ is the [0001] optical axis. All data are for "allowed" α_{zx} or α_{zy} polarizability components ($\hat{z} \parallel \hat{H}$).

The samples were large melt-grown boules, In-doped, from Eagle Picher. The carrier concentration and mobility of each were determined by Hall measurements at ambient temperatures. The carrier concentrations thereby obtained were typically lower than the supplier's estimates by a factor of two to four. (The supplier's nominal values were from percentage In; however, some of the In is apparently compensated or simply does not go into solution.)

For samples with carrier concentrations above 2×10^{18} cm^{-3} the collision lifetimes could also be obtained from plasmon linewidths. The lifetimes obtained in this way agreed within 30% with those from the Hall measurements, indicating that the collision lifetimes do not change appreciably between $\omega = 0$ and $\omega \approx 500$ cm^{-1}, i.e., $\tau(\omega)$ can be treated as independent of frequency.

Table I. Carrier Concentration, Mobility, and Reciprocal Spin-Flip Linewidths for Right-Angle Scattering from CdS and ZnSe in the Diffusion-Dominated Regime

Sample	Concentration n(300 K), cm^{-3}	Mobility μ(300 K), cm^2/V-sec	Collision time τ(300°K), sec	Reciprocal spin-flip width τ_s(100°K) for $\theta = 90°$, sec
CdS	2×10^{19}	55	6×10^{-15}	8×10^{-12}
CdS	4×10^{18}	90	1×10^{-14}	1×10^{-11}
CdS	5.3×10^{17}	280	3×10^{-14}	1.5×10^{-11}
ZnSe	5×10^{17}	100	1×10^{-14}	1.5×10^{-11}

Our most recent experiments on CdS showed that the spin-flip lines were inhomogeneously broadened. This was manifest in the dependence of linewidth upon scattering angle, i.e., momentum transfer. Of the inhomogeneous broadening mechanisms considered by Auyang and Brueck for InSb, we can ignore nonparabolicity of conduction bands, since they are essentially parabolic in CdS and ZnSe. The remaining effects are summarized below.

Brueck's Theory[6]

In Brueck's analysis of spin-flip linewidths in InSb two lifetimes are explicitly included. The first of these is τ_s, a spin lifetime, which should approximate T_2 or T_1, the usual spin–lattice relaxation times. Brueck estimates this as 10^{-10} sec in InSb. This extremely short time is reasonable in narrow-bandgap III–V semiconductors like InSb, because the strong spin–orbit coupling makes spin lifetimes small. T_1 will vary as[11]

$$T_1 \propto [1/(2 - g)]^2 \tag{1}$$

where $2 - g$ varies with spin–orbit coupling as[12]

$$2 - g = \left(\frac{m}{m^*} - 1\right) \frac{2\Delta}{3E_g + \Delta} \tag{2}$$

Here Δ is the spin–orbit splitting of the valence band, E_g is the band gap, and m^* is the effective mass. Thus from (1) and (2) we have

$$T_1 \propto (E_g/\Delta)^2 \tag{3}$$

Note that (3) predicts long T_1 times in wide-gap semiconductors such as CdS.

The second characteristic time introduced in Brueck's theory is τ_p, which describes relaxations in which the electron momentum is changed but the spin direction *and phase* are unaffected. Only processes in which momentum is changed along the field direction $\hat{z}$ are included, but this time τ_p should nevertheless be approximately the same as the Hall mobility lifetime.

Brueck's linewidth formula is (ignoring the k_z dependence of matrix element F)

$$\frac{d^2\sigma}{d\Omega\, d\omega} = |F|^2 \operatorname{Im} \frac{\langle 1/[\omega - \omega_s + h(k_z) - i/\tau]\rangle}{1 + (i/\tau_p)\langle 1/[\omega - \omega_s + h(k_z) - i/\tau]\rangle} \tag{4}$$

where $1/\tau = (1/\tau_p) + (1/\tau_s)$; $\omega_s = \mu_B g H/\hbar$ is the spin splitting; and $h(k_z)$ is, ignoring nonparabolicity, the Doppler contribution

$$h(k_z) = \frac{-\hbar k_z q_z}{m^*} - \frac{\hbar q_z^2}{2m^*} \tag{5}$$

Brueck has shown by numerical calculation that Eq. (4) yields a narrow Lorentzian line in the limit in which τ_p and τ become small. For such cases the apparent linewidth becomes proportional to τ_p, rather than the usual $1/\tau_p$. This linear dependence of Γ upon τ is a characteristic of motional narrowing. Brueck explicitly considers the case of $\tau = 10^{-12}$ sec, $\mathbf{q} \cdot \mathbf{H} = 0$, and nonparabolicity included. The result is $\Gamma \approx 1$ cm^{-1}.

Thus we see that Brueck's formulation is successful in describing free-electron behavior in CdS and ZnSe inasmuch as it predicts a *narrow* Lorentzian for $\tau_p \leq 10^{-12}$ sec, as observed. Brueck's theory is incomplete for CdS in three respects: (1) It does not predict an explicit T dependence, (2) it predicts no q dependence for momentum transfer normal to H, and (3) it predicts $\Gamma \propto \tau_p$, whereas $\Gamma \propto 1/\tau_p$ is observed in Table I. The complementary theory of Auyang helps to remedy these defects, but has some difficulties of its own, as discussed below.

Auyang's Theory[8]

In the diffusion-dominated limit in which nonparabolicity is neglected Auyang's theory predicts a Lorentzian line shape with full width

$$\Gamma_{\parallel} = 2q_z^2 D \tag{6}$$

where q_z is the momentum transfer parallel to the magnetic field. For $\mathbf{q}\cdot\hat{\mathrm{H}} = 0$ the diffusion width is reduced[8] by a factor

$$\Gamma_{\perp} = \Gamma_{\parallel}/[1 + (\omega_c\tau)^2] \tag{7a}$$

i.e.,

$$\Gamma_{\perp} = 2q^2 D[1 + (\omega_c\tau)^2]^{-1} \tag{7b}$$

where τ is the collision time. For our CdS and ZnSe samples $\omega_c\tau < 1$ and $\Gamma_{\perp} \approx 0.8\Gamma_{\parallel}$. This is in extreme contrast with InSb, where $\omega_c\tau \gg 1$.

The diffusion constant D is given in the high-temperature, nondegenerate case by

$$D = (k_B T/m^*)\,\tau \tag{8}$$

assuming a Maxwellian velocity distribution, and by

$$D = (2E_F/3m^*)\,\tau \tag{9}$$

in the degenerate case. Since both Eqs. (8) and (9) predict $\Gamma \propto \tau$ and not $1/\tau$, they are characteristic of motional narrowing theories. Equations (6–8) predict several dependences which we have verified experimentally[10]: (1) Γ is proportional to q_i^2 even for the $\mathbf{q}\cdot\hat{\mathrm{H}} = 0$ geometry, (2) Γ is proportional to T for a wide temperature range, and (3) a Lorentzian line shape is observed.

Most important, Auyang's theory for degenerate semiconductors [Eq. (9)] reconciles an apparent contradiction: All motional narrowing theories predict linewidth Γ proportional to collision time τ, but in Table I our data show Γ roughly proportional to τ^{-1}. This is explained by noting that the low-mobility CdS samples in Table I have higher carrier density. The increased mean velocity of the conduction electrons more than compensates for the reduced mobility and increases the diffusion coefficient. (Note that this holds only for degenerate semiconductors, where $\overline{v^2} = 2E_F/3m^*$ is dependent upon n and not where $\overline{v^2} = KT/m^*$; all three CdS samples in Table I are very nearly degenerate at 100°K.)

As a numerical example of the effect of degenerate statistics, consider the CdS samples in Table I with $n = 5 \times 10^{17}$ and 2×10^{19} cm^{-3}. The $\bar{v}^2$ varies as $n^{2/3}$ and is thus 12 times greater for the high-concentration sample. However, μ (or τ)

is five times *less* for this sample. Hence the observed width should be 2.4 times greater; and, indeed, a value of about 2.0 is observed experimentally for the relative widths. A similar agreement holds for the 4×10^{18} cm^{-3} sample.

Thus we see that motional narrowing theories do apply to CdS. The reason Γ *appears* to vary *inversely* with τ in Table I is that samples of different carrier concentration n are being compared. This reconciles the primary difficulty in applying both Brueck's and Auyang's theories to CdS.

It remains only to explain why the effective collision time τ entering Auyang's theory [Eq. (10)] should be so much shorter than the Hall mobility lifetime or inverse plasmon linewidth. The time τ in Eqs. (8) and (9) fitting our observed width Γ is too short at temperatures $\lesssim 30°K$. The implication is that a narrowing mechanism in addition to the spin diffusion must be considered, e.g., exchange narrowing. Thus a full synthesis of Brueck's motional narrowing theory, which derives from the earlier work of Davies and Blum on electron transport,[13] and Auyang's spin-diffusion expressions, which derive from Kubo's collision-modified densities of states, is not yet available. The CdS data show how complementary these theories are and indicate points, particularly in the definition of effective collision times, where connection between the two theories is not apparent.

References

1. R.E. Slusher, C.K.N. Patel, and P.A. Fleury, *Phys. Rev. Lett.* **18**, 77 (1967).
2. C.K.N. Patel and R.E. Slusher, *Phys. Rev.* **167**, 413 (1968).
3. C.K.N. Patel and R.E. Slusher, *Phys. Rev.* **177**, 1200 (1969).
4. C.K.N. Patel and E.D. Shaw, *Phys. Rev. Lett.* **24**, 451 (1970).
5. A. Mooradian, S.R.J. Brueck, and F.A. Blum, *Appl. Phys. Lett.* **17**, 481 (1970).
6. S.R.J. Brueck, Ph.D. Thesis, MIT 1971.
7. N. Bloembergen, E.M. Purcell, and R.V. Pound, *Phys. Rev.* **73**, 679 (1948).
8. Y.C.S. Auyang, Ph.D. Thesis, MIT, 1971.
9. R. Kubo, S.J. Miyake, and N. Hashitsume, *Sol. St. Phys.* **17**, 270 (1965).
10. P.A. Fleury and J.F. Scott, *Phys. Rev. B* **3**, 1979 (1971); J.F. Scott, P.A. Fleury, and T.C. Damen, *J. Quant. Elect.* **QE8**, 531 (1972); J.F. Scott, P.A. Fleury, and T.C. Damen, *Phys. Rev. B* **6**, 3856 (1972).
11. Y. Yafet, *Solid State Phys.* **14**, 1 (1963).
12. L.M. Roth, B. Lax, and S. Zwerdling, *Phys. Rev.* **114**, 90 (1959).
13. R.W. Davies and F.A. Blum, *Phys. Rev. B* **3**, 3321 (1971).

Fine-Particle Size Effects in Metallic Al and Cu

S. Kobayashi

Department of Physics, Faculty of Science
University of Tokyo, Tokyo, Japan

Introduction

In a fine metallic particle the energy levels of the electrons are quantized, having adjacent level spacings δ of the order of E_F/N, E_F being the Fermi energy and N the number of electrons. At temperatures $kT < \delta$ this discreteness of the levels causes anomalous behavior of the electronic magnetic properties.[1-3] When $kT \ll \delta$ a particle with an even number of electrons (hereafter referred to as an even particle) may have a vanishingly small spin susceptibility χ, because all levels below E_F are occupied by electrons with up and down spins and a finite energy is required to get a difference between up and down populations. On the other hand, χ of an odd particle (with an odd number of electrons) may be Curie-like, since there is one unpaired spin. When the spin–orbit interaction η is large χ of an even particle is modified; when $\eta \gg \delta$, $\chi_{\text{even}} = \chi_{\text{bulk}}$ even when $T \to 0$.

Detailed NMR experiments in Al and Cu fine particles with diameters less than 100 Å have been carried out. It was shown in Refs. 4 and 5 that the observed Knight shifts in the temperature range between 4.2 and 1.03°K are consistent with the above theoretical expectations. In this report we concentrate mainly on the temperature range between 1.0 and 0.5°K.

Experimental

The samples were prepared by a gas evaporation technique[6] and the sizes were determined by electron microscopy. The exposure of the samples to air provides oxide surfaces which ensure electrical insulation between the particles. Some typical samples are listed in Table I, together with the calculated δ.

NMR measurements were performed by using a spin-echo method and a crossed-coil steady-state method at the temperatures of liquid ^{4}He and ^{3}He, between

Table I

Sample	Mean diameter, Å	Half value width of distribution, Å	δ/k, °K
Al # 1	90	50	2.6
Al # 2	65	30	7.0
Cu # 3	65	40	9.1

4.2 and 0.5°K. The line shape is almost the same as that of bulk Al (> 1 μm) at 4.2°K. When the temperature is lowered the peak position shifts in the direction of a smaller Knight shift, and at the same time a very long tail in the direction of a larger Knight shift appears. At the lowest temperature, as shown in Fig. 1, the peak position almost reaches to zero Knight shift and a tail extends over 30 G, which greatly exceeds the bulk Knight shift. The shape in Fig. 1 is strongly H_1 dependent. With a high H_1 (12 mG) a part with a smaller Knight shift is easily saturated and the shape becomes as shown by the dashed line in the figure. The temperature dependence of T_1 of Al (# 1) measured by the spin-echo method is given in Fig. 2. The H_1 of rf pulses is about 200 G, so that most of the spectrum in Fig. 1 contributes to the echo, and consequently T_2^* is shorter (10 μsec) than that of the bulk. Because of a rather poor signal-to-noise ratio, the deviation of the spin–lattice relaxation from a single exponential is not clear and the values of T_1 in Fig. 2 are approximate ones with single exponentials. The T_1 value at the lowest temperature in Fig. 2 agrees with that obtained by a steady-state saturation method at the same temperature in the region where the Knight shifts are larger than in the bulk. The T_1 value of the smaller Knight shift region is one order of magnitude longer than that of the rest. The superconductivity of Al (#2) was checked by a susceptibility measurement using a high-sensitivity SQUID magnetometer at 20 G, and diamagnetism was found below 1.3°K. This indicates that some of the particles are superconductive below 1.3°K.

Figure 3 shows the NMR absorbtion line shape of Cu particles (#3). The character is somewhat different from that of Al. It also has a long tail on the larger Knight shift side, while the peak does not shift. A T_1 measurement shows that T_1 can roughly be expressed by $1/T_1 = aT + b$, with $a = 1/1.27$ sec^{-1} °K^{-1} and $b = 1.5$ sec^{-1}, over the measured temperature range. Because of the second term, it is difficult to conclude that a is really independent of the temperature at low temperatures. The magnetic susceptibility of Cu (# 4) is Curie-like, showing that there exists a considerable amount of paramagnetic oxide on the surfaces of particles.

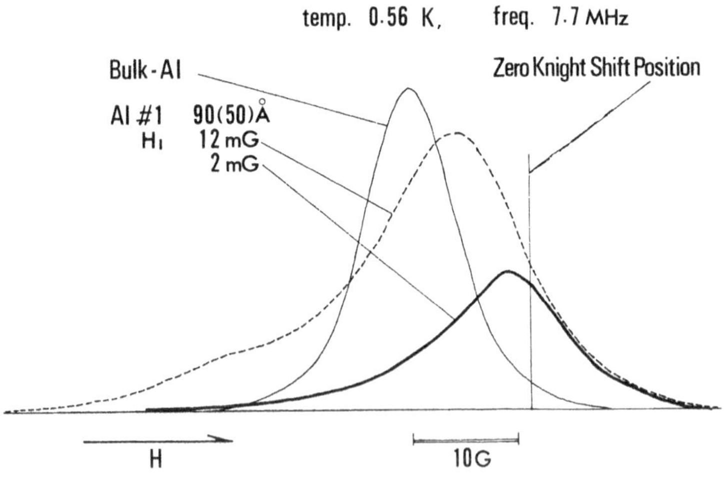

Fig. 1. NMR absorption line shape of Al # 1.

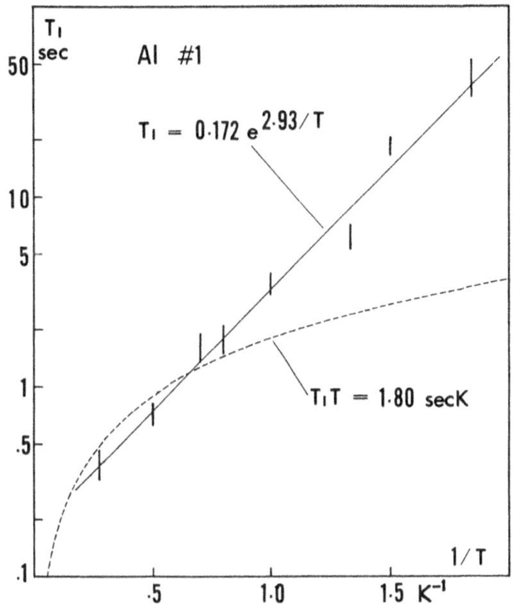

Fig. 2. NMR spin–lattice relaxation time of Al # 1.

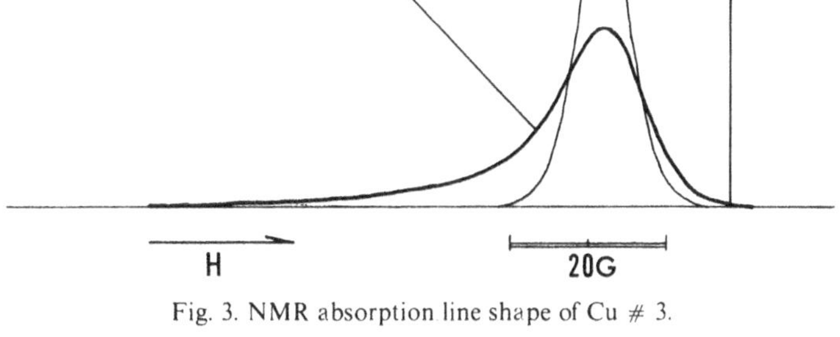

Fig. 3. NMR absorption line shape of Cu # 3.

Discussion

The line shape anomalies observed in Al and Cu particles are, as discussed in Refs. 4 and 5, consistent with what the theories lead us to expect. The tails in the high Knight shift side are to be attributed to odd particles which have a Curie-like susceptibility. The peak shift in Al (at temperatures above 1.3°K) is due to even particles with decreasing susceptibility, and the peak in Cu which does not shift is explained by a nonvanishing susceptibility of even particles because of a large spin–orbit coupling.

Remaining problems in Refs. 4 and 5 are as follows. (i) What happens in Al particles when the temperature is lower than the superconducting transition temperature T_c? (ii) Why are the T_1's of Al and Cu particles not much different from those in the bulk? The present measurements at liquid ^{3}He temperature provide some clues.

For Al the line shape still has a long tail with a large Knight shift at 0.5°K. The major part of the line of # 1 has a T_1 which varies approximately as $T_1 \propto \exp(2.9/T)$, as shown in Fig. 2. The part of the line with a very small Knight shift has a much longer T_1 than the above. (The values of T_1 of Al particles in the liquid ^{4}He range do not differ very much from that of the bulk, which raised the above question in the earlier reports.) Although the exponential variation of T_1 with $1/T$ is akin to that in superconductors, the coefficient in the exponential is too large to be attributed to the superconductivity of Al. Masuda and Redfield[7] reported that in 1000-Å Al, $T_1 \propto \exp(1.3/T)$ in 396 G, and that the coefficient in the exponential decreases with decreasing particle size. If we try to explain our T_1 with superconductivity, it is necessary to take T_c as high as 1.7°K, which disagrees with the susceptibility result and with a recent measurement of T_c in Al particles by Saxena *et al.*[8] It should also be mentioned that the part of the resonance with smaller Knight shift still has a longer T_1 than the above at the lowest temperature.

Based on these considerations, the following may be concluded. Aluminum particles under the condition $\delta > kT$ show even–odd anomalies even below the expected superconducting transition temperature. The spin–lattice relaxation is inhomogeneous in the line and has a character as if there is some finite excitation energy which cannot be wholly attributed to the superconductivity. This excitation energy may have some connection with the discreteness of the energy levels, though the detailed mechanism is not clear.

In the case of Cu the results of the line shape and the spin–lattice relaxation time measurements at temperatures below 1°K are simple extrapolations of those obtained at ^{4}He temperature.[5] The differential peak-to-peak width and the tail in the larger Knight shift side increase with the decreasing temperature. T_1 can still be expressed as $1/T_1 = aT + b$ within the experimental errors. The first and second terms are to be considered to correspond to relaxation via conduction electrons and via spin diffusion, respectively. Because there is a relaxation by-pass b, the expectation that a increases at low temperatures as observed in Al particles is very hard to verify. It may be necessary to choose metals that do not have paramagnetic oxides for the investigation of relaxation in nonsuperconducting metals.

Acknowledgments

The author would like to thank Profs. W. Sasaki and R. Kubo, and Mr. T. Takahashi for valuable discussions, and Messrs. T. Takahashi and K. Mizuguchi for technical assistance.

References

1. R. Kubo, *J. Phys. Soc. (Japan)* **17**, 976 (1962).
2. L.P. Gor'kov and G.M. Eliashberg, *Soviet Phys.—JETP* **21**, 940 (1965).
3. R. Denton, B. Muhlschlegel, and D.J. Scalapino, *Phys. Rev. Lett.* **26**, 707 (1971).
4. S. Kobayashi, T. Takahashi, and W. Sasaki, *J. Phys. Soc. (Japan)* **31**, 1442 (1971).
5. S. Kobayashi, T. Takahashi, and W. Sasaki, *J. Phys. Soc. (Japan)* **32**, 1234 (1972).
6. N. Wada, *Japan J. Appl. Phys.* **6**, 553 (1967).
7. Y. Masuda and A.G. Redfield, *Phys. Rev.* **133**, A944 (1964).
8. A. Saxena, J.E. Crow, and M. Strongin, *Bull. Am. Phys. Soc.* **17**, 333 (1972).

Ultrasonic Shear Wave Attenuation and the Electronic Structure of Cadmium and Zinc*

D. P. Almond, A. R. Birks, M. J. Lea, and E. R. Dobbs

Department of Physics, University of Lancaster
Lancaster, England

Measurements of the electronic attenuation of ultrasonic shear waves have been made down to 0.1°K in a ^3He–^4He dilution refrigerator on high-purity single crystals of Cd and Zn with residual resistivity ratios of 26,000 and 6000 respectively. Sound waves were generated at frequencies ν from 10 to 800 MHz by thin-film CdS transducers grown directly onto the surface of the samples. The total electronic attenuation α in the normal state was measured by applying the necessary magnetic field at temperatures below $0.25T_c$, where the electronic attenuation in the superconducting state is negligible. The acoustic mode studied was the fast shear mode with both the acoustic wave vector $\mathbf{q}$ and its polarization direction $\boldsymbol{\varepsilon}$ in the basal plane. This is the only shear mode in these hexagonal metals which does not interact with their disclocation networks.

The attenuation parameter α/ν is shown as a function of frequency in Fig. 1 for fast shear waves propagating along the $11\bar{2}0$ and $10\bar{1}0$ directions in Cd and Zn. The data were fitted to the equations of Pippard[1,2]

$$\alpha/\nu = \lim_{ql \to \infty} (\alpha/\nu)\, F(ql) \tag{1}$$

where l is the electron mean free path and $F(ql)$ is the free-electron Pippard function for shear waves:

$$F(ql) = \frac{3\pi}{4ql}\left[\frac{2}{3}\frac{(ql)^3}{[1+(ql)^2]\tan^{-1} ql} - 1\right] \tag{2}$$

For pure shear waves in real metals

$$A \equiv \lim_{ql \to \infty} (\alpha/\nu) = (hI^2/4\pi^4 \rho s^2)/(\oint R \cos^2\psi\, d\psi) \tag{3}$$

where I is a Fermi surface integral, ρ is the density, and s is the sound velocity. In the denominator R is the reciprocal of the Gaussian curvature, ψ is the angle between the electron velocity $\mathbf{v}$ and $\boldsymbol{\varepsilon}$, and the integral is taken around the effective zone defined by $\mathbf{q}\cdot\mathbf{v} = 0$. This integral is related to the transverse conductivity σ^q in the anomalous limit by

$$\sigma^q = (e^2/4\pi^2\hbar q) \oint R \cos^2\psi\, d\psi \tag{4}$$

* Work supported by the Science Research Council, London, England.

Most of the data in Fig. 1 are in the $ql \gg 1$ region and the arrows show the high-frequency limits of the attenuation parameter. In Table I it can be seen that agreement with the free-electron value is only good to an order of magnitude and that there is considerable anisotropy in both metals.

Figure 2 shows the free-electron sphere and the second Brillouin zone for Cd (those for Zn are almost identical). The sphere overlaps into the third and fourth zones at the points L and these portions of the Fermi surface form the "butterflies" and "cigars" predicted by the nearly-free-electron (NFE) model. The band structures of these metals have been computed by Stark and Falicov[3] and the resultant Fermi surfaces are very similar to the NFE model except that there is now no overlap into either third or fourth zones at L, and hence the "butterflies" and "cigars" should not exist. Although much experimental data have confirmed the Fermi surfaces derived by Stark and Falicov, there is still some controversy[4] about the existence, or otherwise, of the "butterflies" and "cigars." However, analysis of both the shear wave attenuation data presented here and previous measurements of the attenuation of longitudinal waves [5-7] clearly indicates that these do not exist in Cd or Zn.

Calculations using a simple model Fermi surface indicate the general features that lead to the above conclusion. The two integrals in Eq. (3) have been numerically integrated over a free-electron Fermi sphere with 12 segments missing, which correspond to the "butterflies" and "cigars" as shown in Fig. 2. The integral I is found to be isotropic for propagation directions in the basal plane for this model, and symmetry considerations probably lead to the same conclusion for the Stark–Falicov Fermi surface, though the actual magnitude of I will depend on the detailed Fermi surface structure. Thus any anisotropy in the measured limiting attenuation parameter should result from anisotropy in the conductivity σ^q which involves a purely geometric integral around the effective zone. Since the effective zone for propagation

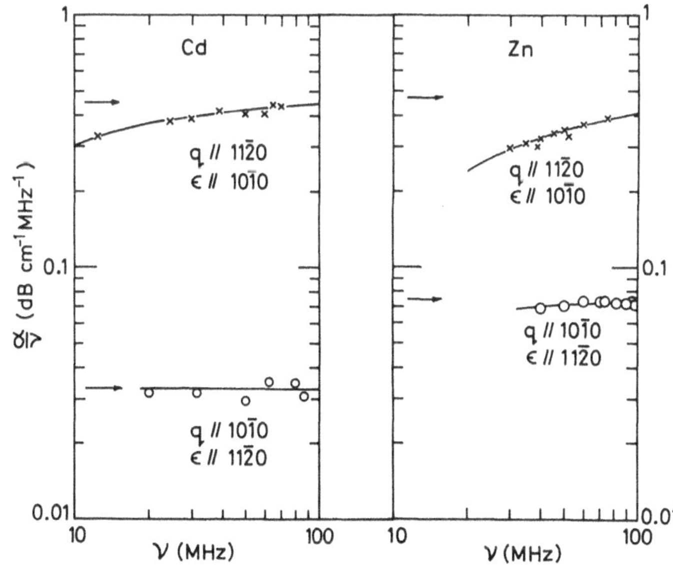

Fig. 1. Pippard plots for fast shear waves in Cd and Zn, fitted to the experimental points. The arrows indicate the values of A.

Table I. Attenuation of Fast Shear Waves in Cd and Zn

Metal	**q**	ε	$A = \lim_{ql \to \infty} (\alpha/v)$, dB cm^{-1} GHz^{-1} Expt.	Free-electron theory	$A_{11\bar{2}0}/A_{10\bar{1}0}$	$\sigma^q_{10\bar{1}0}/\sigma^q_{11\bar{2}0}$*	A,† dB cm^{-1} GHz^{-1}	ν ($q\delta = 1$),* GHz
Cd	$11\bar{2}0$	$1\bar{1}00$	450 ± 20	370	13.6 ± 2	11.5	460	0.35
	$10\bar{1}0$	$1\bar{2}10$	33 ± 5				41	1.21
Zn	$11\bar{2}0$	$1\bar{1}00$	470 ± 20	360	6.4 ± 0.4	11.5	410	0.64
	$10\bar{1}0$	$1\bar{2}10$	74 ± 3				36	2.2

* Estimated from Fermi surface of Stark and Falicov.[3]
† Estimated from model Fermi surface (see text).

along $11\bar{2}0$ lies, as shown in Fig. 2, across the "butterflies," while the $10\bar{1}0$ effective zone does not, it follows from Eq. (4) that $\sigma^q_{11\bar{2}0}$ is much less than $\sigma^q_{10\bar{1}0}$. Since $A \propto 1/\sigma^q$, the shear wave attenuation for propagation along $11\bar{2}0$ should be much larger than for the $10\bar{1}0$ direction, as is observed. The ratios of the attenuations in these two directions are shown in Table I and the similarity with the ratios of the conductivities, as estimated from the Fermi surfaces of Stark and Falicov, must be regarded as strong evidence that the "butterflies" do not exist in Cd or Zn. A similar conclusion can be drawn from the anisotropy of the longitudinal wave attenuation, which also involves a geometric integral around an effective zone.

The absence of the "butterflies" also drastically reduces the value of the isotropic integral I (due to the large decrease in Fermi surface area) and hence the attenuation for the $10\bar{1}0$ direction is much less than the free-electron value. The calculated values of A using the simple model described above, in which 12 segments are removed from the Fermi sphere, are given in Table I. The segments we have removed are slightly larger than suggested by the NFE model, to correspond more

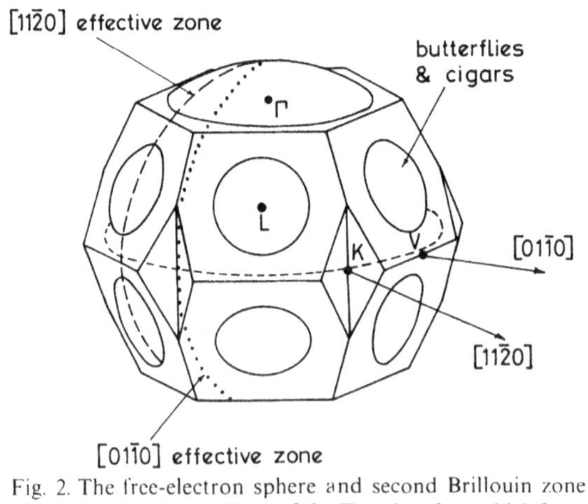

Fig. 2. The free-electron sphere and second Brillouin zone of Cd, showing the portions of the Fermi surface which form the "buterflies" and "cigars." The limiting effective zones for acoustic wave propagation along ⟨$11\bar{2}0$⟩ (- - -) and ⟨$01\bar{1}0$⟩ (···) directions are marked.

closely to the Stark–Falicov Fermi surface. The agreement between the calculated and measured values of A is better for Cd than for Zn. This result is probably related to the presence of the "arms" on the second zone "monster," which occur on the Stark–Falicov $11\bar{2}0$ effective zone in Zn but are absent in Cd. We conclude that both the anisotropy and the magnitude of the shear wave attenuation in Cd and Zn confirm the result of Stark and Falicov that there is no overlap of the Fermi surface into the third zone at L.

The effective skin depth δ is related to σ^q by

$$\delta^2 = 1/4\pi\omega\sigma^q \tag{5}$$

It has been predicted [1,8] that when $q\delta \simeq 1$ the current neutrality condition should break down and consequently the shear wave attenuation in metals should decrease with increasing frequency. When $ql \gg 1$ we can write, from Eq. 3,

$$\alpha/\nu = A/[1 + (q\delta)^4] \tag{6}$$

The skin depth δ is independent of the sample purity and estimates of the frequency at which $q\delta = 1$ are given in Table I for fast shear waves in Cd and Zn. Because of the low value of σ^q for propagation along $11\bar{2}0$ in Cd, as explained above, this frequency is very much less than the free-electron value for this acoustic mode, and so lies in a frequency range which is easily reached experimentally. Preliminary measurements of the attenuation of the $11\bar{2}0$ fast shear wave in Cd are presented in Fig. 3. At the higher frequencies the attenuation parameter α/ν are seen to

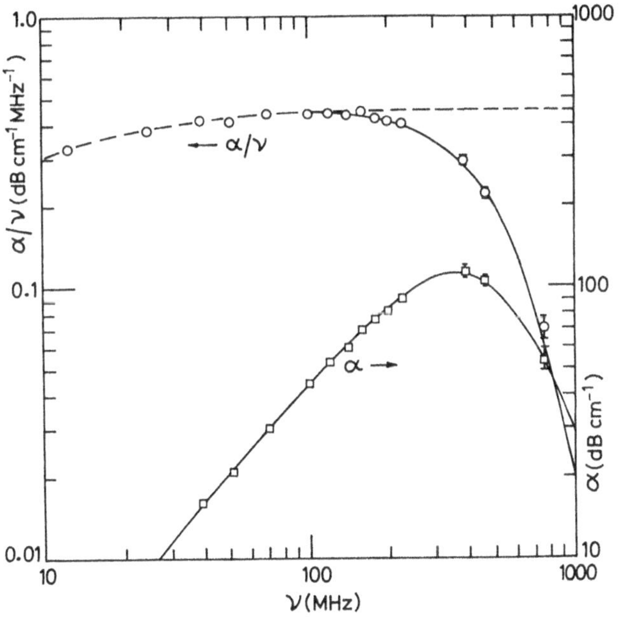

Fig. 3. Upper curves: Pippard plot for fast shear waves along $11\bar{2}0$ in Cd showing the decrease due to the breakdown of the current neutrality condition. The theoretical curves show the expected behavior for $q\delta \ll 1$ (- - -) [Eq. (1)] and for $q\delta = 1$ at 470 MHz (——) [Eq. (6)]. Lower curve: Total electronic attenuation below 0.15 K along $11\bar{2}0$ in Cd for fast shear waves.

be decreasing, as expected, thus showing directly that the neutrality condition for electron–transverse phonon interactions does break down when $q\delta \simeq 1$. The upper solid curve (α/v) in Fig. 3 is of Eq. (6) with $q\delta = 1$ fitted at 470 MHz. Previous ultrasonic measurements on tin[9] and rhenium[10] have demonstrated that electron screening does break down for high-frequency shear waves, but in this paper we have been able to compute the frequency of breakdown for our model Fermi surface to be 350 MHz, and this compares quite well with the experimental value of 470 MHz.

References

1. A.B. Pippard, *Phil. Mag.* **46**, 1104 (1955).
2. A.B. Pippard, *Proc. Roy. Soc. A* **257**, 165 (1960).
3. R.W. Stark and L.M. Falicov, *Phys. Rev. Lett.* **19**, 795 (1967).
4. A.P. Cracknell, *The Fermi Surfaces of Metals*, Taylor and Francis (1971).
5. M.J. Lea and E.R. Dobbs, *Phys. Lett.* **27A**, 556 (1968).
6. D.R. Peck and E.R. Dobbs, *Phys. Lett.* **35A**, 196 (1971).
7. M.J. Lea, J.D. Llewellyn, D.R. Peck, and E.R. Dobbs, *Proc. Roy. Soc. (London)* **A334**, 357 (1973).
8. R. Sandström, preprint of Royal Institute of Technology, Sweden, 1971.
9. E.A. Page, Ph.D. Thesis, Univ. of Maryland, 1969.
10. D.A. Robinson and M. Levy, *Austral. J. Phys.* **24**, 333 (1971).

Electronic Structure of Nickel and Cobalt–Carbon Interstitial Alloys

M. C. Cadeville, F. Gautier, and C. Robert

Laboratoire de Structure Electronique des Solides
Strasbourg, France

In this paper we report transport properties and nuclear magnetic resonance data for *Ni*C and *Co*C alloys and we discuss these results in relation to a model for the electronic structure of these alloys.

The interstitial solubility of carbon in nickel and cobalt is well known.[1-4] We defined previously[4] the solubility range of carbon in cobalt (0–1.1 at.%) and studied the corresponding stabilization of the fcc phase. Moreover, we showed that the saturation magnetization and high-field susceptibilities decrease rapidly with carbon content.[5] These results and the new measurements of transport properties and NMR that we present here allow a simple description for the properties of these alloys, which can be briefly summarized as follows: (1) The interaction effects are small for $c < 1\%$ and the residual resistivities are large ($\rho_0 \sim 5\ \mu\Omega$-cm/at.%); (2) the resistivity associated with carbon in the minority-spin band (↓) is about twice as large as in the majority-spin band (↑); (3) in *Co*C alloys the carbon atoms lie preferentially in octahedral sites of the fcc lattice; (4) for alloys with a virtual bound state, carbon introduces a strong perturbation in the matrix which extends far beyond the impurity and at least up to the third neighbors; (5) in *Ni*C alloys the decreases of magnetization and high-field susceptibility are linear versus concentration: $\partial\mu/\partial c|_{c=0} = -3.6\ \mu_B$/atom and $\partial\chi/\partial c|_{c=0} = -4.2 \times 10^{-6}$emu/g/atom. In *Co*C the results are similar ($\partial\mu/\partial c|_{c=0} \sim -3.5\ \mu_B$/atom).

We first summarize new experimental data for transport phenomena and NMR. Then we discuss these results in relation to the electronic structure of an interstitial impurity in transition metals.

Experimental Results

Details of the specimen preparation are given elsewhere.[4]

Transport Properties. We measured the electrical resistivities and thermoelectric powers of *Ni*C and *Co*C binary alloys and the resistivities of *Ni*CrC and *Co*RuC binary alloys. The measurements we present here were made at low temperatures ($T \leqq 4.2$°K) on rods 2 mm in diameter and about 70 mm in length.

The lattice parameters[4] and the residual resistivities ρ_0 are linear with carbon concent c_C, which shows that the interaction effects are small (Fig. 1); we obtain

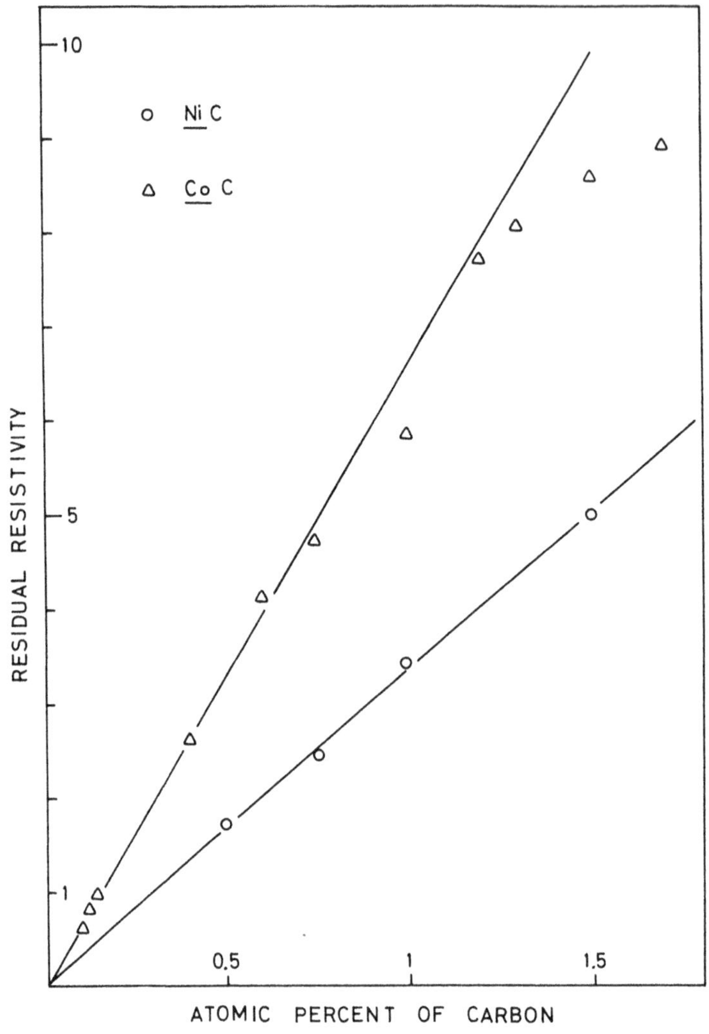

Fig. 1. Electrical residual resistivities of *Ni*C and *Co*C alloys.

$\rho_0 \simeq 3.4\ \mu\Omega$-cm/at.% in nickel and $\rho_0 \simeq 6.6\ \mu\Omega$-cm/at.% in cobalt; this last value for *Co*C is much higher than the published values.[3] In the two-band model for conduction in ferromagnetic alloys we determined the resistivities associated with carbon impurity in each subband σ ($\sigma = \uparrow, \downarrow$) by using the separation method based on the study of ternary alloys. Chromium (or ruthenium) was chosen as auxiliary impurity in nickel (or cobalt) because the resistivities associated with these impurities in each subband were well known from a previous analysis.[7-9] The experimental values of ρ/c_C are a function of only $\lambda = c_{Cr}/c_C$ (or c_{Ru}/c_C) and can be fitted with the theoretical expressions for ρ/c in the two-band model.[8] We obtain for the carbon resistivities in each "σ" band (in $\mu\Omega$-cm/at.%).

In Ni	$\rho_{C\uparrow} = 5.0 \pm 0.5$,	$\rho_{C\downarrow} = 10.6 \pm 1$
In Co	$\rho_{C\uparrow} = 9.3 \pm 1$,	$\rho_{C\downarrow} = 22.4 \pm 1$

For both *Ni*C and *Co*C alloys the carbon residual resistivity is about twice as large in the minority-spin (↓) band as in the majority-spin band (↑) ($\alpha_C = \rho_\downarrow/\rho_\uparrow = 2.1$ in nickel and 2.4 in cobalt). The large value for the resistivity $\rho_{C\downarrow}$ follows from the $s_\downarrow d_\downarrow$ transitions and is of the same order of magnitude as in corresponding substitutional alloys; the value for $\rho_{C\uparrow}$ is much larger than that observed in substitutional alloys without virtual bound states (*Ni*Fe, for example) and has to be explained by the local screening of the carbon impurity on its own cell (as discussed in the next section).

The good agreement between experimental data and the computed values shows that even for two impurities of different types (interstitial and substitutional) the interaction effects between impurities are negligible. Magnetization measurements carried out by Loegel[10] on *Ni*CrC alloys confirm this assumption; the superposition law of moments

$$\bar{\mu} = [1 - (c_1 + c_2)]\,\mu_{\mathrm{Ni}} + c_1(\partial\mu/\partial c_1)_{c_1=0} + c_2(\partial\mu/\partial c_2)_{c_2=0} \tag{1}$$

is obeyed within an accuracy of better than 1%, as seen in Table I. The average magnetizations for the ternary alloys ($\bar{\mu}_{\mathrm{theor}}$) were computed using the following $\partial\mu/\partial c$ values: $-3.6\ \mu_B$/atom for *Ni*C alloys[5] and $-5.2\ \mu_B$/atom for *Ni*Cr alloys.[11,12]

At low temperature ($T < 4.2$°K) the *Ni*C and *Co*C thermoelectric power is linear with temperature and independent of concentration. It is well described by the Mott's formula[13]

$$S_0 = -\frac{\pi^2 k^2 T}{3e}\,\frac{\partial \ln \rho_0}{\partial E}\bigg|_{E=E_F} \tag{2}$$

where ρ_0 is the alloy residual resistivity. Such an expression is not valid at very low concentration when the matrix resistivity ($\rho_{m\sigma}$) is no longer negligible as compared with the impurity resistivity ($\rho_{0\sigma}$), as can be seen in Table II for *Co*C alloys. In both *Ni*C and *Co*C alloys S_0/T is small and positive; the sign of the thermoelectric power impurity is opposite to that of the matrix.

Table I

Alloy composition	$\bar{\mu}$(exp) in μ_B	$\bar{\mu}$(theor) in μ_B	$\Delta\mu = \bar{\mu}_{\mathrm{exp}} - \bar{\mu}_{\mathrm{theor}}$
$Ni_{98.75}Cr_{0.25}C_1$	0.556	0.558	−0.002
$Ni_{98}Cr_1C_1$	0.518	0.515	+0.003
$Ni_{97}Cr_2C_1$	0.462	0.457	+0.005

Table II

*Ni*C alloys		*Co*C alloys	
c, at.%	S_0/T, $\pm 0.005\ \mu$V/deg^2	c, at.%	S_0/T, $\pm 0.005\ \mu$V/deg^2
0.25	+0.017	0.12	0.006
0.75	+0.023	0.4	0.023
1.5	+0.017	1	0.028

NMR Experimental Results in *Co*C Alloys. We used the spin-echo technique for investigating the distribution of hyperfine fields in *Co*C alloys; the details of the experimental apparatus and procedure are given by Durand.[14]

The measurements were made at 4.2°K without external field; we used powders obtained from the rods; these powders were annealed at 1050°C for 1 min and quenched in brine. The fcc lattice parameter determination of a portion of this powder is a good test of the interstitial carbon content.[4]

For each investigated sample we observe (1) the Co main line, which is the same as that of Co in pure cubic Co metal [$\nu_0(4°K) = 217.5$ MHz] and (2) three satellite lines in the low-frequency side at 212.5, 203.5, and about 185 MHz (Fig. 2). The distributions above 220 MHz are believed to come from the hexagonal close-packed

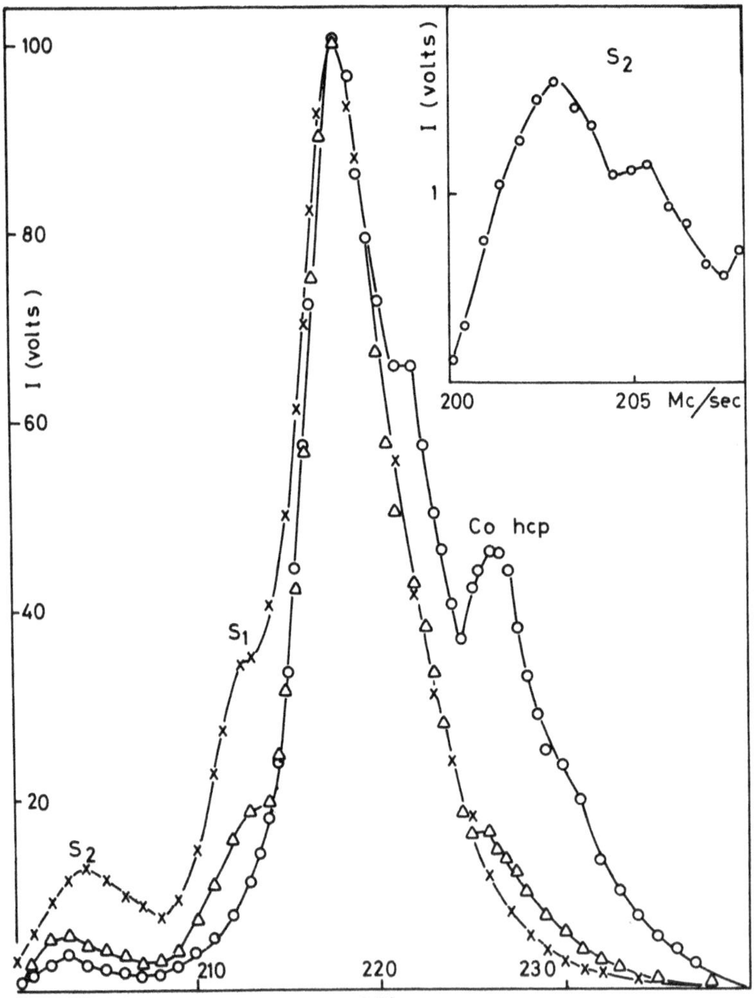

Fig. 2. NMR spectrum of *Co*C alloys. The symbols represent: (○) 0.2 % C, (△) 0.6 % C, (×) 1.1 % C. Inset: details of the second satellite line S_2 for *Co*C with 0.2 % C.

phase, as shown by the rapid decrease in the intensity of these lines with increasing carbon content, in the same way as for the hcp X-ray lines;[4] the hcp lines have quite disappeared at about 1 at.% carbon. The positions of Co fcc satellite lines are independent of carbon concentration.

The integrated relative intensities (I_s/I_0) of both satellite lines S_1 ($\Delta\nu = 5$MHz) and S_2 ($\Delta\nu = 14$ MHz) show a linear behavior versus effective carbon concentration (Fig. 3); the slopes of the straight lines are about $24c$ for I_{S_1}/I_0 and $14c$ for I_{S_2}/I_0. Compared with the numbers of first, second, and third cobalt nearest neighbors of an interstitial atom in a tetrahedral site (respectively 4, 12, 12) or in an octahedral site (respectively 6, 8, 24) the first satellite line S_1 is believed to come from the third Co neighbors of an octahedral carbon atom, and the second satellite line S_2 from both its nearest and next-nearest neighbors (6 + 8). A detailed analysis of this line (S_2) shows that it is in fact the superposition of two near lines at 203 and 204 MHz (Fig. 2), which have about the same intensity and which are difficult to identify. The third satellite line S_3 ($\Delta\nu = 34$ MHz) (not shown in Fig. 2) is wide, with a double structure; it is believed to come from the first and second neighbors of carbon pairs ($\Delta\nu_{S_3} \sim 2\Delta\nu_{S_2}$).

In conclusion, the positions of the satellite lines correspond to a widespread decrease of magnetization around the carbon impurity. The order of magnitude of this decrease is large and of the same order of magnitude as that observed in alloys with virtual bound states (*Co*Os and *Co*Re, for example[14]).

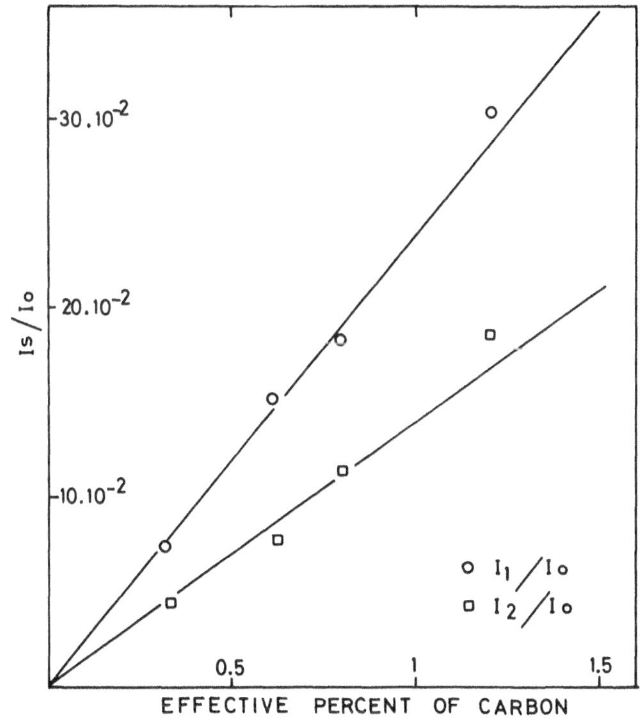

Fig. 3. Dependence of relative intensities I_S/I_0 of satellite lines S_1 and S_2 versus carbon concentration. The straight lines correspond to slopes of $14c$ and $24c$.

Discussion of the Experimental Results—Electronic Structure of Interstitials in Transition Metals

We can easily obtain the screening charge Z_σ introduced by one carbon impurity in each group of spin subbands σ ($\sigma = \uparrow, \downarrow$) from the magnetization measurements:

$$d\mu/dc = \mu_B(Z_\uparrow - Z_\downarrow)$$

and from the Friedel sum rule

$$Z = Z_\uparrow + Z_\downarrow$$

Here μ_B is the Bohr magneton and Z is the impurity excess charge ($Z = 4$ for the carbon impurity). These formulas are valid in the Hartree–Fock approximation when relativistic effects are negligible; they do not assume any particular model for the band structure of the alloy considered. In the present case

$$Z_\uparrow = 0.2, \qquad Z_\downarrow = +3.8 \qquad (NiC)$$
$$Z_\uparrow = 0.3, \qquad Z_\downarrow = +3.7 \qquad (CoC)$$

the screening comes mainly from the minority-spin bands as observed in NiAl alloys. The rigid band model that assumes that the carbon loses nearly all its valence electrons in nickel (or cobalt) and fills the host $d_\downarrow$ holes is not sufficient and a local picture for the screening is necessary for the explanation of NMR and neutron diffuse scattering experiments. This is provided by the extension of the concepts introduced for the study of localized states in normal metals[15] and of the screening in NiAl alloys.[16,17] By the extra orbital model, which describes the wave function for the alloy $|\Psi_{\mathbf{k}}\rangle$ as a linear combination of the host Bloch functions $|\phi_{\mathbf{k}}\rangle$ and of auxiliary orbitals $|\phi_\alpha\rangle$,

$$|\Psi_{\mathbf{k}}\rangle = |\phi_{\mathbf{k}}\rangle + \sum_{\mathbf{k}'} b_{\mathbf{k}\mathbf{k}'}|\phi_{\mathbf{k}'}\rangle + \sum_{\alpha} c_\alpha|\phi_\alpha\rangle$$

The orbitals $|\phi_\alpha\rangle$ are localized around the impurity and represent approximately the crystal wave functions for the energies considered in the interstitial region, whereas the host Bloch functions $|\phi_{\mathbf{k}}\rangle$ describe the wave function in the host cells (the relaxation is neglected here for simplicity). The states $|\phi_\alpha\rangle$ are classified according to the local point symmetry and can be determined in a first approximation as the eigenfunctions (with eigenvalues ε_α) of the perturbed Hamiltonian in the impurity's Wigner–Seitz cell, and zero on its surface (and outside) in order to avoid discontinuities for the description of $|\Psi_{\mathbf{k}}\rangle$. In the present case and for the octahedral symmetry considered it is sufficient to consider the states $|\phi_s\rangle$, $|\phi_p\rangle$ associated with the atomic states 2s, 2p. The scattering equations are easily solved in such a model.[18-20] Another boundary condition can be used, but our qualitative discussion is independent of this choice.

For the study of interstitials in transition metals it is useful to describe briefly two extreme physical situations which can occur according to the values of $\varepsilon_\alpha - \varepsilon_F$ where ε_F is the Fermi energy. Let us summarize briefly the various cases (with one state α, for simplicity).

1. $\varepsilon_\alpha - \varepsilon_F$ is positive and large (say of the same order as the d band width) ($c_\alpha = 0$). The interstitial atom can be only slightly screened in its own cell by the tails of host Bloch functions; large local deviations from neutrality are then observed

and the screening charge occurs mainly from the neighboring shells of $d_\downarrow$ states; the screening in the up bands is small for the small density of states for these bands at the Fermi level—in such a case a large and widespread magnetization *decrease* is observed which can be measured by NMR and diffuse neutron scattering. For example, if such an assumption were valid for *Ni*C alloys, nearly all the *d* holes of the six nearest neighbors of each carbon would be filled ($6 \times 0.55 = 3.3 < Z_\downarrow = 3.8$).

2. When $\varepsilon_\alpha - \varepsilon_F$ is negative and large, localized α states appear below the bottom of the $d\sigma$ bands and can overscreen the impurity excess charge. For example, in the case of *Ni*C, if both $s\sigma$ and $p\sigma$ ($\sigma = \uparrow, \downarrow$) were such localized states, the impurity excess charge Z ($Z = 4$) would be overscreened by these Z_l ($Z_l = 2 + 6$) localized states and the charge $Z - Z_l < 0$ would be screened in the host cells as in the previous case; however, a large and widespread magnetization increase should be observed around the impurity.

3. In intermediate situations $\varepsilon_\alpha - \varepsilon_F$ is small and in such cases the local neutrality in the impurity cell can be satisfied at least partially from the local occupancy of the states $|\phi_\alpha\rangle$ ($c_\alpha \neq 0$). For example, when $\varepsilon_\alpha - \varepsilon_F > 0$ local states can appear for high energies ($\varepsilon_l > \varepsilon_F$ antibonding α-d states) but the total number of up states remains the same in the filled spin-up bands ($Z_\uparrow \simeq 0$, $Z_\downarrow \simeq Z$); however, the local increase of electronic density in the interstitial region is accompanied by a corresponding decrease of the $d\uparrow$ density in the neighboring cells so that we also observe in this case a *negative* widespread magnetic disturbance around each impurity (cf. case 1).

This general discussion is valid for all the interstitials in transition metals (H, B, C, N). In the present case of carbon impurities case (2) is unreasonable based on (i) the observed magnetization decrease around the interstitial in cobalt (NMR) and (ii) the atomic values for ε_s and ε_p. The first case (ionized state) is in agreement with magnetic and NMR data. Moreover, it can explain the large high-field susceptibility decrease ($d\chi/dc \sim -20_{\chi_{Ni}}$). This susceptibility, which arises mainly from the Van Vleck mechanism in the down bands, is quenched at least in the six nearest-neighbor cells of the carbon impurity for which the $d\downarrow$ states are completely occupied. However, from the atomic values of ε_s and ε_p and from the high increased energy required by large deviations from local neutrality, we think it is more reasonable to assume that the electronic structure of the carbon impurity in nickel (or cobalt) is characterized by the last situation 3 with $\varepsilon_p - \varepsilon_F \gtrsim 0$ and $\varepsilon_s - \varepsilon_F < 0$; the local neutrality in the interstitial region can then be (partially) recovered, and the essential features of the model 1 are preserved. Impurity hyperfine field measurements and diffuse neutron scattering will thus allow one to obtain important information concerning the screening and the magnetization in the interstitial region.

References

1. L. Zwell, E.J. Fasiska, Y. Nakada, and A.S. Keh, *Trans. Met. Soc. AIME* **242**, 765 (1968).
2. G. Mah and C. Wert, *Trans. Met. Soc. AIME* **242**, 1211 (1968).
3. J.C. Swartz, *J. Appl. Phys.* **42**, 1334 (1971).
4. M.C. Cadeville and M.F. Lapierre, *Scripta Met.* **6**, 399 (1972).
5. M.C. Cadeville and J. Deportes, *Phys. Lett. A.* **41A**, 237.
6. I.A. Campbell, A. Fert, and A.R. Pomeroy, *Phil. Mag.* **15**, 977 (1967); A. Fert and I.A. Campbell, *Phys. Rev. Lett.* **21**, 1190 (1968).
7. J. Durand and F. Gautier, *J. Phys. Chem. Sol.* **31**, 2773 (1970).

8. P. Leonard, M.C. Cadeville, J. Durand, and F. Gautier, *J. Phys. Chem. Sol.* **30**, 2169 (1969).
9. B. Loegel and F. Gautier, *J. Phys. Chem. Sol.* **32**, 2723 (1971).
10. B. Loegel, Private communication, Strasbourg, 1972.
11. R. Chiffey and T.J. Hicks, *Phys. Lett.* **34A**, 267 (1971).
12. T. Wakiyama and J.P. Rebouillat, *J. Phys. Soc. (Japan)* (to be published).
13. N.F. Mott and H. Jones, *The Theory of the Properties of Metals and Alloys*, Dover, New York (1958).
14. J. Durand, *J. de Physique* **32**, 823 (1971).
15. P.W. Anderson, *Phys. Rev.* **124**, 41 (1961).
16. N.F. Mott, *Adv. Phys.* **13**, 325 (1964).
17. J. Kanamori and K. Terakura, Intermag. Conf., *J. de Phys. Suppl.* **32**, (C1) 282 (1971); K. Terakura and J. Kanamori, *Progr. Theor. Phys.* **46**, 1007 (1971).
18. F. Gautier, *J. Phys. F: Met. Phys.* **1**, 382 (1971).
19. R. Riedinger, *J. Phys. F: Met. Phys.* **1**, 392 (1971).
20. J. Kanamori, K. Terakura, and K. Yamada, *Progr. Theor. Phys.* **41**, 1426 (1969).

10

Transport—Electrical

Induced Torque Anisotropy in Aluminum*

R. J. Douglas, F. W. Holroyd, and W. R. Datars

Department of Physics, McMaster University
Hamilton, Ontario, Canada

The induced torque technique[1] for obtaining information on the galvanomagnetic properties of metals has proved to be a useful adjunct to conventional four-probe experiments. The chief attractions of this technique are its leadless nature and the attendant ease with which it can probe the galvanomagnetic properties of a single sample with the magnetic field in any crystallographic orientation. In these experiments a monocrystalline spherical sample is suspended from a nulling torque transducer in a magnetic field, which is rotating about the torque transducer axis with an angular velocity Ω. This induces eddy currents in the sample which in turn produce a retarding torque which is measured by the torque magnetometer. The torque is an involved, but exactly known, function of the resistivity tensor ρ.[2,3] If we define a rotating coordinate system such that $\mathbf{B} = B\mathbf{z}$ and $\boldsymbol{\Omega} = \Omega\mathbf{y}$, then to first order in Ω the induced torque is

$$N_y = \frac{4\pi R^5 \Omega B^2}{15c^2}\left[(\mathrm{tr}\,\rho)I - \rho^t\right]^{-1}_{xx} \tag{1}$$

where R is the sample radius. If the longitudinal-transverse terms are neglected, the explicit expression in terms of the resistivity components is

$$N_y = \frac{4\pi R^5 \Omega B^2}{15c^2}\,\frac{\rho_{xx} + \rho_{zz}}{(\rho_{xx} + \rho_{zz})(\rho_{yy} + \rho_{zz}) - \rho_{xy}\rho_{yx}} \tag{2}$$

A typical set of torque rotation diagrams for aluminum at 4.2°K is shown in Fig. 1. Here the field was rotated in a (010) plane at 40°/min. The sample was a 4.6-mm diameter sphere (with asphericity less than 1%) spark-cut from a nominal 5-9's purity single crystal. The induced component of the total torque was obtained by digitally recording the total torque in both rotation directions and selecting the component that is odd in Ω. The even-Ω torques are generally small, except at the higher fields, where the de Haas–van Alphen torque oscillations become important. The broad minima in the high-field induced torque evident in Fig. 1 at $\langle 100 \rangle$ directions were also observed as the field was rotated through $\langle 100 \rangle$ directions in the (110) plane. Similar minima were seen at $\langle 111 \rangle$ directions in the (110) and (112) rotation planes. These high-field induced torque minima were also found in spheres that were spark-cut from single crystals that had been strained along a low-symmetry axis by

* This research was supported by the National Research Council of Canada.

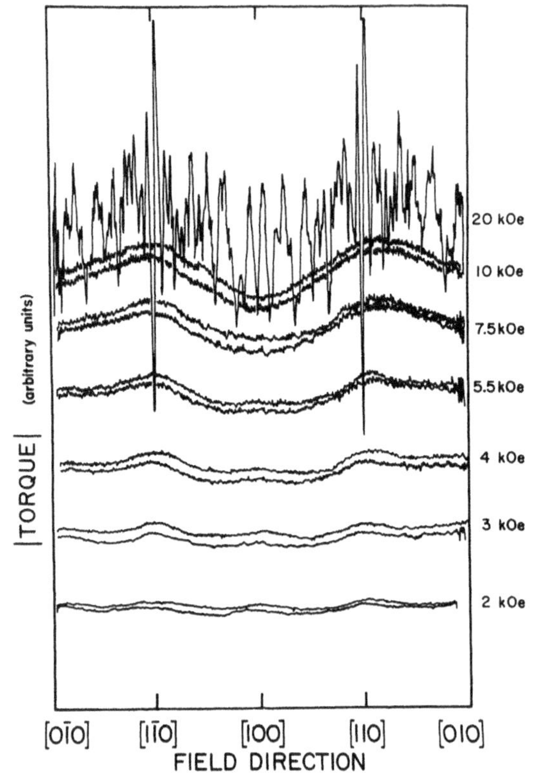

Fig. 1. Absolute value of the measured torque for field rotations in the (010) plane of an unstrained aluminum sphere at 4.2°K. The two traces at each field (except 20 kOe) correspond to the two rotation directions.

1%, 5%, and 8%. These minima persisted as the temperature was raised from 1.2°K to 25°K, where the minima gradually disappeared as the increased phonon scattering removed the sample from the high-field condition. The induced torque tended toward high-field saturation for all field orientations, and showed no evidence of open orbits induced by magnetic breakdown in fields up to 20 kOe. There was no evidence of linear magnetoresistance up to $\omega\tau = 7$.

To interpret these results, the magnetoconductivity tensor of aluminum was calculated using a path integral model (see, e.g., Ref. 4). This model explicitly includes the effects of the variations of the cyclotron frequency for the different orbits and the carrier velocity variations along these orbits. Magnetic breakdown was not included in the calculations. To simplify the path integral evaluation, we used a uniform relaxation time, isotropic mass enhancement of 1.6, and a modified single-OPW Fermi surface consisting of the second band hole surface and toroidal composites of the electron pieces.

Some results of the calculations are of general interest. The computed Hall coefficient reproduces the observed field dependence of the Hall coefficient.[5] It is isotropic to within 1% in both the high- and low-field limits, corresponding to one hole and three electrons per atom, respectively. The computed longitudinal-transverse resistivity saturates at high fields for some field directions at values as high as 14% of the zero-field resistivity. Neglecting the longitudinal-transverse conductivity terms enhances the transverse magnetoresistance by up to 5%, and decreases the calculated induced torque by up to 3%. Throughout this work the longitudinal-transverse terms were explicitly included.

The computed anisotropy of the saturation magnetoresistance is shown in Fig. 2 for field directions in (100), (110), and (112) planes. The dominant anisotropy is that of the longitudinal magnetoresistance. The deep ρ_{zz} minima for the field near the $\langle 100 \rangle$ and $\langle 111 \rangle$ directions arise from the more nearly free-electron behavior of the z component of the Fermi velocity around the cyclotron orbits. Away from symmetry directions more orbits undergo Bragg reflections which reverse the z component of the carrier velocity. These reflections reduce the z component of the effective mean free path in the high-field limit, and thus increase the saturation value of ρ_{zz}. The much smaller computed anisotropies of ρ_{xx} and ρ_{yy} partially counteract the effect of the ρ_{zz} anisotropy on the induced torque; in particular they tend to equalize the induced torque minima at $\langle 100 \rangle$ and $\langle 111 \rangle$.

The computed magnetoresistivity tensor was used with Eq. (1) to obtain theoretical induced torque curves for the (100), (110), and (112) planes. These were least-squares-fitted to our induced torque data using two parameters which scaled $\omega\tau$ and the torque magnitude independently. The fit to the (112) plane is shown in Fig. 3. The agreement is somewhat better for the (100) plane, and somewhat worse in the (110) plane. The two-parameter torque model fits both the field dependence and the

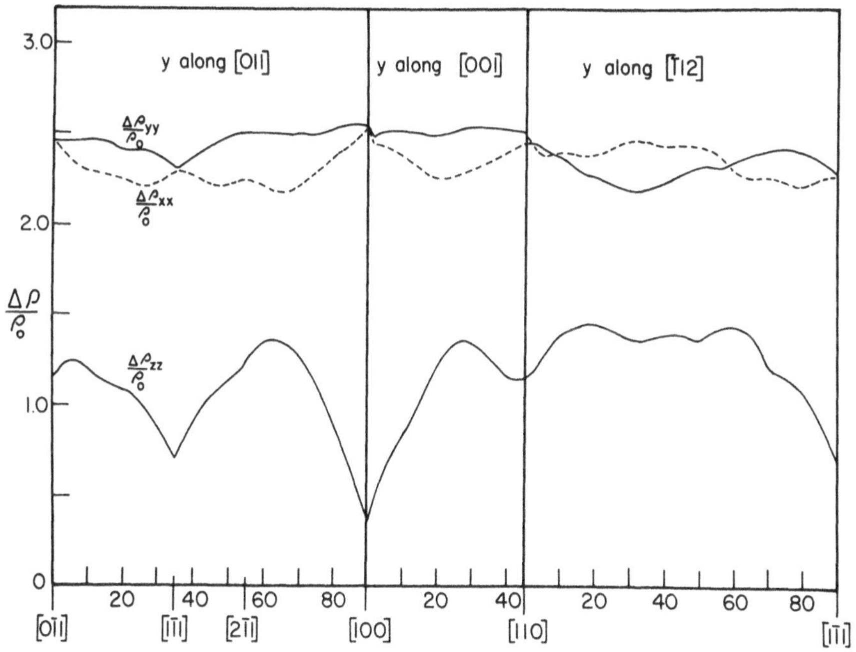

Fig. 2. Theoretical saturation magnetoresistance of aluminum for fields in the (100), (110), and (112) planes. The upper and lower solid curves are respectively the yy transverse and longitudinal (zz) magnetoresistance components. The xx component is indicated by the broken curve.

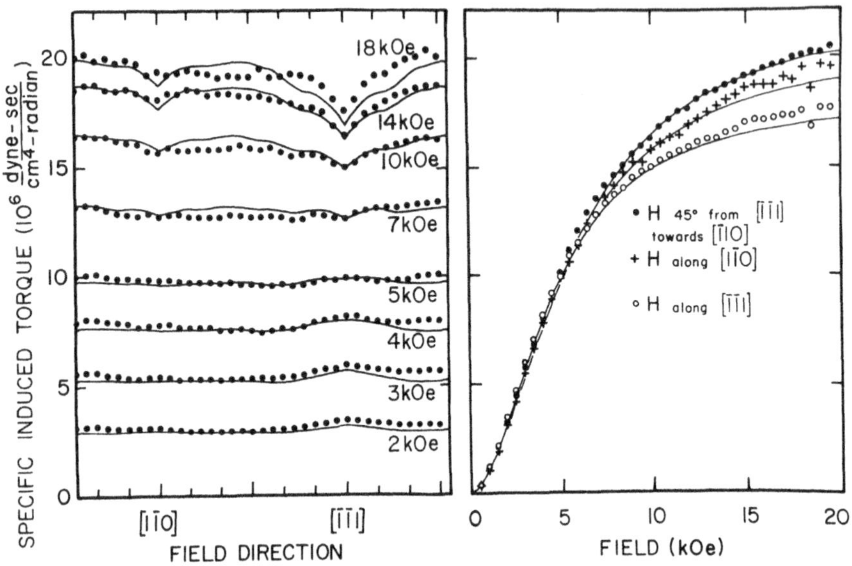

Fig. 3. The points are measured induced torque for fields in the (112) plane of aluminum, the the smooth curves are the two-parameter theoretical fit. Induced torque rotation diagrams are shown at the left for selected fields. At the right are shown the field dependences of the induced torque for three field directions in this plane. The fit gives a relaxation time of 3.07×10^{-11} sec, using an electron mass enhancement factor of 1.6.

anisotropy of the experimental induced torque. At symmetry directions the low-field torque maxima as well as the high-field minima are reproduced. These torque anisotropies are both just reflections of the longitudinal magnetoresistance anisotropy. In the low-field limit, where the Hall terms are negligible in Eq. (2), the lower values of ρ_{zz} near symmetry directions produce torque maxima there; while in the high-field limit, where the denominator of Eq. (2) is dominated by the Hall terms, the lower values of ρ_{zz} yield torque minima near symmetry.

In conclusion, we have shown the complimentary nature of induced torque measurements and path integral conductivity calculations. We have used them in concert to attribute the induced torque anisotropy to the large ρ_{zz} anisotropy which arises from the Bragg reflections of the carriers within the crystal.

Acknowledgment

We are pleased to acknowledge useful communication with L. Falicov and P. Visscher about the path integral method.

References

1. J. Moss and W.R. Datars, *Phys. Lett.* **24A**, 630 (1967).
2. P.B. Visscher and L.M. Falicov, *Phys. Rev. B* **2**, 1518 (1970).
3. J.S. Lass and A.B. Pippard, *J. Phys. E: J. Sci. Instr.* **3**, 136 (1970).
4. J.M. Ziman, *Principles of the Theory of Solids*, Cambridge (1964), Section 9.3.
5. K. Forsvoll and I. Holwech, *Phil. Mag.* **10**, 921 (1964).

Magnetoconductivity Formulation for Arbitrary Fermi Surfaces and General Scattering

Frank E. Richards

The Department of Physics and The James Franck Institute
The University of Chicago, Chicago, Illinois

The goal of magnetoconductivity theory may be defined as the explicit expression of the magnetoconductivity tensor elements σ^{ij} in terms of the quasiparticle parameters (such as velocity $\mathbf{v}_k$) and the basic scattering transition probabilities $Q(k, k')$ which are given for the particular material. The usual approach is to seek a solution to the Boltzmann equation (B.Eq.). With the full collision integral, however, this is such a difficult task that only the simplest cases have been solved to date. Solving the magnetoconductivity problem in the general case is widely regarded as an extemely formidable if not impossible task.[1,2] In this paper we present solutions to this problem, in the general case, derived by using the B.Eq. in two unconventional ways.

We do use the conventional expression for the current,

$$J^i = \int dN_k \, q_k v_k^i g_k$$

where $g_k = (-\partial f_k^\circ/\partial \varepsilon)\, \delta\varepsilon_k$ is the transport part of the distribution function $f_k = f_k^\circ + g_k$, q_k is the carrier charge $\pm e$, and dN_k is written for $dk^3/4\pi^3$, the number of states per cm^3 in the k-space volume element dk^3. The problem is to find g_k or the effective energy shift $\delta\varepsilon_k$, each of which is linear in the electric field E_j while the quasiparticle (qp) parameters and $Q(k, k')$ take their equilibrium values independent of E_j, as in any linear transport theory. At high magnetic fields such that the carrier traverses its orbit several times before relaxing, $\omega\tau \gg 1$, we expand g_k or $\delta\varepsilon_k$ in powers of $1/\omega\tau$,

$$g_k = g_k^{(0)} + g_k^{(1)} + g_k^{(2)}$$

keeping only the lowest order contribution to any tensor element σ^{ij}.

Calculation of σ^{zz}

We find $\delta\varepsilon_k$ by the cancellation of two opposing rates of energy change, in the steady state. The rate of change due to the electric field is $\dot{\varepsilon}_k = q_k E_z \overline{v_k^z}$, where $\overline{v_k^z}$ is the orbit average of the velocity component along the electric field E_z. (For $\omega\tau \gg 1$ and parallel electric and magnetic fields the effect of the magnetic field is to whirl the carrier rapidly through its orbit values of v_k^z, τ_k^{-1}, and $Q(k, k')$, producing an orbit-averaged form of zero-field conduction.) In the absence of scattering this energy change would continue without limit. Scattering causes an opposing rate of energy change, and the two opposing rates of change must be equal in the steady state. Therefore the rate due to scattering has to be proportional to E_j, and to $\delta\varepsilon_k$ as well. And since the desired rate is a scalar, we can always write it as $\delta\varepsilon_k$ times a quantity with the dimensions of inverse time, which we write as $\overline{\tau_k^{-1}}$ and which remains to be determined. That is, $q_k E_z \overline{v_k^z} = \delta\varepsilon_k \overline{\tau_k^{-1}}$, which yields

$$\delta\varepsilon_k = q_k E_z \overline{v_k^z} / \overline{\tau_k^{-1}} \tag{1}$$

$$g_k = (-\partial f_k^\circ/\partial\varepsilon)\, q_k E_z \overline{v_k^z} / \overline{\tau_k^{-1}} \tag{2}$$

$$\sigma^{zz} = e^2 \int dN_k\, (-\partial f_k^\circ/\partial\varepsilon)\, v_k^z \overline{v_k^z} / \overline{\tau_k^{-1}} \tag{3}$$

Thus the existence and uniqueness of the quantity $\overline{\tau_k^{-1}}$ are guaranteed by the postulated existence and uniqueness of g_k itself. This derivation, though simple, is nonetheless perfectly rigorous. In fact, from $\dot{\varepsilon}_k = -\delta\varepsilon_k \overline{\tau_k^{-1}}$ it follows that the effect of scattering upon conduction is to relax it, as if J^z decayed to zero when E_z was turned off—and $\overline{\tau_k^{-1}}$ is the (orbit-averaged) relaxation rate. The results (1)–(3) have also been derived by Chambers,[3] by assuming the fact that we have proved: that scattering relaxes this type of conduction.

Although our g_k is derived by a kinetic calculation independent of the B.Eq., it is a valid g_k and must satisfy the B.Eq.: In effect, the B.Eq. acts as a condition on $\overline{\tau_k^{-1}}$ which determines the explicit expression for $\overline{\tau_k^{-1}}$. This method of determining $\overline{\tau_k^{-1}}$ automatically ensures that the final explicit expression for g_k *is* a solution of the Boltzmann equation.

Expressed in terms of the conventional phase variable ϕ, the B.Eq. is[1]

$$q_k E_z v_k^z \left(-\frac{\partial f_k^\circ}{\partial\varepsilon} \right) - \omega \frac{\partial g_k}{\partial\phi} = \int dN_{k'}\, (g_k - g_{k'})\, Q(k, k') \tag{4}$$

with the familiar collision integral for arbitrary elastic scattering. For g_k and $g_{k'}$ given by (2), and neglecting contributions of higher order in $1/\omega\tau$, the appropriate (orbit-averaged) B.Eq. yields

$$\overline{\tau_k^{-1}} = \frac{\int dN_{k'}\, \overline{Q(k, k')}}{1 + (1/q_k \overline{v_k^z}) \int dN_{k'}\, (q_{k'} \overline{v_{k'}^z} / \overline{\tau_{k'}^{-1}})\, \overline{Q(k, k')}} \tag{5}$$

A similar (but more complicated) result is obtained when the collision integral is expanded to include arbitrary inelastic scattering.[4]

Calculation of σ^{xx}

For the electric field E_x perpendicular to the magnetic field it is well known that usually $\overline{v_k^z} = 0$, so that $g_k^{(0)} = 0$. (The case of zeroth-order conduction, $g_k^{(0)} \neq 0$ due to open orbits, is expressed by analogy with σ^{zz}.[3,4]) Also, it is well known that $g_k^{(1)}$

gives no contribution to σ^{xx}, so the leading contribution to σ^{xx} $(= E_x^{-1}J^x)$ is

$$\sigma^{xx} = E_x^{-1} \int dN_k q_k v_k^x g_k^{(2)}$$

Using the relationship $q_k v_k^x = -e\hbar(\partial k_y/\partial\phi)/m^*$, which can be found by elementary geometry at any point on the carrier's orbit, and integrating by parts with respect to the orbit variable ϕ, one obtains

$$\sigma^{xx} = E_x^{-1} \int dN_k \frac{e\hbar}{m^*}(k_y - \bar{k}_y)\frac{\partial g_k^{(2)}}{\partial\phi} \tag{6}$$

Here the orbit average $\bar{k}_y$ (the orbit center) is the integration constant which is appropriate for each orbit. When $g_k = 0 + g_k^{(1)} + g_k^{(2)}$ is substituted into the B.Eq., the result is that the B.Eq. itself separates according to powers of $1/H$, and the desired quantity $\partial g_k^{(2)}/\partial\phi$ is explicitly expressed by the B.Eq., completing the explicit expression of σ^{xx}. The B.Eq. separates into the two equations,

$$q_k E_x v_k^x\left(-\frac{\partial f_k^\circ}{\partial\varepsilon}\right) - \omega\frac{\partial g_k^{(1)}}{\partial\phi} = 0 \tag{7a}$$

$$-\omega\frac{\partial g_k^{(2)}}{\partial\phi} = \int dN_{k'}(g_k^{(1)} - g_{k'}^{(1)})Q(k,k') \tag{7b}$$

with the collision integral expressed for arbitrary elastic scattering. The explicit expression for $g_k^{(1)}$ has appeared in the literature,* and we have also derived it,[4]

$$g_k^{(1)} = (\partial f_k^\circ/\partial\varepsilon)\, c\hbar E_x H^{-1}(k_y - \bar{k}_y) \tag{8}$$

The expression for σ^{xx} is, then,

$$\sigma^{xx} = \frac{c^2\hbar^2}{H^2}\int dN_k\left(-\frac{\partial f_k^\circ}{\partial\varepsilon}\right)(k_y - \bar{k}_y)\int dN_{k'}\,[(k_y - \bar{k}_y) - (k'_y - \bar{k}'_y)]\,Q(k,k') \tag{9}$$

which is just

$$\sigma^{xx} = (c^2\hbar^2/H^2)\int dN_k(-\partial f_k^\circ/\partial\varepsilon)\,(k_y - \bar{k}_y)^2\, t_k^{-1} \tag{10}$$

where

$$t_k^{-1} = \int dN_k\, \{1 - [(k'_y - \bar{k}'_y)/(k_y - \bar{k}_y)]\}\, Q(k,k') \tag{11}$$

Again, a similar (but more complicated) result is obtained for the quantity t_k^{-1} when the collision integral is expanded to include completely general scattering, while expression (10) is left unchanged.[4] This unconventional application of the B.Eq. was introduced by Adams and Keyes in 1962,[5] but they assumed the relaxation time approximation for the collision integral, so that their solution was only another approximate solution and their work has not been widely recognized.

An understanding of the physical nature of this type of conduction is also of importance. Toward that end, we consider the various terms of the B.Eq., i.e., the

* See Eq. 9.32 of Ref. 1 or Adams and Keyes.[5]

various rates of change of g_k in Eqs. (7a) and (7b). In Eq. (7a) the electric field E_x acts to generate g_k, just as in zeroth-order conduction (σ^{zz}), but the opposing rate of change is due to the whirling caused by the magnetic field (which would quickly smear out $g_k^{(1)}$ if E_x were turned off), rather than to scattering. Equation (7a) is effectively a differential equation for $g_k^{(1)}$, and first-order conduction is therefore independent of scattering. In Eq. (7b) the whirling caused by the magnetic field smears out $g_k^{(2)}$ as well, balanced by the generation of $g_k^{(2)}$ arising in the collision integral. That is, $g_k^{(2)}$ is *generated by* scattering, rather than being relaxed by scattering as $g_k^{(0)}$ is, and there is no relaxation of second-order conduction, no relaxation rate, and no relaxation time approximation. To emphasize the physical nature of this conduction, the quantity t_k^{-1} [Eq. (11)] may be called the generation rate.

This completes the general solution of the magnetoconductivity problem for these two types of conduction. These results should be useful, on the one hand for understanding the experimental data, and on the other hand for extracting information about quasiparticle parameters and especially about scattering.

References

1. J.M. Ziman, *Principles of the Theory of Solids*, Cambridge (1964).
2. R.G. Chambers, in *Solid State Physics, 1. Electrons in Metals*, J.F. Cochran and R.R. Haering, eds., Gordon and Breach, New York (1968); pp. 310–314.
3. R.G. Chambers, in *The Fermi Surface*, W.A. Harrison and M.B. Webb, eds. Wiley, New York (1960), p. 112.
4. F.E. Richards (to be published).
5. E.N. Adams and R.W. Keyes, in *Progress in Semi-Conductors*, A.F. Gibson, ed., Heywood, London (1962), Vol. 6, pp. 85–142.

The Thermal Resistivity and Thermal Magnetoresistivity of Potassium*

R. S. Newrock and B. W. Maxfield

Laboratory of Atomic and Solid State Physics
Cornell University, Ithaca, New York

Recent experimental measurements and calculations have shown that the zero-field electrical resistivity of potassium can be well understood using semiclassical ideas, provided Umklapp scattering processes are considered.[1,2] On the other hand, no suitable, widely acceptable explanation has been found that enables one to understand the observed linear high-field electrical magnetoresistance.† We have undertaken a study of the thermal resistance W and thermal magnetoresistance (TMR) $W(H)$ in an attempt to determine if similar problems exist in the thermal case. Both W and $W(H)$ have been measured in single-crystal and polycrystalline potassium specimens having residual resistance ratios RRR ranging from 1800 to 6500, with measurements of W extending down to a RRR of 195. W was measured between 1.5 and 15°K and $W(H)$ was measured in fields up to 18 kG at many temperatures between 2 and 9°K.

Many experimental problems are encountered when one attempts to measure the thermal resistance of a soft, reactive material such as potassium. A detailed description of the experimental apparatus including mounting procedures, sample preparation, and methods of data acquisition and analysis are given elsewhere.[4] As a general check of our apparatus and measurement technique, we measured the thermal resistance of indium. Using an experimental procedure identical to that used for potassium, we obtain excellent agreement with previous measurements even though the thermal contacts between the sample and thermometers were, by choice, very bad.

The expression generally used to describe the electronic thermal resistivity of a metal is

$$W = (A/T) + BT^2 \tag{1}$$

where A/T is the electron–impurity scattering term and BT^2 represents the electron–phonon scattering contribution. Figure 1(a) shows the temperature dependence of W, plotted as WT vs. T^3, for three potassium samples of purity ranging from RRR = 195 to RRR = 6100. A smooth extrapolation of the data to $T = 0$ gives $A = (WT)_{T=0}$.

* This work was supported partly by the United States Atomic Energy Commission under contract number AT(11-1)-3150, technical report number COO-3150-6, and partly by the Advanced Research Projects Agency through the Materials Science Center at Cornell University, MSC Report #1839.

† For references to experimental and theoretical magnetoresistance work that has been done relating to potassium see Refs. 3.

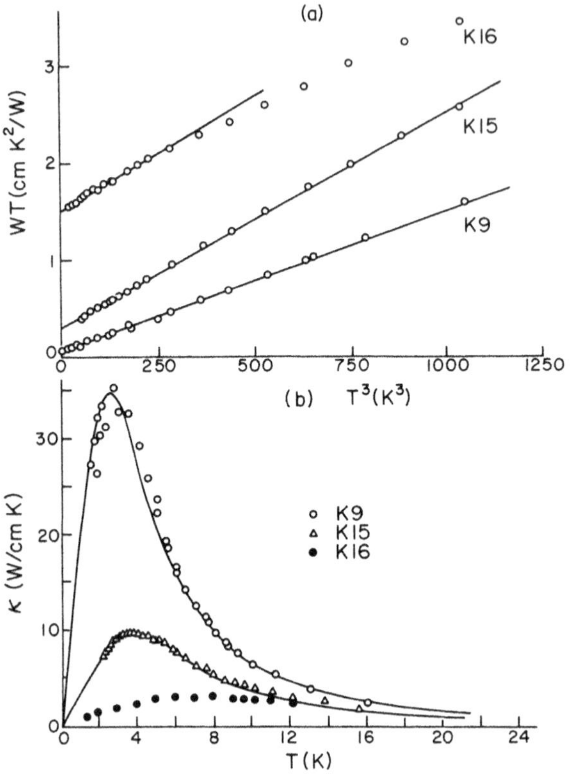

Fig. 1. The temperature dependence of (a) the thermal resistance and (b) the thermal conductivity of three potassium specimens: K-16, polycrystalline, RRR = 195; K-15, single crystal, RRR = 1095; K-9, polycrystalline, RRR = 6130. The curves in (b) are calculated using Eq. (1) ($\kappa = 1/W$), where A and B are determined, respectively, from the intercept and slope of the data in (a).

For comparison with other work on potassium, the thermal conductivity $\kappa = 1/W$ for the same three samples is shown in Fig. 1(b).

The temperature-dependent portion of the thermal resistance times temperature can be written as $W_T T = WT - (WT)_{T=0}$; the last term is temperature independent. We can compare W_T with recent theoretical work.[5] Figure 2 shows W_T/T^2 as a function of temperature for both an impure (RRR = 195) and a pure (RRR = 6100) specimen. The solid and dashed curves represent the theoretical thermal resistance due to electron–phonon scattering as calculated by Ekin[5]; as indicated in the figure legend, each curve corresponds to a different pseudopotential. Effects of electron–phonon Umklapp processes are included specifically and phonon drag effects are neglected. The Kohler variational method using the standard relaxation-time trial function is used to calculate the curves in Fig. 2(a), while a modified trial function containing a second-order energy-dependent term is used to calculate the curves in Fig. 2(b).

Data plotted in this manner emphasize deviations from the simple Bloch theory of thermal conductivity. If only normal processes are important, then the thermal resistivity should increase as T^2 at low temperatures; such behavior corresponds to a straight, horizontal line in Fig. 2.

Good quantitative agreement is obtained between our measurements on the impure specimen shown in Fig. 2(a) and the calculated values, even though a very simple trial function is used to calculate the electron–phonon contribution to the thermal resistivity. This is expected, since it can be shown that if a particular scattering mechanism is dominant and can be described by a relaxation time, then the relaxation-time trial function can be used to obtain the temperature dependence of those scattering mechanisms that cannot be described by a relaxation time.[4] In the impure specimen impurity scattering is dominant and can be described by a relaxation time. One expects calculations using the simple trial function to become increasingly less accurate as electron–phonon scattering becomes more important in relation to impurity scattering. Figure 2(b) shows results on a pure specimen; at low temperatures

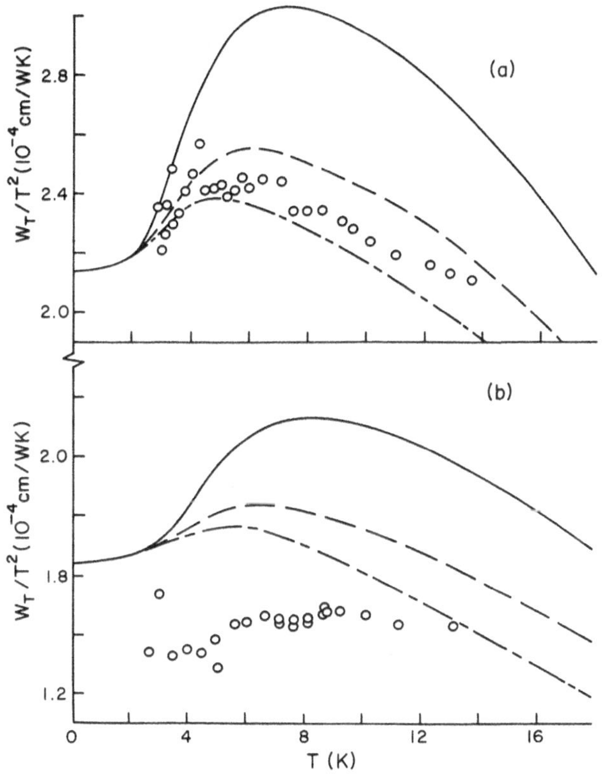

Fig. 2. The temperature dependence of W_T for the three specimens shown in Fig. 1. The curves are theoretical values calculated by Ekin using different pseudopotentials: (——) Bardeen pseudopotential; (—-—) Ashcroft pseudopotential; (— —) Lee-Falicov pseudopotential. The relaxation-time trial function was used for calculations in (a) and the modified trial function for curves in (b).

the experimental values lie 10–20% below the calculated values even though a potentially better trial function was used.

What appears to be excellent agreement between theory and experiment at high temperatures for the impure specimen is not quite that good. Sample K-15 is sufficiently impure so that above 6°K the lattice thermal conduction is appreciable. This makes the thermal resistance lower than that due to electronic processes alone. When lattice conduction is taken into account the measured temperature dependence for the impure sample at high temperatures is in poor agreement with the calculated dependence shown in Fig. 2(a).

The experimental results for pure specimens at high temperatures also show a temperature dependence that differs considerably from the calculated values. Possible reasons for this behavior in both the pure and impure specimens are discussed in detail elsewhere.[4] Briefly, it may be due to the choice of the trial function and/or deviations from Matthiessen's rule (over the range of purity studied here about a 50% deviation from Matthiessen's rule is observed in the low-temperature data).

In the temperature interval 4–7°K, Umklapp scattering contributes as much as 50% to the total thermal resistivity.[5] The observed temperature dependence at low temperatures and the reasonably good agreement as to the magnitude of the thermal resistivity show quite clearly that Umklapp scattering plays an important role in determining the thermal resistivity. Even so, the resistivity is represented quite closely by a T^2 dependence. It is a curious coincidence that just when the thermal resistivity due to normal processes begins to deviate below a T^2 dependence the resistivity due to Umklapp processes increases in nearly the correct proportion to give a total thermal resistivity having a quadratic temperature dependence. This should be contrasted with the electrical resistivity, where no simple power law is sufficient to describe the low-temperature results.

Our results are in reasonable quantitative agreement with those of MacDonald *et al.*,[6] whose measurements were done on specimens encapsulated in glass capillaries. Contrary to the observations of Stauder and Mielczarek,[7] we find no evidence for any anomalous contribution to the thermal resistance of potassium in any specimens that we have measured. Possible reasons are discussed elsewhere[4]; we conclude that the anomalous behavior reported by Stauder and Mielczarek is not a property of potassium.

Preliminary data on the thermal magnetoresistivity of potassium have been obtained in fields up to 18 kG. Figure 3 shows the TMR at a number of temperatures between 2.5 and 6.0°K. $W(H)$ is nearly linear in H at the higher temperatures. The curvature increases rapidly for temperatures below 4°K. As shown in Fig. 3(b), TMR curves for different temperatures cross. That is, at some high field the low-temperature TMR will decrease initially with increasing temperature. Higher-temperature data show a tendency to cross at fields above 20 kG. Such effects are observed in all pure (RRR $>$ 1800) specimens.

It is interesting to compare these results with the electrical magnetoresistance (MR). Under conditions for which the data in Fig. 3(a) were taken the electrical MR would increase nearly linearly with H above a few kilogauss. If the Wiedemann–Franz law were obeyed, one would expect a similar behavior for the TMR. Clearly this is not the case. This is quite surprising since it is at low temperatures where the scattering is most likely to be elastic that one expects the Wiedemann–Franz law to be

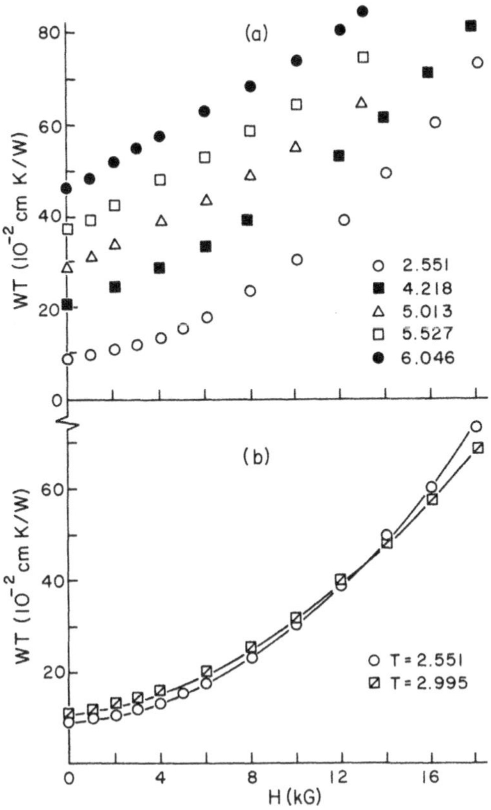

Fig. 3. The thermal magnetoresistance for a single-crystal specimen of potassium having a RRR = 4140; note the crossing of the TMR curves in (b). Specimen size is 1 × 3 × 40 mm.

obeyed. It is difficult to draw any firm conclusions from our magnetic field results until the work is extended to higher fields.

References

1. J.W. Ekin and B.W. Maxfield, *Phys. Rev. B* **4**, 4215 (1971).
2. P.N. Trofimenkoff and J.W. Ekin, *Phys. Rev. B* **4**, 2392 (1971).
3. H. Taub, R.L. Schmidt, B.W. Maxfield, and R. Bowers, *Phys. Rev. B* **4**, 1134 (1971); J.A. Schaefer and J.A. Marcus, *Phys. Rev. Lett.* **27**, 935 (1971); A.W. Overhauser, *Phys. Rev. Lett.* **27**, 938 (1971).
4. R.S. Newrock and B.W. Maxfield, *Phys. Rev. B* **7**, 1283 (1973).
5. J.W. Ekin, *Phys. Rev. B* **6**, 371 (1972).
6. D.K.C. MacDonald, G.K. White, and S.B. Woods, *Proc. Roy. Soc. (London)* **A235**, 358 (1956).
7. R.F. Stauder and E.V. Mielczarek, *Phys. Rev.* **158**, 630 (1967).

Magnetoresistance Anisotropy in Potassium

T. K. Hunt and S. A. Werner

Scientific Research Staff, Ford Motor Company
Dearborn, Michigan

Introduction

The recent experiments of Schaefer and Marcus[1] (SM) showing a large, field-dependent anisotropy in the induced torque on small single-crystal potassium spheres do not seem to permit an explanation which preserves the conventional ideas about the Fermi surface of potassium metal. The anisotropy they observe does not have the symmetry of the bcc lattice and thus indicates the presence of another "preferred" direction in the crystal. These results disagree with the earlier torque measurements of Lass and Pippard[2] on a large single-crystal potassium sphere.

Overhauser[3] has offered an explanation of these anomalies using a charge-density-wave (CDW) model for the electronic ground state of potassium. He assumed the CDW wave vector $\mathbf{Q}$ to be oriented along one of the $\langle 110 \rangle$ directions, forming a "single-Q" state throughout the small spherical crystal, thus creating the preferred direction. In agreement with the SM data, a minimum in the predicted torque pattern develops at low fields for orientations which bring this preferred $\langle 110 \rangle$ direction into the plane containing the applied field $\mathbf{B}_{\mathrm{dc}}$ and the axis of rotation $\boldsymbol{\Omega}$.

An alternative explanation by Lass[4] shows that an ellipsoidal shape anisotropy leads to a transition from twofold to fourfold anisotropy of the induced torque as the magnetic field is increased. For shape anisotropy to account quantitatively for the SM results, however, requires a 10% asphericity and a correlation of this asymmetry with specific crystallographic axes in each of the many SM samples that showed the torque anisotropy. Such a large shape anisotropy would have led to an easily observed twofold anisotropy in the induced torque at 77°K and at room temperature, which was not seen.[5] The model requires a linear magnetoresistance with a coefficient 5–10 times larger than is usually observed in potassium (see, e.g., Ref. 6).

Uncertainty as to the physical origin of the anomalous torque results led us to investigate similar samples ($\sim$ 4-mm-diameter spheres) using another inductive technique[7]* which still preserves the basic measurement geometry of the SM experiments.

* In many respects the technique is similar to the techniques used to observe helicons; see, e.g., the review by Maxfield.[8]

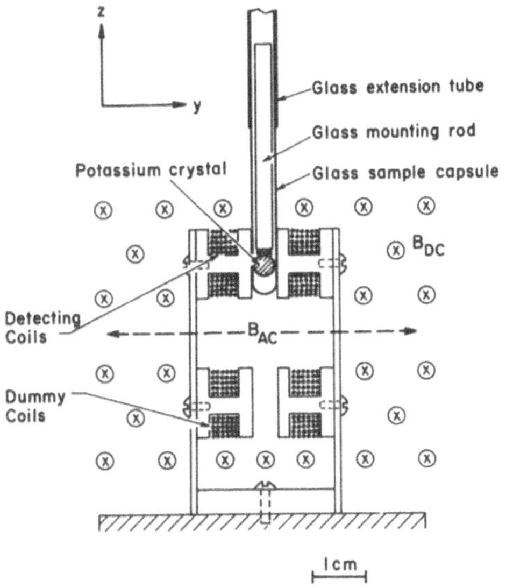

Fig. 1. Schematic diagram of the experimental system showing the relative orientations of the applied ac and dc magnetic fields. The ac field of ~25 G is produced by a pair of rectangular coils. The dc field is produced in the x direction by an 11-in. ADL electromagnet. The angular rotation is performed using a stepping motor.

Experimental Method

The experimental sample and coil system are shown schematically in Fig. 1. Use of a separately excited ac magnetic field rather than the motion of a large dc field to induce the probing current allows measurement in very small dc fields and over a wide range of frequencies. For our potassium spheres at 4.2°K we are able to make measurements at 1.5 Hz, well below the eddy current decay rate, and up to 2400 Hz in magnetic fields from a few gauss up to 9 kG. The field of the induced currents in the sample is sensed by detector coils and analyzed for amplitude and electrical phase using a PAR HR-8 lock-in amplifier.

The detected voltage is proportional to the projection of the induced ac magnetic moment along the detecting coil y axis. Our "in-phase" signal occurring at a 180° phase angle with respect to the exciting field is the main component at low frequencies. The ratio of the "out of phase" or 90° component to this "in-phase" component is the ratio of the frequency ω to the eddy current decay rate. In the low-frequency regime, where the skin depth is larger than the sample size, the y component of the induced moment is (in MKS units) approximately

$$M_y = \frac{2\pi R^5 (\sigma/q)\,\omega B_{ac}}{15\mu_0}\left[1 + \frac{1}{4}\left(\frac{\omega_c \tau}{q}\right)^2\right]^{-1} \qquad (1)$$

Here σ is the zero-field conductivity, assumed isotropic, $1/\tau$ is the mean scattering rate appropriate to σ, R is the radius of the sphere, B_{ac} is the amplitude of the exciting field at angular frequency ω, and μ_0 is the vacuum permeability. The purely empirical linear magnetoresistance seen at high fields,[6] for which there is as yet no widely accepted theoretical explanation,[3,9,10] is reflected in the quantity $q \cong (1 + \alpha\omega_c\tau)$. α is a small number, of order 0.002–0.006, and is quite sample dependent.[6] For $B_{dc} \neq 0$ the Lorentz force causes the induced-current loops to tip out of the plane normal to the exciting field and for $\omega_c\tau \gg 1$ they lie nearly in the plane containing *both* the ac and dc fields. The bracketed term in Eq. (1) reflects the fact that as the current loops tip, both the magnitude of the induced moment and its relative projection along its original axis decrease with the increasing tipping angle $u = \tan^{-1}[\frac{1}{2}(\omega_c\tau/q)]$. This rapid decrease in the in-phase signal is always observed in the present experiments.

Our spheres were prepared by injecting molten droplets of potassium at about 75°C beneath the surface of a room-temperature bath of degassed mineral oil using a simple syringe. The starting material was the standard MSA high-purity potassium which we find commonly yields residual resistivity ratios of 3000–5000. The 3–4-mm-diameter spheres produced by this method have uniformly bright, shiny surfaces as observed microscopically. Upon thorough examination, however, a very large fraction of the specimens are seen to contain deep holes, or large dimples, either of which could constitute a serious shape anisotropy. Our samples were carefully selected to be free of such defects and comprised perhaps 1% of the specimens originally produced. We have also attempted solidification of spheres by slow cooling in a warm oil bath maintained with a temperature gradient. This method produces specimens with much rougher, nonuniform surfaces and with the same general character and frequency of dimples and holes.

The sample is mounted with petroleum jelly or silicone grease on the end of a short glass rod and inserted into a thin-walled glass capsule. The short rod is then inserted into a 1-m-long glass extension tube, permitting the sample to be rotated about the z axis and positioned in the coils from outside the dewar.

For a number of samples the crystallographic orientation was determined with X rays using the transmission Laue precession method. We have never observed a potassium sample of this size and purity to be other than a single crystal (whether containing a hole or not). X-ray analysis of several samples with holes and dimples reveals no particular correlation between the orientation of the defects and the crystallographic axes.

Experimental Results

We have investigated a number of samples having various crystallographic orientations. In every case we observe a small (usually less than 3%) twofold, two-peaked anisotropy at room temperature. For $B_{dc} = 0$ the magnitude of the anisotropy $A \equiv$ (max. signal − min. signal)/(max. signal + min. signal) is independent of temperature from 300 to 1.4°K and of frequency over our available range.

The low-temperature conductivity was obtained from the ratio of the "in-phase" to "out-of-phase" signals at $B_{dc} = 0$. Expressions for both are given by Zimmerman.[7] This method gives the correct conductivity at room temperature and tests on Cu and Al spheres also support its reliability. For increasing B_{dc} we observe a growing and ultimately very large twofold anisotropy while the mean signal decreases as predicted by Eq. (1). The progression of this growth for a particular sample is shown in Figs. 2 and 3. At 9.1 kG, corresponding to $\omega_c\tau \approx 23$ in this sample, the anisotropy A approaches unity. We have no theoretical explanation for this effect. We have seen no evidence of a four-peaked anisotropy in any of our samples even in fields of 9 kG, where it is clearly visible in the SM experiments. At 77°K and above B_{dc} has no effect on A.

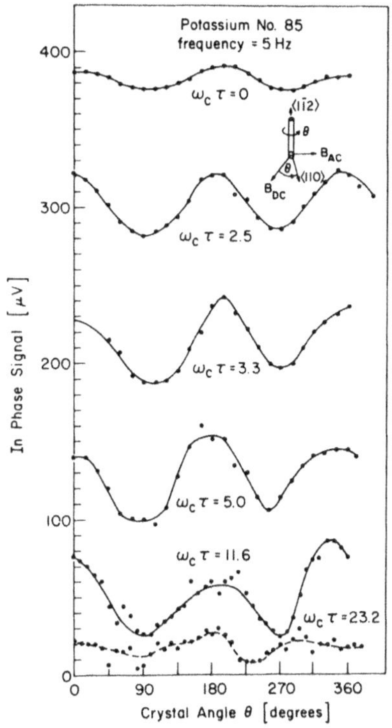

Fig. 2. Voltage generated "in phase" with the exciting field as a function of crystal angle θ for various magnetic fields. The residual resistivity ratio for this sample was 3780.

Equation (1) predicts a distinctive behavior of the mean signal as the sample is cooled from 4.2 to 1.4°K, thereby freezing out the large-angle phonon Umklapp scattering and increasing the conductivity.[11] For small $\omega_c\tau$ we expect the signal to increase at 1.4°K, for the signal is then proportional to σ. At large $\omega_c\tau$ we expect the signal to decrease at 1.4°K since it should then be proportional to $1/\sigma$. We have confirmed this expected behavior, observing an increase in σ of about 10% at 1.4°K. We find that A as a function of B_{dc} is temperature independent down to 1.4°K. We have also run single-crystal samples having large holes and observed no substantial differences in the anisotropy behavior from that of "good" spherical samples.

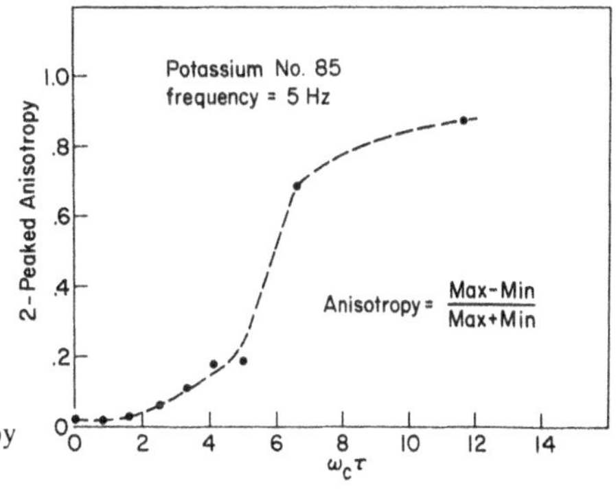

Fig. 3. The rapid increase of the two-peaked anisotropy with magnetic field.

Discussion

The torque experiments of SM are sensitive to the same component of the induced moment as the present measurements since their $\tau = M_y B_{dc}$. Because their induced moment also depends on B_{dc}, we expect that our signal $V(\theta)$ is related to the torque by $\tau(\theta) = \text{const} \times [B_{dc}^2 V(\theta)]$. The two experiments should therefore exhibit the same symmetry and, if scaled by B_{dc}^2, the same dependence on B_{dc}. They do not. It is clear that effects occurring close to the sample surface are important in all such measurements, for the signals are heavily weighted by the current loops at large radii. It appears inescapable, however, that the magnetic field has a large effect on the symmetry of the resistivity tensor and that the effect is probably dependent on the method of sample preparation.

In no sample have we seen any large zero-field anisotropy. In the CDW model this would require that if our samples are single-**Q** domains, the anisotropy parameter γ giving the ratio of the resistivities parallel to and perpendicular to **Q** must be considerably less than the value 4 which Overhauser proposed to explain the SM results.

We find that for some samples the angular location of the anisotropy pattern does not shift with B_{dc}, while for others it does. In one particularly carefully oriented sample (with $\langle 001 \rangle$ vertical) the maximum and minimum signals shifted from $\langle 100 \rangle$ along $\mathbf{B}_{ac}$ to $\langle 110 \rangle$ along $\mathbf{B}_{ac}$ as the field was increased from 0 to 7.6 kG.

We have made a large number of measurements at higher frequencies (10–2400 Hz) and it appears that the general behavior at high frequencies, including the helicon modes, grows continuously out of the low-frequency results. A more complete discussion of these continuing higher-frequency experiments and the anisotropic effects observed in them will be given in a future publication.

Acknowledgments

We are pleased to acknowledge many helpful discussions with Dr. A. W. Overhauser and are indebted to Prof. G. W. Ford for his derivation of the shape anisotropy results, from which the simpler result given in Eq. (1) was obtained.

References

1. J.A. Schaefer and J.A. Marcus, *Phys. Rev. Lett.* **27**, 935 (1971).
2. J.S. Lass and A.B. Pippard, *J. Phys. E: J. Sci. Instr.* **3**, 137 (1970).
3. A.W. Overhauser, *Phys. Rev. Lett.* **27**, 938 (1971).
4. J.S. Lass, *Phys. Lett.* **39A**, 343 (1972).
5. J.A. Marcus, Private communication.
6. H. Taub, R.L. Schmidt, B.W. Maxfield, and R. Bowers, *Phys. Rev. B* **4**, 1134 (1971).
7. J.E. Zimmerman, *Rev. Sci. Instr.* **32**, 402 (1961).
8. B.W. Maxfield, *Am. J. Phys.* **37**, 241 (1969).
9. J.R. Reitz and A.W. Overhauser, *Phys. Rev.* **171**, 749 (1968).
10. L.M. Falicov and Henrik Smith, *Phys. Rev. Lett.* **29**, 124 (1972).
11. J.W. Ekin, *Phys. Rev. Lett.* **26**, 1550 (1971).

Low-Temperature Magnetoresistance of a Cylinder of Crystalline Lead from Inductive Permeability Measurements

R. C. Callarotti and P. Schmidt

Instituto Venzolano de Investigiciones Científicas
Department de Ingeniería, Caracas, Venezuela

Introduction

At 4°K lead is a type 1 superconductor with surface sheath superconductivity. We discuss measurements in cylindrical samples under the effect of a dc magnetic field H and an ac magnetic field h $[h = h_0 \exp(-i\omega t)]$. Both fields are aligned along the axis of the cylinders (Fig. 1 shows the geometry considered). The field $H \gg h$. As H is increased past H_{c3} the samples become normal and their resistance will increase since lead shows anisotropic magnetoresistivity.

Here we will be concerned mainly with the normal lead magnetoresistance as measured by inductive methods. We have measured the real part μ' and the imaginary part μ'' of the relative magnetic permeability μ of our samples as a function of H and ω ($0 < H < 20$ kG, $30 < f < 1000$ Hz, where $f = \omega/2\pi$).

The transverse magnetoresistance of lead has been measured by Alekseevskii and Gaidukov[1-3] in the usual four-terminal method. These results were interpreted according to the theory of Lifshitz and Peschanskii.[4,5] (See also the review article by Fawcett.)[6] According to the direction of the magnetic field, different types of orbits are traversed by the electrons, thus causing different types of magnetoresistive behavior. Three different behaviors can be observed in lead[2] (let α be the angle between a crystal direction and the direction of the measuring current density **J**, and ρ the resistivity):

(a) At constant H, $\rho(\alpha, H) \rightarrow$ independent of α; at constant α, $\rho(\alpha, H) \sim H^2$. This behavior implies compensated closed orbits, and it occurs in lead for a wide range of directions for the magnetic field, centered about the [001] and [111] axes, or their equivalents.

(b) At constant H, $\rho(\alpha, H) \rightarrow$ independent of α; at constant α, $\rho(\alpha, H) \sim$ saturates with increasing H_0. This behavior implies noncompensated closed orbits, or open orbits without a mean direction. In lead this only occurs for H parallel to the [110] axis.

(c) $\rho(\alpha, H) = AH^2 \cos^2\alpha + B$; A and B are constants. This behavior implies open orbits with a mean direction and was observed for all those measured directions where H lay in the (111) plane (in lead).

For our measurements we chose $H \parallel [100]$, corresponding to case (a). Then we can express the magnetoresistance in a power expansion up to second order in the

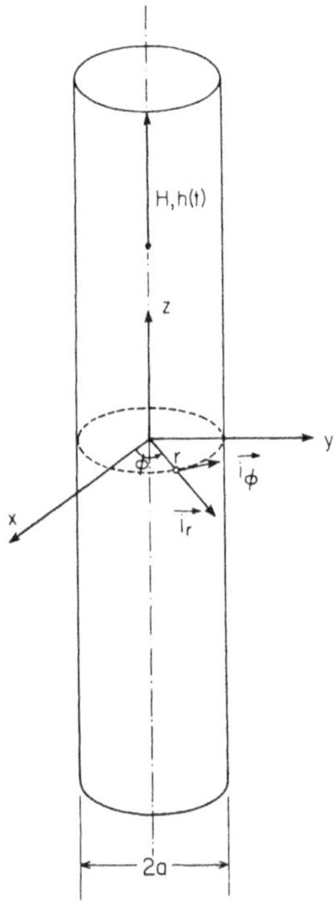

Fig. 1. Geometry considered in the experiments.

magnetic field. In x,y,z coordinates the resistivity tensor ρ_{ij} is[7]

$$\rho_{ij} = \rho_{ij} + \rho_{ijk}H_k + \rho_{ijkl}H_kH_l \tag{1}$$

where ρ_{ij} is the resistivity for zero applied fields, H_i is a component of H, and ρ_{ijk} and ρ_{ijkl} are third- and fourth-order tensors.

Since lead is a face-centered crystal ($m3m$, O^h), we have (for $\mathbf{H} = H\mathbf{i}_z$)

$$|\rho| = \begin{vmatrix} \rho_0 + \rho_{1133}H^2 & +\rho_1 H & 0 \\ -\rho_1 H & \rho_0 + \rho_{2233}H^2 & 0 \\ 0 & 0 & \rho_0 + \rho_{3333}H^2 \end{vmatrix} \tag{2}$$

where

$$\rho_{1133} = \rho_{2233} \equiv b \tag{3}$$

Notice that the usual transverse magnetoresistance coefficient (in the x or y direction) would be

$$[\rho_{11}(H) - \rho_0]/\rho = bH^2 \tag{4}$$

while ρ_1 represents the Hall effect.

Inductive Measurements

From Maxwell's equations, neglecting displacement currents, we can easily show that for the geometry of Fig. 1, the field inside the cylindrical sample of radius a

must satisfy

$$\nabla \times \mathbf{Z} = i\omega\mu_0 h(r, \theta)\,\mathbf{i}_z \tag{5}$$

where, for convenience, we have defined

$$\mathbf{Z} \equiv \rho[\nabla \times h(r, \theta)\,\mathbf{i}_z] \tag{6}$$

and where the other symbols have their usual meaning.

By transforming the tensor ρ_{ij} given by Eq. (1) to a cylindrical coordinate system $(r\phi z)$ we can show that the components of the vector $\mathbf{Z}$ are

$$Z_r = (\rho_0 + bH^2)\left(\frac{1}{r}\frac{\partial h}{\partial \phi}\right) - (\rho_1 H)\left(\frac{\partial h}{\partial r}\right) \tag{7}$$

$$Z_\phi = -(\rho_1 H)\left(\frac{1}{r}\frac{\partial h}{\partial \phi}\right) - (\rho_0 + bH^2)\left(\frac{\partial h}{\partial r}\right) \tag{8}$$

$$Z_z = 0 \tag{9}$$

The substitution of Eqs. (7)–(9) into (5) and the evaluation of the curl of $\mathbf{Z}$ yields the following differential equation:

$$\frac{1}{r}\frac{\partial}{\partial r}\left[r\frac{\partial h}{\partial r}\right] + \frac{1}{r^2}\frac{\partial^2 h}{\partial \phi^2} + k^2 h = 0 \tag{10}$$

where

$$k^2 = \left[\frac{i\omega\mu_0}{\rho_0 + bH^2}\right] \equiv \frac{i\omega\mu_0}{\rho(H)} \tag{11}$$

Equation (10) is easily solved and its solution (applying the boundary condition $h(a, \phi) = h_0$ for all ϕ's) is

$$h = h_0 J_0(kr)/J_0(ka) \tag{12}$$

and μ can then be calculated to be

$$\mu(ka) = (2/ka)\,J_1(ka)/J_0(ka) \tag{13}$$

This, of course, is the same relative permeability as that of an isotropic cylinder[8] with effective resistivity $\rho(H) = \rho_0 + bH^2$.

Samples and Measurements

Our samples were made of Cominco 6-9's lead, whose resistivity ratio was greater than 10^4. They were oriented by X-ray diffraction. The detection system has been described elsewhere.[9]

Experimental curves for μ' vs. H at different frequencies are shown in Fig. 2. By scaling the frequency in accordance with Eq. (11) one can derive all the different curves from one of them. In Fig. 2 we have used the 250-Hz curve as the reference. The agreement obtained indicates that Eq. (13) should be correct.

Equation (13) can be expanded in terms of Kelvin functions[9] so as to obtain

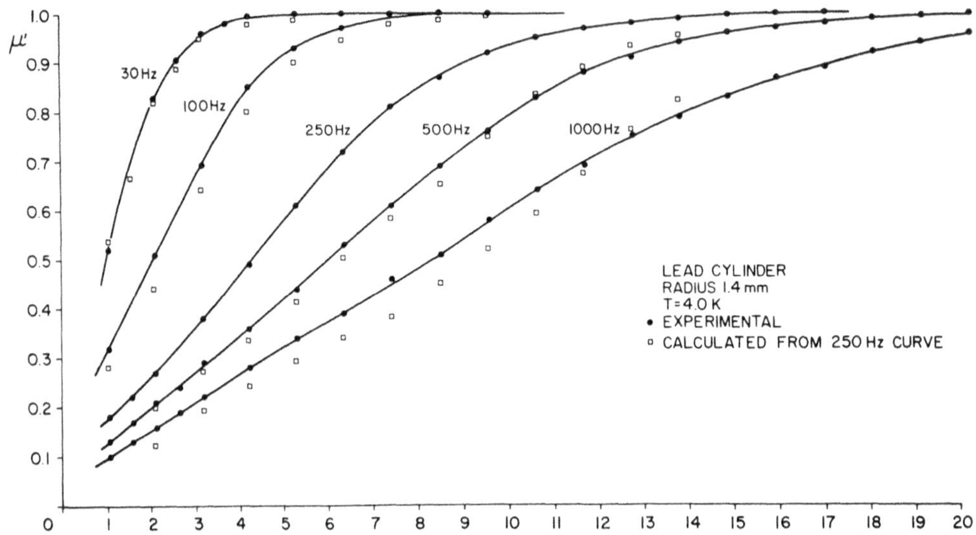

Fig. 2. Experimental μ' vs. H curves ($H \parallel [100]$ axis).

theoretical expressions for $\mu'(|k|a)$ and for $\mu''(|k|a)$. Then by comparison with experimental curves one obtains an effective resistivity for every value of H.

The effective resistivity values thus obtained and then plotted on a reduced Kohler's diagram are shown in Fig. 3. We find

$$\frac{\Delta\rho}{\rho(H=0)} \equiv \frac{\rho(H) - \rho_0}{\rho_0} \propto H^{+1.85} \tag{14}$$

Measurements of the temperature and angular dependence of the ratio H_{c3}/H_{c1} were also obtained from the inductive measurements, but they will be reported elsewhere.

Conclusions

We have examined the possibility of using inductive measurements in cylinders to determine transverse magnetoresistance data. For the case of cubic crystals, where the magnetic field has been oriented so that only closed compensated orbits are traversed, the magnetoresistance measured inductively can be easily interpreted. Of course, in these cases one could just as well measure the effective resistance by transient methods [10-12] instead of the steady-state permeability measurements discussed here.

The advantage of using a superconductor for inductive magnetoresistance measurements resides in the fact that, for $H < H_{c1}$, $\mu' = \mu'' = 0$. Thus the zero reference level of the curves is clearly indicated.

Acknowledgments

We express our thanks to Dr. R. M. Rose for supplying the samples, and to Dr. J. J. Engelhardt for his assistance with the experimental system.

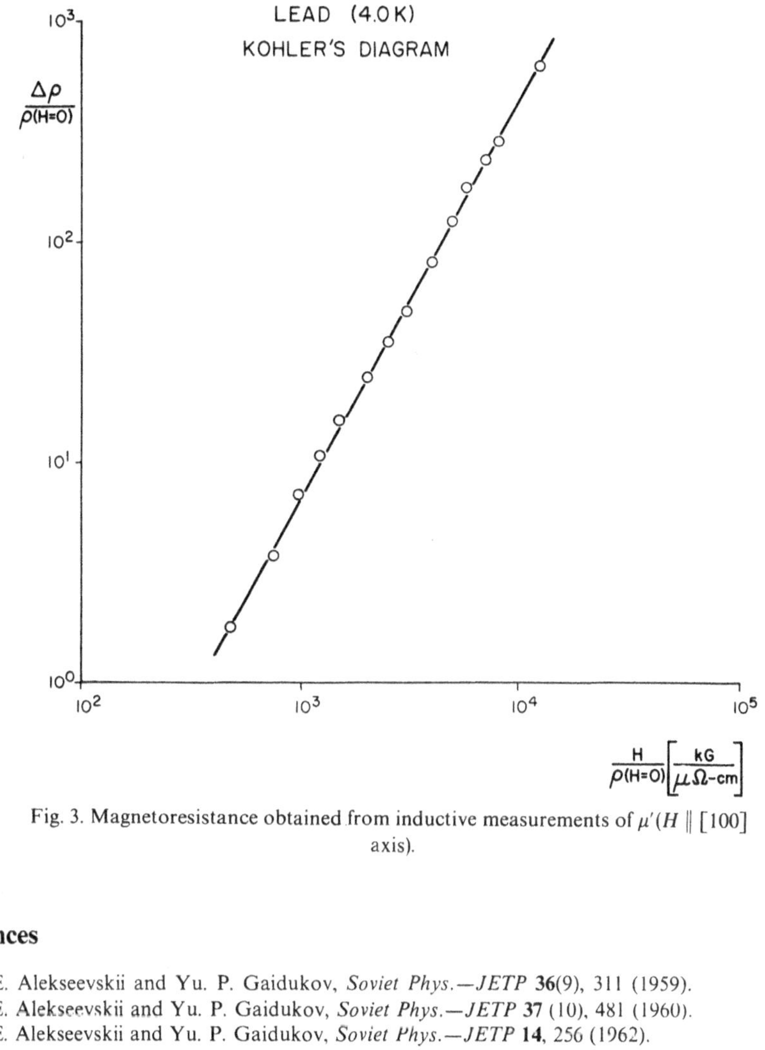

Fig. 3. Magnetoresistance obtained from inductive measurements of $\mu'(H \parallel [100]$ axis).

References

1. N.E. Alekseevskii and Yu. P. Gaidukov, *Soviet Phys.—JETP* **36**(9), 311 (1959).
2. N.E. Alekseevskii and Yu. P. Gaidukov, *Soviet Phys.—JETP* **37** (10), 481 (1960).
3. N.E. Alekseevskii and Yu. P. Gaidukov, *Soviet Phys.—JETP* **14**, 256 (1962).
4. I.M. Lifshitz and V.G. Peschanskii, *Soviet Phys.—JETP* **11**, 137 (1960).
5. I.M. Lifshitz and V.G. Peschanskii, *Soviet Phys.—JETP* **35**(8), 875 (1959).
6. E. Fawcett, *Adv. Phys.* **13**, 139 (1964).
7. A.C. Smith, J.F. Janak, and R.B. Adler, *Electronic Conduction in Solids*, McGraw-Hill, New York (1969).
8. L.D. Landau and I.M. Lifshitz, *Electrodynamics of Continuous Media*, Addison-Wesley, Reading, Massachusetts (1960), p. 194.
9. R. Callarotti, J.J. Engelhardt, and M. Vecchi, *Phys. Rev.* (to be published).
10. C.P. Bean, R.W. De Blois, and L.B. Nesbitt, *J. Appl. Phys.* **30**, 1976 (1959).
11. R.C. Callarotti and M. Alfonzo, *J. Appl. Phys.* **43**, 3040 (1972).
12. R. Callarotti, P. Schmidt, and H. Arqué, *J. Appl. Phys.* **43**, 3952 (1972).

Quantum Limit Studies in Single-Crystal and Pyrolytic Graphite

John A. Woollam

NASA Lewis Research Center
Cleveland, Ohio

David J. Sellmyer*

Materials Science Center
Massachusetts Institute of Technology
Cambridge, Massachusetts

and

Rodney O. Dillon† and Ian L. Spain†

Department of Chemical Engineering
University of Maryland
College Park, Maryland

Introduction

The occupied regions of momentum space in graphite are confined to very elongated volumes near the hexagonal Brillouin zone edges. The elongation is along the hexagonal axis (c axis), so that cross-sectional areas perpendicular to that axis are very small. Hence, when magnetic fields are applied parallel to the c axis only the $n = 0$ Landau levels remain below the Fermi energy for fields above 7 T.[1]

The energy bands in graphite can be described by a set of seven parameters[2] (including the Fermi energy as a parameter). In a quantizing magnetic field the Landau levels are described by a secular equation determined by diagonalizing the Hamiltonian.[3] A major unusual feature for the levels along the Brillouin zone edge is that the $n = 0$ Landau level for electrons is independent of magnetic field, while that for holes is field dependent.

A second notable feature of graphite is that it is compensated (electron and hole densities equal) and the theoretical quantum limit (QL) results of Kubo *et al.*[4] and Adams and Holstein[5] cannot be applied directly.

* Supported by the National Science Foundation, Grant No. GP-21312. Present address: Department of Physics, University of Nebraska, Lincoln, Nebraska.

† Supported by the Atomic Energy Commission, Grant No. AT-(40-1)-3782.

Theory

McClure and Spry[6] have chosen a simple model which closely approximates graphite and assumes that scattering is by randomly placed, screened ionized impurities, where the screening is magnetic field dependent. The results of this theory give

$$\sigma_{xx} = K(\pi/16)(\Delta Z)^2 ec(n_s + p_s) B^{-1} \quad (1)$$

$$\sigma_{xy} = (p_s - n_s) ecB^{-1} \quad (2)$$

for the conductivity σ_{xx} and Hall conductivity σ_{xy}, where two types of scatterers are assumed: ionized acceptors with density p_s and ionized donors of density n_s. In Eqs. (1) and (2) ΔZ is the valence difference from pure graphite, B is the magnetic field and K is a constant (considered equal to unity in Ref. 6).

Refinements of this theory were made by Barnes,[7] who showed that K in Eq. (1) depended on the type of potential assumed, but that σ_{xx} was still proportional to B^{-1}.

Recently Sugihara[8] has calculated σ_{xy} using details of the band structure of graphite for fields above 7 T, and finds deviation from the classical dependence on B^{-1}, as given in Eq. (2). This deviation was observed experimentally over a wide magnetic field range above 7 T, and is explained by Sugihara as due to the presence of neutral scattering centers.

Experimental Results and Discussion

A magnetic field was applied along the c axis of several pressure-annealed pyrolytic graphite samples,[9] as well as in a natural single crystal. The magnetoresistivities ρ_{xx} and Hall resistivities ρ_{yx} were measured as functions of magnetic field and σ_{xx} and σ_{xy}, calculated from

$$\sigma_{xx} = \rho_{xx}/(\rho_{xx}^2 + \rho_{yx}^2) \quad (3)$$

$$\sigma_{xy} = \rho_{yx}/(\rho_{xx}^2 + \rho_{yx}^2) \quad (4)$$

which are valid for the field along the c axis. Temperatures ranged between 4.2 and 1.5°K, and dc methods were used. Leads were attached with either conductive epoxy or copper plating and results were found to be independent of method.

Samples of pyrolytic graphite* were designated by PG followed by a number. PG9I was a neutron-irradiated sample. Only one single crystal, designated SC-1, was studied. In Table I the resistivities at 300 and at 4.2°K are given, together with their ratios. These ratios are indicative of the purity/crystallinity of the samples. Samples PG9 and SC-1 have the highest ratios reported for any pyrolytic or single-crystal, respectively. At 4.2°K the average conductivity mobility for PG9 is 8 m^2/V sec, the highest reported for any pyrolytic material.

In Fig. 1 selected results of ρ_{xx} vs. B are given.† Of interest are the field dependences above 7 T and the magnitude of ρ_{xx} at high field compared with the value in zero field. From results of ρ_{xx} and ρ_{yx} measurements plots of σ_{xx} and σ_{xy} vs. B on log-log scales were made for each sample to determine the field dependences for

* Kindly supplied by Dr. A. Moore and the Union Carbide Corporation.

† Preliminary results of measurements on a single crystal were reported earlier.[10]

Table I. Quantities Obtained from Experiment for Several Pyrolytic Graphite Samples, the Single Crystal of Ref. 6, and a Natural Single Crystal, All at 4.2°K.

Sample	ρ_{xx}(300°K) (B = 0), $\mu\Omega$-m	ρ_{xx}(4.2°K) (B = 0), $\mu\Omega$-m	ρ_{xx}(300)/ ρ_{xx}(4.2) (B = 0)	k in $\sigma_{xx} \propto B^k$ (10–15 T)	$n_s + p_s$,* 10^{24} m^{-3}	$\lvert n_s - p_s \rvert$,† 10^{24} m^{-3}	n_s, 10^{24} m^{-3}	p_s, 10^{24} m^{-3}
PG3	0.5	0.31	1.6	S‡	1.3	0.02	0.64	0.66
PG4	0.51	0.06	8.5	− 1.0§	0.9	0.01	0.46	0.45
PG5	0.47	0.05	9.1	− 1.1§	1.2	0.04	0.62	0.58
PG7	0.53	0.11	5.0	− 1.1§	1.9	0.09	1.0	0.91
PG9	0.52	0.04	12.3	− 1.1§	1.1	0.005	0.54	0.55
PG9I	0.65	0.48	1.3	− 0.5	1.8	#	#	#
Ref. 6	#	0.03	#	− 1.0	0.7	0.01	0.36	0.35
SC-1	0.37	0.008	46	S‡	0.6	0.01	0.32	0.31

* σ_{xx} data, 10–15 T.
† σ_{xy} data at 10 T.
‡ S: Saturates to constant value
§ Deviates above 15 T.
# Not measured.

comparison with Eqs. (1) and (2). Results on all samples are summarized in Table I. It is found that σ_{xx} is approximately proportional to B^{-1} up to about 15 T in most samples, as predicted by McClure *et al.*[6,7]

In Fig. 2 σ_{xx} is plotted vs. B^{-1} and from Eq. (1) the slope should be proportional to $p_s + n_s$, the density of scattering centers, and these are listed in Table I. Since Eq. (1) is not followed over all field ranges, only the slopes below 15 T are used. From our Hall coefficient results and Eq. (2) we have calculated $n_s - p_s$ and these numbers are listed in Table I. The values of n_s and of p_s for the various samples are found to be

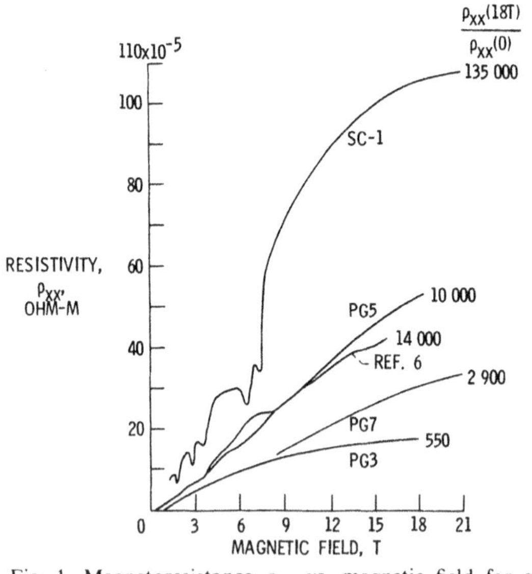

Fig. 1. Magnetoresistance ρ_{xx} vs. magnetic field for a selected set of pyrolytic and single-crystal samples at 4.2°K. Numbers at right of curves indicate ratio of resistance at 18 T to the value at zero field.

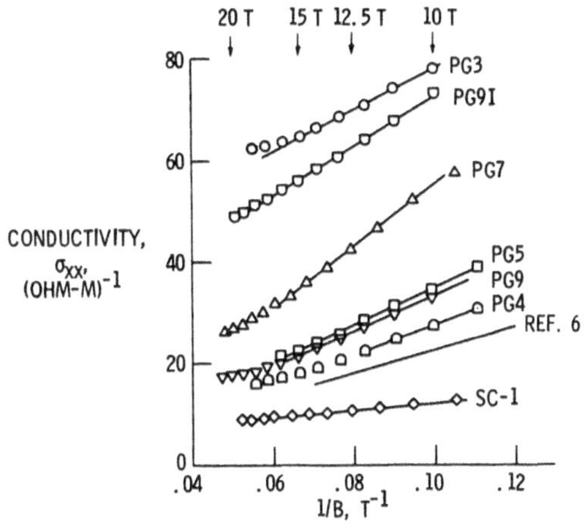

Fig. 2. Conductivities σ_{xx} vs. magnetic field at 4.2°K for several pyrolytic samples and one single crystal. McClure and Spry's data are shown for comparison.

nearly equal, as shown in Table I. Such a result is very surprising, since it is unlikely that several samples would each have equal numbers of donor and acceptor impurities. It is more likely that only one type exists. Since the theoretical coefficient K in Eq. (1) depends on the scattering potential assumed by McClure and Spry[6] and Barnes,[7] it seems reasonable to use our data to adjust the coefficient. We thus assume that one type of scattering center dominates, and that the density is given by the Hall effect results. This assumption gives reasonable values for the relaxation time using the equations developed in Ref. 6. If the value of K in Eq. (1) is increased to 50, the densities determined by Eqs. (1) and (2) will be nearly the same. This value of K is closer to that found by Barnes and McClure than found previously by McClure and Spry. Further differences between $n_s + p_s$ and $|n_s - p_s|$ are then explainable as being due to differing amounts of more than one scattering type.

Another major result is a direct comparison with a recent theory by Sugihara[8] of the deviation[10] of σ_{xy} from the B^{-1} dependence for SC-1. The Sugihara calculation uses the high-field wave functions and energy levels unique to graphite. Our experimental results are explained by the Sugihara theory if a density of $\sim 10^{24}/m^3$ of neutral impurity centers is assumed.

Finally, we would like to mention two experimental results which are unusual and as yet unexplained. The deviation from the B^{-1} dependence of σ_{xx} above 15 T (lower in PG3 and SC-1) has not been theoretically explained. A second unusual result is a decrease in the magnetoresistivity with decreasing temperature. Both of these results will be discussed more completely elsewhere.

References

1. J.A. Woollam, *Phys. Rev. B* **3**(4), 1148 (1971).
2. J.W. McClure, in *The Physics of Semimetals and Narrow Gap Semiconductors*, Carter and Bates, eds., Pergamon Press, New York and Oxford (1971), pp. 127–137.
3. J.W. McClure, *Phys. Rev.* **119**(2), 606 (1960).

4. R. Kubo, S. Miyake, and N. Hashitsume, in *Solid State Physics*, Vol. 17, F. Seitz and D. Turnbull, eds., Academic Press, New York (1965), pp. 269–364.
5. E.N. Adams and T.D. Holstein, *J. Phys. Chem. Solids* **10**, 254 (1959).
6. J.W. McClure and W.J. Spry, *Phys. Rev.* **165**(3), 809 (1968).
7. D. Barnes, Ph. D. Thesis, Univ. of Oregon, 1967.
8. K. Sugihara, to be published.
9. A.W. Moore, A.R. Ubbelohde, and D.A. Young, *Proc. Roy. Soc.* **A280**, 153 (1964).
10. J.A. Woollam, *Phys. Lett.* **32A**(6), 371 (1970).

Acoustic Open-Orbit Resonance Line Shape Under Diffusive Scattering of the Conduction Electrons*

J. Roger Peverley

The Catholic University of America
Washington, D.C.

The Diffusion Equation

When applied fields are present the distribution function $f_{\mathbf{k}}(\mathbf{r})$ for the conduction electrons in a metal differs from the equilibrium value f_0 by a small amount which can be defined by $f = f_0 - \chi\, \partial f_0/\partial \varepsilon$, so that χ is just the extra energy imparted to the electrons by the fields. If the Boltzmann transport equation is formulated in terms of this variable, the collision term takes the form

$$(\partial\chi/\partial t)_{\text{coll}} = \int [\chi(\mathbf{k}') - \chi(\mathbf{k})]\, P(\mathbf{k}, \mathbf{k}')\, d\mathbf{k}' \tag{1}$$

where $P(\mathbf{k}, \mathbf{k}')$ is the rate for transitions between points $\mathbf{k}$ and $\mathbf{k}'$ *on the Fermi surface*. We assume that the scattering angle is small, so that $P(\mathbf{k}, \mathbf{k}')$ is appreciable only for small values of $\mathbf{q} = \mathbf{k}' - \mathbf{k}$. In this case the surface integration (1) is for all practical purposes an integration over a plane tangent to the Fermi surface at $\mathbf{k}$ and we can use Cartesian coordinates to describe the components of $\mathbf{q}$ on this plane. If χ varies slowly on the Fermi surface, we can expand it about the point $\mathbf{k}$ by means of Taylor's theorem. Using tensor notation, we have

$$\chi(\mathbf{k}') = \chi(\mathbf{k}) + q_i\left(\frac{\partial\chi}{\partial q_i}\right)_{\mathbf{k}} + \frac{1}{2}\, q_i q_j \left(\frac{\partial^2\chi}{\partial q_i\, \partial q_j}\right)_{\mathbf{k}} + \cdots \tag{2}$$

and substituting into (1) we obtain the diffusion equation

$$\partial\chi/\partial t = D_{ij}\, \partial^2\chi/\partial q_i\, \partial q_j \tag{3}$$

where the tensor diffusion constant is given by

$$D_{ij} = \tfrac{1}{2} r \langle q_i q_j \rangle \tag{4}$$

with

$$\langle q_i q_j \rangle = (1/r) \int q_i q_j P(\mathbf{k}, \mathbf{q})\, d\mathbf{q} \tag{5}$$

and

$$r = \int P(\mathbf{k}, \mathbf{q})\, d\mathbf{q} \tag{6}$$

* Research supported by the U.S. Office of Naval Research, Contract No. N00014-67A-0377-0007.

The vanishing of the linear term in (2) is ensured by the minimal symmetry requirement $P(\mathbf{k}, \mathbf{q}) = P(\mathbf{k}, -\mathbf{q})$ and this property (valid for small $\mathbf{q}$) also ensures that the 2×2 tensor D_{ij} is symmetric and hence that it can be made diagonal by a suitable rotation of coordinate axes on the tangent plane. The parameter $r(\mathbf{k})$ defined by (6) is just the total collision rate and has a familiar significance. The quantities $\langle q_i q_j \rangle$, on the other hand, are the second moments of the distribution $P(\mathbf{k}, \mathbf{q})$ for the random variable $\mathbf{q}$. The mean square step size $\langle q^2 \rangle = \langle q_x^2 \rangle + \langle q_y^2 \rangle$ is, by symmetry, an invariant equal to the trace of $\langle q_i q_j \rangle$ and its square root, denoted q_{rms}, is clearly another important parameter of the system.

Generalized Relaxation Time

If an electron known to be at the point $\mathbf{k}$ at time $t = 0$ is found at the point $\mathbf{k}''$ after N steps, then the distribution of the quantities $\mathbf{Q} = \mathbf{k}'' - \mathbf{k}$ (in the planar approximation) has the property

$$\langle Q_i Q_j \rangle = N \langle q_i q_j \rangle = r \langle q_i q_j \rangle t \tag{7}$$

so that

$$\langle Q^2 \rangle = \operatorname{Tr} \langle Q_i Q_j \rangle = r \langle q^2 \rangle t. \tag{8}$$

If we assign a distance k_0 as the criterion for randomization of the electron (which will vary from one transport problem to another), then the randomization rate is just

$$1/\tau_0 \sim r \langle q^2 \rangle / k_0^2 \tag{9}$$

where τ_0 is a kind of relaxation time for the problem. (The approximation sign takes care of effects due to Fermi surface curvature, finite q_{rms}, etc.) The lower limit for (9) occurs for the case of dc electrical resistivity, where k_0 is of the same order as the Fermi radius k_F; Pippard[1] has, in fact, shown by exact calculation that the effective relaxation time for a free-electron gas under diffusive scattering is given by $\tau_0 = k_F^2/2D$, where D is the (isotropic) diffusion constant. The upper limit for (9) occurs when $k_0 \sim q_{\text{rms}}$, because then randomization can be achieved in a single collision. This situation can arise when χ is large only in a very limited region on the Fermi surface, so that the expansion (2) can no longer be stopped at the second term and the diffusion equation (3) becomes invalid. In such a case it is better to go back to (1) and rewrite it, neglecting the first term in the integrand by comparison to the second, so that

$$\partial \chi / \partial t = -\chi / \tau \tag{10}$$

where the relaxation time τ is just the reciprocal of the total collision rate r defined in (6). This definition of τ is appropriate for all $k_0 \lesssim q_{\text{rms}}$ and joins smoothly with the definition (9) for τ_0 when $k_0 = q_{\text{rms}}$.

Selectivity

Transport phenomena can be graded according to the magnitude of the ratio k_0/q_{rms}. Highly selective phenomena are such that this ratio is less than unity, so that (10) is obeyed. Such phenomena can be used to determine the parameter $r(\mathbf{k})$, which for electron–phonon scattering varies as T^3. Probably the best example is the magnetic surface-state resonance, which has been used by Koch and Doezema[2] to determine $r(\mathbf{k})$ at points on the copper Fermi surface. Less selective phenomena have $k_0/q_{\rm rms}$ greater than unity, so that several steps are required for randomization. Such methods (in the limit $q_{\rm rms} \to 0$) obey (3) and hence can, in principle, be used to determine the product $r\langle q^2 \rangle$, which for electron–phonon scattering varies as T^5 (as in the Bloch–Grüneisen law for electrical resistivity). Thus, to determine the parameter $\langle q^2 \rangle$ experimentally, *both* selective and unselective methods are required.

Acoustic and magnetoacoustic phenomena, of which there is a rich variety, tend to be intermediate in selectivity; for this reason we consider it a worthwhile exercise to reexamine some of these phenomena subject to the assumption of diffusive scattering.

The Acoustic Open-Orbit Resonance

Many metals are known to exhibit open electron orbits for certain orientations of an applied magnetic field. The acoustic resonance occurs when the period of the open orbit in real space matches the lattice disturbance due to an impressed sound wave. The resonance condition is $n\omega_c = \mathbf{q} \cdot \bar{\mathbf{v}} \pm \omega$, where the integer n is the order of the resonance, ω_c is the cyclotron frequency, $\mathbf{q}$ is the sound wave vector, $\bar{\mathbf{v}}$ is the mean velocity of the open orbit electrons, and ω is the sound frequency (typically much smaller than $\mathbf{q} \cdot \bar{\mathbf{v}}$). In the conventional theory[3,4] there is a Lorentzian resonance in the attenuation coefficient α given by

$$\alpha \sim |A_n|^2 \tau/[1 + \Omega^2\tau^2] \tag{11}$$

where $\Omega = n\omega_c - \mathbf{q} \cdot \bar{\mathbf{v}} \pm \omega$ and τ is the conventional relaxation time. The quantity A_n in the numerator is an intrinsic amplitude factor which need not concern us here.

The width Δk of the band of electrons undergoing open orbit is typically somewhat smaller than the overall Fermi surface dimensions; this fact accounts for the intermediate selectivity of the method. The assumptions underlying (11) are only expected to be valid in the limit $q_{\rm rms} \gg \Delta k$. The opposite limit, $q_{\rm rms} \ll \Delta k$, is equivalent to the assumption of a diffusion model, which we now examine.

Boundary Conditions

We assume that the randomization of electrons takes place in two stages. First, an electron known to be on the open orbit at some instant of time diffuses under the influence of small-angle collisions until this motion carries it across the boundary plane separating open orbits from neighboring closed orbits. According to our previous argument, this will, on the average, take a time of order $(\Delta k)^2/D$. Second, the presence of a strong magnetic field ensures that the trajectory of the electron deviates drastically from those of its fellows which remained on the open orbit. In general, there will be a discontinuity in the cyclotron mass at the boundary plane, and unless the masses at the two sides of the boundary happen to be in a simple ratio, an electron which crosses the boundary will be effectively randomized in a further time $2\pi/\omega_c$, which will be much less than the time for diffusion if the magnetic field is strong.

According to this argument, we are able to establish boundary conditions for the problem, namely that variations of χ along the open direction vanish at the boundaries. With this assumption the problem can be solved by the usual trajectory method. The full calculation has been presented elsewhere[5] and we give here only the result. The nth attenuation resonance is given by

$$\alpha \sim \sum_{m=1}^{\infty} \frac{(2m-1)^{-2} |A_n|^2 \tau_m}{1 + \Omega^2 \tau_m^2} \tag{12}$$

where

$$\tau_m = (\Delta k)^2/(2m-1)^2 \pi^2 D \tag{13}$$

and where Ω is defined as in (11). The first term in (12) is by far the largest, since even the second term is 81 times smaller at resonance. This first term is a Lorentzian, having exactly the same form as (11) but with an effective relaxation time $\tau_1 = (\Delta k)^2/\pi^2 D$, which is of the general form (9) with $k_0 \sim \Delta k$.

Electron–Phonon Scattering

The choice of (11) or (12) to describe the line shape when the scattering is due to electron–phonon collisions depends on the magnitude of the parameter q_{rms} in comparison with the open orbit width Δk. At sufficiently high temperatures one might have $q_{\text{rms}} \gg \Delta k$, so that the line shape would follow (11) with $1/\tau \sim T^3$. In the opposite limit $q_{\text{rms}} \ll \Delta k$ the scattering would be diffusive and one would expect (12) to apply with $1/\tau_1 \sim T^5$. Thus while the line shape is expected to be Lorentzian under all conditions, the temperature dependence should give an indication of the magnitude of the parameter q_{rms} for electron–phonon scattering. This parameter is expected to be of the order of $kT/\hbar v_s$, where v_s is the phonon velocity, so that we have the possibility of being able to distinguish between longitudinal and transverse phonons by means of their velocity separation.

Experiment

Electron–phonon scattering rates for the [111]-directed open orbit in copper have been measured by Talaat and Peverley[6] on the basis of the linewidths of the acoustic open-orbit resonance. Over the range 7–18°K the scattering rate is well represented by the equation $1/\tau = (1.6 \pm 0.2) \times 10^5 T^4$. This temperature dependence is displayed in Fig. 1. Also shown is the rate obtained by a highly selective method (magnetic surface-state resonance[2]) and by an unselective method (dc electrical resistivity[7]). The former rate is for a Fermi surface point 16° from [100] in the (001) plane; we select this value for comparison because it is the lowest rate given by the authors and the effects due to anisotropy of $r(\mathbf{k})$ are thereby minimized.

The intermediate scattering rate obtained acoustically is consistent with the intermediate selectivity expected on *a priori* grounds. The apparent T^4 dependence, however, is probably due to the fact that the typical step size q_{rms} is comparable with the open-orbit width, so that the real situation is between the limiting cases (11) and (12). The analysis carried out so far[6] suggests that transverse or quasitransverse phonons are mainly responsible for the scattering; a definitive answer to this

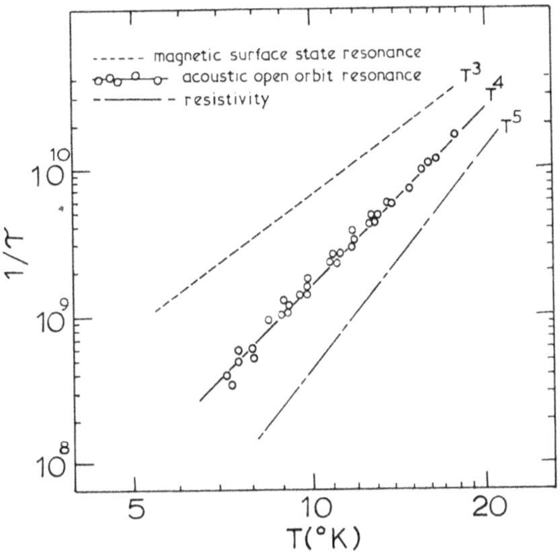

Fig. 1. Electron-phonon scattering rates in copper, illustrating the dependence on selectivity.

question, however, must probably await a proper Monte Carlo treatment of the scattering problem.*

References

1. A.B. Pippard, *Proc. Roy. Soc. (London)* **A305**, 291 (1968).
2. J.F. Koch and R.E. Doezema, *Phys. Rev. Lett.* **24**, 507 (1970).
3. E.A. Kaner, V.G. Peschanskii, and I.A. Privorotskii, *Zh. Eskperim. i Teor. Fiz.* **13**, 147 (1961).
4. J. Mertsching, *Phys. Stat. Sol.* **14**, 3 (1966).
5. J.R. Peverley, *Phys. Rev.* (to be published).
6. M.H. Talaat and J.R. Peverley, *Phys. Rev.* (to be published).
7. R. Berman and D.K.C. MacDonald, *Proc. Roy. Soc. (London)* **A255**, 135 (1962).

* **Note Added in Proof:** This has now been accomplished, see D.S. Khatri and J.R. Peverly, *Phys. Letters* **46A**, 36 (1973).

Resistivity of n-Type Bi_2Te_3 from 1.3 to 300°K*

W. T. Pawlewicz and J. A. Rayne

Carnegie-Mellon University, Pittsburgh, Pennsylvania

and

R. W. Ure, Jr.

University of Utah, Salt Lake City, Utah

The resistivity of tellurium-doped, n-type Bi_2Te_3 has been measured from 1.3 to 300°K on single crystals with carrier concentrations varying between 2.5×10^{17} and 1.9×10^{19} cm^{-3}. Corresponding mobilities at 4.2°K range from 2×10^5 to 10^4 cm^2 V^{-1} sec^{-1}, respectively. Data have been obtained by a standard four-probe method, using an integrating digital voltmeter to read the potential difference along the specimen. The sample is maintained in close thermal contact with a vacuum-jacketed copper block, the temperature of which is maintained electronically to better than 0.01°K from 1.5 to 20°K. Between 300 and 30°K the sample temperature is measured by a platinum resistance thermometer, while below 30°K a calibrated germanium thermometer is used.

At low temperatures all the specimens are degenerate, with corresponding Fermi degeneracy temperatures ranging from 50°K for the lowest carrier concentration to about 400°K for the highest. For carrier densities n_H less than 4×10^{18} cm^{-3} it has been demonstrated[1] that the conduction band is described by a six-valley ellipsoidal model. The behavior of the sample with lowest carrier density, $n_H = 2.5 \times 10^{17}$ cm^{-3}, is consistent with this model. At temperatures below 20°K the sample exhibits metallic behavior with an ideal resistivity which varies as T^5. This dependence, which agrees with the Bloch–Grüneisen theory for $T \ll \theta_D$, where θ_D is the relevant Debye temperature, strongly suggests that the dominant scattering mechanism is the electron–phonon interaction. Figure 1(a) shows a fit of the data to the expression

$$\rho - \rho_0 = \beta T^5 \tag{1}$$

where $\beta = 3.10 \times 10^{-12}$ Ω-cm°K^{-5}. The maximum temperature over which this equation is valid is approximately $\theta_D/8$ (θ_D = 165°K for Bi_2Te_3). Beyond this value the resistivity gradually assumes a T^3 dependence, until the onset of intrinsic semiconducting behavior at 200°K.

For higher carrier density specimens the ideal resistivity at low temperatures varies as T^2. This behavior is shown in Fig. 1(b) for a sample with $n_H = 4.1 \times 10^{18}$

* Work supported in part by the National Science Foundation.

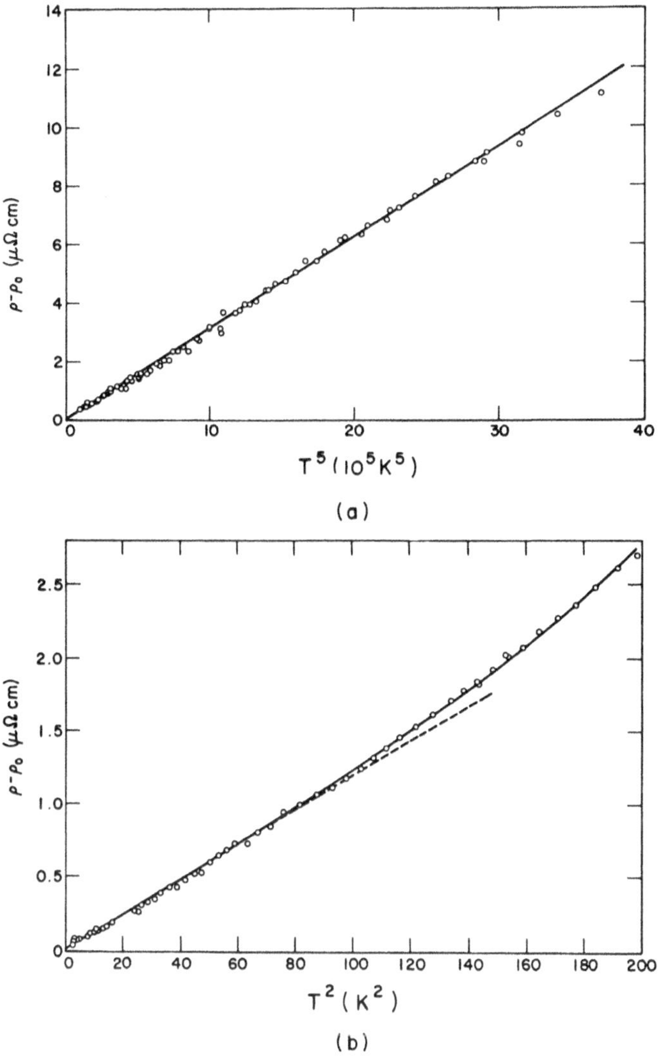

Fig. 1. (a) Fit of resistivity data for sample with $n = 0.25 \times 10^{18}$ cm^{-3} to equation $\rho - \rho_0 = \beta T^5$; (b) Fit of resistivity data for sample with $n = 4.1 \times 10^{18}$ cm^{-3} to equation $\rho - \rho_0 = \alpha T^2$.

cm^{-3}. The dashed line is given by the equation

$$\rho - \rho_0 = \alpha T^2 \tag{2}$$

where $\alpha = 1.21 \times 10^{-8}$ Ω-cm°K^{-2}. Above 10°K the resistivity deviates significantly from this relation and assumes a stronger dependence on temperature.

At carrier densities in excess of 4×10^{18} cm^{-3} it is known that a second heavy-mass conduction band begins to be populated. According to Baber,[2] the existence of this populated second band allows the possibility of electron–electron scattering,

Table I. Variation with Carrier Concentration of Coefficient α in a Fit of Resistivity Data for n-Type Bi_2Te_3 to the Equation $\rho - \rho_0 = \alpha T^2 + \beta T^5$

Sample	Carrier concentration, 10^{18} cm^{-3}	α, 10^{-8} Ω-cm °K^{-2}
1	0.252	0.02 ± 0.20
2	0.766	0.70 ± 0.15
3	2.57	1.19 ± 0.07
4	4.10	1.21 ± 0.06
5	18.4	1.04 ± 0.04

which contributes a T^2 term to the resistivity. Since there is a large difference between the effective masses for the two groups of carriers in Bi_2Te_3, this term might be expected to dominate the temperature dependence of the resistivity in agreement with experiment. Using the scheme for interband scattering proposed by Baber and the known energy band parameters for Bi_2Te_3, the behavior of the magnitude of the T^2 contribution can be calculated as a function of carrier density. Assuming the number of carriers in each band remains constant with temperature, α should be negligible for n_H less than 4×10^{18} cm^{-3}. For n_H greater than 4×10^{18} cm^{-3}, α

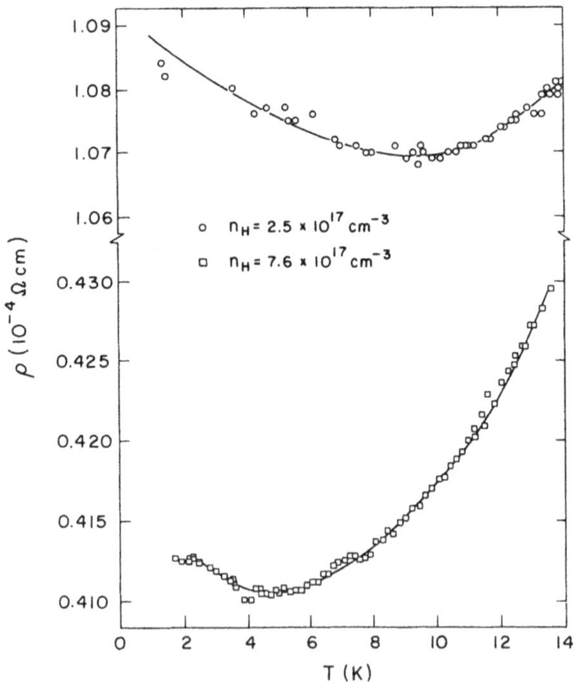

Fig. 2. Variation of resistivity for samples with low carrier concentration as a function of temperature. The existence of a resistance minimum in each case is clearly evident.

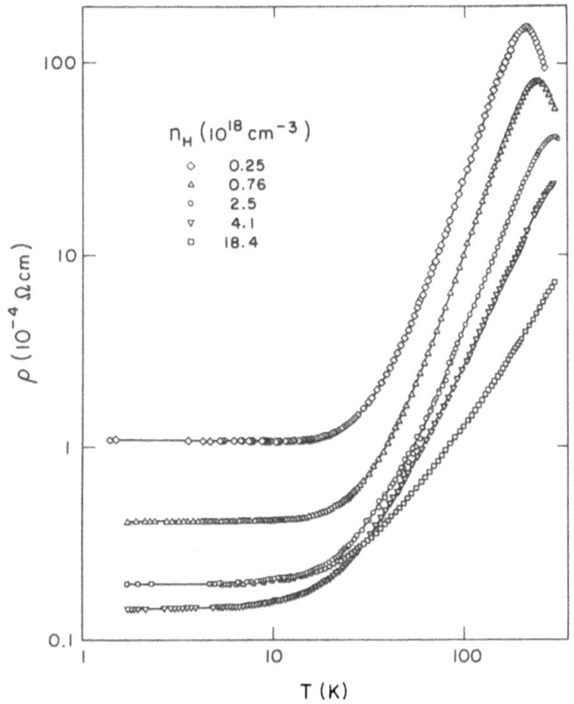

Fig. 3. Logarithmic plot of resistivity of all samples in the temperature range 1–300°K.

should increase abruptly, reach a maximum, and then fall off slowly with increasing n_H. The values of α for the five specimens examined thus far are given in Table I. The experimental variation of α with n_H is as expected, with the exception of the lower-carrier-density samples, for which the effect of thermal excitation of carriers from the first band to the second is evident.

Values of α for the two lowest carrier-density specimens have been deduced by resolving the data into T^2 and T^5 components by means of a computer least-squares analysis. The relatively large uncertainties in α reported for these samples are the result of an interesting phenomenon that occurs at the lowest temperatures. Figure 2 shows that the resistivity in each case exhibits a shallow minimum, which occurs between 9 and 10°K for $n_H = 2.5 \times 10^{17}$ cm^{-3} and near 5°K for $n_H = 7.6 \times 10^{17}$ cm^{-3}. In both cases the resistivity at 1°K is a few percent higher than that at the minimum. The observation of these minima only in the lowest carrier-density samples suggests that carrier freezeout is the explanation of the effect. For these lightly doped specimens there may still exist a small energy gap between the impurity band and the conduction band edge. If this were the case, the resistivity would rise exponentially as T approaches absolute zero. Experiments below 1°K are being planned to investigate this phenomenon.

Figure 3 is a logarithmic plot of the resistivity of each sample from 1 to 300°K. At high temperatures the resistivity of each sample varies in accordance with the power law

$$\rho = \text{const} \times T^x \tag{3}$$

Table II. Values of Exponent and Temperature Range for Fit of Resistance Data to the Equation ρ = const $\times$ T^x

Sample	Carrier concentration, 10^{18} cm^{-3}	Power law exponent	Temperature range
1	0.252	3.08	35–130
2	0.766	2.96	35–140
3	2.57	2.55	30–185
4	4.10	2.42	40–205
5	18.4	1.67	50(> 300)

beginning near 30°K in samples with relatively low carrier densities and near 50°K for the highest carrier densities. The power law dependence is observed until the onset of intrinsic semiconducting behavior. It is found that x decreases monotonically from 3.1 to 1.7 with increasing carrier concentration. Values of x and the approximate temperature regions in which the power law is applicable are given in Table II for the samples investigated.

References

1. R.B. Mallinson, J.A. Rayne, and R.W. Ure, Jr., *Phys. Rev.* **175**, 1049 (1968).
2. W.G. Baber, *Proc. Roy. Soc. (London)* **A158**, 383 (1937).

Effect of Surface Condition on the Lorenz Number and Other Electron Transport Properties of Size-Limited Single-Crystal Tungsten Rods*

D. R. Baer and D. K. Wagner

Laboratory of Atomic and Solid State Physics
Cornell University, Ithaca, New York

It is well known that when the electron mean free path l of a pure metal becomes comparable to the smallest specimen dimension surface scattering of electrons plays an important role in determining the transport properties of the metal. The effects of the surface scattering on observable characteristics of the specimen are termed size effects and are responsible for a number of interesting phenomena, some of which include: a resistivity that depends upon specimen size; a negative component of the longitudinal magnetoresistance; and an oscillating component of the transverse magnetoresistance.[1-3] Size effects have been observed for both electrical and thermal conduction and the same basic theories are used to explain both.

Theoretical treatments of size effects have been hampered, however, by a lack of knowledge about the nature of the scattering of electrons at the surface. Consequently, it has been assumed customarily that a fraction of the electrons striking the surface are specularly reflected and that the remainder are diffusely scattered. Generally, the results of most size-effect experiments have supported the view that the surface scattering is mainly diffuse. On the other hand, there is some evidence that specular scattering can be important, the most notable of which is the existence of magnetic surface states.[4] Still other experiments have proved difficult to interpret in terms of either specular or diffuse scattering, suggesting that perhaps this description of the scattering at the surface is oversimplified.

Recent measurements by Wagner *et al.*[5] have indicated that the Lorenz number of tungsten may be affected by the surface scattering. Normally in a bulk metal the Lorenz number L is considered to be a good indicator of the amount of inelastic scattering present. At low temperatures L is usually reduced below the standard value L_0 of $2.445 \times 10^{-8}\ W\Omega/{}^\circ K^2$ by the inelastic scattering of electrons by various excitations of the metal crystal (e.g., phonons). At still lower temperatures such scattering mechanisms are frozen out, leaving only elastic processes (e.g., scattering by defects, impurities, etc.) and L returns to L_0. This behavior has been observed in many metals. In the measurements by Wagner *et al.*[5] the sample (sample W-8 of Ref. 5) with the largest ratio of mean free path to sample diameter produced a Lorenz number about 10% lower than L_0 upon extrapolation to zero temperature, while

* Work supported by the U.S. Atomic Energy Commission under Contract Number AT(11-1)-3150, Technical Report No. COO-3150-7.

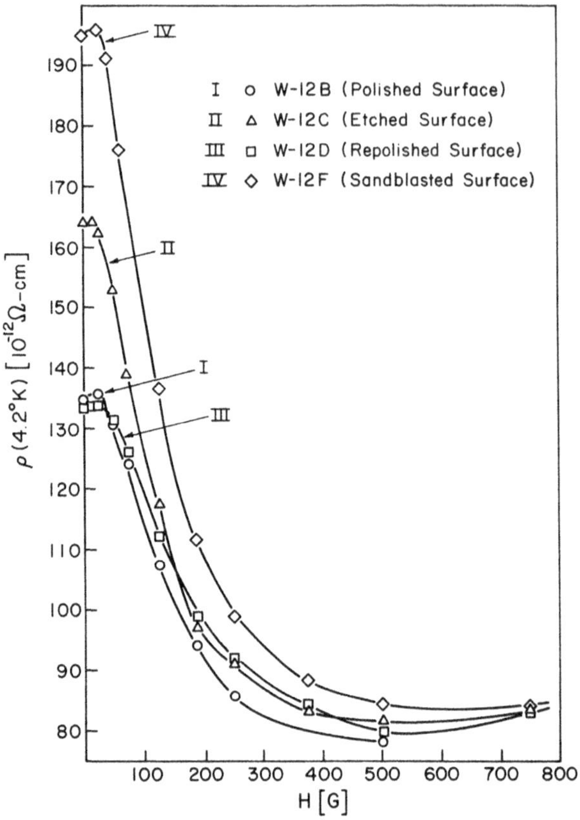

Fig. 1. The longitudinal magnetoresistivity as a function of magnetic field at 4.2°K for various conditions of the sample surface.

for the bulk specimens the Lorenz number was observed to approach L_0. After removing the polished surface of this sample by electroetching they observed an increase in the electrical and thermal resistivities and the Lorenz number approached L_0. The present study was undertaken to explore in greater detail the effect of surface condition on the electrical and thermal resistivities and on the Lorenz number.

Measurements of the electrical and thermal resistivities ρ and W as a function of temperature from 1.5 to 5.5°K and of the longitudinal magnetoresistance at 4.2°K were made on variously oriented, single-crystal tungsten rods ranging from 0.13 to 3 mm in diameter. The electrical and thermal measurements were made as described by Wagner *et al.*[5] The precision of the measurements was roughly $\frac{1}{2}$% for the electrical resistivity and 1% for the thermal resistivity. The longitudinal magnetoresistance measurements were made in a superconducting solenoid homogeneous to about 0.1% and the samples were aligned in the field to about 1.5°. Experimental checks showed no evidence that current jetting[6] had any effect on the longitudinal magnetoresistance measurements.

The effect of surface scattering on the electrical resistivity can be seen clearly

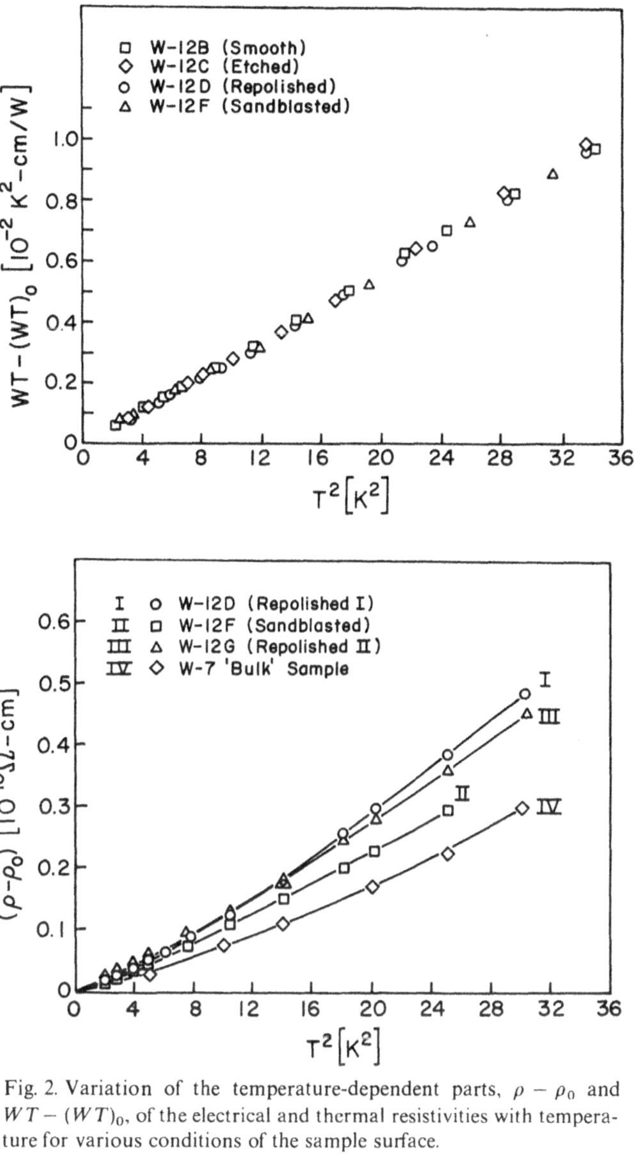

Fig. 2. Variation of the temperature-dependent parts, $\rho - \rho_0$ and $WT - (WT)_0$, of the electrical and thermal resistivities with temperature for various conditions of the sample surface.

in the measurements of the longitudinal magnetoresistance shown in Fig. 1. This figure shows the behavior of the longitudinal magnetoresistance of sample W-12 with the surface in the initial polished state, after etching, after repolishing, and finally after sandblasting. Note that the zero-field resistivity is changed *reversibly* by altering the condition of the sample surface; in particular, it was increased by electroetching, decreased by repolishing, and again increased by sandblasting. In all cases, however, the resistivity approaches the same value as the magnetic field is increased. Since the fraction of electrons striking the surface is reduced by a longitu-

Table I. Characteristics of the Tungsten Samples

Sample*	d, mm	$\rho(299)/\rho(0)$	$(WT)_0$, 10^{-2} Kcm/W	ρ^0 10^{-10} Ω-cm	$L(0) = \rho_0/(WT)_0$, 10^{-8} W Ω °K^2	Surface treatment	Surface condition
W-12B	0.9	52,000	0.495 ± 0.004	1.04 ± 0.015	2.10 ± 0.05	Electropolished	Mirror
W-12C	0.9	38,500	0.663 ± 0.005	1.395 ± 0.005	2.10 ± 0.025	H_2O_2 etched	Slightly and selectively etched
W-12D	0.8	47,500	0.555 ± 0.006	1.13 ± 0.005	2.03 ± 0.025	Electropolished	Mirror
W-12E	0.8	42,000	0.601 ± 0.004	1.275 ± 0.006	2.120 ± 0.025	Electroetched	Selectively etched
W-12F	0.8	30,000	0.750 ± 0.005	1.790 ± 0.007	2.39 ± 0.025	Sandblasted	Matt
W-12G ($H = 0$)	0.74	43,000	0.569 ± 0.004	1.260 ± 0.005	2.21 ± 0.025	Electropolished	Mirror
W-12G ($H = 480$ G)	0.74	—	0.285 ± 0.005	0.690 ± 0.005	2.42 ± 0.05	Electropolished	Mirror
W-12H	0.74	37,000	0.657 ± 0.005	1.445 ± 0.006	2.20 ± 0.025	Goldplated	Gold
W-12J	~ 0.48	17,000	1.31 ± 0.01	3.12 ± 0.02	2.38 ± 0.035	Sandblasted	Matt
W-4	1	30,000	0.721 ± 0.005	1.780 ± 0.004	2.47 ± 0.02	Electroetched	Matt
W-15A	0.28	14,000	1.57 ± 0.01	3.85 ± 0.01	2.46 ± 0.03	Electropolished	Mirror
W-15B	0.28	9,000	2.27 ± 0.01	5.66 ± 0.015	2.49 ± 0.03	Sandblasted	Matt
W-16	0.23	23,000	1.07 ± 0.006	2.62 ± 0.005	2.45 ± 0.01	Electropolished	Mirror

* All orientations were [110], except for W-16, for which it was [111].

dinal magnetic field, this suggests that the bulk properties of the specimen have not been altered by our treatment of the surface.

The temperature dependence of the electrical resistivity is also observed to change reversibly by altering the condition of the sample surface, as can be seen in the lower graph in Fig. 2. Unlike its temperature-independent counterpart, the temperature-dependent part of the electrical resistivity is increased by polishing, decreased by sandblasting, and increased again by repolishing. Note, however, that in all cases the temperature dependence of the resistivity is greater than that of a typical bulk sample. If these effects are present in the thermal resistivity, they are clearly considerably smaller, as can be seen by referring to the upper graph in Fig. 2.

The Lorenz number is also observed to depend upon the surface condition. Table I lists the zero-temperature extrapolated values of the electrical and thermal resistivities ρ_0 and $(WT)_0$ and their quotient $\rho_0/(WT)_0$, which is the zero-temperature Lorenz number $L(0)$. By referring to the data for sample W-12 in the table, it is seen that a roughened surface yields a value of $L(0)$ approximately equal to L_0, while a smooth (electropolished) surface yields a value of $L(0)$ which is 10–20% lower than L_0. We obtained further evidence that the condition of the surface is responsible for this effect by measuring the electrical and thermal resistivities of the repolished sample (W-12G) in a longitudinal magnetic field strong enough to remove most surface scattering. The results of these measurements are plotted in Fig. 3. From the graph it can be seen that at a field of 480 G the zero-temperature Lorenz number $L(0)$ approaches L_0, while at zero field it approaches a value about 10% lower. Since it is well known for a bulk metal that the Lorenz number will equal L_0 if the scattering is elastic (this is also valid in a magnetic field), these results demonstrate directly that the observed deviations from the Wiedemann–Franz law must be ascribed to the polished condition of the surface.

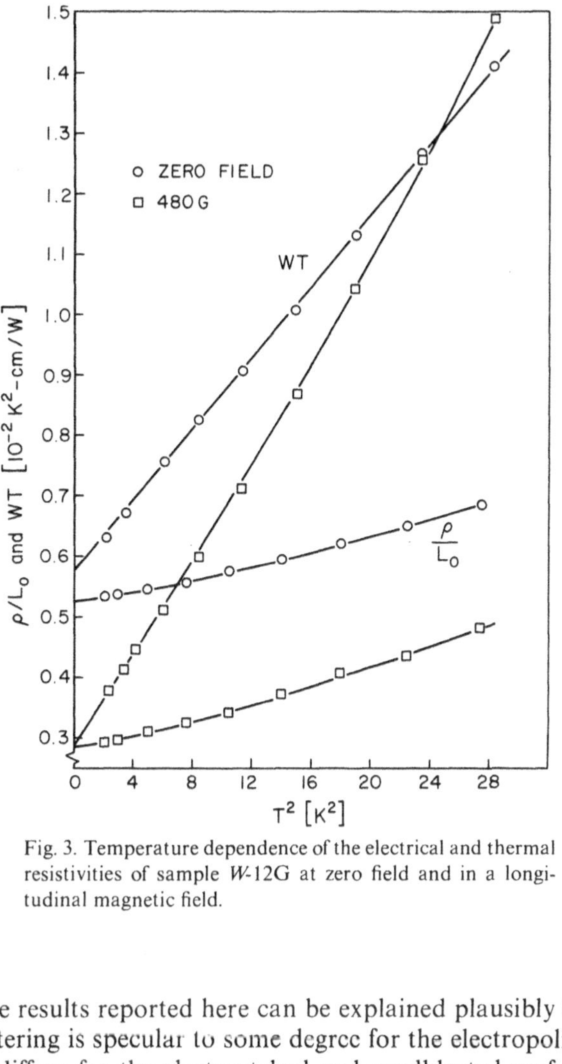

Fig. 3. Temperature dependence of the electrical and thermal resistivities of sample *W*-12G at zero field and in a longitudinal magnetic field.

Some of the results reported here can be explained plausibly by assuming that the surface scattering is specular to some degree for the electropolished surface and predominantly diffuse for the electroetched and sandblasted surfaces.* This would readily explain the decrease of the resistivities upon electropolishing of the sample surface. It may also explain the variation of the temperature-dependent part of the electrical resistivity with the condition of the surface; however, because this is a much more subtle effect, there is at present no sound basis for drawing such a conclusion. Whatever the cause of this behavior, the lack of a similar variation of the temperature dependence of the thermal resistivity is probably due to the fact that the electron mean free path for thermal conduction is shorter than that for electrical conduction in the temperature range of our measurements. On this basis one expects that the thermal resistivity would be less sensitive than the electrical resistivity to the presence of the surfaces, and this is consistent with all our observations.

* Because our surfaces are not ultraclean, surface coatings may also play a role. For a study of the effect of surface coatings on the electrical resistivity of thin films see Ref. 7.

The deviation of the Lorenz number from L_0 that is associated with the electropolished surface is not easily understood, however. If the temperature dependence of ρ or WT changes dramatically below 1.5°K, our extrapolation to zero temperature would not be valid and the deviation of the Lorenz number from L_0 might only be apparent. An inelastic surface scattering mechanism yet to be frozen out at 1.5°K might possibly cause such a behavior. If, on the other hand, the deviation of the Lorenz number from L_0 does persist to zero temperature as our extrapolations indicate, then our present understanding of the size effect is inadequate. Both bulk and surface scattering processes should be elastic at zero temperature, in which case one would expect the Lorenz number to equal L_0.* Measurements at lower temperatures could provide useful additional information about these deviations from the Wiedemann–Franz law.

Acknowledgments

We are grateful to B. Addis for zone-refining and electropolishing the tungsten samples used in this study and to B. W. Maxfield and T. N. Rhodin for their interest in this work.

References

1. E.H. Sondheimer, *Adv. Phys.* **1**, 1 (1952).
2. R.G. Chambers, in *Physics of Metals: I. Electrons*, J.M. Ziman, ed., Cambridge University Press, Cambridge (1969), Chapter IV.
3. G. Brandli and J.L. Olsen, *Mat. Sci. Eng.* **4**, 61 (1969).
4. J.F. Koch, in *Solid State Physics, Vol. I, Electrons in Metals*, J.F. Cochran and R.R. Haering, eds., Gordon and Breach, New York (1968), p. 253.
5. D.K. Wagner, J.C. Garland, and R. Bowers, *Phys. Rev.* **B3**, 3141 (1971).
6. W.A. Reed, E.F. Blount, J.A. Marcus, and A.J. Arko, *J. Appl. Phys.* **42**, 5453 (1971).
7. K.L. Chopra and M.R. Randlett, *J. Appl. Phys.* **38**, 3144 (1967).
8. R.F. Greene, in *Solid State Surface Science*, M. Green, ed., Dekker, New York (1969), Vol. I, Chapter II.

* A formal treatment of the problem using the formulation of the surface boundary condition developed by Greene[8] indicates that the Wiedemann–Franz law is obeyed when the scattering is elastic.

Observation of a Negative Conduction in Antimony in High Magnetic Fields

T. Morimoto and K. Yoshida

Institute of Atomic Energy, Kyoto University
Kyoto, Japan

Several years ago we found a new phenomenon of a negative resistance (in a total, not differential sense) in that the potential drop in bismuth crystals in high magnetic fields seemed to be opposed to the original current flow,[1-3] although the term "negative resistance" was not appropriate since it has been pointed out by Morimoto[2,4] that it might in fact be a self-induction phenomenon arising from the interaction of the collective flow of an electron–hole plasma with the applied magnetic field, and that it is a characteristic phenomenon of semimetals having an equal number of electrons and holes as well as an extremely high velocity of surface recombination.[5-7] Since antimony is a semimetal having an electronic structure similar to bismuth,[8-12] it might be expected that antimony would also indicate a negative conduction under the appropriate conditions of high magnetic fields and low temperatures. Here we report such a phenomenon observed in antimony crystals with different crystallographic orientations.

Single crystals of antimony were grown by zone-leveling ingots of 99.9999% purity. Several rectangular samples with cross sections ranging from (2.3 mm)^2 to (7 mm)^2, lengths ranging from 8 to 12 mm, and three different crystallographic orientations were cut out. After repeated polishing and etching they were annealed for 2 hr at 300°C in a vacuum of 2×10^{-6} mm Hg to remove the stress. The resistance ratio $R(293)/R(1.5)$ of the samples was about 3700. All these samples indicated negative conduction at low temperatures. Here we describe a typical example. As shown in Fig. 1(a), measurement of the potential drop was made by using a pair of probes, x and z probes of 0.1-mm-diameter silver wire, attached respectively at the middle of x and z surfaces of the rectangular sample by condenser discharge. A constant current J was passed along the long direction by using a YEW' type 2853 constant-current supply. Antimony # 2, which will be described here, has the long direction along the y axis and a length of 7.96 mm. It has a relatively large cross section of 6.46 mm (x) by 5.87 mm (z), and the distances of the x and z probes are 1.9 and 1.59 mm, respectively.

Figure 2 shows the angular dependence of the potential drop measured by the x probe for Sb # 2 at 1.6°K when the magnetic field direction is rotated in the xy plane; the field strength applied and the total current passed through the sample were 20 kG and 100 mA, respectively. As can be clearly seen, the hatched region around the y axis indicates negative potential drop against the original current flow, while the potential drop measured by the z probe indicates normal, positive

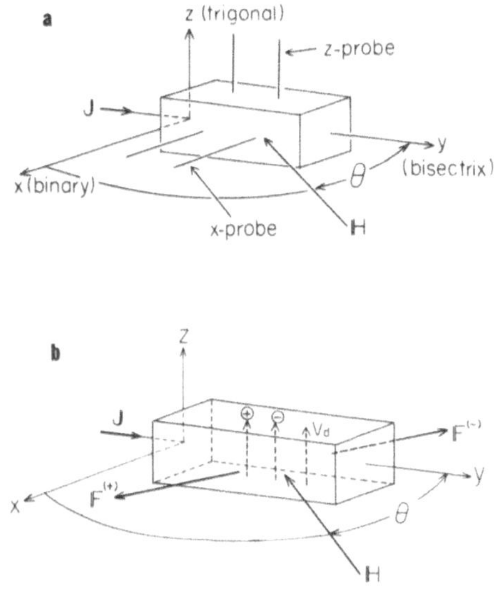

Fig. 1. (a) Arrangement of sample and leads. (b) Intuitive explanation of self-induction caused by the collective flow of an electron–hole plasma across the magnetic field. $F^{(+)}$ and $F^{(-)}$ indicate the forces acting on holes and electrons, respectively.

magnetoresistance. In the figure the shift of the zero level of voltage comes from the inclusion of the induced emf associated with the rotation of the electromagnet.

Figure 3 shows the magnetic field dependence of the potential drop at $\theta = 29°$ where the maximum of the negative potential drop is observed at 1.6°K. It can be seen that the x probe indicates a negative voltage which increases with increasing magnetic field in proportion to H^2, with Shubnikov–de Haas-like oscillations. On the other hand, the z probe indicates a positive voltage, showing the normal magnetoresistance. In the above setting the magnetic field direction was rotated through the xy plane. However, when the magnetic field is applied in the yz plane, the z probe now indicates a negative potential drop and the x probe shows a positive potential drop in the new geometry. Here it should be noted that the negative potential drop is observed locally in the middle of the sample; however, the positive potential drop is always set up between the two ends of the sample so as to pass a constant current through the sample.[2,3]

The same phenomenon was also observed in the ac experiment. In contrast to bismuth, however, a remarkable decrease in the negative voltage was observed

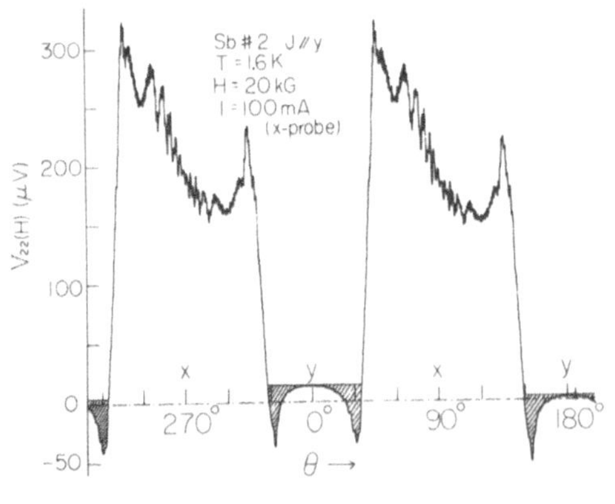

Fig. 2. Angular dependence of the voltage $V_{22}(H)$ measured by the x probe at 1.6°K under a magnetic field of 20 kG. Negative conduction is seen in the hatched region.

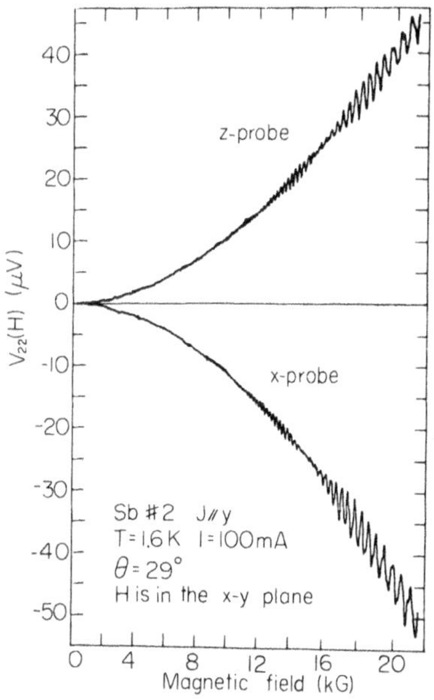

Fig. 3. Magnetic field dependence of voltage $V_{22}(H)$ at 1.6°K and at $\theta = 29°$.

in antimony at frequencies higher than 10 kHz. This may suggest a difference in the contribution of the classical skin effect to this phenomenon in antimony and bismuth, since the former has 100 times higher concentration of free carriers than the latter.

In the earlier investigation of this phenomenon in bismuth the possibility of the inclusion of spurious voltage coming from the misalignment of the leads was carefully checked by reversing the setting of the sample, and it was shown that the quadratic Hall voltage did not play an essential role in the phenomenon.[2] In addition, observation of negative conduction in Sb # 4, which has a rather small cross section of 2.3×2.4 mm^2 and a rather long length of 12 mm along the binary direction, so that a reversed flow of the current in the sample seems unlikely, seems to support strongly a hypothesis of self-induction of semimetals under high magnetic fields,[2,4] whose intuitive explanation is given in Fig. 1(b) and the following.

In semimetals having an equal number n of electrons and holes, such as bismuth and antimony, there exists a drift motion of both carriers with the same velocity $\mathbf{v}_d$ perpendicular to the current and magnetic field directions, since the Hall electric field is vanishingly small at high magnetic fields and the velocity of the surface recombination of electrons and holes is high enough to justify neglect of the counter-diffusion. Hence such a collective flow of charge carriers across the magnetic field will induce an emf given by

$$V = -(1/c) \int (\mathbf{v}_d \times \mathbf{H})\, ds \tag{1}$$

This is equivalent to assuming the induction of an apparent electric field E'_y along the original current direction as follows:

$$E'_y = (1/c)(\mathbf{v}_d \times \mathbf{H})_y \tag{2}$$

The direction is opposite to the original current flow as shown in Fig. 1(b), so that it could give rise to a negative conduction in high magnetic fields, where the condition $\mathbf{v}_d \gg |J/en|$ is satisfied, J and e being the current density and the electronic charge, respectively. If we assume a simplified two-band model of electrons and holes in the geometry of Fig. 1, $\mathbf{v}_d$ is given by the relation

$$-\mathbf{v}_d/(J/|e|n) = (\mu_z^*/c)\,H_x = \langle \omega_c \tau \rangle \tag{3}$$

where μ_z^* is the effective ambipolar mobility,[6,7] ω_c is the cyclotron frequency, and τ is the relaxation time. Hence we can estimate the order of the magnitude of the induced negative electric field to be

$$E_y' = -J\mu_z^*\,H^2(\sin^2\theta)/c^2|e|n \tag{4}$$

which agrees with the experimental value.[2] On the z plane, the induction voltage vanishes from the boundary condition that $\mathbf{v}_d = 0$ at the surface. However, the effect is usually masked by the larger effect of the normal magnetoresistance of the material, because we actually measure the sum of the resistance term plus the above induction term. Therefore we can observe such negative conduction only around the longitudinal field direction, where the normal magnetoresistance vanishes at high fields. Needless to say, the mechanism of such self-induction of semimetals is quite different from that of resistance. The former comes from the collective motion of charge carriers across the magnetic field and may be reversible, while the latter results from the scattering, and thus the individual motion, of charge carriers, i.e., the relaxation process, and is irreversible.

To date such negative conduction has not been observed in antimony samples at 77°K, in contrast to bismuth. However, except for such minor differences, we may conclude that semimetals such as bismuth and antimony could exhibit a self-induction under high magnetic fields in addition to the well-known resistance.

Acknowledgments

We are grateful to Y. Tashiro and T. Minato for their technical assistance.

References

1. T. Morimoto, *J. Phys. Soc. Japan* **23**, 126 (1967).
2. T. Morimoto and K. Yoshida, Tech. Repts. Eng. Res. Inst. Kyoto Univ. No. 150 (1970).
3. T. Morimoto and K. Yoshida, in *Proc. 12th Intern. Conf. Low Temp. Phys., 1970*, Academic Press of Japan, Tokyo (1971), p. 557.
4. T. Morimoto, *Bull. Eng. Res. Inst. Kyoto Univ.* **36**, 1 (1969).
5. T. Hattori, *J. Phys. Soc. Japan* **23**, 19 (1967).
6. T. Yamada, *J. Phys. Soc. Japan* **20**, 1424 (1965).
7. A.A. Lopez, *Phys. Rev.* **145**, 823 (1968).
8. L.M. Falicov and P.J. Lin, *Phys. Rev.* **141**, 562 (1966).
9. L.R. Windmiller, *Phys. Rev.* **149**, 472 (1966).
10. S. Epstein and H.J. Juretschke, *Phys. Rev.* **129**, 1148 (1963).
11. G.N. Rao, N.H. Zebouni, C.G. Grenier, and J.M. Reynolds, *Phys. Rev.* **133A**, 141 (1964).
12. M.S. Bresler and N.A. Red'ko, *Soviet Phys.—JETP* **34**, 149 (1972).

11
Transport—Thermal

Transport Properties Near Magnetic Critical Points: Thermopower*

S. H. Tang

State University of New York, Stony Brook, New York and Brookhaven National Laboratory, Upton, New York

T. A. Kitchens

Brookhaven National Laboratory, Upton, New York

F. J. Cadieu

Queens College, Flushing, New York

and

P. P. Craig

Brookhaven National Laboratory, Upton, New York and National Science Foundation, Washington, D. C.

Introduction

One of the most amazing characteristics of physics has been the ability of a few simple arguments to explain many complicated effects. No field of research illustrates this fact so clearly as the study of critical phenomena. Here the static properties of many diverse systems experimentally show similar characteristics. This universality indicates that these properties are not very dependent on the detailed interactions between the elements which undergo the ordering, and has led to the rather general yet phenomenological theory of scaling which has been quite successful in correlating most of the experimental data (see, e.g., Ref. 1).

In the hope of learning more of the elemental interactions near the critical point, the dynamic properties have received increasing attention. Predominant in importance are neutron scattering experiments on the long- and short-range order effects in order–disorder and magnetic systems (see, e.g., Ref. 2). Laser light scattering experiments have elucidated liquid–gas systems near the critical points (see, e.g., Ref. 3) and acoustic attenuation techniques also have been utilized. During the past few years theoretical activity has primarily followed two different approaches: one assuming that the same universality seen in the static properties will obtain,[1,4] and

* Work supported by the U.S. Atomic Energy Commission. Invited paper.

the second assuming that more detailed information of the interactions is necessary to explain the dynamic behavior (see, e.g., Ref. 5).

The experimental techniques discussed above probe the critical system by an external disturbance; it is also possible to study a critical system by utilizing the elements of the system itself as probes. Such measurements of mass and spin diffusion, electrical and thermal conductivity, and viscosity tend to be experimentally simpler and do not introduce an external disturbance. On the other hand, these properties are average dynamic properties, averaged over some of the states of the probe element. Thus the price one pays for the experimental simplicity is that the observable quantities are often complicated convolutions of the variables of fundamental interest. In fact, the unfolding of the experimental data to obtain the quantities of interest may not even be unique.

In this paper we will present some of the results from our study of one of the transport properties, the thermopower, and review the work of other investigations of this property. Two different analyses have been used. One is a phenomenological approach based on the assumption of local equilibrium. This approach is primarily justified by empirical correlation of the temperature derivative of the resistivity and also of the thermopower to the specific heat. The beauty of this analysis is that it provides some information on the nature of the elements of the system. The other analysis is much more fundamental, beginning with the general Boltzmann transport equation, but becomes rather bollixed by the details of the calculation. In the following discussion we attempt to provide a justification of the first analysis in terms of the second. Then we shall summarize the experimental data on the thermopower in the vicinity of the second-order phase transition in several systems which we believe to be typical of particular genres: nickel as the itinerant ferromagnet; gadolinium as the localized moment ferromagnet; some comments on the antiferromagnets as characterized by the heavy rare earths; and β-brass as the order–disorder system.

Critical Behavior of Thermopower

The theoretical discussions will be clearer after a brief outline of the experimental situation, as illustrated by the data for the thermopower and resistivity in the itinerant ferromagnet nickel. Data for a wide range of temperature T about the Curie point T_C are shown in Fig. 1. In the vicinity of T_C the qualitative similarity between ρ and thermopower Q is striking. Upon differentiation with respect to temperature, and critical behavior of both $d\rho/dT$ and dQ/dT is reminiscent of the specific heat anomaly. This is seen by writing dQ/dT in the form

$$dQ/dT = (A/\lambda)(\varepsilon^{-\lambda} - 1) + B, \qquad \varepsilon = |(T - T_C)/T_C| \tag{1}$$

where A and B are constants, or at most are very weakly temperature dependent in the neighborhood of T_C, and λ is the exponent which characterizes the critical behavior; a similar expression can be written for $d\rho/dT$. The values of λ and hence the type of power law dependence can be analyzed by a log-log plot of experimentally determined d^2Q/dT^2 vs. ε. It is found that for the thermopower of nickel $\lambda^+ = 0.14 \pm 0.1$ and $\lambda^- = -0.3 \pm 0.1$, and for the resistivity $\lambda^+ = 0.1 \pm 0.1$ and $\lambda^- = -0.3 \pm 0.1$, where the superscript plus refers to the region $T > T_C$ and minus to $T < T_C$. These values are consistent with that of the specific heat anomaly in nickel.

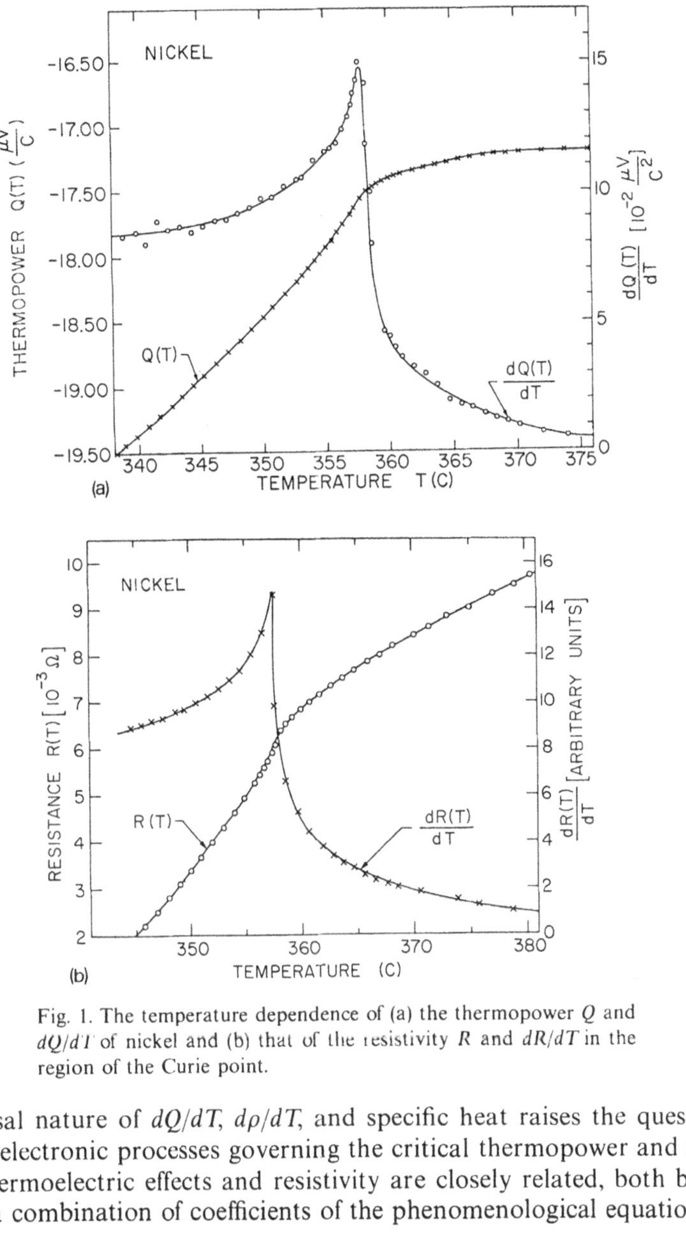

Fig. 1. The temperature dependence of (a) the thermopower Q and dQ/dT of nickel and (b) that of the resistivity R and dR/dT in the region of the Curie point.

The universal nature of dQ/dT, $d\rho/dT$, and specific heat raises the question as to the type of electronic processes governing the critical thermopower and resistivity.

The thermoelectric effects and resistivity are closely related, both being coefficients or a combination of coefficients of the phenomenological equations[6]

$$J_e = -L_{ee}\nabla\mu - L_{es}\nabla T \tag{2a}$$

$$J_s = -L_{es}\nabla\mu - L_{ss}\nabla T \tag{2b}$$

which relate the particle flow density J_e, the entropy flow density J_s, and the generalized conjugate driving forces, namely the electrochemical potential gradient $\nabla\bar{\mu}$ and the temperature gradient ∇T. The thermopower $Q = 1/e\,\nabla\bar{\mu}$, with $J_e = 0$, is the fundamental quantity from which all thermoelectric effects can be derived. Here e is, by definition, the charge of an electron. Recalling Eqs. (2a) and (2b), we define

$$S^* \equiv [J_s/J_e]_{\nabla T=0} = L_{es}/L_{ee}$$

which has units of entropy per charge carrier. Thus the thermopower can be expressed in terms of this "transport entropy," namely $Q = -(1/e)\,S^*$. In practice, Q can be determined from one of three experimental quantities, the Seebeck coefficient, the Peltier heat, or the Thompson heat, via the Onsager–Kelvin relations.[6]

The coefficients in Eqs. (2a) and (2b) must in general be evaluated by means of statistical mechanics. Assuming a system of noninteracting charge carriers and a single scattering time, solution of the Boltzmann equation is possible, as exemplified by the Mott equation,[7]

$$Q = \frac{\pi^2 k^2 T}{3e}\left[\frac{d\ln N}{dE} + \frac{d\ln v^2}{dE} + \frac{d\ln \tau}{dE}\right]_{E=E_F} \tag{3}$$

Here the bracketed derivatives with respect to energy are to be evaluated at the energy of the actual Fermi surface of the metal, a calculation which is difficult in practice. N, v, and τ are the density of states, velocity, and lifetime of the electrons, respectively. The lifetime τ represents the various effects of scattering by phonons, impurities, localized d (or f) spins and spin fluctuations, etc. It appears that in the vicinity of T_C the most likely sources of anomalous Q must be the fluctuations of the density of states at the Fermi surface as well as the spin-disorder scatterings.

Effects of Fluctuations on Thermopower

In 1936 Mott[8] proposed an s-d electron interaction model for nickel. He considered the probability of interband scattering of s electron into a hole in the d band,

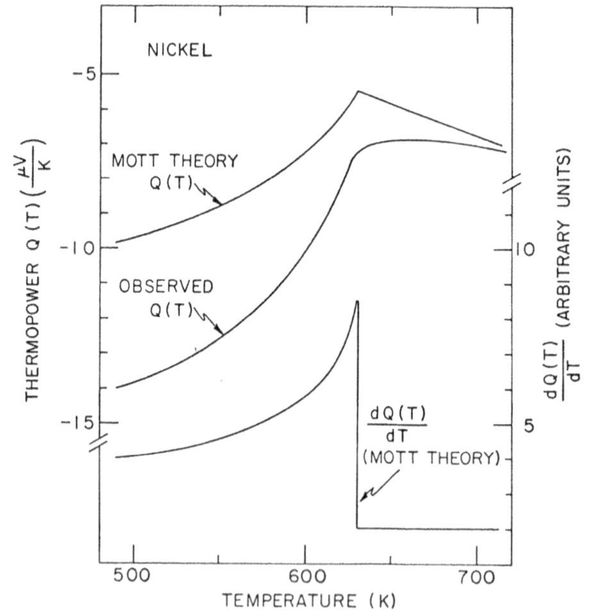

Fig. 2. The thermopower anomaly in nickel as predicted by Mott on the assumption of s-d interband scattering.

and computed the thermopower shown in Fig. 2. Just below T_C the Zeeman splitting of the spin-up and spin-down *d* bands changes rapidly with temperature. The resultant density-of-states variations will probably be observed as a decrease in thermopower, as well as in the resistivity; the contour of Q vs. T due to such effects essentially follows an inverted magnetization curve immediately below T_C. The qualitative feature of Fig. 2 suggests that the model has some merit for $T < T_C$. This calculation does not account for critical fluctuations for $T > T_C$ and is probably insufficient for the rare earths, where the magnetization is due to the unpaired 4*f* spins.

Mott had also recognized the nonvanishing contribution from spin-disorder scattering. However, only recently have such concepts been applied to resistivity with some success. The situation in resistivity has been summarized by Kawatra and Budnick[9] and by others.[10] In essence there are two approaches: the mean field theory of de Gennes and Friedel[11] and the scaling theory of Fisher and Langer[12] (hereafter referred to as FL). The point of departure in the two approaches has been the choice of the Fourier transform of the static spin-spin correlation function $\hat{\Gamma}(K, T)$, which enters into the relaxation time τ as

$$\tau_0/\tau = \Gamma(0, T) + (1/8k_F{}^4)\int_0^{2k_F} \hat{\Gamma}(K, T)\, K^3\, dK \tag{4}$$

where the Born approximation of the scattering of electrons by a lattice of localized spins is assumed, and k_F and K are the Fermi momentum vector and momentum transfer vector, respectively. The static spin–spin correlation function is legitimately used here since near T_C the lifetime of a fluctuation is long compared to the time an electron takes to traverse a correlation length.

For large values of K there is a correspondence between the functional form of $\hat{\Gamma}(K, T)$ and that of thermopower and resistivity, as shown by FL. The formal argument for the transport coefficients to diverge as the specific heat can now be seen easily. Explicitly, following FL, Eq. (4) is rewritten as

$$\tau_0/\tau = \sum_{s=0} \nu_s f(R_s)\, \Gamma(R_s, T) \tag{5}$$

which is closely reminiscent of the energy for the spin system*

$$U(T) = -\sum_{s=1} \nu_s J(R_s)\, \Gamma(R_s, T) \tag{6}$$

Here the sum is over successive shells of spins and ν_s is the number of spins in the s shell. The exchange interaction $J(R_s)$ and the integral $f(R_s)$ are temperature independent. The structure factor $f(R_s)$ is given by FL to be

$$f(R) = \frac{1}{4k_F^4 R}\frac{d^2}{dR^2}\frac{\cos 2k_F R}{R}$$

Sufficiently close to T_C it may be expected that the same single term will dominate both the series, so that $1/\tau$ may be expected to diverge in the same way as does the spin energy.

* This is known to be correct for two-dimensional Ising models; see, e.g., Ref. 13.

The situation regarding thermopower is a little more complex than that involving resistivity. Supposing that spin–spin fluctuations dominate the critical region, combining Eqs. (3) and (5) and neglecting for the moment the effect of density of states and effective mass variations, one has from the spin-disorder scattering contribution

$$Q = -\frac{\pi^2 k^2 T}{3e}\frac{1}{k_F}\left[-4 + \frac{1}{k_F}\frac{\sum_{s=0}\Gamma(R_s, T)\,v_s \sin 2k_F R_s/k_F R_s}{\sum_{s=0}\Gamma(R_s, T) f(R_s)\,v_s}\right]\left[\frac{\partial k_F}{\partial E}\right]_{E=E_F} \tag{7}$$

In particular, assuming the leading term to be $R_s = R$, then substituting $f(0) = 1$ and $\Gamma(0) = 1$ for $T > T_C$, one has

$$\frac{dQ}{dT} = -\frac{\pi^2 k^2 T}{3e}\frac{1}{k_F}\frac{1}{[1 + vf(R)\Gamma(R, T)]^2}\frac{\sin 2k_F R}{k_F R}\left[\frac{\partial k_F}{\partial E}\right]_{E=E_F}\frac{d\Gamma(R, T)}{dT} \tag{8}$$

According to FL the leading term satisfies $2k_F R \sim 1$; thus $dQ/dT > 0$ for electrons. Alternatively, Q can be expressed in terms of the Fourier transform of the correlation function; combining Eqs. (4) and (3), one has[14]

$$Q = -\frac{\pi^2 k^2 T}{3e}\frac{1}{k_F}\left[\frac{\tau}{\tau_0}\hat{\Gamma}(2k_F, T) + \frac{4\tau}{\tau_0}\Gamma(0, T) - 4\right]\left[\frac{\partial k_F}{\partial E}\right]_{E=E_F} \tag{9}$$

Therefore dQ/dT is governed by the temperature dependence of Eqs. (5) and (6), namely a specific heat type of singularity. Thus the similarity between dQ/dT and $d\rho/dT$ is contained in the Mott equation. To recapitulate, the specific heat and the temperature derivatives of resistance and thermopower are shaped by the same spin–spin correlation function near T_C for the nickel system; this may be a more general condition. Consequently, it is the static and not the dynamic properties of the critical fluctuations that govern the anomalous component of the thermopower and the resistivity. We have illustrated this fact empirically in a previous paper.[15] This calculation and our earlier experimental verification for the nickel system strongly suggest that equilibrium thermodynamics can be usefully invoked in consideration of the anomalous electron transport properties near Curie points. We believe this justifies the thermodynamic analyses used many years ago by many others* and most recently by us.

We shall now briefly review the explicit forms of the spin–spin correlation function. In the treatment of de Gennes and Friedel[11] the Ornstein–Zernike (OZ) form of $\Gamma(R, T)$, which is a mean-field approximation, is employed,

$$\Gamma(R, T) \sim e^{-\kappa R}/R \tag{10}$$

where $\kappa = \xi^{-1} \sim \varepsilon^{\nu}$ is the inverse correlation length. This leads to a cusplike maximum in resistivity[9,10] (e.g., as observed in $GdNi_2$) and also in thermopower.[20] Fisher and Langer had pointed out, however, that even if the spins are strongly

* See Ref. 16 for a critical review.

correlated, they will not scatter coherently if R is appreciably greater than the electron mean free path. It is conceived that short-range spin fluctuations must dominate when T is sufficiently close to T_C; the static spin–spin correlations are chosen as[12]

$$\Gamma(R, T) \sim D(\kappa R)/R^{1+\eta} \tag{11}$$

where $D(\kappa R)$ is of such a form that $d\Gamma/dT$ diverges as the specific heat and η is a small positive number. When T moves away from T_C in either direction the OZ behavior must eventually obtain. The criteria for the switching from OZ behavior to FL behavior have not yet been determined. To date there are no measurements on the thermopower of systems which are believed to exhibit clearly the dual nature of spin-disorder scatterings.

For $T < T_c$ Eq. (11) or (10) for $\Gamma(R, T)$ leads to wrong predictions for resistivity[9,10] and for thermopower as well.[20] This is possibly due to the effects of d-hole density fluctuations in addition to the long- and short-range spin fluctuations. The difficulty for constructing such a model lies in our present inability to determine the relative importance of the various mechanisms. However, if we imagine one single mechanism to dominate, then (9) suggests that critical thermopower provides a probe for that pair correlation function with fluctuations of K equal to $2k_F$.

Thermodynamics of Critical Thermopower

Upon the justification given above, we return to the rather ancient thermodynamic analysis of the thermopower.[16] From the relation $Q = -S^*/e$ the temperature derivative of thermopower is

$$\frac{dQ}{dT} = -\frac{1}{e}\frac{dS^*}{dT} \tag{12}$$

a form reminiscent of the equilibrium heat capacity $C_e = TdS/dT$. The pertinent question to raise at this juncture is whether or not the transport entropy S^* is formally equal to the static entropy S. In the linear region of thermopower, where diffusion dominates, the thermopower is surely determined by the dynamic properties and there is no justification to assume that $S^* = S$. In the critical region, where fluctuations govern, the situation warrants reconsideration. We have already shown that the collective effects on the electron transport properties from the thermal spin–spin fluctuations (or any random fluctuations) are static rather than dynamic properties. Moreover, this is reasonable since close to the critical point the fluctuation length $\xi \sim \varepsilon^{-\nu}$ is long compared to the electronic mean free path; the fluctuation contribution may very well be treated in terms of equilibrium theory. Superimposed upon the fluctuation contribution are the nonequilibrium contributions, which do not exhibit singular behavior. Also, the values of C_e include the effect of electron–phonon interaction, namely the electrons are constantly clothed by phonons independent of the static or dynamic nature of the phenomena.

In reality, most metals suffer from volume dilation, which is formally expressed in the Gibbs–Duhem equation $d\bar{\mu} = -TdS + v\,dp$. From our previous analysis it is clear that we have neglected the $v\,dp$ term here. This point will be discussed in more detail later.

Experimental Work

Reviews of the critical resistivity are available,[9,10] and therefore we shall be concerned only with the critical thermopower, which is less well understood, for the iron group, rare earths, and order–disorder metals.

Iron Group. The critical thermopower of polycrystalline nickel provides a nearly ideal test system of an itinerant ferromagnet, and the data have been cited earlier (Fig. 1). For $T > T_C$ short-range spin fluctuations appear to be appropriate, while for $T < T_C$ the roles of spin fluctuations and changes in the electronic density of states are less certain. The universal nature of critical behavior in thermopower and resistivity is best illustrated by comparing dQ/dT with $d\rho/dT$ (Fig. 3a); the straight line through all the data points completes the proof.

From the thermodynamic point of view

$$Q = Q_{\text{crit}} + Q_{\text{diff}}, \qquad C_e = C_{\text{crit}} + C_{\text{diff}} \tag{13}$$

where crit and diff refer to the contributions due to critical fluctuations and diffusion, respectively. We must have $T dQ_{\text{crit}}/dT$ proportional to $-(1/e)\, C_{\text{crit}}$. The constant of proportionality is a measure of the density of charge carriers. A graph of dQ/dT vs. C_{crit} is shown in Fig. 3(b). The slopes of the straight lines for the branches $T > T_C$ and $T < T_C$ are equal, within the experimental tolerance. Accounting for the factor of T, these data give a value of 0.85 electron per atom, whereas Hall effect measurements obtain a value of 0.82 electron per atom.[17] The values of the electronic density derived from thermopower or Hall effect have been subject to modification in perhaps different ways due to appreciable hole conduction; in comparing thermopower with specific heat, we have assumed that both 4s electrons and 3d holes are itinerant. However, since the band structure of Ni is so complex, the evidence of itinerancy in nickel drawn from such a comparison must be viewed with caution. There is no discrepancy in the sign of dQ/dT as suggested by Stoner.[18] The different intercepts in Fig. 3(b) perhaps indicate a variation of the diffusion thermopower as caused by the onset of ferromagnetism; a similar effect also appears in the analysis of resistivity.

In recent work on iron Pak and Zelenskiy[19] report agreement in the magnitude of C_{crit} with the specific heat deduced from $T dQ/dT$, within the uncertainty of charge-carrier density in iron. They have not, however, compared the form of divergence nor derived the values of the critical exponents. Also, a survey of the literature[20] shows that the thermopowers of dilute alloys of nickel all appear to inherit the qualitative features of Ni, with the magnitude of dQ/dT decreasing as the other component is increased.

Heavy Rare Earths. The Seebeck effect near the Curie points in heavy rare earths allows crystalline anisotropy and critical fluctuation effects to be studied in a single system of highly localized spins, as well as ferromagnetism and antiferromagnetism in a class of not dissimilar metals. A classic example is the systematic study by the Iowa group of the metals gadolinium, terbium, dysprosium, holmium, erbium, and yttrium.[21] Recently the resistivity,[24] thermal expansion,[22] and thermopower[23] of gadolinium have been reexamined in greater detail where pertinent to the study of critical phenomena.

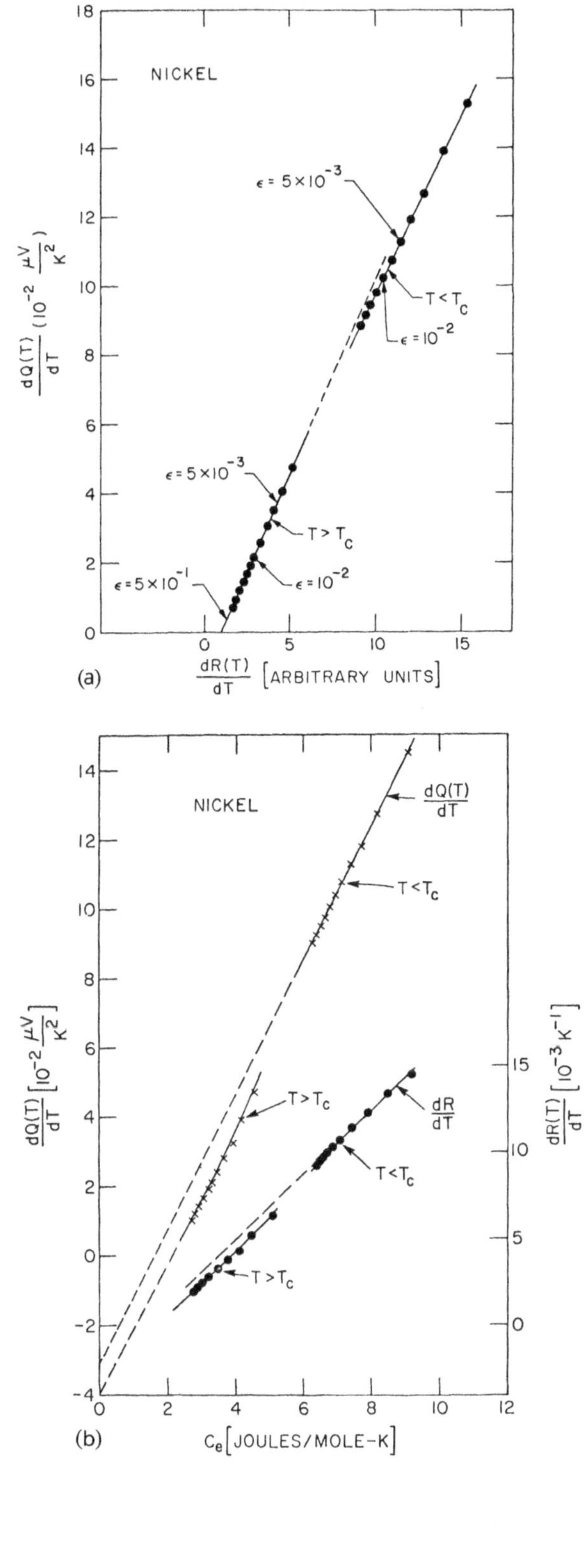

Fig. 3. Graph of dQ/dT and dR/dT vs. the magnetic specific heat of nickel (b), and comparison of dQ/dT with dR/dT as measured on the same nickel specimen (a). The specific heat data were measured by Connelly *et al.*

The disparate behavior of critical thermopower along the a and c axes of Gd is shown in Fig. 4. The a-axis behavior conforms to the general temperature dependence as observed in the isotropic nickel, while the c-axis behavior shows an anomalous temperature dependence. It is possible to reconcile the a- and c-axis thermopowers by ascribing the humpback anomaly to the effect of anomalous negative lattice expansion unique along the c axis of Gd; similar spin–lattice couplings in resistivity along the c axis of Gd have been considered by Zumsteg *et al.*[24] For thermopower, in particular, the coefficient of linear expansion α and C_e are both second derivatives of the Gibbs potential and their contributions to the thermopower have in principle the same footing, as suggested by the Gibbs–Duhem relation $d\bar{\mu} = -s\,dT + v\,dp$. Assuming that the various mechanisms are independent, the lattice term is, phenomenologically,

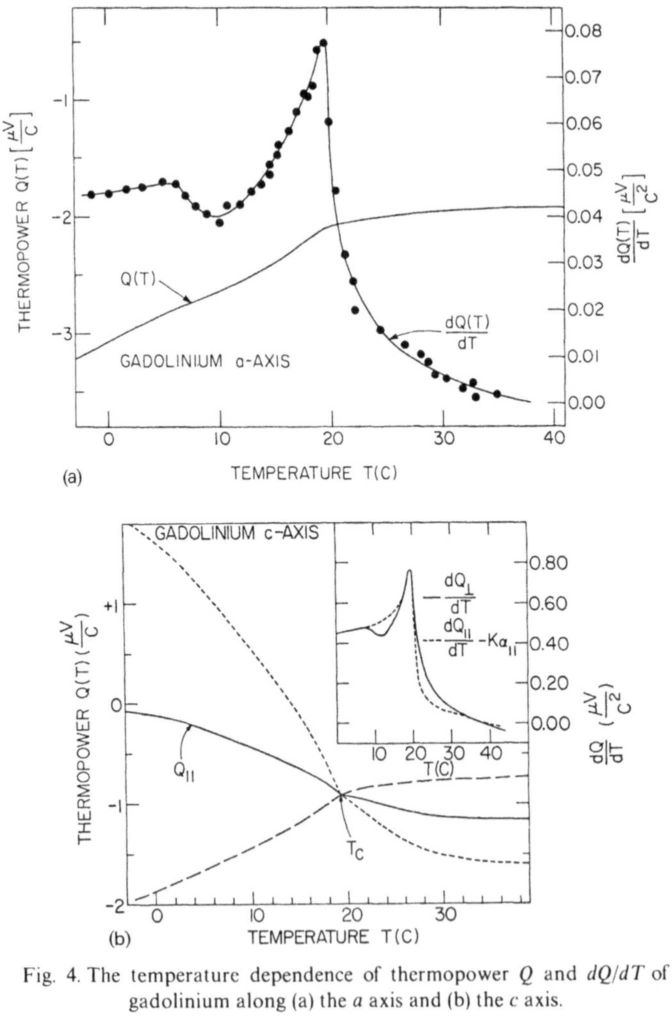

Fig. 4. The temperature dependence of thermopower Q and dQ/dT of gadolinium along (a) the a axis and (b) the c axis.

$$\left[\frac{dQ}{dT}\right]_{\text{latt}} = \left[\frac{dQ}{dL_c}\right]\left[\frac{dL_c}{dT}\right] = L_c\alpha_c\left[\frac{dQ}{dL_c}\right] \tag{14}$$

where L_c and α_c are, respectively, the c-axis lattice parameter and coefficient of linear expansion. Recalling Eq. (4), we see that the thermopower is closely bound up with the shape of Fermi surfaces and in particular with the proximity of Fermi surfaces to the Brillouin zone boundary. These effects may account for the spin–lattice couplings in thermopower.

After subtracting the lattice contribution to the c axis the remaining part consists of spin-disorder scattering and diffusion contributions. The resulting data resemble those observed along the a axis, as illustrated in Fig. 4(b). An estimate of $L_c dQ/dL_c$ can also be effected by writing $L_c dQ/dL_c = L_c(d\rho/dL_c)\, dQ/d\rho$, where the derivatives $L_c d\rho/dL_c$ and $dQ/d\rho$ can be estimated from the work of Zumsteg *et al.* and our own; one finds that

$$[dQ/dT]_{\text{latt}} = 650\alpha_c \quad \mu\text{V}/°\text{C}^2$$

which gives roughly the magnitude of negative divergence observed.

Thermodynamic analysis can be carried out for the a-axis measurements. By assuming a reasonable number of three electrons per atom the specific heat associated with the ordering of conduction electrons can be determined, which is about 1/7 of the total magnetic specific heat. Comparing with the value of the itinerant magnetic moment, which is about 1/15 of the total,[25] it appears that the conduction electrons possess about twice the entropy per electron of the localized f electrons.

All the other heavy rare earths exhibit spiral antiferromagnetism. Two additional factors are involved: (a) The spin-pair fluctuations can occur with a finite exchange momentum of the order of the antiferromagnetic reciprocal lattice vector K_A, leading to anisotropy in the transport properties with respect to the direction of K_A; (b) below the Néel temperature T_N the formation of magnetic superzone boundaries may destroy a large part of the Fermi surfaces and thus have an important bearing on the thermopower. The problem has been considered by Suezaki and Mori[26] for resistivity with the help of scaling theory (Fig. 5). Recently the thermopower has been considered by Ausloos and Lo,[27] who claim that the temperature coef-

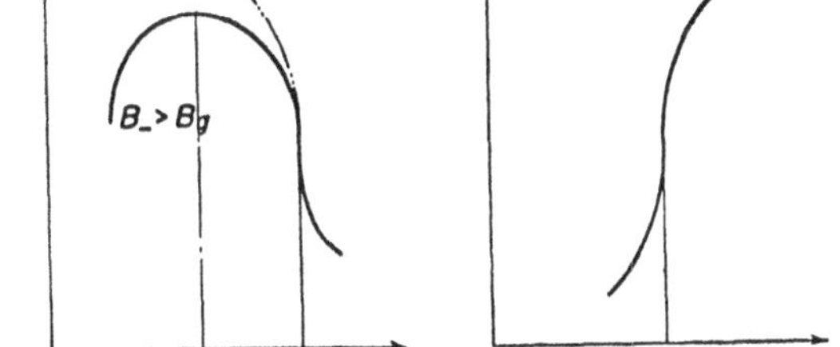

Fig. 5. Qualitative illustration of the prediction of Suezaki and Mori for the anomaly in ρ near the Néel temperature of an antiferromagnet. A similar qualitative behavior for the thermopower has been claimed by Ausloos and Lo.

ficient of Q diverges as the resistivity. That is, along the direction perpendicular to K_A the divergence is expected to correspond to the case of ferromagnets, though much stronger because of the large-angle scattering implied by K_A, while along the direction parallel to K_A the formal prediction takes the form

$$dQ/dT = -B_+\varepsilon^{-(\alpha+\gamma-1)}, \qquad T > T_N$$
$$dQ/dT = B_-\varepsilon^{-(\alpha+\gamma-1)} - B_g\varepsilon^{-(\alpha+\gamma)/2} \qquad T < T_N \tag{15}$$

where the B_+ and B_- terms represent the effects of spin fluctuations and the B_g term the effect of reducing electron density due to superzone energy gaps. In the isotropic Heisenberg model for the localized spins the values are $\alpha = 0$ and $\gamma = \frac{4}{3}$. For temperatures sufficiently close to T_N the B_g term will overwhelm the B_- term, due to a stronger temperature dependence of the exponents.

Thermopowers of antiferromagnetic Tb, Dy, Er, and Yt, but not of Ho, apparently exhibit some resemblance to that of ferromagnetic Gd.[2] There are some indications, however, that the anomalies are stronger than those of Gd along the *a* and *c* axes. All the heavy rare earths possess the quality of anomalous lattice contraction near T_N, the effects of which may not be negligible. The predictions of Eq. (15) have not been verified experimentally.

Order–Disorder Systems. Order–disorder systems which have pure second-order transitions are, from the theoretical point of view, equivalent to the localized moment antiferromagnets. However, from the point of view of the electron transport phenomena there is an essential difference: The ordering is in the ionic system and affects the electronic properties somewhat more indirectly than in the magnetic systems. For example, in β-brass, perhaps the best known of the order–disorder systems, $2k_F$ is smaller than K_0, the reciprocal lattice vector of the ordered phase, so the effects of the superzone and large-angle scattering (momentum transfers of K_0) are expected to be negligible. However, the indirect effects can still provide the same type of similarity between the transport properties and the specific heat via a theory analogous to that of Fisher and Langer. That such a theory is reasonable has been demonstrated in the simultaneous measurement of the temperature derivative of the resistivity and the specific heat carried out by Simons and Salamon[28] on β-brass. This elegant experiment showed the two quantities to have the same temperature dependence in the ordered and also in the disordered phases for $10^{-5} \leqq \varepsilon = |(T - T_0)/T_0| \leqq 10^{-1}$.

Parks,[10] in his recent review article on electron transport phenomena, has argued that β-brass is a test case for the validity of the old thermodynamic view of thermopower. As discussed above, the electronic system is only secondarily affected by the ordering and thus the thermopower will only be weakly governed by the anomaly, while Parks expects "just as strong an anomaly in β-brass as in nickel" from the scattering theory. Thus a small anomaly in β-brass is seen by Parks as offering support for the thermodynamic viewpoint.

Recent measurements by the Rochester group[29] of the thermopower in β-brass show that the anomaly in this quantity is over an order of magnitude smaller than in nickel. The Rochester group has not been able to characterize the anomaly; indeed, they have used a very large temperature gradient of $\Delta T = 1$ and $5°$K (corre-

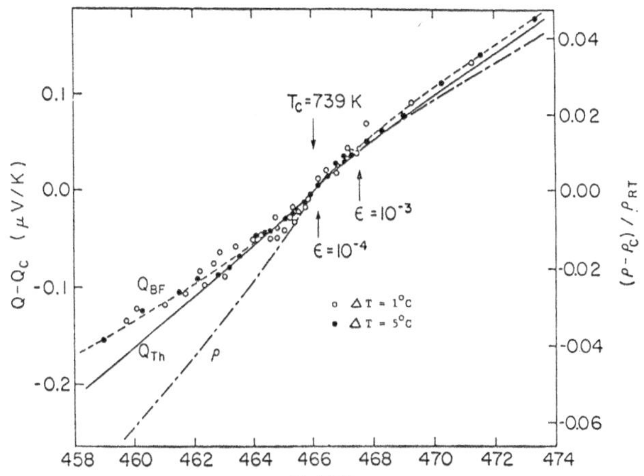

Fig. 6. The temperature dependence of the thermopower of β-brass relative to chromel in the region of the order–disorder transition. Also shown for comparison is the resistivity.

sponding to $\varepsilon = 1.4 \times 10^{-2}$ and 0.7×10^{-1}!), as shown in Fig. 6. Contrary to the conclusions drawn in that paper, we see the smallness of the effect as in fact consistent with the old thermodynamic view.

Thermopower measurements in order–disorder systems provide a unique probe of a conduction system embedded in a system undergoing a nonmagnetic ordering. It appears unlikely that the density–density, concentration–concentration, and density–concentration correlation functions used by the Rochester group provide a complete explanation. If these correlation functions are averaged over the ΔT used in their experiments, there is not enough deviation from $Q \propto T$ to characterize the data, using their estimated parameters. The anomaly may well be traced to the ion density–electron density correlations or effects due to the lattice expansion, since the thermopower appears to have a temperature dependence of similar shape near T_0. Clearly this possible lattice expansion mechanism cannot be dismissed by a comparison of the magnitude of the temperature coefficients of these properties as was done by the Rochester group, since the coupling constant is unknown. There are other possible effects which also need to be taken into account. The number of vacancies is drastically increased just above T_0 and, at least in other systems (e.g., Cu_3Au), has given rise to hysteretic effects even in the transport properties.[30] Further investigations on these systems should be very informative.

Summary

In conclusion, we have reviewed the experimental studies of thermopower of systems undergoing second-order phase transitions. The similarity between the critical behavior of thermopower and resistivity appears to be universal. The anomalous thermopower can be attributed to the effects of spin–spin fluctuations in magnetic systems and electron density fluctuations in order–disorder alloys. The old thermodynamic treatment is justified in the light that the anomalous thermopower provides a good measure of static electron entropy in metals.

Acknowledgments

The authors wish to acknowledge many useful discussions with Dr. J.I. Budnick and Dr. D.F. Brewer.

References

1. L.P. Kadanoff, W. Gotze, and D. Hamblen, R. Hecht, E.A.S. Lewis, V.V. Palciauskas, M. Royl, J. Swift, D. Aspnes, and V. Kane, *Rev. Mod. Phys.* **39**, 395 (1967).
2. R. Nathans, *Current Problems in Neutron Scattering*, CNEN (1970).
3. H.L. Swinney and H.Z. Cummins, *Phys. Rev.* **171**, 152 (1969).
4. B.I. Halperin and P.C. Hohenberg, *Phys. Rev.* **177**, 952 (1969).
5. H.E. Stanley, *Introduction to Phase Transitions and Critical Pheñomena*, Oxford, New York (1971), chapter 15.
6. C.A. Domenicali, *Rev. Mod. Phys.* **26**, 237 (1954).
7. J.M. Ziman, *Electrons and Phonons*, Oxford, New York (1960), p. 397.
8. N.F. Mott, *Proc. Roy. Soc. (London)* **A156**, 368 (1936).
9. M.P. Kawatra and J.I. Budnick, *Dynamical Aspects of Critical Phenomena*, Plenum, New York (1971), p. 257; P.P. Craig and W.I. Goldberg, *J. Appl. Phys.* **40**, 964 (1969).
10. R.D. Parks, *Magnetism and Magnetic Materials* (AIP Conf. Proc. No. 5), Am. Inst. of Physics, New York (1971), p. 630.
11. P.G. de Gennes and J. Friedel, *J. Phys. Chem. Solids* **4**, 71 (1958).
12. M.E. Fisher and J.S. Langer, *Phys. Rev. Lett.* **20**, 665 (1968).
13. M.E. Fisher, *Rep. Progr. Phys.* **30**, 615 (1967).
14. K. Levin, private communication.
15. S.H. Tang, T.A. Kitchens, and P.P. Craig, *Phys. Rev. Lett.* **27**, 593 (1971).
16. C. Herring, in *Magnetism*, Vol. IV, G.T. Rado and H. Suhl, eds., Academic Press, New York (1968), p. 129.
17. E. Fawcett and W. Reed, *Phys. Rev. Lett.* **9**, 336 (1962).
18. E.C. Stoner, *Phil. Mag.* **12**, 737 (1931).
19. V. Pak and Y.A. Zelenskiy, *Fiz. Metal. Metalloved.* **30**, 328 (1970).
20. S.H. Tang, Ph. D. Thesis, SUNY at Stony Brook, New York, 1972.
21. L.R. Silland and S. Legvold, *Phys. Rev.* **137**, A1139 (1965).
22. F.J. Cadieu and D.H. Douglass, *Phys. Rev. Lett.* **21**, 680 (1968).
23. S.H. Tang, F.J. Cadieu, T.A. Kitchens, and P.P. Craig, *Magnetism and Magnetic Materials* (AIP Conf. Proc. No. 5), Am. Inst. of Physics, New York (1971), p. 1265.
24. F. Zumsteg, F.J. Cadieu, S. Marcelja, and R.D. Parks, *Phys. Rev. Lett.* **25**, 1204 (1970).
25. R.M. Moon, W.C. Koehler, J.W. Cable, and H.R. Child, *Phys. Rev. B* **5**, 997 (1969).
26. Y. Suezaki and H. Mori, *Progr. Theor. Phys.* **41**, 1177 (1969).
27. M. Ausloos and S. Lo (to be published).
28. D.S. Simons and M.B. Salamon, *Phys. Rev. Lett.* **26**, 750 (1971).
29. G.A. Thomas, K. Levin, and R.D. Parks (to be published).
30. A.C. Damask, Z.A. Fuhrman, and E. Germagnoli, *J. Phys. Chem. Solids* **19**, 265 (1961).

A New Method for the Investigation of Thermoelectric Phenomena in Pure Metals*

J. C. Garland

Department of Physics
The Ohio State University, Columbus, Ohio

Although studies of the thermodynamic power of semiconductors and metal alloy systems have been made for many years, there has not been a concomitant degree of activity in studies of pure metals. The reason for this relative inactivity has not been so much a lack of interest, as the experimental difficulty in obtaining accurate data, particularly at liquid helium temperatures, where thermopower information would be most valuable. The magnitude of the low-temperature thermoelectric power of most pure metals is typically 10^{-7}–10^{-8} V/°K, so that if differences in temperature along a specimen are not to exceed a few millidegrees, the resulting voltages are of the order of 10^{-10} V. Although the discovery of the dc Josephson effect has made it possible in the past few years to measure voltages of this magnitude with great precision, there remains the considerable thermometry problem of resolving very small temperature gradients with microdegree accuracy. In this paper an alternate method for investigating the thermoelectric properties of metals is proposed which does not rely upon the measurement of a temperature gradient. This method has been applied to very pure tungsten with the following results: (a) Overall data resolution was at least an order of magnitude better than comparable thermoelectric power data, and (b) the data were found to be substantially more sensitive than the thermopower to low-temperature scattering effects.

This method involved the measurement of a thermoelectric ratio $G = eK_1/K_2$, where K_1 and K_2 are general transport coefficients defined by[1]

$$\mathbf{j} = e^2 K_0 \mathbf{E} + (e/T) K_1(-\nabla T) \tag{1}$$

$$\mathbf{u} = eK_1 \mathbf{E} + (1/T) K_2(-\nabla T) \tag{2}$$

($\mathbf{j}$ is the electric current density and $\mathbf{u}$ is the heat current density). The coefficients K_i are given by Fermi surface integrals of the form

$$K_i = (1/4\pi^3) \int \lambda_k \mathbf{v}_k (\varepsilon - \mu)^i (-\partial f_k^0/\partial\varepsilon)\, d^3\mathbf{k} \tag{3}$$

where the scattering processes in the metal have been represented by an energy-dependent and $\mathbf{k}$-dependent scattering length λ_k. If we assume that λ_k has the same

* Work supported in part by the National Science Foundation under Grant No. GH-33385.

functional form for all of the K_i, then G may easily be shown to be given by

$$G = e[d \log \sigma(\varepsilon)/d\varepsilon]_\mu \tag{4}$$

where $\sigma(\varepsilon)$ is the (energy-dependent) electrical conductivity and μ is the chemical potential. Under the same circumstances the thermoelectric power S, defined by $S = K_1/eTK_0$, is given by

$$S = eL_0T[d \log \sigma(\varepsilon)/d\varepsilon]_\mu \tag{5}$$

where L_0 is the classical Lorentz number. The assumption that the scattering function λ_k is common to each of the transport coefficients K_i is reasonable for "catastrophic" elastic scattering, but is likely to be in serious error for the small-angle inelastic scattering characteristic of electron–phonon collisions at low temperatures; this type of scattering has a pronounced effect on K_1 and K_2 but a relatively small effect on K_0. From an experimental point of view it is possible to use measurements of the Wiedemann–Franz (WF) ratio as a guide to the validity of Eqs. (4) and (5); when the WF ratio approximates the Lorentz number to within a few percent, then

$$S(T) = L_0TG(T) \tag{6}$$

which shows that $S(T)$ and $G(T)$ differ principally by a factor of temperature. For a free-electron metal and a simple relaxation time scattering approximation G is independent of temperature and equal to $3e/2\mu$.

While simultaneous measurements of the electrical resistance, thermal resistance, and thermoelectric power may be used to obtain the transport coefficients K_1, K_2, and K_3 and all combinations of these [including $G(T)$], this is in practice a rather cumbersome procedure. It is possible, however, to obtain $G(T)$ directly by a simple means which does not involve the measurement of a temperature gradient. According to this technique, a heat current $\mathbf{u}$ and electric current $\mathbf{j}$ are simultaneously applied to the specimen in such a way as to cancel the electric field within the metal. Dividing Eq. (1) by Eq. (2) then yields

$$G = (j/u)_{E=0} \tag{7}$$

Intuitively, the zero-field condition simply means that the "thermoelectric field" $S\nabla T$ is equal in magnitude but opposite in direction to the ordinary field $\mathbf{j}/\sigma$, and this balance condition will be reached when a voltmeter shows there to be no potential gradient along the direction of current flow.* Because the voltage null must be measured under nonisothermal conditions, it is important that the connections to the sample be made with superconducting wire to avoid spurious thermoelectric voltages.

The simple potentiometric arrangement used to measure $G(T)$ in this experiment is shown in Fig. 1(a). A heater is attached to one end of a rod-shaped sample, while the other end is maintained at a constant temperature T_0. A section of the sample is short-circuited by a superconducting Josephson junction galvanometer[2] which indicates a current flow whenever the heater is turned on. Ideally, this current would be equal to the open-circuit thermoelectric voltage $S\,\Delta T$ divided by the resistance R of the sample, although in practice it is reduced substantially by the contact

* The "observed" electric field actually differs from the real electric field by a small term equal to $-\nabla\mu$.

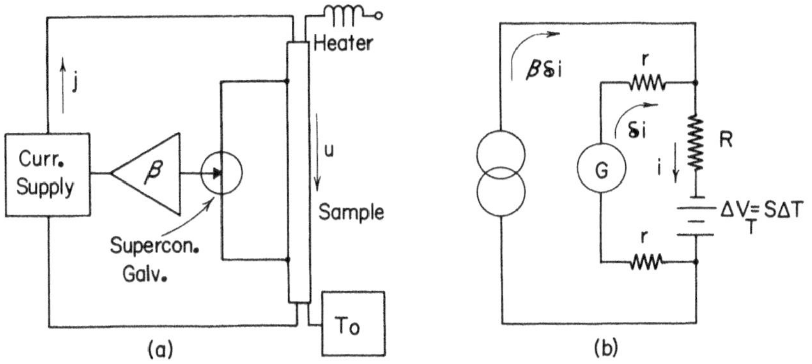

Fig. 1. (a) Schematic outline of the experiment used to measure $G(T)$. (b) Equivalent circuit representation of the experiment.

resistance between the superconducting loop and the sample. The "off-balance" indication of the galvanometer is amplified by a circuit with an effective current gain β and used to supply a feedback current $\mathbf{j}$ which restores the galvanometer to equilibrium. The measured quantities in the experiment are therefore $\mathbf{j}$, $\mathbf{u}$, and T_0. Figure 1(b) shows an equivalent circuit representation of the experiment. Here the sample is treated as a thermoelectric "battery" with internal resistance R and open-circuit voltage $\Delta V_T = S\,\Delta T$, and the feedback circuit is approximated by an ideal source of current $\beta\delta i$, where δi is the differentially small equilibrium current through the galvanometer. Summing the voltages around the loop formed by the galvanometer and the sample to zero, and supposing that the current entering or leaving a junction is conserved, we find

$$-\Delta V_T + iR + 2(\delta i)\,r = 0 \tag{8}$$

$$(\delta i)(\beta + 1) - i = 0 \tag{9}$$

where r is the contact resistance between the sample and the superconducting galvanometer leads, and i is the current through the sample. Combining Eqs. (8) and (9) then yields

$$\delta i = \Delta V_T/[2r + (\beta + 1)R] \approx \Delta V_T/\beta R \qquad \text{for} \quad \beta \gg 1 \tag{10}$$

In order for the measurement accuracy to be limited by the inherent noise of the galvanometer rather than by feedback error, we then require $\beta \geqq \Delta V_T/\varepsilon R_1$, where ε is the rms noise current in the galvanometer—in this experiment about 10^{-7} A. Choosing $\Delta V_T \approx 10^{-11}$ V and $R \approx 10^{-8}\ \Omega$ as typical values for the tungsten samples used in this experiment, we find that β must have a minimum value of 10^4 in order for the measurement accuracy of G to be noise limited. We may estimate the corresponding fractional error in G as follows: From Eq. (10) the measured feedback current $i_0 = \beta\,\delta i$ can be written as

$$i_0 = \left(\frac{\beta}{\beta + 1}\right)\frac{\Delta V_T/R}{1 + 2r/(\beta + 1)R} \approx \frac{\Delta V_T}{R}\left(1 - \frac{2r}{\beta R}\right) \tag{11}$$

The fractional error in G is then given by

$$\frac{\delta G}{G} = \frac{(\Delta V_T/R) - i_0}{\Delta V_T/R} \approx \frac{2r}{\beta R} \tag{12}$$

In this experiment $r \approx 10^{-7}\,\Omega$ and $R \approx 10^{-8}\,\Omega$, so that $\delta G/G \approx 0.1\%$, a value which corresponds closely to the measured scatter of the data.

Figure 2 shows the temperature dependence of G for three single-crystal tungsten samples over the temperature range 1.2–7.0°K. The residual resistance ratio of these crystals varied between 23,000 and 77,000. Comparative measurements of the absolute thermoelectric power over the same range of temperatures are shown in Fig. 3. The thermoelectric power data were obtained using conventional techniques; voltages were measured using a Josephson junction quantum interferometer in a potentiometric circuit, while temperature gradients were obtained by using matched bare germanium chips as thermometers in an ac differential bridge arrangement. The measurements of both $S(T)$ and $G(T)$ were made with the direction of the currents parallel to the [110] crystal axis. Although a detailed discussion of these data will be published elsewhere, there are two observations which seem pertinent here.

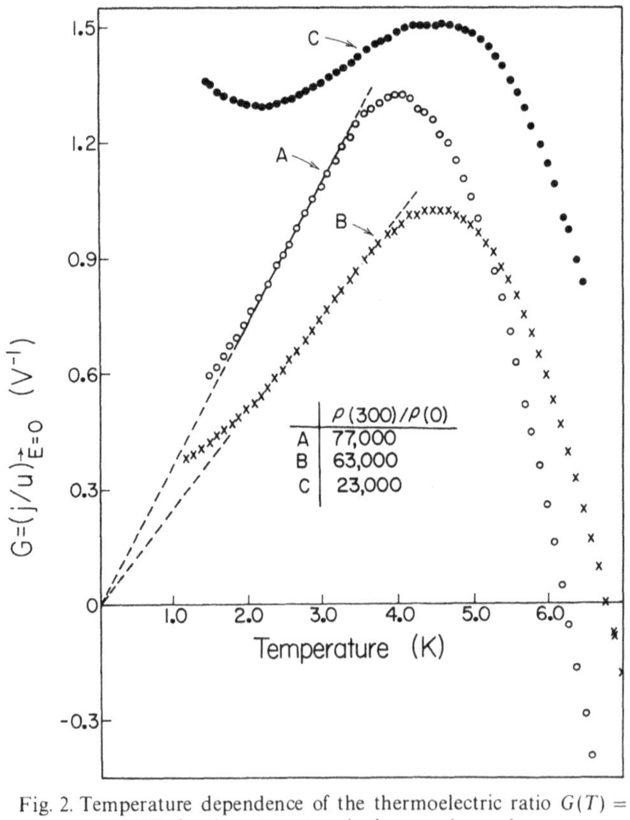

Fig. 2. Temperature dependence of the thermoelectric ratio $G(T) = eK_1/K_2$ for three tungsten single-crystal samples.

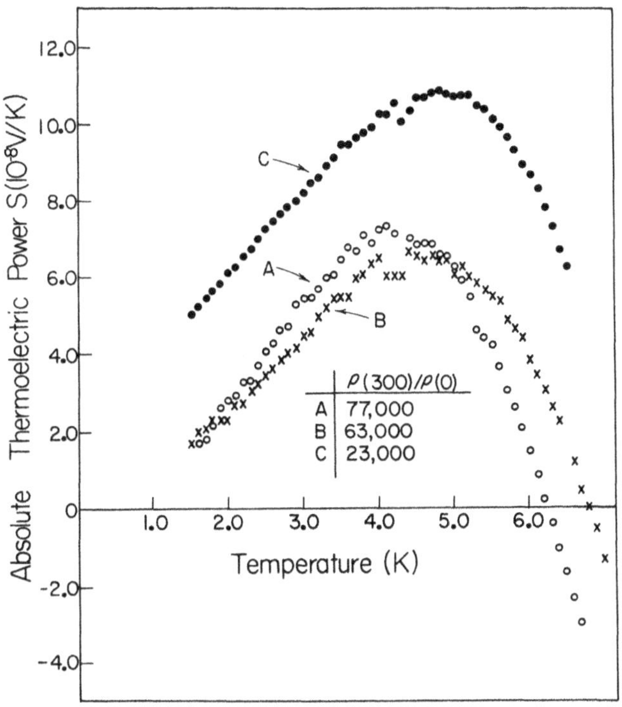

Fig. 3. Temperature dependence of the absolute thermoelectric power $S(T)$ for three tungsten single-crystal samples.

First, the resolution of the measured values of $S(T)$ is clearly inferior to the corresponding values of $G(T)$, particularly for those samples which had the largest residual resistance ratio. Although the specimens used here were of relatively modest dimensions (diameters were either 1.5 or 3.0 mm), the increasingly common use of large-diameter samples to avoid the complications of boundary scattering could make conventional thermopower measurements all but impossible. Second, the temperature dependence of $G(T)$ shows a pronounced dependence on impurity concentration below 4°K. This effect is evident as only a slight positive curvature of $G(T)$ for samples A and B at the lowest temperatures, but appears as an increase in $G(T)$ below 2°K for sample C. The comparative measurements of $S(T)$ do not show an equivalent sensitivity to different scattering rates; this is in part a result of the decreased resolution of the data, but also because the extra factor of T [see Eq. (6)] tends to dominate $S(T)$ at low temperatures.

References

1. J.M. Ziman, *Principles of the Theory of Solids*, Cambridge Univ. Press, Cambridge, England (1964), p. 194.
2. J. Clarke, *Phil. Mag.* **13**, 115 (1966).

The Thermoelectric Power of Iron*

F. J. Blatt

Physics Department, Simon Fraser University
Burnaby, British Columbia, Canada

Recently MacInnes and Schröder[1] suggested that the large maximum in the thermoelectric power of iron near 200°K could be the result of a magnetothermoelectric effect which derives from the large effective internal magnetic field associated with anisotropic (skew) scattering induced by spin–orbit interaction. They reported results of calculations based on the expression derived by Sondheimer[2] for the thermoelectric power of an ideal two-band conductor in a magnetic field. The calculated curves are in very satisfactory agreement with the measurements of Blatt *et al.*[3] In performing these calculations, MacInnes and Schröder made a number of simplifying assumptions, but used reasonable values for the various parameters that enter into Sondheimer's expression. The effective field was used as an adjustable parameter, with values that ranged between 5×10^5 and 10^7 Oe.

Sondheimer's expression, on which the results of MacInnes and Schröder rest, was derived for two bands of ideal form, i.e., spherical energy surfaces, and is valid only if the magnetic field is transverse to the temperature gradient. In a longitudinal field such a model yields no field dependence of the transport coefficients,[4] except for quantization effects which are, however, outside the range of applicability of Sondheimer's formulas.

Consequently, an examination of the thermoelectric power of iron as a function of magnitude and direction of an external magnetic field can serve as a critical test of the Sondheimer–MacInnes–Schröder theory. If it is assumed that in a polycrystalline sample in the unmagnetized state the domains are randomly oriented, a transverse field condition exists within two-thirds of the sample. If the transverse effective field is indeed crucial to the appearance of the thermopower maximum, application of a transverse external field of sufficient magnitude to orient all domains normal to the temperature gradient should result in an increase of the thermopower by roughly 50%, whereas application of a corresponding longitudinal field should greatly reduce the thermoelectric power. The measurements reported here were motivated by these considerations.

Experimental Details

Measurements were made using a cryostat of fairly simple design. The temperature of one copper block was measured using a gold–0.07% iron versus chromel

* Research supported in part by the National Research Council, Canada, and by the National Science Foundation.

thermocouple, and the temperature difference across the sample, mounted between two insulated copper blocks, was measured with a differential thermocouple of the same materials. The thermocouple was calibrated against a calibrated germanium resistance thermometer between 4.2 and 100°K, and found to differ slightly from the NBS table only below 90°K. The thermocouple readings below 20°K were corrected for the field dependence of the thermocouple using the results of von Middendorff.[5]

A superconducting solenoid was employed to generate the field, and data were obtained for eight field values between 0 and 50 kOe. The samples came from the same roll of wire which had been used by Blatt *et al.*[3] and had a nominal purity of 99.996%. Measurements were made on the as-drawn wire as well as on wire annealed for 1 hr just above the α–γ transition.

Data were taken at temperature intervals of about 2–5°K, and the temperature difference across the sample was maintained between 1 and 5°K, the larger steps and gradients being employed at the higher temperatures. The raw data were then analyzed using a least-squares fit routine. In every instance the standard deviation of the calculated curve was less than 10^{-2} μV/°K, and none of the individual data points deviated from the corresponding smoothed curve by more than 0.15 μV/°K.

Results and Discussion

The results of these measurements are summarized in Figs. 1–3. Figures 1 and 2 show only curves for fields of 0 and 2.0 kOe, transverse and longitudinal, since

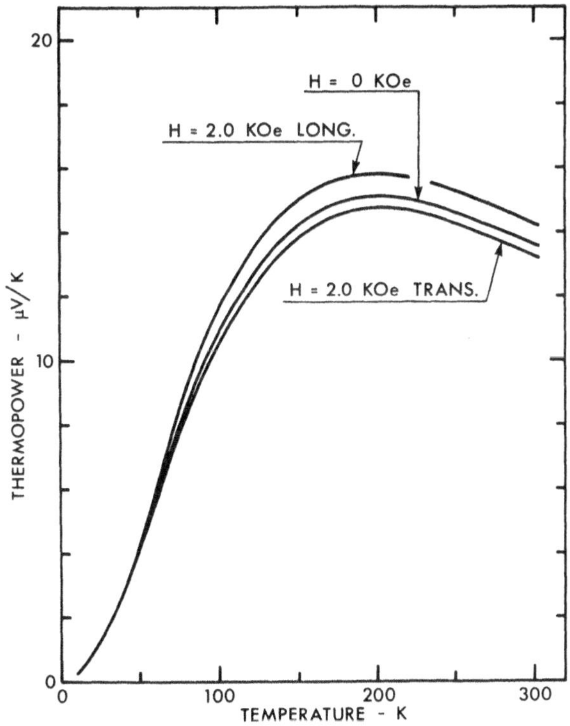

Fig. 1. The thermoelectric power of unannealed iron at zero field and in transverse and longitudinal fields of 2.0 kOe.

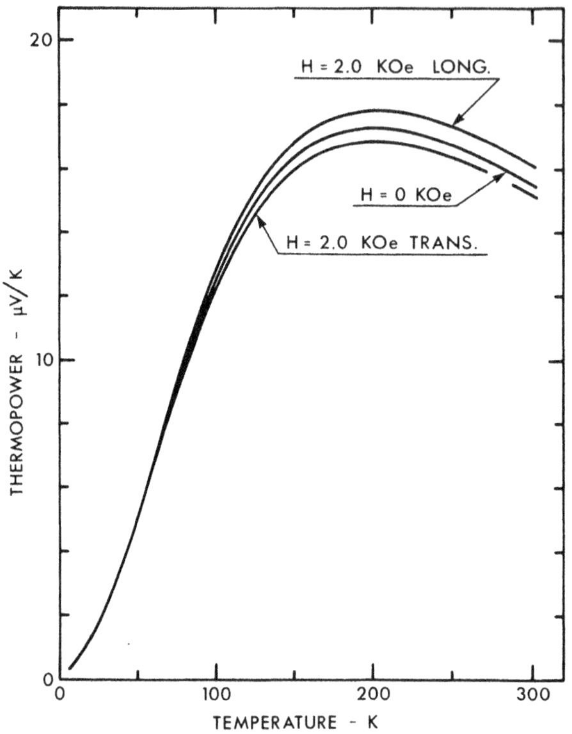

Fig. 2. The thermoelectric power of annealed iron at zero field and in transverse and longitudinal fields of 2.0 kOe.

the curves for the other field values practically overlap those for 2.0 kOe. That this is so at 200°K can be seen from Fig. 3, and it is true over the entire temperature range studied. The zero-field results are in good agreement with previous measurements[3] and the field dependence of the thermopower is in satisfactory agreement with earlier results of Kousmine,[6] who made measurements at room temperature.

Contrary to what might have been anticipated according to the Sondheimer–MacInnes–Schröder argument, the effect of domain orientation is rather small and, moreover, is in the wrong direction. That is, a transverse field results in a slight reduction of the thermopower instead of a large increase, and a longitudinal field causes a small increase instead of a substantial reduction. The small changes in the thermopower that are observed are clearly associated with domain alignment; once technical saturation has been attained further increase of the magnetic field produces no significant change in the thermopower.

Evidently, the MacInnes–Schröder theory appears to be unsatisfactory, at least in its present form. Our results do not, of course, argue against spin–orbit-induced scattering as a possibly important mechanism in producing the large thermoelectric peak in iron. They do show, however—and this comes as no great surprise—that a calculation based on the assumption of spherical energy surfaces cannot yield results in agreement with observation. Whether the use of a more realistic Fermi surface[7] might lead to results consistent with our data remains to be seen.

Since the thermoelectric behavior of iron appears to be still unresolved, it is not inappropriate to speculate about other possible explanations. Blatt *et al.*[3] suggested magnon drag as one possibility, but this appears rather unlikely in light of theoretical calculations.[8] In transition metals electron–electron-induced *s-d* transitions can play an important role in transport, and in the case of the thermopower lead to anomalous behavior.[9] Colquitt *et al.*[9] did not allow for a change in the Fermi wave vector of *d* electrons with temperature, and since their results are sensitive to the ratio k_s/k_d and since in iron k_d does change significantly with temperature, these calculations can also yield curves resembling that of iron. However, since the work of Colquitt *et al.* is also based on a spherical energy surface model, further elaboration which retains that model is hardly warranted.

Blatt *et al.* dismissed phonon drag on the grounds that generally the phonon-drag maximum appears at a temperature of about $\theta_D/5$, where θ_D is the Debye temperature, whereas in iron the maximum appears at about $\theta_D/2$. Moreover, the dependence of the peak on cold-work and alloying did not appear to conform to expectations for phonon drag, and the temperature dependence of the thermopower at low temperatures followed more closely than expected for magnon drag than phonon drag. However, the T^3 dependence of phonon drag rests on the Debye

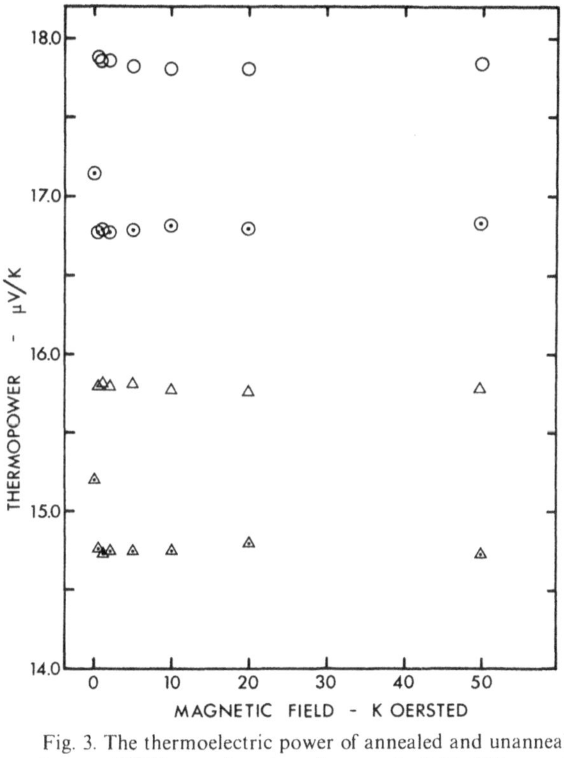

Fig. 3. The thermoelectric power of annealed and unannealed iron at 200°K as a function of magnetic field. (○) annealed iron, longitudinal field; (⊙) annealed iron, transverse field; (△) unannealed iron, longitudinal field; (◬) unannealed iron, transverse field.

approximation and the assumption of a Fermi surface that is, if not spherical, at least of roughly spherical form; the dependence of phonon drag on impurity content also depends on the Fermi surface, and at low temperatures phonon drag may be enhanced or, as is more common, diminished due to alloying;[10] finally, there is no *a priori* reason which prohibits the appearance of a phonon-drag maximum at $\theta_D/2$ rather than $\theta_D/5$. Hence, in view of the better understanding gained in the intervening years, perhaps phonon drag should not be dismissed so lightly.

Acknowledgments

I wish to thank the Physics Department of Simon Fraser University for the hospitality extended to me during 1972, and especially Prof. D.J. Huntley, who gave me free access to his laboratory and shared his equipment with me.

References

1. W.M. MacInnes and K. Schröder, *Phys. Rev. B* **4**, 4091 (1971).
2. E.H. Sondheimer, *Proc. Roy. Soc.* **A193**, 484 (1948).
3. F.J. Blatt, D.J. Flood, V. Rowe, P.A. Schroeder, and J.E. Cox, *Phys. Rev. Lett.* **18**, 395 (1967).
4. F.J. Blatt, *Physics of Electronic Conduction in Solids*, McGraw-Hill, New York (1968).
5. A. von Middendorff, *Cryogenics* **11**, 318 (1971).
6. T. Kousmine, *Helv. Phys. Acta* **7**, 732 (1934).
7. A.V. Gold, L. Hodges, P.T. Panousis, and D.R. Stone, *Intern. J. Magnetism* **2**, 357 (1971).
8. L.E. Gurevich and I.Y. Korenblit, *Soviet Phys.—Solid State* **6**, 1960 (1964); M. Roesler, *Phys. Stat. Sol.* **7**, K75 (1964); C.M. Bhandari and G.S. Verma, *Nuovo Cimento* **60B**, 249 (1969).
9. L. Colquitt, H.R. Fankhauser, and F.J. Blatt, *Phys. Rev. B* **4**, 292 (1971).
10. C. van Baarle, *Physica* **33**, 424 (1967); R.S. Crisp and J. Rungis, *Phil. Mag.* **22**, 217 (1970).

Low-Temperature Thermoelectric Power of *n*-Type GaAs in a Strong Magnetic Field*

J. P. Jay-Gerin

Centre de Recherches sur les Très Basses Températures
Grenoble—Cedex, France

Introduction

Askenazy *et al.*[1] have shown that a sample of *n*-type GaAs with $N = 3.8 \times 10^{16}$ electrons/cm^3 at $T = 2°$K experiences a large increase in both longitudinal, $\rho_{\parallel}$ and transverse, $\rho_{\perp}$, magnetoresistivity starting at $\sim$ 50 kG, reaching a maximum at $\sim$ 100 kG, and decreasing again to a relatively small and slowly varying value by about 150 kG. This behavior is associated with the transition to a nondegenerate electron state which takes place when the lowest Landau level is driven through the Fermi level.

Jay-Gerin and Wallace[2] have recently studied this transition to nondegeneracy by calculating the electronic specific heat of an isotropic semiconductor (GaAs) in the extreme quantum limit. With increasing field the curve of C_v vs. H rises rapidly to a maximum, then decreases toward the value $\frac{1}{2}Nk_B$ (N is the number of electrons per unit volume, k_B is Boltzmann's constant) for large fields, indicating that the electrons simulate a one-dimensional non-degenerate gas.

In this work we wish to study, under certain simplifying assumptions, the thermoelectric power (TEP) of a semiconductor (*n*-type GaAs) at very low temperatures when subjected to a magnetic field strong enough so that only the $n = 0$ Landau level is occupied. We consider a sample of high purity so that the inequality $\omega_c\tau \gg 1$ ($\omega_c = eH/m^*c$ is the cyclotron frequency and τ is the electron relaxation time) is verified. GaAs is chosen because of its very small g-factor ($g \approx 0.32$), so that Zeeman splitting can be ignored. As is well known,[3] the TEP of a semiconductor at low temperatures is accounted for by the sum of two contributions: S_e, the electronic diffusion component, and S_{ph}, the phonon-drag component. For low electron concentrations these two parts can be computed independently of each other. We report here the calculation of both S_e and S_{ph} in the extreme quantum limit on the basis of a parabolic band model for electrons. Details of this calculation will be published elsewhere.

Calculation of S_e

Using the π-approach method to thermoelectric problems (∇T, the thermal gradient, is assumed to be always zero),[4] we consider the electron transport phenom-

* This work was performed at the Eaton Electronics Research Laboratory, McGill University, Montreal, Canada.

ena in the presence of a strong magnetic field $\mathbf{H} \parallel Oz$ and a vanishingly small electric field $\mathbf{E} \parallel Ox$. In sufficiently strong transverse magnetic fields the purely electron diffusion TEP is independent of the electron scattering parameters and can be calculated accurately from knowledge of both the electron concentration and the band structure. For the parabolic band S_e is found to be

$$S_e = -\frac{k_B}{e}\left[\frac{3}{2}\frac{\mathscr{F}_{1/2}(\xi)}{\mathscr{F}_{-1/2}(\xi)} - \xi\right] \tag{1}$$

with

$$\xi = (\zeta - \tfrac{1}{2}\hbar\omega_c)/k_B T$$

Here, e is the magnitude of the electronic charge, m^* is its effective mass, c is the velocity of light, and the $\mathscr{F}$'s are the Fermi–Dirac integrals.[5] The Fermi energy ζ strongly depends upon the magnetic field strength in the extreme quantum limit.[2,6]

Calculation of S_{ph}

Consideration of the electronic motion in the presence of crossed electric ($\mathbf{E} \parallel Ox$) and magnetic ($\mathbf{H} \parallel Oz$) fields shows that electrons mainly drift along the y direction with the Hall velocity $\bar{v}_y$ which is independent of the electron-scattering mechanisms and much greater than the mean electron velocity parallel to the electric field, v_x. To a zero order approximation with respect to scattering, $\bar{v}_x = 0$ and $\bar{v}_y = -cE/H$. In this case, because of the strong electron–phonon interaction, the electron flow drags phonons along the y direction, inducing the Peltier heat $(U_{ph})_y$. The phonon-drag TEP is then given by

$$S_{ph} = (1/T)\left[(U_{ph})_y/E\right]\rho_{xy}(H) \tag{2}$$

where $\rho_{xy}(H) = H/Nec$ is the element of the magnetoresistance tensor.

Now, the usual Boltzmann transport equation for phonons of mode $\mathbf{q}, j$ leads to

$$\mathrm{U}_{ph} = \sum_{\mathbf{q},j} \hbar\omega_{qj}\,\mathbf{v}_{qj}\,\tau_{ph}\,(\partial N_{qj}/\partial t)_{e\text{-}ph} \tag{3}$$

Since the electron constant-energy surfaces are assumed to be isotropic, only the longitudinal phonons with $j = l$ are effective in the phonon-drag phenomenon. For the acoustic modes of vibration we take the dispersion relation $\omega_{\mathbf{q}} = sq$, where s is the velocity of sound, q is the magnitude of $\mathbf{q}$, and $\mathbf{v}_{\mathbf{q}} = s(\mathbf{q}/q)$ is the phonon group velocity. The phonon relaxation processes other than those on electrons are approximated by a relaxation time τ_{ph}. In the very low temperature range we consider here, τ_{ph} is due only to boundary scattering and therefore is independent of q and T. On the other hand, we also assume τ_{ph} to be independent of H. $(\partial N_{\mathbf{q}}/\partial t)_{e\text{-}ph}$ is the time rate of change of the phonon distribution $N_{\mathbf{q}}$ due to the electron–phonon collisions, and is calculated correct up to the first order in the electric field and to the second in the electron–phonon interaction. In the calculation we suppose that all electrons are in the ground oscillator state with the quantum number $n = 0$ and that n does not change during scattering. The results we obtain are given by the following equations:

$$S_{ph} = -\frac{k_B}{e}\frac{1}{4\pi^2 N}\int_0^{q_{max}} dq\, q^2 C_{ph}\left(\frac{\hbar\omega_q}{k_B T}\right)[Y_1(q) - Y_2(q)] \tag{4}$$

where $q_{max} = k_B\Theta_D/\hbar s$, with Θ_D the Debye temperature; $C_{ph}(x) = x^2 e^x/(e^x - 1)^2$ is the Einstein function; and

$$Y_1(q) = [\alpha(q)]^{-1}\int_0^{\alpha(q)} dz\, \eta(q, z)$$

$$Y_2(q) = [\alpha(q)]^{-3}\int_0^{\alpha(q)} dz\, z^2\eta(q, z)$$

with

$$\alpha(q) = \tfrac{1}{2}(\hbar^2 q^2/2m^* k_B T)^{1/2}$$

$$\eta(q, z) = \Gamma(q, z)/[\Gamma(q, z) + \tau_{ph}^{-1}]$$

$$\Gamma(q, z) = \frac{\gamma(q)\exp(4k_B T z^2/\hbar\omega_c)}{|z|\{\cosh[z^2 + (\hbar\omega_q/4zk_B T)^2 - \xi] + \cosh(\hbar\omega_q/2k_B T)\}}$$

and

$$\gamma(q) = \frac{\Omega}{4\pi}\frac{\alpha(q)}{q}|C(q)|^2\omega_c\left(\frac{2m^*}{\hbar^2}\right)^2 \sinh\left(\frac{\hbar\omega_q}{2k_B T}\right)\exp\left(-\frac{\hbar^2 q^2}{2m^*\hbar\omega_c}\right)$$

Here Ω is the volume of the crystal and $C(q)$ describes the mode of electron coupling to acoustic phonons.

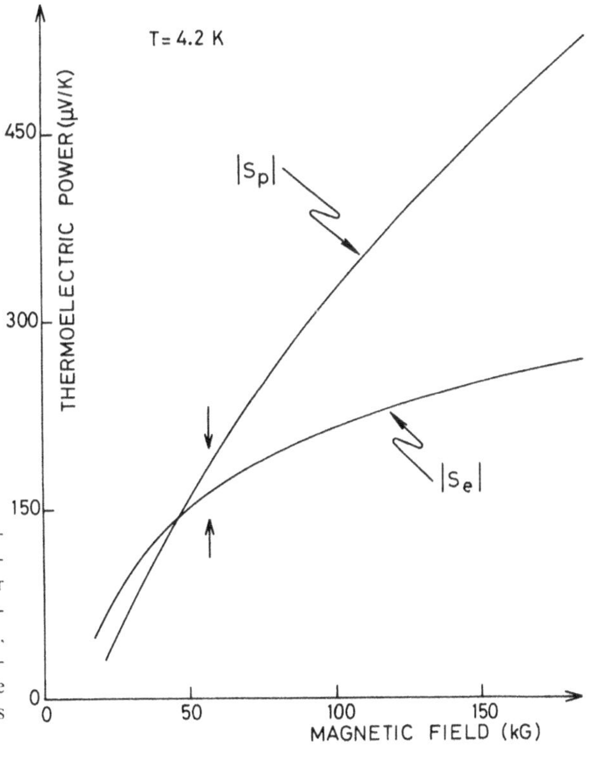

Fig. 1. Electron-diffusion and phonon-drag components of the TEP as a function of magnetic field, calculated for n-type GaAs with $N = 1.2 \times 10^{16}$ electrons/cm^3, at a temperature $T = 4.2$ K, and for $\tau_{ph}^{-1} = 5 \times 10^7$ sec^{-1}. The arrows indicate the field at which the bottom of the $n = 0$ Landau coincides with the Fermi level.

Numerical Results for GaAs

Numerical calculation of both S_e and S_{ph} for GaAs with $N = 1.2 \times 10^{16}$ electrons/cm^3 at $T = 4.2°$K is carried out in the case of scattering due to the deformation potential, for which we have[3]

$$|C(q)|^2 = C^2 \hbar q / 2\rho s \Omega$$

where C is the deformation potential constant and ρ is the density of the crystal. The following numerical constants are used: $C = 7.0$ eV, $\rho = 5.31$ g/cm^3, $s = 5.24 \times 10^5$ cm/sec, $\Theta_D = 347°$K, and $m^* = 0.07\ m_0$ (m_0 is the free-electron mass). The results are shown in Fig. 1.

Acknowledgments

The author wishes to thank Prof. P.R. Wallace, Dr. O.P. Gupta, and Prof. J.D.N. Cheeke for many helpful and stimulating discussions.

Note Added in Proof:

There are three major conclusions from this study: (1) The change in the electron distribution from degenerate to nondegenerate, which takes place at ≈ 56 kG, has a large effect on the variation of S_e and S_{ph} with magnetic field. This effect is essentially characterized by a large and rapid increase in both $|S_e|$ and $|S_{ph}|$. (2) Unlike $\rho_{\parallel}$ and $\rho_{\perp}$, no anomalous behavior occurs at the transition to nondegeneracy. (3) For large magnetic fields, $|S_e|$ and $|S_{ph}|$ increase monotonically with H, although the rate of increase of $|S_e|$ is much slower than that of $|S_{ph}|$.

References

1. S. Askenazy, J.P. Ulmet, and J. Leotin, *Sol. St. Commun.* **7**, 989 (1969).
2. J.P. Jay-Gerin and P.R. Wallace, *J. Low Temp. Phys.* **7**, 525 (1972).
3. S.M. Puri, *Phys. Rev.* **139**, A995 (1965).
4. C. Herring, T.H. Geballe, and J.E. Kunzler, *Phys. Rev.* **111**, 36 (1958).
5. J.S. Blakemore, *Semiconductor Statistics*, Pergamon Press, Oxford (1962), p. 346.
6. P.R. Wallace and O.P. Gupta, *Phys. Stat. Sol.* **50**, 619 (1972).

Lattice Thermal Conductivity of Cold-Worked *Cu*–10 at.% Al between 0.5 and 4°K*

J. A. Rowlands,† A. Kapoor and S. B. Woods

Department of Physics, University of Alberta
Edmonton, Alberta, Canada

Thermal conductivity measurements were made on a cold-worked copper alloy to test various theories[1-3] which have been put forward to describe the effect of dislocation scattering on the lattice thermal conductivity. A general characteristic of these theories is that below some temperature where the dominant phonon wavelength becomes larger than a characteristic length associated with some feature of the dislocation structure the dislocation scattering of phonons decreases and consequently the lattice thermal conductivity increases. Two groups of workers[2,3] have made several series of measurements between 4.2 and 1.2°K on *Cu*–Al alloys. However, despite rather similar experimental results, their interpretations differ markedly. It has been suggested that further measurements at lower temperatures might clarify the situation.

Therefore we made measurements in an apparatus similar to that described by Jericho,[4] and took precautions, particularly with the calibration of the germanium thermometers, so that we could, we believe, measure κ to $\pm\frac{1}{2}\%$ over the whole range from 0.5 to 4°K. To test the reliability of the apparatus, we made measurements on a nominally pure Pd specimen with a thermal conductance very similar to that of *Cu*–10 at.% Al alloy. The experimental results for the total thermal conductivity κ were combined with the measured electrical resistivity ρ to obtain the Lorenz ratio, defined by $L = \kappa\rho/T$. Values of L for the Pd specimen are plotted against temperature T in Fig. 1. Since the Pd specimen has a residual resistivity ratio ρ_{293}/ρ_0 of 170, the Lorenz ratio at helium temperatures can be expected to *decrease* with increasing T due to the introduction of "inelastic" electron–electron scattering.[5] Measurements above 2.5°K by other authors[6] support this conclusion. Our measurements can be empirically fitted by one straight line between 0.5 and 4°K, within the experimental uncertainty. Although the extrapolated value L_0 lies above the Sommerfeld value L_S, this may not be significant because of the uncertainty of the method of extrapolation. These results verify our belief that the apparatus is accurate to $\pm\frac{1}{2}\%$.

High-purity copper and aluminum were melted in a silica crucible in vacuum by induction heating to make a homogeneous ingot, which was melted again and cast into an evacuated $\frac{1}{4}$-in.-diameter quartz tube. The rod was then reduced to $\frac{1}{8}$-in. diameter by cold-swaging. The electrical and thermal conductivities of the specimen were first measured with the specimen in this cold-worked state. The specimen was

* Work supported in part by the National Research Council of Canada.

† I. W. Killam post-doctoral fellow.

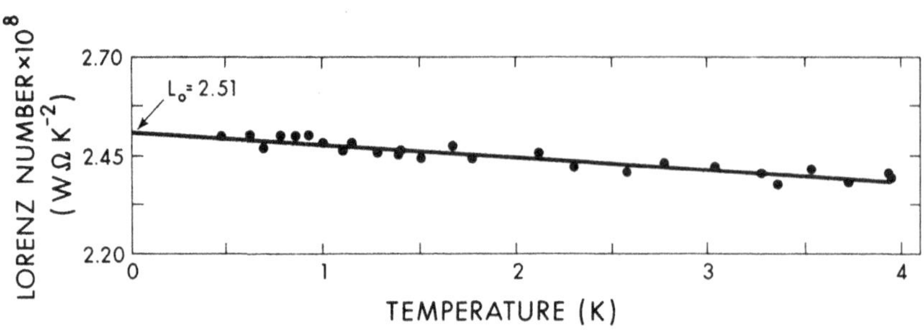

Fig. 1. Lorenz number versus temperature of "pure" palladium.

then annealed for 12 hr at 600°K and remeasured. Two further annealing stages at 675 and 1000°K, both for 12 hr, were each followed by conductivity measurements. The temperature of the final anneal at 1000°K was high enough to cause visible recrystallization. In Fig. 2 the Lorenz ratio obtained from the conductivities is shown. Since for this alloy below 4°K ρ is constant, κ_g can be easily extracted from these data. We assume the validity of the Wiedemann–Franz law for the electronic part of the thermal conductivity and subtract L_S, the Sommerfeld value for the Lorenz number, from the measured values, which is conveniently done by shifting the origin of Fig. 2 vertically. The lattice conductivity κ_g can then be recovered simply by multiplying $L - L_S$ by T/ρ. Since ρ for the most highly annealed specimen is only 10% less than that of the unannealed specimen, Fig. 2 is almost equivalent to κ_g/T vs. T plots, hence a

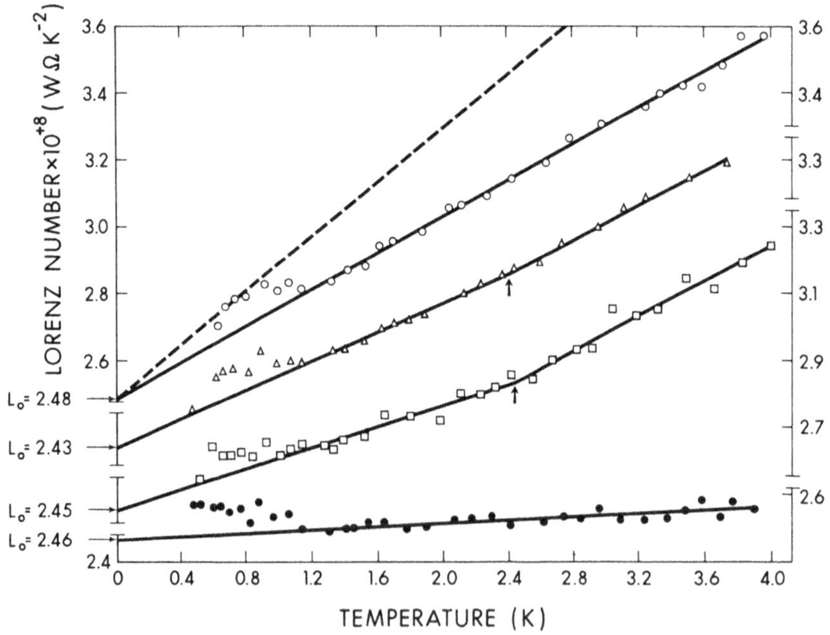

Fig. 2. Lorenz number versus temperature of *Cu*–10 at.% Al at various stages of anneal: (●) cold-worked specimen; (□) same specimen after 600°K anneal; (△) same specimen further annealed at 675°K; (○) same specimen finally annealed at 1000°K; (– – –) an estimate of $\kappa_{ge}/T \equiv 1/W_{ge}T$.

straight-line dependence of the data reveals a T^2 dependence of κ_g. This temperature dependence is characteristic of both electron and dislocation scattering of phonons.[7] If we represent the thermal resistivity due to dislocation scattering of phonons by W_{gd} and that due to electrons by W_{ge} and assume that the total lattice conductivity is related to these two resistivities by the formula[2] $1/\kappa_g \equiv W_g = W_{gd} + W_{ge}$, then $T/\kappa_g \equiv W_g T = W_{gd}T + W_{ge}T$.

In the curves where the specimen was partially annealed a "kink," designated by small arrows, is apparent at 2.5°K. This feature is absent in the fully cold-worked and fully annealed states. Although the other authors have observed a similar kink, its explanation is uncertain. Mitchell *et al.*[2] attribute the kink to a change in the electron scattering of phonons, i.e., in W_{ge} at 2.5°K. However, (1) the kink diminishes as the dislocations are annealed out, reducing W_{gd}, and (2) in our results and those of Leaver and Charsley[3] the kink is absent for the fully annealed specimen where W_{ge} would be expected to dominate. Although Mitchell *et al.*[2] see a slight kink in their annealed specimen, we believe it is consistent with a dislocation scattering mechanism, or possibly some other defect scattering, which is decreased by annealing. In Fig. 2 the apparent absence of a kink for the unannealed specimen, where W_{gd} is dominant, is simply due to the slope being so small when compared to the experimental uncertainty that the presence or absence of a kink is unobservable.

The other remarkable feature in Fig. 2 is the appearance of "humps" in κ_g/T below 1°K. If we had assumed that W_{ge} entirely dominated the phonon scattering in our most annealed specimen, as Mitchell *et al.*[2] and more recently Friedman *et al.*[8] do for theirs, we would be in the difficult position of having to explain an actual decrease in W_{ge} below 1.2°K. Instead we assume that W_{gd} becomes negligible below $\sim$ 0.7°K in this specimen and the broken line in Fig. 2 then represents κ_{ge}/T for this alloy. It is readily demonstrated that W_{gd} is not negligible at any temperature within the measurement range, by drawing lines parallel to the broken line through each L_S value. Although $W_{gd}T^2$, which according to standard theory should be constant, decreases below 1°K in each specimen, and by more than an order of magnitude in the unannealed specimen, where the effect is most marked, some defect scattering remains. It may also be noted that defect scattering sufficiently limits the conductivity of the highly annealed specimen that sensitive measurements might detect a kink near 3°K if it existed. This is a possible explanation of other experimental results.[2,8]

Preliminary measurements similar to these, on *Cu*–30 at.% Zn, indicate the presence of an even larger hump below 1.2°K, in accordance with qualitative predictions.[9] In order to determine whether the effect is due to dislocations or other defects introduced by swaging, we are attempting measurements on a well-annealed single crystal. Possibly electron microscopy can be used to determine the defect concentrations associated with conductivity changes as the crystal is cold-worked in stages.

References

1. M. W. Ackerman and P. G. Klemens, *Phys. Rev. B* **3**, 2375 (1971).
2. M. A. Mitchell, P. G. Klemens, and C. A. Reynolds, *Phys. Rev. B* **3**, 1119 (1971).
3. A. D. W. Leaver and P. Charsley, *J. Phys. F: Metal Physics* **1**, 28 (1971).

4. M. H. Jericho, *Phil. Trans. Roy Soc.* **A257,** 385 (1965).
5. J. M. Ziman, *Electrons and Phonons,* Oxford Univ. Press, New York (1960).
6. J. T. Schriempf, *Phys. Rev. Lett.* **19,** 1131 (1967).
7. P. G. Klemens, in *Solid State Physics,* Academic Press, New York (1958), Vol 7, p. 1.
8. A. J. Friedman, T. K. Chu, P. G. Klemens, and C. A. Reynolds, *Phys. Rev. B* **6,** 356 (1972).
9. P. G. Klemens, *J. Appl. Phys.* **39,** 5304 (1968).

AC Measurement of the Seebeck Coefficient of Nickel Near Its Curie Temperature*

Chester Piotrowski,† Craig H. Stephan,‡ and Jack Bass

Department of Physics, Michigan State University
East Lansing, Michigan

There has been recent interest in the behavior of the Seebeck coefficient S of Ni in the vicinity of its Curie temperature T_C. Nagy and Pal[1] reported observing anomalous behavior in S within a few degrees of T_C. Their data were not precise enough to allow any conclusions concerning the form of the "nonanomalous" portion of S. Tang *et al.*[2] on the other hand, found no anomalous behavior. They reported that dS/dT, the derivative of S with respect to temperature, had very nearly the same form as the heat capacity of Ni in the vicinity of T_C. The measurements of both Nagy and Pal[1] and Tang *et al.*[2] were made using dc temperature gradients. In this note we report the first ac measurements of the Seebeck coefficient of Ni in the vicinity of T_C. We first briefly discuss the advantages and disadvantages of ac measurements of S and describe the measuring system and procedures used. We then compare our experimental results with those obtained by Nagy and Pal[1] and Tang *et al.*,[2] and with results obtained earlier by Grew.[3]

The analysis by Tang *et al.* suggests that the quantity of primary interest in these measurements will be the first derivative of S in the vicinity of T_C. In Ni, to obtain values of this derivative having a noise level of only a few percent, it is necessary to have values for S with a noise level of less than 0.1%. With normal dc measurements it is difficult to reduce noise arising from slowly varying thermal voltages to less than about 10^{-8} V. For measurements with Ni–Pt or Ni–Au thermocouples a noise level in S of 0.1% thus requires a temperature difference of at least 1°K. Nagy and Pal used a temperature difference of 0.2°K and were only able to achieve a noise level of a few percent. Tang *et al.* used a temperature difference of 1°K, allowing a smaller noise level. With ac techniques involving phase-sensitive detection it is possible to reduce noise voltages to less than 10^{-9} V, allowing data with a noise level of 0.1% to be obtained with ac temperature differences of less than 0.1°K—ten times smaller than with dc techniques.

Our measuring system was an improved version of that previously used for measurements of the Seebeck coefficient of Al.[4] In the present experiment the sample was a 0.025-mm-thick Ni foil. The hot junction was formed by spot-welding onto this foil, as closely together as possible, a 0.025-mm-diameter Au wire and a 0.05-mm-

* Supported in part by the National Science Foundation and in part by the U.S. Atomic Energy Commission.

† Present address: Department of Physics, Cornell University, Ithaca, New York.

‡ Present address: Xerox Corp., Rochester, New York.

diameter Pt wire. This produced a parallel combination of two thermocouples, a Au–Pt thermocouple, and either a Ni–Au or a Ni–Pt thermocouple, depending upon whether Au or Pt was used as the common arm. The sample and leads were placed in an oven, the temperature of which was measured using a chromel–alumel thermocouple whose output voltage was displayed on a digital voltmeter. An ac temperature gradient was produced along the Ni sample by heating it with a chopped light beam, focused on the spot-welded region but on the opposite side of the foil from the spot welds. The resulting ac voltages from the two parallel thermocouples were phase-sensitively detected and their ratio measured using a Hewlett-Packard No. 3450A digital voltmeter. Taking this ratio eliminated the necessity of measuring the ac temperature gradient. The Seebeck coefficient of Ni was obtained by multiplying the measured ratio by the Seebeck coefficient of Au–Pt (which we measured independently using standard dc techniques), and then subtracting the Seebeck coefficient of Pt.[5]

The primary difficulty in these measurements arises from the fact that the ac temperature difference decays rapidly with distance from the point at which the ac heat is applied; for Ni in the vicinity of T_C ($\sim 631°K$) the ac temperature excursion becomes small in a distance of only about 0.5 mm. It is thus necessary to use small wires as leads and to spot-weld them closely together to ensure that the ac temperature excursions of the two thermocouples are very nearly the same. In addition, ac heating produces along the sample a dc temperature difference which must be kept small over the distance between the spot-welded Au and Pt leads. This was achieved by using samples of appropriate geometry.

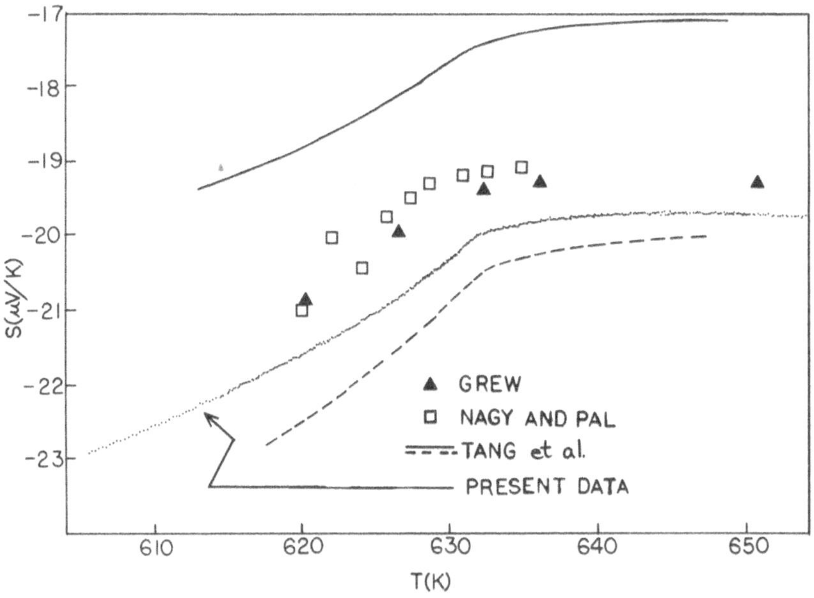

Fig. 1. The Seebeck coefficient S of nickel. The solid curve represents S for Ni as given by Tang *et al.*[6] The dashed curve represents S for Ni as determined using data by Tang *et al.*[2] for Ni–Pt and our assumed values of S for Pt.[5] The data of Nagy and Pal[1] are for a sample on which they observed no anomaly.

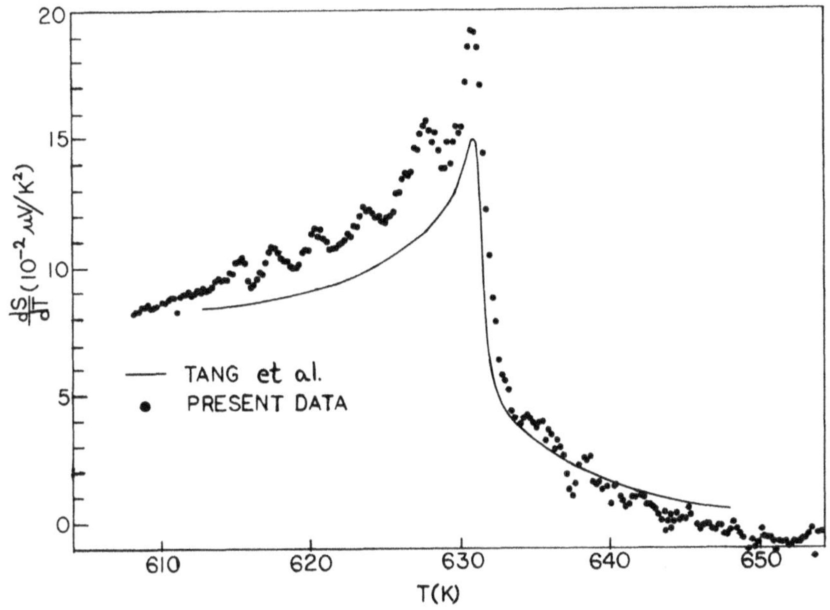

Fig. 2. dS/dT for nickel. Each of our values for dS/dT was obtained by fitting a group of 21 data points to a quadratic function and then taking the derivative of this function at the midpoint of the temperature range used. This procedure smoothed the data over temperature intervals decreasing from about 5°K well away from T_C to 1°K in the vicinity of T_C. The solid line represents dS/dT for Ni as given by Tang *et al.*[6] Both sets of data have been normalized to the same T_C. A portion of the difference between our data and theirs probably arises from different values used for the Seebeck coefficient of Pt.

A measuring run yielded approximately 500 data points, each consisting of a simultaneous measurement of the Seebeck coefficient ratio and the voltage output of the chromel–alumel thermocouple. The measurements extended over a temperature range of 50°K. Within $\pm$ 15°K of T_C we used $\frac{1}{2}$-min measuring intervals and a rate of change of the ambient temperature varying from 0.1 to 0.2°K/min for different samples.

In Fig. 1 we compare the Seebeck coefficient measured using an ac temperature excursion of 0.1°K with values obtained by other investigators. We see that our data are in good agreement with these other results. We also see that our data exhibit no anomalous behavior in the vicinity of T_C of the type reported by Nagy and Pal.[1] In a number of runs on a series of different samples we never observed any such anomalous behavior, even though we used ac temperature differences as small as 0.05°K. Thus it cannot be argued, as it might have been concerning the data of Tang *et al.*,[2] that the use of too large a temperature difference obscured the anomaly.

In Fig. 2 we compare values of dS/dT for this same sample with those found by Tang *et al.*[6] Each value for dS/dT was obtained by fitting a group of 21 data points with a quadratic function and then taking the derivative of this function at the midpoint of the temperature interval used. The complete set of data was obtained by repeating this procedure at 0.2°K intervals over the entire temperature range studied. This procedure smoothed the data over a temperature interval which decreased from about 5°K well away from T_C to 1°K in the vicinity of T_C. The use of an equal number of data points for each fit was intended to yield data having similar scatter in all temperature regions.

We see that our values for dS/dT are in semiquantitative agreement with those obtained by Tang *et al.* A portion of the difference probably arises because we used different values of S for Pt than those used by Tang *et al.*

We have now made quantitative measurements on two independent samples, with both increasing and decreasing temperature, using Pt and Au as alternative standards. In all cases we obtained results in semiquantitative agreement with those shown. As indicated by the data of Fig. 2, our values for dS/dT were generally larger than those reported by Tang *et al.*

To summarize, we have shown that it is possible to obtain accurate values for S and dS/dT using ac heating with temperature differences as small as 0.1°K. We obtained values for S and dS/dT in satisfactory agreement with those previously reported. We found no evidence of anomalies in S such as those reported by Nagy and Pal.[1]

Acknowledgments

We thank Dr. S. H. Tang for a prepublication draft of the paper by Tang *et al.*[6] We also thank Profs. P. S. Signell and J. T. Holdeman, as well as J. F. Reiser, for assistance with the computer analysis of our data.

References

1. I. Nagy and L. Pal, *Phys. Rev. Lett.* **24**, 894 (1970).
2. S. H. Tang, P. P. Craig, and T. A. Kitchens, *Phys. Rev. Lett.* **27**, 593 (1971).
3. K. E. Grew, *Phys. Rev.* **41**, 3561 (1932).
4. R. H. Freeman and J. Bass, *Rev. Sci. Instr.* **41**, 1171 (1970).
5. N. Cusack and P. Kendall, *Proc. Phys. Soc. (London)* **27**, 898 (1958).
6. S. H. Tang, T. A. Kitchens, F. J. Cadieu, and P. P. Craig, this volume.

12

Energy States and Heat

Optical and Heat Capacity Studies of the Solid Solutions $Ti_xTa_{1-x}S_2$*

J. A. Benda, C. N. King, K. R. Pisharody,† and W. A. Phillips

Department of Applied Physics
Stanford University, Stanford, California

Introduction

The layered transition metal dichalcogenides have recently been the subject of considerable interest. This is due in part to the intriguing intercalation properties of these materials, where alkali metals or organic molecules may be inserted between the layers; but in addition, the elucidation of the band structure of these quasi-two-dimensional materials presents interesting problems. In particular, the $1T$ phase of TiS_2 and TaS_2 is not understood. This phase has octahedral coordination of the metal by the chalcogen atoms, giving a single-layer repeat unit. This is in contrast to the phases with trigonal or mixed coordination, where multilayer repeat units occur. In terms of the band structure developed for octahedral coordination[1] in these compounds, $1T$-TiS_2 should be a semiconductor and $1T$-TaS_2 a metal. The resistivity of TiS_2, however, is metallic, and recent studies[2] have shown that impurities cannot account for this behavior. The required impurity concentration is greater by at least an order of magnitude than the measured value. On the other hand, TaS_2 has a resistivity which increases with decreasing temperature.[3] Wilson and Yoffe[1] have proposed the existence of a superlattice to explain this semiconducting behavior, while Geertsma *et al.*[4] have suggested that spin–orbit splitting is responsible.

In an attempt to throw light on the unusual behavior of both $1T$-TaS_2 and $1T$-TiS_2, Thompson *et al.*[2] measured the resistivity of solid solutions of $1T$-$Ti_xTa_{1-x}S_2$. They found that the resistivity of the solid solutions varied monotonically between the limits TiS_2 and TaS_2. For $x < 0.4$ the resistivity increased with decreasing temperature, but the derived activation energy was not constant. At very low temperatures the decreasing activation energy suggested impurity conduction.

In this paper we extend the studies of this solid solution through measurements of heat capacity and optical absorption at liquid helium temperatures. The results of these two measurements are presented separately in the next two sections, and the results are discussed in the final section.

* Research supported by U. S. Air Force Office of Scientific Research, Office of Aerospace Research, under Grant No. AFOSR 68-1510C. Invited paper.

† Present address: Indian Institute of Technology, Kanpur, India.

Optical Absorption

The technique of measuring optical absorption between 0.089 and 3.0 eV at liquid helium temperatures has been described in detail previously,[5] and only an outline will be given here. The radiation absorbed by the sample is measured directly, as a temperature rise, by means of a surface-doped silicon bolometer attached to the sample. Samples were attached to the bolometers by grease. The total intensity of the incident radiation is monitored by a highly absorbing surface attached to a second bolometer, and the ratio of powers gives the absolute absorption A. The absorption measurements are believed accurate to about 2–3% from 0.5 to 2.4 eV and about 4% outside of this range. The energy resolution is about 2%.

In the present experiments samples were grown by iodine vapor transport. Crystals, at intervals of $\Delta x = 0.1$, were used. For $0.1 < x < 0.9$ the crystals were the same as those used in Ref. 2. The crystals formed good optical surfaces perpendicular to the c axis.

The results of the optical absorption measurement are plotted in Figs. 1–4. By convention the reflectivity R ($R = 1 - A$ for these optically thick samples) is plotted as a function of energy. The most noticeable feature is the presence of a plasma edge which shifts upward in photon energy, but which becomes increasingly damped as one progresses from TiS_2 to TaS_2. In TaS_2 only a remnant of the plasma edge exists. Extension of the measurements (at room temperature) shows a continuously increasing reflectivity with decreasing energy.[6]

An attempt was made to fit the plasma edges using the Drude model for free-carrier absorption with three adjustable parameters: ω_p^2, the square of the plasma frequency; τ, the optical scattering time; and ε_{1b}, the core electronic contribution to the static dielectric constant. Fits, within the error of the experimental points, were

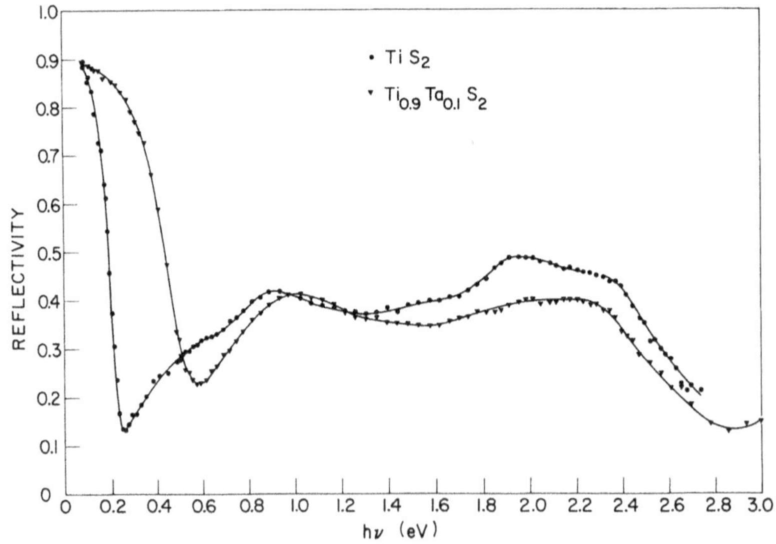

Fig. 1. Reflectivity of $Ti_xTa_{1-x}S_2$ between 0.1 and 0.3 eV; $x = 1.0$ and 0.9.

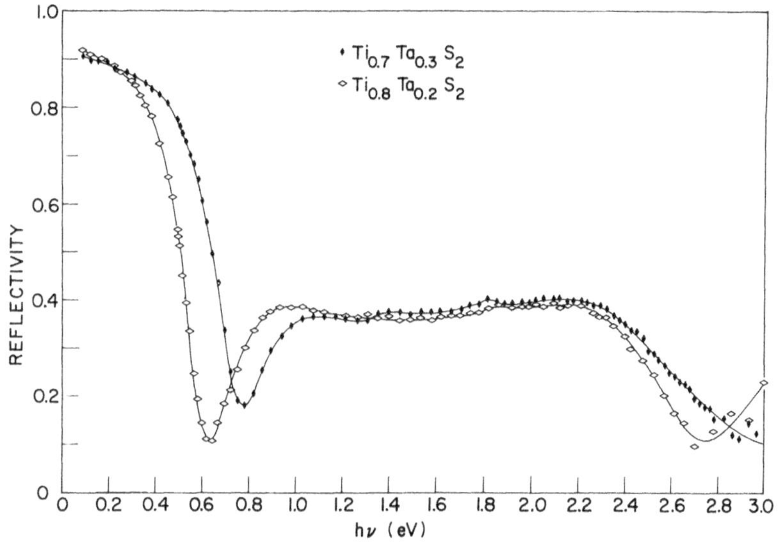

Fig. 2. Reflectivity of $Ti_xTa_{1-x}S_2$ between 0.1 and 0.3 eV; $x = 0.8$ and 0.7.

made for samples with between 0% and 40% Ta; at higher concentrations the interband transitions overlap the plasma edges.

This model assumes that interband transitions do not contribute to the dielectric constant in the frequency range of the curve fits except through the constant ε_{1b}. The first interband transitions appear to commence at about 0.95 eV, which, although above the plasma edges fitted, may still influence the derived parameters. The following scheme, using a frequency-dependent $\varepsilon_{1b}(\omega)$ to take interband transitions into account,

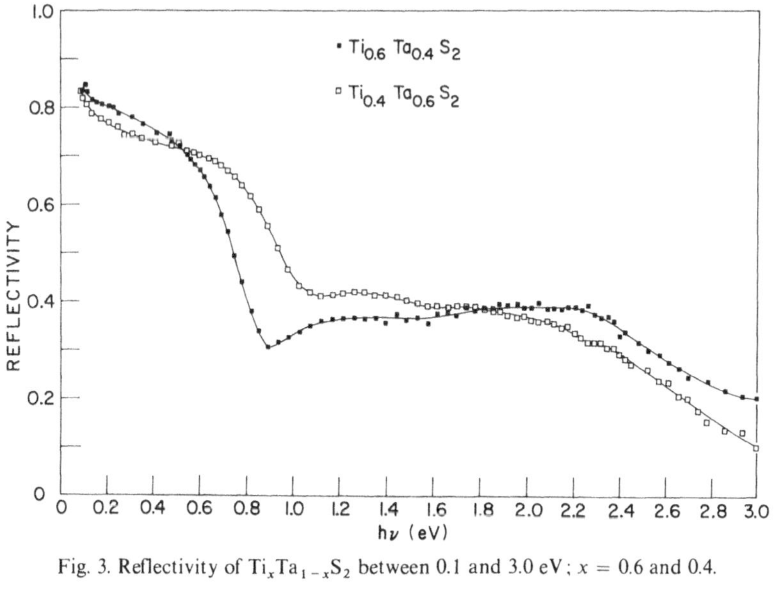

Fig. 3. Reflectivity of $Ti_xTa_{1-x}S_2$ between 0.1 and 3.0 eV; $x = 0.6$ and 0.4.

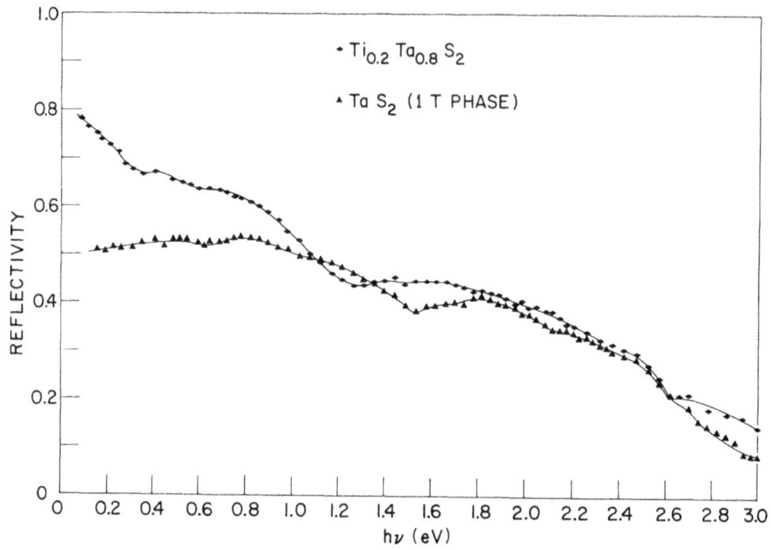

Fig. 4. Reflectivity of $Ti_xTa_{1-x}S_2$ between 0.1 and 3.0 eV; $x = 0.2$ and 0.

was based on the work of Lee *et al.*[7] They measured the index of refraction of TiS_2 directly, and after correcting for free-carrier absorption, claimed the interband portion could be fitted to a simple oscillator relation up to 0.7 eV,

$$\varepsilon_{1b}(\omega) = \frac{n_\infty^2 - 1}{1 - \omega^2/\omega_0^2}$$

where n_∞ is the electronic contribution to the long-wavelength index of refraction, and $\hbar\omega_0 \cong 1.0$ eV, the position of the first interband structure. The same plasma edges were then fitted with this $\varepsilon_{1b}(\omega)$ using n_∞ as a parameter in the fit. This second set of scattering times and plasma frequencies were all within 10% of those derived above, indicating that interband transitions do not seriously affect these values. The parameters obtained by both fitting procedures are presented in Table I and plotted

Table I

x in $Ti_xTa_{1-x}S_2$	Fit with $\varepsilon_{1b}(\omega) = \text{const}$			Fit with $\varepsilon_{1b}(\omega) = (n_\infty^2 - 1)/(1 - \omega^2/\omega_0^2)$		
	$(\hbar\omega_p)^2$, eV^2	τ, 10^{-15} sec	ε_{1b}	$(\hbar\omega_p)^2$, eV^2	τ, 10^{-15} sec	n_∞^2
1.0	0.74	12.3	13.5	.65	12.3	12.3
0.9	5.46	4.10	20.1	4.82	3.90	15.2
0.8	4.71	5.33	13.5	4.12	4.99	10.3
0.7	9.94	3.73	18.1	8.15	3.36	10.9
0.6	15.2	1.48	16.3	13.6	1.36	9.3

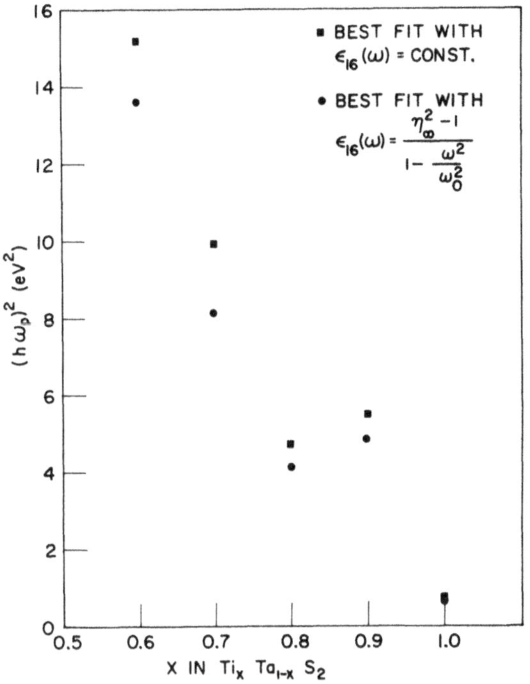

Fig. 5. The variation of the plasma frequency ω_p as a function of composition for $1.0 > x > 0.5$.

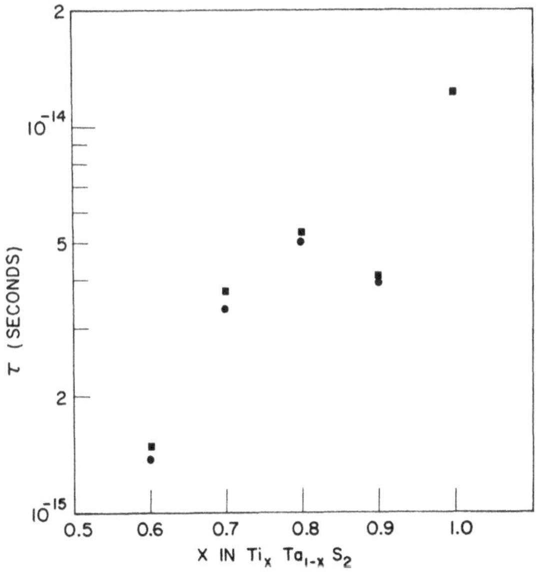

Fig. 6. The variation of the optical scattering time τ as a function of composition for $1.0 > x > 0.5$.

in Figs. 5 and 6. For $x > 0.6$ the frequency-dependent $\varepsilon_{1b}(\omega)$ does not give a better fit to the experimental data.

It will be noted from Table I that the dielectric constant ε_{1b} does not remain constant through the series. This is probably reasonable since the effect of replacing Ti by Ta is certain to change the interband transitions; in addition, the core states of Ta are likely to have a different polarizability from Ti. The value of n_∞ derived for TiS_2 is 3.5, in good agreement with the value of 3.8 reported by Lee *et al.*[7]

The position and magnitude of the peaks in reflectivity between 1.0 and 3.0 eV in TiS_2 agree well with room-temperature results.[8] The peak at 1.9 eV in TiS_2 disappears on adding Ta, while the peak at 1.0 eV shows an upward shift in energy with decreasing x. A third peak, at 2.3 eV, shows little variation with x. A possible explanation will be given in the final section.

Heat Capacity

The heat capacity of six samples has been measured between 1 and 10°K, using a technique which has been described in earlier publications.[9] Samples in the form of pressed powder pellets or single crystals are attached to a silicon bolometer. Typically the mass of the sample is 100 mg. The bolometer is attached to a large thermal mass by a link of thermal conductivity k, and the heat capacity c is determined from measurements of the time constant τ, where $\tau = c/k$. The thermal conductivity k is measured directly during the experiment. This method gives an estimated accuracy of $\pm 2\%$.

The results on four samples are presented in Figs. 7 and 8. From Fig. 7 we see

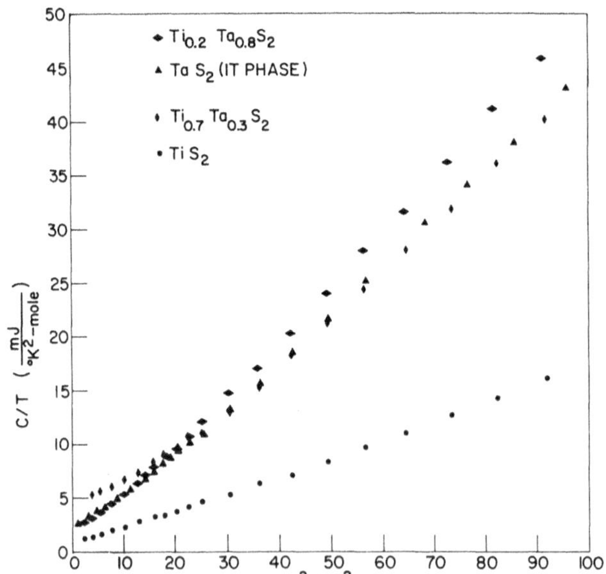

Fig. 7. The specific heat of four samples of the $Ti_xTa_{1-x}S_2$ system between 1.2 and 10°K.

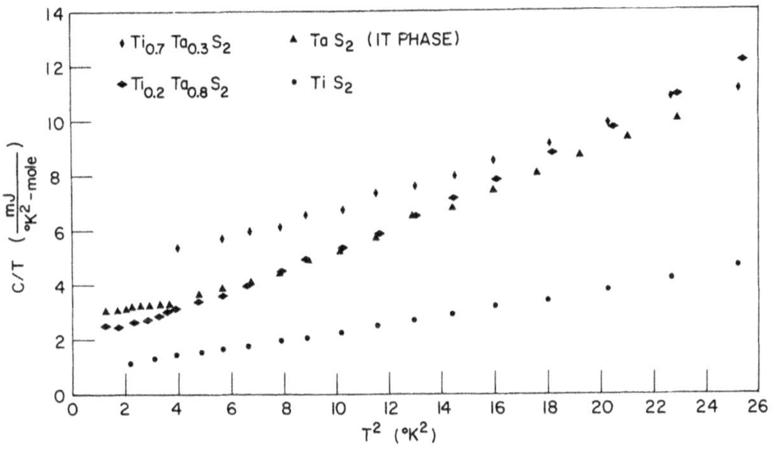

Fig. 8. The specific heat of four samples of the $Ti_xTa_{1-x}S_2$ system between 1.2 and 5°K.

the heat capacity in TiS_2 below 7°K can be written in the form

$$C = \gamma T + \beta T^3$$

where γ and β are the coefficients of the electronic and lattice contributions, respectively. This form of heat capacity is characteristic of metals (and degenerate semiconductors), although recent results in disordered systems have shown that the identification of γT with free electrons is not always correct.[10] In view of the metallic resistivity of TiS_2,[2] however, an electronic contribution to the heat capacity is probable. This result is independent of any specific model; further analysis will be left until the next section. From β we obtain a Debye temperature θ_D of 234°K. The T^3 behavior is a low-temperature approximation to the lattice heat capacity, and it appears in TiS_2 that this model can describe the data, within the scatter of the experimental points, up to 7°K, or about $\theta_D/30$. In view of the small effective modulus for transverse sound waves propagating perpendicular to the layers, this is slightly surprising, but similar results have been seen in other metallic layered structures.[11] The small sound velocity is expected only for a small range of orientation, and so these low-frequency modes may make a negligible contribution to the heat capacity.

This analysis into T and T^3 terms is not obviously applicable to all the samples, but we can use the TiS_2 results as a guide in subtracting the lattice contribution. From Fig. 7 we see that the lattice heat capacity is about three times larger at the Ta end of the composition range, which is expected on the basis of the larger mass of Ta. Therefore θ_D is smaller by a factor of 0.7, and since the form of the lattice heat capacity should be similar in all these layered structures, we expect the Debye approximation to be good to the same fraction of θ_D, 5°K rather than 7°K. We therefore fit a straight line to the results in Fig. 8, giving more weight to the high-temperature points. The derived values of γ, β, and θ_D are given in Table II. Notice that this value of γ will be an underestimate of the zero-temperature intercept.

In $Ti_{0.7}Ta_{0.3}S_2$ this procedure leads to a well-defined γ, since the experimental points appear to fall on the line, whereas in $Ti_{0.2}Ta_{0.8}S_2$ the low-temperature points are systematically high, although the deviation is small. This deviation is part of a

Table II

x in $Ti_xTa_{1-x}S_2$		γ, mJ/mole-°K^2	β, mJ/mole-°K^4	θ_D, °K
1.0	TiS_2	0.8 ± 0.1	0.15 ± 0.01	234 ± 5
0.7	$Ti_{0.7}Ta_{0.3}S_2$	4.0	0.28	190 ± 5
0.2	$Ti_{0.2}Ta_{0.8}S_2$	1.0	0.43	165 ± 4
0.2*	$Ti_{0.2}Ta_{0.8}S_2$	1.4	0.46	161 ± 3
0	TaS_2	1.4	0.38	172 ± 4
0†	TaS_2 with excess S	2.7	0.44	165 ± 4

* Crystal.
† Quenched with excess S.

systematic trend: TaS_2 shows large deviations from the straight line at low temperatures. The value of θ_D in TaS_2, however, agrees well with earlier measurements in the $2H$ phase.[11]

This type of deviation from linearity in plot of C/T against T^2 is quite common and is often identified with a Schottky anomaly due to impurities. Since the $1T$ phase is not stable at room temperature, the impurity level might be appreciable, and further details of sample preparation might be appropriate at this point. This sample was prepared by heating $2H$ powder to 900°C and then quenching to room temperature. Since all the materials appeared to convert, the impurity level should be comparable to a $2H$ sample. A second sample was prepared by quenching in the presence of excess sulfur, which tends to stabilize the $1T$ phase. The results are given in Table II. The heat capacity of the second sample could be represented by almost the same lattice contribution, but by a larger γ. The deviation of the points from the straight line at low temperatures was essentially the same. A transported crystal of the $x = 0.2$ sample gave a similar but less drastic effect. Therefore impurities certainly influence the low-temperature heat capacity in TaS_2, but at this point we have no way of determining how much of the total γ is due to this effect. Further studies are required to give a definitive answer.

A magnetic field of 6 kG did not change the heat capacity of $1T$-TaS_2 at 1.2°K (to 2%), giving some evidence that the effect is not magnetic in origin. Since only the magnitude, and not the form of the nonlattice contribution, changes with impurity level, this low-temperature heat capacity may not be resolvable into separate electronic and impurity contributions.

The most important general result of these measurements may be summarized as follows: In all the compositions studied the heat capacity indicates the presence of excitations with a density of states which is almost independent of energy. In TiS_2 this is probably electronic in origin.

Discussion

Two models have been proposed to explain the metallic resistivity of TiS_2. Wilson and Yoffe suggested that the tendency of the titanium dichalcogenides to form nonstoichiometric compounds with an excess of metal atoms led, in TiS_2, to the formation of a degenerate semiconductor. This idea has been convincingly disproved by Thompson *et al.*,[2] who show that TiS_2 as prepared by them contained no more than 0.2% excess titanium. Even in their samples, however, they observe a metallic resistivity with a carrier concentration of at least 10% of the number of titanium atoms. They argue that TiS_2 is therefore a semimetal. There appears, however, to be a slight contradiction in their discussion of the transport properties of the $TiTaS_2$ system. If TiS_2 is a semimetal, then it should contain equal numbers of holes and electrons, and the transport properties cannot be completely described in terms of a single-band model. The negative Hall coefficient indicates that the electrons have a higher mobility than the holes, even though the band structure model indicates that the electron band has *d*-like character, while the hole band is *s*- or *p*-like.

The optical results described in the present paper give additional evidence for the semimetallic character of TiS_2. We assume that the effect of adding small amounts of Ta to TiS_2 may be adequately described in terms of the rigid band model. There is some evidence for this from the interband structure in the reflectivity: The peaks at 1.9 and 2.3 eV are unchanged in position on adding tantalum. From Fig. 5 we see that the effect of adding Ta is to increase ω_p^2, and from the slope of this plot we deduce an effective mass (for the "conduction band" in the semimetal) approximately equal to the free-electron mass m_0. We are also taking ω_p^2 to reflect only the free carriers, either electrons or holes, and to be determined by some weighted average of the two.

The optical relaxation time in TiS_2 is 1.2×10^{-14} sec. Allowing for a factor of two between the dc and the optical relaxation time, we obtain a value of 2.4×10^{-14} sec for the relaxation time. Combining this with the Hall mobility,[2] and assuming single-band conduction, we find in this case a drift effective mass of $5m_0$. In view of the different definitions of effective mass, this factor of five is not unreasonable, although an analysis of the Hall data in terms of two-band conduction would tend to decrease the magnitude of the drift effective mass. The exact value depends on the relative mobilities of the electrons and holes. This Hall data, interpreted in terms of a single band, give a carrier concentration of about 2×10^{21} in TiS_2, while ω_p^2 gives a slightly lower value of 5×10^{20} cm^{-3}. This difference probably lies in the simplified analysis of the two effects, since ω_p has the same value at both low and high temperatures,[12] indicating little change of the carrier concentration with temperature. This result, that the carrier concentration is about 20% of the number of titanium atoms, is also consistent with the heat capacity data.

The optical spectra of TiS_2 is very similar to that of ZrS_2 and HfS_2 except for the peak at 0.95 eV.[8] This shifts upward in energy as Ti is replaced by Ta. All of this suggests it is a transition to the Fermi level from the valence band. In the band calculation by Bromley *et al.*[13] for zirconium and hafnium dichalcogenides the peak corresponding to 1.9 eV in TiS_2 is associated with a transition to the bottom of the conduction band, and this is consistent with its disappearing on replacing some Ti by Ta. The peak corresponding to 2.3 eV in TiS_2 is associated with a transition to a higher point in the conduction band and so remains unchanged in position or magnitude as Ti is replaced by Ta.

The most striking feature of the mixed-system optical properties is the rapid decrease of scattering time with increasing Ta content (see Fig. 6), which follows closely the behavior of the Hall mobility. The spectra qualitatively indicate that this trend continues beyond $x = 0.6$ with time constants probably in the 10^{-16} sec range near TaS_2, although no quantitative fit to the results was attempted. A lifetime of this magnitude implies a large broadening, about 10 eV, of the electron states. This may be due to correlation or localization effects, but irrespective of the exact nature of this scattering, it would appear that a free-electron model is not directly applicable to $1T$-TaS_2 or to $Ti_xTa_{1-x}S_2$ for small x. This result is consistent with the model proposed by Geertsma *et al.*, although it provides no evidence as to the origin of this localization. On the other hand, if a superlattice is responsible for the semiconducting behavior of $1T$-TaS_2, then the addition of a small amount of Ti should give metallic behavior, in contrast to the systematic trends observed in this system. However, until the effect of impurities on the heat capacity and electrical resistivity of TaS_2 is better understood, we cannot draw a more detailed picture of the electron states in this material.

Acknowledgments

It is a pleasure to thank Prof. T. H. Geballe for many stimulating discussions and much encouragement during the course of this work. We also thank J. F. Revelli for supplying us with several of the samples used in this study.

References

1. J. S. Wilson and A. D. Yoffe, *Adv. Phys.* **18,** 193 (1969).
2. A. H. Thompson, K. R. Pisharody, and R. F. Koehler, Jr., *Phys. Rev. Lett.* **29,** 163 (1972).
3. A. H. Thompson, F. R. Gamble, and J. F. Revelli, Jr., *Sol. St. Commun.* **9**, 981 (1971).
4. W. Geertsma, C. Haas, R. Huisman, and F. Jellinek, *Sol. St. Commun.* **10**, 75 (1972).
5. R. Bachmann, H. C. Kirsch, and T. H. Geballe, *Sol. St. Commun.* **9**, 57 (1971).
6. G. Lucovsky, private communication.
7. P. A. Lee, G. Said, R. Davis, and T. H. Lim, *J. Phys. Chem. Solids* **30,** 2719 (1969).
8. D. L. Greenaway and R. Nitsche, *J. Phys. Chem. Solids* **26,** 1445 (1965).
9. R. Bachmann, F. J. Di Salvo, Jr., T. H. Geballe, R. L. Greene, R. E. Howard, C. N. King, H. C. Kirsch, K. N. Lee, R. E. Schwall, H.-U. Thomas, and R. B. Zubeck, *Rev. Sci. Instr.* **43**, 205 (1972).
10. R. C. Zeller and R. O. Pohl, *Phys. Rev. B* **4**, 2029 (1971).
11. F. J. Di Salvo, R. Schwall, T. H. Geballe, F. R. Gamble, and J. H. Osiecki, *Phys. Rev. Lett.* **27,** 310 (1971).
12. G. Lucovsky, R.M. White, J.A. Benda, and J.F. Revelli, Jr., *Phys. Rev. B* **7**, 3859 (1973).
13. R.B. Murray, R.A. Bromley, and A.D. Yoffe, *J. Phys. C* **5**, 746 (1972).

Anomalous Specific Heat of Disordered Solids

Herbert B. Rosenstock

Naval Research Laboratory
Washington, D. C.

Twelve years ago, Anderson[1] noted that the specific heat in disordered solids at low temperatures was larger than elastic measurements would have implied. A simple explanation for this apparent discrepancy was offered shortly thereafter.[2] More recently the specific heat of many disordered solids has been found to be linear in the temperature at temperatures below 1°K.[3,4] This paper attempts to provide a possible explanation for this phenomenon.* We argue that cavities in a solid, each containing a weakly bound particle, can provide the proper temperature dependence.

We first observe[2] that low-frequency vibrations and elastic vibrations are not necessarily one and the same thing in disordered solids (whereas they are indeed the same thing in ordered solids). To understand this, let us visualize a cavity of somewhat large size, as will form naturally at a number of places in a disordered solid. Imagine further that inside that cavity there is located one atom, weakly bound to only one of the atoms that are bounding that cavity. The vibrations of that single atom in the cavity will be controlled by that one force constant, and will be low in frequency since the force constant is weak. Therefore, like all low frequencies, it will make an appreciable contribution to the specific heat of the glass [see Eq. (1)]. On the other hand, being essentially isolated from the rest of the lattice, it will make no contribution to its elastic properties. Such cavities, and consequently such vibrational modes which are of low frequency but not elastic ("acoustic"), cannot exist in ordered solids. In ordered solids experiments on elastic ("acoustic") vibrations will therefore detect all the low-frequency modes which contribute to the low-temperature specific heat; in disordered solids they will not. The nature of the atom in the cavity may differ in different glasses: it could be an impurity, or a "nonbridging oxygen," or some other intrinsic or external species.

To put the preceding on a more quantitative basis, we note that at temperature T the specific heat of one quantum oscillator of frequency ω is

$$c(\omega) = k(z/\sinh z)^2, \qquad z = h\omega/4\pi kT \tag{1}$$

where k is Boltzmann's constant and h is Planck's constant. An assembly of such oscillators will contribute an amount

$$C = \int c(\omega)G(\omega)d\omega \tag{2}$$

* See Refs. 5–7 for earlier attempts at an explanation.

to the specific heat of the entire solid, if the distribution of frequencies is denoted by $G(\omega)$. The preceding is true for any set of oscillators. Here we are referring specifically to the local oscillators inside a cavity as described in the preceding paragraph; the frequency of any such oscillator is given simply by

$$\omega = (\alpha/m)^{1/2} \tag{3}$$

where α is the force constant by which the mass m is bound to the cavity. We may perhaps assume that m is the same for all particles, but we do not actually know the magnitude of the "weak" force constant α or its distribution. We therefore do not know $G(\omega)$ either. Accordingly, a reasonable description of the limit of a very weakly bound particle is that of a free particle in a box. The specific heat of such a system can be written as[8]

$$C(\xi) = (3k/2)f(\xi) \tag{4}$$

where

$$\xi = \sigma/kT \tag{5}$$

$$\sigma = h^2/8mL^2 \tag{6}$$

$$f(\xi) = \begin{cases} f_1(\xi), & \text{small } \xi \\ f_2(\xi), & \text{large } \xi \end{cases} \tag{7}$$

with f_1 and f_2 given in Ref. 8, Eqs. (4) and (5). The function f_1 is actually exact for all ξ, though cumbersome to evaluate when ξ is large; f_2 gives five-figure precision for all $\xi > 2$. The expression for the contribution to the low-temperature specific heat can, instead of (2), be written as

$$C = \int_0^\infty c(\xi)n(\sigma)\,d\sigma \tag{8}$$

$n(\sigma)$ being the distribution of the parameter σ related to the size of the cavity by Eq. (6), or as

$$C/\tfrac{3}{2}k = kT\int_0^\infty f(\xi)\,d\xi\,n(\sigma) \tag{9}$$

Since σ depends on T, L, and m, Eq. (9) relates the temperature dependence of the specific heat to the distribution of cavity sizes and masses. If, in particular,

$$n(\sigma) = \text{const} \tag{10}$$

then we find that the specific heat of the substance is linearly dependent on the temperature, as has recently been observed[3,4] in silica, germania, selenium, polystyrene, and other substances.

To relate the distribution $n(\sigma)$ to the cavity size distribution $N(L)$, we use $N(L)\,dL = n(\sigma)d\sigma$ and assume m = const (only one kind of particle is involved); then Eq. (6) yields

$$N(L) = h^2n(\sigma)/4mL^3$$

and the condition for a linear T dependence, $n(\sigma)$ = const, becomes

$$N(L) \propto L^{-3} \tag{11}$$

To what extent is the assumption of an L^{-3} size distribution of cavities, and the description of an atom vibrating in it by a free particle, physically realistic? It cannot be valid for $L < 2$ Å; this would be no cavity at all, and forces would be strong. Nor is $L > 20$ Å acceptable; few, if any, cavities of such size would be expected to persist without collapse in a normal solid. However, all that is actually required is that $n(\sigma)$ be (reasonably) constant in that range for which $f(\xi)$ in Eq. (9) is appreciable. To investigate for what values of the physical parameters this will be satisfied, we therefore assume a simple $n(\sigma)$: It is constant where finite, but has an upper and lower cutoff,

$$n(\sigma) = 1/(\sigma_2 - \sigma_1)\begin{cases} 1, & \sigma_1 < \sigma < \sigma_2 \\ 0, & \text{otherwise} \end{cases} \tag{12}$$

Evaluation of Eq. (9) with Eqs. (12) and (7) has given us five figures, one each for an atomic species (atomic masses $A = 8, 16, 32, 48, 72$) and each figure showing several curves, some for "cutoffs" on cavity size that give good linearity over the entire temperature region of experimental interest, $0.1 < T < 1$, others showing substantial deviation. For brevity, only one of these figures, for $A = 16$, is exhibited here (Fig. 1), but the results of all of them are summarized in Table I.

A reasonable way to interpret the table is as follows. For a given species of atomic mass A the specific heat is sensitive to the cavity size distribution only in the range of cavity sizes from L_2 to L_1. In particular, if in that range the distribution obeys Eq. (11), a term linear in the specific heat will appear from 0.1 to 1°K.

We conclude that a linear specific heat term can be produced by cavities containing atoms of mass from about 8 to 40. Lower masses would demand the existence of

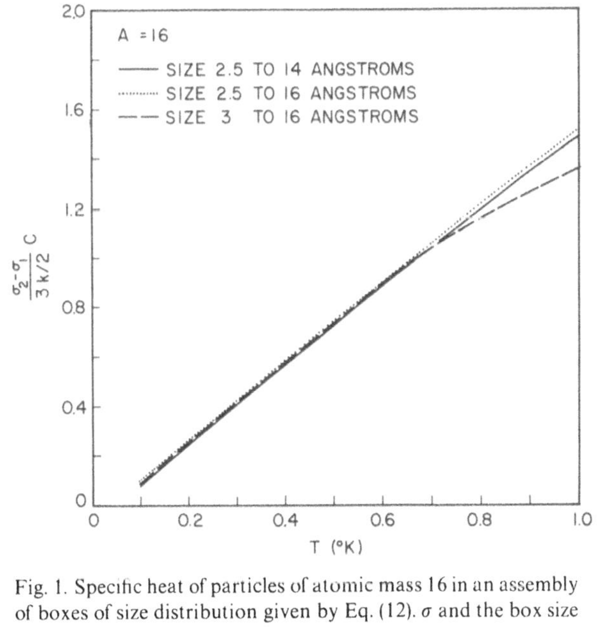

Fig. 1. Specific heat of particles of atomic mass 16 in an assembly of boxes of size distribution given by Eq. (12). σ and the box size L are related via Eq. (6).

Table I*

Atomic mass A	Permissible cutoffs on cavity size: Largest allowable lower cutoff L_2, Å	Smallest allowable upper cutoff L_1, Å
8	3	20
16	2.5	14
32	2.0	10
48	1.6	8
72	1.5	8

* Listed for several atomic species is the range of cavity sizes over which Eq. (12) for the cavity size distribution must be obeyed in order to give rise to a specific heat that is linear in the temperature range 0.1–1°K. L_2 is the largest allowable lower cutoff; L_1 is the smallest allowable upper cutoff.

cavities larger than 20 Å in length (though, on account of the L^{-3} distribution, not very many); larger masses demand the existence of cavities so small that the assumption on which our calculation is based, that the particles in them is effectively free, is no longer realistic.

References

1. O. L. Anderson, *J. Phys. Chem. Solids* **12,** 41 (1959).
2. H. B. Rosenstock, *J. Phys. Chem. Solids* **23,** 659 (1961).
3. R. B. Stephens, G. S. Cieloszyk, and G. L. Salinger (in press).
4. R. B. Stephens, R. C. Zeller, and R. O. Pohl, *Bull. Am. Phys. Soc.* **16,** 377 (1971).
5. H. B. Rosenstock, *J. Non-Cryst. Solids* **7,** 123 (1972).
6. P. Fulde and H. Wagner, *Phys. Rev. Lett.* **27,** 1180 (1971).
7. P.W. Anderson, B.I. Halperin, and C.M. Varma, *Phil. Mag.* **25,** 1 (1972).
8. H. B. Rosenstock, *Am. J. Phys.* **30,** 38 (1962).

Specific Heat of Strongly Disordered Crystalline and Amorphous Films*

W. Buckel and Ch. Ohlerich

Physikalisches Institut der Universität Karlsruhe
Karlsruhe, West Germany

Superconducting films produced in a strongly disordered state by quench-condensation of the vapor onto a cold substrate show remarkable changes (mostly enhancement) of their superconducting transition temperature T_c.[1] These interesting experimental results have been known for a long time,[2] but there was no possibility of explaining them within the framework of our knowledge of superconductivity.

Only a few years ago the extension of the BCS theory to strong coupling superconductors gave us a chance to understand the influence of disorder on T_c. In an approximation of this theory one gets

$$T_c \propto \langle \omega_D \rangle \exp[-(1+\lambda)/\lambda]$$

where the parameter λ describing the attractive electron–electron interaction via phonons is defined by

$$\lambda \equiv 2\int_0^\infty \alpha^2(\omega)\, F(\omega)\, (d\omega/\omega)$$

where ω is the phonon energy, $F(\omega)$ is the phonon density of states, and $\alpha^2(\omega)$ is the electron–phonon interaction parameter. The term $\alpha^2(\omega)F(\omega)$, which determines λ and therefore also T_c, is called the Eliashberg function.

From tunneling experiments we know that the Eliashberg function is increased at low frequencies for disordered films.[3] This means an increase of λ due to the disorder. However, it is still not clear whether the change of the Eliashberg function is mainly due to a change of $\alpha^2(\omega)$ or of $F(\omega)$. An answer can be found by measuring the specific heat of such disordered films, since the specific heat is essentially determined by $F(\omega)$.

The very small mass of such films and the necessity of measuring the specific heat immediately after the condensation at temperatures as low as possible cause some difficulties which can be solved only by special experimental arrangements. For these experiments an ac method discussed by Sullivan and Seidel[4] has been used. A special effort has been put into constructing sample holders which approximate an ideal RC component as closely as possible. This was achieved by an arrangement of several films on a thin collodion substrate. The temperature was measured by Bi–Ag thermocouples. The ac heating source was the radiation from a luminescent diode.

* This work has been supported by the Deutsche Forschungsgemeinschaft.

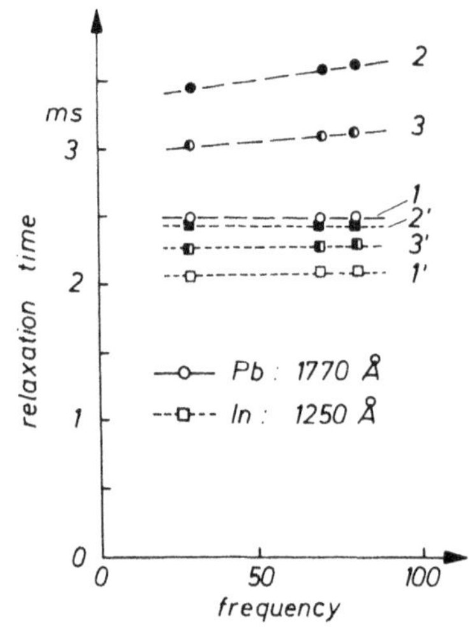

Fig. 1. Frequency dependence of the relaxation time for a Pb and an In film. Primed curves are for In: 1, 1′: τ values of the substrate. 2, 2′: τ values measured immediately after depositing the film at 4.2°K. 3, 3′: τ values measured at 4.2°K after annealing to about 200°K.

The relaxation time τ_{RC} of an ideal *RC* component should be frequency independent. Some results for a Pb and an In film are shown in Fig. 1. Only weak frequency dependences, as shown here for the Pb film, have been observed. Therefore the data can be analyzed using the simple formulas for an ideal *RC* component.

Figure 1 also shows the influence of the disorder on the heat capacity of the film. The curves marked by 1 and 1′ give the τ values for the substrate. After condensing the material at 4.2°K the larger τ values of curves 2 and 2′ were observed for Pb and In, respectively. Annealing to about 200°K decreased the τ values measured at 4.2°K for both materials. Since the thermal resistance of the sample holder remains constant during the whole procedure of condensing and annealing, the τ values immediately reflected the change of the heat capacity. The curves 3 and 3′ give the τ values after annealing. The heat capacity of the samples decreased.

Figure 2 shows the annealing behavior of the excess heat capacity due to the disorder for an In film 520-Å thick. In Fig. 2(a) the normalized specific heat, measured at 4.2°K, is plotted versus the highest annealing temperature seen by the film. The excess heat capacity decreased during annealing in a way similar to the drop in the residual resistance. This is shown in Fig. 2(b). Here the same values of the normalized specific heat are plotted versus the residual resistance of the film after the different annealing steps. These results, which have also been observed for many other films, confirm that the excess heat capacity measured immediately after the condensation is really due to the disorder.

Four different materials, pure In, pure Pb, Pb + 33 at.% Bi, and Bi + 12 at.% Sb, have been studied in the present work by this method. At the substrate temperature of 4.2°K the alloy systems grow in an amorphous state.[5,6] Some typical results are collected in Table I. For disordered Pb films and especially for the amorphous films the ratio of the heat capacities c/c_0 tends to increase with decreasing temperature. Measurements at lower temperatures could decide whether this increase is the onset of a specific heat anomaly as observed in materials like amorphous SiO_2 or GeO_2.[7]

Ewert[8] measured the specific heat of quenched condensed films in the temperature

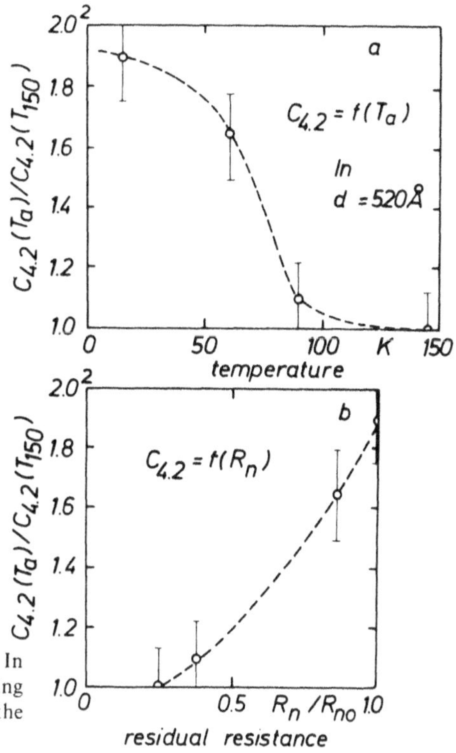

Fig. 2. Annealing behavior of the specific heat of an In film: (a) measured at 4.2°K as a function of the annealing temperature; (b) measured at 4.2°K as a function of the residual resistance.

Table I. Relative Change of the Specific Heat after Quench-Condensation*

Material	Thickness, Å	C/C_0			
		1.8 K	3 K	4.2 K	6 K
In†	520	—	—	1.9 ± 0.2	—
	1250	1.7 ± 0.2	—	1.8 ± 0.2	1.4 ± 0.2
	2350	1.5 ± 0.2	1.5 ± 0.2	1.5 ± 0.2	—
Pb†	1770	2.1	1.9	1.7	—
Pb‡	2900	3.1	2.7	2.3 ± 0.1	—
	2800	—	—	2.0 ± 0.1	—
Bi§	~1000	6(!) ± 1	—	4.7 ± 0.5	3.3 ± 0.5

* C = specific heat immediately after condensation, C_0 = specific heat after annealing and recooling.

† Crystalline.

‡ Amorphous, with 33 at.% Bi.

§ Amorphous, with 12 at.% Sb.

range from 5 to 8°K. He did not observe any change for the disordered crystalline films. Since the films were thicker and the substrate temperature during the condensation was substantially higher (~ 30°K) than in these experiments, the different results may be due to different degrees of frozen-in disorder. For the amorphous films our results extrapolated to the higher temperatures are in fairly good agreement with those of Ewert and Sander.[9]

With regard to superconductivity, the results can be compared with specific heat data calculated by Happel *et al.*[10] from tunneling experiments, assuming that $\alpha^2(\omega)$ does not depend on the disorder. From these calculations the specific heat of quench-condensed pure Pb films measured at 4.2°K should be about five times larger than after annealing to 77°K. Our experiment gave only a factor of 1.7. Also, for amorphous Pb + 33 at.% Bi the calculated enhancement is larger than the observed one. Therefore the change of the Eliashberg function determined by an analysis of tunneling experiments cannot at present be quantitatively understood by only a change of $F(\omega)$.

Acknowledgments

We gratefully acknowledge very helpful discussions with N. Barth, S. Ewert, and W. Sander.

References

1. W. Buckel and R. Hilsch, *Z. Physik* **138**, 109 (1954).
2. A. Shalnikov, *Nature* **142**, 74 (1938).
3. K. Knorr and N. Barth, *J. Low Temp. Phys.* **4**, 469 (1971).
4. P. F. Sullivan and G. Seidel, *Phys. Rev.* **173**, 679 (1968).
5. P. Baier, *Z. Physik* **213**, 89 (1968).
6. W. Sander, *Z. Physik* **147**, 361 (1957).
7. R. C. Zeller and R. O. Pohl, *Phys. Rev. B* **4**, 2029 (1971).
8. S. Ewert, *Z. Physik* **237**, 47 (1970).
9. S. Ewert and W. Sander, *Z. Physik* **247**, 21 (1971).
10. H. Happel, K. Knorr, and N. Barth, *Z. Physik* **249**, 185 (1971).

Surface Effects in the Low-Temperature Heat Capacity of Ionic Crystals*

G. P. Alldredge, T. S. Chen, and F. W. de Wette

Department of Physics, The University of Texas
Austin, Texas

The presence of surface-localized vibrational modes[1,2] is reflected in the frequency distribution as an enhancement at lower frequencies with an accompanying depletion of the bulk bands at higher frequencies.† For samples of high specific surface area this should give an observable surface-proportional enhancement of the heat capacity, which is inherently a low-temperature effect that appears most strongly in the temperature regime in which the (bulk) Debye temperature varies strongly. In this paper we apply the results of realistic lattice calculations to the prediction of the surface specific heat for four ionic crystals, with three aims in mind: (1) to compare with existing data for NaCl[3-5]; (2) to encourage similar experiments on several other crystals; and (3) to call attention to some subtle implications of the surface-phonon spectra of real crystals.

For ionic crystals of the rocksalt structure we may focus on the {100} faces, since their predominance in small-particle samples is strongly favored by the anisotropy of surface energy. We have already briefly reported applications of the shell model to the study of surface phonons for the (001) faces of NaCl[6], RbF and RbCl,[7] and LiF.[8] Following our earlier application of a rigid-ion model of NaCl to calculate the vibrational surface thermodynamic functions,[9] we use the shell model to calculate the surface specific heat for these crystals. For a film of 15 (001) layers with both surfaces free we calculate a specific heat per particle c_v. Next the bulk specific heat per particle c_v^b is calculated independently, and the surface specific heat C_v^s is determined as

$$C_v^s = (N/A)(c_v - c_v^b) \qquad (1)$$

where N is the total number of particles in the film and A is the total surface area for both sides of the film. In this work no corrections are made either for static relaxation near the surfaces or for effects of thermal expansion.

In Fig. 1 we display shell model predictions for the surface specific heat of the (001) surface of the four crystals; the temperature scale is based on the Θ_0 derived from the elastic constants of the models. (For reference, Table I gives the Θ_0 derived

* Research sponsored in part by the U.S. Air Force Office of Scientific Research under Grant No. AF OSR-71-1973, and in part by the Robert A. Welch Foundation.

† References to most work through 1969 are given in Chapter 9 of Ref. 1; further references are given in Refs. 2 and 5–9.

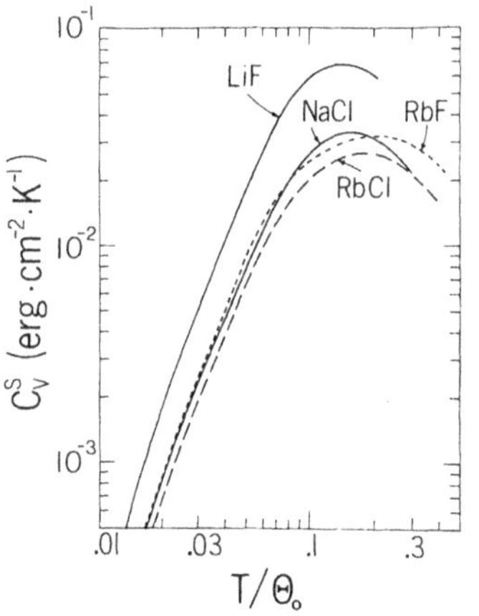

Fig. 1. Surface specific heat for the (001) surface of NaCl, LiF, RbF, and RbCl. The Θ_0 used to scale the temperature is that obtained from the elastic constants derived from the models.

from experimental elastic constants, as well as several other specific heat parameters.) Two principal features are apparent: (1) For T between $0.02\Theta_0$ and $0.05\Theta_0$, C_v^s appears approximately proportional to T^α, and (2) at a higher, characteristic temperature T_m, C_v^s reaches a characteristic maximum $C_{v,\max}^s$. In contrast to the results of isotropic continuum theories,[10-12] which predict $\alpha = 2$, the straight portions in Fig. 1 yield α in the range of 2.42 (NaCl) to 2.56 (RbF). (The T^α regime probably extends below $0.02\Theta_0$, but lack of a sufficient number of low frequencies in the root sampling makes the results unreliable there.) Furthermore, although the C_v^s curves approximately parallel each other in the T^α regime, the maximum values are attained at different values of T_m/Θ_0 (from 0.14 to 0.21), and there are some possibly experimentally observable differences in the shapes of the curves near $C_{v,\max}^s$.

Table I. Summary of Specific Heat Parameters for the Four Ionic Crystals Treated in This Study*

	NaCl	LiF	RbF	RbCl
Θ_0(expt),[15] °K	322	734	222	169
Θ_0(model), °K	318	727	240	185
$\Delta\Theta/\bar{\Theta}$, %	13[16]	19[17]	30[16]	20[16]
$T(\Theta_{\min})$,°K	35[16]	85[17]	17[16]	12[16]
$T_m = T(C_{v,\max}^s)$,°K	50	105	50	35
T_m/Θ_0(model)	0.16	0.14	0.21	0.19
$C_{v,\max}^s$, erg cm^{-2}°K^{-1}	0.0333	0.0677	0.0322	0.0268
Specific area, m^2/g	218	250	121	142

* Θ_0(expt) is the zero-temperature (bulk) Debye temperature derived from experimental elastic constants; Θ_0(model) is the analogous quantity derived from the (bulk) elastic constants obtained with the shell models used in this paper; $\Delta\Theta/\bar{\Theta}$ is the ratio of the range of variation of the bulk Θ (from $T = 0$ to $\sim\frac{1}{3}\Theta_0$) to the median value $\bar{\Theta}$ over the same temperature domain; $T(\Theta_{\min})$ is the absolute temperature at which the (bulk) Θ is a minimum (in the low-temperature domain); T_m is the absolute temperature at which each of the curves in Fig. 1 attains its maximum $C_{v,\max}^s$; the last entry gives the specific areas of the 15-layer films of the present work.

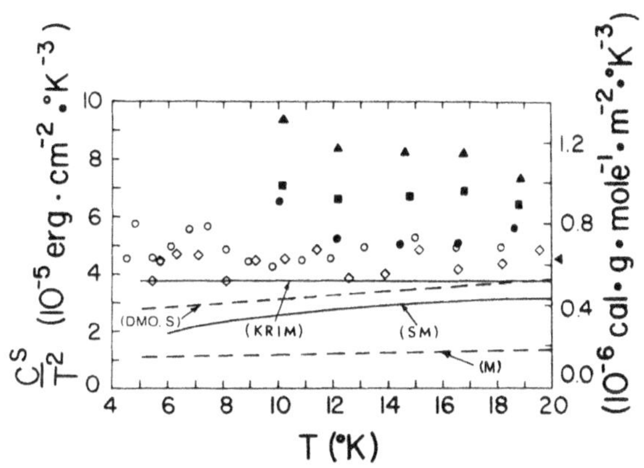

Fig. 2. Surface specific heat divided by the square of the absolute temperature. The solid lines labeled KRIM and SM are results obtained from a rigid-ion model[9] and from the present model, respectively. The dashed lines represent the continuum theories of Refs. 12 and 11 (DMO, S) and of Ref. 10 (M), respectively, as applied in Ref. 5. Experimental points: (■) 59 m^2/g, (●) 38 m^2/g, (▲) 39 m^2/g, Refs. 3 and 4; (○) 78.3 m^2/g, (◇) 41.8 m^2/g, Ref. 5.

In Fig. 2 we show a comparison in the T^α regime among the experimental results, the results of continuum theories, and the lattice results of Ref. 9 and the present work. (The apparent lack of proportionality to surface area in the data of Refs. 3 and 4 may be due to early experimental difficulties in determining surface area.[9]) The data of Ref. 5 seem more consistent with a value of $\alpha = 2$ (obtained also with the rigid-ion model in Ref. 9) than with the value $\alpha \approx 2.4$ obtained here, but with the residual scatter in the data the point remains somewhat uncertain. The fact that the present results lie significantly below the data may be due in part to the neglect of surface relaxation and, perhaps, in part to residual systematic difficulties in surface area determination.

We conclude with some remarks as to why one might expect in the lower temperature regime appreciable deviations from $\alpha = 2$ due to lattice effects. To illustrate the point, Fig. 3 for the 15-layer film of LiF displays both the dispersion curves and the surface excess frequency distribution $f^s(\omega)$.[8] First, it is obvious that surface-wave dispersion appears at significantly lower frequencies than for the bulk, and when one recalls that the bulk Debye temperature is varying rapidly in just this temperature regime, one should not be surprised by strong deviations from Debye-like behavior for the surface specific heat. Second, there is a class of long-wavelength, low-frequency surface modes—the shear-horizontal (SH) surface modes—which have *no* dispersion-free branch along the $\langle 110 \rangle$ directions for the (001) surface.[13] Hence no Debye-like assumption of a constant phase velocity can be made for the SH surface waves along the $\langle 110 \rangle$ directions. Space limitations do not permit a discussion of the SH surface waves away from the $\langle 110 \rangle$ directions, where one must take account of whether the anisotropy ratio $\eta = 2c_{44}/(c_{11} - c_{12})$ is greater or less than unity (see, e.g., Ref. 14), but it appears that their general effect may give significant additional deviations from Debye-like behavior in the surface specific

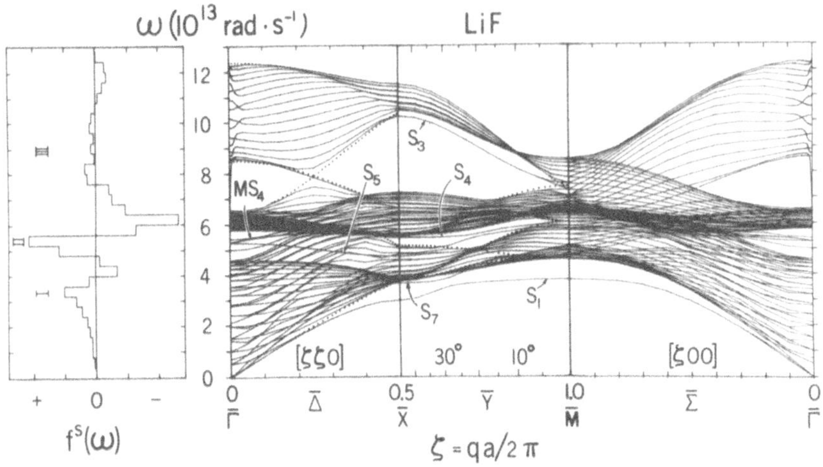

Fig. 3. Dispersion curves and surface excess distribution of frequencies $f^s(\omega)$ for a 15-layer film of LiF with free (001) surfaces. $f^s(\omega)$ is the difference between equally normalized histograms for frequency distributions obtained for the film and the bulk ; $f^s(\omega)$ has been smoothed somewhat by means of a staggered-bin procedure.

heat. Thus it may be that Debye approximations are inadequate for describing the surface specific heat of cubic $\{100\}$ surfaces and any other surfaces for which lattice (and anisotropy) effects are important at long wavelengths and low frequencies.

References

1. A.A. Maradudin, E.W. Montroll, G.H. Weiss, and I.P. Ipatova, *Theory of Lattice Dynamics in the Harmonic Approximation*, 2nd ed., Academic Press, New York (1971).
2. R.E. Allen, G.P. Alldredge, and F.W. de Wette, *Phys. Rev. B* **4**, 1648, 1661 (1971).
3. J.A. Morrison and D. Patterson, *Trans. Faraday Soc.* **52**, 764 (1956).
4. D. Patterson, J.A. Morrison, and F.W. Thompson, *Can. J. Chem.* **33**, 240 (1955).
5. J.H. Barkman, R.L. Anderson, and T.E. Brackett, *J. Chem. Phys.* **42**, 1112 (1965).
6. T.S. Chen, G.P. Alldredge, F.W. de Wette, and R.E. Allen, *Phys. Rev. Lett.* **26**, 1543 (1971).
7. T.S. Chen, G.P. Alldredge, and F.W. de Wette, *Sol. St. Commun.* **10**, 941 (1972).
8. T.S. Chen, G.P. Alldredge, and F.W. de Wette, *Phys. Lett.* **40A**, 401 (1972).
9. T.S. Chen, G.P. Alldredge, F.W. de Wette, and R.E. Allen, *J. Chem. Phys.* **55**, 3121 (1971).
10. E.W. Montroll, *J. Chem. Phys.* **18**, 183 (1950).
11. R. Stratton, *Phil. Mag.* **44**, 519 (1953); *J. Chem. Phys.* **37**, 2972 (1962).
12. M. Dupuis, R. Mazo, and L. Onsager, *J. Chem. Phys.* **33**, 1452 (1960).
13. G.P. Alldredge, *Phys. Lett.* **41A**, 281 (1972).
14. G.P. Alldredge, R.E. Allen, and F.W. de Wette, *J. Acoust. Soc. Am.* **49**, 1453 (1971).
15. C.R. Cleavelin, D.O. Pederson, and B.J. Marshall, *Phys. Rev. B* **5**, 3193 (1972).
16. G. Raunio and S. Rolandson, *Phys. Rev. B* **2**, 2098 (1970).
17. G. Dolling *et al.*, *Phys. Rev.* **168**, 970 (1968).

Heat Capacity Measurements on Isotopes of Molybdenum

R. Viswanathan* and H. L. Luo*

Department of Applied Physics and Information Science
University of California, San Diego, La Jolla, California

and

J. J. Engelhardt†

Institute for Pure and Applied Physical Sciences
University of California, San Diego, La Jolla, California

Introduction

Little is known concerning the heat capacity of transition metal isotopes. One likely reason is the prohibitive cost of large isotopic samples for conventional methods. The steady-state calorimetry technique, which requires samples less than 50 mg, is most suitable for such measurements. As a general test of this recently developed technique[1] and to determine its accuracy, we have chosen to measure the low-temperature heat capacity of Mo isotopes between 1.5 and 20°K. The reasons for choosing Mo as a first attempt are several: (1) The phonon spectrum of Mo is well established to be quite normal.[2] (2) Several of the Mo isotopes are relatively evenly abundant, and high-purity samples are easily available. (3) The purification process of Mo is well known and can be easily controlled.[3,4] (4) No complication of phase transition occurs in Mo above 1°K, and hence the electronic heat capacity coefficient γ and the Debye temperature θ_D can be quite accurately extrapolated. (5) The γ and θ_D of Mo are well recorded,[5] so that measured values of Mo isotopes can be critically compared.

Sample Preparation and Measurements

The heat capacity apparatus is based on the steady-state calorimetry technique[1] with a laser beam as a heat source,[6] the details of which will be published elsewhere.

The enriched Mo isotopes of masses 92, 96, and 100 were obtained from Oak Ridge National Laboratory. Spectroscopic analysis provided by the supplier only gave upper limits of metallic impurities. These samples were melted under intense arc for 9 min to remove all gaseous and most metallic impurities. The superconducting

* Supported by the U.S. Atomic Energy Commission under contract AT(04-3)-34.
† Supported by the U.S. Air Force Office of Scientific Research, Office of Aerospace Research, under contract AFOSR F44620-72-C-0017.

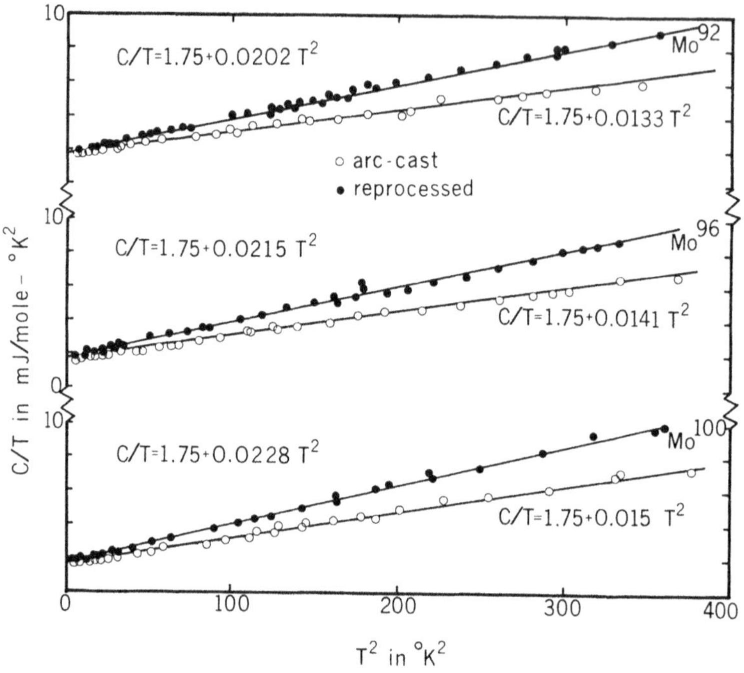

Fig. 1. Low-temperature heat capacity of Mo isotopes.

transition temperatures T_c were measured inductively on these samples. The transitions were very sharp and an isotope effect of -0.38 was deduced, which was in good agreement with previous measurements[7,8] and served as a stringent test for the purity of the samples. However, our heat capacity measurements on these samples indicated a systematic deviation from the published data. In an attempt to understand the discrepancy, the samples were reprocessed in the following steps: (1) Oxidized completely by heating at 650°C under pure oxygen for two days; (2) reduced in hydrogen atmosphere at 600°C for 2 hr and at 1000°C for two additional hours; (3) again intense arc-melting for 9 min. Then heat capacity measurements were repeated on the reprocessed samples. As a control, measurement was also made on three-pass zone-refined natural Mo obtained from Materials Research Corp.

Results and Discussion

The heat capacity data between 1.5 and 20°K are given in Figs. 1 and 2 in the conventional C/T vs. T^2 plots. Table I lists the parameters deduced from these graphs, in comparison with the best available literature data[5,9] on natural Mo.

The low and consistent values of γ (Fig. 1 and Table I) indicate that all our samples were quite pure, and the isotopic mass does not have any effect on γ, and hence the density of states, as expected. However, in all cases, θ_D was found to be much higher for the arc-cast samples. It was very intriguing to observe that θ_D was reduced drastically by the reprocessing. It is reasonable to believe that the

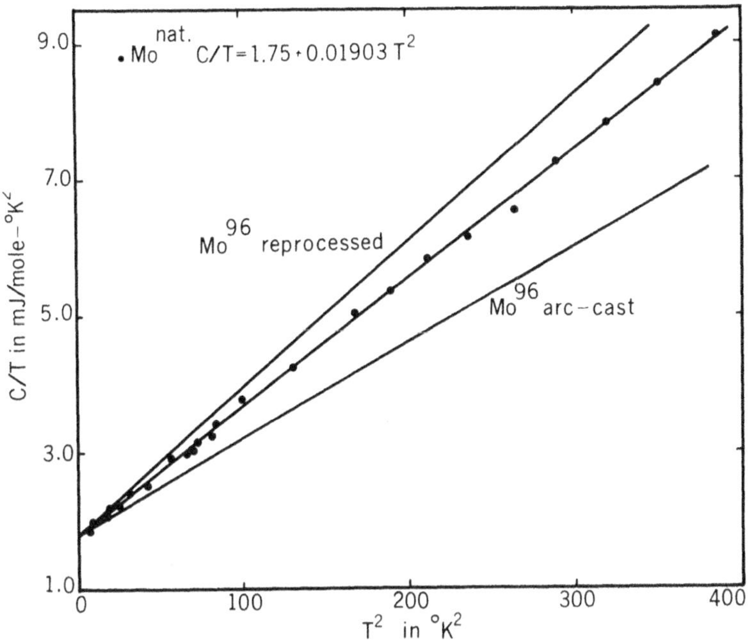

Fig. 2. Low-temperature heat capacity of natural Mo and the isotope ^{96}Mo.

decreases in θ_D were due to the effective removal of carbon impurities by the oxidation and reduction process.

Figure 2 indicates the large change in slope of the C/T vs. T^2 plot (and hence θ_D) of the isotope ^{96}Mo by the reprocessing, in relation to the data on natural Mo. The data on the latter were carefully analyzed by least-square fit, giving the values of $\gamma = 1.73 \pm 0.008$ mJ/mole °K^2 and $\theta_D = 467.1 \pm 1.7$°K, indicating that the precision of the measurement is about $\pm\frac{1}{2}\%$ for both γ and θ_D. The absolute accuracy, however, is liberally estimated to be $\pm 2\%$.

The two sets of θ_D values of the arc-cast and reprocessed samples were analyzed

Table I. Electronic Heat Capacity Coefficient γ and Debye Temperature θ_D of Natural Mo and Isotopes of Mo

Sample		γ, mJ/mole °K^2	θ_D, °K
^{92}Mo	Arc-cast	1.75	527
	Reprocessed	1.75	458
^{100}Mo	Arc-cast	1.75	506
	Reprocessed	1.75	440
^{96}Mo	Arc-cast	1.75	516
	Reprocessed	1.75	449
Mo^{nat}	Arc-cast	1.75	467
Mo^{nat}		1.83[5]	460,[5] 474[9]*

* Elastic constant data.

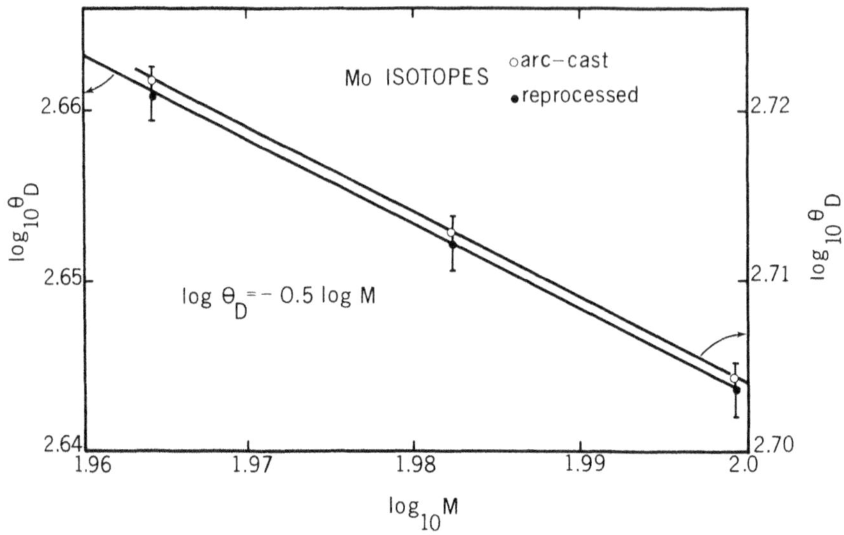

Fig. 3. Log θ_D as a function of log M for Mo isotopes.

separately by least squares as a function of the isotopic mass M. The θ_D was found to be proportional to $M^{-\alpha}$, where $\alpha = -0.503 \pm 0.005$ for the arc cast and $\alpha = -0.497 \pm 0.008$ for the reprocessed samples, exactly satisfying the Debye theory of specific heats (Fig. 3).

Acknowledgments

We have benefited greatly by discussions with Prof. L. J. Sham and Dr. A. C. Lawson.

References

1. P.F. Sullivan and G. Seidel, *Phys. Rev.* **173**, 679 (1968).
2. A.D.B. Woods and S.H. Chen, *Sol. St. Commun.* **2**, 233 (1964).
3. T.H. Geballe, B.T. Matthias, E. Corenzwit, and G.W. Hull, *Phys. Rev. Lett.* **8**, 313 (1962).
4. J.J. Engelhardt, *Phys. Rev.* **179**, 452 (1969).
5. F. Heiniger, E. Bucher, and J. Muller, *Phys. Kond. Materie* **5**, 243 (1966).
6. R. Viswanathan and H.L. Luo, *Sol. St. Commun.* **9**, 1733 (1971).
7. B.T. Matthias, T.H. Geballe, E. Corenzwit, and G.W. Hull, *Phys. Rev.* **129**, 1025 (1963).
8. E. Bucher and C. Palmy, *Phys. Lett.* **24A**, 340 (1967).
9. F.H. Featherstone and J.R. Neighbours, *Phys. Rev.* **130**, 1324 (1963).

Specific Heat and Debye θ of CdS:0.45% Mn*

M. M. Kreitman, G. J. Svenconis, J. J. Kosiewicz, and J. T. Callahan

University of Dayton, Dayton, Ohio

Introduction

Although the first research articles on the II–VI compounds date back to the middle of the nineteenth century and extensive literature on II–VI compounds has been accumulated, there are few reports of experimental specific heat and Debye temperature data. Shaulov and Kostina,[1] using an adiabatic calorimeter, have reported experimental specific heat values for CdS over the temperature range 55–300°K. Theoretical heat capacity results have been obtained by Goel and Singh[2] and Nusimovici *et al.*[3-5] The former employed a simple model involving a nearest-neighbor approximation, while the latter introduced more sophisticated models with additional force parameters and a long-range Coulomb potential, and treated higher-order neighbors. The purpose of this paper is to introduce the results of heat capacity measurements on CdS:0. 45% Mn over a temperature range extending down to approximately 3°K and to compare these experimental data and calculated Debye temperatures with theoretical results.

Apparatus and Approach

The specific heat measurements were carried out on a 209-g mass of 16 CdS crystals between 3 and 300°K in an adiabatic calorimeter. The equipment used, except for minor changes, is essentially the same as that described by Kosiewicz.[6] The crystals (Eagle Pitcher Corp., Norman, Oklahoma) were melt-grown and doped 0.45% Mn by weight. Because of the small values of the specific heat predicted by the Debye theory at liquid helium temperatures, the crystals were bonded together with G.E. varnish and tied with nylon string in order to form the 209-g mass.

The calorimeter was equipped with a specimen shield and a guard ring, both of which, in addition to the test specimen, were instrumented with platinum and carbon resistor thermometers and noninductively wound manganin heaters. Thermometer readings were obtained through the use of an ac resistance bridge.

Results and Discussion

Experimental heat capacity measurements are shown in Fig. 1 along with the results of other workers. Comparison with the experimental data of Shaulov and

* Work supported by Aerospace Research Laboratories, Air Force Systems Command, Contract F33615-71-C-1877.

Kostina indicates good agreement below 75°K. From 200 to 300°K the increase in the difference between the two sets of specific heat measurements grows from 5% to approximately 9%, which is beyond the limits of experimental error. It is not known to what this difference may be attributed.

Much larger differences exist in the theoretical results for the heat capacities. Not only do they disagree among themselves, but each falls far away from the experimental results. Goel and Singh[2] used the fact that each atom is bonded to its neighbor by covalent forces which could be represented by springs. Although other forces have to be accounted for, as a first approximation their assumption is feasible and its accuracy is dependent on the relative magnitudes of the omitted forces. Their specific heat model is reasonable since the major contribution to the heat capacity is usually due to lattice vibrations. They determined the normal mode frequencies for 2400 wave vector values from which the frequency distribution was drawn. Nusimovici and Birman[3] repeated the calculations while introducing more force parameters. They included first, second, and third neighbors, introduced a long-range Coulomb potential, and employed 40,000 vector values to generate the frequency distribution. The two frequency distributions obtained by the two different methods were quite different. In spite of this, the specific heat is known to be rather insensitive to large changes in the frequency distribution[7] and fair agreement between the two models should be expected. Subsequent calculations by Nusimovici *et al.*[4,5] based on a more refined model have yielded essentially the same Debye curve (see Fig. 2). It is not known why there should be such large differences in the theoretical results, especially at low temperatures.

In considering the generally large discrepancy between theory and experiment, it might be offered that the theory ignores the effect of the expansion of the crystal on the force constants. For CdS some idea of the magnitude of this effect can be obtained from the consideration of the variations of θ_{elastic} with temperature[8] (small: ~ 1.2%). Also, in spite of the fact that the theoretical calculation is essentially a low-temperature calculation, the $k = 0$ values for the dispersion curves were fit to high-temperature data in order to determine parameter values and the generation of acoustic modes. It might have been better to determine the parameters by a fit

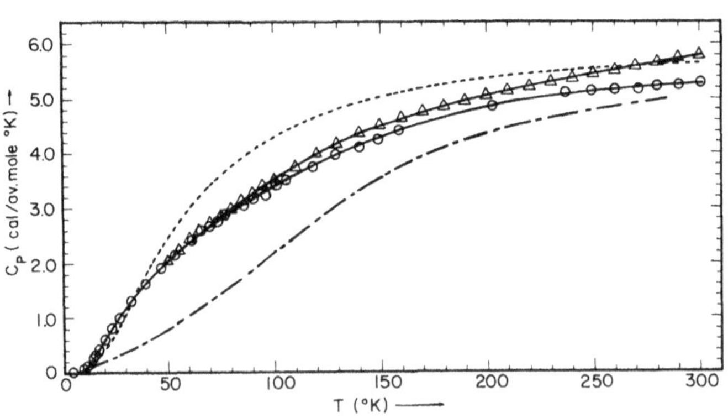

Fig. 1. Comparison of present specific heat results with other experimental and theoretical results. (○) Present work; (△) Shaulov and Kostina; (- - - - -) Goel and Singh; (- - -) Nusimovici.

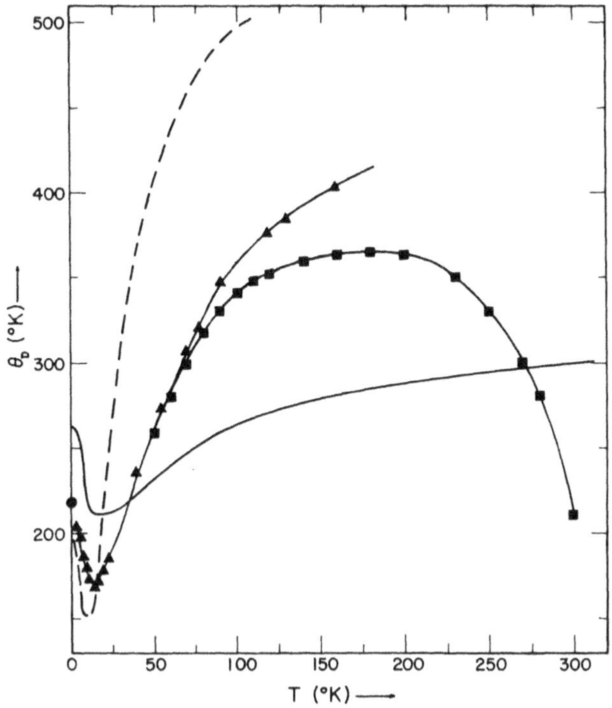

Fig. 2. Experimental and theoretical Debye temperatures as a function of temperature for CdS. (▲) Present work (specific heat); (■) Shaulov and Kostina (specific heat); (●) Gerlich (elastic constants); (– – –) Goel and Singh (*nn* theory); (———) Nusimovici and Birman (theoretical).

to low-temperature elastic constant data instead of optical data, and thus confine the comparison with experimental specific heat data at low temperatures.

The experimental specific heat values permitted the value of θ_{thermal} to be tabulated down to 3.2°K using Debye tables[9] and assuming the approximation $C_p = C_v$. The validity of this assumption is based on the fact that the calculated correlation between C_p and C_v at 300°K ($\sim 0.19\%$) is well within experimental error, and at lower temperatures the difference is expected to decrease further.[10] Therefore $C_p = C_v$ was assumed over the entire temperature range for CdS.

The temperature range of usefulness of the θ–T curve is limited to $0 \to \theta/2 \approx 0 \to 110$°K since for $T \to \theta$ the specific heat tends toward the Dulong–Petit value of 6 cal/av mole°K and is not sensitively dependent on the value of θ_{thermal}. This causes difficulties in calculating θ temperatures from specific heat data since small changes in the heat capacity will produce large changes in the θ value. For example, a change of 0.01 cal/av mole°K (well within the experimental error) in the value of C_p at 300°K tends to produce a change of 6°K in the θ value. At high temperatures where the vibrations are not harmonic, there are theoretical difficulties in interpreting specific heats. The choice of 110°K, which is approximately $\frac{1}{2}\theta_{\text{thermal}}$, is a standard, chosen for consistency.[8,11]

The closed circle on the graph at $T = 0$ (see Fig. 2) indicates the result obtained by Gerlich's low-temperature elastic constant data[12,13] ($\theta_{0,\,\text{elastic}} \approx 215°\text{K}$). This result was based on two assumptions: that the errors produced by the use of room-temperature values for the peizoelectric and adiabatic dielectric constants and the errors produced by the use of the Debye model to estimate the correction for the change in path length traversed by sound waves used in the measurements at low temperatures are small. These assumptions are reasonable.

Our $\theta_{0,\,\text{thermal}}$ value was obtained by using a computer fit of our data to the equation $C_p = \gamma T + 464.4(T/\theta_{0,\text{thermal}})^3$. From a plot of C_p/T vs. T^2 it appears that the T^3 region lies between 3.2 and 5.2°K. The lower limit is produced by an anomaly in the heat capacity caused by the Mn impurity. Above 3.2°K the effect of the Mn impurity appears to be negligible. The limit of the T^3 region determined is in approximate agreement with Blackman's estimate, $T \approx \theta/50$.

Our low-temperature limiting value of $\theta_{0,\,\text{thermal}} = \theta_{\text{thermal}}(T = 0) = 216 \pm 4°\text{K}$ was found to correspond satisfactorily with the value of 215°K obtained by Gerlich. This result differed by 7.4% with the theoretical value of 200°K obtained by Goel and Singh[2] and by 19.2% with the theoretical value of 263°K obtained by Nusimovici *et al.*[3-5]

Acknowledgments

We are grateful to Drs. John G. Daunt, Eugene Frank, and David Gerlich for help in this work.

References

1. Y.K. Shaulov and E.N. Kostina, *Russ. J. Phys. Chem.* **42**, 1007 (1968).
2. N.S. Goel and R.P. Singh, *Ind. J. Pure Appl. Phys* **1**, 343 (1963).
3. M.A. Nusimovici and J.L. Birman, in *II–VI Semiconducting Compounds*, D.J. Thomas, ed., W.A. Benjamin, New York (1967), p. 1204.
4. M.A. Nusimovici, Ph.D. Thesis, University of Paris, 1968.
5. M.A. Nusimovici, M. Balkanski, and J.L. Birman, *Phys. Rev.* **1**, 595 (1970).
6. J.J. Kosiewicz, M.S. Thesis, University of Dayton, Ohio, 1968, unpublished.
7. J. DeLaunay, *Sol. St. Phys. Ser.* **2**, 219 (1956).
8. M. Blackman, in *Handbuch der Physik* Vol. 7/1 (1955), p. 325.
9. P.A. Giguere and M. Boisvert, *Tables des Fonctions Thermodynamiques de Debye*, Les Presses de l'Université Laval, Quebec (1962).
10. J.K. Roberts and A.R. Miller, *Heat and Thermodynamics*, Blackie and Son, London (1961), p. 209.
11. F.H. Herbstein, *Phil. Mag.* **10**, 313 (1961).
12. D. Gerlich, *J. Phys. Chem. Solids* **28**, 2575 (1967).
13. D. Gerlich, Personal communication.

INSTRUMENTATION AND MEASUREMENT

13

Plenary Topics

New Techniques in Refrigeration and Thermometry*

John C. Wheatley

Physics Department, University of California—San Diego
La Jolla, California

Introduction

The purpose of this paper is to discuss in a context of specific scientific experiments a group of instruments, techniques, and phenomena which are important in the study of the properties of matter in the millidegree range or below. These are (1) pressure sensing via capacitors, (2) application of superconducting devices to measurement, (3) thermal contact via magnetic coupling to ^{3}He, (4) the adiabatic compression device, (5) adiabatic demagnetization of CMN/CDP, and (6) the dilution refrigerator. This group is certainly not comprehensive, for I have left out important subjects like nuclear demagnetization, applications of NMR, and orientation of radioactive nuclei. My purpose is not so much to describe the instruments themselves, as to mention what I think are interesting scientific problems connected with them or introduced by them or what kinds of new problems have been solved by them. Here again I will be specific and confine my discussion to a few selected experiments of current interest.

Since they are fundamental to the principal experiments to be discussed, I begin by mentioning the first three items on the above list. We owe development of the pressure-sensitive capacitor as applied to very low temperatures to Dwight Adams and his colleagues. A version of the basic Straty–Adams[1] design we use in La Jolla is shown in Fig. 1. This device, when used with a General Radio capacity bridge, has a sensitivity in the vicinity of 34 atm of about 10^{-5} atm using a 0.1 sec averaging time. Devices like it have been used extensively by Adams' group, particularly to obtain information on the properties of solid ^{3}He, and we may expect them to play an important role in measurement and thermometry in the future.

There is an extensive literature for the second item, the application of superconducting devices to highly sensitive electric and magnetic measurements, and it is dealt with elsewhere.[22] For present purposes we will regard them as sensitive detectors of magnetic flux (typical sensitivity 10^{-4} flux quanta per root cycle[2]) which when coupled to external circuits sense flux, current, or voltage as required by an experiment.

The third item, magnetic coupling of energy into ^{3}He, is a phenomenon rather than an instrument, but its use constitutes an important technique for utilizing or measuring low temperatures. The phenomenon was discovered in the contact of cerous magnesium nitrate (CMN) with pure liquid ^{3}He and has been studied in

* Work supported by the U.S. Atomic Energy Commission under Contract No. AT(04-3)-34, P.A. 143.

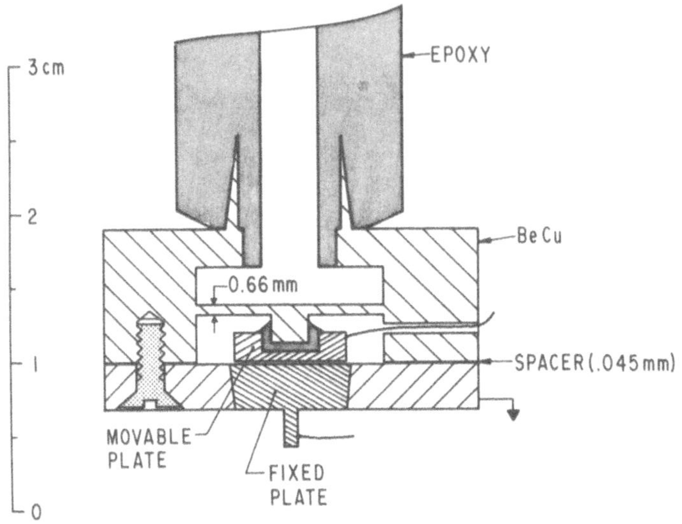

Fig. 1. Scale drawing of a Straty–Adams type capacitor used at La Jolla for measurements of the melting pressure of ^{3}He.

this system most recently by Bishop *et al.*[3] and was first discussed theoretically by Leggett and Vuorio.[4] Briefly, a Curie law paramagnet in contact across an interface via an electron–nuclear magnetic interaction with a Fermi liquid will have a thermal contact resistance qualitatively like that between the paramagnetic nuclei and Fermi liquid electrons in a metal. The latter thermal resistance is linear in T (contrast the T^{-3} dependence of the phonon resistance), and this does appear to be the temperature dependence of the surface CMN–^{3}He thermal resistance at low temperatures when the ^{3}He is pure enough. A ^{4}He impurity tends to condense preferentially on the CMN surface, however, and about one monolayer of ^{4}He is sufficient to increase seriously the surface thermal resistance.[5] Bishop and Mota have found a similar problem in contact to metallic platinum. A further snag[3] in thermal contact to CMN is an internal resistance. We have found recently with Richard Webb that $Na_3Ce(C_7H_3NO_4)_3 \cdot 15H_2O$, introduced by Wolf and collaborators[6] and called CDP, is qualitatively like CMN in low thermal contact resistance, but no quantitative measurements are available. It would be very useful to achieve good surface thermal contact to a metal. We have not yet succeeded, but we hope that the near future will bring some interesting papers in this area.

Adiabatic Compression of ^{3}He: The Pomeranchuk Effect

Surely one of the main reasons for developing the millidegree technology is to allow experimentation on ^{3}He. I begin with adiabatic compression, with reference to the discoveries, principally by the Cornell group,[7,8] of the properties of solid and liquid ^{3}He. When a two-phase mixture of low-entropy liquid ^{3}He and higher-entropy solid ^{3}He are compressed the mixture cools; routinely at present, so far as is known, to temperatures of about 2 m°K. The ^{3}He is self-cooling, the source of refrigeration being located at the solid–liquid interface, and hence thermal contact

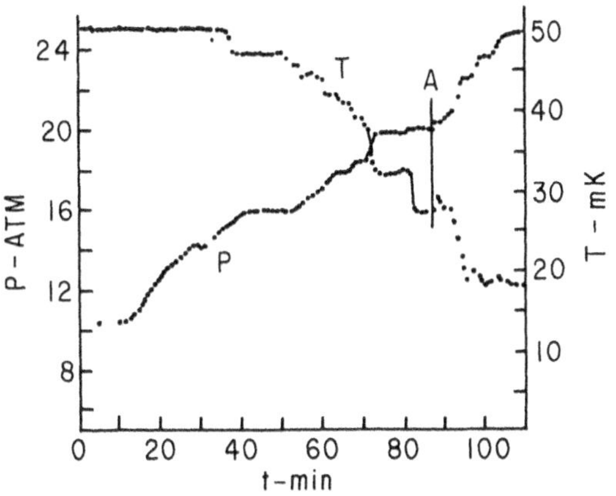

Fig. 2. First demonstration of cooling of ^3He by adiabatic compression. From Anufriyev.[10]

problems are much reduced. In a practical device the ^3He is contained in a flexible-walled cell. Mechanical compression of the cell is achieved by means of liquid ^4He under pressure either directly in contact with the outer walls[9] or via a mechanical linkage.[7]

The first demonstration of cooling by compression was given by Anufriyev[10] in 1965. We show his published results in Fig. 2. He used an external CMN thermometer; in all probability his ^3He cooled to much lower temperatures than those indicated. In addition to thermometry, friction must also have been a problem with Anufriyev's compression cell, which had been precooled with ferric alum. Present-day devices are precooled using dilution refrigerators. Temperatures comparable to the lowest presently available were achieved in late 1968 in experiments at La Jolla,[11] and we have frequently said of these experiments that the principal technical

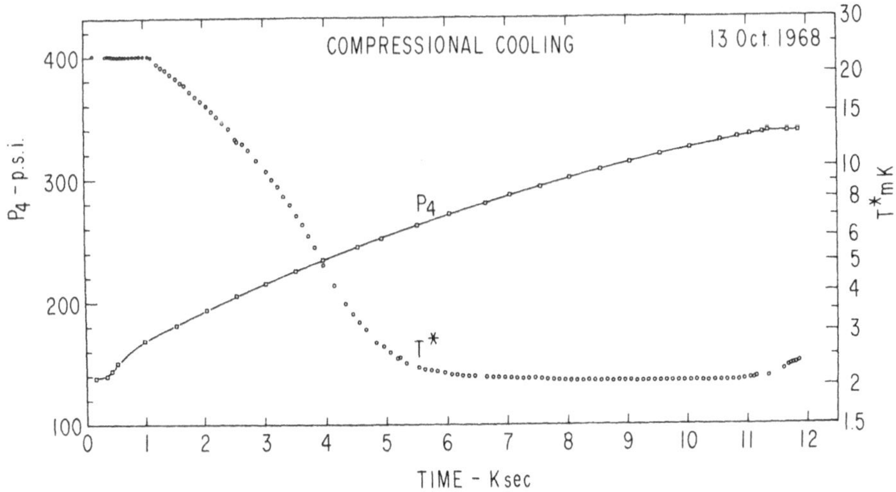

Fig. 3. Cooling of ^3He by adiabatic compression. The compression is actuated by increasing the liquid ^4He pressure P_4. T^* is the CMN magnetic temperature measured in the ^3He.

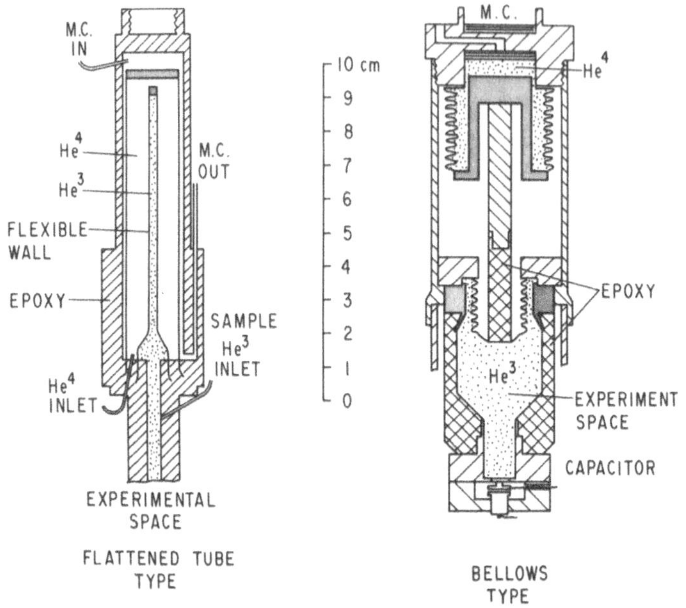

Fig. 4. Schematic drawings of practical compressional cooling devices. The flattened tube type is from Ref. 9 and the bellows type from Ref. 7. In both there is precooling by a dilution refrigerator, from which thermal isolation is achieved via the Kapitza resistance.

discovery was the very low frictional heating of the compressional cell. An example of these early results is shown in Fig. 3. Since then a variety of experiments on solid/liquid ^{3}He along the melting curve have been performed, including the bulk nuclear polarization of solid ^{3}He; and the Cornell group has developed a large-volume compressional cell using a bellows system. Schematic drawings of two types of cell are shown in Fig. 4. A great advantage of the bellows type device is its large volume, which permits both a large refrigeration capacity and, perhaps more importantly, introduction of entire experiments inside the cooled ^{3}He. A principal disadvantage of compressional cooling devices for studying ^{3}He properties is that the cooled mixture of solid and liquid ^{3}He is heterogeneous so that experiments on either solid only or liquid only are difficult to arrange. For NMR experiments this difficulty has been overcome by superposing a gradient field on a uniform magnetic field so that different parts of the cell come into resonance at different uniform field excitation currents. What is found empirically is that solid forms quite inhomogeneously, so that there may be substantial regions which are either mostly solid or mostly liquid.

Recently the Cornell group,[7] combining the compressional cooling and pressure-sensitive capacitor techniques, reported several new features on the pressurization curve (not the melting curve) of ^{3}He. Since they may figure importantly in the area of thermometry, we introduce them here. To give a concrete notion of what they are like, we show in Fig. 5 chart recorder tracings of capacity (pressure) vs. time obtained recently in La Jolla with Richard Johnson and Douglas Paulson, showing the three features, called points A, B, and B′ by the Cornell group. Although point B is somewhat irreproducible, point A is quite reproducible, as is, to a lesser extent, point B′. Point A is observed on both compression and decompression, point B on compression, and point B′ on decompression. The melting curve[12] and dP_m/dT[13] are already used for thermometry in experiments using compressional cooling. The former is subject to inaccuracy due to uncertainty in the value of the total pressure, while the latter is inaccurate due to uncertainties in the molar volume change on melting, the liquid entropy, and the solid entropy. For millidegree thermometry the minimum in the melting curve is too far away to serve as a useful fixed point. On the other hand, point A can be identified readily to within 0.01 psi, corresponding roughly to an uncertainty in T_A of about 0.02 m°K, or possibly somewhat more, so that possibly it would be a useful fixed point for millidegree thermometry, at least as a means of comparing experiments done at different laboratories. In this connection we anticipate our discussion of noise thermometry to present in Fig. 6 the melting curve of ^{3}He as determined[13] both using CMN thermometry calibrated via dP_m/dT and using noise thermometry. The latter measurements are highly preliminary, but we present them here in order to discuss the possibilities. To measure a temperature, the pressure is stabilized for several thousand seconds while averaging of thermal noise from a resistor in contact with the helium proceeds. The Johnson noise thermometer, calibrated at one point only, 4.2°K, read 2.81 m°K at point A and 2.37 m°K at point B. However, the lowest noise temperature obtained even on compressing past point B was 2.34 m°K, leading us to suspect that some residual heat leak into the Johnson noise sensor might have held its temperature above the surrounding ^{3}He. If one were to assume that the lowest temperature

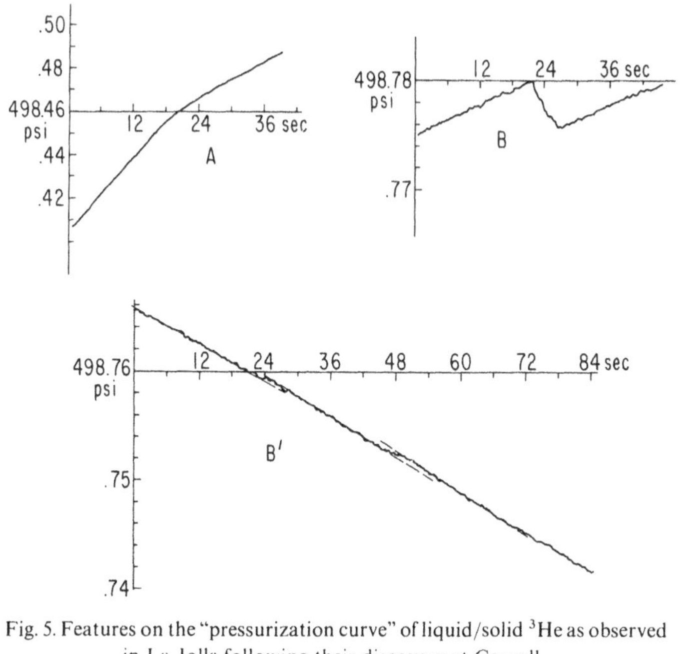

Fig. 5. Features on the "pressurization curve" of liquid/solid ^{3}He as observed in La Jolla following their discovery at Cornell.

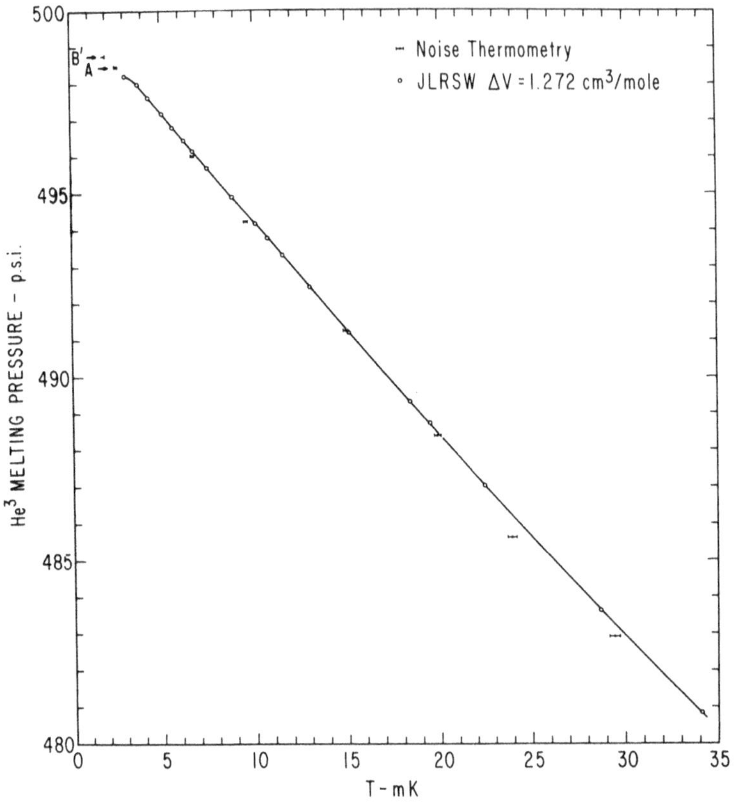

Fig. 6. Melting curve of ^{3}He as given in Ref. 13 and as measured in preliminary work using noise thermometry. Magnetic temperatures for CMN have been corrected from Ref. 13 to be in accord with the recent measurements of Ref. 15.

were 2.0 m°K, then one can estimate that a likely temperature for point A is 2.6 $\pm$ 0.1 m°K, but further experiments are required. The statistical problems with noise thermometry are indicated by the error bars, but the systematic deviations of the noise temperatures from those CMN temperatures obtained via dP_m/dT indicate possible inaccuracies both in the calibration of the noise thermometer and in the entropy and molar volume differences. We note that use of Grilly's[14] new value for Δv_m makes the agreement worse. It is clear both that a definitive work needs to be done and that the possibilities are great.

The pressure-sensitive capacitor is also an important tool for the study of liquid ^{3}He and may be important in connection with fixed points for thermometry. Recently in La Jolla with Richard Webb and Richard Johnson we were studying high-pressure (32.2 atm) liquid ^{3}He which had been cooled by adiabatic demagnetization of finely powdered CMN mixed with it. The cell geometry was somewhat complicated but included both a tube leading to a pressure-sensing capacitor and a CMN thermometer which had been located in an appendix to allow measurements of thermal conductivity. Reproducibly on warming a "feature" appeared on the chart recorder giving capacity (^{3}He pressure) as function of time. A smooth curve through the recorder trace is shown in Fig. 7, where capacity has been converted to pressure to improve interpretation. Although thermal equilibrium within the cell was poor due to the very low liquid thermal diffusivity, it is suggestive that the feature occurred at a magnetic temperature of 2.7 m°K. It appears that the pressure-sensitive capacitor should become a standard sensing instrument for work with ^{3}He.

Advances in Millidegree Thermometry Using Superconductive Devices

Here I draw attention to recent work[2] on low-frequency ac susceptibility measurements and noise temperature measurements with a low-temperature limit near 2 m°K. I omit discussion of static magnetization measurements,[17,18] although they usually have higher sensitivity, since they have been discussed adequately earlier. In the present work the technique was used of cooling pure low-pressure liquid ^{3}He by mixing it with finely powdered CMN, precooled using a dilution refrigerator and then adiabatically demagnetized. The ^{3}He temperatures could be monitored during the demagnetization process to ensure that the time scale for reduction of the field was satisfactory, and magnetic temperatures below 3 m°K could be held for many hours. The cell used[15] is shown in Fig. 8. Two types of thermometer were placed in contact with the ^{3}He: a 16-Hz magnetic susceptibility thermometer using a powdered electron paramagnetic, and a thin copper sheet with superconducting leads which served as a Johnson noise thermometer.

The paramagnetic sample is placed in a miniature superconducting mutual inductance coil set with an astatic secondary. Unbalance signals are nulled out using a simple ac bridge,[2] unbalance being observed via a SQUID serving as null detector. In such a system a "memory" is not required for the SQUID system, so flux slippage in the detector is not a problem. The advantage of the measuring system is that very small samples (1–10 mg) and small measuring fields (0.1 G rms) are possible.

The noise thermometer is of the "current-actuated" type in which the noise currents are coupled to a SQUID using a coil.[2] A "voltage-actuated" noise thermo-

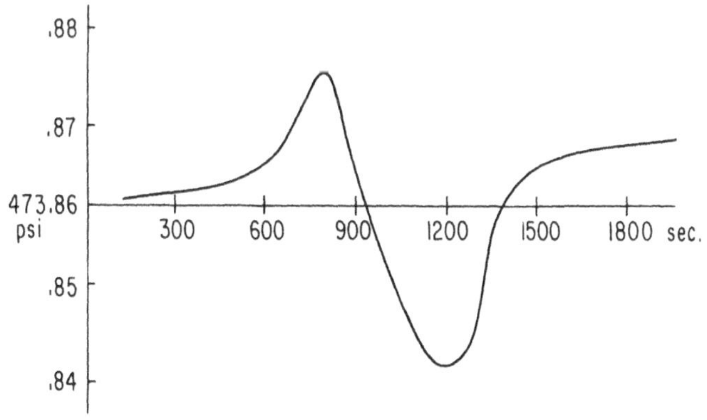

Fig. 7. A pressure feature which appears reproducibly on the warming curve of a mixture of CMN and pure liquid ^{3}He at 32.2 atm after adiabatic demagnetization. The graph is a tracing of a chart recording plotting pressure (capacity) as function of time.

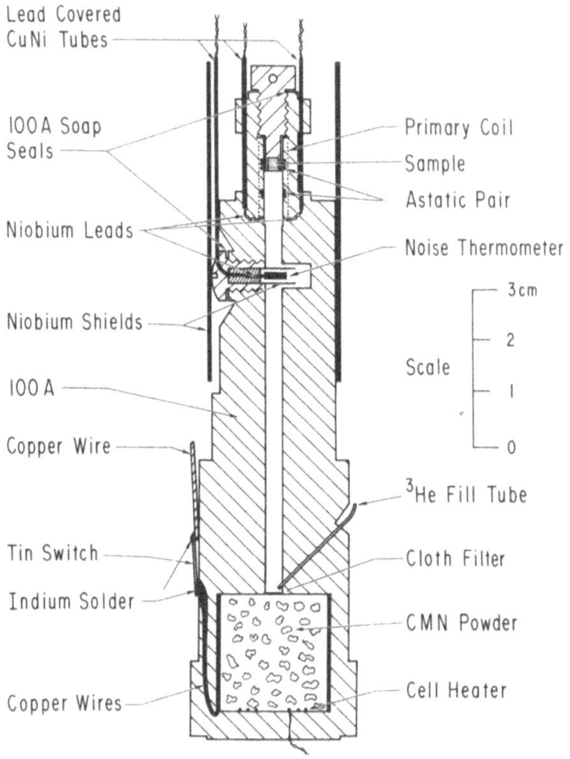

Fig. 8. Adiabatic demagnetization cell for intercomparing a Johnson noise thermometer and a magnetic thermometer using pure liquid ^{3}He as thermal contact agent.

meter[19] is discussed elsewhere.[20] In our noise thermometer the noise currents modulate the average field in an rf-biased SQUID operating in a feedback mode with current fed back into the rf coil to keep the average SQUID flux constant. The feedback voltage, which exactly mirrors the input noise currents, is amplified in a band-pass amplifier and then electronically squared and averaged. The mean square noise currents are proportional to the absolute temperature. The noise is very substantial, as may be seen in Fig. 9, in which the output of an rf-biased SQUID is shown (above) with a noise source in the SQUID and (below) when there is no noninstrumental noise source. The noise bandwidth is 2.5 Hz and the trace width is 5 sec. In our experiments in noise thermometry at millidegree temperatures the contribution of the SQUID and electronics to the noise temperature is about 50 μ°K. Details of the method and results for CMN are given elsewhere,[21] but we show results for CDP (supplied by Prof. W. P. Wolf) in Fig. 10. The temperature insensitivity of χ' is not believed to be due to poor thermal contact.

Noise thermometry appears to be a very powerful tool. At this point we feel that it has the potential for much lower temperatures and that the electron noise temperatures are probably quite accurate absolute temperatures. However, a noise thermometer sensor requires a substantial averaging time to achieve precise results. Furthermore, in common with all other metallic thermometers used for sensing the temperature of ^{3}He, the noise-producing resistance must be put in contact with the ^{3}He via a thermal boundary resistance. We have unfortunately not yet found a way to reduce drastically the thermal boundary resistance to a metal. In the present thermometer near 2 m°K its time constant for thermal equilibrium was 8×10^2 sec, indicating both a high thermal resistance and a higher than expected heat capacity. Our experiments suggest that millidegree sensors of the future should be of pure metals with low volume and high surface area, although this could be relaxed if a means for magnetic coupling of the metal to ^{3}He could be found.

These experiments emphasize the effectiveness with which salts like CMN can cool liquid ^{3}He because of their high refrigerating capacity and the low thermal boundary resistance. They also emphasize the need for better metal–^{3}He thermal contact, both for thermometry and for refrigeration using nuclear demagnetization.

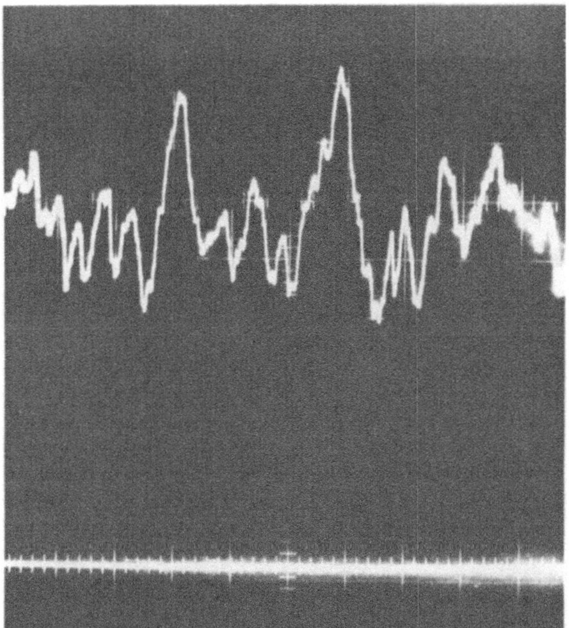

Fig. 9. Example of Johnson noise at 4.2°K as opposed to instrumental noise. In the upper trace (5 sec across) the output of an rf-biased SQUID system is shown when a copper rod at 4.2°K (specially arranged with lead plating and solder-covered copper foil to have a time constant of about 1 sec) is inserted in a symmetric SQUID. The lower trace shows the essentially negligible noise from an empty SQUID. The noise bandwidth is 2.5 Hz.

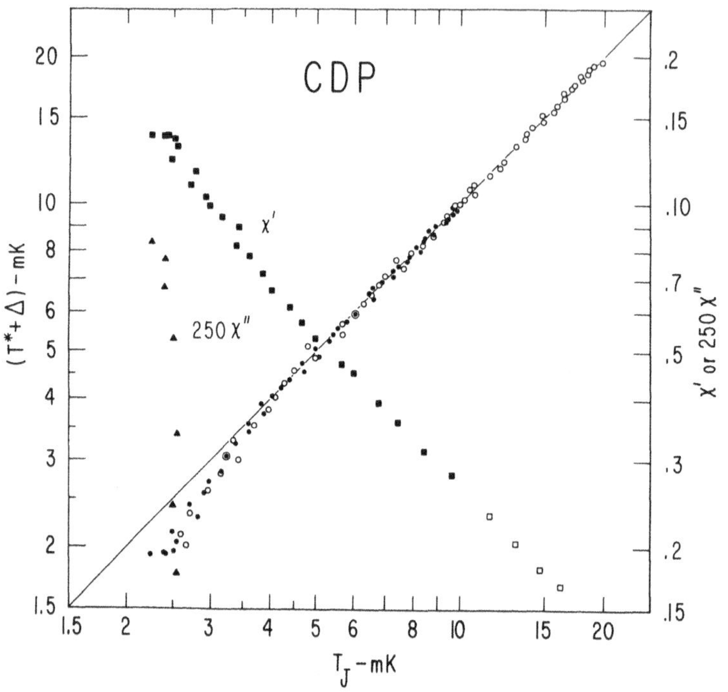

Fig. 10. Magnetic temperature and 16-Hz susceptibility of [CDP^6 $Na_3Ce(C_7H_3NO_4)_3 \cdot 15H_2O$] as function of the Johnson noise temperature.

References

1. G.C. Straty and E.D. Adams, *Rev. Sci. Instr.* **40**, 1393 (1969).
2. R.P. Giffard, R.A. Webb, and J.C. Wheatley, *J. Low Temp. Phys.* **6**, 533 (1972).
3. J.H. Bishop, D.W. Cutter, A.C. Mota, and J.C. Wheatley, *J. Low Temp. Phys.* **10**, 379 (1973).
4. A.J. Leggett and M. Vuorio, *J. Low Temp. Phys.* **3**, 359 (1970).
5. W.C. Black, A.C. Mota, J.C. Wheatley, J.H. Bishop, and P.M. Brewster, *J. Low Temp. Phys.* **4**, 391 (1971).
6. J.C. Doran, U. Erich, and W.P. Wolf, *Phys. Rev. Lett.* **28**, 103 (1972).
7. D.D. Osheroff, R.C. Richardson, and D.M. Lee, *Phys. Rev. Lett.* **28**, 885 (1972).
8. D.D. Osheroff, W.J. Gully, R.C. Richardson, and D.M. Lee, *Phys. Rev. Lett.* **29**, 920 (1972).
9. R.T. Johnson and J.C. Wheatley, *J. Low Temp. Phys.* **2**, 423 (1970).
10. Yu. D. Anufriyev, *Soviet Phys.—JETP Lett.* **1**, 155 (1965).
11. R.T. Johnson, R. Rosenbaum, O.G. Symko, and J.C. Wheatley, *Phys. Rev. Lett.* **22**, 449 (1969).
12. R.A. Scribner, M.F. Panczyk, and E.D. Adams, *J. Low Temp. Phys.* **1**, 313 (1969).
13. R.T. Johnson, O.V. Lounasmaa, R. Rosenbaum, O.G. Symko, and J.C. Wheatley, *J. Low Temp. Phys.* **2**, 403 (1970).
14. E.R. Grilly, *J. Low Temp. Phys.* **4**, 615 (1971).
15. R.A. Webb, R.P. Giffard, and J.C. Wheatley, *Phys. Lett.* **41A**, 1 (1972).
16. R.A. Webb and J.C. Wheatley, *Phys. Rev. Lett.* **29**, 1150 (1972).
17. R.A. Buhrman, W.P. Halperin, S.W. Schwenterly, J. Reppy, R.C. Richardson, and W.W. Webb in *Proc. 12th Intern. Conf. Low Temp. Phys., 1970,* Academic Press of Japan, Tokyo (1971), p. 831.
18. E.C. Hirschkoff, O.G. Symko, L.L. Vant Hull, and J.C. Wheatley, *J. Low Temp. Phys.* **2**, 653 (1970).
19. R.A. Kamper, J.D. Siegwarth, R. Radebaugh, and J.E. Zimmerman, *Proc. IEEE* **59**, 1368 (1971).
20. H. Marshak and R.J. Soulen, this volume.
21. R.A. Webb, R.P. Giffard, and J.C. Wheatley, this volume.
22. B.H. Taylor, this volume; J. Clarke, this volume.

Cryogenic Metrology*

B. N. Taylor

Institute for Basic Standards
National Bureau of Standards, Washington, D.C.

Introduction: Metrology—What Is It?

The dictionary definition of metrology is "the science of weights and measures." Traditionally, the word is associated with the type of work carried out by the National Bureau of Standards (NBS) and its sister institutions throughout the world. It brings to mind the base units of the Système International d'Unités (SI): the metre,† second, kilogram, ampere, kelvin, candela, and mole. Furthermore, metrology is usually associated with measurements of very high accuracy, i.e., at the part per million (ppm) level, and often with determinations of fundamental physical constants. In the broader sense, however, the word metrology may be associated with any measurement problem which places demands on the state of the art. An example might be the measurement or even detection of some quantity which under more favorable environmental conditions could be determined to very high accuracy.

It is the purpose of this paper not only to concisely review the application of low-temperature (cryogenic) techniques and phenomena to selected areas of "traditional" metrology, but also to touch upon some applications which may only be considered metrological in the broader sense. Throughout the discussion will be rather general, i.e., the focus will be primarily on the basic principles. It is suggested that the references be consulted for details. The topics to be covered are as follows.

I. Past applications: a very brief summary of some past applications of cryogenics to metrology, pre-1962 (the year 1962 is chosen as the dividing line since cryogenic metrology has only come of age during the last decade with the discovery of the Josephson effects).

II. Present applications. (a) Fundamental constants: $2e/h$, hN, h/m_e, and c, where e is the electron charge, h is the Planck constant, N is the Avogadro constant, m_e is the free-electron rest mass, and c is the speed of light in vacuum; (b) Basic measurements and standards: volt maintenance, thermometry, time and frequency, and rf attenuation; (c) Novel instrumentation: A selection of specific state-of-the-art and highly promising instruments requiring cryogenics for their operation.

III. Comments on the future of cryogenic metrology.

† Internationally accepted spelling.

Past Applications of Cryogenics to Metrology

Past applications of cryogenics to metrology fall under its broader definition as given above. Typical applications are as follows.*

Superconducting Galvanometer. Probably the earliest cryogenic instrument was the superconducting galvanometer of Smith *et al.*[2] built in the mid-1930's. The device, which was a more or less conventional galvanometer with a superconducting moving coil, was to be used to test the Hall effect and to measure thermoelectric effects in superconductors. Pippard and Pullan[3] later improved on these early designs and constructed a single-turn galvanometer capable of detecting 10^{-12} V.

Superconducting Chopper Amplifier. Such a device was built by Templeton[4] in the mid-1950's. An 800-Hz magnetic field superimposed on a dc field was used to drive a tantalum wire in and out of the superconducting state, thereby "chopping" or converting to ac the dc emf to be measured which was placed in series with the tantalum wire. The ac signal was then detected by conventional means.

Superconducting Reversing Switch. Templeton[5] also constructed a superconducting reversing switch with tantalum coils which were driven normal by means of a magnetic field. It was used for distinguishing between the desired signal and thermal emf's in cryogenic low-level dc measuring circuits.

Superconducting Bolometer. Several workers have used a dc-biased thin superconducting film held just below its transition temperature as a bolometric detector of infrared radiation (see, e.g., Ref. 6). The radiation to be detected is chopped at a convenient frequency and allowed to fall upon the bolometer. The chopped radiation gives rise to oscillations in the resistance of the bolometer (via oscillations in its temperature) and consequently an ac voltage across it. This ac signal is then detected by a suitable narrowband amplifier.

Masers. The use of microwave maser amplifiers cooled to liquid helium temperatures for low-level microwave signal detection, e.g., in radio astronomy, was well developed by the early 1960's.[7]

Parametric Amplifiers. Cooled parametric amplifiers for low-level microwave signal detection was also commonplace by the early 1960's. Of particular interest, however, is a liquid-nitrogen-cooled varactor parametric amplifier built by Cutkosky[8] at NBS, which operates in the audiofrequency range, i.e., 1592 Hz. The device is particularly useful as a null detector for capacitance bridges. At 1592 Hz (10^4 rad/sec) a signal current of 10^{-14} A through a 1000-pF capacitance can be detected in less than 1 sec. Minimum noise figures below 0.001 dB (referred to room temperature) are obtained. This null detector will play a key role in the NBS experiment to determine the absolute ohm via the calculable capacitor method.

Present Applications of Cryogenics to Metrology

Fundamental Physical Constants. One of the goals of metrology is the development of a measurement system based on invariant properties of nature rather than variable man-made artifacts such as metre bars and kilograms. Consequently, metrology and the determination of fundamental physical constants have been

* See Ref. 1 for a fairly complete review of superconducting instrumentation as of the early 1960's.

closely associated through the years. Indeed, the determination of fundamental constants to ever greater levels of accuracy has provided some of the most difficult metrological problems known. It is the one field which continually pushes at the limits of the measurement system. For example, the value of the speed of light is presently limited by the ^{86}Kr length standard.[9]

From the point of view of pure science, the attempt to improve our knowledge of the values of the fundamental constants continues unabated for at least two reasons. First, to make quantitative predictions from the basic theories of physics requires accurate numerical values for the constants which appear in the theories. Second, and more important, the careful examination of the numerical values of these constants as determined from experiments in different fields of physics can in turn test the overall consistency and correctness of the basic theories themselves.

A case in point is the measurement of $2e/h$ via the ac Josephson effect first reported in 1967 by Parker *et al.*[10] and its subsequent impact on quantum electrodynamics (QED).[11] The importance of this measurement stemmed from the fact that a value of the fine structure constant α independent of QED theory could be derived from $2e/h$. This α value, in turn, permitted critical comparisons between QED theory and experiment which did not require *a priori* information from quantum electrodynamics itself. Because of its importance, we begin our discussion of cryogenic metrology as applied to the determination of fundamental constants with the ac Josephson effect measurement of $2e/h$.

2e/h. The determination of $2e/h$ via the ac Josephson effect is rather straightforward in comparison with most fundamental constants experiments. A Josephson device, either a tunnel junction or point contact, is irradiated with microwave radiation of frequency ν_0, typically 10 GHz. The radiation modulates the time rate of change of the quantum mechanical phase difference between the two superconductors, resulting in constant-voltage current steps in the current–voltage curve of the Josephson device. (These steps, first observed by Shapiro,[12] are shown in Fig. 1). The voltages V_n at which the steps occur are related to ν_0 by the so-called (generalized)

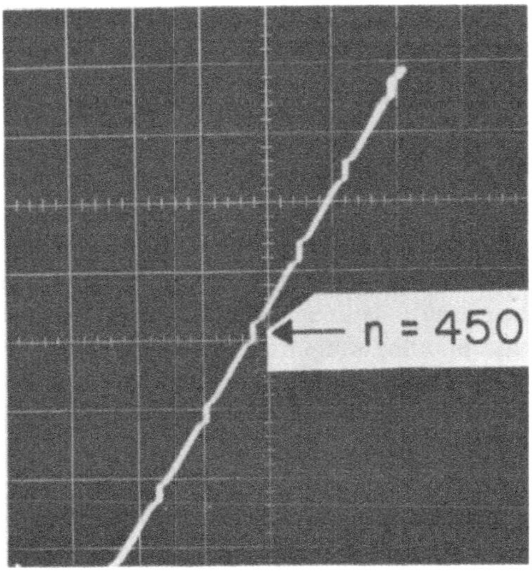

Fig. 1. Oscilloscope trace showing the 450th microwave induced constant voltage current step in a Pb–Pb oxide–Pb Josephson tunnel junction. The frequency of the incident radiation is 11 GHz. The vertical (current axis) scale is 50 μA/cm and the horizontal (voltage axis) scale is 25 μV/cm. (From Finnegan *et al.*[17])

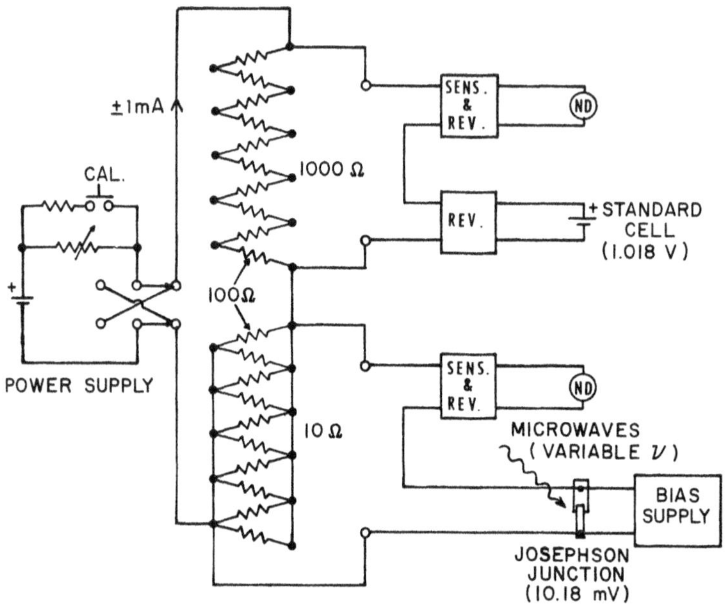

Fig. 2. Schematic diagram showing how at NBS 10-mV Josephson voltages are compared to the emf of a standard cell via a fixed 100:1 Hamon network. (The compensating resistors for the networks are not shown.)

Josephson frequency voltage relation, $n\nu_0 = (2e/h)\,V_n$, where n, an exact integer, is the step number.[12] (The magnitude of $2e/h$ is such that for 10-GHz radiation each step is separated by about 20 μV.) A Josephson device thus behaves as a frequency-to-voltage converter with the constant of proportionality being the invariant fundamental physical constant $2e/h$. Obviously, the simultaneous determination of ν_0 and V_n (and counting the step number n) will yield a value for this constant.

Several groups throughout the world have now carried out such measurements,[13-15] including the National Bureau of Standards. In the recent work at the Bureau by Finnegan *et al.*,[16] which is a continuation of work at the University of Pennsylvania,[17] two Pb–Pb Josephson tunnel junctions on a single substrate were placed in series in order to increase the total junction voltage to be measured to 10 mV. This voltage is then compared to the emf of a standard cell (which is known in terms of the U.S. legal or as-maintained volt) by means of a special 100:1 fixed-ratio potentiometer utilizing two interchangeable Hamon networks. (See Fig. 2.) The latter permit the establishment of the 100:1 ratio to an accuracy of one or two parts in 10^8. By adjusting both the current through the Hamon networks and the frequency of the applied microwave radiation, it is possible to directly compare the standard-cell voltage (approximately 1.018 V) with the 10 mV (actually 10.18 mV) Josephson-step voltage with an uncertainty of only a few parts in 10^8. (Thermal emf's and Hamon network uncertainties are eliminated by suitable reversals and interchanges.)

Figure 3 shows a series of $2e/h$ measurements recently taken at NBS over a 3-hr period against a single standard cell. Each point consists of two Hamon network interchanges and two junction polarity reversals. The statistical standard deviation of the mean of the five points σ_M is 0.014 ppm. The systematic uncertainties in the apparatus are of the order of 0.03 ppm (one standard deviation estimate). Thus $2e/h$ in terms of a particular unit of voltage can be determined with an accuracy of a few parts in 10^8. However, α cannot be obtained to this accuracy until the other quantities appearing in the equation which relates α to $2e/h$ are also determined to a few parts in 10^8.[2]

hN. Although $2e/h$ is the only fundamental constant to date which has been successfully measured to high accuracy using cryogenics, other low-temperature fundamental constants experiments are now underway which have sufficient promise and are of sufficient interest to warrant brief mention.

One such experiment is the determination of the product hN via the analog of the ac Josephson effect in superfluid helium. Prior to this conference, it was believed that this effect is observed when the superfluid flow between two "weakly" connected baths of superfluid helium at different pressure heads is modulated by an applied sound wave field. (The weak coupling is achieved via an approximate 10-μm-diameter orifice connecting the two baths; and the sound field is established via a quartz transducer near the orifice driven at, say, 100 kHz.) The chemical potential difference between the two baths and the frequency of the sound field ν_0 are supposedly related by the Josephson-like equation $nh\nu_0 = M_{He}g(\Delta z)_0$ where n is an integer, M_{He} is the mass of the helium atom, g is the local value of the acceleration due to gravity, and $(\Delta z)_0$ is the level difference corresponding to stationary flow between the two baths. (Such stationary level differences were first reported by Richards and Anderson in 1965.[18,19]) Noting that $M_{He} = M^*_{He}/N$, where M^*_{He} is the atomic mass of the helium atom, it becomes apparent that the product hN can be determined by measuring ν_0 and the level differences since the atomic mass of the helium atom is known via mass spectroscopy to very high accuracy (parts in 10^8), and a value of g at any

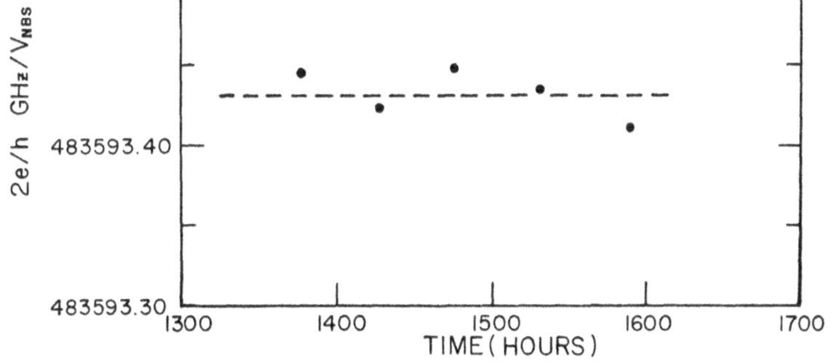

Fig. 3. Measurements of $2e/h$ obtained at NBS against a single standard cell during an approximate 3-hr period.

particular site can be determined to comparable accuracy.[11] (For $\nu_0 = 100\text{kHz}$ the distance between adjacent levels is about 1 mm.) Indeed, Khorana and Douglass[20] have reported a 1% measurement of hN. A measurement accurate to a few ppm, which might possibly be achieved experimentally, would be of great interest because it would improve significantly our present knowledge of N as well as other constants.

On the other hand, in spite of the fact that several groups[18-22] have observed these stationary levels supposedly arising from the ac Josephson effect, recent observations reported at this conference by Musinski and Douglass strongly indicate that they may be due to acoustic resonances in the apparatus. Indeed, these authors believe that no one has yet observed the analog of the ac effect in superfluid helium! Clearly, much additional work is called for.

h/m_e. A high-accuracy measurement of this quantity is of importance because of its relationship to the fine structure constant α. It can readily be shown that $\alpha = (2R_\infty \lambda_C)^{1/2}$, where R_∞ is the Rydberg constant for infinite mass and λ_C is the electron Compton wavelength, which is equal to $h/m_e c$. Since the speed of light is known to an accuracy of better than $1/10^8$ and R_∞ to an accuracy of $1/10^7$,[9,11] a measurement of h/m_e to, say, 1 ppm will yield a 0.5-ppm value of α. Such a value would serve to check the result derived from $2e/h$ and would be relevant to critical tests of quantum electrodynamics.

That a measurement of h/m_e could be made using low-temperature physics was first demonstrated by Zimmerman and Mercereau[23] in 1965. Imagine a thin-film superconducting ring (e.g., evaporated on a fused quartz rod) containing a Josephson-type weak link. The quantum mechanical phase difference ϕ across the weak link is given by

$$\phi = \hbar^{-1} \int_1^2 \mathbf{p} \cdot d\mathbf{r}$$

where $\mathbf{p}$, the canonical momentum, is given by the usual equation, $\mathbf{p} = 2m_e\mathbf{v} + 2e\mathbf{A}$. Assuming the current–phase relation for the weak link is given by the Josephson equation $I = I_1 \sin \phi$, one obtains

$$I = I_1 \sin\left[2/\hbar \oint (m_e\mathbf{v} + e\mathbf{A}) \cdot d\mathbf{r}\right]$$

where the integral is to be taken around the ring. If now the ring is rotated with angular frequency ω, then $\mathbf{v} = \mathbf{r} \times \omega$ and the maximum supercurrent through the weak link is periodic in ω with period $h/4m_e S$, where S is the area of the ring. (This assumes that any externally applied magnetic flux is constant with time.) Thus, measuring the maximum supercurrent as a function of rotation rate can yield a value for h/m_e, provided the area S of the ring is known. (The supercurrent in the ring may be monitored via the periodic dependence of the rf impedance of a coil coupled to the ring.) In the original work of Zimmerman and Mercereau h/m_e was determined with an accuracy of 4%. Subsequently, Parker and Simmonds[24] refined the method and were able to report a 0.04% result in 1970. Falco and Parker report a 0.02% value at this conference, but more important, they find that rings up to 1 cm in diameter can be successfully used. This should significantly simplify future determinations of the effective area of the ring. It should be noted that the rotating superconductor experiment is not as clean theoretically as the measurement of $2e/h$ via microwave-induced steps because of relativistic corrections to m_e. One must eventually understand the role of the electron binding energy (which depends on the work function of the metal used) and Fermi velocity before a 10-ppm value of h/m_e can be obtained.[24]

c. An accurate value of the speed of light in vacuum is of importance because (1) *c* plays a key role in geodesy and lunar and planetary ranging, and (2) if *c* can be measured to an accuracy comparable to the present ^{86}Kr length standard, then serious consideration could be given to replacing the present length standard with an exact adopted value of *c*. The result would be a unified standard of length, time, and frequency. The more obvious method to determine *c* is to measure both the wavelength λ and frequency ν of the same laser line. While the novel modulation technique of Bay *et al.*[25] can achieve the desired accuracy, the direct frequency counting method, developed to its most advanced state by Evenson and co-workers[26,27] at NBS, is inherently a more accurate technique. Indeed, Evenson *et al.*[9] have recently reported a six parts in 10^{10} measurement of the frequency of the methane-stabilized (via saturated absorption) He–Ne laser oscillating at 88 THz (3.39 μm). When combined with the measurement of the wavelength of this line by Barger and Hall[28] a $3.6/10^9$ value for *c* was obtained.[9]

The frequency counting technique is simple in principle but difficult in practice. The idea is to mix the radiation to be measured with the radiation from a klystron of known frequency and to look for the beat note. This requires a mixer capable of generating very high harmonics and possessing the highest possible frequency response. To date the best mixer appears to be a tungsten-on-nickel diode. However, although its frequency response extends to at least 88 THz (3.39 μm), its ability to generate harmonics is somewhat limited. Consequently, in the measurement of the frequency of the 3.39-μm line a rather complex laser frequency chain had to be employed.[27]

On the other hand, the Josephson junction is unique in its ability to generate high harmonics of applied signals. In early work by McDonald and co-workers[29] a 60-MHz beat signal was obtained between the 100th harmonic of an *X*-band ($\sim$ 10 GHz) klystron and the 891-GHz (337-μm) radiation from an HCN laser. In more recent work McDonald and colleagues[30] obtained a beat note between the 401st harmonic of an *X*-band signal and the 3.8-THz (78-μm) radiation from a water vapor laser. McDonald (this conference) is now attempting to refine his techniques in order to observe the beat signal between the 1100th harmonic of an *X*-band source and the 10-THz (28-μm) water vapor emission. If successful, this result would greatly simplify the measurement of infrared frequencies and perhaps future determinations of *c*. Thus the Josephson mixer work not only comes under the general heading of cryogenic determinations of fundamental constants, but is really the first entry in the next general area of discussion: basic measurements and standards.

Basic Measurements and Standards

Volt Maintenance via 2e/h. The unit of voltage is generally maintained in the national standards laboratories of the various countries as well as in primary calibration laboratories via large groups of saturated standard cells kept at constant temperature and for which the mean emf of the group is assumed constant with time. However, standard cells are not without their problems. Their emf's are highly temper-

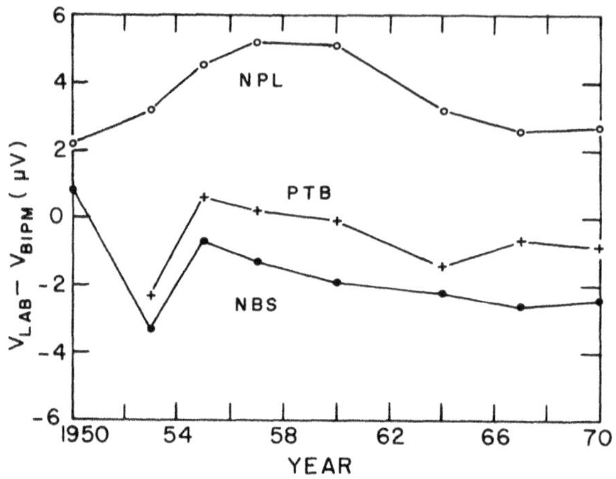

Fig. 4. Relationship over the last twenty years among the units of voltage maintained by three national laboratories and the Bureau International des Poids et Mesures (BIPM) at Sèvres. France. (NPL = National Physical Laboratory, U.K.; PTB = Physikalisch-Technische Bundesanstalt, Germany; and NBS = National Bureau of Standards, U.S.)

ature dependent (about 55 ppm/°C), subject to shock and vibration, and change with time (aging effect). Figure 4 shows the relationship over the last twenty years between the units of voltage maintained by three national laboratories and the unit of voltage maintained by the Bureau International des Poids et Mesures.[11,31] Large relative changes in the units are apparent. (Such volt differences are obtained by interlaboratory shipment of standard cells.)

It follows from the earlier discussion of the ac Josephson effect measurements of $2e/h$ that their reliability is now such that they can be used to eliminate variations of the type apparent in Fig. 4; or in other words, to monitor and maintain a unit of voltage. Indeed, Fig. 5 shows the temporal variation of $2e/h$ as measured in terms of the NBS volt over the past year.[32] Since the $2e/h$ measurements are accurate to a few parts in 10^8 and $2e/h$ is an invariant constant of nature, the apparent drift in $2e/h$ must be ascribed to a drift in the unit of voltage. In order to eliminate this drift in the future, NBS has decided as of July 1, 1972 to adopt an exact value for $2e/h$; and to periodically adjust the emf's assigned the standard cells used to maintain the NBS volt so that V_{NBS} is always consistent with the adopted number.[33] (The value adopted by NBS is $2e/h = 483593.420$ GHz/V_{NBS} and is indicated by the dashed line in Fig. 5.) The amount of adjustment will of course be determined by periodic $2e/h$ measurements. It is believed that in this way the NBS unit of voltage can be maintained constant indefinitely to within $1/10^7$.

Recently the values of $2e/h$ measured at several national laboratories in terms of their respective as-maintained units of voltage were compared by means of a shippable, temperature-regulated, volt-transport standard.[34] This standard enabled the relationships between the units of voltage maintained by the four laboratories involved to be determined to within 1–2 parts in 10^7. Since the values of $2e/h$ obtained at these laboratories all agreed to within this amount (when converted to a common unit of voltage), it can be concluded that the Josephson effect measurements of $2e/h$ are well in hand and that serious consideration should be given to adopting a single international value of $2e/h$ for use in maintaining units of voltage.

Thermometry. There have been a rather significant number of applications of cryogenic phenomena to thermometry over the past decade, especially phenomena associated with superconductivity (For a recent review, see Ref. 35.) The more important of these applications include the following.

(a) Heat capacity. The monotonic and exponential temperature dependence of the heat capacity of a superconductor below its transition temperature T_c can serve quite well to detect the lack of smoothness in a temperature scale below T_c. Indeed, early measurements of the heat capacity of vanadium[36] uncovered substantial errors in the T48 temperature scale.

(b) Critical field. The temperature dependence of the critical magnetic field of a superconductor can be used as a secondary thermometer after calibration against a suitable primary thermometer. Indeed, Ries and Mapother[37] have suggested that 1 m°K accuracy between 0.3 and 1.2°K can be obtained via a critical field thermometer based on Al.

(c) Transition temperatures. Schooley and Soulen[38] have demonstrated that the transition temperatures of pure superconductors can be used as fixed points in the cryogenic range, i.e., 0.5–7°K. Transition widths of a variety of superconductors (Pb, In, Al, Zn and Cd) were examined and found to be less than 1 m°K, and the transitions reproducible to 1 m°K. It may also be possible to extend the range to $0.01 < T < 20$°K using other suitable superconductors.

(d) Superconductive tunneling. That the exponential-like temperature dependence

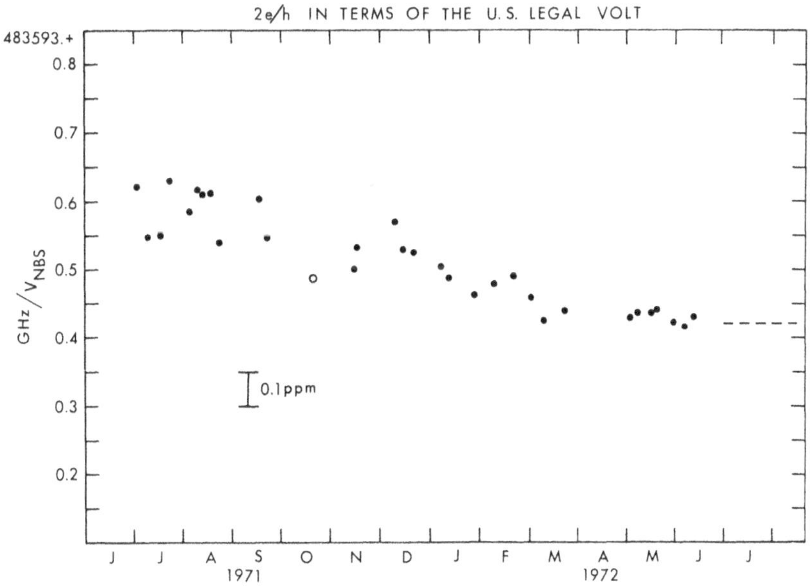

Fig. 5. Plot showing the apparent temporal variation of $2e/h$ as measured in terms of the NBS volt over the past year. The dashed line indicates the value of $2e/h$ adopted for maintenance of the NBS volt after July 1, 1972 (see text).

of the tunnel current in a superconducting tunnel junction can be used as a secondary thermometer was first demonstrated by Giaever *et al.*[39] in 1962. The useful range of such a thermometer would appear to be $0.2T_c < T < 0.9T_c$.

(e) Noise thermometry. One of the more unique applications of the ac Josephson effect is the absolute noise thermometer. First proposed by Kamper in 1967,[40] it was demonstrated experimentally by Kamper and Zimmerman in 1970[41] after earlier work by Silver *et al.*[42] The principle of the device is quite straightforward. A Josephson junction (usually a point contact) is connected in a low-inductance circuit across a small resistance $R(\sim 10^{-5}\Omega)$. A direct current through the resistor biases the junction at a voltage such that it oscillates at a Josephson frequency of order 30 kHz. [Recall the Josephson equation $\nu = (2e/h)\,V$, where $2e/h \approx 484$ MHz/μV.] Due to the thermal (Johnson) noise in the resistor, the voltage across the Josephson junction fluctuates, and the voltage fluctuations are converted to frequency fluctuations according to the Josephson relation. If now the variance $\sigma^2(\nu)$ of the frequency fluctuations are measured, then it can be shown that

$$\sigma^2(\nu) = [2k_B R/\tau(2e/h)^2]T$$

where k_B is the Boltzmann constant, τ is the counter gate time, and T is the absolute temperature. Since all of the quantities appearing in the coefficient of T are known or can be readily determined, the Josephson noise thermometer can serve as a means of measuring thermodynamic temperature (i.e., in kelvins). Kamper *et al.*[43] have shown that noise temperatures in the m°K range can be observed. Recently Soulen and Marshak[44] (and this conference) have compared a Josephson noise thermometer with two low-temperature scales below 1°K: the ^{3}He vapor pressure scale (extrapolated to 0.02°K via cerium magnesium nitrate) and the scale derived from the γ-ray anisotropy of ^{60}Co in single-crystal ^{59}Co. In both cases the agreement was quite satisfactory over the temperature range covered (from 0.02 to 0.7°K for the former and from 0.02 to 0.05°K for the latter). For a discussion of other methods of determining temperatures in the m°K range, see Wheatley.[63]

Time and Frequency. Quite recently significant advances have been made in the utilization of external high-Q superconducting cavities for the stabilization of microwave oscillators. Benard and co-workers[45] have stabilized a 10-GHz reflex klystron and have reported fractional frequency standard deviations of about 2×10^{-12} for 10-sec averaging times. Stein and Turneaure[46] (and this conference) stabilized an 8.6-GHz Gunn effect oscillator with phenomenal results: fractional frequency fluctuations of 1.5×10^{-14} for 10-sec averaging times and a fractional frequency drift of 1.6×10^{-14}/hr for a 32-hr interval. (The short-term result exceeds the state of the art for any method for 10-sec averaging times.) Stein and Turneaure (this conference) point out that there are many applications for such oscillators, including: (1) attempting to measure the temporal variation of the fine structure constant; (2) a compact clock for use in experiments to test general relativity; (3) measuring gravitational radiation (gravity waves); (4) improving atomic time standards, e.g., the cesium atomic clock and the proposed hydrogen storage beam atomic clock; and (5) providing an ultrastable X-band source of radiation for measuring the frequencies of infrared lasers. (See the discussion on the measurement of the speed of light.)

RF Attenuation. Recently Kamper and Simmonds[47] described a broadband superconducting quantum interference device (SQUID) designed to operate at 9 GHz. (We shall say a bit more about SQUIDs later.) The device, which can be used to measure radiofrequency currents in much the same way as the more conventional 30-MHz SQUID can be used to measure dc currents, consists of a point-contact Josephson junction mounted in a reduced-height X-band waveguide inductively coupled to a 50-Ω coaxial line. As demonstrated by Kamper and co-workers,[48] one of the more useful applications of the device is the measurement of rf attenuation. It can be shown that if an rf current of amplitude I_1 is passed through the coaxial line, then the average reflected microwave power from the SQUID varies with I_1 as $J_0(2\pi I_1/I_0)$, where J_0 is the zeroth-order Bessel function and I_0 is the current required to change the magnetic flux in the SQUID by one quantum ($I_0 \sim 70\mu\text{A}$). Thus, by using the appropriate electronics, the reflected power may be made to trace out the Bessel function (its null points in particular) as the rf current is varied. Since values of the argument at which the Bessel function goes to zero have been tabulated, the current attenuation ratios may be readily calculated. In a test of the method using a 30-MHz piston attenuator it was found that the computed attenuation ratios were in excellent agreement with the values assigned by the NBS Calibration Service. Although this accuracy makes the device useful as presently developed, its great advantage is its broadband response; the SQUID device can operate from dc to 1 GHZ, while the conventional piston attenuator is a very narrow band and can only be used at a discrete frequency.

Novel Instrumentation. During the past ten years many rather novel instruments requiring cryogenics for their operation have been constructed. In most cases the metrological problems they were designed to solve fall under the broader definition of metrology rather than the more traditional one. While many of these instruments take advantage of the Josephson effects, some quite useful devices do not. Here we very briefly discuss some of the cryogenic instruments which we believe to be the most unique.

SQUID Technology. The use of the SQUID (superconducting quantum interference device) in a variety of circuit configurations for determining current, resistance, and voltage is now commonplace. As a voltmeter, the SQUID's sensitivity of 10^{-17} V is unprecedented. Similarly, its ability to easily detect magnetic fields of less than 10^{-14} T cannot be matched by any other device. In this regard, we show in Fig. 6 human magnetocardiograms taken in a magnetically unshielded environment by Zimmerman and Frederick.[49] They were obtained using a portable magnetic gradiometer utilizing a pair of differentially connected flux transformers and a SQUID detector biased at 30 MHz. Such magnetocardiograms (MCG's) compare quite favorably with ECG's and may eventually provide a new tool for the diagnosis of heart defects.

However, it is not our purpose here to discuss the large body of literature which now exists on the technology and application of SQUIDs, but to simply note their existence and some of their uses. We refer the reader to Refs. 50–52 for complete details and additional references.

Cryogenic Dividers and Ratio Devices. Any Josephson device used to carry out $2e/h$ measurements for the purpose of maintaining units of voltage must of course be cooled to liquid helium temperatures. It is therefore natural to consider the possibility of constructing an all-cryogenic potentiometer for comparing the Josephson step voltages to the emf of a standard cell. Indeed, Sullivan[53] at NBS has designed and constructed a prototype all-cryogenic apparatus similar to that schematically depicted in Fig. 2. A silicon-bronze alloy wire of low temperature coefficient is used to wind the resistors, and a mechanical superconducting switch is used to connect them either in series or parallel as required. The low-level null detector is a SQUID. (Because of the superconducting switches and superconducting connecting leads, the usual compensating resistors of the room-temperature version are not required. The result is a much simpler device.) The only portions of the apparatus which are at room temperature are the 1-mA Hamon network power supply, the standard cell, and the standard cell null detector. (An overall view of the apparatus is shown in Fig. 7.) Tests of the prototype against a similar room-temperature apparatus have been very encouraging. A similar device has been constructed by Gallop[54] at the National Physical Laboratory, U.K., using resistors wound from phosphor bronze wire.

Sullivan[55] has also constructed a rather clever apparatus for digitally measuring dc current (and voltage) ratios. In this method the change in the number of flux

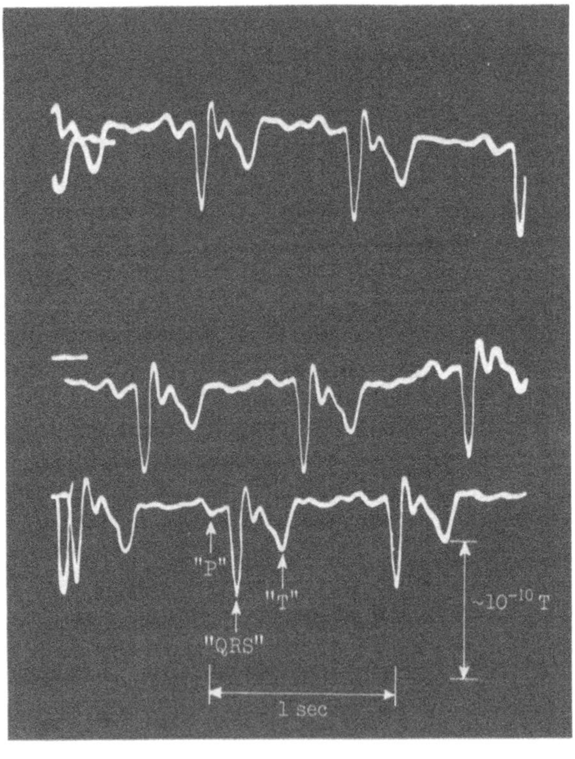

Fig. 6. Typical magnetocardiograms taken in an unshielded environment with a SQUID gradiometer magnetometer (from Zimmerman and Frederick.[49])

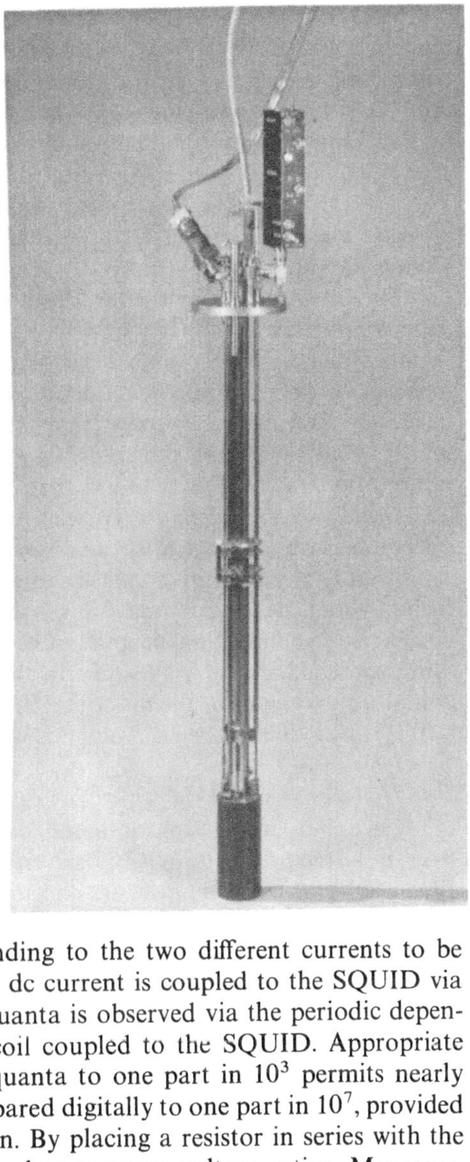

Fig. 7. Photograph of the low-temperature voltage divider and null detector of Sullivan. The box at the top contains the circuitry for the SQUID null detector readout. The divider resistors are wound on the large brass spool at the bottom. The thin-film Josephson tunnel junction, SQUID, and superconducting switches are inside the spool. (From Sullivan.[53])

quanta in a multihole SQUID corresponding to the two different currents to be measured is determined by counting. The dc current is coupled to the SQUID via a multiturn coil and the change in flux quanta is observed via the periodic dependence of the rf impedance of a second coil coupled to the SQUID. Appropriate circuitry for interpolating between flux quanta to one part in 10^3 permits nearly equal currents in the mA range to be compared digitally to one part in 10^7, provided the circuit parameters are properly chosen. By placing a resistor in series with the dc current coil, the device can also be used to measure voltage ratios. Moreover, by operating the device in a feedback mode, it can be used as a current regulator. Regulation of a 3-mA current to $1/10^9$ has been achieved.

Another promising ratio device has recently been constructed by Harvey[56] of the National Standards Laboratory, Australia. It is an all-cryogenic dc ratio transformer similar to the "classical" room-temperature dc ratio transformer developed by MacMartin and Kusters[57] in which the magnetic flux in a high-permeability magnetic core with several windings is sensed in order to determine the ampere-turn balance condition. In the cryogenic version this sensing is done via a SQUID.

In Harvey's prototype unit the equality of the 11 ratio windings wound on the magnetic core is ensured by winding them with a multistrand cable having a superconducting sheath. A test of the transformer showed that the ratio of seven of the individual windings and the sheath to one of the windings used as a reference differed by less than 0.01 ppm. The largest deviation of the remaining three was 0.025 ppm. The noise in the system is such that 0.002-ppm ratio measurements can be made. The practicality of the transformer was also tested by connecting the windings in series to obtain a known 10:1 ratio and using it to calibrate cryogenic resistors having the same nominal ratio.

Superconducting Gravimeter. Prothero and Goodkind[58] have built an ingenious superconducting gravimeter and have used it in extensive studies of earth tides. It consists of a 2.54-cm-diameter superconducting sphere levitated in the magnetic field generated by two superconducting coils. Any changes in the vertical gravitational or inertial force are measured as changes in the vertical position of the sphere. These are detected via a capacitance displacement transducer. Fractional changes in the acceleration due to gravity of less than $1/10^{10}$ can be observed with this apparatus.

Gravity Wave Detectors. At least two groups are currently carrying out work on cryogenic gravity wave detectors which, if successful, should yield instruments significantly more sensitive than the apparatus used by Weber in his original experiments. Hamilton *et al.*[64] are using a 5000-kg cylinder supported via a persistent current superconducting magnet. (The cylinder is coated with a high-field, thin-film superconductor.) A superconducting accelerometer operated as a gravimeter will be used to monitor the motion of the bar and the entire apparatus can be cooled to 0.1°K by a dilution refrigerator. A somewhat similar apparatus is being fabricated by Opfer *et al.*[65]

Non-Josephson Cryogenic Electrical Instruments. A significant number of cryogenic but non-Josephson-based electrical instruments have been developed over the last ten years, primarily for the detection of low-level dc signals in the picovolt range. Of particular interest are the magnetic amplifier device of Poerschke and Wollenberger[59] and the vibrating mutual inductor instrument of Ries and Scatterthwaite.[60]

Whither Cryogenic Metrology

We conclude this review with some comments on the future of cryogenic metrology. However, prognostications beyond the next five years are not given because of the impossibility of extrapolating the present rapid rate of progress so far ahead.

Fundamental Constants

2e/h. It is quite likely that the capability for measuring $2e/h$ to an accuracy of $1/10^9$ in terms of some fixed voltage at the 1-V level will soon become available via cryogenic potentiometers and null detectors. However, it does not seem likely that one will find a voltage reference with sufficient long- or even short-term stability (i.e., a few hours) to take advantage of this capability. It might just be possible to achieve five parts in 10^9 short term by selecting a particularly stable standard cell and maintaining its temperature constant to 100 μK.

hN. The cryogenic determination of this quantity is now a completely open question since, as noted previously, recent work indicates that the ac Josephson effect in liquid helium has yet to be observed. Even if the stable head differences found in the past are in fact related to *hN* as formerly believed, it is unlikely that any useful number (e.g., a value accurate to 10 ppm) would be forthcoming in the next several years, if at all. The reason for this is the many difficulties associated with the experiment, not the least of which is the influence of mechanical vibrations.

h/m_e. A sufficiently accurate measurement of this quantity to be of interest (e.g., 5 ppm) should be available within the next few years via the rotating superconducting experiment of Parker and co-workers. The fact that interference patterns from 1-cm-diameter rings have now been observed is the principal source of this optimism. Maintaining the ambient magnetic flux sufficiently constant (i.e., to within 10 fT) and understanding the motional and binding corrections to m_e still remains to be achieved, as does the accurate determination of the area of the ring at the operating temperature. However, the latter seems amenable to present dimensional metrology technology.

c. Josephson device mixers should soon play a significant role in the measurement of the frequency of a variety of laser lines in the infrared. Whether they can be used to directly measure the frequency of laser lines in the optical region is still an open question, but one probably requiring some pessimism according to the work of McDonald and co-workers.[29,30] In any event, Josephson mixers should soon be of use in future measurements of *c*.

Other Fundamental Constants and Related Quantities. Cryogenics will no doubt play an increasing role in future determinations of other fundamental constants. For example, we are now considering at NBS how a liquid-nitrogen-cooled solenoid could be used in the experiment to measure the ratio of the NBS ampere to the absolute ampere via the current balance method. Rich[61] at the University of Michigan is contemplating an improved determination of the anomalous moment of the electron using a high-field superconducting solenoid. And of course, there remain to be done the experiments outlined in the article by Taylor *et al.*[11] and not discussed here, i.e., measurement of the flux quantum $h/2e$ and the quantity h/m_e by methods alternate to those previously given.

Basic Measurements and Standards

Volt Maintenance via $2e/h$. As noted previously, NBS is already on the "Josephson volt standard." Moreover, the good agreement between $2e/h$ values determined at several national laboratories (also noted previously) should lead to the international adoption of a single exact value for $2e/h$ for use in maintaining units of voltage by all those laboratories desiring to do so. This may well take place as early as 1975.

Thermometry. Most of the applications of cryogenic phenomena to the measurement of temperature previously discussed are currently being used and no doubt will find wider acceptance in the future. The Josephson junction noise thermometer looks extremely promising at this point in time and may become the method to use for absolute temperature measurements in the range $T < 10°K$.

Time and Frequency. The use of high-Q superconducting cavities to stabilize various kinds of oscillators is an extremely promising area. Stein and Turneaure[66] predict that fractional frequency fluctuations of 10^{-16} for 10-sec averaging times are possible as well as fractional frequency drift rates of 1×10^{-14}/yr. As these authors point out, such oscillators may be useful for improving time standards and for carrying out a variety of unique physics experiments, e.g., tests of the temporal variation of the fine structure constant and general relativity.

RF Attenuation. The broadband calibration of rf attenuators via the SQUID device of Kamper *et al.*[48] may well become the standard technique in the field within the next few years.

It is impossible to predict what other applications of cryogenics to the basic measurements and standards field will be found in the near future. However, one can certainly be sure that there will be others. The use of cryogenic dividers and ratio devices (see below) for intercomparing standard resistors is one very real possibility.

Novel Instrumentation

Special-Purpose Apparatus. The stability of persistent superconducting currents and the low thermal and mechanical noise characteristic of materials cooled to cryogenic temperatures will ensure the continued design and construction of such unique apparatus as the gravimeter and gravity wave detector. Also in this category are Fairbank's[62] superconducting linear accelerator and his superconducting gyroscope for testing general relativity.

SQUID Technology. SQUID-based instrumentation in all of its various forms will no doubt become even more sophisticated during the next few years and will find its way into most research laboratories. The commercial manufacture of such instrumentation should speed up this widespread utilization. Indeed, at least three small companies are now marketing a variety of SQUID instruments and it is even rumored that one of the largest companies in the instrumentation field is considering undertaking a major program to develop commercial SQUID instruments.

Classical dc and Low-Frequency Instruments. In the author's opinion, this field will be the "new wave" of the next five years. Cryogenic dividers, potentiometers, ratio transformers, etc, will prove to be of great use in a variety of "classical" applications such as standard resistor comparisons, volt boxes, and current comparators. The all-cryogenic $2e/h$ volt comparator is, of course, the best example to date.

In conclusion, it is probably safe to say that in the not too distant future, alongside the traditional drum of oil found in every metrology laboratory will be a storage dewar of liquid helium.

References

1. D.H. Parkinson, *J. Sci. Instr.* **41**, 68 (1964).
2. H.G. Smith and F.G.A. Tarr, *Trans. Roy. Soc. Can.* Sect. 3, **29**, 23 (1935); H.G. Smith, K.C. Mann, and J.O. Wilhelm, *Trans. Roy. Soc. Can.* Sect. 3, **30**, 13 (1936).
3. A.B. Pippard and G.P. Pullan, *Proc. Cambr. Phil. Soc.* **48**, 188 (1952).
4. I.M. Templeton, *J. Sci. Instr.* **32**, 314 (1955).
5. I.M. Templeton, *J. Sci. Instr.* **32**, 172 (1955).
6. J.S. Hulbert and G.O. Jones. *Proc. Phys. Soc. (London)* Sect. B, **68**, 801 (1965).
7. W.J. Tabor and J.T. Sibilia, *Bell Syst. Tech. J.* **42**, 1863 (1963).
8. R.D. Cutkosky, *IEEE Trans. Instr. Meas.* **IM-17**, 232 (1968).
9. K.M. Evenson, J.S. Wells, F.R. Peterson, B.L. Danielson, G.W. Day, R.L. Barger, and J.L. Hall, *Phys. Rev. Lett.* **29**, 1346 (1972).

10. W.H. Parker, B.N. Taylor, and D.N. Langenberg, *Phys. Rev. Lett.* **18**, 287 (1967).
11. B.N. Taylor, W.H. Parker, and D.N. Langenberg, *Rev. Mod. Phys.* **41**, 375 (1969).
12. S. Shapiro, *Phys. Rev. Lett.* **11**, 80 (1963); S. Shapiro, A.R. Janus, and S. Holly, *Rev. Mod. Phys.* **36**, 223 (1964).
13. J.C. Gallop and P.W. Petley, in *Proc. 4th Intern. Conf. Atomic Masses and Fundamental Constants,* J.H. Sanders and A.H. Wapstra, eds., Plenum Publishing Corp., New York (1972), p. 398.
14. I.K. Harvey, J.C. Macfarlane, and R.B. Frenkel, *Metrologia* **8**, 114 (1972); *Phys. Rev. Lett.* **25**, 853 (1970).
15. V. Kose, F. Melchert, H. Fack, and H.-J. Schrader, *PTB Mitt.* **81**, 8 (1971).
16. T.F. Finnegan, T.J. Witt, B.F. Field, and J. Toots, in *Proc. 4th Intern. Conf. Atomic Masses and Fundamental Constants,* J.H. Sanders and A.H. Wapstra, eds., Plenum Publishing Corp., New York (1972), p. 403.
17. T.F. Finnegan, A. Denenstein, and D.N. Langenberg, *Phys. Rev. B* **4**, 1487 (1971).
18. P.L. Richards and P.W. Anderson, *Phys. Rev. Lett.* **14**, 540 (1965).
19. P.L. Richards, *Phys. Rev. A* **2**, 1532 (1970).
20. B.M. Khorana and D.H. Douglass, Jr., in *Proc. 11th Intern. Conf. Low Temp. Phys. 1968*, St. Andrews University Press, St. Andrews, Scotland (1969), Vol. 1, p. 169.
21. B.M. Khorana and B.S. Chandrasekhar, *Phys. Rev. Lett.* **18**, 230 (1967); B.M. Khorana, *Phys. Rev.* **185**, 299 (1969).
22. J.P. Hulin, C. Laroche, A. Libchaber, and B. Perrin, *Phys. Rev. A* **5**, 1830 (1972).
23. J.E. Zimmerman and J.E. Mercereau, *Phys. Rev. Lett.* **14**, 887 (1965).
24. W.H. Parker and M.B. Simmonds, in *Precision Measurement and Fundamental Constants.* D.N. Langenberg and B.N. Taylor, eds., NBS Special Publication 343, USGPO, Washington, D.C. (1971), p. 243.
25. Z. Bay, G.G. Luther, and J.A. White, *Phys. Rev. Lett.* **29**, 189 (1972).
26. K.M. Evenson, G.W. Day, J.S. Wells, and L.O. Mullen, *Appl. Phys. Lett.* **20**, 133 (1972).
27. K.M. Evenson, J.S. Wells, F.R. Peterson, B.L. Danielson, and G.W. Day, *Appl. Phys. Lett.* **22**, 192 (1972).
28. R.L. Barger and J.L. Hall, *Appl. Phys. Lett.* **22**, 196 (1972).
29. D.G. McDonald, A.S. Risley, J.D. Cupp, and K.M. Evenson, *Appl. Phys. Lett.* **18**, 162 (1971).
30. D.G. McDonald, A.S. Risley, J.D. Cupp, and K.M. Evenson, *Appl. Phys. Lett.* **20**, 296 (1972).
31. E.R. Cohen and B.N. Taylor, in *Proc. 4th Intern. Conf. Atomic Masses and Fundamental Constants,* J.H. Sanders and A.H. Wapstra, eds., Plenum Publishing Corp., New York (1972), p. 543.
32. B.F. Field, T.F. Finnegan, and J. Toots, *Metrologia* **9**, 155 (1973).
33. *Nat. Bur. Std. Tech. News Bull.* **56**, 159 (1972).
34. W.G. Eicke, Jr. and B.N. Taylor, *IEEE Trans. Instr. Meas.* **IM-21**, 316 (1972).
35. J.F. Schooley, in *The Science and Technology of Superconductivity,* W.D. Gregory, W.N. Mathews, Jr. and E.A. Edelsack, eds., Plenum Publishing Corp., New York (1973), Vol. 2, p. 625.
36. W.S. Corak, B.B. Goodman, C.B. Satterthwaite, and A. Wexler, *Phys. Rev.* **102**, 565 (1956).
37. R.P. Ries and D.E. Mapother, in *Temperature, Its Measurement and Control in Science and Industry,* H.H. Plumb, ed., ISA, Pittsburgh, Pa. (1972), Vol. 4, p. 885.
38. J.F. Schooley and R.J. Soulen, in *Temperature, Its Measurement and Control in Science and Industry,* H.H. Plumb, ed., ISA, Pittsburgh, Pa. (1972), Vol. 4, p. 169.
39. I. Giaever, H.R. Hart, Jr., and K. Megerle, *Phys. Rev.* **126**, 941 (1962).
40. R.A. Kamper, in *Proc. Symp. on Physics of Superconducting Devices,* Charlottesville, Va., 1967 (Office of Naval Research Report).
41. R.A. Kamper and J.E. Zimmerman, *J. Appl. Phys.* **42**, 132 (1971).
42. A.H. Silver, J.E. Zimmerman, and R.A. Kamper, *Appl. Phys. Lett.* **11**, 209 (1967).
43. R.A. Kamper, J.D. Siegwarth, R. Radebaugh, and J.E. Zimmerman, *Proc. IEEE* **59**, 1368 (1971).
44. R.J. Soulen and H. Marshak, in *Proc. 1972 Applied Superconductivity Conf.*, IEEE Pub. No. 72 CHO682-5-TABSC (1972), p. 588.
45. J. Benard, J.J. Jimenez, P. Sudraud, and N.T. Viet, *Electron. Lett.* **8**, 117 (1972).
46. S.R. Stein and J.P. Turneaure, *Electron. Lett.* **8**, 321 (1972).
47. R.A. Kamper and M.B. Simmonds, *Appl. Phys. Lett.* **20**, 270 (1972).
48. R.A. Kamper, M.B. Simmonds, R.T. Adair, and C.A. Hoer, in *Proc. 1972 Applied Superconductivity Conf.*, IEEE Pub. No. 72 CHO682-5-TABSC (1972), p. 696.
49. J.E. Zimmerman and N.V. Frederick, *Appl. Phys. Lett.* **19**, 16 (1971).

50. R.P. Giffard, R.A. Webb, and J.C. Wheatley, *J. Low Temp. Phys.* **6**, 533 (1972).
51. W.W. Webb, *IEEE Trans. Magn.* **MAG-8**, 51 (1972).
52. J.E. Zimmerman, *Cryogenics* **12**, 19 (1972).
53. D.B. Sullivan, *Rev. Sci. Instr.* **43**, 499 (1972).
54. J.C. Gallop, National Physical Laboratory Report Qu22, April 1972.
55. D.B. Sullivan, in *Proc. 1972 Applied Superconductivity Conf.*, IEEE Pub. No. 72 CHO682-5-TABSC (1972), p. 631.
56. I.K. Harvey, *Rev. Sci. Instr.* **43**, 1626 (1972).
57. M.P. MacMartin and N.L. Kusters, *IEEE Trans. Instr. Meas.* **IM-15**, 212 (1966).
58. W.A. Prothero and J.M. Goodkind, *J. Geophys. Res.* **77**, 926 (1972).
59. R. Poerschke and H. Wollenberger, *Cryogenics* **10**, 333 (1970).
60. R.P. Ries and C.B. Satterthwaite, *Rev. Sci. Instr.* **38**, 1203 (1967).
61. A. Rich, private communication.
62. W.M. Fairbank, in *Proc. Symp. on Physics of Superconducting Devices*, Charlottesville, Va., 1967 (Office of Naval Research Report).
63. J.C. Wheatley, this volume.
64. W.O. Hamilton, P.B. Pipes, and P.S. Nayar, this volume.
65. J.E. Opfer, S.P. Boughn, W.M. Fairbank, M.S. McAshan, H.J. Paik, and R.C. Taber, this volume.
66. S.R. Stein and J.P. Turneaure, this volume.

Macroscopic Quantum Devices*

J. Clarke†

Department of Physics, University of California, Berkeley
and Inorganic Materials Research Division
Lawrence Berkeley Laboratory, Berkeley, California
and Royal Society Mond Laboratory, Cambridge, England

Either one or two Josephson junctions may be inserted into a superconducting ring to form a superconducting quantum interference device (SQUID). The dc SQUID consists of two junctions mounted on a ring. The critical current of the device is an oscillatory function of a magnetic flux applied at right angles to the plane of the ring, the period being the flux quantum Φ_0. Changes in flux of much less than Φ_0 can be detected: In a device of inductance 10^{-9} H the resolution is typically $10^{-4}\Phi_0/\sqrt{\text{Hz}}$. This sensitivity is perhaps two orders of magnitude below the theoretical noise limit. The rf SQUID consists of a single junction mounted on a ring, and is weakly coupled to the inductance of a tank circuit. The tank circuit is excited at its resonant frequency by an rf current of sufficient amplitude that the peak induced current in the rf SQUID exceeds the critical current of the junction. At an appropriate rf current bias level the voltage across the tank circuit is a periodic function of the flux applied to the SQUID, the period again being Φ_0. In a device of inductance 10^{-9} H and with a tank circuit resonant frequency of 30 MHz the resolution is typically $10^{-4}\Phi_0/\sqrt{\text{Hz}}$. This figure is close to the theoretical noise limit.

Either type of SQUID may be used in conjunction with a flux transformer to measure magnetic fields (with a resolution of 10^{-14} T/$\sqrt{\text{Hz}}$), magnetic field gradients (with a resolution of 10^{-12} T/m$\sqrt{\text{Hz}}$), and magnetic susceptibilities. SQUIDs may also be used as voltmeters. The resolution is limited by Johnson noise developed in the resistance of the low-temperature circuit, provided that this resistance is not too large: The upper limit in the liquid ^{4}He temperature range appears to be a few tens of ohms. In the Johnson-noise-limited measurement the power resolution is $4kTB \simeq 6 \times 10^{-23}$ W at 1°K in a 1-Hz bandwidth.

* Supported in part by the U.S. Atomic Energy Commission. A detailed account appears in the January 1973 issue of *Proc. IEEE*.

† Alfred P. Sloan Foundation Fellow.

14

Thermometry

Refrigeration and Thermometry in the Microkelvin Range*

R. G. Gylling†

Department of Technical Physics
Helsinki University of Technology, Otaniemi, Finland

Introduction

In every experiment at low temperatures there are three main difficulties that have to be overcome: (1) how to produce and maintain the cold; (2) how to cool the specimen to be investigated; and (3) how to measure the low temperature.

With the advent of the dilution refrigerator the number of people doing research in the millikelvin range has vastly increased, and much interest has been focused on these problems. Today the low-temperature experimentalist in general masters the difficulties arising at temperatures above 10 m°K, but there remains the more exotic temperature range below 10 m°K and especially the microkelvin range, which is now becoming accessible to experimental work. Here new solutions to the above-mentioned problems have to be found since old methods generally do not work satisfactorily.

Refrigeration

The only method of reaching equilibrium temperatures below 1 mK that has been put into practice in the past five years has been nuclear refrigeration.[1,2] The lowest lattice temperature that has been produced is 370 μ°K; this was achieved using nuclear refrigeration of copper.[3,4] However, recently a Czechoslovak group reached 860 μ°K by demagnetizing a sample of $Ce_2Mg_3(NO_3)_{12} \cdot 24H_2O$ (CMN) diluted to 10% with $La_{1.8}Ce_{0.2}Mg_3(NO_3)_6 \cdot 24H_2O$(LMN).[5] This temperature was inferred by measuring the susceptibility of the sample. Although this method has a cooling capacity vastly inferior to nuclear refrigeration, it may come into use due to the much morc easily attainable initial conditions of the demagnetization and because of the anomalously low boundary resistance between CMN and pure ^{3}He.

Except for the fact that nuclear moments are about 2000 times smaller than electronic magnetic moments, the method of nuclear demagnetization resembles demagnetization of a paramagnetic salt to a large extent. Many of the relations valid for this well-known technique may be directly applied. One of the most striking differences is, however, the occurrence of two separate energy baths in a nuclear refrigerant: the lattice including the conduction electrons, and the nuclei. The

* Invited paper.
† Present address: Instruments for Technology Ltd., Rvomela, Finland.

energy exchange within these baths is good, which leads to good internal equilibrium, while the energy exchange between the baths may be quite poor. This makes it possible to define two distinct temperatures, the lattice and the nuclear spin temperatures. At low temperatures only metals have a good enough energy exchange between these baths to be useful in nuclear refrigeration experiments. The relation governing the energy exchange is

$$\frac{dE}{dt} = \frac{\lambda N B^2}{\mu_0 K}\left(\frac{T_e}{T_n} - 1\right)$$

where λ is the nuclear Curie constant, K is the Korringa constant of the material used as refrigerant, N is the number of moles, and B is the applied magnetic induction.

When performing nuclear refrigeration experiments we want to cool the electrons and the lattice using the demagnetized nuclei. It is then of prime importance that the energy exchange between nuclei and electrons be as good as possible. This is achieved by using a refrigerant with a large value of λ/K and leaving a substantial final magnetic induction on the demagnetized nuclei. We then have to compromise between a low final temperature and a large refrigeration capability.

The experimental setup in Helsinki[3] differs from earlier experimental arrangements mainly as to the use of a strong superconducting magnet wound of filamentary wire and a powerful dilution refrigerator for producing the necessary large initial value of B/T. The starting conditions of demagnetization that have been realized so far are B_i = 5T(50 kG) and T_i = 16 m°K. This corresponds to an entropy reduction of only 1.3% for copper, and it is thus clear that there is still room for vast improvement.

A typical result of our demagnetization runs is shown in Fig. 1. This particular experiment was performed using a 12-mole sample of copper and a final magnetic induction of 0.072 T (720 G).

A number of different metals suitable for nuclear refrigeration are compared in Fig. 2. Here the time dependence of the conduction electron temperature after

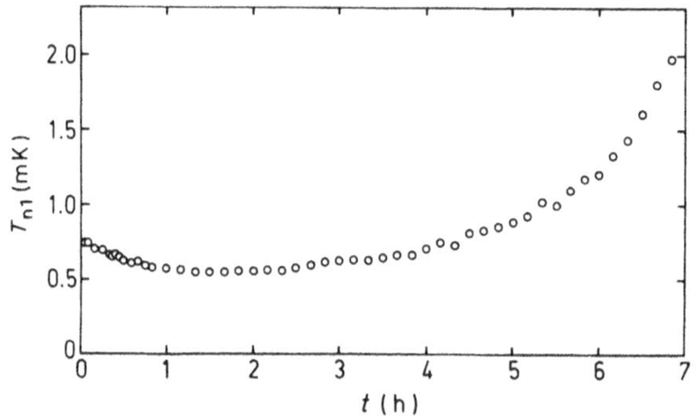

Fig. 1. Observed time behavior of the nuclear spin temperature in the thermometer region of a 12-mole copper nuclear refrigeration stage in an applied magnetic induction of 0.072 T. The initial conditions of demagnetization were T_i = 16 m°K and B_i = 5T (50 kG), and the heat leak was 15 nW.

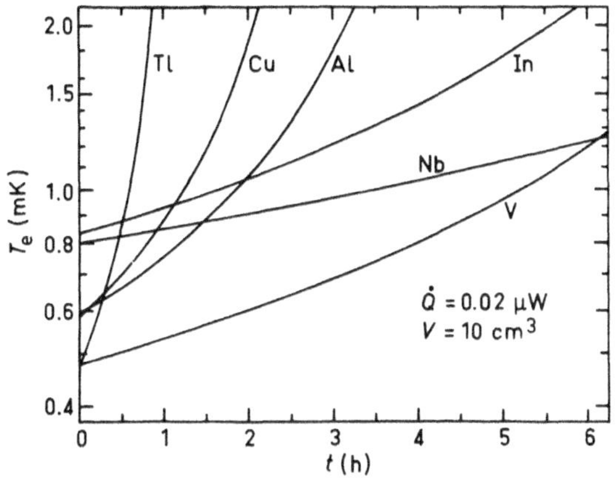

Fig. 2. Comparison of the calculated time behavior of the electron temperature in six metals. It has been assumed that the volume of the specimen is 10 cm^3 and that the initial conditions of the demagnetization are the same in each case: $T_i = 16$ m°K and $B_i = 5$ T. The heat leak is assumed to be 20 nW. The final induction is 0.15 T for Tl, Cu, Al, and V, and 0.25 T for In and Nb. See text for further explanations.

demagnetization is plotted. It has been assumed that the initial conditions are the same as those achieved by us, $B/T = 310$ T/°K, that the volume of the sample is 10 cm^3, and that the heat influx from the surroundings is quite large, 20 nW. In order to have a large cooling capacity below 1 m°K, the final induction is assumed to be 0.15 T in the case of Tl, Cu, Al, and V. In the case of In the strong quadrupole interaction corresponds to an effective internal magnetic induction of 0.25 T and no external induction is needed, except for a 0.03-T induction which is necessary for keeping the metal in the normal state. In Nb the critical induction is about 0.25 T, which makes it impossible to use lower final magnetic inductions in demagnetization experiments.

We see that it is possible to achieve microkelvin temperatures even in the electron and lattice systems for prolonged periods although the sample is rather small and the heat leak is considerable. It is also evident from Fig. 2 that copper is by no means the best refrigerant. Symko[6] has experimentally demonstrated nuclear refrigeration using indium.

According to Fig. 2, vanadium is one of the most promising metals. The entropy removed per unit volume in V is, even for these modest initial conditions, about five times larger than the entropy reduction that can be achieved with CMN. If sufficiently good thermal contact could be achieved, it should be entirely possible to cool, for instance, ^{3}He below 1 m°K by using this refrigerant.

We may thus conclude that the problem of producing and maintaining microkelvin temperatures is now solved also in practice. In the future there may well be several cooling methods in this temperature range at the disposal of the low-temperature physicist, such as hyperfine enhanced nuclear refrigeration[7] and Pomeranchuk cooling in a magnetic field.

Transfer of Heat

Because all heat transfer mechanisms in metals become weaker when the temperature is lowered, heat transfer problems become very serious in the microkelvin range.

Figure 3 shows the thermal resistances encountered in an experiment involving a nuclear refrigeration stage and a metallic or a liquid sample. A thermometer used to measure the temperature of the sample is also included in the figure.

Metallic Samples. In the case of a metallic sample the two thermal impedances encountered during cooling are associated with the electronic thermal conduction κ and the nuclear spin–lattice relaxation time τ_1. Provided the thermal contact between the two electron systems is perfect, the sample nuclei will be cooled at a rate determined by the spin–lattice relaxation time of the sample, which is thus the optimum cooling rate that can be achieved. Effectively this means that only metals, where the spin–lattice relaxation is fast and obeys the Korringa relation $\tau_1 T_e = K$, can be cooled in this type of experiment.

When the temperature of a metallic sample is measured by means of a metallic thermometer material which is in electronic contact with the sample one is faced with exactly the same problems as above. Assuming again that perfect thermal contact can be achieved between the electron systems, the response time of the thermometer is equal to the spin–lattice relaxation time of the thermometric material. In platinum this time is 300 sec at 100 μ°K; in thallium it is only 60 sec. Thermometry should thus be possible in metallic systems.

The thermal time constant of the electron system in a metal is usually short enough to be neglected (around 0.1 sec) and ideally it is independent of temperature. The absolute value of the thermal resistance of the electron system, which is proportional to T^{-1}, can be substantial in the microkelvin range. As an example, the cross section needed in a copper sample to keep the temperature drop below 10% over a distance of 10 cm, at 100 μ°K, and in the presence of a 10-nW heat leak is 50 cm^2, which corresponds to a diameter of 8 cm.

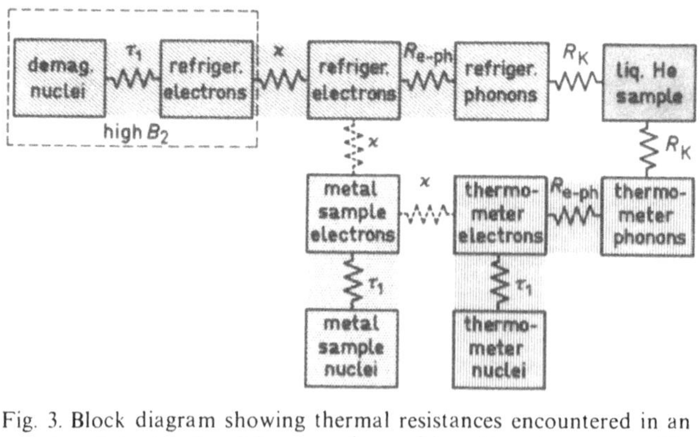

Fig. 3. Block diagram showing thermal resistances encountered in an experimental setup involving a nuclear refrigeration stage, a metallic and/or a liquid helium sample, and a metallic thermometer to measure the sample temperature.

Liquid Helium Samples. Thermal contact between a nuclear refrigeration stage and a liquid helium sample is worse. Here two additional thermal resistances are encountered, the Kapitza boundary resistance and the electron–phonon resistance. The Kapitza resistance R_K has a T^{-3} temperature dependence and it is inversely proportional to the surface of the boundary; thus it may be somewhat reduced by using a finely divided material. In bulk matter the electron–phonon resistance $R_{\text{e-ph}}$ is proportional to T^{-4} and inversely proportional to the volume of the material. Due to its steeper temperature dependence, $R_{\text{e-ph}}$ will always dominate below a certain temperature, which usually is of the order of 10 m°K.

Thermal contact to a metallic thermometer material immersed in liquid helium suffers from the same two thermal resistances. It is worth noticing that the thermal time constant $R_{\text{e-ph}}C_e$, which at low enough temperatures will determine the response time of the thermometer, is a constant characteristic of the thermometric material used. This is similar to the spin–lattice relaxation time but is in contrast to the thermal time constants that involve R_K, which can be affected by altering the surface-to-volume ratio of the thermometric sample. Due to the steep temperature dependences of these time constants, it is found that the limiting temperature below which the response time of the thermometer becomes too long for practical applications (say $\tau \gtrsim 300$ sec), is quite similar for most materials. This temperature is about 1 m°K.

As far as the electron–phonon resistance is concerned, it has been assumed that the relation valid for bulk material can be applied to systems where one or more dimensions become smaller than the phonon wavelength. It is not clear how the resistance will be changed in such systems; it might be that the resistance is increased at higher temperatures but that the temperature dependence at the same time becomes less steep, which would improve the situation considerably at microkelvin temperatures.

The preceding discussion has clearly demonstrated the importance of discovering a metallic system where a direct coupling exists between the electron spins and the nuclear spins of a liquid ^{3}He sample. The important advantage of such a coupling would be that it bypasses not only the Kapitza boundary resistance but also the electron–phonon resistance. The impetus for this search was derived from the discovery that such a coupling exists between CMN and ^{3}He.[8] Although many laboratories, including ours, are investigating this problem, there have been no reports of success so far.

Experimental. In order to illustrate the problems associated with heat transfer in the microkelvin range, the results of one nuclear demagnetization to a final magnetic induction of 0.036 T will be examined. The sample was 12 moles of copper and the initial conditions were $B/T = 290$ T/°K. The time behavior of the temperature of different parts of the nuclear stage is illustrated in Fig. 4. The shape of the nuclear stage is shown in the insert. The temperatures indicated are T_{e1} and T_{n1}, electron and nuclear spin temperatures in the thermometer region, respectively, and T_{e2} and T_{n2}, electron and nuclear spin temperatures in the demagnetized region of the nuclear stage. Due to a 12-nW external heat leak, there is a considerable difference between T_{e1} and T_{e2} (see Fig. 4), and since τ_1 is about 2200 sec for copper at 0.5m°K, the minimum nuclear spin temperature recorded with the NMR thermometer will always be considerably higher than the electron temperature that is achieved in the demagnetized region. The actual minimum temperature recorded in any experiment was 520 μ°K, which corresponded to $T_{e2} = 370$ μ°K and $T_{n2} =$ 300 μ°K.

In another experiment the thermal contact between platinum powder of 3 μm average grain size and a dilute solution of ^{3}He in ^{4}He was investigated by immersing the platinum powder in the liquid helium in a mixing chamber. The expected different thermal time constants are plotted in Fig. 5. Above 10 m°K τ_1 is the dominant time constant for such a fine powder; we were able to verify that there is no other time constant present which is larger than about 0.1 τ_1 in this temperature range. This means that, although the grain size is about 2–3 times smaller than the phonon wavelength, the electron–phonon thermal resistance is not enhanced at 10 m°K.

In summary we can say that good thermal contact is very difficult to achieve in the microkelvin range. To cool liquid samples with nuclear refrigeration seems

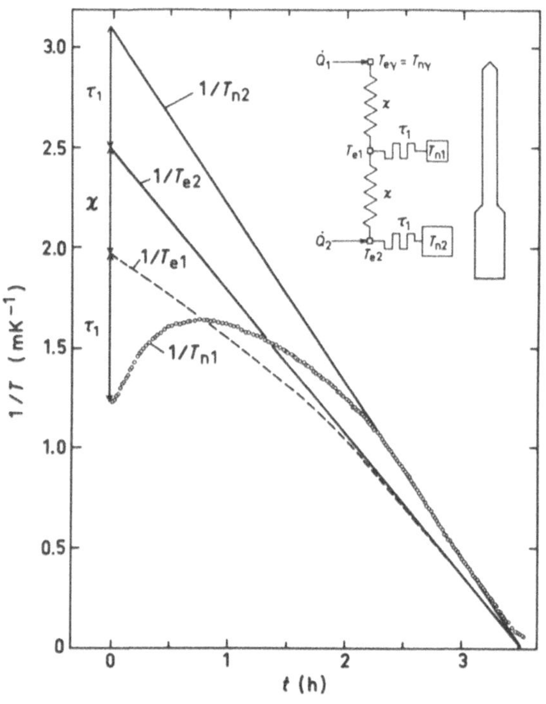

Fig. 4. Diagram showing the observed warming up of a 12-mole nuclear refrigeration stage in an applied magnetic induction of 0.036 T. The insert shows a block approximation and the actual shape of the nuclear stage. T_{e1} and T_{n1} are electron and nuclear spin temperatures in the pulsed NMR thermometer region, and T_{e2} and T_{n2} are the corresponding temperatures in the demagnetized region. Circles denote actual temperature recordings, while other curves are calculated from the heat leak of 12 nW and using the measured values of the electronic thermal resistance κ and the spin–lattice relaxation time τ_1.

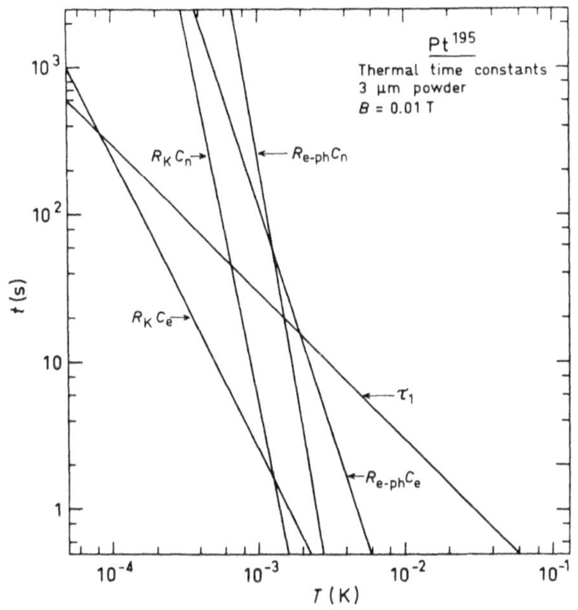

Fig. 5. Calculated thermal time constants of 3-μm average grain size platinum powder immersed in liquid helium. The applied magnetic induction is assumed to be 0.01 T. For explanation of symbols see text.

to become impossible below 1 m°K, and the same applies to thermometry in liquid samples. The only hope of improvement lies with the discovery of new materials in which more efficient heat transfer mechanisms are present. In the case of metallic samples the situation is better; temperatures as low as 100 μ°K should be possible both to realize and to measure provided favorable experimental conditions can be achieved.

Thermometry

In the future new methods, such as noise thermometry, may become practical in the microkelvin range, but at present the only reliable secondary thermometers in this temperature range are those based on the nuclear Curie law.

The nuclear susceptibility may be measured using static or resonant methods. Lately the static method has become as sensitive as resonant methods by the use of superconducting magnetometers based on the Josephson effect (SQUIDs). The drawback of this method, however, is the sensitivity to residual flux or to magnetic impurities. A possible combination of the static and resonant methods would be to employ the SQUIDs for detection of the nuclear magnetic resonance.[9]

Pulsed NMR Thermometry. Nuclear magnetic resonance (NMR) methods, including both continuous wave and pulsed NMR, do not suffer from this disadvantage since they are selective. The NMR signal is enhanced over the static signal by a factor $B/\Delta B$, which may be as large as 1000 for some materials. Here

Table I. Properties of Materials Suitable for Pulsed NMR

Nucleus	Abundance, %	Critical inductance of superconductor, T	Korringa constant K, sec°K	Spin–spin relaxation time τ_2, μsec	χ/χ_{Pt} (ω = const)
^{63}Cu	69.1	—	1.24	80	30
^{65}Cu	30.9		1.08		
^{117}Sn	7.6	0.03	0.035	100	0.8
^{119}Sn	8.6				
^{133}Cs	100	—	0.13	1500	3
^{195}Pt	33.8	—	0.03	1000	1
^{203}Tl	29.5	0.02	0.006	30	11
^{205}Tl	70.5				

B is the applied induction and ΔB is the linewidth of the resonance. There are a number of advantages that make pulsed NMR preferable to continuous wave NMR: (1) The temperature which is recorded corresponds to the situation immediately prior to the measurement, and thus no disturbance from the measuring pulse will affect the result; (2) the NMR signal observed is the nuclear free precession (NFP) signal alone, which eliminates the problem of separating the signal from a large coherent background; (3) the time required for a measurement is shorter, about 1 msec as compared to about 10 sec for continuous wave NMR; and (4) the measuring procedure is easily made automatic with digital output, and the heat input of one measurement is easily controlled.

In Table I the properties of five materials that are well suited for pulsed NMR thermometry are listed.

The requirements that must be met by a good thermometer for the microkelvin range are: (1) fast thermal response, (2) small heat input caused by the measurement, (3) small heat capacity, and (4) ease of calibration. A pulsed NMR thermometer consisting of a finely divided thermometric sample in a weak polarizing magnetic induction ($B < 0.05$ T) meets these requirements. The material must be finely divided in order to allow the transmitter pulse to penetrate the metal and to reduce eddy-current heating. The steady induction cannot be too large since the heat capacity and the eddy-current heating both are proportional to B^2. The calibration of the thermometer is simple in principle since only one measurement at a known temperature is needed.

The block diagram of an automated pulsed NMR thermometer is shown in Fig. 6. The recording apparatus, which provides digital readout of the NFP signal, consists of a rectifying unit, an analog gate determining the integration interval, and an RC integrator. The thermometric materials that have been used so far are copper and platinum; the corresponding NFP signals are shown in Figs. 7 and 8. In copper, where only a few periods of the NFP signal are integrated, a phase-sensitive detector is used to rectify the signal. In platinum, where one may integrate over more than 100 periods, an ordinary linear rectifying method is chosen. The short-time resolution achieved at temperatures below 0.5°K is 0.2 % using platinum.

When a pulsed NMR thermometer is used in the microkelvin range but calibrated against the vapor pressure of helium the necessary extrapolation over three

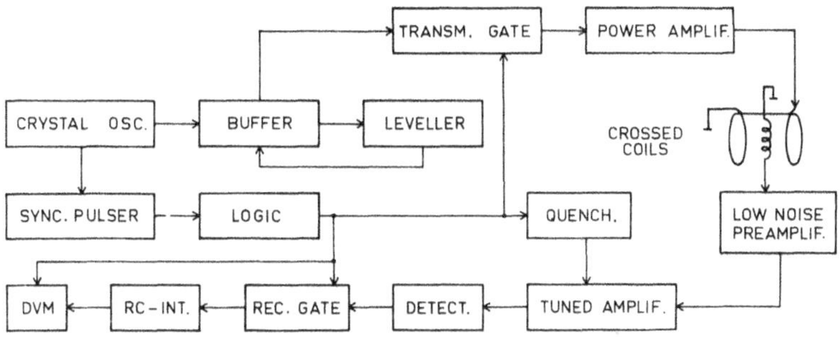

Fig. 6. Block diagram of a pulsed NMR thermometer with digital readout.

decades in temperature impairs the accuracy of the calibration. This drawback is considerably reduced when using a nuclear orientation (NO) gamma-ray thermometer to calibrate the NMR thermometer in the temperature range from 3 to 30 m°K. On the other hand, it is more difficult to achieve thermal equilibrium between the two thermometers at these temperatures.

Korringa Law Thermometry. Another promising method of calibration has been used by us lately,[10,11] which consists of employing the Korringa relation to deduce the temperature by measuring τ_1. The advantage of this method is that thermal equilibrium is much easier to achieve, because only internal equilibrium in the thermometric sample is needed. In this manner the pulsed NMR thermometer becomes self-contained, and provides both a secondary and a primary thermometric output—NMR signal and τ_1, respectively.

In copper, where the relaxation is slow, τ_1 was measured by the field sweep method. The dc magnetic induction acting on the thermometer nuclei was altered, causing a change in the nuclear spin temperature. After waiting sufficiently long

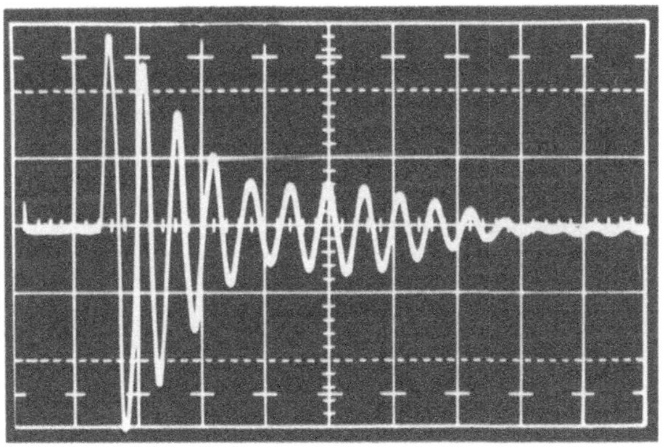

Fig. 7. The nuclear free precession signal of copper in 0.0073 T applied magnetic induction. The beat in the signal is due to the two copper isotopes ^{63}Cu and ^{65}Cu with abundances of 69% and 31%, respectively. The Larmor frequencies of the isotopes differ by 7%. The horizontal scale of the figure is 20 μ sec/division.

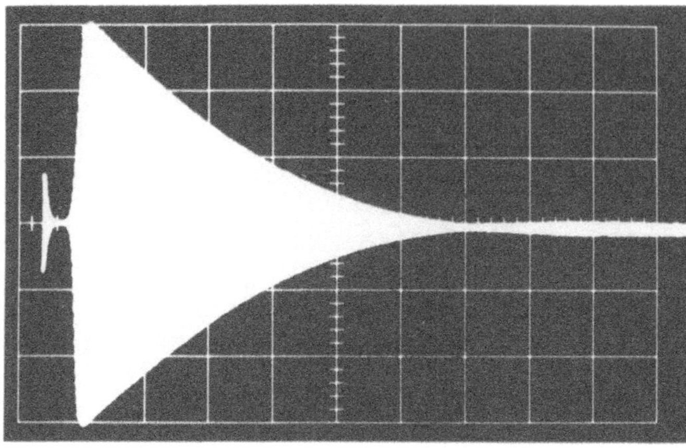

Fig. 8. The nuclear free precession signal of platinum in 0.045 T applied magnetic induction. The horizontal scale of the figure is 200 μ sec/division.

for the nuclei to come into equilibrium with the lattice, the field was swept back to the original value and the subsequent relaxation toward the lattice temperature was recorded. The temperature dependence of the spin–lattice relaxation time between 1 and 10 m°K is shown in Fig. 9. It is seen that τ_1 is accurately proportional to T^{-1} over the whole temperature interval and that the Korringa constant is

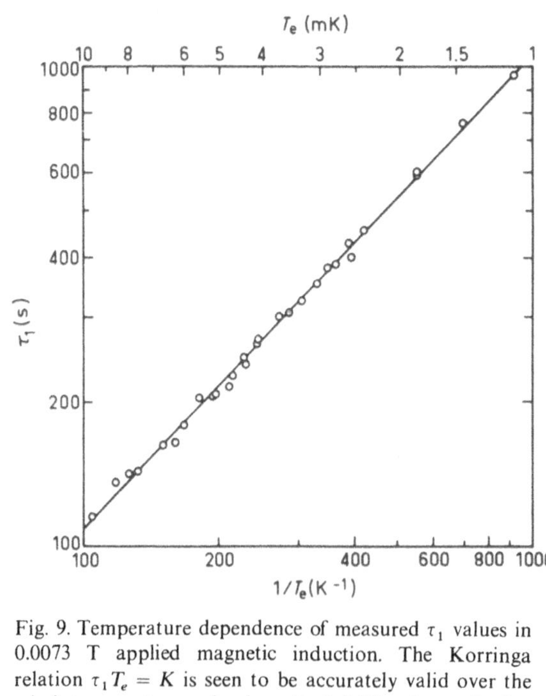

Fig. 9. Temperature dependence of measured τ_1 values in 0.0073 T applied magnetic induction. The Korringa relation $\tau_1 T_e = K$ is seen to be accurately valid over the whole temperature region investigated. The observed value of the Korringa constant is $K = 1.09$ sec °K.

1.09 sec °K. The applied magnetic induction in which the measurements were performed was 0.0073 T and it was found that the limiting high-induction value of the Korringa constant, 1.17 sec °K, also measured by others,[12] was not attained before the applied induction was increased to about 0.025 T. This was attributed to the effect of 2-ppm Mn impurities found in our copper specimen. Thus one must be quite careful in choosing the value of the Korringa constant to be used for calibration purposes. Provided this value is accurately known, about 4% absolute accuracy may be achieved by this method of calibration. In comparison, the accuracy achieved by calibrating against an NO thermometer is about 2%.

Similar measurements on platinum are described elsewhere.[10] The spin-lattice relaxation in platinum is so fast that the experimental measurement procedure must be automated, but on the other hand it seems that here τ_1 is less sensitive to impurities and thus constitutes a better thermometric standard.

We may thus conclude that accurate thermometry in the microkelvin range is feasible using the well-established method of NMR. There is reason to believe that the Curie law nuclear susceptibility will provide a reliable thermometric variable which is linear in T^{-1} down to about 10 μ°K, provided the external dc magnetic induction is kept small enough. With careful calibration of the NMR thermometer the attainable accuracy should be better than 2%, while the resolution will be an order of magnitude better.

References

1. R.H. March and O.G. Symko, *IIR Bull. Annexe* **1965**(2), 57.
2. E.B. Osgood and J.M. Goodkind, *Cryogenics* **6**, (1966).
3. P.M. Berglund, G.J. Ehnholm, R.G. Gylling, O.V. Lounasmaa, and R.P. Søvik, in *Proc. 1970 Ultralow Temperature Symp.* R.A. Hein, D.U. Gubser, and E.H. Takken, eds., Naval Research Laboratory Report No. 7133, Washington, D.C. (1970), p. 113.
4. R.G. Gylling, *Acta Polytech. Scand.*, **81**, 1 (1971); P.M. Berglund, G.J. Ehnholm, R.G. Gylling, O.V. Lounasmaa, and R.P. Søvik, *Cryogenics* **12**, 297 (1972).
5. R.S. Safrata, Private communication.
6. O.G. Symko, *Phys. Lett.* **25A**, 385 (1967).
7. K. Andres and E. Bucher, *J. Appl. Phys.* **42**, 1522 (1971).
8. W.R. Abel, A.C. Anderson, W.C. Black, and J.C. Wheatley, *Phys. Rev. Lett.* **16**, 273 (1966).
9. T. Day, this volume.
10. M.I. Aalto, H.K. Collan, R.G. Gylling, and K.O. Nores, to be published, and this volume.
11. M.I. Aalto, P.M. Berglund, H.K. Collan, G.J. Ehnholm, R.G. Gylling, and O.V. Lounasmaa, *Physica* **173**, (1972).
12. A.G. Anderson and A.G. Redfield, *Phys. Rev.* **116**, 583 (1959).

The Temperature Scale Defined by ^{60}Co γ-Ray Anisotropy and Noise Thermometry

H. Marshak and R. J. Soulen, Jr.

National Bureau of Standards
Washington, D.C.

Introduction

The development of new techniques to obtain low temperatures and the subsequent proliferation of experiments necessitates improvements in both the technology of temperature measurements and the temperature scale itself. Kamper and Zimmerman[1] have demonstrated that a Josephson junction connected in parallel to a resistor converts the Johnson-noise voltage fluctuations into frequency fluctuations, and that the variance of these is then simply related to the absolute temperature. If the temperature determined by the noise thermometer is truly absolute, it should be particularly useful since its range extends four orders of magnitude upward from 1 m°K.

In order to test the noise thermometer, it should be compared with another thermometer whose temperature is also derived from a fundamental relation. At low temperatures there are only a few thermometers that meet this criterion. Of these, one stands out as having considerable potential, namely a nuclear γ-ray anisotropy thermometer. If all the parameters of the nuclear spin system are known along with the decay properties, a measurement of the γ-ray anisotropy would yield the absolute temperature via the Boltzmann factor. The purpose of this paper is to evaluate one γ-ray anisotropy thermometer, ^{60}Co in hcp ^{59}Co single crystal, and report the results of our comparison with the noise thermometer.

^{60}Co γ-Ray Anisotropy Thermometer

Cobalt-60 in hcp ^{59}Co single crystal was chosen because no magnetic field is necessary to achieve ordering of the nuclear spin system. The presence of a magnetic field would perturb the noise thermometer. In addition, the decay of ^{60}Co is well known and there are numerous papers on its use as a thermometer in hcp ^{59}Co single crystal.[2-5] In the present work we are not concerned with its use as a general-purpose low-temperature thermometer, but rather with the temperature scale it defines.

The normalized angular distribution of γ rays from an oriented radioactive nucleus is given by

$$W(\theta) = \sum_{k=0}^{k_{\max}} B_k(T)\, U_k R_k Q_k P_k(\cos\theta) \tag{1}$$

where θ is the angle between the direction of emission of the γ ray and the orientation axis—in our case the c axis of the crystal. The $B_k(T)$ are statistical tensors which contain all the temperature dependence. The U_k are the angular momentum reorientation parameters, the R_k are the angular correlation coefficients, the Q_k are the solid angle correction factors, and the P_k are the Legendre polynomials. In the following subsections we shall evaluate the errors in each of these quantities and their effect on the absolute temperature obtained using (1).

1. $B_k(T)$. The only uncertainty in the statistical tensors is due to the uncertainty in the values of the energy levels E_m which enter into the Boltzmann factor. Cobalt has a large hyperfine interaction and there are indications of a small quadrupole interaction in hcp ^{59}Co.[6] If there were also a small quadrupole interaction in hcp ^{60}Co, the values for E_m would be given by

$$E_m = \pm \tfrac{1}{2}Am + p[m^2 - \tfrac{1}{3}I(I+1)] \tag{2}$$

where A is the hyperfine coupling constant and p is the quadrupole coupling constant. Although the value of A can be obtained directly by doing a nuclear magnetic resonance—oriented nuclei (NMR-ON) experiment, none has been carried out.* To obtain the value of A for our case, we assume that the effective fields are the same for both ^{60}Co and ^{59}Co; i.e., there are no isotope effects. Thus

$$H'_{\text{eff}}(^{60}\text{Co}) = H_{\text{eff}}(^{59}\text{Co}), \qquad \frac{h\nu' I'}{\mu'_I} = \frac{h\nu I}{\mu_I}, \qquad \nu' = \nu \frac{I}{I'} \frac{\mu'_I}{\mu_I} \tag{3}$$

where ν is the frequency corresponding to the energy separation of adjacent levels and $\mu_{\pm}$ is the nuclear magnetic moment. The value of ν has been measured by NMR techniques[7] and is 227.9 ± 0.1 MHz. The ratio μ'_I/μ_I has been measured[8] quite accurately and is 0.8191 ± 0.0016. Values for ν' and its uncertainty can be readily calculated since I and I' are known. The value of $A/2$ and its error are 6.2715 ± 0.0126 m°K.†

Ignoring the quadrupole term for the present, we can calculate the effect of the uncertainty in $A/2$ on the final temperature. $W(\theta)$ is computed for $\theta = 0$ since this corresponds to the largest change in intensity with temperature. The values used for the other quantities in Eq. (1) are: $U_2 = 0.9393744$, $U_4 = 0.7977240$, $R_2 = -0.4477023$, $R_4 = -0.3043790$, $Q_2 = 0.9965$, $Q_4 = 0.9886$, and $P_2 = P_4 = 1$. The calculations show a constant error of $\pm 0.2\%$ in temperature from 2.0 to 40.0 m°K.

Since a nuclear quadrupole splitting of $\pm 0.182 \pm 0.008$ MHz has been measured for hcp ^{59}Co, we should consider the effect of including a similar interaction for ^{60}Co. If we assume that the electric field gradient and the quadrupole moment are the same for both isotopes,‡ then the quadrupole splitting would be $\pm 0.085 \pm 0.004$ MHz for hcp ^{60}Co. This corresponds to ± 4.08 μ°K. Putting this into Eq. (2) and computing $W(0)$, we obtain the error as a function of temperature. At 20 m°K the uncertainty is ± 0.08 m°K.

* We have recently attempted such a measurement with Dr. D. B. Utton of our laboratory, without success; most likely our sample was too thick (7.5×10^{-2} mm).

† Deriving the splitting this way, rather than by just using the values for the effective field and the magnetic moment, circumvents any cumulative or accidental errors in these quantities.

‡ This assumption is probably good to about 10 or 20%.

2. U_k. In the β decay of ^{60}Co to ^{60}Ni the forces tending to reorient the nucleus can be neglected, and an exact value for U_k can be calculated.

3. R_k. If we set the energy response of our detector to count only the higher-energy γ ray (1.33 MeV), we can calculate R_k exactly since it is pure E2. If we want to count both it and the lower energy γ ray (1.17 MeV), we will have an error introduced due to the possible small admixture of M3 radiation. The M3/E2 amplitude has been measured for this transition and is found to be $\delta = -0.003 \pm 0.005$.[9] Calculating the modified R_k for this transition and weighting both transitions with equal probability of being detected, we can calculate the effect of this on the temperature. At 20 m°K the error is +0.106 to −0.026 m°K.

4. Q_k. Solid-angle correction factors have been calculated for our particular geometry (3 × 3 in. NaI detectors 20 in. from rectangular source), using Monte Carlo techniques.* The correction due to the finite source size was negligible, as we expected. The values for Q_2 and Q_4 were 0.9965 ± 0.0005 and 0.9886 ± 0.0020, respectively. The errors introduced into the temperature by these uncertainties were quite small. At 20 m°K the error is ±0.007 m°K.

5. $P_k(\cos\theta)$. There are three sources of error that enter into the angular term. First there is the misalignment of the *c* axis of the crystal with the zero-degree position of the detector. Second there is the mosaic spread of the crystal. Finally there are closure domains perpendicular to the *c* axis. The uncertainty in alignment can be determined by measuring the angular distribution at low temperatures. This was found to be ±1°, which results in an error of ±0.007 m°K at 20 mK. The mosaic spread of our crystal was found to be less than 1°, and the error due to this is negligible. The error due to closure domains for a rectangular sample can be calculated from the equation

$$W'(0) = W(0)\,f + W(\pi/2)/(1 - f) \tag{4}$$

where f is the fraction of correctly aligned domains, $1 - f$ is the fraction of closure domains, and $W(0)$ and $W(\pi/2)$ are the effects we would have for no closure domains. The value of f can be calculated for our sample and was found to be 0.9935. At 20 m°K this leads to a 0.12 m°K higher temperature.

6. Combined error. If we consider all the errors discussed above, we can calculate the maximum error using (1). The results of this calculation are given in Table I. These errors can be decreased considerably by reducing the amount of closure domains† and only counting the 1.33-MeV γ ray.

Having discussed the uncertainties in calculating the temperature from Eq. (1), we now turn to the measurement of $W(0)$ itself. The precision with which we can measure $W(0)$ depends upon the stability of the counting system and how well we can determine the background. In order to achieve a 1% accuracy in temperature, we would have to measure $\Delta W(0)/W(0)$ to 0.08% at 40 m°K, 0.14% at 30 m°K, and 0.28% at 20 m°K. In this temperature region (20–40 m°K) a measurement of $\Delta W(0)/W(0)$ to 0.2% is possible (using standard counting equipment) in a reasonable length of time (our 2-μCi sample gives us 60 counts/sec). More precise measurements can be made with more sophisticated counting equipment.

* These were kindly done by Mr. W. Snow of Argonne National Laboratory.

† One could reduce the effect of closure domains by increasing the length of our rectangular sample or by following a suggestion of Prof. D. A. Shirley to activate only the center part of our sample.

Table I. Maximum Calculated Error in the Temperature

T(Actual), m°K	T'(Calc.), m°K	% Error
40.0	40.719	+ 1.80
30.0	30.527	1.76
20.0	20.361	1.80
10.0	10.219	2.19
5.0	5.247	4.93
2.0	2.150	7.48

Experimental Results and Conclusions

The ^{60}Co and noise thermometers were compared at discrete temperatures from 20 to 40 m°K using a ^3He–^4He dilution refrigerator. The lower limit was set by the operation of the refrigerator and the upper limit by the insensitivity of the ^{60}Co thermometer. Both thermometers were mounted directly to the copper mixing chamber through gold-plated mechanical joints. Data were taken at stabilized temperatures for periods of approximately 6 hr. Typically, two different cold runs were taken in sequence, and then the refrigerator was warmed to above 0.3°K for the warm counts. This procedure was then repeated for other temperatures. Background data were taken both before and after a series of runs was made. Some of the temperature points were repeated many times in the course of the experiment using a reproducible secondary thermometer. The results of this comparison are shown in Fig. 1. The error bars shown indicate the combined uncertainties of both thermometers, and in several cases reflect the larger scatter of noise temperature. From this figure it appears that there is a discrepancy of approximately 1 m°K between the thermometers. Although some of this might be explained by the results in Table I, we feel that these data are too preliminary for us to draw any definite conclusions.

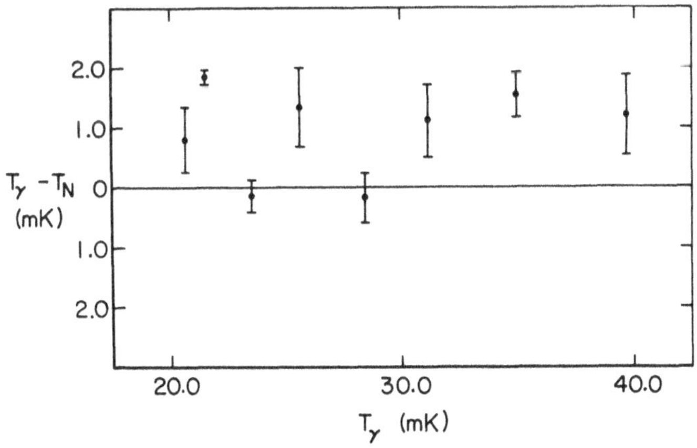

Fig. 1. The temperature differences between the ^{60}Co γ-ray anisotropy thermometer (T_γ) and the Josephson junction noise thermometer (T_N).

Acknowledgments

It is a pleasure to acknowledge the technical assistance of Mr. R. B. Dove. We thank Mr. W. Snow of Argonne National Laboratory for calculating the solid-angle correction factors.

References

1. R.A. Kamper and J.E. Zimmerman, *J. Appl. Phys.* **42**, 132 (1971).
2. T.L. Thorp, B.B. Triplett, W.D. Brewer, N.E. Phillips, D.A. Shirley, J.E. Templeton, R.W. Stark, and P.H. Schmidt, *J. Low Temp. Phys.* **3**, 589 (1970).
3. P.M. Berglund, H.K. Collan, G.J. Ehnholm, R.G. Gylling, and O.V. Lounasmaa, *J. Low Temp. Phys.* **6**, 357 (1972).
4. M.F. Cracknell and G.V.H. Wilson, *Proc. Roy. Soc.* **A296**, 71 (1966).
5. G.C. Handra, V.R. Pandharipande, and T.S. Radhakrishnan, *Phys.Lett.* **25A**, 754 (1967).
6. P.C. Riedi and R.G. Scurlock, *Phys. Lett.* **24A**, 42 (1967).
7. R.F. Jackson, R.G. Scurlock, D.B. Utton, and T.H. Wilmshurst, in *Proc. Intern. Conf. on Magnetism, Nottingham, 1964*, Institute of Physics and the Physical Society, p. 384.
8. W. Dobrowolski, R.V. Jones, and C.D. Jeffries, *Phys. Rev.* **101**, 1001 (1956).
9. J.R. Sites, H.A. Smith, and W.A. Steyert, *J. Low Temp. Phys.* **4**, 605 (1971).

Accurate NMR Thermometry Below 1°K

D. Bloyet, A. Ghozlan, P. Piéjus, and E. J.-A. Varoquaux

Institut d'Electronique Fondamentale
Université Paris, Orsay, France

Temperature Measurement by Nuclear Magnetic Resonance

Nuclear magnetic resonance (NMR) techniques can provide an accurate and convenient determination of temperatures below 1°K through nuclear susceptibility measurements. NMR thermometry was used at first[1,2] for the measurement of temperatures below 10 m°K. Its range of application has been extended more recently up to liquid helium temperatures.[3-5]

We use a pulsed NMR method,[3] and most of our measurements have been performed on ^{63}Cu at a frequency of 1.5 MHz. The basic accuracy of the amplitude measurement is one part in one thousand. We shall first describe the general layout of the electronic equipment capable of such accuracy and then discuss the various factors that might impair the low-temperature measurement.

Pulsed NMR Apparatus

The pulsed NMR experiment is by itself fairly conventional: the setup is shown in Fig. 1. The problems associated with the design and operation of the electronic equipment involve (a) the signal-to-noise ratio, (b) the overall gain calibration procedure, (c) the long-term stability, and (d) the system linearity.

(a) An operating frequency of 1.5 MHz has been chosen to yield a 60 dB signal-to-noise ratio at 1°K in 0.5 cm^3 of copper.

(b) Special attention has been paid to the long-term stability of the electronics and the gain calibration of the whole system. This has led us to incorporate in the system a calibration unit to allow the in situ calibration of the gain and linearity. The calibration unit consists of a high-quality Otecna-300 Ω attenuator with a buffer output amplifier. The attenuation factors, specified to only 0.7% by the manufacturer, are measured in situ with a laboratory-made peak detector and a set of A-55 J. Fluke bolometers. These steps are essential since attenuator calibration depends on external circuitry and cannot be specified independently of the source and load parameters.

(c) The master oscillator is controlled by quartz; the output amplitude is stabilized to better than 5×10^{-4} on the long term. The short-term noise is such that the maximum scatter of the calibration signal read directly by the receiver is less than 2×10^{-4}.

(d) The receiving channel includes a low-noise FET preamp (1.5 dB noise figure for a 1-kΩ source resistance), a two-stage selective amplifier with 200 kHz band-

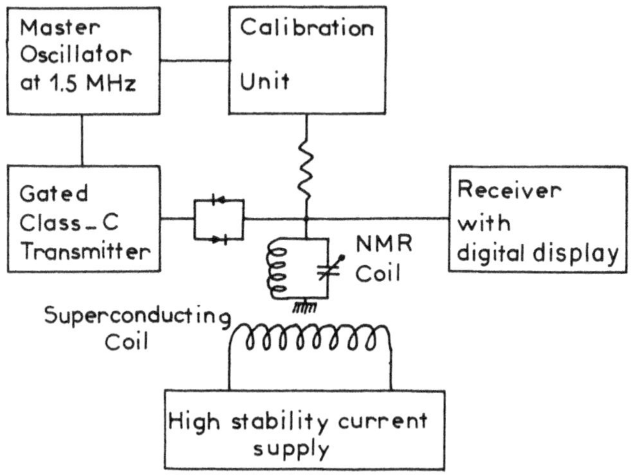

Fig. 1. General layout of the NMR appararus.

width, a sampling linear detector, and a digital display. The detector, which has been described elsewhere,[6] is linear to better than 0.1% over a 30 dB dynamic range and is not sensitive to the phase of the signal. It should be appreciated that a lock-in detector is not well suited to precise amplitude measurements. The signal is sampled for 66.7 μsec (100 periods) with a delay time after the pulse which is usually 53.3 μsec. The receiver has fully recovered from blind-off 40 μsec after the pulse. The receiver gain is calibrated immediately after each measurement.

(e) A single resonance coil is connected directly to the receiver, to the transmitter through a set of back-to-back diodes, and to the calibration unit through a high-stability resistor. The spin excitation transverse field H_1 is 35 Oe. The effect on the signal of changes in the pulse duration is shown in Fig. 2. A $\pi/2$ nutation angle is obtained for a pulse duration of ten periods ($\tau = 6.7$ μsec).

(f) The H_0 field, whose value is 1330 Oe for ^{63}Cu at 1.5 MHz, is produced by a compensated superconducting coil ($\Delta H \sim 0.5$ Oe). The effect on the signal of scanning the field is shown in Fig. 3. The two peaks for ^{65}Cu and ^{63}Cu are in the ratio of the gyromagnetic constants (7% apart). The central peak is an admixture of the two isotopes and its free precession decay exhibits a strong beat pattern. The apparatus is usually tuned on ^{63}Cu. The long-term field stability, better than 5×10^{-5}, is seen in Fig. 3 to be adequate.

High-Temperature Calibration and Ultimate Accuracy

The temperature calibration is done over the range 1.2–2°K against two calibrated resistances. The reference temperature is known to ± 2 m°K. The rms deviation from Curie's law for our two copper specimens is of the order of 5×10^{-4} over this temperature range, which is a check on both the samples and the electronics. The combined errors in the NMR temperature calibration amount to $2\text{–}3 \times 10^{-3}$. The calibration constant reproduces to 2×10^{-3} after warmup to room temperature.

Taking into account a possible systematic error of 10^{-3} in the attenuator calibra-

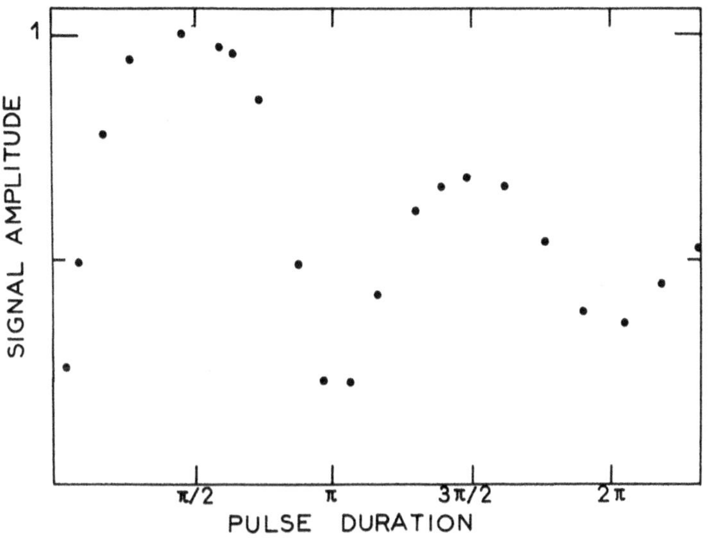

Fig. 2. Signal amplitude (in arbitrary units) in terms of the rf pulse duration expressed as a nutation angle $\theta = \gamma H_1 \tau$. It is seen from the $3\pi/2$ peak that H_1 is fairly homogeneous over the sample.

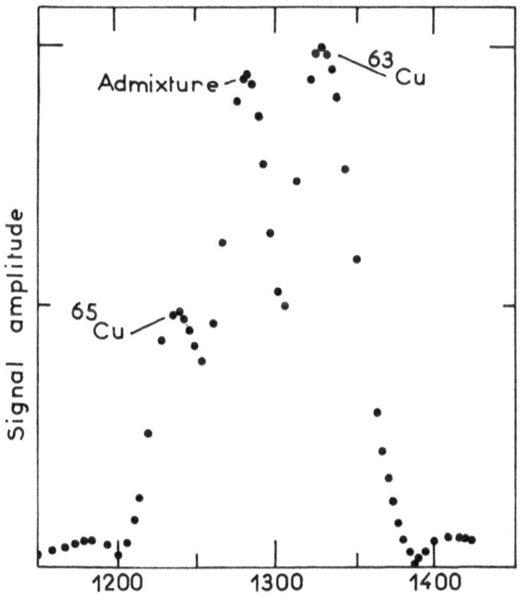

Fig. 3. Signal amplitude (in arbitrary units) in terms of the H_0 field in Oe.

tion, the ultimate accuracy of a low-temperature measurement is $3\text{–}4 \times 10^{-3}$, under the assumption that the measured amplitude is inversely proportional to the temperature.

Checks on the Sample NMR Parameters

As long as the Zeeman energy is the dominant energy, Curie's law is expected to hold for the static susceptibility of the nuclear spins in the temperature range of interest. This will also be the case for the dynamic susceptibility observed in a pulsed NMR experiment if the shape of the free precession decay does not change with temperature. If such a change occurs, it might mean that some nuclei are shifted off resonance into regions where they are not observed any more. The signal amplitude will no longer vary as $1/T$ if the number of nuclei involved is changing with T. Three mechanisms can give rise to such an effect.

(1) A change in the skin depth due to resistivity variations. This effect is not important for finely divided samples of pure metals below 2°K.

(2) Quadrupolar interactions: The copper specimens are badly strained, so that a fairly strong quadrupolar broadening[7] is to be expected. The total intensity of the resonance line should be down by about 40%, but again the effects should show no temperature dependence below 2°K as the lattice becomes very nearly steady.

(3) Magnetic impurities[8]: The local field about a magnetic impurity shifts the neighboring host nuclei off resonance. Since this field decreases away from the impurity as $1/r^3$ and is proportional to the impurity magnetization $\langle S_z \rangle$ along the H_0 direction, the number of host nuclei that are wiped out should be changing with temperature as $\langle S_z \rangle$. As reported by Nagasawa and Steyert,[9] the wipeout number in *Cu*Fe (380 ppm) changes from 1000 at 4.2°K to 2000 at 1.4°K. This temperature variation is much faster than the one expected from $\langle S_z \rangle$. Besides, no such variation was observed in the more recent experiments by Potts and Welsh.[10] It is likely that far below the Kondo temperature ($T_K \sim 15$°K for *Cu*Fe) the wipeout number becomes temperature independent. A small amount of iron impurity will not spoil the NMR thermometry on copper. However, this will not be so for *Cu*Cr ($T_K \sim 0.5$°K) and *Cu*Mn ($T_K \sim 10\text{–}50$ m°K). Both Cr and Mn can be found even in fairly pure (99.999%) copper with ppm concentrations: with typical wipeout numbers of 1000, such impurities are on the verge of being harmful.

Since the effects described above should be manifested in the relaxation times, we have checked T_1 with fair precision and T_2 with higher precision in our samples. The Korringa constant is found to be equal to 1.27 sec °K. The time T_2, which is 54 μsec, exhibits a rather small but systematic decrease by 6×10^{-3} between 1°K and 35 m°K. Part of this decrease is due to the onset of radiation damping as the nuclear magnetization increases.[11] This type of broadening can be corrected for, since it does not reveal any departure from the $1/T$ law. However, we cannot rule out the possibility that the residual effect is due to magnetic impurities. If such is the case, it is not clear how the broadening variation should be related to the intensity deviation from the ideal $1/T$ law. It is expected that this deviation would not exceed $2\text{–}3 \times 10^{-3}$. A direct comparison between various metallic samples is required in order to take full advantage of the accuracy of the NMR apparatus.

References

1. R.E. Walstedt, E.L. Hahn, C. Froidevaux, and E. Geissler, *Proc. Roy. Soc.* **284A**, 499 (1965).
2. E.B. Osgood and J.M. Goodkind, *Cryogenics* **6**, 54 (1966).
3. O. Avenel, P. Piéjus, and E. J.-A. Varoquaux, *IEEE Trans. Instr. Meas.* **IM-19**, 408 (1970).
4. Yu. D. Anufriev and V.P. Peshkov, in *Proc. 12th Intern. Conf. Low Temp. Phys. 1970*, Academic Press of Japan, Tokyo (1971), p. 835.
5. M.I. Aalto, P.M. Berglund, H.K. Collan, G.J. Ehnholm, R.G. Gylling, M. Krusius, and G.R. Pickett, *Cryogenics* **12**, 184 (1972).
6. D. Bloyet, P. Piéjus, E.J.-A. Varoquaux, and O. Avenel, *Rev. Sci. Inst.* **44**, 383 (1973).
7. T.S. Rowland, *Progr. in Mat. Sci.*, Vol. 9.
8. J. Winter, *Magnetic Resonance in Metals*, Oxford (1971), p. 121.
9. H. Nagasawa and W.A. Steyert, *J. Phys. Soc. Japan* **28**, 1202 (1970).
10. J.E. Potts and L.B. Welsh, *Phys. Rev. B* **5**, 3421 (1972).
11. N. Bloembergen and R.V. Pound, *Phys. Rev.* **95**, 8 (1954).

A Hyperfine Thermometer for Use at Low Temperature*

F. R. Szofran† and G. Seidel

Department of Physics, Brown University
Providence, Rhode Island

Introduction

This paper describes a primary thermometer that is potentially useful in the millikelvin temperature region. The temperature is determined absolutely, with no calibration required, by measuring the relative thermal populations of an ensemble of systems among several energy levels. For reasons described elsewhere[1] the system most likely to be useful as a thermometer employing this approach is a paramagnetic ion possessing a hyperfine interaction (nuclear spin $\neq 0$) located at a site of cubic symmetry.

Such a system is characterized conveniently and adequately by the spin Hamiltonian

$$\mathcal{H} = g\beta \mathbf{H} \cdot \mathbf{S} + A\mathbf{I} \cdot \mathbf{S} - g_n' \beta_n \mathbf{H} \cdot \mathbf{I} \tag{1}$$

The first two terms in the Hamiltonian are the usual isotropic electronic Zeeman and isotropic hyperfine interactions. The third term is the nuclear Zeeman interaction, where g_n' includes the effect of the enhancement of the magnetic field at the nucleus by the polarization of the electronic charge distribution.

For $S = 1/2$ this Hamiltonian gives rise to energy levels which are given by the well-known Breit–Rabi equation,

$$E(F, m_F) = -\frac{A}{4} - g_n' \beta_n H m_F + \frac{A}{2}(2I + 1)(F - I)[1 - 2\delta_{m_F,p}\theta(x)] \times \left[1 + \frac{4m_F}{2I + 1}x + x^2\right]^{1/2}$$

where

$$F = I \pm S, \qquad m_F = F, F - 1, \ldots, -F$$

$$x = (g\beta + g_n'\beta_n) H/(A/2)(2I + 1)$$

$$\delta_{m_F,p} = 1 \quad \text{if} \quad m_F = p$$

$$= 0 \quad \text{if} \quad m_F \neq p$$

* Work supported in part by the National Science Foundation.

† Present address: Department of Physics, University of Nebraska, Lincoln, Nebraska.

$$p = -(I + \tfrac{1}{2})(Ag/|Ag|)$$

$$\theta(x) = 0 \quad \text{for} \quad x < 1$$
$$= 1 \quad \text{for} \quad x > 1$$

At low fields ($x < 1$) the states in a given F multiplet are split approximately by $g\beta H/(2I + 1)$. However, due to the form of the energy dependence on the field, the energy splittings are not exactly equal and under appropriate conditions the various rf transitions $\Delta F = 0$, $\Delta m_F = \pm 1$ can be resolved.

The rf transition probabilities W_i can be calculated with accuracy in a straightforward manner from the Hamiltonian. The intensities are given by

$$I_i(T) = NW_i \frac{\exp(-E_{i+1}/kT) - \exp(-E_i/kT)}{\sum_n \exp(-E_n/kT)} \tag{2}$$

where $E_{i+1} < E_i$. Notice that at a given temperature the *relative* intensities of the various lines are completely determined by the transition probabilities and the energy levels of the states:

$$\frac{I_i}{I_j} = \frac{W_i[\exp(-E_{i+1}/kT) - \exp(-E_i/kT)]/\sum_n \exp(-E_n/kT)}{W_j'[\exp(-E_{j+1}'/kT) - \exp(-E_j'/kT)]/\sum_n \exp(-E_n'/kT)} \tag{3}$$

The primes are necessary because the field and therefore the transition probabilities and energy levels are different for different transitions at the same frequency.

Since all quantities on the right-hand side of Eq. (3) except T are known, and since the transition probabilities are temperature independent, T may be found by an iterative solution of this equation. In practice however, it is better to write $M_i = cI_i(T)$, where M_i is the measured value of the intensity and c depends on the sensitivity of the spectrometer and other experimental parameters. The quantity $\sum_i[M_i - cI_i(T)]^2$ is then minimized by allowing T and c to be the free parameters. For a system in which more than two transitions can be observed the temperature is overdetermined, thereby permitting a check in various possible sources of experimental error such as differential saturation.

Of the several different samples investigated experimentally for possible use as a low-temperature thermometer, only ThO_2 doped with 1% Er^{3+} will be discussed below.

The ^{167}Er isotope has a natural abundance of ~23% and nuclear spin 7/2, and is the only isotope of Er having a nuclear moment. The Er^{3+} electronic configuration is $4f^{11}$; in ThO_2 the ground state is a Γ_7 doublet. Thus a negative hyperfine interaction constant was predicted.[2] The electronic g value (6.752) and magnitude of the hyperfine interaction ($|A| = 0.02326$ cm^{-1}) have been measured by Abraham *et al.*[3] The authors measured $g_n'(-3.39)$ and experimentally confirmed the negative sign of A. Therefore all of the spin Hamiltonian parameters are known and calculations of transition probabilities and energy levels can be carried out as needed. A plot of the energy levels as a function of magnetic field is shown in Fig. 1; resonance transitions at a frequency of 250 MHz are indicated. Notice that the maximum magnetic field required is ~500 G.

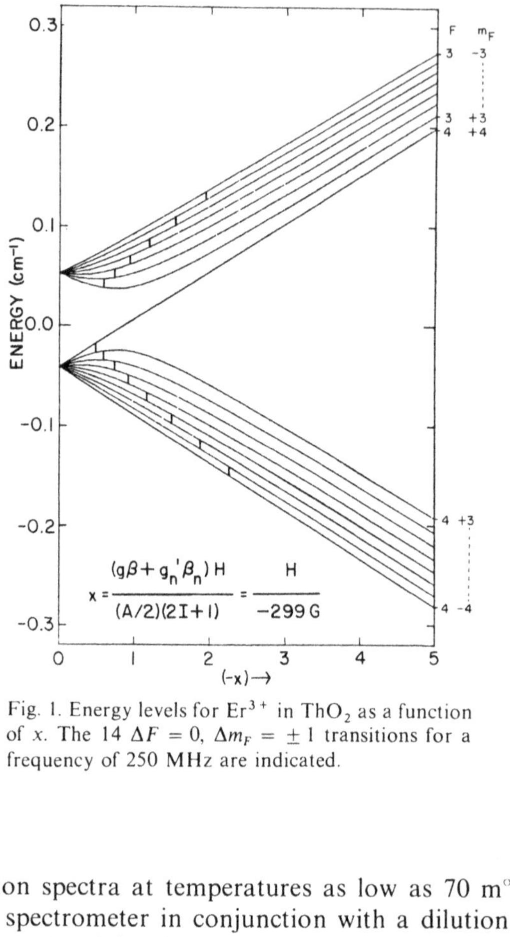

Fig. 1. Energy levels for Er^{3+} in ThO_2 as a function of x. The 14 $\Delta F = 0$, $\Delta m_F = \pm 1$ transitions for a frequency of 250 MHz are indicated.

Experiment

Direct absorption spectra at temperatures as low as 70 m°K were obtained using a bridge-type spectrometer in conjunction with a dilution refrigerator. The powdered sample was placed inside the epoxy mixing chamber of the refrigerator for optimum thermal contact, while the sample coil was placed outside the mixing chamber. With this arrangement the frequency of the resonant tank circuit could be changed without opening the mixing chamber. Consequent loss of fill factor was not a problem, due to the size of the signal.

The signal oscillator was locked to the resonant frequency of the sample tank circuit so that a pure absorption line shape was obtained. While the signal was large enough to obviate the need for lock-in detection, it was not so large that it introduced nonlinearities into the measurements. Typically the fractional change in Q of the tank circuit on passing through resonance was less than 3%. The signal was detected by a superheterodyne receiver followed by a crystal detector which was operated in the square law region. Under these conditions the change in dc voltage out of the detector was proportional to the rf power absorbed by the sample.

The dc signal was amplified and recorded on an x–y recorder as a function of magnetic field. A digitizer was then used to determine the area under each resonance peak. This determined the relative intensities because each area was proportional to the intensity of the corresponding resonance transition.

The temperature of the mixing chamber was measured independently with a Ge resistance thermometer calibrated against the susceptibility of CMN.

Results and Discussion

A plot of the field derivative of the absorption spectrum at 190 MHz is shown in Fig. 2. The resonance lines are a few gauss wide and resolution is satisfactory at frequencies as low as 150 MHz. In the neighborhood of 80 m°K, where most of the measurements were made, the temperature determined from the relative intensities was in agreement with the Ge thermometer to within 5 m°K. The precision of the present measurements is limited principally by the fact that the relative intensities of the resonance lines are not very sensitive to temperature for $T \gg h\nu/k$. Since our refrigerator was not designed for operation at much lower temperatures, the data were limited to the region $T > 7\ h\nu/k$. The accuracy of the measurement should increase by greater than a factor of ten for $T \simeq (1/2)\,h\nu/k$ (4.6 m°K for 190 MHz), where the temperature derivative of the relative intensities is a maximum.

One of the main considerations of any thermometer operating in the millikelvin range is thermal relaxation rates. Although difficulty in establishing thermal equilibrium was not encountered in the present experiment, principally because of cross-relaxation within the spin system and relatively strong coupling to the lattice, the problem becomes much more severe at low temperatures. Rather than

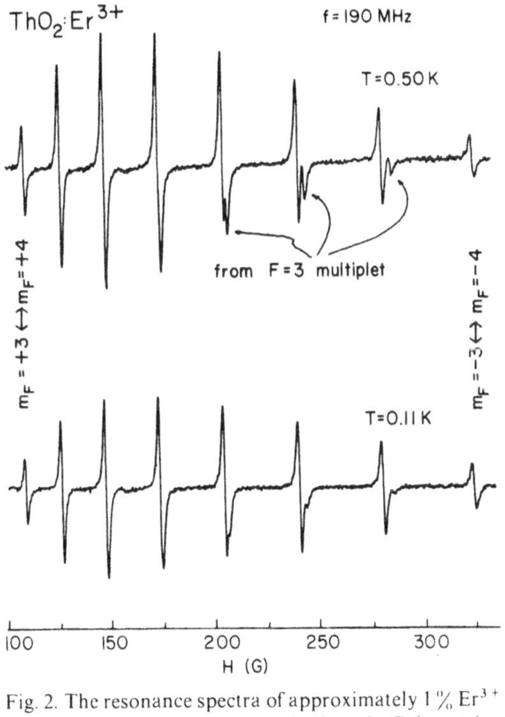

Fig. 2. The resonance spectra of approximately 1 % Er^{3+} in ThO_2 at the temperatures indicated. Gain setting of the spectrometer is lower for the lower trace.

using paramagnetic ions in dielectrics to try to solve this problem, more promising systems are $4f$ ions in metals. Narrow ESR lines are observed in powdered cubic metals (see, e.g., Ref. 4), and the relaxation rates of the $4f$ ions, because of the large interaction with the conduction electrons, are many orders of magnitude greater than in dielectrics. Work is now in progress to convert our refrigerator for use at much lower temperatures and to investigate such factors as the linewidth and signal strength of $Au:Er^{3+}$.

Acknowledgments

We thank M. Abraham of ORNL for the loan of the $ThO_2:Er^{3+}$ sample and B. Gregory for helpful discussions.

References

1. F.R. Szofran, B. Gregory, and G. Seidel, in *Proc. Fifth Symp. on Temperature*, Washington, 1971 (to be published).
2. A. Abragam and B. Bleaney, *Electron Paramagnetic Resonance of Transition Ions*, Clarendon, Oxford (1970).
3. M. Abraham, R.A. Weeks, G.W. Clark, and C.B. Finch, *Phys. Rev.* **137**, A138 (1965).
4. E.P. Chock, D. Davidov, R. Orbach, C. Rettori, and L.J. Tao, *Phys. Rev. B* **5**, 2735 (1972); D. Davidov, R. Orbach, C. Rettori, D. Shaltiel, L.J. Tao, and B. Ricks, *Phys. Rev. B* **5**, 1711 (1972), and references contained therein.

Pulsed Platinum NMR Thermometry below 2°K

M. I. Aalto, H. K. Collan, R. G. Gylling,

and K. O. Nores

Helsinki University of Technology
Otaniemi, Finland

Several requirements can be set for a good secondary thermometer. The relationship between the measured quantity and temperature should be simple, the thermometer should have a fast response, and it should be accurate, easily calibrated, and simply automated. It is useful if some kind of internal consistency check is available.

The pulsed platinum NMR thermometer described in this report fulfills all these requirements; it is intended for rapid temperature and spin–lattice relaxation time measurements. This type of thermometer makes use of the fact that the nuclear magnetization follows the Curie law down to well below 1 m°K. Therefore the temperature dependence is so simple that only one calibration point is needed. The response time is short in platinum because this metal has a short nuclear spin–lattice relaxation time τ_1. Furthermore, platinum has a long spin–spin relaxation time τ_2, which facilitates the accurate recording of the thermometer signal.

The nuclear magnetization is measured by first exciting the nuclear spins by a short rf pulse at the nuclear Larmor frequency and then recording the subsequent nuclear free precession (NFP) signal, which is proportional to the magnetization.

The general arrangement of our thermometer is closely similar to that described earlier.[1] The transmitter is crystal controlled at 413 kHz. The transmitter pulse is first cut to a suitable length and subsequently amplified and fed to the transmitter coil in the cryostat. The NFP signal is observed by means of a receiver coil and then passed through a low-noise preamplifier and a tuned amplifier. A quenching circuitry shunts the inputs of the two amplifiers during the transmitter pulse. The ac NFP signal is converted into a dc voltage by a rectifying detector followed by an *RC*-integrator. An analog gate between the detector and the integrator is kept open during the integration time. When the integration has been completed the analog gate is turned off and the DVM is triggered to read the stored voltage. This reading is proportional to the amplitude of the NFP signal, which, in turn, is proportional to $1/T$.

The time base for the above operations is provided by a pulser, which is synchronized with the crystal oscillator and drives a digital counter. The counter logic gives command pulses for the transmitter and receiver signal gates, the quenching circuitry, and the DVM.

The overall precision of the recording apparatus is better than 0.1% over a

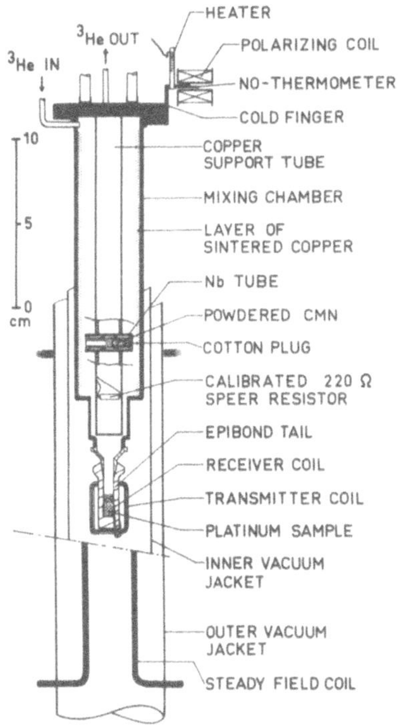

Fig. 1. Experimental arrangement.

wide dynamic range. The accuracy in the actual measurement depends on many factors other than the performance of the electronics. With careful elimination of external disturbances we can achieve an accuracy of 1 %, with a short-time resolution of 0.2 %.

The installation of the thermometer is shown in Fig. 1. The thermometric specimen was a 3.7-g sample of powdered platinum with average grain size of 3 μm.* The powder was in direct contact with liquid helium inside the mixing chamber of a dilution refrigerator. In order to allow the rf signals to reach the platinum sample it was located in an Epibond tail. The cylindrical receiver coil was wound on the inside surface of the tail to maximize the filling factor. The transmitter and the steady field coils were both of saddle shape and wound of superconducting wire.

In order to check the $1/T$ dependence of the NFP signal, several other thermometers were installed. For temperatures above 30 m°K we relied on an accurately calibrated Speer carbon resistor. For the low-temperature end a nuclear orientation (NO) thermometer[2] consisting of ^{54}Mn nuclei in nickel was used. Furthermore, to check the overall consistency of all thermometers, we had a 20-mg sample of powdered CMN in direct contact with liquid helium in the mixing chamber. The static magnetic susceptibility of CMN was measured with a superconducting magnetometer.†

* Engelhard Industries, Newark, New Jersey. The impurity analysis given by the manufacturer was: Mg, 2 ppm; Fe, 40 ± 10 ppm; Ni, 15 ppm; Co, 10 ppm; Au and Ag, 0.2 %.

† SHE Manufacturing Corp., San Diego California.

The relation between $1/T$ and the output voltage of the integrator of the pulsed NMR thermometer is shown in Fig. 2. The temperature is determined by the carbon resistor (filled circles) and by the nuclear orientation thermometer (open circles). The nuclear magnetization seems to follow the Curie law, as expected. The consistency of the various temperature determinations and previous experience lead us to believe that our temperature scale below 0.5°K is correct within 2%.

In a thermometer based on the measurement of nuclear magnetization a possibility of internal calibration is provided by the nuclear spin–lattice relaxation process; according to the Korringa law, $\tau_1 T = K$, where K is the Korringa constant, the spin–lattice relaxation τ_1 is proportional to $1/T$. Furthermore, τ_1 is a material constant and thus thermometry based on the measurement of the relaxation time gives the absolute temperature without recourse to external temperature standards. This method of calibration has the additional advantage that only internal equilibrium must be achieved in the thermometric specimen, in contrast to the usual case where equilibrium must be maintained between two separate thermometers.

Our thermometer was designed for investigating the feasibility of this calibration method. In the following we describe briefly the τ_1 measurement.

The principle of measuring τ_1 is first to destroy the static part of the nuclear magnetization with a 90° transmitter pulse and then to follow the transient recovery of the nuclear magnetization back to the equilibrium value. This was done by measuring the nuclear spin temperature using 10° pulses at four instants during the relaxation. An electronic circuitry automatically fires the pulses with preset time intervals after the 90° pulse.

Since the platinum powder was inside liquid helium, it was not in metallic

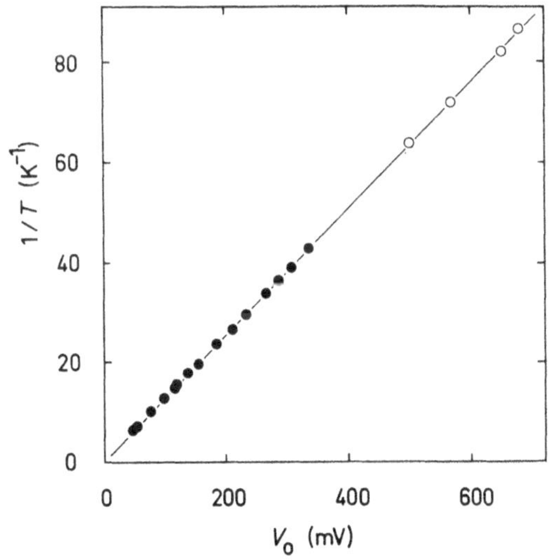

Fig. 2. A calibration plot for the pulsed NMR thermometer. Here $1/T$, as determined by the calibrated Speer resistor (filled circles) and by the NO thermometer (open circles), is plotted against the integrator output voltage V_0.

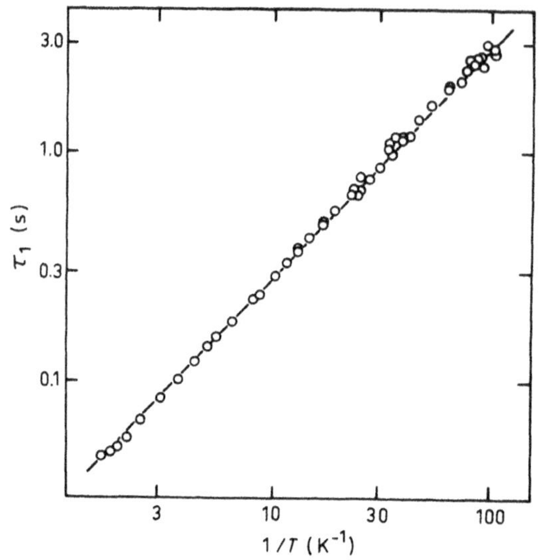

Fig. 3. Results of τ_1 measurements in a field of 450 G. The solid line through the points corresponds to the relation $\tau_1 T = 29.6$ msec °K.

contact with the surroundings. Especially at the lowest temperatures of our τ_1 measurements this caused some problems.

The main complication is that the conduction electron system momentarily warms up considerably when the pulses are applied. This is due to eddy current heating of the platinum powder. As a result the relaxation is not strictly exponential. However, if we apply the four measuring pulses without the preceding 90° pulse, then we directly obtain the correction factors by which the observed values should be multiplied to make them follow an exponential relaxation.

Our τ_1 results are shown in Fig. 3. The temperature was derived from the amplitude of the equilibrium NFP signal, using the calibration shown in Fig. 1. We see that between 1°K and 10 m°K τ_1 is inversely proportional to temperature. The Korringa constant is 29.6 msec °K. This agrees well with earlier measurements at higher temperatures.[3-5]

In conclusion, our thermometer has been found very satisfactory. Its accuracy in $1/T$ is 1% and its response time is 3 sec at 10 m°K. Internal calibration is possible by the measurement of τ_1, which can be done easily with the help of automated electronics. The temperature calibration obtained in this way has an accuracy of 3%.

References

1. M.I. Aalto, P.M. Berglund, H.K. Collan, G.J. Ehnholm, R.G. Gylling, M. Krusius, and G.R. Pickett, *Cryogenics* **12**, 184 (1972).
2. P.M. Berglund, H.K. Collan, G.J. Ehnholm, R.G. Gylling, and O.V. Lounasmaa, *J. Low Temp. Phys.* **6**, 357 (1972).
3. R.E. Walstedt, M.W. Dowley, E.L. Hahn, and C. Froidevaux, *Phys. Rev. Lett.* **8**, 406 (1962).
4. J. Butterworth, *Phys. Rev. Lett.* **8**, 423 (1962).
5. L.E. Drain, *J. Phys. Chem. Solids* **24**, 379 (1963).

Relationship Between Noise Temperatures and Powdered CMN Magnetic Temperatures: Application to the Properties of ^{3}He*

R. A. Webb, R. P. Giffard,† and J. C. Wheatley

Physics Department, University of California—San Diego
La Jolla, California

Johnson noise coupled to a Josephson effect device has recently been used to measure the absolute temperature of a resistor by Kamper *et al.*[1] down to a temperature of 23 m°K and by Giffard *et al.*[2] in the temperature range 4.2°K to 12 m°K. We report here the results of a more detailed and systematic investigation of Johnson noise thermometry using an rf-biased point-contact superconducting device (SQUID) to sense the Johnson noise currents from a resistor.

The noise resistor is placed in the very low-temperature part of our apparatus and is connected via superconducting leads to a superconducting coil located in one hole of a two-hole symmetric SQUID at 4.2°K.

In one set of experiments, the details of which are described elsewhere,[3] a beryllium copper noise resistor was placed in a dilute solution appendix of a mixing chamber of a dilution refrigerator. A cerium magnesium nitrate (CMN) thermometer was located in a second, similar appendix. From 4.2°K to 5.4 m°K the mean square Johnson noise currents were measured as a function of the magnetic temperature T^*. Above 16 m°K the data were used to determine Boltzmann's constant to 1%. Below 16 m°K we observed systematic deviations of the noise currents, all toward a lower T^*, with the deviations reaching a maximum at 10 m°K and then returning to zero at 5.4 m°K, our lowest temperature.

To determine whether or not these deviations were a property of the CMN or simply a real temperature difference between the two thermometers, an adiabatic demagnetization cell was built. A new noise resistor made from copper foil and a 16-mg powdered CMN sample in the shape of a right-circular cylinder were placed in a 3.2-mm-I.D. column located above a larger (10.7 g) powdered CMN cooling salt. The thermal contact was achieved by means of pure liquid ^{3}He cooled by adiabatic demagnetization of the CMN. The susceptibility of the CMN thermometer was measured using a 16-Hz ac bridge using a SQUID as a null detector.[2] A single temperature point (4.2°K) was used to provide the calibration for the output of our electronics in terms of the Johnson noise temperature. In a separate experiment the resistance of the sensor was found to change less than 0.4% from 4.2°K to 3 m°K.

The measured noise is greater than the actual noise by an amount which is

* Supported by the U.S. Atomic Energy Commission under Contract No. AT (04-3)-34, P.A. 143.
† Present address: Clarendon Laboratory, Oxford, England.

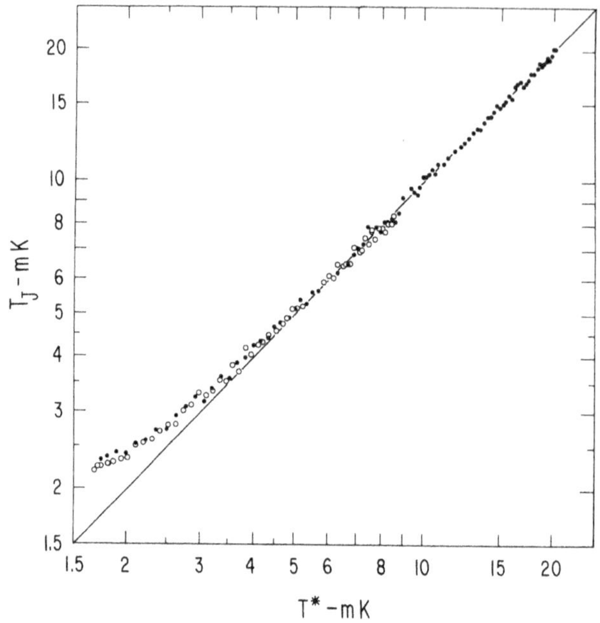

Fig. 1. Johnson noise temperature plotted against the magnetic temperature of CMN. Solid circles are data taken on the third demagnetization; open circles are data taken on the fourth demagnetization. A 0.05-m°K device noise has been subtracted from each noise temperature. The straight line is $T_J = T^*$.

called the "device noise," and when referred to temperature this noise was found to be 0.05 m°K for this experiment. In equilibrium a 2000-sec averaging interval was necessary to obtain a precision estimated at $\pm 1\%$.

The results for the third and fourth demagnetizations[4] are shown compositely in Fig. 1, where ln T_J is displayed as a function of ln T^*. Below 11 m°K a 2000-sec averaging interval was used for each T_J; above 11 m°K this time was 4000 sec. The measured 0.050-m°K "device noise" has been subtracted from each T_J. In the temperature range 8–20 m°K we fitted the data to the equation $T_J = T^* + \Delta$ and found $\Delta = 0 \pm 0.12$ m°K.

We now discuss those observations that have increased our confidence in the noise thermometer results. First, above $T^* = 2$ m°K no systematic difference is observed between the third and fourth demagnetizations even though the initial temperatures and the warmup rates were different. Second, the dilution refrigerator results were quantitatively predicted by the Johnson noise formula to within the overall accuracy of the measurement. In both the dilution refrigerator work and the demagnetization work the noise power spectrum consistently gave the expected result of a $(1 + \omega^2\tau^2)^{-1}$ part and a part due to the white spectrum[2] of "device noise." To within experimental error the spectrum was constant over the entire measuring range. In particular, no $1/f$ noise was observed. Hence it is likely that T_J accurately measures the temperature of the copper sensor.

There are at least two possibilities why the noise resistor might not measure the actual temperature of the ^3He. One is the existence of an RC-type time constant describing the response of the thermometer. A time constant of $\sim$ 800 sec was measured at our lowest temperature, much less than the 35,000 sec the cell stayed below $T^* = 3$ m°K. The second and most serious possibility for error is that a tiny heat leak to the copper sensor would elevate its temperature above the ^3He via Kapitza resistance.

We can attempt to make an independent estimate of the absolute temperature T_i using the known heat capacity of ^3He and the latest data from Hudson[5] for the CMN heat capacity to obtain the total heat capacity of our cell vs. T. Using dT_J/dt and the above heat capacities, we find for the fourth demagnetization a residual heat leak of $1.4_8 \times 10^{-2}$ erg/sec for $T_J = 2.4$–7 m°K. Assuming this heat leak was constant back to the first measurement of T_J and assuming the Hudson heat capacity data, we find an initial temperature of $T_i = 2.0$ m°K. Using the heat capacity data of Huntzicker and Shirley,[6] we find $T_i = 2.1$ m°K. These are to be compared with the measured Johnson noise temperature of $T_J = 2.2$ m°K. Assuming that the absolute temperature of the ^3He was 2.0 m°K when $T_J = 2.2$ m°K and that this difference is caused by a heat leak directly to the thermometer, we find, assuming a T^{-3} Kapitza resistance, that the temperature difference between the two would be less than 0.1 m°K at $T_J = 2.6$ m°K.

In a later experiment the CMN thermometer was replaced by a powdered $Na_3[Ce(C_7H_3NO_4)_3] \cdot 15H_2O$ (CDP) sensor, a substance magnetically similar to CMN.[7] The same Johnson noise thermometer was used and at the lowest temperature behaved the same as in the CMN work.

Finally, measurements[8] of the critical field curve of tungsten provided a T–T^*

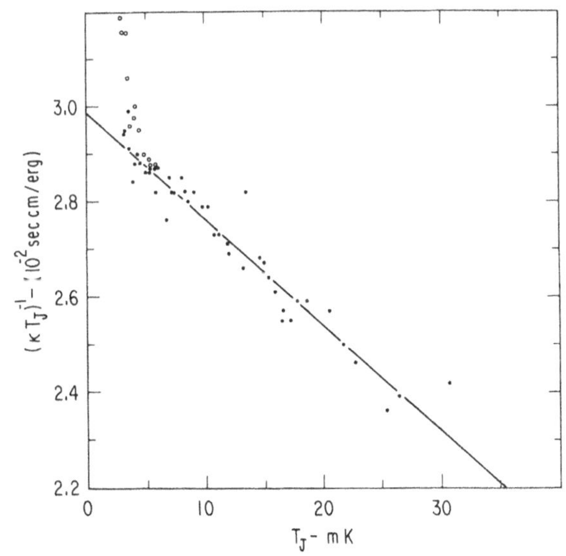

Fig. 2. Thermal conduction in pure liquid ^3He at saturated vapor pressure. $(\kappa T_J)^{-1}$ is plotted against the Johnson noise temperature T_J (solid circles). κ is the thermal conductivity as measured in Ref. 9. The solid line is the best fit to the current data. The open circles are $(\kappa T^*)^{-1}$ vs. T^*.

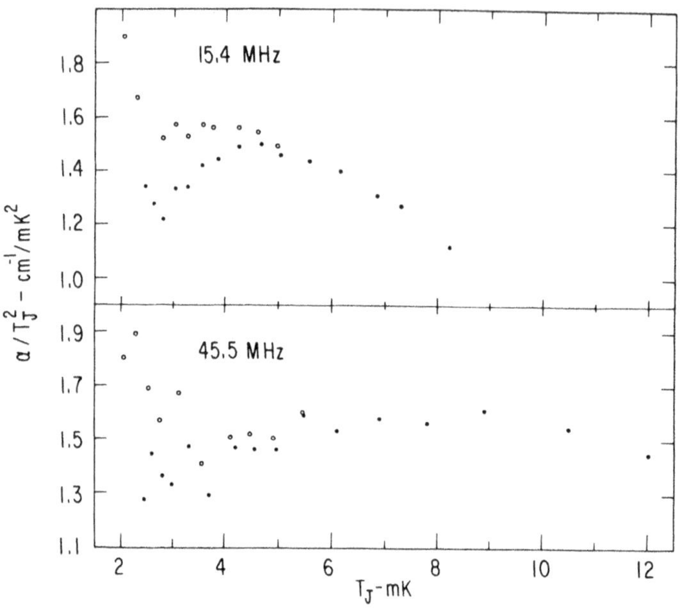

Fig. 3. Sound attenuation in pure ^3He at 0.32 atm for two different frequencies. α is the amplitude attenuation coefficient as given in Ref. 10. α/T_J^2 is plotted against T_J, the Johnson noise temperature (solid circles). Open circles are the original data $\alpha/(T^*)^2$ vs. T^*.

relation for powdered CMN similar to the one in Fig. 1, including $T_{BCS} = 2.4$ m°K when $T^* = 2.0$ m°K.

We have applied the results of the Johnson noise correction to the CMN temperature scale to reanalyze some of the properties of pure ^3He. It appears that only the thermal conductivity at low pressure and the attenuation of sound data will be affected by the Johnson noise correction. In Fig. 2 we present a plot of $(\kappa T_J)^{-1}$ vs. T_J (solid circles), where κ is the thermal conductivity of pure ^3He at saturated vapor pressure as measured by Abel *et al.*[9] The open circles are the original $(\kappa T^*)^{-1}$ vs. T^* data. It appears that the Johnson noise correction, although quite large below 4 m°K, improves the original data. There is one more correction to the conductivity data that might be significant. We plan to reexamine the raw data and recompute conductivities on the basis of a ΔT_J as opposed to the ΔT^* which was originally used. However, the accuracy of dT_J/dT^* is not good.

The amplitude attenuation coefficient of sound α in pure ^3He at 0.32 atm will be changed slightly at low temperatures. Abel *et al.*[10] found $\alpha \simeq A(T^*)^2$ from their lowest temperatures up to the point where viscous attenuation became important. In Fig. 3 we present a plot of α/T_J^2 vs. T_J (solid circles) for both the 15.4- and 45.5-MHz data. The open circles represent the original data plotted as $\alpha/(T^*)^2$ vs. T^*. Here it would appear that the Johnson noise correction is too large, although the size of the correction would be reduced if there were a difference of only 0.1 m°K in the Δ between the CMN used in this experiment and the CMN used in the Johnson noise experiment.[11]

References

1. R.A. Kamper, J.D. Siegwarth, R. Radebaugh, and J.E. Zimmerman, *IEEE Proc.* **59**, 1368 (1971).
2. R.P. Giffard, R.A. Webb, and J.C. Wheatley, *J. Low Temp. Phys.* **6**, 533 (1972).
3. R.A. Webb, R.P. Giffard, and J.C. Wheatley, *J. Low Temp. Phys.* **13**, 383 (1973).
4. R.A. Webb, R.P. Giffard, and J.C. Wheatley, *Phys. Lett.* **41A**, 1 (1972).
5. R.P. Hudson, private communication.
6. J.J. Huntzicker and D.A. Shirley, *Phys. Rev. B* **2**, 4420 (1970).
7. R.A. Webb and J.C. Wheatley, *Phys. Rev. Lett.* **29**, 1150 (1972).
8. W.C. Black, R.T. Johnson, and J.C. Wheatley, *J. Low Temp. Phys.* **1**, 641 (1969).
9. W.R. Abel, R.T. Johnson, J.C. Wheatley, and W. Zimmermann, *Phys. Rev. Lett.* **18**, 737 (1967).
10. W.R. Abel, A.C. Anderson, and J.C. Wheatley, *Phys. Rev. Lett.* **17**, 74 (1966).
11. W.R. Abel and J.C. Wheatley, *Phys. Rev. Lett.* **21**, 597 (1968).

15
Superconducting Instruments

Magnetochemical Measurements of Biochemical Compounds with a Superconducting Magnetometer*

M. Cerdonio,† R. H. Wang, G. R. Rossman, and J. E. Mercereau

California Institute of Technology
Pasadena, California

Introduction

Over the past decade Josephson's predictions[1] have been extensively examined and verified in minute detail by many experimenters. Out of this study has come a scientific base upon which a new superconducting quantum electronics is being developed in many laboratories. This paper is a report of the development of a magnetometer instrument, based on principles of quantum superconductivity, which is now in routine use in our laboratory for magnetochemical measurements. For this purpose the high sensitivity of the superconducting magnetic sensor must be matched by very high standards of reproducibility, reliability, and precision of measurement over a temperature range from room temperature to below 4.2°K in magnetic fields up to 800 G.

We first outline the nature of biochemical problems and relate it to the requirements on a magnetometer system. A description of the apparatus and its methods of use are given. Data taken with this instrument on the protein-digestion enzyme carboxypeptidase and the oxygen carrier hemerythrin are presented. Finally, we discuss the relevance of susceptibility measurements to the problem of the relationship between the structure and oxygen affinity in hemoglobin.

Transition metal ions usually play a central role in many metal-containing proteins as the active sites of biological molecules.[2] The metal ions can participate in oxygen and electron transfer, bring about structural changes, and affect enzymatic activities. Iron and copper can act as oxygen carriers in nonenzymic proteins such as hemoglobin, hemerythrin, and hemocyanin. There are at least 40 known metalloproteins which contain either iron or copper and still others with molybdenum, manganese, cobalt, zinc, vanadium, etc.[3] They are involved in various important metabolic pathways such as the biosynthesis of DNA, electron transfer in photosynthesis, and protein digestion.

The magnetic moment of a metal ion in any chemical substance is sensitive to the geometric arrangement of the ion's nearest neighbors. Often if there is more than one metal ion in a molecule, intramolecular antiferromagnetic coupling may

* Work supported in part by the Office of Naval Research, Contract No. N00014-67-A-0094-0013. Invited paper.

† Supported in part by CNR. Permanent address: Physics Department, University of Rome, Rome, Italy.

be present. An accuracy of $0.1\mu_B$/ion or better is usually needed for meaningful correlation between the geometric arrangement and the magnetic moment as well as for an estimate of the magnitude of the exchange constant (usually of the order of 150°K) in the case of antiferromagnetism. The temperature range of interest lies between room temperature and liquid helium temperature, around which contributions from zero-field splitting to paramagnetism show up. Between these two limits a temperature-dependent magnetic moment may arise due to either incomplete quenching of orbital angular momentum or low-spin–high-spin thermal equilibrium or antiferromagnetic coupling between neighboring ions.[4,5] A metalloprotein may often contain only one or a few metal ions in a molecule of molecular weight typically of 10^5 and is usually available in quantities of less than a few tens of milligrams.

To avoid the paramagnetic saturation effect, the externally applied magnetic field should not exceed a few kilogauss at helium temperatures. In addition, a reliable calibration standard independent of temperature and magnetic field is needed. In the temperature range of interest an accuracy of about 0.1 μ_B/ion for a change of a few Bohr magnetons amounts to an accuracy of a few parts of 10^{-9} emu/cm^3 for the corresponding change of susceptibility of 10^{-7} emu/cm^3.

We have developed an instrument which meets all these specifications by using different properties of superconductors: perfect diamagnetism with the well-known volume susceptibility $-(4\pi)^{-1}$ emu/cm^3 for calibration standard, persistent current to provide stable magnetic field and shielding as well as the basis for flux transformer, and flux quantization as the basis for high-sensitivity magnetic sensor. One useful feature of this superconducting magnetometer is the capability of nulling the large diamagnetic contribution from the protein host, which is often orders of magnitude larger than the contribution from the extremely dilute transition metal ions, by measuring only the temperature-dependent part of susceptibility. This feature will be illustrated by data taken on hemerythrin.

Instrumentation

A simplified schematic diagram of the experimental arrangement is shown in Fig. 1. The specimen is placed inside a thin extension of a glass dewar surrounded by a superconducting solenoid immersed in a liquid helium bath. The solenoid is enclosed by a superconducting shield and the dewars are mounted inside a mu-metal shield.

A copper foil of high purity and thermal conductivity surrounds the sample, providing a uniform temperature distribution and thermal contact with the calibrated Ge and GaAs thermometers* as well as the heater; to further improve and ensure a fast thermal equilibrium, helium exchange gas is introduced into the sample dewar. By control of helium exchange gas in the vacuum jacket of the sample dewar and with a proper use of the heater, the temperature of the sample can be made to drift slowly around any desired value between about 2 and 250°K with a typical rate of drift of 1°K/min at 100°K. It is also possible to achieve temperatures close to room temperature at the cost of a larger drift rate.

Any change in the magnetic flux in the sample compartment is detected by two

* Thermometers provided by Lake Shore Cryogenics, Hamburg, New York.

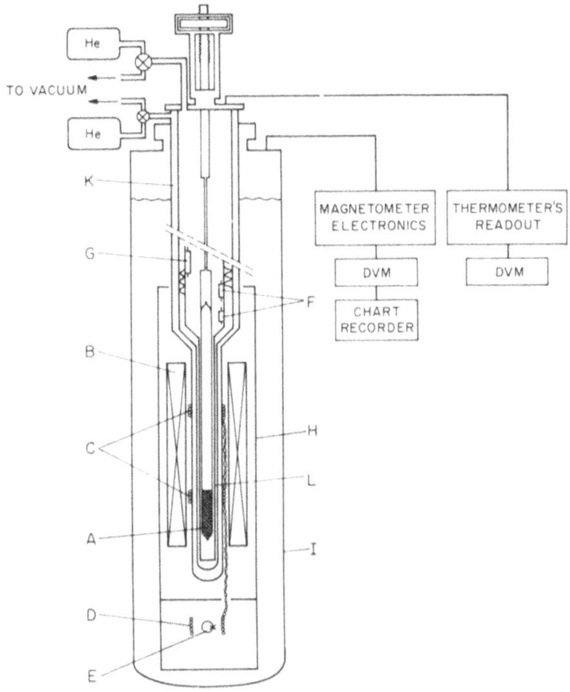

Fig. 1. A schematic diagram of the magnetometer. (A) Sample. (B) superconducting solenoid. (C) flux transformer pickup coils, (D) secondary coil of flux transformer, (E) superconducting sensor. (F) GaAs and Ge thermometers. (G) heater. (H) superconducting shield, (I) He dewar, (K) sample dewar, (L) copper shield.

superconducting pickup coils which are constructed to automatically compensate for magnetization changes due to the sample holder and its surroundings to within 0.15%. These coils and the secondary coil comprise a dc superconducting flux transformer which transfers any flux change in the sample to the superconducting sensor. A heat switch is provided in the flux transformer line to open up the transformer circuit and isolate the sensor during large changes in the magnetic field of the solenoid. A flux transfer ratio of up to 20% has been achieved. The superconducting sensor consists of a thin, cylindrical, superconducting film with one weak link deposited on a 3-mm diameter aluminum oxide rod. Details of the operation of the sensor and the electronic detection system have been described elsewhere.[6,7] For the present purpose we recall that the output signal is a periodic function of the sample-related flux with period $\varphi_0 = hc/2e = 2.07 \times 10^{-7}$ G-cm^2, the flux quantum, and that the flux sensitivity is independent of the background field. The system is linearized by using a feedback loop which compensates for any flux change in the sensor due to sample-related flux, and the sensor effectively "locks on" to any desired flux value.[8] Typical noise in operation with a flux transfer ratio of 10% and an applied field of 200 G is $10^{-3}\varphi_0$ with a 1-sec response time. In terms of the magnetization of the sample this amounts to a change of about 10^{-8} G in the quoted ambient field.

Two modes of operation are possible: a "total" susceptibility mode in which the sample is moved from one coil to the other at a fixed temperature and the output of the magnetometer is recorded; and a "relative" susceptibility mode in which the sample is kept with its magnetic center in one of the two pickup coils and the output of the magnetometer is recorded as a function of temperature as this is made to drift slowly from an arbitrary chosen value. Contribution from the sample holder is effectively compensated since it extends through both pickup coils while the sample is only in one coil. The output N of the magnetometer in units of the flux quantum is related to the volume susceptibility χ by the linear relation

$$N = \alpha(H)S\chi F^{-1} \tag{1}$$

where at typical fields of 150 G and for such dilute magnetic samples we assume that the volume susceptibility is field independent, S is the cross-sectional area of the sample, $\alpha(H)$ is a linear function of the applied field and is determined by calibrating the instrument with a superconducting sample, and the geometric factor F has a value of unity if the calibrating material has the same dimensions as the sample, otherwise it can be calculated. Although the measurements are made for a sample of constant mass, the volume may change over a wide range of temperature. The significant quantity is the gram susceptibility $\chi_g = \chi/d$, where d is the density. So relation (1) becomes

$$N = \alpha(H)\, m\chi_g (Fl)^{-1} \tag{2}$$

where l appears as a characteristic length of the sample in the direction of the applied field. For a cylindrical sample with the axis parallel to the field and of length much longer than the diameter D of the pickup coils, the F factor is independent of l and l itself is just the length of the cylinder. For a spherical sample the geometric factor F is proportional to Dl^{-1}, so that the length cancels out and the constant D can be included in the calibration function of $\alpha(H)$.

It is evident that the appearance of l in Eq. (2) for a cylindrical sample is going to limit the accuracy of measurements over a wide temperature range unless the linear expansion coefficient is known. This effect may be as large as a few parts in 10^{-9} emu/cm^3.

The present sensitivity and reproducibility of this magnetometer for a typical 0.05-cm^3 sample in a magnetic field ranging from 12 to 800 G are as follows: The *sensitivity* to susceptibility changes at any temperature between 1.7 and 300°K is $\delta\chi = 3 \times 10^{-11}$ emu/cm^3 for sample volumes as small as 0.05 cm^3. In the temperature range of 2–120°K the *reproducibilities* in relative changes of the volume susceptibility $\Delta\chi(T)$ and the magnetic moment $\Delta\mu(T)$ are $\pm 1.5 \times 10^{-9}$ emu/cm^3 and $\pm 1 \times 10^{-9}$ emu, respectively, for a sample whose total change in susceptibility $\Delta\chi$ is $\lesssim 1.2 \times 10^{-5}$ emu/cm.^{3} For a sample whose total change in volume susceptibility is $\gtrsim 1.2 \times 10^{-5}$ emu/cm^3 the reproducibility in $\Delta\chi(T)$ is $\pm 4 \times 10^{-4}$ $\Delta\chi(T)$. At higher temperature the reproducibility is mainly limited by the background magnetism contribution due to construction materials. For *total* susceptibility measurements between 2 and 200°K, where the sample is moved from one coil into another, the reproducibilities in $\chi(T_0)$ and $\mu(T_0)$ are $\pm 1.5 \times 10^{-8}$ emu/cm^3 and $\pm 9 \times 10^{-9}$ emu, respectively, for $\chi(T_0) < 4 \times 10^{-5}$ emu/cm^3; and are $\pm 4 \times 10^{-4}$ $\chi(T_0)$ and $\pm 4 \times 10^{-4}$ $\mu(T_0)$, respectively, for $\chi(T_0) > 4 \times 10^{-5}$ emu/cm^3. The reproducibility is highest in the determination of relative susceptibility changes $\Delta\chi(T)$ and is near the limit given for the accuracy for a cylindrical sample. A more detailed comparison with other instruments is difficult to make because no data are available about their reproducibilities. Usually only the sensitivity is reported (see, e.g., Ref. 9), which in a few cases is comparable to our reproducibility, but in this context we would like to mention that our sensitivity is at least one order of magnitude better than our reproducibility.

The present limitations are by no means intrinsic. Mechanical vibrations and background paramagnetism of construction materials are the main limiting factors in the present performance. These limitations can be reduced by a more rigid construction, e.g., assembling the solenoid, the flux transformer, and the sample holder as a single unit and a proper choice of construction materials. While the system was designed for use with sample volume of about 0.05–0.2 cm^3, there is no fundamental limit to the size since one can easily maximize the sensitivity by optimizing the flux transfer ratio.

Measurements

Carboxypeptidase *A*. The enzyme carboxypeptidase *A* (CPA, mol. wt. 34,900) is present in the pancreas of mammals, where it removes the C-terminal amino acid from peptide chains ingested as food. Each native protein molecule contains a zinc ion which is essential for its enzymatic activity.[10] Three positions of the distorted tetrahedral coordination around the zinc ion are occupied by ions from the protein, and the fourth position is the binding site for a substrate or a water molecule.[11]

The "entatic" site hypothesis proposes that the catalytic action of CPA results at least in part from the strained or distorted coordination sphere of the metal ion. Such distortion is said to produce a high-energy configuration at the metal ion resembling a transition state.[12] One can substitute the diamagnetic Zn^{2+} ion by Ni^{2+}, Co^{2+}, and Cu^{2+}, among others.[11] While Co-CPA and Ni-CPA retain enzymatic activity, Cu-CPA has none. This loss of enzymatic activity has been taken as evidence in favor of the "entatic" site hypothesis since Cu^{2+} is known to prefer a different geometric arrangement, i.e., a square planar coördination.

We have determined the magnetic moments of Ni-CPA and Co-CPA. The magnetic susceptibility of 14 mg of Ni-CPA containing 16.8 μg of Ni was measured in the range 2.5–75°K at a field of 150 G using the "relative" mode of operation to remove the temperature-independent contributions of susceptibility. The resulting temperature-dependent paramagnetic susceptibility versus inverse temperature is shown in Fig. 2. The corresponding magnetic moment in units of the Bohr magneton is determined to be $\mu_e = (2.67 \pm 0.10)\ \mu_B$/ion, assuming that the deviation from Curie behavior below 10°K is due to a possible contribution from zero-field splitting (Fig. 3). A "total" susceptibility measurement was carried out for 24.4 mg of Co-CPA containing 29.7 μg of Co at fields of about 40 G between 2 and 130°K. After subtracting out the temperature-independent contributions the susceptibility was plotted against inverse temperature (Fig. 4) and the magnetic moment is $\mu_e = (4.77 \pm 0.15)$ μ_B/ion.

In light of ligand field calculations and overall magnetochemical evidence, both Ni-CPA and Co-CPA fall in the ranges for an octahedral coordination, with

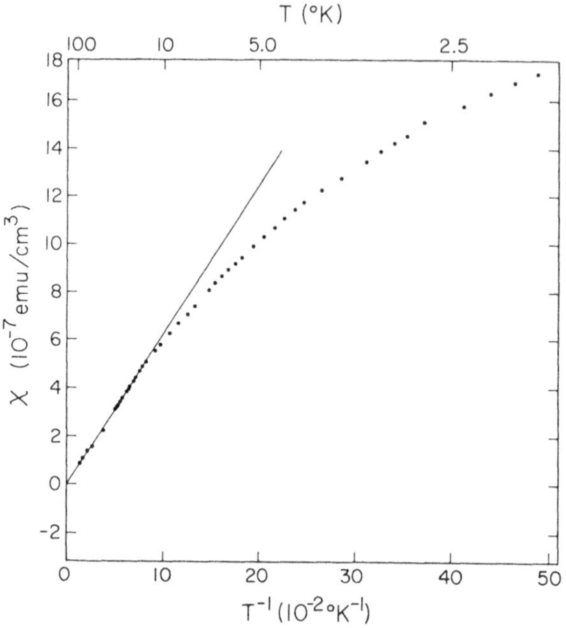

Fig. 2. Magnetic susceptibility versus inverse temperature for 14 mg of Ni-CPA containing 16.8 μg of Ni in a magnetic field of 150 G after subtracting out the temperature-independent contributions of susceptibility.

magnetic moments of $(2.8-3.2)\mu_B$ and $(4.7-5.2)\mu_B$, respectively, whereas in a tetrahedral coordination their respective moments would be $(3.2-4.0)\mu_B$ and $(4.3-4.7)\mu_B$. Thus, contrary to the entatic site hypothesis, the Ni(II) ion is seen to be in a six-coordinate environment and does not necessarily have a high-energy structure. The "irregular tetrahedral" coordination of the cobalt(II) ion proposed in this hypothesis[13] is not supported by the measurement of magnetic susceptibility. The spectroscopic results also support our conclusion.*

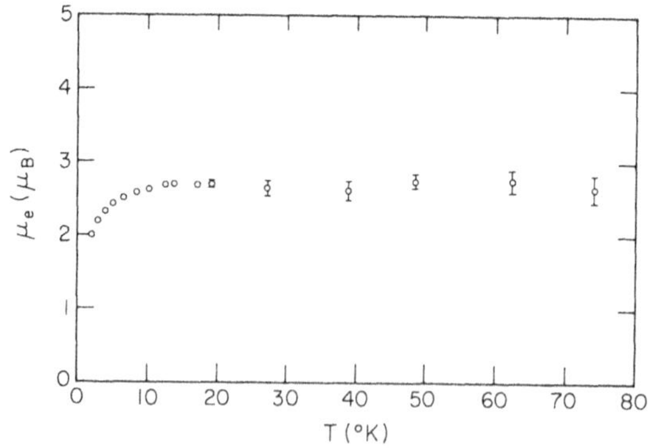

Fig. 3. Magnetic moment per Ni ion versus temperature for the Ni-CPA sample determined from the data of Fig. 2.

* Root and Rosenberg.[14] We are grateful for their preparation of the Ni-CPA and CO-CPA samples.

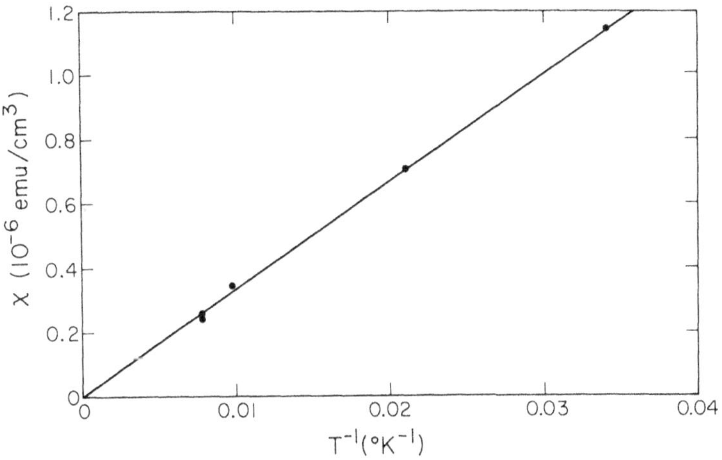

Fig. 4. Magnetic susceptibility versus inverse temperature for 24.4 mg of Co-CPA containing 29.7 μg of Co at fields of about 40 G after subtracting out the temperature-independent contributions of susceptibility.

Oxy- and Metaquo-hemerythrin. Hemerythrin, a non-heme iron protein, is an oxygen carrier in lower invertebrates. This protein (mol. wt. 108,000) consists of eight subunits, each containing two iron atoms. Previous attempts to measure the magnetic susceptibility of oxy-hemerythrin failed to show any temperature dependence, leading to postulation of strong antiferromagnetic coupling between the two iron ions in each subunit.

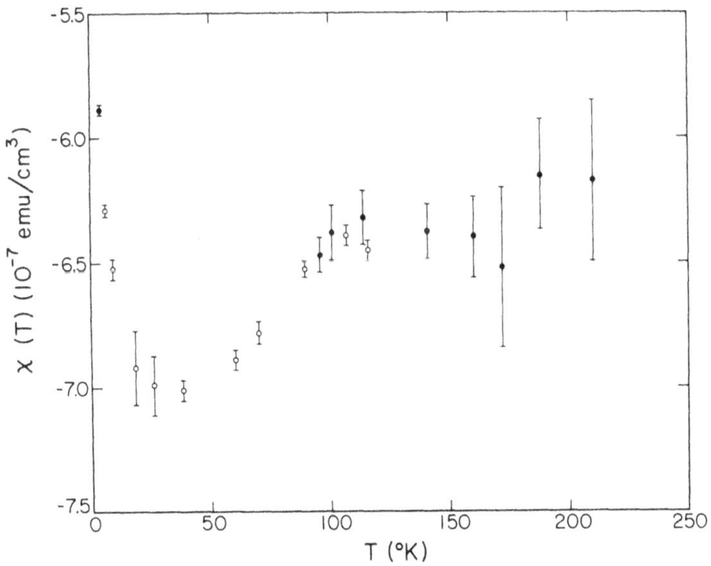

Fig. 5. Susceptibility versus temperature for 0.162 ml of an alcoholic suspension of oxy-hemerythrin containing 460 μg of iron at 47 G. Data shown as solid and open circles are obtained from measurements of total susceptibilities and susceptibility changes, respectively.

The temperature dependence of the magnetic susceptibility of 0.162 ml of oxy-hemerythrin containing 460 μg of iron measured in a field of 47 G is presented in Fig. 5. The solid data points correspond to total susceptibility measurements, and the others to relative susceptibility. The corresponding temperature dependence of the magnetic moment per iron ion is shown in Fig. 6, assuming that the rise in susceptibility below 40°K is due to paramagnetic impurities. A similar antiferromag-

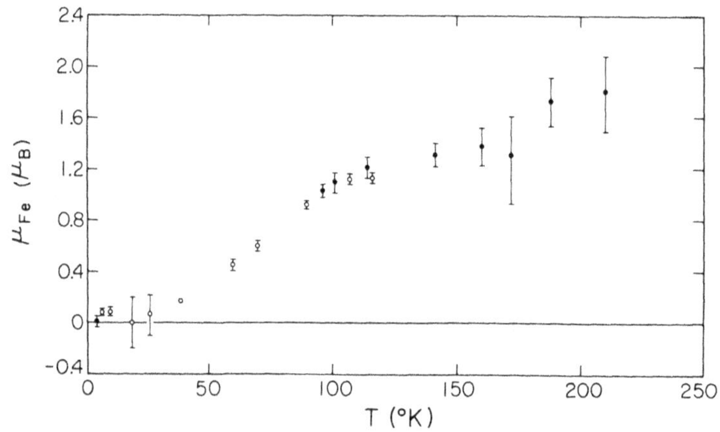

Fig. 6. Magnetic moment per Fe ion versus temperature for 0.162 ml of an alcoholic suspension of oxy-hemerythrin containing 460 μg of iron at 47 G, assuming paramagnetic impurities with a Curie constant $C = 4.94 \times 10^{-7}$°K-emu/cm^3.

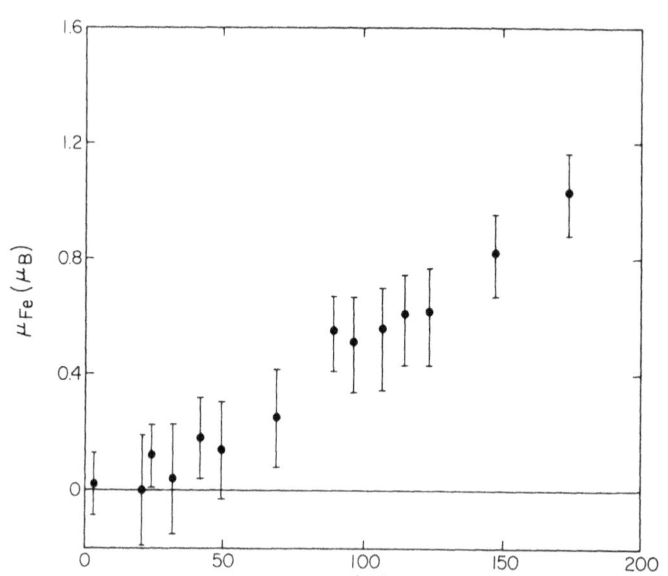

Fig. 7. Magnetic moment per Fe ion versus temperature for 0.122 ml of an alcoholic suspension of metaquo-hemerythrin containing 350 μg of iron, assuming paramagnetic impurities with a Curie constant $C = 6.04 \times 10^{-7}$°K-emu/cm^3.

netic behavior of a fully oxidized ferric form of hemerythrin, metaquo-hemerythrin, was observed (Fig. 7) for a 0.122-ml sample containing 350 μg of Fe measured at 19 G.

The results can be fitted by the standard spin–spin interaction model[5] to give the spin states of the pairs of iron (III) atoms as $S_1 = S_2 = 5/2$ for both oxy- and metaquo-hemerythrin and the magnitudes of the exchange coupling constant J as -116 and -200°K, respectively. The magnetic susceptibility data also confirm the model of oxy- and metaquo-hemerythrin as suggested by the electronic spectral results. A full discussion and interpretation of the hemerythrin results in connection with optical data appear elsewhere.[15]

Hemoglobin. Hemoglobin (mol. wt. 64,000) consists of four protein subunits, two α and two β subunits. In each subunit a heme group, comprised of a divalent iron bound to a nitrogen-rich porphyrin group, is embedded in a crevice of the protein. The binding of oxygen to hemoglobin is said to be cooperative in that the affinity to oxygen is increased with increasing degree of oxygenation (see Ref. 16 for a review).

Perutz[17] has proposed a mechanistic model which connects the cooperative effect to the structural change of the protein. Upon oxygenation, the iron ion of one subunit goes from the high-spin, low-oxygen-affinity state to the low-spin, high-affinity state, and moves closer to the porphyrin plane. This movement of the iron ion in one subunit produces a structural change of neighboring subunits, which in turn pushes the other iron ions toward the porphyrin planes, and induces a transition to the low-spin state. This model can explain how oxygen affinity increases when hemoglobin is partially oxygenated. Among other models, there is one which ascribes the cooperative effect to the opening and closing of the protein crevice independent of the spin state of the iron ion.[18] While this is not the place to discuss the relative merits of these models, a systematic study of the spin state of the iron ion by magnetic susceptibility measurements offers a relevant test of some of the models.

As pointed out earlier, to avoid the paramagnetic saturation effect, the externally applied magnetic field should be limited to a few kilogauss near helium temperature.[19] While this field level would seriously limit the sensitivity of conventional techniques applied to this problem, it has no bearing on the operation of our magnetometer, which is operated typically at a few hundred gauss. We have begun measurements on the susceptibility of hemoglobins with the superconducting magnetometer but as yet have not been able to obtain reliable results, due to the difficulties in the sample preparation and analysis. In the magnetic resonance technique a signal due to impurities which lies far away from the frequency of interest can be ignored. However, in the static magnetic susceptibility measurement all impurity moments become important and a 0.1-μg trace of paramagnetic impurity or nanograms of ferromagnetic impurity can render the sample useless. Molecular oxygen must be removed by careful degassing of solutions and solids. Reagents and sample solutions must be filtered through fine filters to remove the small quantities of gelatinous hydrous oxides which persist in most transition metal solutions. These new levels of sensitivity will demand a new degree of sophistication in sample preparation. Conceivably it is relatively easier to prepare a high-quality sample of a few milligrams than the relatively large amount usually needed for accurate measurements in conventional methods.

Conclusion

A new experimental instrument for measurements of temperature-dependent magnetic susceptibility over the temperature range of 2°K to room temperature has been developed. The high sensitivity and reproducibility of the superconducting magnetometer open a new realm of magnetic measurements on weakly magnetic materials. Already data have been obtained on biochemical compounds of unknown coordination environment, some of which were unmeasurable with previous techniques. Unlike the conventional techniques, this instrument shows great promise for further improvement in sensitivity, response time, and convenience.

Acknowledgments

We would like to thank Prof. Harry B. Gray, Dr. H. A. Notarys, Dr. Wray H. Huestis, Dr. Charles Root, and Mr. Steve Decker for many helpful discussions as well as criticisms. We thank Dr. H. E. Hoenig for his valuable contribution to the early stage of this work.

References

1. B.D. Josephson, *Phys. Lett.* **1**, 251 (1962).
2. B.L. Vallee and W.E.C. Wacker, *The Proteins*, 2nd ed., Vol. V, *Metalloproteins*, H. Neurath, ed., Academic Press, New York (1970).
3. E. Frieden, The Chemical Elements of Life, *Sci. Am.* (July 1972).
4. B.N. Figgis, *Introduction to Ligand Fields*, Interscience, New York (1966).
5. A. Earnshaw, *Introduction to Magnetochemistry*, Academic Press, New York (1968).
6. J.E. Mercereau, *Rev. de Physique Appliqué* **5**, 13 (1970).
7. H.A. Notarys, R.H. Wang, and J.E. Mercereau, to be published in IEEE special issue on superconducting electronics.
8. M. Nisenoff, *Rev. de Physique Appliqué* **5**, 21 (1970).
9. B.N. Figgis, Magnetochemistry, in *Technique of Inorganic Chemistry*, H.B. Janassen and A. Weissberger, eds., Interscience, New York (1965), Vol. IV, p.137.
10. B.L. Vallee, J.A. Rupley, T.L. Coombs, and H. Neurath, *J. Biol. Chem.* **235**, 66 (1960).
11. J.A. Hartsmith and W.N. Lipscomb, in *The Enzymes*, P.D. Boyer, ed., *Hydrolysis: Peptide Bonds*, Vol. 3, 3rd ed., Academic Press, New York (1971), p. 1.
12. B. L. Vallee and R.J.P. Williams, *Proc. Nat. Acad. Sci. U.S.* **59**, 498 (1968).
13. S.A. Latt and B.L. Vallee, *Biochem.* **10**, 4263 (1971).
14. C. Root and R. Rosenberg, Private communication.
15. J.D. Dawson, H.B. Gray, H.E. Hoenig, G.R. Rossman, J.M. Schredder, and R.H. Wang, *Biochem.* **11**, 461 (1972).
16. E. Antonini and M. Brunori, *Hemoglobin and Myoglobin in their Reactions with Ligands*, North-Holland, Amsterdam (1971).
17. M.F. Perutz, *Nature* **237**, 495 (1972).
18. R.G. Schulman, S. Ogawa, K. Wüthrich, T. Yamane, J. Peisach, and W.E. Blumberg, *Science* **165**, 251 (1969).
19. A. Tasaki, in *Probes of Structure and Function of Macromolecules and Membrane*, Vol. II, *Probes of Enzymes and Hemoproteins*, B. Chance, T. Yonetani, and A.S. Mildvan, eds., Academic Press, New York (1971), p.247.

Superconducting-Cavity-Stabilized Oscillators of 10^{-14} Stability*

S. R. Stein and J. P. Turneaure

High Energy Physics Laboratory and Department of Physics
Stanford University, Stanford, California

The use of superconducting cavities to stabilize klystrons and Gunn effect oscillators to produce fractional frequency stabilities of 10^{-12}–10^{-13} has been recently reported.[1,2] We have made progress in developing this technique and have achieved an order of magnitude improvement in short-term stability for stabilized Gunn effect oscillators at 8.6 GHz. Our best measured value for the standard deviation of the fractional frequency fluctuations $\sigma_y(N = 2, \tau = 10^2 \text{ sec})$ is 1.5×10^{-14}. There has also been two orders of magnitude improvement in long-term stability, with a measured average fractional frequency drift of 1.6×10^{-14} hr^{-1} during a 32-hr interval. Our most immediate interest in these oscillators, other than the development of the oscillators themselves, is to attempt a measurement of the temporal change in the fine structure constant. We are also interested in the development of a compact clock for the measurement of various time effects related to general relativity. Others are interested in such oscillators to make measurements of gravitational radiation,[3] to improve atomic time standards, such as the cesium atomic clock[4] and the proposed hydrogen storage beam atomic clock,[5] and to provide an ultrastable X-band source for infrared-microwave frequency synthesis between the frequency of methane and the frequency of cesium.[5]

The superconducting-cavity-stabilized oscillator (SCSO) technique for utilizing a superconducting cavity in attaining high frequency stability was chosen for a number of reasons: (1) Almost all of the circuit adjustments are at room temperature; (2) the room-temperature circuit elements can be made from standard waveguide components; and (3) the stabilized Gunn effect oscillator can deliver 25 mW of useful power, while the superconducting cavity may only have a few μW of incident power. The SCSO system schematic is shown in Fig. 1. Power from the Gunn effect oscillator enters one port of a four-port magic tee, where the power is split between the symmetric arms of the tee. In one of the symmetric arms the crystal modulator reflects two 1-MHz sidebands with a suppressed carrier. In the other symmetric arm a variable short reflects the carrier such that the carrier enters the fourth port 90° out of phase with the 1-MHz sidebands. The power incident on the cavity now appears approximately as frequency modulation with no amplitude modulation. When the carrier is close to the resonant frequency the phase and

* This work was supported in part by the U.S. Office of Naval Research, under contract N00014-67-A-0112-0061. Invited paper.

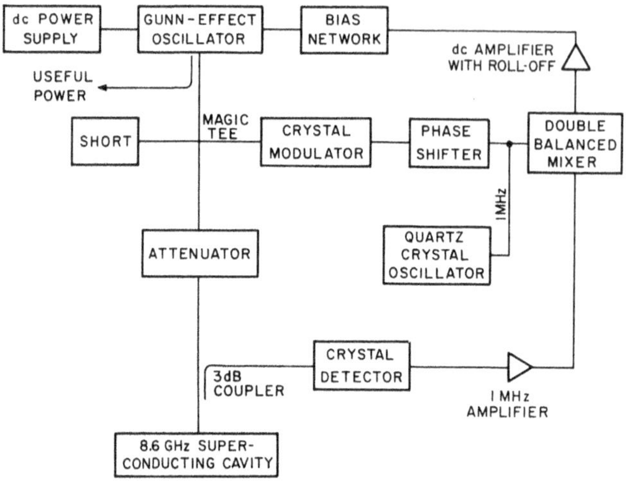

Fig. 1. A simplified block diagram for a single superconducting-cavity-stabilized oscillator.

amplitude of the reflected carrier depend on the difference between the oscillator and cavity frequencies. Under the same condition, however, the two 1-MHz sidebands are 10^5 bandwidths away from the cavity resonant frequency and are reflected exactly as if the transmission line were terminated by a short. Thus, as the phase of the carrier shifts relative to the sidebands, the 1-MHz frequency modulation is transformed into amplitude modulation which can be detected at the crystal detector. The 1-MHz component of the detected signal is ac-amplified and then product-detected against the 1-MHz reference with a double balanced mixer. When the phase shifter is adjusted to the optimum value the output of the double balanced mixer has the necessary characteristics of a discriminator voltage. After appropriate dc amplification, frequency response shaping, and biasing, the signal drives a varactor diode in the Gunn effect oscillator cavity. This particular SCSO system has the advantage that since the carrier and sidebands travel over the same rf path to the cavity and then to the crystal detector, the production of spurious 1-MHz amplitude modulation due to changes in this path length is negligible.

The superconducting cavities are the most important element in the SCSO: To obtain high frequency stability, the cavity must have a very high Q and a very stable resonant frequency. For these measurements the superconducting TM_{010} mode cavities were machined from solid niobium and were processed using chemical polishing and ultrahigh-vacuum firing.[6] The thickness of the cavity walls is about one-half the cavity inside diameter (the inside diameter is about 2.5 cm). This relatively large wall thickness reduces the sensitivity of the cavity resonant frequency to changes in acceleration and external pressure. The two cavities were mounted in a single vacuum can to isolate the cavities from pressure fluctuations of the helium liquid and to thermally isolate the two cavities so they could be independently temperature-regulated. The two cavities and their surrounding vacuum can were placed in a top-loading liquid helium dewar which could maintain 1.3°K for a period of somewhat over 24 hr. The region of the cavities was magnetically shielded to reduce the ambient magnetic field to about 10 mOe. Room-temperature thermal radiation coming down the waveguides to the cavities was greatly attenuated by a series of four copper baffles extending five-eighths of the way across each of the waveguides, which were tuned to have a VSWR of less than 1.2 at the operating frequency. The temperature of each cavity was regulated by a temperature controller whose temperature-sensing element was a germanium thermometer.

Frequency stability measurements were made in the time domain using two independent SCSOs. The schematic of the system used to measure $\sigma_y(N, r, \tau, \omega_B)$, which characterizes the frequency stability and is defined in Ref. 7, is shown in Fig. 2. The outputs of the SCSOs, which were isolated from each other by more than 100 dB, were mixed to a lower frequency f_1, which was about 25 MHz. The 25 MHz was amplified and then mixed with a synthesized signal derived from a 5-MHz quartz crystal oscillator to produce a signal f_3, which for most measurements was about 1 kHz. The signal f_3 was passed through a -6 dB/octave low-pass filter to limit the noise bandwidth and then used to gate a period time counter. A 5-MHz signal was used for the clock input to the period time counter. Data were collected on paper tape in a form that was suitable for computer analysis. The frequency-stability measuring system did not appear to contribute significant additive noise to σ_y.

The quantity $\sigma_y(N, r, \tau, \omega_B)$ was measured for averaging times τ of 10^{-3}–10^2 sec with a repetition time T of 10^{-1} sec for $\tau < 10^{-1}$ sec and with an $r = T/\tau$ of 1.04 for $\tau > 10^{-1}$ sec and a noise bandwidth $f_B = \omega_B/2\pi$ of 2.5 kHz. The operating conditions of the SCSOs during these measurements are given in Table I. σ_y was calculated both for $N = 2$ and $N = 100$ using 100 samples, and it was subsequently divided by $\sqrt{2}$ to account for the fact that the measured value of σ_y is for two oscillators. Figure 3 shows $\sigma_y(N = 2)$ plotted as a function of τ; it is seen that σ_y is 8×10^{-11} for $\tau = 10^{-3}$ sec and 1.5×10^{-14} for $\tau = 10^2$ sec. For τ between 10^{-3}

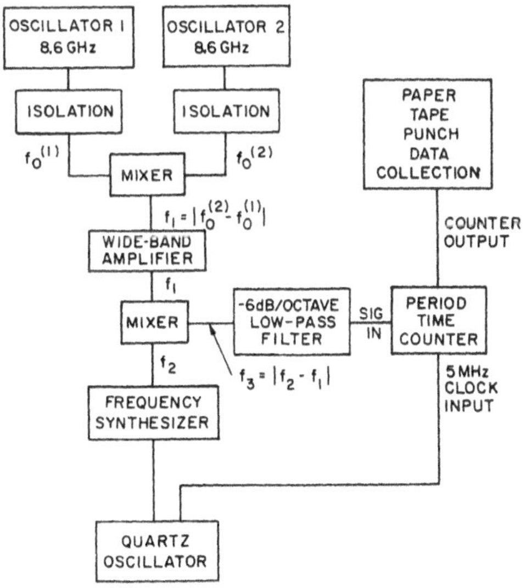

Fig. 2. A simplified block diagram of the system used to measure the frequency stability in the time domain of a pair of superconducting-cavity-stabilized oscillators.

Table I. Operating Conditions of the Superconducting Cavity Stabilized Oscillators

	SCSO system 1	SCSO system 2
System bandwidth, kHz	200	200
System dc loop gain,* dB	220	200
Cavity operating temperature, °K	1.4	1.4
Cavity frequency, GHz	8.650	8.625
Cavity loaded Q	7×10^8	7×10^8
Cavity coupling factor	40	0.5
Cavity incident power, μW	84	51
Cavity static fractional frequency shift	7×10^{-11}	2×10^{-11}

* System dc loop gain (dB) = open-loop frequency shift/closed-loop frequency shift.

and 10^1 sec $\sigma_y(N = 2)$ is nominally proportional to $1/\tau$. Although this dependence is indicative of white or flicker phase noise, measurements of the noise on the beat frequency f_3 (cf. Fig. 2) using an oscilloscope with f_3 set equal to less than 10^{-1} Hz indicate that the primary noise for $\tau < 10^1$ sec is frequency modulation of the cavity resonant frequency due to vibration. For τ between 10^1 and 10^2 sec $\sigma_y(N = 2)$ is independent of τ, which is probably related to random phase shifts of the carrier with respect to the 1-MHz sidebands. For $\tau < 10^1$ sec $\sigma_y(N = 100)$ is essentially equal to $\sigma_y(N = 2)$, but it is larger than $\sigma_y(N = 2)$ for larger averaging times [for $\tau = 10^2$ sec $\sigma_y(N = 100)$ is about 2.5 $\sigma_y(N = 2)$].

Measurements were also made to determine the long-term stability of the SCSOs. These long-term measurements were made over a time period of 32 hr with an

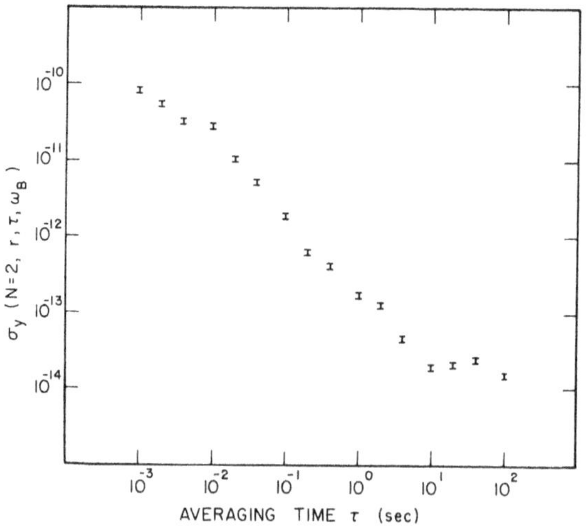

Fig. 3. Plot of $\sigma_y(N = 2, r, \tau, \omega_B)$ as a function of the averaging time τ. For $\tau < 10^{-1}$ sec the measurement repetition time T is 10^{-1} sec and for $\tau > 10^{-1}$ sec $r = T/\tau = 1.04$. The measurement noise bandwidth $f_B = \omega_B/2\pi$ is 2.5 kHz. Each point on the graph is computed from 100 samples.

averaging time τ of just under 10^2 sec and a repetition time of 10^2 sec. Figure 4, which is a plot of every tenth fractional frequency drift measurement versus time, shows the relative long-term stability of the two SCSOs. Examination of Fig. 4 shows that the maximum peak-to-peak variation of the fractional frequency drift is 5.5×10^{-12}. The fractional frequency drift with time appears to consist of random frequency drift and sudden jumps in the fractional frequency of up to 10^{-12}. Although linear frequency drift is a small part of the relative frequency drift observed with time, it was nonetheless calculated and is equal to $1.6 \times 10^{-14}\ \mathrm{hr}^{-1}$ for the 32-hr time interval.

These measurements have been principally made to evaluate the SCSO performance and to decide what are the most important contributions to the frequency noise and drift. The sources of frequency noise and drift can be divided into two categories: The stability of the superconducting cavity resonant frequency, and the instability introduced by the rest of the SCSO system. Only some of the factors that contribute to frequency noise and drift are discussed.

The superconducting cavity resonant frequency depends on a number of factors, any one of which can lead to frequency noise and drift.

1. Vibrations can introduce frequency modulation of the resonant frequency. Qualitatively, measurements indicate that vibrational noise was the principal limiting factor for $\tau < 10^1$ sec. The dewar was supported by the laboratory floor, with some damping material between the floor and the dewar bottom. Some improvement should be obtained by supporting the helium dewar on springs with a natural mechanical resonant frequency of less than 1 Hz and with appropriate damping.

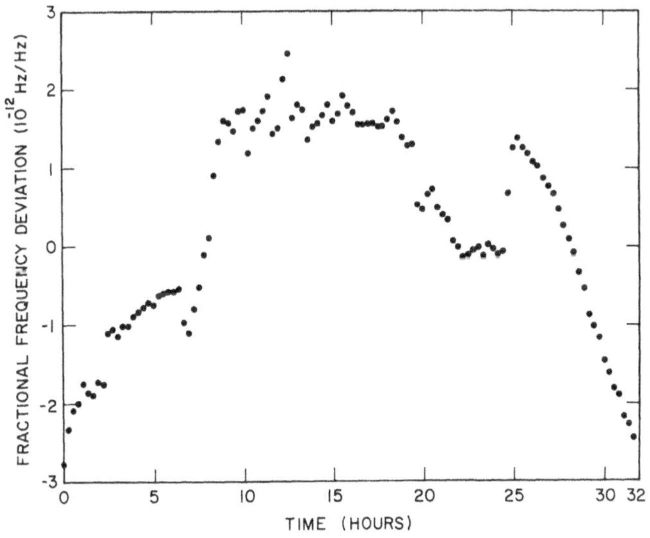

Fig. 4. The relative fractional frequency drift of two superconducting-cavity-stabilized oscillators as a function of time. These long-term measurements were made over a time period of 32 hr with an averaging time τ of just under 10^2 sec and a repetition time of 10^2 sec. Only every tenth measurement is plotted.

2. A static frequency shift of the cavity resonant frequency is produced by the electromagnetic radiation pressure[6] and by a nonlinear surface reactance.[8] The static fractional frequency shift is approximately proportional to the power incident on the cavity and was 7×10^{-11} for SCSO system 1. Since measurements on the Gunn effect oscillator power output stability show that the power occasionally jumps by 0.5%, the static fractional frequency shift can be expected to also occasionally jump by 3.5×10^{-13}. Two improvements can be made: the cavity incident power can be regulated, and the cavity incident power can be reduced so the static fractional frequency shift is less. The latter alternative, of course, leads to a lower signal-to-noise ratio for the SCSO system.

3. Temperature changes of the cavity produce a resonant frequency shift due to the temperature dependence of the surface reactance and to the thermal expansion. Improvement in temperature control of the cavities was the principal improvement made since the previously reported work.[2] The temperature dependence of the resonant frequency has been measured,[9] and at 1.4°K the fractional frequency temperature coefficient is about 1.0×10^{-9} °K^{-1}. With the improved temperature control the temperature dependence of the resonant frequency does not appear to be a limit in the measurements presented here; however, temperature control will become very important if long-term drift is substantially decreased.

4. Penetrating radiation can result in resonant frequency shifts due to the production of charged particles in the cavity volume. Earlier measurements indicate this source of resonant frequency shift is not important for fluctuations in naturally occurring radiation.[2]

5. Acceleration changes produce resonant frequency shifts. An earlier estimate[2] gives a fractional frequency shift of 10^{-9}–10^{-10} sec^2/m for the cavities used in these experiments. For cavities that are geographically fixed this source of resonant frequency shift should not be important.

6. Low-temperature creep and dislocation jumps can result in resonant frequency changes. Infrequent dislocation jumps are not serious since with an ensemble of three or more SCSOs one can determine which cavity suffered a frequency jump. The magnitude of low-temperature creep probably cannot be estimated, but it may be roughly measured by observing the relative drifts among a number of SCSOs after all other effects have presumably been removed.

Additional frequency noise and drift are introduced by the rest of the SCSO system which are not the direct result of resonant frequency shifts.

1. The phase stability of the carrier relative to the phase of the 1-MHz sidebands limits the frequency stability. The fractional frequency was measured to change 4.6×10^{-12} per degree of relative phase change. Thus the 5.5×10^{-12} fractional frequency drift observed in the long-term drift data corresponds to only a 1.2° relative phase shift in the 32-hr measurement period. We have observed much larger relative phase shifts in the modulator over similar times and can thus conclude that long-term frequency drift is dominated by the drift of the carrier phase relative to the phase of the 1-MHz sidebands. It should be possible to improve the relative phase stability in the modulator. Also, it is possible to increase the superconducting cavity Q by a factor of 10^2, which will in turn reduce the phase stability requirement by a factor of 10^{-2}.

2. The Gunn effect oscillator can have rather large drifts, of the order of 100

kHz. The very large loop gain of 220 dB reduces this drift to an insignificant fractional frequency drift of about 10^{-16}.

3. The signal-to-noise ratio of the SCSO system has been adequate for the present measurements. However, in future measurements we would like to decrease the signal by a factor of about 18 dB, so that the static frequency shift is about 10^{-12}. The signal-to-noise ratio of the crystal detector appears to be adequate for present values of σ_y even with this reduced signal level, but any further reduction in σ_y will require a detector with a lower noise figure.

We believe there is a good possibility of reducing $\sigma_y(N = 2, \tau < 10^3 \text{ sec})$ to the range of 5×10^{-15} to 5×10^{-16} in the near future. However, because of signal-to-noise limitations and other factors, the superconducting-cavity-stabilized oscillator will probably not be capable of achieving a short-term fractional frequency stability of better than 1×10^{-16}. The long-term stability will be limited by a number of factors whose magnitude can not be estimated. Nevertheless, we feel that a fractional frequency drift rate of 1×10^{-14} per year is not an unreasonable possibility for an ensemble of oscillators. Achieving this level of performance will make the superconducting-cavity-stabilized oscillator not only useful for making a measurement of the temporal change of the fine structure constant, but also as a useful device for systems and physical measurements that require ultrastable frequencies and times.

References

1. J. Benard, J.J. Jimenez, P. Sudraud, and N.T. Viet, *Electron. Lett.* **8**, 117 (1972).
2. S.R. Stein and J.P. Turneaure, *Electron. Lett.* **8**, 321 (1972).
3. J.B. Braginskii and M.B. Menskii, *Soviet Phys.—JETP Lett.* 13, 417 (1971).
4. D. Allan and J.A. Barnes, private communication.
5. D. Halford, private communication.
6. J.P. Turneaure and N.T. Viet, *Appl. Phys. Lett.* **16**, 333 (1970).
7. J.A. Barnes, A.R. Chi, L.S. Cutler, D.J. Healey, D.B. Leeson, T.E. McGunigal, J.A. Mullen, Jr., W.L. Smith, R.L. Sydnor, R.F.C. Vessot, and G.M.R. Winkler, *Trans. IEEE on Instr. Meas.* **IM-20**, 105 (1971).
8. J. Halbritter, Externer Bericht 3/68-8, Kernforschungszentrum, Karlsruhe (1968).
9. C.M. Lyneis, private communication.

The Relationship of Josephson Junctions to a Unified Standard of Length and Time*

D. G. McDonald, A. S. Risley, and J. D. Cupp

National Bureau of Standards
Institute for Basic Standards,
Boulder, Colorado

We have been working on infrared frequency synthesis using Josephson junctions with a goal of substantially simplifying existing techniques in this field and thereby contributing to the possible realization of a unified standard. To describe the technical context of our work, it is necessary to leave the area of low-temperature physics and discuss saturated absorption spectroscopy, developments in infrared frequency synthesis, and their relationship to standards of time and length.

Saturated Absorption Spectroscopy

Laser saturated absorption spectroscopy was discovered[1] in 1967 and is equal in importance to the discovery of molecular beam resonance techniques in the late 1930's. It is central to our discussion of a unified standard of time and length, but it is not yet a part of the cultural lore of low-temperature physicists, so a brief discussion is in order.

From the point of view of frequency-stabilizing lasers the most prominent example of saturated absorption spectroscopy is the absorption by methane molecules of the 88-THz ($\lambda = 3.39\ \mu$m) emission of a He–Ne laser.[2†] It is the most thoroughly investigated example and has resulted in the most highly stabilized laser system yet developed. We shall consider only this example.

The experimental arrangement (Fig. 1) consists of a He–Ne gain cell (containing the lasing medium, electrically pumped by a power supply which is not shown) and a methane absorption cell placed between mirrors forming the optical laser cavity. A small amount of laser power can be coupled out of the cavity through a partially transmitting end mirror and the detector gives a measure of that power.

To begin with, consider the absorption cell to be evacuated. Then, if one of the end mirrors is translated slowly along the length of the cavity, it is found that the laser can be continuously tuned from zero output power to a maximum output and then off again, as shown in Fig. 2(a). Changing the length of the cavity simply changes

* Contribution of the National Bureau of Standards, not subject to copyright. Invited paper.

† The He–Ne laser is misnamed, for historical reasons. It should be called a Ne laser since the laser output is from a Ne transition. That the very slightly tuneable emission of the laser coincides in frequency with a resonant transition in methane is simply an accident of nature.

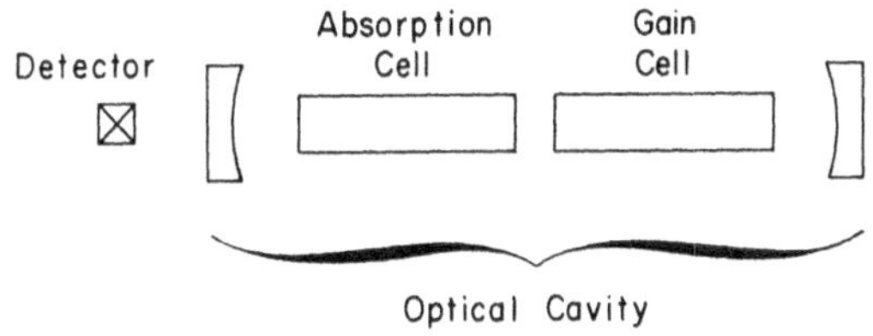

Fig. 1. Schematic representation of apparatus for saturated absorption spectroscopy. The gain cell is filled with a mixture of He and Ne, and the absorption cell with methane. The end mirrors define the optical cavity and the detector responds to the laser output power from a partially transmitting mirror on the left end of the cavity.

its resonant frequency and tunes the spectrally narrow cavity resonance across the Doppler-broadened emission line of the lasing Ne.

If a suitable pressure of methane is introduced into the absorption cell and the laser is again tuned across its gain curve, the detector will give a response as illustrated in Fig. 2(b), if the laser power level is properly adjusted. The sharply increased power output at the center of the figure is due to the fact that methane molecules have been optically pumped to saturation (i.e., the relevant upper and lower levels are nearly equally populated) and therefore the gas becomes more transparent. The narrowness of this induced transparency line, or saturated absorption line as it is called, can be understood by considering the interaction of electromagnetic traveling waves in the laser cavity with the Doppler-shifted resonances of methane molecules. A wave traveling to the right in the laser cavity can, in general, be resonant with only a very restricted group of molecules which happen to have a suitable longitudinal velocity component to Doppler-shift their resonance to coincidence with the frequency of the electromagnetic wave. When that wave is reflected from the end mirror and propagates to the left through the gas it cannot, in general, interact with the same longitudinal velocity group as before, because the Doppler shift is now in the opposite direction. The key point is that only molecules with zero longitudinal velocity can

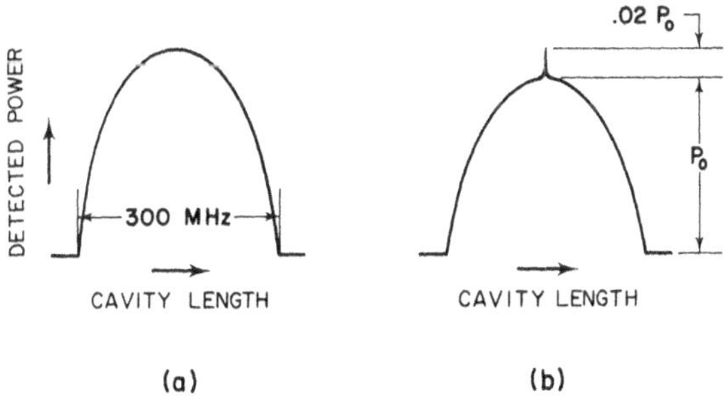

Fig. 2. (a) Detected power versus laser cavity length with the absorption cell evacuated. This curve gives the laser power output as a function of frequency. (b) With methane in the absorption cell the small but very narrow saturated absorption line appears near the center of the laser emission curve. The height of the saturated absorption line is not drawn to scale.

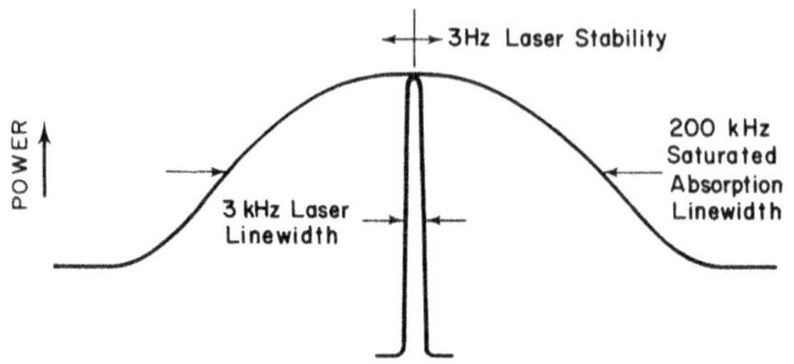

Fig. 3. Illustration of the saturated absorption linewidth, the laser emission linewidth, and, at the top, the stability of the servo-controlled laser with 1000-sec integration time. The linewidths are not drawn to scale.

absorb the same frequency wave traveling in both directions, and therefore this particular group of molecules becomes saturated at a different (lower) laser power level than any other group. Thus the sharp line of Fig. 2(b) is from saturated absorption by methane molecules with very nearly zero longitudinal velocity, and consequently is very narrow compared with the Doppler-broadened gain curve of the laser. The narrow line provides a highly suitable characteristic for frequency-stabilizing the laser.

The basic problem in frequency-stabilizing lasers is that the output frequency of a laser is directly proportional to the length of its optical cavity, and consequently the frequency is subject to thermal and vibrational changes. Earlier attempts at frequency-stabilizing lasers used servo control of the laser cavity length to maintain the laser emission at the top of the gain curve of Fig. 2(a) or of some similarly broad line. The advantage of using the much sharper saturated absorption line of Fig. 2(b) is obvious and has resulted in several orders of magnitude of improvement in frequency stability.

A summary of the relevant spectral linewidths for the stabilized system is shown in Fig. 3. The scales have been changed, so the saturated absorption line no longer appears as a narrow feature. The laser power spectrum is much narrower in comparison, and when it is servo-controlled to the methane line center the fractional stability is 3×10^{-14}, or 3 Hz at 88×10^{12} Hz, using a 1000-sec integration time.[3]*

Length Standards

The presently accepted definition of the meter is based on the 6058-Å emission from a ^{86}Kr electrical discharge tube used in conjunction with an interferometer. Paradoxically, although this system defines the meter, it cannot be used to measure

* A clear distinction should be made between stability and accuracy. Stability refers to the performance of a particular system as a function of time. Accuracy refers to an absolute comparison of the frequencies, for example, of two independently constructed methane-stabilized lasers. The accuracy can be no better than the stability and is generally less, due to systematic errors. The accuracy of the methane system is presently a few times 10^{-11}. The optimistic view is that the systematic errors can be reduced to make the accuracy approximately equal to the stability. A primary standard must have high accuracy as well as stability.

a meter, at least not in one step. Due to the small degree of coherence of the Kr source, it does not produce interference fringes if the optical path is more than a few tens of centimeters. This is the principal practical defect of the the Kr standard and it would be eliminated if the methane wavelength, defined by a stabilized laser, were adopted as the length standard.

Frequency Standards and Frequency Synthesis

The very great stability of the methane-stabilized He–Ne laser is far more than is required for length metrology (which is limited by diffraction corrections to accuracies of $\sim 1 \times 10^{-10}$) and is comparable with that of the present primary frequency standard. This standard is a ^{133}Cs atomic beam system oscillating at 9.192631770 GHz.* To be considered as a frequency standard, however, an obvious prerequisite for the 88-THz oscillator is that one must be able to measure its frequency relative to the 9-GHz standard, a frequency $\sim 10^4$ times lower.

The conceptual framework for measuring one oscillator frequency relative to another involves the process of harmonic mixing. Consider two oscillators of frequency ν_1 and ν_2, with ν_1 known, ν_2 unknown, and $\nu_1 \ll \nu_2$. In harmonic mixing, the two signals are applied to a nonlinear element (i.e., a device with a nonlinear current–voltage characteristic, e.g., a diode) and a difference frequency or beat frequency is observed according to the relation

$$\nu_b = \nu_2 - n\nu_1 \tag{1}$$

where n is the relevant harmonic number for ν_1. Usually ν_2 is known to a fair degree of accuracy and ν_1 is adjusted to make ν_b fall in the band pass of the preamplifier attached to the nonlinear element. The harmonic number n associated

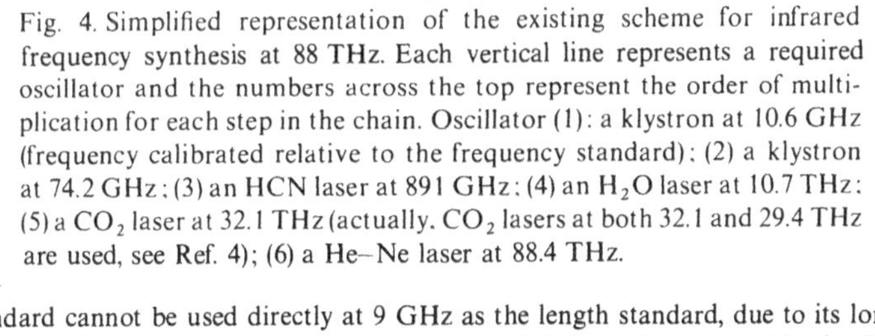

Fig. 4. Simplified representation of the existing scheme for infrared frequency synthesis at 88 THz. Each vertical line represents a required oscillator and the numbers across the top represent the order of multiplication for each step in the chain. Oscillator (1): a klystron at 10.6 GHz (frequency calibrated relative to the frequency standard); (2) a klystron at 74.2 GHz; (3) an HCN laser at 891 GHz; (4) an H_2O laser at 10.7 THz; (5) a CO_2 laser at 32.1 THz (actually, CO_2 lasers at both 32.1 and 29.4 THz are used, see Ref. 4); (6) a He–Ne laser at 88.4 THz.

* The Cs standard cannot be used directly at 9 GHz as the length standard, due to its long wavelength. Diffraction corrections to the wavelength are unacceptably large for reasonable size interferometers. The emphasis in our discussion is on the development of the methane-stabilized laser as a unified standard. An equally logical development would be to use the Cs atomic beam system, in conjunction with a frequency multiplication chain, as a unified standard. Ultimately the choice depends on the relative accuracies of the 9-GHz and 88-THz oscillators. At the present time the accuracy of the Cs system is about two orders of magnitude better than that of the methane system.

with the observed beat is determined by shifting ν_1 a known amount and measuring the corresponding shift of ν_b.

To measure the 88-THz frequency relative to the primary standard requires $n \approx 10^4$, and this is probably impossible, due to the decrease of harmonic power with increasing n. To overcome this problem in the recent first absolute measurement of frequency as high as 88 THz[4] a stepwise scheme was used, as illustrated in Fig. 4.

Each of the arrows in the figure represents harmonic multiplication by the indicated number so as to either measure or control the frequency of the next higher oscillator in the chain. The best result obtained so far for the methane frequency is 88.376181627($\pm$ 50) THz.[5] Incidentally, this frequency measurement, when combined with a wavelength measurement,[7] gives the best available value for the velocity of light.[7]

It has been admirably demonstrated that accurate frequency measurements can be made in the infrared, but it is a complex procedure because of the large number of oscillators required.

Josephson Junctions

Josephson junctions offer the possibility of eliminating some of the intermediate oscillators. The theoretical basis for using Josephson junctions for this purpose can readily be derived from the usual Josephson equations.[8,9] With applied voltage amplitudes v_1 and v_2 at frequencies ν_1 and ν_2 ($\nu_1 \ll \nu_2$) the current amplitude at the beat frequency of Eq. (1) is

$$i(\nu_b) \propto J_n(2ev_1/h\nu_1)\, J_1(2ev_2/h\nu_2)$$

where $J_1(x)$ is an integer-order Bessel function and e and h are the usual fundamental constants. The important conclusion is that as the order of harmonic mixing n is increased, the beat frequency current amplitude decreases only slowly (maximum value of $J_n \propto n^{-1/3}$), and therefore very high-order mixing should be observable. However, as a note of caution, it should be added that this theory omits resistive losses and shunt capacitance effects of real junctions, so $n^{-1/3}$ is a *minimum* rate of fall as the harmonic number is increased.

Objections have been raised that Josephson junctions cannot be expected to function at frequencies much above the Cooper-pair binding energy. Nearly all of our experiments have been with Nb, which has a threshold frequency for pair-breaking of 675 GHz. We have obtained[10] a measurable Josephson response at frequencies as high as 12 times the pair-breaking frequency, and this is consistent with the microscopic theory,[11] which predicts only a ν^{-1} rate of fall of the current at high frequencies.[12]

Point-contact junctions are used in order to facilitate the coupling of the infrared radiation into the junction. The application of antenna theory has proven useful in this connection.[13]

Our first harmonic mixing experiments[9] were for 84th-order mixing of an X-band signal with the 891-GHz ($\lambda = 337\ \mu$m) output of the HCN laser. They verified the high harmonic generation capability of Josephson junctions. The experimental arrangement is represented schematically in Fig. 5 and can be thought of as an infrared heterodyne receiver. The significant difference as compared with more

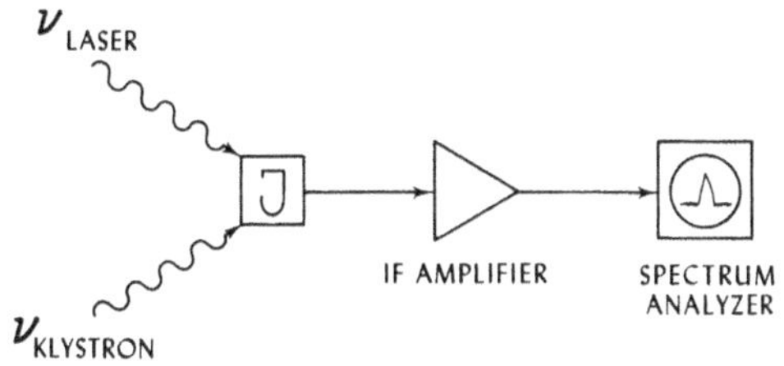

Fig. 5. Josephson junction frequency synthesis scheme. The junction generates high-order harmonics of the klystron signal and mixes them with the applied laser signal. The appropriate beat signal is amplified and fed to the spectrum analyzer for measurement of its frequency.

common heterodyne receivers is that local oscillator power is generated in the Josephson junction as a high harmonic of the applied klystron frequency. The beat frequency of Eq. (1) is generated in the same junction as the harmonics and, after passing through the i.f. amplifier, is displayed as a spectral line on the analyzer.

The klystron frequency can be determined by conventional electronic counter methods and, since the beat frequency, as observed on the spectrum analyzer, is a relatively low frequency compared with that of the laser, only a relatively crude calibration of the spectrum analyzer is required to give a high absolute accuracy for the laser frequency.

In order to extend these experiments to the next available higher frequency laser line at 3.8 THz ($\lambda = 78\ \mu$m), a systematic study was made to more nearly optimize the system. An important result of this study was that for a given order of harmonic mixing, raising the i.f. or beat frequency (by changing the klystron frequency slightly) from 18 MHz to 9 GHz raised the available beat signal power by a factor of 74.[14] This is consistent with expectations for a parametric device.

At 3.8 THz it was again possible to observe a beat with an X-band source[14]* this time with harmonic number 401 and a signal-to-noise ratio of 17 dB with a noise bandwidth of 8 kHz. We also studied the systematics of the dependence of the available beat signal power on the order of harmonic mixing n required to reach the fixed upper frequency. The empirical result is that the power decreases approximately as n^{-2}.

This result is useful for assessing the chances of extending harmonic mixing experiments to the next higher frequency laser line, at 10.7 THz ($\lambda = 28\ \mu$m). It requires generation of approximately the 1100th harmonic of the X-band source. An auxiliary experiment has shown that mixing between two fundamentals, both near 10.7 THz, produces a signal-to-noise ratio of 60 dB. Using the n^{-2} rule, then, we estimate that 1100th order harmonic mixing will yield a signal-to-noise ratio of $\lesssim 1$, a difficult situation. Improvements in the equipment are required.

To this end, the apparatus is presently being rebuilt as illustrated in Fig. 6. The 9-GHz traveling wave maser preamplifier, which was also used for the 3.8-THz experiments,[14] is being incorporated in a more nearly optimum way with the

* The development of a suitable X-band source will be discussed elsewhere.

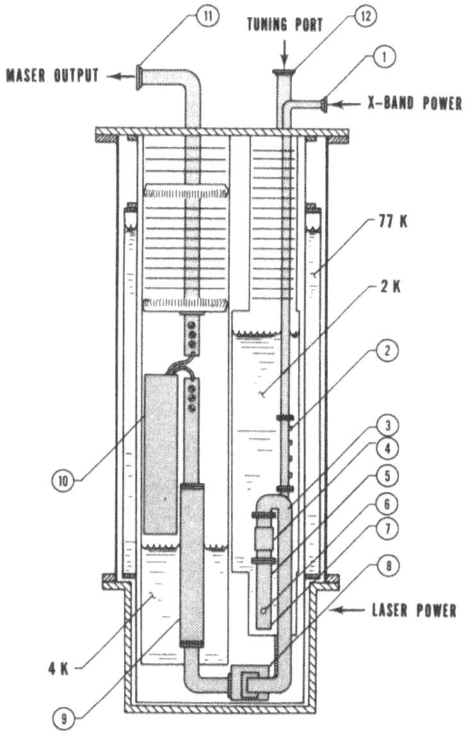

Fig. 6. Josephson junction–ruby maser, infrared laser, harmonic mixing system. The system uses a Josephson junction both for generating high harmonics of the applied X-band power and for mixing the harmonics with the applied infrared power. Maximum sensitivity is obtained by using a low-noise maser preamplifier for the beat frequency signal. X-band power enters the waveguide at port 1 and passes through the band-pass filter (2) (band pass 9.365–9.635 GHz), which excludes most of the unwanted noise. A 10-dB Schwinger directional coupler directs the X-band power through a sliding stub tuner (4) and a tapered waveguide section (5), to the Josephson junction (6). Additional tuning is provided by a sliding short (7), below the junction. Laser power impinges on the junction through a 3-mm-diameter iris in the waveguide wall. The waveguide system is tuned and impedance-matched for a beat frequency of 9 GHz. Beat frequency power propagates past the tuner (4), straight through the directional coupler (3), and through the waveguide switch (8). Next it passes through a low-pass microwave filter (9), which stops maser pump power at 23 GHz from propagating backward through the waveguide to the Josephson junction. Finally, the signal is coupled into and out of the maser through coaxial cable and then to a conventional room-temperature heterodyne receiver at port 11. The sliding stub and sliding short tuners are adjusted by observing the waveguide system through port 12 and the waveguide switch (8).

Josephson junction mixer. The transmission line and impedance transformer (sliding stub tuner) between the Josephson device and the maser will now be at liquid He temperature and consequently we will be able to take full advantage of the 2°K noise temperature of the maser. These changes will lower the system noise temperature by about a factor of 30. In addition, the 10.7-THz laser has now been frequency-stabilized[5] so that digital time-averaging of the spectrum analyzer output signal can be used to further improve the signal-to-noise ratio. We therefore have a reasonable hope of observing the beat from the 1100th harmonic.

In summary, we note that the very great stability of the methane-stabilized laser has already greatly improved length metrology and is competitive with the present frequency standard. The development of a sufficiently simple method of infrared frequency synthesis will make a unified standard practical. Since no other possibilities for simplification have progressed measurably, the Josephson junction experiments should be pushed to their limit.

Acknowledgments

We want to acknowledge stimulating and informative discussions on these topics with Donald Halford, Helmut Hellwig, and John Hall. Furthermore, we express our thanks to K. M. Evenson, J. S. Wells, F. R. Petersen, B. L. Danielson, G. W. Day, and R. L. Barger for supplying data prior to publication.

References

1. P.H. Lee and M.L. Skolnick, *Appl. Phys. Lett.* **10**, 303 (1967).
2. R.L. Barger and J.L. Hall, *Phys. Rev. Lett.* **22**, 4 (1969); H. Hellwig, H.E. Bell, P. Kartaschoff, and J.C. Berquist, *Appl. Phys. Lett.* **43**, 450 (1972).
3. J.L. Hall, G. Kramer, and R.L. Barger, in *Conf. on Precision Electromagnetic Measurements, 1972, Boulder, Colorado,* IEEE Cat. No 72 CHO 630-4-PREC.
4. K.M. Evenson, G.W. Day, J.S. Wells, and L.O. Mullen, *Appl. Phys. Lett.***20**, 133 (1972).
5. K.M. Evenson, J.S. Wells, F.R. Petersen, B.L. Danielson, and G.W. Day (to be published).
6. R.L. Barger and J.L. Hall (to be published).
7. K.M. Evenson, J.S. Wells, F.R. Petersen, B.L. Danielson, G.W. Day, R.L. Barger, and J.L. Hall (to be published).
8. C.C. Grimes and S. Shapiro, *Phys. Rev.* **169**, 397 (1968).
9. D.G. McDonald, A.S. Risley, J.D. Cupp, and K.M. Evenson, *Appl. Phys. Lett.* **18**, 162 (1971).
10. D.G. McDonald, V.E. Kose, K.M. Evenson, J.S. Wells, and J.D. Cupp, *Appl. Phys. Lett.* **15**, 121 (1969).
11. N.R. Werthamer, *Phys. Rev.* **147**, 255 (1966).
12. D.G. McDonald, K.M. Evenson, J.S. Wells, and J.D. Cupp, *J. Appl. Phys.* **42**, 179 (1971).
13. L.M. Matarrese and K.M. Evenson, *Appl. Phys. Lett.* **17**, 8 (1970).
14. D.G. McDonald, A.S. Risley, J.D. Cupp, K.M. Evenson, and J.R. Ashley, *Appl. Phys. Lett.* **20**, 296 (1972).
15. A.S. Risley, J.H. Shoaf, J.R. Ashley, and F.M. Palka (to be published).

Static Susceptibility of Biological Molecules and SQUID Detection of NMR*

E. P. Day

Department of Physics, Stanford University
Stanford, California

A static susceptibility apparatus capable of detecting changes of 10^{-10} cgs in a 0.25-cm^3 room-temperature sample has been developed, using the technology of superconductivity. The instrument has been used to detect nuclear magnetic resonance by following rf-induced changes in the longitudinal nuclear magnetization. It has also been used to search for changes in the diamagnetic susceptibility of biological molecules as they undergo conformational changes. The static susceptibility instrument will be discussed first, to be followed by a discussion of the nuclear magnetic resonance experiment and a discussion of the search for changes in the diamagnetism of biological macromolecules.

A direct measurement of changes in magnetic susceptibility can be made with a sensitive superconducting magnetometer as a detector of magnetic flux changes and a persistent mode superconducting magnet as the source of a large, stable magnetic field. A sample, located in the high-field region of the magnet, is placed in a pickup coil coupled to the magnetometer. Changes in magnetic flux through the pickup coil are caused by changing sample position relative to the coil, by changing sample temperature, or by applying radio or microwave frequency appropriate to cause nuclear or electronic paramagnetic resonance.

A schematic drawing of the apparatus is given in Fig. 1. The 0.25-cm^3 sample cylinder is located in a double-walled, room-temperature, copper dewar finger, making it possible to vary the sample temperature from 0 to 100°C with forced N_2 gas flow. Two opposing, matched pickup coils are used to allow the cancellation of effects common to sample and control while detecting events unique to the sample. The large magnetic field of the persistent mode magnet is confined to the vicinity of the sample by surrounding the magnet with a superconducting cylinder capped with ferromagnetic end pieces. This shield configuration was chosen to enhance the magnet's homogeneity and to isolate the high-field region at the sample from the low-field region at the magnetometer. The double weak-link magnetometer is shielded from the earth's field with two mu-metal shields and from fields within the apparatus with a superconducting shield.

The crucial design problem in the development of this instrument was that of creating a stable magnetic field environment at the pickup coil. Flux creep, vibration,

* Work supported in part by the Center for Materials Research, Stanford, California and Office of Naval Research.

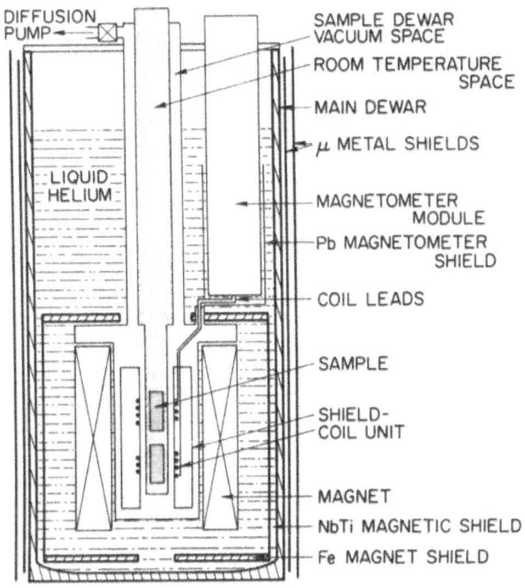

Fig. 1. Schematic diagram of the static susceptibility apparatus.

and paramagnetism are three sources of noise which have been dealt with in achieving the present sensitivity of 10^{-10} cgs. Flux creep is an intrinsic instability of a high-field magnet. To shield against this noise and to cut down on the effects of vibration, a superconducting shield capable of being heated normal was built into the apparatus between the magnet and the pickup coil. The effects of vibration are reduced by attaching the pickup coil rigidly to this shield, forming a shield–coil unit. Vibration of this unit relative to the magnet does not lead to noise at the pickup coil as long as there is no relative motion of the coil and the shield. Aluminum and OFHC copper were used as the sole construction materials within the shield to eliminate the presence of all but diamagnetic materials from the vicinity of the pickup coil. The nuclear paramagnetism of both aluminum and copper is undesirable. Temperature regulation of the shield–coil unit has been used in the present apparatus to eliminate noise from this source. The apparatus response time of less than a second is set by the eddy current behavior in the dewar walls between the sample and the pickup coil. The noise level with the magnet charged to approximately 8 kOe is typically less than 10^{-10} cgs.

An unusual experiment made possible with the present sensitivity is the detection of nuclear magnetic resonance by following rf-induced changes in the longitudinal magnetization. The basic idea for this type of experiment originated with Fairbank.[1] It was first discussed in the literature by Pierce[2] and later by others.[3] What had not been anticipated is the fact that broad nuclear magnetic resonance lines can be detected at full sensitivity using low levels of rf power. This fact makes it likely that this technique will find application in isotope trace analysis in solids, in thermometry, in geological dating, in the detection of broad lines in biological molecules, and possibly in the detection of chemical shifts in solids.

Adiabatic fast passage of a nuclear resonance line results in the inversion of the longitudinal magnetization. This will cause a change of $2M_0$ in the magnetometer reading of the susceptibility apparatus, where M_0 is the equilibrium magnetization of the resonant nuclei. The signal will be $2M_0$ independent of the rf power level as long as the conditions of adiabatic fast passage have been met.

The conventional means of detecting adiabatic fast passage uses a transverse pickup coil as opposed to the longitudinal coil used in the present method. A transverse coil detects a pulse as the magnetization swings past the transverse plane on its way toward inversion. For broad resonance lines of width H_L the pulse height is reduced by the factor (H_1/H_L) because the spins are not in phase in the transverse direction. H_1 is the amplitude of the rf field. The only way to overcome this incoherence is to increase the rf field amplitude until H_1 equals H_L. This requires large amounts of rf power with the 1–10 G linewidths typical of solids. Note that with the present technique employing a longitudinal detection coil the incoherence of the transverse pulse does not enter into consideration. The full signal of $2M_0$ is detected upon passing from completely below the broad line to completely above it, since the entire spin population undergoes inversion in the process of passage.

The configuration for the nuclear magnetic resonance experiment is identical to that shown in Fig. 1 with the addition of an rf coil located inside the sample holder and oriented at right angles to the sample cylinder axis. The rf energy is coupled into the coil by a coaxial line brought in along the sample stick.

The results of an adiabatic fast passage experiment on the protons in water is shown in Fig. 2. The curve shown is a superposition of sweeps in opposite directions through resonance. The linewidth is indicated by the distance between rise slopes of the opposing sweeps. The broad line is due to the magnet's inhomogeneity of 3.1×10^{-3} as measured with a Hall probe. The longitudinal spin relaxation time of the protons in water is shown by the exponential decay from maximum signal in either direction. The value of 3.5 ± 0.4 sec derived from this curve is consistent with the known value. The resonant frequency of 31.81 ± 0.005 MHz at the crossover point of opposing sweeps is consistent with the magnetic field value of 7.47 kOe measured with the Hall probe.

The ordinate of the trace was calibrated against the bulk diamagnetic susceptibility of water. It should be possible to reverse this calibration procedure and use the well-known nuclear susceptibility to calibrate the units of static susceptibility which are uncertain at present (Ref. 4, pp. 85–87).

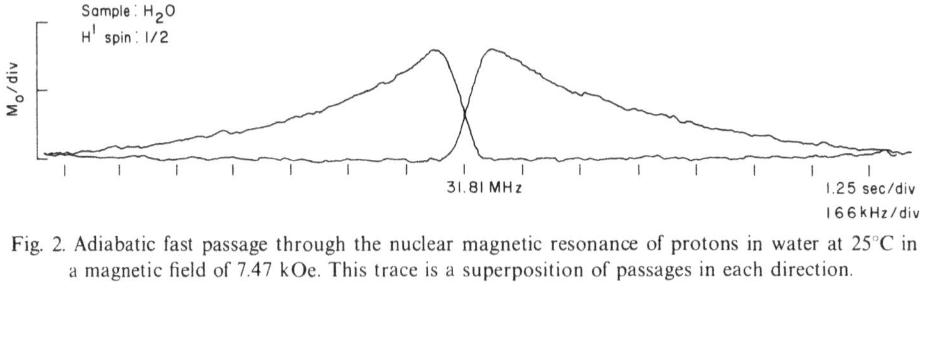

Fig. 2. Adiabatic fast passage through the nuclear magnetic resonance of protons in water at 25°C in a magnetic field of 7.47 kOe. This trace is a superposition of passages in each direction.

A separate publication[5] describes the detection of ^7Li and ^{19}F nuclei in room-temperature LiF powder. These broad resonances were detected at full sensitivity using 40 dB less rf power than required by conventional detection.

A static susceptibility instrument of greater sensitivity than that presently available is needed in the search for possible changes in the diamagnetism of biological macromolecules in aqueous solution. The present sensitivity of 10^{-10} cgs corresponds to the ability to detect changes of the order of 10^{-4} of the diamagnetic susceptibility of water. A typical concentration for a biological molecule in aqueous solution is 10^{-3}, making it possible to see 10% effects with the present sensitivity. Clearly, greater sensitivity is needed since susceptibility changes, if they exist, are to be expected at the 1% level or less.

What reason is there to expect small changes in the diamagnetism of biological macromolecules upon denaturation? Denaturation refers to the uncoiling of one of these large, linear, threadlike molecules from the specific three-dimensional globular configuration which it assumes in its natural, functional state. This unique three-dimensional shape arises from the specific intramolecular interactions of weak bonds between the amino acid building blocks of the linear threadlike molecule. Because this unique configuration arises from weak bonds such as van der Waals forces, electrostatic forces, and hydrogen bonds, it can be disrupted with slight changes in temperature or *p*H.

Changes in the diamagnetic susceptibility of biological molecules upon denaturation are expected because there is some (admittedly disputed) evidence that there are changes in the diamagnetic susceptibility of small organic molecules with changes of physical state (Ref. 4, pp. 89–91 and Ref. 6). The dramatic change in conformation of a biological macromolecule upon denaturation is a phase transition similar in many respects to a change of physical state for smaller molecules.

Preliminary experiments have been done with molecules which can be put in aqueous solution at much higher concentration than the 10^{-3} figure stated earlier in this discussion. Lysozyme, a globular protein which acts as a digestive enzyme, was run at a concentration of 15% w/v. At this concentration we can detect changes at better than 10^{-3} of the molecular diamagnetism. No changes in static susceptibility were seen upon denaturation of lysozyme with heating. Likewise, no changes in static susceptibility were seen upon thermal denaturation of ribonuclease (run at a concentration of 10% w/v).

Conclusive evidence on this issue of possible changes of diamagnetic susceptibility with denaturation of biological macromolecules will await the development of a static susceptibility apparatus of greater sensitivity than the present 10^{-10} cgs.

Acknowledgments

This research benefited from the active support of Dr. William M. Fairbank. The suggestion of using adiabatic fast passage was made by Dr. James E. Opfer. John M. Pierce contributed the design of the room-temperature dewar and John Philo contributed throughout all phases of this work.

References

1. M. Bol, B.S. Deaver, and W.M. Fairbank, U.S. Patent 3,454,875.
2. J.M. Pierce, A Persistent Current Magnetometer with Novel Applications, Symp. on Physics of Superconducting Devices, April 28–29, 1967.

3. R.A. Kamper, L.O. Mullen, and D.B. Sullivan, Some Applications of the Josephson Effect, National Bureau of Standards Technical Note No. 381, National Bureau of Standards, Washington, D.C., October 1968.
4. P.W. Selwood, *Magnetochemistry*, 2nd ed., Interscience Publishers, New York (1956).
5. E.P. Day, *Phys. Rev. Lett.* **29**, 540 (1972).
6. Ya. G. Dorfman, *Diamagnetism and the Chemical Bond,* American Elsevier, New York (1965), pp. 57, 104.

Gravitational Wave Detection at Low Temperatures*

W. O. Hamilton, P. B. Pipes, and P. S. Nayar

Department of Physics, Louisiana State University
Baton Rouge, Louisiana

We have constructed an apparatus at the NASA Mississippi Test Facility which makes use of low-temperature techniques to detect gravitational waves.[1] This is part of an experiment being conducted jointly between LSU and Stanford.[2] The final construction and assembly are now underway and the preliminary cryogenic testing of the apparatus is scheduled to begin during the next several months. We will describe the apparatus design and calculated performance, and then briefly discuss the instrumentation which should enable us to detect the effects of gravitational radiation on a massive test body.

A schematic of the apparatus is shown in Figs. 1 and 2. These line drawings are scale tracings from the assembly blueprints for the apparatus. The gravity wave antenna consists of an aluminum bar 10 ft long and 3 ft in diameter weighing approximately 12,000 lb. No mechanical support is shown for the antenna since it will be supported on a magnetic field created by pancake coils lying in the surrounding shell. For this purpose the aluminum bar will be coated with a high-field superconductor. Through the efforts of the Stanford group we have already constructed and tested magnetic supports to the pressures required to support a 36-in.-diameter aluminum bar. The support is stable with the coils in the persistent mode.

The shell surrounding the antenna and holding the magnet will be cooled by a dilution refrigerator which is under construction at LSU. The refrigerator is of conventional design and is designed to remove 500 ergs/sec at 0.05°K. It is being designed to circulate $5-10 \times 10^{-4}$ mole/sec. The apparatus will be similar to that reported by Ehnholm and Gylling.[3]

The next outer shell will contain ^{4}He at about 2°K. Helium reservoirs on the end caps will contain approximately 100 liters of helium each. Thermal uniformity along the shield will be maintained by a manifold of tubes running along the shell and connecting the two reservoirs. This design has been copied from the very successful cryogenic design for the Stanford superconducting accelerator. A minimum length of weld is exposed to the vacuum by such a design, so the problems attendant upon leak testing are reduced. This shell is fabricated entirely of 5083 aluminum and seamless aluminum tubing. The shell construction is heavy enough to allow exchange gas to be used inside this shell.

The boil-off gas from the helium temperature shell is circulated through the next shell, which is constructed of light-weight 1100 aluminum sheet and seamless tubing. The construction technique is the same as that for the previously described

* Research supported by the National Science Foundation and the Air Force.

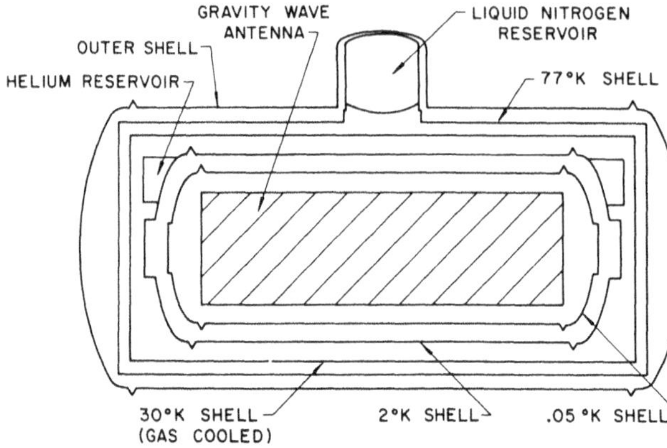

Fig. 1. Side view of the gravitational radiation detector. The bar length is 10 ft and its resonant frequency is 800 Hz.

2°K shell, but the shell is not made as an individual vacuum container. The aluminum body of the shell serves to conduct heat to the skeleton of tubing through which gas circulates. This shell will run at 40°K when the dilution refrigerator is not running and near 30°K when it is.

The nitrogen-cooled shell is of the same construction, using 1100 aluminum sheet and tubing. The nitrogen reservoir is contained in a tower situated on the top of the apparatus. Superinsulation is planned for use between the nitrogen shell and the main vacuum shell.

The outer vacuum shell is constructed of 5083 aluminum plate $\frac{3}{4}$ in. thick. It is $17\frac{1}{2}$ ft long and $7\frac{1}{2}$ ft in diameter. All of the shells except the innermost 0.050°K shell are suspended from this shell. The innermost shell is separately suspended, as shown in Fig. 2.

The vibration filters shown in Fig. 2 consist of alternate layers of Isomode neoprene padding and 1-in.-thick steel plate. This is the type of filter successfully used by Weber in isolating his apparatus from vibrational excitation. Weber reports that vibrations at acoustic frequencies are reduced by a factor of four in each layer.

Weber indicates that the gravitational wave signals he detects cause the length of his bar to change by approximately 10^{-14} cm. This is a length change of approximately the same size as the room-temperature thermal noise displacement of the end of his bar. By reducing the temperature of the bar to below 0.050°K the thermal noise displacements are reduced by a factor of 100 or more and the corresponding power sensitivity is increased by the ratio of the temperatures. Of course, the isolation of the antenna from ground vibration must be increased also. This points out the importance of the vibration filters and of the magnetic support. The uniform magnetic support should provide an unexcelled freedom from vibrational coupling to the antenna.

The magnetic support also makes feasible the use of magnetic cooling to reduce the bar temperature below that of the dilution refrigerator. Very small amounts of paramagnetic salt can be used to obtain sizable temperature reductions. The absence of heat conduction through the supports should allow the antenna to remain at ultra low temperatures for an extended time.

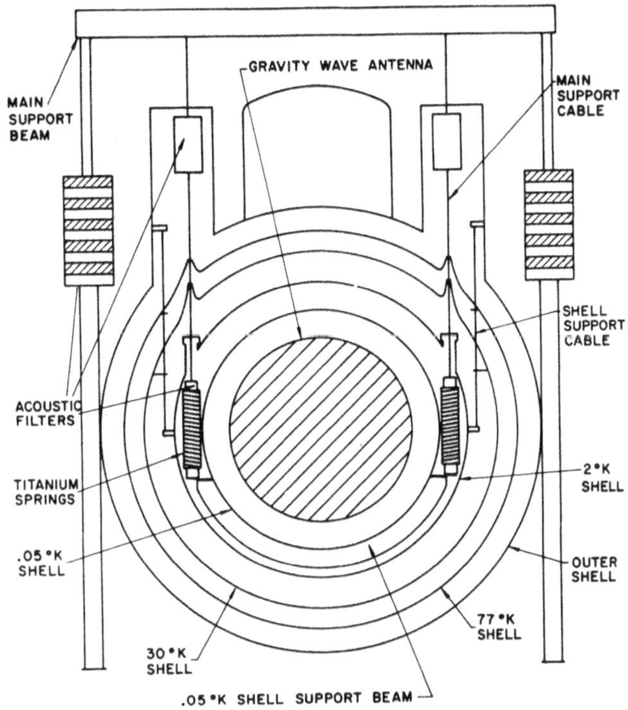

Fig. 2. Support schematic for the gravitational radiation detector.

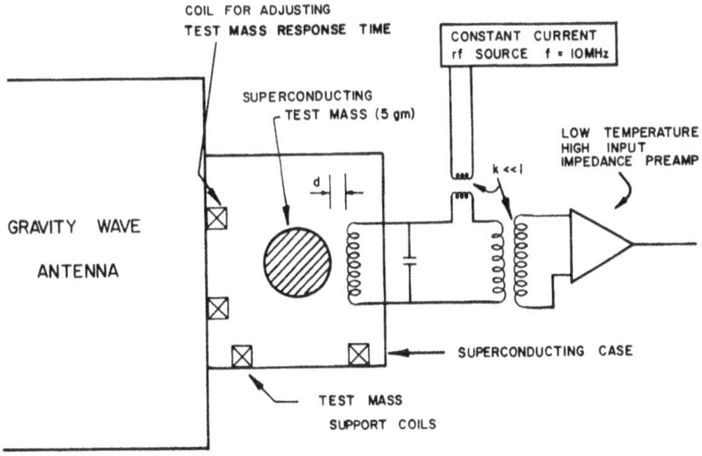

Fig. 3. Schematic of accelerometer readout.

The motion of the bar will be detected by a superconducting accelerometer which we are developing at LSU. An accelerometer of a different design is being developed at Stanford. Our device consists of a superconducting test mass supported on a persistent current magnetic field. The concept is illustrated schematically in Fig. 3. A tuned superconducting inductance is placed in close proximity to the test mass. Motion of the inductance with respect to the test mass will change the value of the inductance and hence the frequency of the tuned circuit. The sensitivity of such a device is proportional to the circuit Q. We have constructed a test of this device in a vertical configuration as a gravimeter. Our preliminary sensitivity, using a number of nonsuperconducting circuit elements and hence a greatly reduced Q, agrees well with calculations. We have at present a sensitivity to motion of several angstroms amplitude at frequencies above 3 Hz. Other workers have obtained Q's of 10^5 in similar circuits using all-superconducting elements, and with such a Q our sensitivity should be better than 10^{-17} cm according to calculation.

We have now constructed two identical sets of apparatus. The first, which we are presently assembling for low-temperature test at MTF, will then be set up at LSU. The second apparatus will be assembled at Stanford. The length changes of the bar will be monitored at both stations and stored on magnetic tape. Correlations between the two bars will be analyzed by computer programs developed by the Mississippi Test Facility. A similar apparatus is also being constructed at the University of Rome and correlations will be investigated among all three stations.

Acknowledgments

It is a pleasure to acknowledge many fruitful conversations with W. Montgomery, S. Kleve, T. Bernat, G. Pizzella, I. Modena, and the entire Stanford group.

References

1. J. Weber, *Phys. Rev. Lett.* **22**, 1302 (1969).
2. W.O. Hamilton, *Bull. Am. Phys. Soc.* **15**, 1361 (1970).
3. G.J. Ehnholm and R.G. Gylling, *Cryogenics* **11**, 39 (1971).

Low-Temperature Gravity Wave Detector*

J. E. Opfer, S. P. Boughn, W. M. Fairbank, M. S. McAshan, H. J. Paik, and R. C. Taber

Department of Physics, Stanford University
Stanford, California

We are engaged, in cooperation with Hamilton *et al.*,[1] in a program to develop and build gravitational radiation detectors which can be cooled to extremely low temperatures. Cooling the detectors, which utilize massive aluminum bars as antennas, improves the sensitivity of detection by reducing thermal noise in the antennas and by allowing the use of low-noise superconducting electronics. In addition, a superconducting system can be used to support the antennas to reduce vibration-induced noise. Because cooling a detector to liquid helium temperatures already gives substantial improvement over performance at room temperature, we are constructing a prototype detector to operate at 1°K. This prototype detector will be a sensitive gravity wave detector as well as an aid in the development of detectors operating at much lower temperatures.

In Fig. 1 the gravity wave antennas and the accompanying support and isolation system for the prototype detector are shown schematically. The antennas are two 750-kg aluminum bars 40 cm in diameter and 210 cm long, each having a cross section for absorption of gravitational radiation approximately half that of the antennas used by Weber.[2] Each bar is coated with a superconducting shield and is supported by a magnetic field maintained by a persistent current in the support coils. Horizontal motion of the bars is reduced by a set of balance coils. The low-temperature system is shown supported by means of a nest of springs and acoustic filters made of Isomode pads and steel blocks. Shielding against electromagnetic radiation is provided by a lead shield surrounding the liquid helium container.

Low-Temperature System

The low-temperature system provides a 21-in. by 20-ft-long space surrounded by an annular-shaped liquid helium container. Radiation heat leaks are reduced to 0.1 W by a helium-gas-cooled shield insulated from a liquid-nitrogen-cooled shield by layers of aluminized Mylar. The support cables contribute an equal heat leak, resulting in a 0.2 W steady-state heat leak into the liquid helium.

Support System

The magnetic support system for the antenna consists of a set of persistent superconducting coils and a layer of superconducting Nb–Ti on the surface of the

* Work supported by the National Science Foundation.

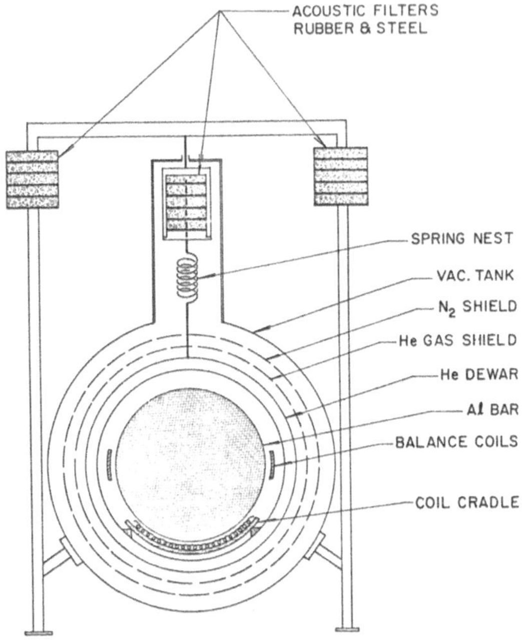

Fig. 1. Schematic diagram of the support and isolation system for a low-temperature gravity wave detector.

aluminum bar. Interaction of the current in the coils with image currents in the Nb–Ti provides the supporting force. Because the coils maintain constant magnetic flux linking them, increasing the height of the bar decreases the currents flowing in the support coils and, consequently, the supporting force, until at the equilibrium height the support force balances the weight of the bar.

Extensive studies of the support system have been made on an aluminum bar 9 in. long and 6 in. in diameter. This bar and the support system are shown in Fig. 2. Piezoelectric crystals are attached to the bar, which is coated with a 0.015-in. layer of plasma-sprayed Nb–Ti. This assembly is inserted into a cradle which is fitted with a persistent levitation coil made of 0.025 × 0.050-in. copper-clad Nb–Ti wire. Another coil is mounted above the bar to simulate the effect of a heavier bar. These studies showed that the system was capable of supporting a 16-in.-diameter bar with a considerable safety margin. Further, the mechanical Q of the bar while supported was observed to be unaffected by the Nb–Ti layer. At low temperature the observed Q in a vacuum environment was 6×10^5.

The magnetic support just described provides a number of important advantages over other means of support. Excellent mechanical isolation from vibrations at the 1200-Hz resonant frequency of the bar is achieved because of the low resonant frequency (approximately 5 Hz) of the support system and the absence of a direct acoustical path for higher frequency vibrations. Because the support forces are distributed along the bar, there is a relatively small coupling between the support coils and the bending modes of the bar. Distributing the support forces also minimizes acoustic emission arising from stresses induced in the bar by the support.

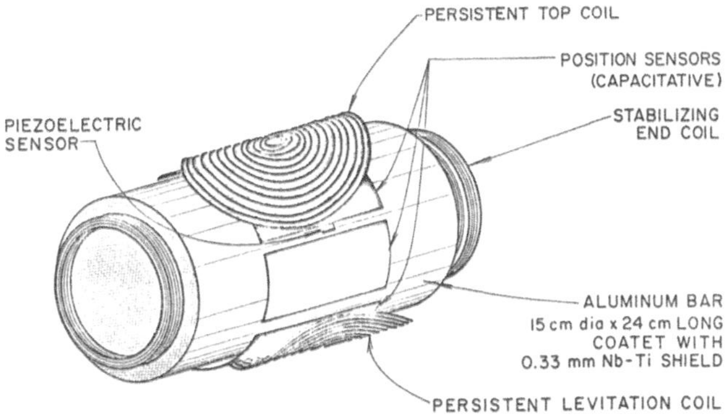

Fig. 2. Magnetic levitation test system.

Finally, the magnetic support system provides thermal isolation to allow further cooling of the detector.

Accelerometer

In order to detect excitation of the aluminum bar antenna by a gravity wave, a resonant accelerometer is used to measure the acceleration of the end of the bar. A schematic diagram of the accelerometer, which makes extensive use of superconducting materials and electronics, is shown in Fig. 3. In the operation of the accelerometer, the position of a circular superconducting niobium diaphragm relative to an outer support ring is indicated by an inductance change of the two flat coils L_1 and L_2 situated near the diaphragm. These coils are connected together and have a large persistent current I_0 flowing through them. Inductance changes in L_1 and L_2 are observed by the accompanying change in magnetic flux in a third coil L_3 connected across L_1 and L_2. In turn this flux change is measured with a superconducting magnetometer. High sensitivity is achieved by storing large amounts of magnetic energy in the region near the diaphragm and by utilizing a sensitive superconducting magnetometer to measure flux changes. Because the persistent current in L_1 and L_2 contributes substantially to the restoring force on the diaphragm, varying this current allows an exact tuning the diaphragm to the frequency of the bar.

Although a complete discussion of the response of the gravity wave detector is beyond the scope of this paper, it is possible to consider two important properties

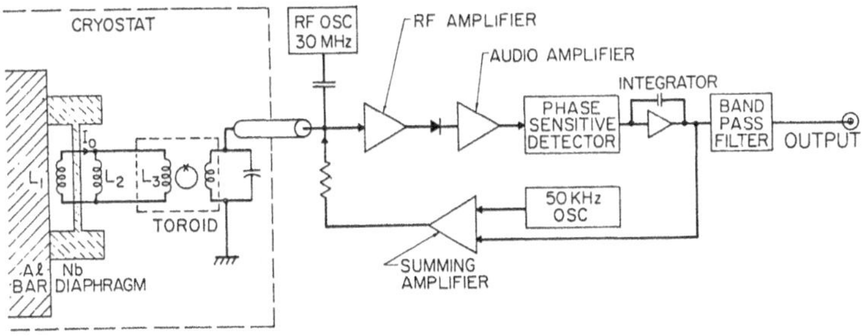

Fig. 3. Accelerometer used to detect motion of a gravitational wave antenna.

of the system. The first of these is the sensitivity of the accelerometer to an energy of excitation of the bar, and the second is the level of excitation of the bar caused by noise in the accelerometer. To determine the first property, it should be noted that the accelerometer and the antenna on which it is mounted together make up a coupled oscillatory system which allows transfer of energy back and forth between the antenna and the diaphragm. If the resonant frequency of each of the two taken separately is exactly the same, then when they are coupled the excitation energy of the antenna can be completely transferred to the diaphragm after a time equal to half the period of the beat frequency of the coupled system. In terms of the uncoupled resonant frequency ν and the effective masses m and M of the diaphragm and antenna, respectively, this beat frequency is $\nu(m/M)^{1/2}$. The corresponding maximum amplitude of motion of the diaphragm is $(M/m)^{1/2}$, or typically 500 times greater than that of the bar. If now a fraction γ of the stored energy of the diaphragm resides in the magnetic field, a proper choice of L_3 allows $\gamma/2$ times the stored energy to be transferred into and out of the coil L_3 at the diaphragm resonant frequency. In terms of the stored energy equal to a thermal energy kT, the corresponding maximum magnetic field in L_3 is $(4\pi\gamma kT/V)^{1/2}$, or 2×10^{-8} G for a coil volume $V = 1\ \text{cm}^3$, $T = 1°\text{K}$, and $\gamma = 0.2$. This value of γ is readily realizable with a persistent current of 5 A in coils L_1 and L_2 spaced 10^{-2} cm from the superconducting diaphragm. Such signal levels are readily detectable by means of a superconducting magnetometer with a sensitivity of 10^{-9} G in a 1-Hz bandwidth.

In order to determine the expected level of excitation of the bar caused by noise in the accelerometer, it is important to note that excitation of the diaphragm results in excitation of the bar by an equal amount at a later time. It is sufficient then to consider only excitation of the diaphragm itself by electronic noise in the magnetometer system. If the effect of operating the magnetometer under closed-loop conditions is to introduce a noise current into the input circuit as indicated by Giffard *et al.*,[3] then it can be shown that excitation of the diaphragm is negligible for values of noise current measured by them. In order to resolve this question directly, we are performing an experiment designed to observe thermally excited motion in the diaphragm, using the magnetometer readout shown in Fig. 3. Except for problems of spurious vibrations and flux noise, this experiment is equivalent to an observation of Johnson noise in an *RLC* circuit resonant at frequency ν that is coupled inductively to a magnetometer.

Conclusion

A prototype low-temperature gravity wave detector is being developed which will have improved performance over room-temperature detectors both because of reduced thermal noise in the antennas and reduced electronic noise. Construction of the detector and continuing studies of the noise properties of the system are expected to give information concerning the limit of performance achievable by further cooling of the detector.

References

1. W.O. Hamilton, P. B. Pipes, and P. S. Nayar, this volume.
2. J. Weber, *Phys. Rev. Lett.* **22**, 1320 (1969).
3. R.P. Giffard, R.A. Webb, and J.C. Wheatley, *J. Low Temp. Phys.* **6**, 533 (1972).

Large-Diameter Superconducting Quantum Interference Devices for a Measurement of h/m*

C. M. Falco and W. H. Parker

Department of Physics
University of California, Irvine, California

We report on progress made toward a precision measurement of the fundamental physical constant h/m, where m is the rest mass of the free-electron, using the macroscopic phase coherence of superconductors. Previous measurements using a 2-mm-diameter superconducting quantum interference device (SQUID) demonstrated the possibility of using a rotating ring to measure h/m.[1] However, our objective of a measurement with an accuracy of a few parts per million will require SQUIDs of much greater diameter. The work reported here is directed toward developing and studying the behavior of large-diameter SQUIDs.

The measurement of h/m is possible because the phase-coherent superconductor wave function is single-valued and leads to the quantization condition

$$\oint \mathbf{p} \cdot d\mathbf{l} = nh \tag{1}$$

where $\mathbf{p} = m^*\mathbf{v} + e^*\mathbf{A}$. The quantities m^* and e^* are, respectively, the mass and charge of the fundamental charged "particle" of the superconducting state. In the absence of any relativistic corrections it can be shown that $m^* = 2\text{m}$. If the quantization condition is applied to a thin superconducting ring (as shown in Fig. 1) and the ring rotated about its axis at an angular frequency ω, Eq. (1) gives

$$nh = 2m^*\omega S + e^*\Phi \tag{2}$$

or

$$\Phi/\Phi_0 = n - (2m^*\omega S/h) \tag{3}$$

where S is the area enclosed within the ring, Φ is the magnetic flux enclosed by the ring, and $\Phi_0 = e^*/h$. In a bulk superconducting ring the integer n is fixed, thus prohibiting a measurement of the spacing between adjacent states and h/m. However, if the superconducting ring is suitably weakened at some point, transitions to adjacent states become possible. These transitions manifest themselves as changes in the effective rf impedance of the ring, which becomes periodic in the quantity $2m^*\omega S/h$. Measurements of the periodicity of this impedance along with S and ω allow a precision determination of h/m. A block diagram of the electronic system used to detect this periodicity is shown in Fig. 1.

* Research supported by the National Science Foundation and National Bureau of Standards.

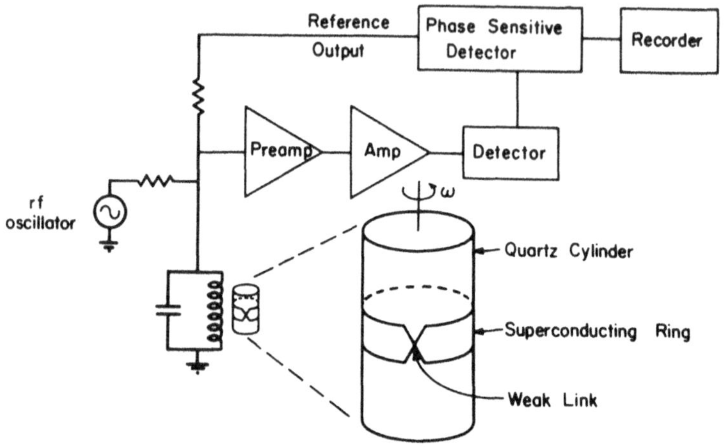

Fig. 1. Block diagram of the electronic system used to sense the current in a superconducting ring containing a weak link.

In order to observe quantum effects in large-diameter SQUIDs, the critical current of the weak link must be small over a relatively broad temperature range. Of the many methods of weakening the superconducting ring we have investigated, the most suitable is the Notarys–Mercereau technique of normal metal underlays.[2] Our work uses Au–Sn weak links typically 5 μm wide and 2 μm long. The advantage of these devices is that the critical current of the weak link remains less than 1 μA over a temperature range as large as 0.35°K, as shown in Fig. 2. The need for small critical currents was demonstrated by the analog computer studies of the circuit in Fig. 1, which predicted the best signal would be obtained when the critical current I_c is approximately Φ_0/L, where L is the inductance of the SQUID.[3] This prediction was tested by measuring the signal from a superconducting ring of diameter 6.9 mm and length 27.8 mm. The ring was then cut, current and voltage leads attached, and the critical current measured as a function of temperature. The result is shown in Fig. 2. From other studies reported here the inductance of a ring of this size is 2.75 $\pm$ nH. The criterion $I_c \cong \Phi_0/L$ implies that the maximum signal should occur at $I_c = 0.76 \pm 0.07$ μA. From Fig. 2, the maximum signal occurs when the critical current of the weak link is 0.77 ± 0.05 μA, in agreement with the predicted value. Using the data of Fig. 2, it is possible to plot the signal against $I_c/(\Phi_0/L)$. The result is in good agreement with the behavior found in the analog studies except that the observed signal decreases for $I_c > \Phi_0/L$ rather than remaining constant as predicted.

A simple argument based on the energy dissipated in the SQUID suggests that the detected signal should be inversely proportional to the inductance of the ring.[4] By observing the maximum signal from one of these devices as the length of the ring was repeatedly reduced (and consequently the inductance increased) the L^{-1} behavior of the signal was demonstrated. In order to obtain the straight line in Fig. 3 the inductance of the SQUID has been taken as $L_{\text{ring}} + L_{\text{weak link}}$, where L_{ring} is the calculated inductance of a circular ring with the dimensions of the SQUID and $L_{\text{weak link}}$ is typically 1.0–1.5 nH. Extrapolation of this curve to

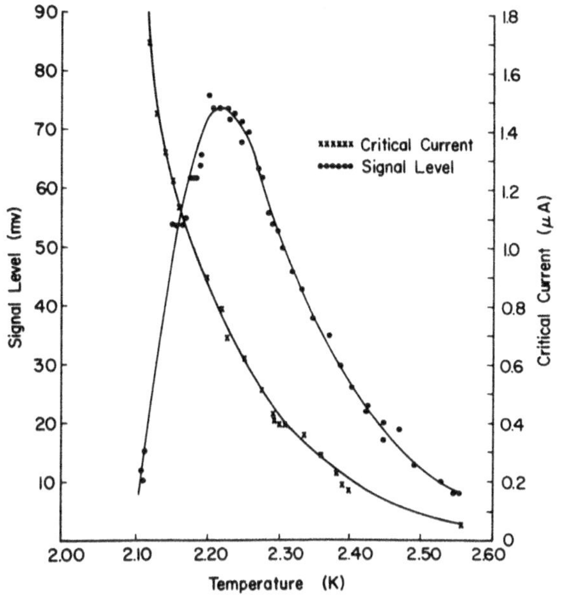

Fig. 2. Detected signal and critical current of a 6.9-mm-diameter SQUID as a function of temperature.

zero signal gives a value for the maximum inductance and hence the maximum diameter of rings that can be used in the h/m measurement.

The maximum inductance from Fig. 3 of approximately 15 nH can be understood on the basis of the intrinsic noise of the weak link. Measurements of the current–voltage characteristic of weak links indicate that the minimum critical current not completely obscured by the intrinsic noise is approximately 2×10^{-7} A.[5] This

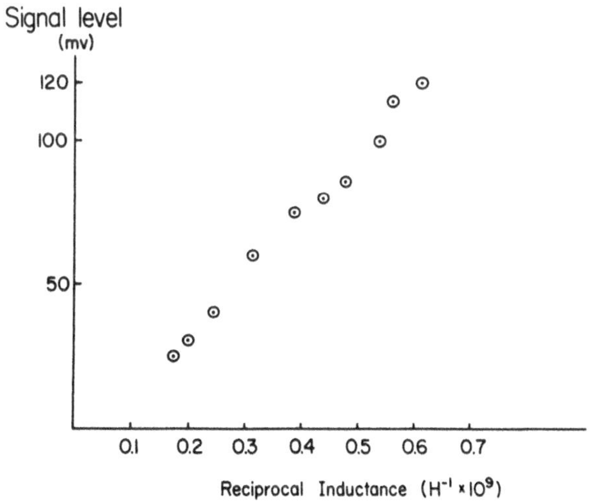

Fig. 3. Detected signal from a 2.9-mm-diameter SQUID as a function of reciprocal inductance.

result, together with the criterion for best signal, $I_c \cong \Phi_0/L$, predicts $L_{max} \cong \Phi_0/I_c(\min) \cong 10^{-8}$H, in agreement with the observed limit.

It would appear that the primary consideration in obtaining useful signals from large-diameter SQUIDs is the inductance. The geometric size of the ring should not be an important factor provided the inductance is less than approximately 10 nH. To date we have successfully operated rings 1 cm in diameter with a signal-to-noise ratio sufficient to resolve a quantum phase change of $2\pi/1000$ with a 1-sec averaging time. This corresponds to a magnetic field resolution of 2.5×10^{-10} G.

We have also investigated methods of reducing the inductance of rings of fixed length. One method which gives increased signals is the insertion of either superconducting or high-conductivity normal metal plugs into hollow tubes on which the rings have been fabricated. Using this technique with a 1.0-cm SQUID fabricated on a tube with 1-mm walls, the signal was increased by 100%. This corresponds well with the expected increase due to the estimated decrease in the inductance. Using a technique such as this, it is possible to envision using a 2-cm diameter (or larger) ring in a measurement of h/m.

Previous studies have indicated that the coupling of the SQUID to the coil of the resonant circuit affects the detected signal level. The analog studies indicated that the maximum signal should be obtained for $k^2Q \approx 1$, where $k = M/(L_{ring}L_{coil})^{1/2}$ and Q is the quality factor of the resonant circuit. Other studies have found $k^2Q \approx \pi/2$ to be the optimal condition.[6] Using a 1-cm-diameter ring, we have measured the signal as a function of k^2Q over the range 0.56–6.5. The parameter k^2Q was varied by partially withdrawing the superconducting ring from the coil of the resonant circuit. The coupling constant k was calculated from the shift of the resonant frequency and Q was calculated from the frequencies of the half-power points. The detected signal monotonically increased with increasing k^2Q, and reached a maximum for $k^2Q \gtrsim 5$. The maximum signal is 1.7 times as large as at $k^2Q = \pi/2$ and 2.4 times as large as at $k^2Q = 1$.

SQUIDs incorporating a Sn–Au underlay weak link are very stable under temperature cycling. We have cycled such devices between room temperature and 2°K more than ten times over a period of several months with no observable effect on the behavior of the device. This durability makes them particularly convenient to use.

The use of larger-diameter rings in a more precise determination of h/m will require the reduction of both spatial and temporal variations in the ambient magnetic field to a lower level than achieved at present. We anticipate that this will not be difficult. In our present system the ambient magnetic field is reduced to less than 3 μG over a 55-cm-long section of the cryostat by use of a single room-temperature μ-metal shield of 93 cm length, 17 cm diameter, and 1.3 mm wall thickness. The current through 24-cm-diameter coils placed coaxially 14 cm from each end of the shield is adjusted for minimum field. The remaining field is "frozen" by a Sn superconducting shield. Tin is used since the temperature can be slowly and uniformly lowered through the superconducting transition. The use of a superconducting shield, however, does not ensure temporal stability of the magnetic field. We have found that magnetic flux occasionally moves within the superconducting shield for unidentified reasons.

For our present determination of h/m we have deposited 400-Å-thick Sn rings incorporating a Sn–Au weak link on 3-mm-diameter fused silica rods. The cross-sectional area at room temperature has been determined for us by the National Bureau of Standards.* The one-standard-deviation uncertainty in the area measurement is 80 ppm. The area at 2°K is obtained by correcting the room-temperature area by the 80 ± 20 ppm total linear thermal expansion of fused silica.[7] The uncertainty arises because the total thermal expansion can vary by 20 ppm between samples manufactured under slightly different conditions. The fused silica rod is rotated by a synchronous motor driven by a power amplifier and a stable oscillator at a frequency of approximately 40 Hz. At this rotation rate the quantum phase change for a 3-mm-diameter ring is approximately 20π. This change is doubled by measuring between $-\omega$ and ω.

The best numerical value for h/m obtained from measurements on a 3-mm-diameter rotating superconducting ring and its one-standard-deviation uncertainty is $h/m = (7.2759 \pm 0.0016) \times 10^{-4}$ J sec kg^{-1}, in agreement with the presently accepted value.[8] The 0.022% uncertainty is the root sum square of 0.002% uncertainty in the rotation rate, 0.02% in the quantum phase change, and 0.009% in the low-temperature ring area.

The ultimate accuracy to which h/m can be measured with rotating superconductors will be determined by the accuracy of the area measurement. The results reported here clearly indicate that rings as large as 1 cm, if not larger, can be used. Techniques under development at the National Bureau of Standards should make it possible to measure the area of a 1-cm-diameter high-quality fused silica cylinder to an accuracy of approximately 4 ppm. We thus expect that the present practical limit to the accuracy of a determination of h/m utilizing the periodic response of the current in a rotating superconducting ring containing a single weak link is 5–6 ppm.

References

1. W.H. Parker and M.B. Simmonds, in *Proc. Intern. Conf. on Precision Measurement and Fundamental Constants*, D.N. Langenberg and B.N. Taylor, eds., National Bureau of Standards, Special Publication 343, U.S. Government Printing Office, Washington, D.C. (1971), p. 243.
2. H.A. Notarys and J.E. Mercereau, in *Proc. Intern. Conf. on the Science of Superconductivity*, F. Chilton, ed., North-Holland, Amsterdam (1971); *Physica* **55**, 424 (1971).
3. M.B. Simmonds and W.H. Parker, *J. Appl. Phys.* **42**, 38 (1971).
4. J.E. Zimmerman, *J. Appl. Phys.* **42**, 4483 (1971).
5. M. Simmonds and W.H. Parker, *Phys. Rev. Lett.* **24**, 876 (1970).
6. R.P. Gifford, R.A. Webb, and J.C. Wheatley, *J. Low Temp. Phys.* **6**, 533 (1972).
7. D.F. Gibbons, *J. Phys. Chem. Solids* **11**, 246 (1959).
8. B.N. Taylor, W.H. Parker, and D.N. Langenberg, *Rev. Mod. Phys.* **41**, 375 (1969).

* We would like to thank Dr. J. A. Simpson of the National Bureau of Standards for arranging the necessary measurements.

Frequency Pulling in Josephson Junctions

B. T. Ulrich

Department of Physics
University of Texas, Austin, Texas

The voltage V across a Josephson junction and the current I through the junction obey the equations $V = (\hbar/2e)\,d\varphi/dt$ and $I = I_0 \sin\varphi$, where φ is the quantum phase difference across the junction. When $V \neq 0$ an oscillating current flows in the junction, and this current emits electromagnetic radiation. The time average of the voltage equation $\langle V \rangle = (\hbar/2e)\langle d\varphi/dt \rangle = (\hbar/2e)\langle \omega \rangle$ relates the average voltage across the junction to the average time derivative of the phase. This paper describes an experiment in which a junction coupled to a microwave cavity locks to a mode of the cavity and emits *most* of its radiation at the cavity resonant frequency ω_c, which may differ from $\langle \omega \rangle$ by as much as 1%. The "frequency pulling" effect described here refers to the observation that the frequency of the emitted radiation differs from the frequency $\langle \omega \rangle$ that would be determined by $\langle V \rangle$. This result is interpreted in terms of a parametric relaxation mechanism in which the phase difference across the junction progresses at the cavity frequency ω_c most of the time and then periodically relaxes rapidly to maintain the average angular frequency, $\langle d\varphi/dt \rangle = (2e/\hbar)\langle V \rangle$, with the relaxation taking place rapidly compared to the time during which the phase progresses at the cavity frequency. During the entire process the detailed voltage–phase relationship $V = (\hbar/2e)\,d\varphi/dt$ is valid at all times. [A frequency pulling effect in which V differs from $(\hbar/2e)\,d\varphi/dt$ has been shown by McCumber[1] not to occur.]

The implications of the present experiments are the following: $2e/\hbar$ is usually measured by impressing microwave radiation of a known frequency on the junction and by measuring the voltage of the induced constant-voltage steps in the current–voltage characteristic. One might imagine that $2e/\hbar$ could be measured by the inverse experiment in which voltage is applied across the junction and the frequency of the emitted radiation is measured. The present experiments show that the inverse experiment would be difficult because the frequency of the emitted power could be influenced by the modes of a surrounding cavity such that the emitted power would be at a cavity frequency rather than at $2e\langle V \rangle/\hbar$. Also, the present experiments show that in the measurement of $2e/\hbar$ in which external microwave radiation is impressed on the junction the amplitude of this radiation must be chosen sufficiently large that phase slip at another frequency is not possible.

In these experiments the wavelength of the primary power component of the radiation emitted by a Josephson junction consisting of a niobium point contact on superconducting tin was measured with a Fabry-Perot interferometer external to the dewar.[2] The point contact was tightly coupled to a surrounding coaxial

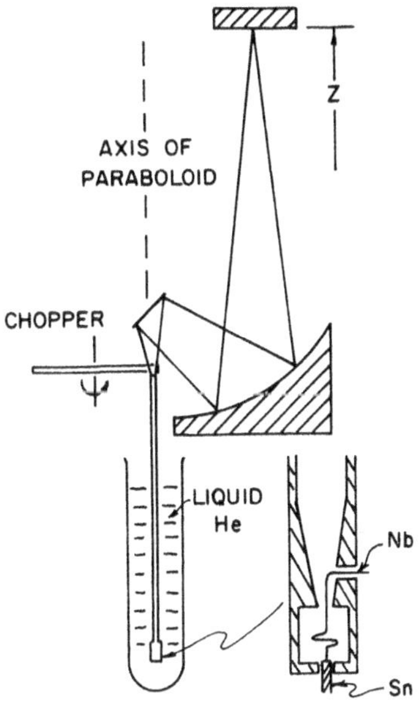

Fig. 1. Optical arrangement.

cavity, which was weakly coupled to an oversize waveguide leading out of the dewar, as shown in Fig. 1. The exit of the waveguide was reimaged by reflecting optics onto a mirror (called the "end mirror") which formed the far end of an external Fabry-Perot cavity. The end mirror reflected the microwave radiation back to the same junction that had emitted the radiation and this junction detected the reflected radiation in a homodyne detection mode.[3] The amplitude and phase of the reflected wave was varied by varying the position Z of the end mirror. The resulting change in the current–voltage characteristic of the junction was detected at dc by biasing the junction through a 19-Ω source impedance. The current–voltage characteristics of the junction showed step structure at the voltages corresponding to cavity resonances. The dynamic resistance at the bias point just above the step at 441 μV was 50 Ω, and the change in the dc bias voltage as a function of the end mirror position was detected by a voltage amplifier. This voltage is shown in Fig. 2 as a function of end mirror position Z. For the data presented the end mirror was at an 1015 mm optical path distance from the junction, and the wavelength of the emitted radiation, as determined from twice the end mirror displacement for one fringe period was 1.41 mm.

The primary power component wavelength was found to be at a cavity mode wavelength, and independent of $2e\langle V\rangle/h$ for variations of 1.7% away from the cavity frequency. This conclusion arises from the following consideration: The peak-to-peak variation in the dc bias voltage across the junction over a fringe period is 15.4 μV out of the bias voltage of 441 μV. If the frequency of the emitted radiation were proportional to the voltage, this bias variation would correspond to a 1.7% variation in the wavelength, and several full wavelengths would sweep past the junction during the movement of the mirror shown. The position of the mirror as

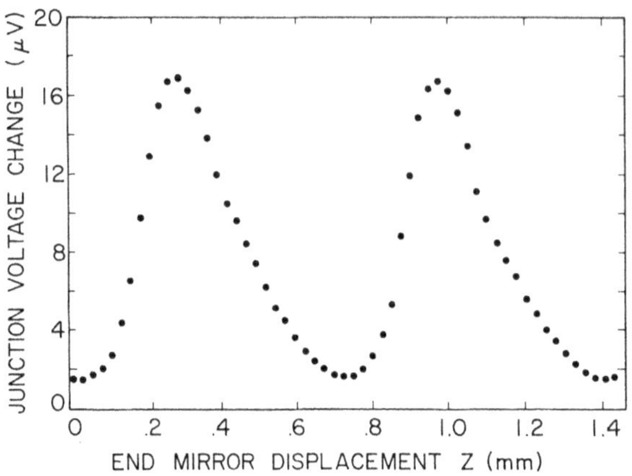

Fig. 2. Interference fringes showing that emitted wavelength is nearly independent of junction bias voltage.

a function of the voltage across the junction would be given by the expression

$$Z = \frac{\lambda_0}{2} \frac{V_0}{V_0 + v} \left[\frac{1}{2\pi} \cos^{-1}\left(\frac{v}{V_0} \right) - \frac{2Z_0}{\lambda_0} \frac{v}{V_0} \right]$$

where Z_0 is the one-way junction-to-end-mirror distance, Z is the deviation from Z_0, V_0 is the junction bias voltage for wavelength λ_0, and v is the deviation from V_0. Since in fact this rapid (and multiple-valued) variation of v as a function of end mirror position Z is not observed, it is concluded that the average wavelength of most of the power emitted by the junction does not vary linearly with the voltage across the junction. The next simplest hypothesis to make is that most of the power is emitted at a cavity mode frequency of the tightly coupled cavity. Microwave emission was observed only at nine discrete wavelengths between 1.1 and 2.6 mm, which could be selected by changing the bias voltage for the cavity geometry employed. No attempt has been made to calculate the wavelengths of these modes because the geometry is complicated by the S bend in the niobium wire.

The following mechanism is proposed for the "frequency pulling": The phase $\varphi(t)$ progresses approximately as $\varphi = \omega_c t$ until it differs significantly from $2e\langle V \rangle t/\hbar$. Then φ changes abruptly to catch up to $2e\langle V \rangle t/\hbar$. This abrupt change takes much less time than the time spent progressing at $\omega_c t$. The abrupt phase change corresponds to a voltage pulse across the junction, which brings the time average of the phase up to $2e\langle V \rangle t/\hbar$. This abrupt "catching up" must occur sufficiently often that the voltage pulse across the junction needed to bring the time average of $(h/2e)\, d\varphi/dt$ to the average bias voltage does not become too large: the voltage must be smaller than the energy gap of the superconductors.

In these experiments the frequency of the voltage pulses causing the abrupt changes in phase was determined by a delay line connected to the junction. The delay line had a high impedance at its other end, so that the voltage pulse reflected back to the junction would have the same sign as the voltage pulse leaving the junction. The frequency of the abrupt changes was 200 MHz, as determined by the round trip transit time of the delay line. The 200-MHz voltage due to this parametric oscillation has been observed directly by connecting a 200-MHz amplifier with high input impedance to the end of the delay line opposite to the junction.[4]

References

1. D.E. McCumber, *Phys. Rev. Lett.* **23**, 1228 (1969).
2. B.T. Ulrich and E.O. Kluth, in *Proc. 1972 Conf. on Applied Superconductivity*, IEEE Conference Record, 72 CHO 682-5 TABSC.
3. B.T. Ulrich and E.O. Kluth, *Proc. IEEE* **61**, 51 (1973).
4. B.T. Ulrich, this volume.

The Flux Period of Josephson Junction Interferometers*

D. L. Stuehm and C. W. Wilmsen

Colorado State University
Fort Collins, Colorado

Introduction

The use of Josephson junction interferometers[1] for the measurement of very small magnetic fields and the absolute value of magnetic fields depends, respectively, on accurately knowing the flux period of the interference pattern and the location of zero applied field. In addition, the shape of the interference pattern becomes complex when multiple-junction interferometers are used. This paper presents a qualitative description of the flux period of two-junction interferometers and numerically computed interference patterns for three-junction interferometers. The numerical technique solves the Josephson equations subject to the boundary conditions and takes into consideration the phase variations across the junction due to the applied and self-induced magnetic fields in the junctions. The details of the technique are described elsewhere.[2]

Two-Junction Interferometer

Why the flux period of a two-junction interferometer is less than the flux quantum can be qualitatively understood from Fig. 1. If a flux quantum threads the superconducting loop between the two junctions, then integration of the vector potential around the loop indicates that there exists a phase shift of 2π rad between the inside edges of the two junctions. If there is no phase change across the junctions, then the current through both junctions is a maximum and a peak in the interference pattern occurs one flux quantum from the zero flux peak. However, for any finite-length junction the pair phase across the junction is not constant but is related to the local field by

$$d\phi(y)/dy = (2ed/\hbar c)\, H(y) \tag{1}$$

where $d = 2\lambda_L + T$, λ_L is the London penetration depth, and T is the insulator thickness. Considering junction phase, the second interferometer current peak occurs at a flux somewhat less than the flux quantum after the first, as indicated in Fig. 1. These facts lead to the conclusion that the period of the interference pattern is a function of the junction lengths in addition to the interference loop area.

The geometry of a Josephson junction interferometer is shown in Fig. 2. If one includes one-half the area of each junction in the interference loop area, the flux

* Supported by NASA, Grant No. NGR 06-002-094

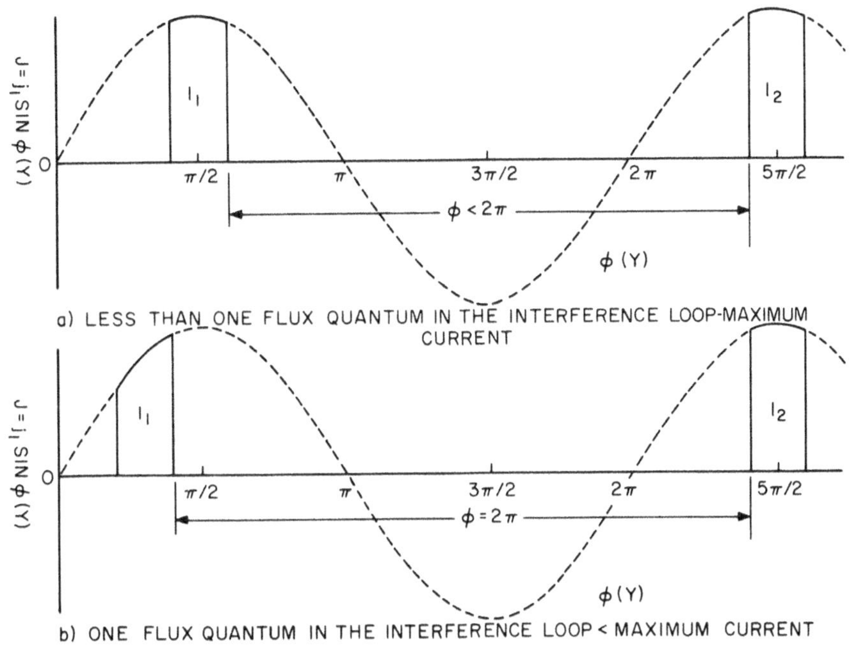

Fig.1. Phase variation for a two-junction interferometer for two values of applied flux.

period of the interferometer is one flux quantum. This result, obtained numerically, takes into consideration the phase change across the junctions due to the applied and self-induced fields in the junctions.

Three-Junction Interferometer

Figure 3 illustrates three computed interference patterns for the three-junction, symmetric-feed interferometer. This demonstrates the features of the three-junction

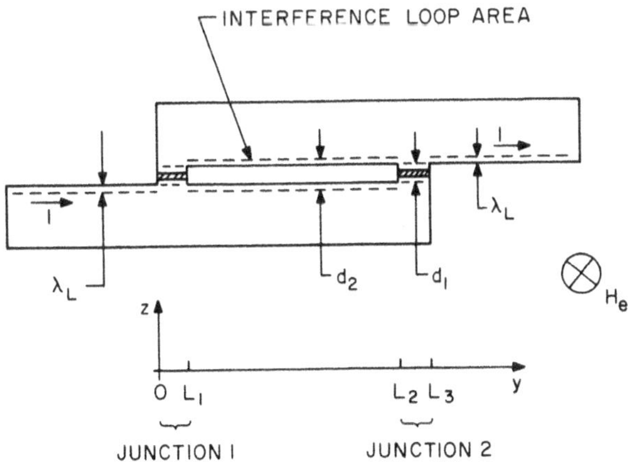

Fig. 2. Geometry of a two-junction, symmetric-feed interferometer.

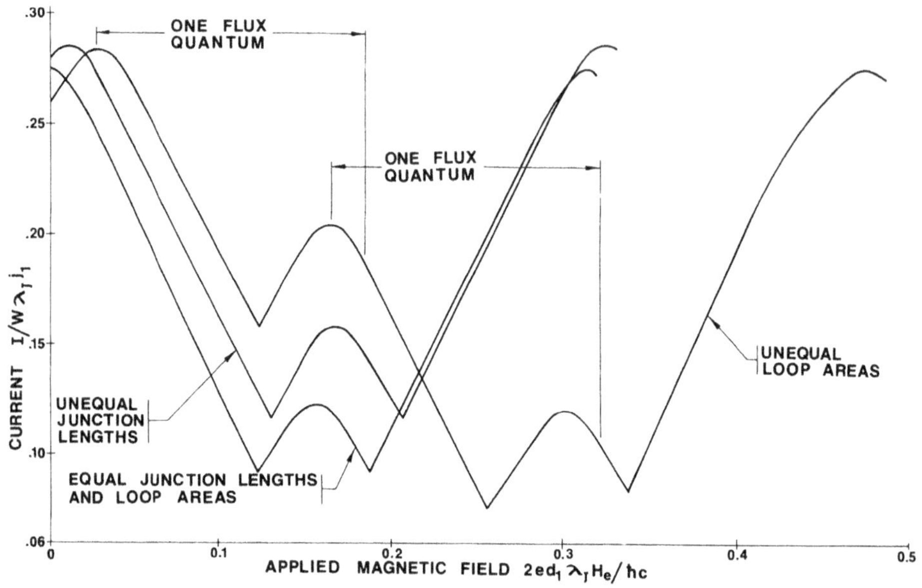

Fig. 3. The interference patterns of three three-junction interferometers.

pattern and how it is affected by unequal junction lengths and unequal loop areas. The high peaks are separated by less than two flux quanta, if one neglects the junction areas. By including the central junction area plus one-half of each end junction area, the flux period is two flux quanta. Note that the first peak occurs at zero applied field.

If the junction lengths are unequal, the first high peak does not occur at zero applied field, because the unequal currents produce a net flux in the interference loops at zero applied field. This net flux must be cancelled by an applied field in order for maximum current to flow. The flux period is not affected by unequal junction lengths.

When the two loop areas are unequal the interference pattern changes considerably. The amplitude of the first peak increases and occurs at nonzero applied field. The second peak (low) increases in amplitude and shifts toward the first peak. The third peak changes from a high peak to a low peak. Thus the pattern repeats after two lower peaks and has a period of approximately $3\Phi_0$.

In summary, the flux period of two- and three-junction interferometers is a multiple of the flux quantum $hc/2e$ if one includes in the interference loop area part of the junction area. Only in the case of unequal interference loop areas does the separation between adjacent current maxima differ from the flux quantum.

References

1. R.C. Jaklevic, J. Lamb, J.E. Mercereau, and A.H. Silver, *Phys. Rev.* **140,** A1628 (1965).
2. D.L. Stuehm and C.W. Wilmsen, *J. Appl. Phys.* **43,** 3588 (1972).

Magnetic Field and Q Dependence of Self-Induced Current Peaks in Resonant Josephson Tunnel Junctions

Klaus Schwidtal and C. F. Smiley*

Electronics Technology and Devices Laboratory (ECOM)
Fort Monmouth, New Jersey

The most general theory of the self-induced current "steps" in the dc I–V characteristics of resonant Josephson tunnel junctions reported thus far has been developed by Kulik.[1] In contrast to previous theories,[2-4] Kulik makes no restrictive assumptions other than that the junction length L is small with respect to the Josephson penetration depth λ_J. This is the very same assumption made in deriving the Fraunhofer diffraction equation for the magnetic field dependence of the maximum dc Josephson current I_0. We will therefore assume that Kulik's theory[1] is applicable whenever the measured $I_0(H_e)$ curve can be described by the Fraunhofer diffraction equation, which is practically the case for all rectangular junctions with a uniform oxide barrier layer, and $L/\lambda_J \leq 2$.

The magnetic field dependence of the nth current step can be written in the form

$$I_n = I_n^{\max} F(n, \Phi/\Phi_0)$$

where $F(n, \Phi/\Phi_0)$ is a semiperiodic function of the magnetic flux Φ in the junction, with a period of Φ_0 and a principal maximum $F(n, \Phi/\Phi_0) = 1$ at $\Phi/\Phi_0 = n/2$. (Φ_0 is the quantum unit of magnetic flux.) $I_n^{\max}$ is a function of the mode number n, the reduced length L/λ_J, and the quality factor $Q_n = \omega_n/\Delta\omega$. The term Q_n has been introduced in the theory to represent the losses in the nth mode; Q_n^{-1} is proportional to the linewidth $\Delta\omega$. But while in previous theories[2-4] $I_n^{\max} = I_0\alpha$, with

$$\alpha = (L/2\pi n\lambda_J)^2 Q_n$$

Kulik's $I_n^{\max}$ is a more complex function of α, with $I_n^{\max} \to 0$ for $n \to \infty$ *and* for $n \to 0$. This function has a maximum at $\alpha = 1$, which is

$$(I_n^{\max})^{\max} = 0.34\, I_0^{\max}$$

To the best of our knowledge, no systematic experimental test of Kulik's theory[1] has previously been made. This may be due to the fact that a quantitative comparison of measured $I_n(H_e)$ curves with the theory requires Q_n to be known, which it is not. But one can look at the problem the other way around. Kulik's theory[1] allows one to obtain Q_n data from measured $I_n(H_e)$ curves; this is much easier to do than to measure $Q_n = \omega_n/\Delta\omega$ directly. It then remains to show that the resulting $Q_n(n)$ functions are reasonable. It is in this way that we have proceeded.

* Present address: Gates Radio Co., Quincy, Illinois.

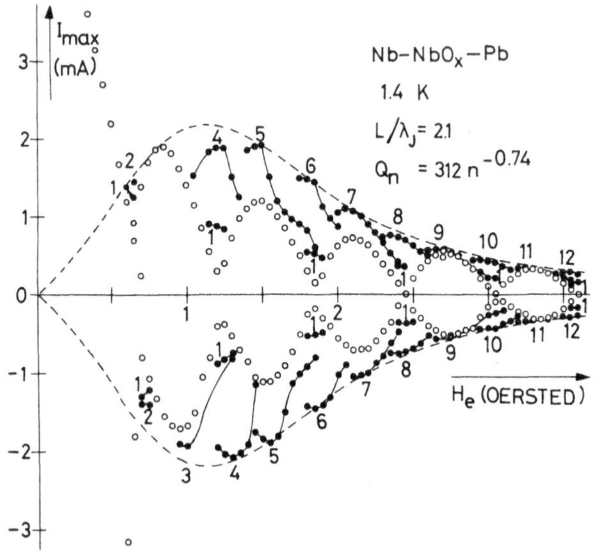

Fig. 1. Magnetic field dependence of I_0 (open circles) and I_n (full circles; the numbers are n) for a Nb–Pb junction, measured "asymmetrically" in the usual four-probe configuration. The dashed line is I_n^{max} vs. H_e, calculated from Kulik's theory[1] with $Q_n = 312n^{-0.74}$, $L = 0.3$ mm, $w = 0.3$ mm.

We have measured the magnetic field dependence of I_0 and I_n for cross-type Pb–PbO_x–Pb and Nb–NbO_x–Pb thin-film Josephson tunnel junctions with $L = 0.3$ mm and L/λ_J systematically varied between 1 and 10. The magnetic field H_e was applied perpendicular to L. Results are shown in Figs. 1 and 2. With the experimentally obtained I_n vs. H_e curves we have plotted theoretical I_n^{max} vs. H_e curves obtained from Kulik's theory,[1] using the respective $Q_n(n)$ functions given in the figure captions.

From the figures it is immediately obvious that Kulik's theory[1] does describe the observed magnetic field dependence pattern very well. The I_n^{max} vs. H_e curve does indeed have a maximum, and this maximum is $(I_n^{max})^{max} = 0.34\, I_0^{max}$ for Nb–Pb junctions, and for Pb–Pb junctions at 4.2°K. Only for Pb–Pb junctions at 1.5°K is $(I_n^{max})^{max} < 0.34\, I_0^{max}$. These features remain essentially the same for junctions with L/λ_J between 1 and 10, even though for junctions with $L/\lambda_J > 2$ the theory is not expected to hold strictly. The highest current step, $(I_n^{max})^{max}$, moves to larger mode numbers with increasing L/λ_J ratios. For Pb–Pb junctions at 4.2°K the mode number of the highest current step n_{max} is related to the reduced length by $n_{max}^3 \approx 6(L/\lambda_J)^2$. For Pb–Pb junctions at 1.5°K $(I_n^{max})^{max}$ is always $< 0.34\, I_0^{max}$. Often we even find that $I_n^{max}(1.5°K) < I_n^{max}(4.2°K)$, that is, the step height decreases with decreasing temperature.* As we shall see, this behavior can be explained by geometric effects. It remains to show that the experimentally obtained $Q_n(n)$ functions are reasonable.

* A similar decrease in step height with decreasing temperature has been reported in Ref. 5.

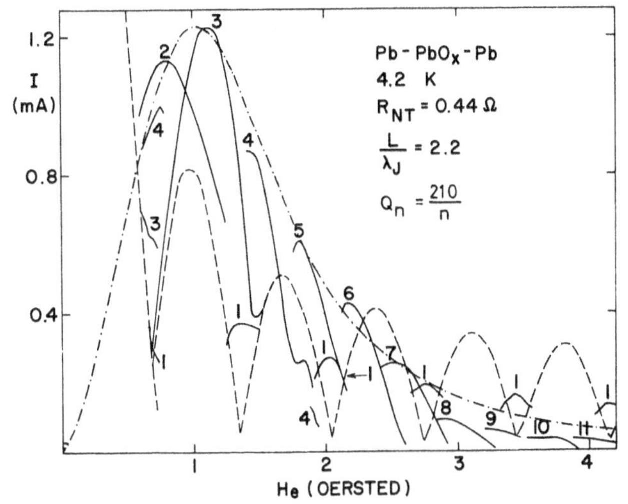

(a)

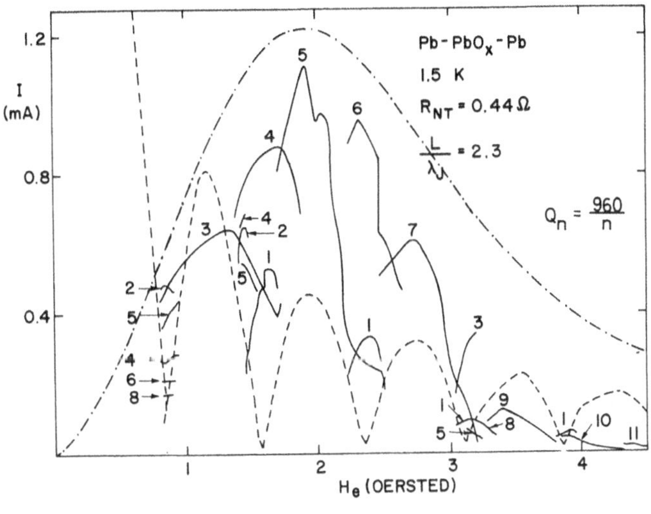

(b)

Fig. 2. Magnetic field dependence of I_0 (dashed line) and I_n (full line) for a Pb–Pb junction, measured "symmetrically" in a six-probe configuration (see Ref. 7). The numbers are the mode numbers n. The dash-dotted line is I_n^{max} vs. H_e, calculated from Kulik's theory, using the indicated Q_n value. $L = 0.3$ mm, $w = 0.1$ mm.

There are three main effects which contribute to the resonance linewidth, i.e., to Q_n^{-1}, and which add as $Q_n^{-1} = \sum_i Q_{ni}^{-1}$.

1. The Q associated with Ohmic losses in the oxide layer, due to a finite single-particle tunnel current I_t at the resonance voltage $V_n = (\hbar/2e)\,\omega_n$. With $R = V_n/I_t$ and C the junction capacitance, this "tunnel current Q" can be written as[3]

$$Q_{nt} = \omega_n RC \propto nRC$$

For $T/T_c \approx 0.2$ and $L/\lambda_J \approx 2$ this leads for our junctions to $Q_{nt} > 10^4 n$, which does not agree at all with the observed Q_n. Only for $T/T_c \approx 0.6$ and for low mode numbers is Q_{nt} of the same order of magnitude as the observed Q_n.

2. The Q associated with rf losses in the superconducting films. Pedersen *et al.*[6] have estimated that this Q is of order 10^3 for Pb–Pb junctions, and that

$$Q_{nf} \propto \omega_n^{-\beta} \propto n^{-\beta}$$

where β lies between 0.7 and 1, depending on whether the superconducting films approach the London or Pippard limit. Q_{nf} agrees very well with the Q_n observed for Nb–Pb junctions (Fig. 1; $Q_n = bn^{-\beta}$ has been fitted to the experimental data by the technique of least mean squares, which yields $b = 312 \pm 23$, $\beta = 0.74 \pm 0.04$). Niobium thin films are superconductors in the London limit. Their surface resistance, which is proportional to Q_f^{-1}, is also expected to be higher than that of Pb thin films. We therefore believe that in the case of Nb–Pb junctions Q_{nf}^{-1} dominates, i.e., Q_{nf}^{-1} is larger, or much larger, than the other contributions to Q_n^{-1}. Lead thin films are superconductors in the Pippard limit, and we therefore expect $\beta \approx 1$ for Pb–Pb junctions. This is in good agreement with the experimental data at 4.2°K (Fig. 2a). This agreement, together with the fact that b in $Q_n = bn^{-\beta}$ does not vary with L/λ_J, leads us to believe that rf losses in the Pb thin films dominate the energy dissipation in Pb–Pn junctions as well. But since $Q_{nf}^{-1}(\mathrm{Pb}) < Q_{nf}^{-1}(\mathrm{Nb})$, the Q_n of Pb–Pb junctions is limited by the geometric effect, at least at 1.5°K.

3. The theory assumes that the junction length L, which determines the resonance frequency ω_n, is a constant over the junction width w. But in real junctions there will always be small statistical variations of L. This effect will broaden the linewidth, and can formally be described by a "geometric Q," Q_{ng}, but which does not contribute to energy dissipation,

$$Q_{ng} = \omega_n/\Delta\omega = \lambda_n/\Delta\lambda = 2L(n\Delta L) \propto n^{-1}$$

The best accuracy with respect to a constant L that can be obtained by deposition of thin films through a stenciled mask is a ΔL of the order $1\,\mu\mathrm{m}$, so that for our junctions, with $L = 0.3$ mm, $2L/\Delta L = nQ_{ng} \approx 600$; i.e., $Q_{ng} \approx 600/n$. If the geometric effect dominates, i.e., if $Q_{ng} < Q_{nt}, Q_{nf}$, it will also reduce I_n, because at the step voltage V_n only part of the junction will be in resonance. This effect is clearly seen for Pb–Pb junctions at 1.5°K (Fig. 2b, with $Q_n > Q_{ng}$). On the other hand, for Nb–Pb junctions (Fig. 1), and for Pb–Pb junctions at 4.2°K (Fig. 2a), the observed Q_n is less than Q_{ng}, so that in these cases Q_{nf} dominates.

We conclude that the measured magnetic field dependence of I_n is in very good agreement with Kulik's theory.[1] At low temperatures, $T/T_c \approx 0.2$, the Q_n of Nb–Pb junctions is limited by rf losses in the Nb thin films; while the Q_n of Pb–Pb junctions is limited by small statistical variations of the junction length over the width of the junction.

References

1. I.O. Kulik, *Zh. Tekhn. Fiz.* **37,** 157 (1967) [*Soviet Phys.—Tech. Phys.* **12,** 111 (1967)].
2. R.E. Eck, D.J. Scalapino, and B.N. Taylor, in *Low Temperature Physics LT-9* (Proc. 9th Intern. Conf., 1964), Plenum Press, New York (1965), p. 415.
3. I.O. Kulik, *Zh. Eksperim. i Teor. Fiz. Pisma* **2**, 134 (1965) [*Soviet Phys.—JETP Lett.* **2**, 84 (1965)].
4. D.D. Coon and M.D. Fiske, *Phys. Rev.* **138,** A744 (1965).
5. J.T. Chen, T.F. Finnegan, and D.N. Langenberg, in *Proc. Intern. Conf. on the Science of Superconductivity*, F. Chilton, ed., North-Holland, Amsterdam (1971), p. 413; *Physica* **55,** 413 (1971).
6. N.F. Pedersen, T.F. Finnegan, and D.N. Langenberg, *Phys. Rev. B* **2**, 4151 (1972).
7. K. Schwidtal, *Phys. Rev. B* **2,** 2526 (1970).

Parametric Oscillations in Superconducting Double Point-Contact Structures

B. T. Ulrich*

École Normale Supérieure, Paris, France

Parametric phase slip oscillations have been observed in double point-contact Josephson junctions, in which the junction critical current could be varied by an external magnetic field. The phase slip in the junction was synchronized parametrically by a delay line connected to the junction. The purpose of these experiments was to study the phase slip process in inductively dominated Josephson junctions where LI_0 was much larger than the flux quantum, and where I_0 could be varied without changing either the junction geometry or the temperature. This paper discusses the bias voltage and magnetic field dependence of the phase slip, and also the relation of parametrically synchronized phase slip processes to frequency pulling in Josephson junction microwave emission experiments.[1]

The parametric oscillations to be described form a class of dynamic phenomena which can occur in a Josephson junction at frequencies lower than the frequency given by the Josephson voltage–frequency relation. These subharmonic parametric phase slip processes occur when a Josephson junction is connected to certain types of external circuits. In the present experiments the phase slip was synchronized to occur at 212 MHz by a delay line connected to the junction. At each slip of the phase across the junction the voltage pulse defined by the voltage–phase relationship $V = (\hbar/2e)\, d\varphi/dt$ traveled up the delay line, was reflected with the same sign by a high impedance, and returned to the junction with a delay of 4.7 nsec to the junction, at which time the phase slip was triggered again by the returning voltage pulse. Such synchronized phase slip processes can occur in junctions which do not have a delay line expressly connected. In these cases the current and voltage leads to the junction can behave as transmission lines, and discontinuities in the lines reflect the pulses which synchronize the phase slip in the junction. In the present experiments the radiofrequency properties of the environment of the junction were controlled in such a way that the phase slip occurred at 212 MHz. Nevertheless, parametric phase slip processes can occur very frequently in experiments with inductively dominated point-contact Josephson junctions. The parametric oscillations may go unnoticed by the experimenter because their frequency would usually be outside the pass band of the amplifiers used to observe the current–voltage (I–V) characteristics of the junction, although the oscillations occur at much lower frequency than that given by the Josephson voltage–frequency relation. It is the thesis of this paper that these subharmonic oscillations play an important role in determining both

* On leave from the University of Texas, Austin, Texas.

the $I-V$ characteristics and the high-frequency electromagnetic behavior of a Josephson junction.

We approach the theoretical description of the parametric oscillations through a sequence of simple models that exhibit the relevant physical processes. First we discuss a single Josephson junction in series with an inductance, then discuss the effects of including the junction shunt conductance, and finally consider the effects of the junction shunt capacitance. In the discussion of the parametric oscillations we represent the double point contact by a single Josephson junction with magnetic-field-dependent critical current $I_0(B)$. It can be shown from the observed magnetic field dependence of the double-junction critical current that effects of the loop inductance of the double junction are small.[2]

The time development of the phase difference φ across a Josephson junction with critical current I_0 is described by the first-order differential equation

$$I = \left(G_s L I_0 \cos \varphi + \frac{\hbar G_s}{2e} \right) \frac{d\varphi}{dt} + I_0 \sin \varphi \tag{1}$$

when the junction in series with inductance L is driven by a current source I shunted by a source conductance G_s. This equation has a well-known analytic solution for $\varphi(t)$. If $2\pi L I_0 > \Phi_0$, where $\Phi_0 = h/2e = 2.07 \times 10^{-15}$ Wb is the flux quantum, then the solution for $\varphi(t)$ is a multiple-valued function of time. As φ increases, time at first increases, and then "time runs backward," between alternate points where $d\varphi/dt$ becomes infinite. In the usual treatments the equation is "turned off" whenever $d\varphi/dt$ becomes infinite, and φ is allowed to jump to one of its values for which $d\varphi/dt$ is finite at the same value of time. In the experiment to be described here the inductance of the niobium point was 6×10^{-9} H and the maximum critical current $I_0(B)$ of the double junction was 24.3 μA. Thus the ratio $2\pi L I_0/\Phi_0 \approx 450 \gg 1$, placing the conditions of the experiment well within the region of multiple-valued solutions of Eq. (1).

The multiple-valued behavior of $\varphi(t)$ is an artifact of simplifying the equivalent circuit surrounding the junction. If the internal shunt conductance G_0 of the junction is included, then the phase obeys

$$I = \frac{\hbar G_s G_0 L}{2e} \frac{d^2\varphi}{dt^2} + \left[G_s L I_0 \cos \varphi + (G_0 + G_s) \frac{\hbar}{2e} \right] \frac{d\varphi}{dt} + I_0 \sin \varphi$$

It can be shown by analytic techniques that a unique periodic limit cycle exists for this equation, and that the frequency of oscillation is given by the Josephson voltage–frequency relation.[3] Specifically, the equation does not possess subharmonic solutions.

When the junction shunt capacitance C is included the autonomous equation describing the phase becomes third order,

$$I = \frac{\hbar G_s L C}{2e} \frac{d^3\varphi}{dt^3} + \left(\frac{\hbar G_s G_0 L}{2e} + \frac{\hbar C}{2e} \right) \frac{d^2\varphi}{dt^2}$$
$$+ \left[G_s L I_0 \cos \varphi + (G_0 + G_s) \frac{\hbar}{2e} \right] \frac{d\varphi}{dt} + I_0 \sin \varphi$$

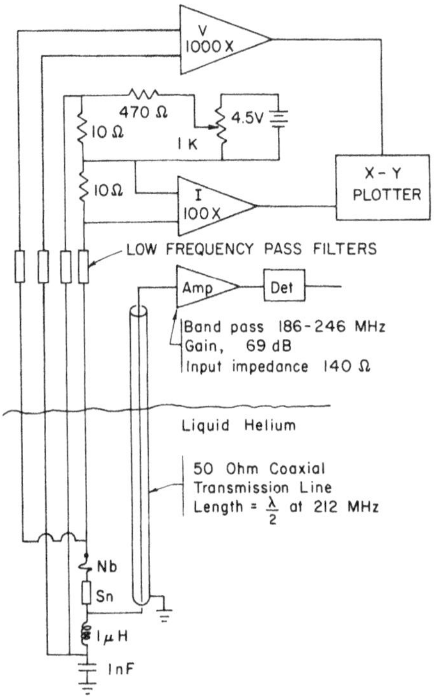

Fig. 1. Block diagram of apparatus used to observe parametric oscillations.

and subharmonic solutions become possible for large enough C. The equation describing a junction coupled to a delay line is presently under investigation.

This experiment on parametric oscillations in Josephson junctions used the same junction mount as that used for experiments on microwave emission and frequency pulling with point-contact junctions.[1] The junction was a chemically etched niobium wire touched to the end of a tin cylinder by a differential screw mechanism, with contact being made in liquid helium at 1.53°K. The point made a double contact to the tin, with enclosed area of about 10^{-6} cm^2 determined by the magnetic field period of the critical current. The junction was surrounded by a cylindrical coaxial cavity which in turn was weakly coupled to a 5-mm circular waveguide leading up the neck of the dewar. The external optics which form the Fabry-Perot cavity used for the microwave emission experiments was removed for this experiment, and the waveguide exit was open and pointed toward the ceiling. A 186–246-MHz rf amplifier with 69 dB gain was connected to the junction by a semirigid coaxial transmission line that was one half-wavelength long at 212 MHz. The amplifier input impedance was 140 Ω. Thus a voltage pulse emitted by the junction during phase slip was conducted by the 50-Ω coaxial line to the amplifier, and partially reflected back to the junction with the same sign after a 4.7-nsec delay. A crystal detector at the output of the rf amplifier gave a dc output voltage proportional to the rf power output from the amplifier. The dc bias circuit and rf circuit are shown in Fig. 1.

The I–V characteristic was typical of a point-contact junction that shows some tunneling behavior. A tunneling break of 38 μA was observed at 2.13 mV, a voltage consistent with the sum of the energy gaps for niobium and for tin. The low-voltage portion of the $I-V$ curve is shown in Fig. 2. The junction was mounted in a stainless steel dewar, with neither mu-metal nor superconducting shield around the junction, and a magnetic field external to the dewar was used to modulate the total critical current from 9.7 to 24.3 μA, corresponding to critical currents $I_1 = 17.0$ μA and $I_2 = 7.3$ μA for each of the two junctions. The double junction was biased with a 19.7-Ω source resistance and the voltage change as a function of the external magnetic field is shown in Fig. 3. Because the $I-V$ curve has nearly zero differential conductivity in the region just above the critical current, this voltage dependence as a function of magnetic field very nearly represents the critical current dependence as a function of magnetic field. From the nearly sinusoidal shape of $I_0(B)$ it follows that the loop inductance of the double junction is small, and with regard to the parametric oscillations, the primary effect of the double junction structure is to modulate the junction critical current.

The junction bias voltage dependence of the power in the parametric oscillation is shown in Fig. 2. Two curves are shown, one for maximum $I_0(B)$ and the other for minimum $I_0(B)$. The parametric oscillation is observed up to 1.5 mV bias, and this oscillation is *not* the oscillation given by the Josephson voltage–frequency relation, which implies that the 212-MHz oscillation occurs at 0.4 μV bias voltage. The 200-MHz oscillation may be regarded as sweeping the $I-V$ characteristic, and the $I-V$ characteristic plotted by the $X-Y$ plotter is thus the time average of the 200-MHz oscillation. This interpretation is supported by the fact that no constant-voltage

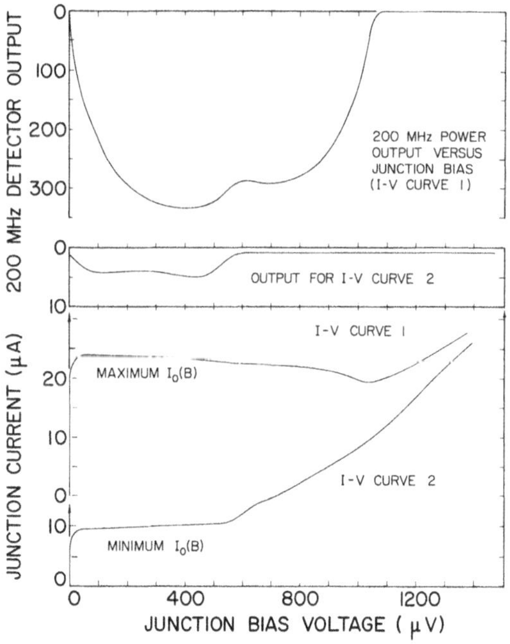

Fig. 2. Double point-contact Josephson junction current–voltage characteristics, and power output of the parametric oscillations at the magnetic fields for maximum and minimum $I_0(B)$. The ordinate voltage scale is common for all graphs.

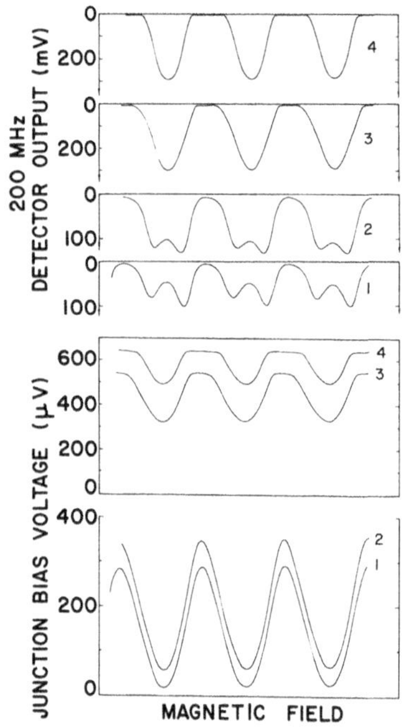

Fig. 3. Magnetic field dependence of bias voltage for the junction biased with 19.7-Ω dc load resistance, and corresponding dependence of the parametric oscillation, for four settings of the load line. The magnetic field scale is common for all graphs.

steps were observed in the $I-V$ curve when 4-mm-wavelength microwave radiation was impressed on the junction.

The magnetic field dependence of the oscillation is shown in Fig. 3, with bias voltage for each value of the magnetic field indicated in the lower curves. The double-peaked structure in the oscillation is due to the bias voltage becoming large enough to pass the maximum of the oscillation versus voltage bias curve for fixed magnetic field.

Acknowledgment

It is a pleasure to thank Dr. Albert Libchaber for his hospitality in the Laboratoire de Physique des Solides of the École Normale Supérieure during this research.

References

1. B.T. Ulrich, this volume.
2. T.A. Fulton, L.N. Dunkleberger, and R.C. Dynes, *Phys. Rev. B* **6**, 855 (1972).
3. B.T. Ulrich and G. Nicolis, in preparation.

Analog Computer Studies of Subharmonic Steps in Josephson Junctions

C. A. Hamilton

Quantum Electronics Division
National Bureau of Standards, Boulder, Colorado

It is well known that a superconducting weak link (Josephson junction) which is exposed to microwave radiation exhibits constant-voltage steps in its V–I curve. These steps occur at the voltages $n(\hbar\omega/2e)$ and are a consequence of phase-locking between the applied radiation ω and the Josephson self-oscillation ω_0 ($\omega_0 = 2eV_{dc}/\hbar$). Certain types of junctions show additional steps at the voltages $(n/m)(\hbar\omega/2e)$, $n \neq m > 1$.[1] These have become known as subharmonic steps. If the shunt conductance seen by the junction is sufficiently high, the Josephson current will significantly affect the junction voltage. This feedback situation results in the generation of harmonics of the self-oscillation. The subharmonic steps occur when one of these harmonics phase-locks with the applied radiation.[2–4] (The term "harmonic steps" would thus be a more appropriate name.)

We have used an analog simulation of a Josephson junction in an appropriate circuit model to study the subharmonic steps. The circuit model is shown in Fig. 1. It consists of an ideal Josephson junction, shunted by a resistance R and a capacitance C, and driven by a constant-current source with both a dc and an rf component. The analog Josephson junction has a critical current of 1 mA and a voltage-to-frequency conversion constant $2e/h = 483$ Hz/V. In Fig. 2 we have used the analog model to plot a series of V–I curves as a function of the capacitance C. In these curves the rf drive current I_{rf} has been adjusted to maintain the principal steps ($V_{dc} = n\hbar\omega/2e$) at constant amplitude. The first important observation is that with $C = 0$ no subharmonic steps are observed. This is true even though the junction is producing harmonics of the self-oscillation with substantial amplitude. To demonstrate this point, we plot in Fig. 3 the frequency spectrum of the Josephson current with and without a small capacitance. The rf drive has been removed and the bias point corresponds to the 0.5 step of Fig. 2. It is apparent that a capacitance which produces large subharmonic steps has only a small effect on the spectrum of the self-oscillation. In fact, the capacitor actually decreases the harmonic content of the self-oscillation. The important effect of the capacitor is to produce a phase lag in the feedback voltage relative to the junction current. This phase shift allows harmonics of ω_0 to phase-lock with the applied frequency over a range of direct current. This argument is further verified by replacing the shunt capacitor with a series inductor. This produces a phase lead in the feedback voltage relative to the junction current. Similar subharmonic steps are observed in this case. Apparently, when the impedance

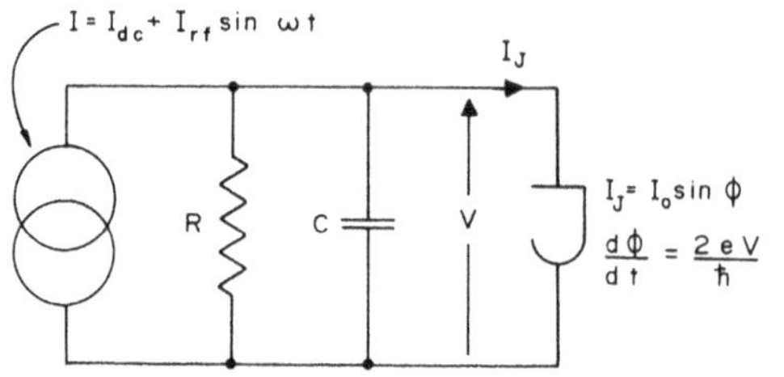

Fig. 1. Circuit model for the analog simulation.

seen by the junction is purely resistive, the relative phases of the harmonics of ω_0 are not free to simply phase-lock with an external signal. This result conflicts with previous numerical calculations which predict the existence of subharmonic steps from a purely resistive circuit.[2] Since we are not aware of the details of these calculations, this discrepancy remains unresolved. Our own numerical calculations are consistent with the analog results.

Figure 2 shows that the subharmonic behavior peaks at a value of capacitance where $R \approx 1/\omega C$. For small values of C there is insufficient phase shift for strong phase-locking; and for large C the shunt admittance across the junction becomes large and harmonic generation is suppressed.

We have also studied the dependence of the subharmonic steps on the amplitude of the applied rf signal. In Fig. 4 we plot a series of V–I curves as a function of the applied rf amplitude. From this and similar data we conclude that all of the subharmonic steps between any two principal steps have about the same rf amplitude dependence. This dependence is roughly proportional to the product of the dependences of the principal steps above and below the group of substeps. Thus, all the substeps between any two principal steps go to zero when either of the principal

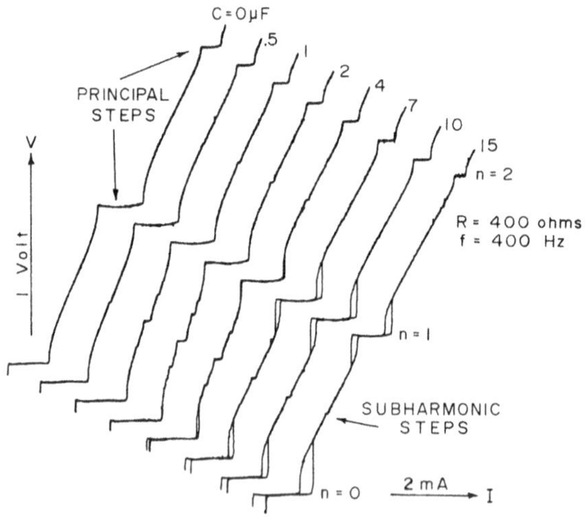

Fig. 2. A series of V–I curves as a function of the capacitance C. At each value of C, I_{rf} is adjusted to maintain the principal steps at constant amplitude.

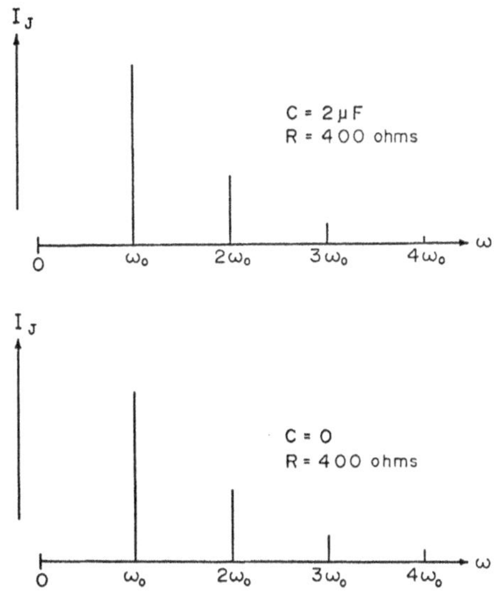

Fig. 3. Frequency spectrum of the Josephson current with and without a small shunt capacitance.

steps go to zero. Since adjacent principal steps tend to oscillate out the phase, the substeps oscillate at twice the rate of the principal steps.

It has been generally assumed that the only requirement for the generation of subharmonic steps is the existence of harmonics of the self-oscillation. The analog results show that the relative phase of these harmonics is crucial and that some reactive element in the circuit is essential. This explains why subharmonic steps are observed in some junctions and not in others. For example, point-contact junctions formed

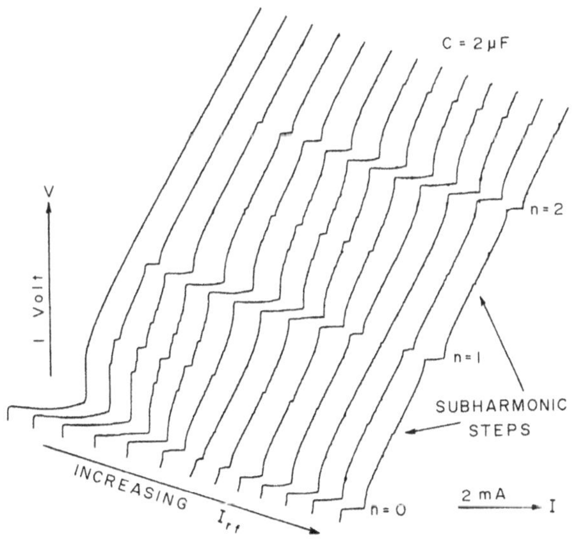

Fig. 4. Subharmonic steps as a function of the applied rf amplitude.

with a very sharp point (low capacitance) generally do not produce subharmonic steps. However, when the contact pressure is increased to substantially deform the point (thus increasing the capacitance) subharmonic steps usually appear. The existence of subharmonic steps can thus serve as an additional probe of the rf impedance seen by the junction.

Acknowledgments

The author is indebted to Dr. Donald McDonald and Dr. Eric Johnson for many helpful discussions on this work.

References

1. A.H. Dayem and J.J. Weigand, *Phys. Rev.* **155,** 419 (1967).
2. J.R. Waldram, A.B. Pippard, and J. Clarke, *Phil. Trans. Roy. Soc.* (*London*) **A 268,** 265 (1970).
3. J. Clarke, *Phys. Rev. B* **4,** 2963 (1971).
4. D.B. Sullivan, R.L. Peterson, V.E. Kose, and J.E. Zimmerman, *J. Appl. Phys.* **41,** 4865 (1970).

16

Cryogenic Instrumentation and Measurement Techniques

Low-Temperature Susceptibility Measurements of Cerium Magnesium Nitrate

G. O. Zimmerman* and D. J. Abeshouse*

Department of Physics, Boston University
Boston, Massachusetts

and

E. Maxwell† and D. R. Kelland†

Francis Bitter National Magnet Laboratory‡
Massachusetts Institute of Technology, Cambridge, Massachusetts

Measurements in Zero Magnetic Field

The susceptibility of a cerium magnesium nitrate (CMN) single-crystal sphere, a powder sphere, and a powder 1 × 1 cm cylinder were measured in zero magnetic field as a function of specific entropy by adiabatically demagnetizing the CMN from a temperature in the vicinity of 1°K.

The procedure was to demagnetize the CMN from a high temperature, about 1°K, and then observe the ambient warmup of the salt so as to be able to extrapolate the susceptibility to $t = 0$, the time when the magnetic field vanished. The apparatus used was the same as that in Ref. 1.

In order to test the adiabaticity of this process, the samples were cycled several times. In the cycling process the sample was isolated with the magnetic field on, and then the field was reduced linearly to zero in a typical time of 30 min. Susceptibility measurements were then made for about 5 min and the field reapplied over a time span of about 30 min. The process was then repeated. During that time the sample remained isolated. Although we could measure the susceptibility to an accuracy of one part in 10^5, there was no evidence of nonadiabaticity in the process except that which could be attributed to the ambient heat leak. A generous limit can thus be placed on the error due to irreversible processes of about 2%.

At 1°K the magnetic susceptibility of CMN was assumed to be that of an assembly of noninteracting dipoles of spin 1/2,

$$S/R = \ln 2 \cosh (g\mu H/2kT) - (g\mu H/2kT) \tanh (g\mu H/2kT) \qquad (1)$$

where the g factor for CMN in the direction of easy magnetization was taken to be 1.84. This assumes that g is a constant and does not change with the applied magnetic

* Supported in part by grant No. AF-AFOSR 1117-66.
† Work supported by U.S. Air Force Office of Scientific Research.
‡ Supported by National Science Foundation.

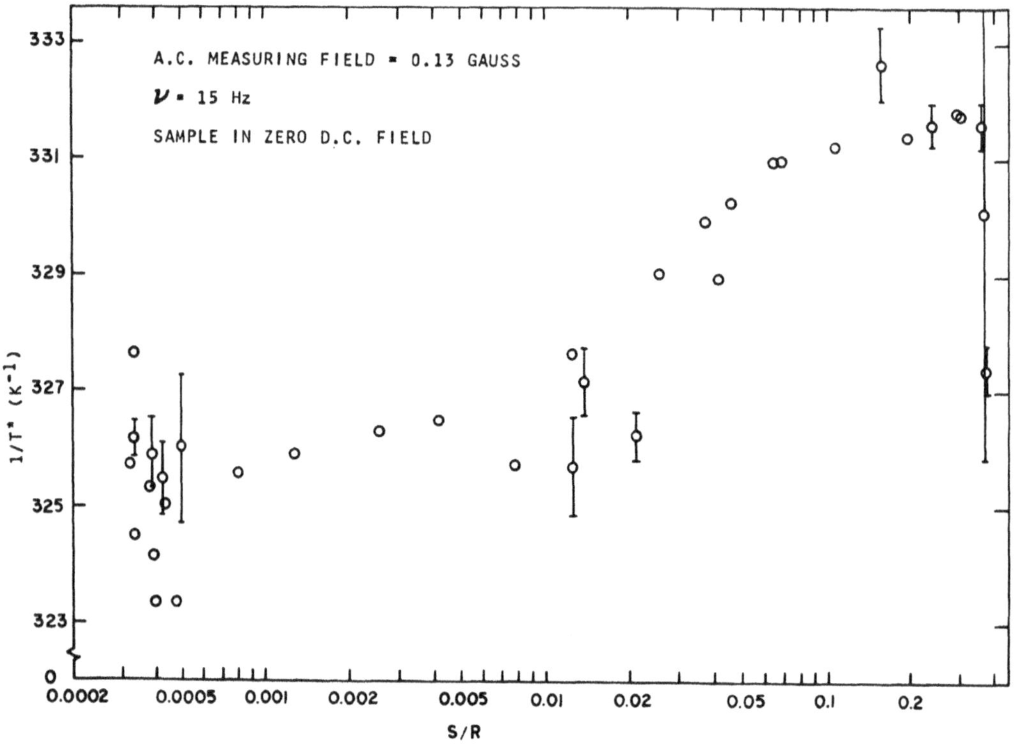

Fig. 1. Low-entropy susceptibility of a single-crystal sphere of CMN in terms of $1/T^* = \chi/C$.

field. This assumption might not be valid in high magnetic fields,[2] where it might be in error by as much as 5 %. The highest magnetic fields used in this investigation were 10^5 Oe. The correction due to the change in the g factor will not be made here, in order to facilitate the comparison of these data with others.

Results of measurements in the low-entropy region on a single-crystal sphere of CMN along the easy magnetization axis are shown in Fig. 1. The data were taken over a period of several months and the reproducibility increased our confidence in this method. The errors are due mainly to the extrapolation of the susceptibility to $t = 0$. One of the failings of our data is that because of apparatus design, our highest entropy was $S/R = 0.55$, which corresponds to a T^* of about 4.5 m°K. The lack of high-entropy data in our experiment presented a problem because the two complete measurements at high entropies, that of Hudson and Kaeser[3] and that of Mess *et al.*,[4] disagree with each other over the entire region of their measurement, and our highest entropy data fall somewhere between the two.

Two least-square fits (to an eight-order polynomial) were then made to our data anchored at high entropy to those of Refs. 3 and 4, respectively. The equation which was fitted is given by

$$\chi/C = \sum_{n=0}^{8} a_n (S/R)^n \tag{2}$$

Table I. Coefficients of Eq. (2) with High Entropy Fitted to Data of Refs. 3 and 4

	Hudson and Kaeser[3]	Mess *et al.*[4]
a_0	3.2521×10^2	3.2536×10^2
a_1	1.5766×10^2	7.4360×10^1
a_2	-1.5138×10^3	1.8108×10^3
a_3	9.5842×10^3	-3.5022×10^4
a_4	-5.9586×10^4	2.1666×10^5
a_5	2.8350×10^5	-6.1015×10^5
a_6	-7.4956×10^5	8.0583×10^5
a_7	9.4740×10^5	-4.2777×10^5
a_8	-4.5370×10^5	2.9397×10^4
RMS residual	0.90	3.858
Maximum deviation	2.60	20.66

with the coefficients and deviations given in Table I. From the standard deviations it can be seen that our data are in somewhat better agreement with those of Ref. 3 than with those of Ref. 4. On the other hand, it should be noted that the least-squares fit is mainly a low-entropy approximation of the data since that is where most of our points were taken, and that at $S = R \ln 2$ the magnetic susceptibility is $\chi/C = 40$, about 13 % of the susceptibility at the maximum.

A similar method to the one above was used in the measurements on the powdered sphere and the powdered cylinder. The powdered sphere, with a diameter of 2 cm, and the powdered cylinder, with diameter and length both equal to 2.54 cm, consisted of ground CMN powder compacted into the respectively shaped nylon containers. In order to increase the thermal bonding between the powder grains, a drop of water was added to each container and subsequently pumped off.

The powder entropy at high temperatures can be computed in two ways, each yielding a different result. In the first instance one can assume an average $g = \frac{2}{3}(1.84)$ and substitute it into (1). On the other hand, if one assumes that the crystallographic axes of the powder grains are distributed uniformly over a sphere, one has

$$S/R = \int_0^{\pi/2} [\ln 2 \cosh (x \sin \theta) - x \sin \theta \tanh (x \sin \theta)] \sin \theta \, d\theta \tag{3}$$

where $x = g_\perp \mu H/kT$. In both cases $g_\parallel$ is assumed to be zero. The error of this assumption is assumed to be negligible since $g_\parallel$ is of the order of 0.03.

The results of this measurement are given in Fig. 2. The open points are those computed using an average g, while the closed points are those computed by means of Eq. (3). For comparison the single-crystal data are also shown. One can see that at the lowest entropies the single-crystal data and those of the cylinder are in excellent agreement, while those of the powder sphere have a significantly higher $T^* = C/\chi$. At higher entropies the powder cylinder exhibits values which are intermediate between those of the single-crystal sphere and those of the powder sphere.

From Fig. 2 one can compare the T^* of the single-crystal sphere, the powder sphere, and the powder cylinder at constant entropy. This is done in Table II. $T^*_{cs} = 4.5$ m°K is the highest temperature at which we could obtain reliable S/R

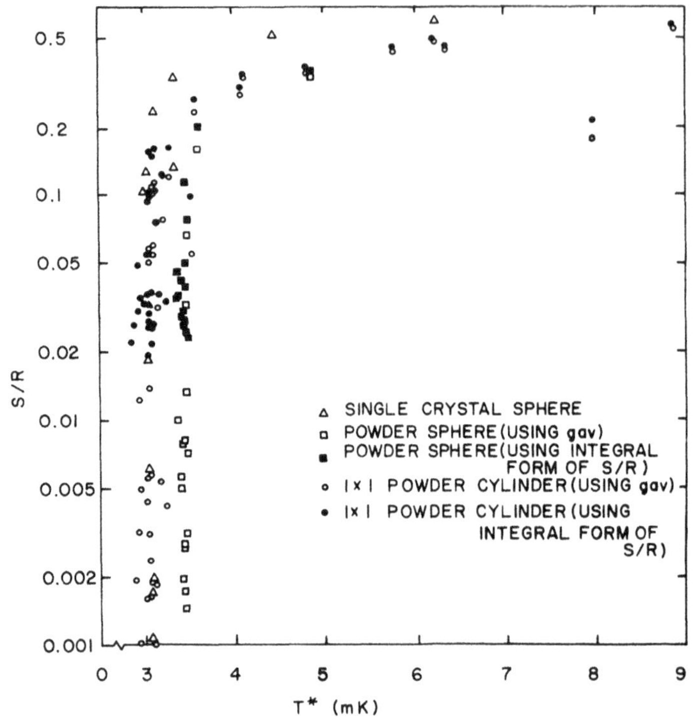

Fig. 2. S/R as a function of T^* for the single-crystal sphere (△), powder sphere (□), and 1 × 1 cm powder cylinder (○). Solid symbols use integral form.

data. One sees that at low temperatures the T^* of the crystal sphere and of the 1 × 1 cm cylinder agree. If we take the single-crystal T^* to be the true temperature, a dubious assumption, then at the lowest T^* the powder cylinder should give a true temperature. At higher temperatures there is a disagreement of as much as 2 m°K, which does contradict the additional 1.7 m°K of Abraham and Eckstein.[5] We thus add another set of data to "the delta campaign."[6]

Measurements in a Magnetic Field

The above measurements were carried out in zero magnetic field with special coils compensating for the ambient field at the apparatus. The reason for this precaution was the observation that the magnetic susceptibility changes markedly in all the temperature regions when a small magnetic field is applied. In order to investigate this phenomenon further, we measured the effect of an externally applied magnetic field on the susceptibility of a single-crystal sphere. Although measurements were made with the field applied both parallel and normal to the measuring field, both of which were normal to the c axis, we shall here report only on the case of the applied field parallel to the measuring field. The results for the other case were similar.

In this measurement the CMN was demagnetized to zero magnetic field and measurements of the susceptibility were made as a function of time as the sample warmed due to ambient heating. Different external magnetic fields were then applied

Table II. Comparison of T^* Values for Crystal Sphere (cs), Powder Sphere (ps), and Powder Cylinder (pc)

T^*_{cs}, m°K	T^*_{ps}, m°K	T^*_{pc}, m°K
3.00	3.50	3.05
3.05	3.55	3.10
3.10	3.60	3.40
3.20	4.10	3.90
3.50	5.00	4.80
4.00	5.75	5.75
4.50	6.50	6.50

adiabatically and the susceptibility in those fields measured. Smooth curves were then drawn through the points at each value of the externally applied field. At each particular time (consequently, constant entropy) the magnetic susceptibility as a function of the applied magnetic field obeyed the equation

$$\chi(H, S)/C = [\chi(0, S)/C] - A(S)\, H^{\eta(S)} \tag{4}$$

where the left-hand side is the magnetic susceptibility in an applied field, the first term on the right is the susceptibility in zero field at that entropy, and A and η are functions dependent on the entropy or susceptibility in zero field. These are shown as a function of T^* for values above the minimum in Fig. 3.

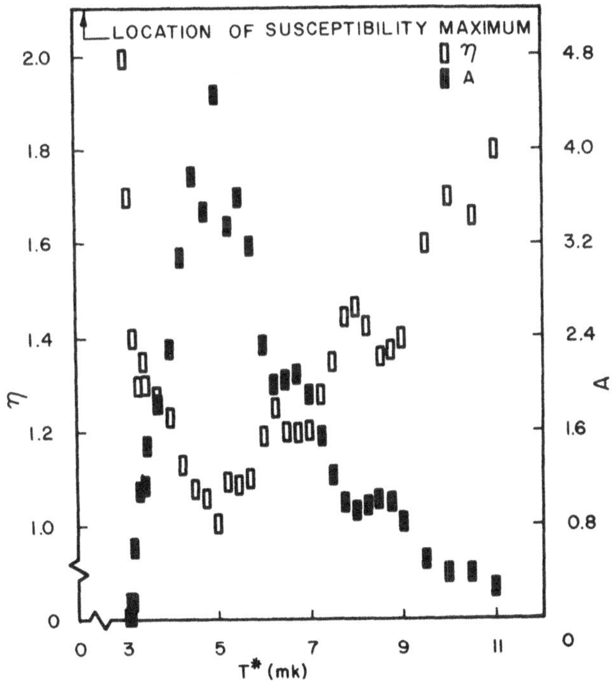

Fig. 3. Exponent η and coefficient A of Eq. (4) as a function of T^*.

As one can see from Fig. 3, A has a maximum at $T^* = 5$ m°K and becomes small, although never zero, at both high and low temperatures. The function η has a minimum of about 1 at that temperature and approaches the classical value of 2 at both high temperature and at the susceptibility maximum. These results are surprising since one would have expected either the value of 2, the classical value, or the value of $-1/5$ or $-1/4$, the scaling law value, for the exponent. Moreover, one would not expect there to be a minimum in the exponent at that high a temperature, a temperature of about 5 m°K as compared with that of about 1 m°K for the susceptibility maximum. We have not been able to come up with a plausible explanation for this behavior since most theories which treat the influence of an applied magnetic field on susceptibility obtain an H^2 dependence.

References

1. D.J. Abeshouse, G.O. Zimmerman, D.R. Kelland, and E. Maxwell, *Phys. Rev. Lett.* **23,** 308 (1969).
2. S.J. Williamson and J.A. Cape, *Phys. Rev. Lett.* **21,** 370 (1968).
3. R.P. Hudson and R.S. Kaeser, *Physics* **3,** 95 (1967).
4. K.W. Mess, J. Lubbers, L. Niesen, and W.J. Huiskamp, *Physics* **41,** 260 (1969).
5. B.M. Abraham and Y. Eckstein, *Phys. Rev. Lett.* **20,** 649 (1968).
6. R.P. Hudson, *Cryogenics* **9,** 76 (1969).

Application of ^{4}He Calorimeter to the Specific Heat of Titanium

K. L. Agarwal and J. O. Betterton, Jr.

Louisiana State University—New Orleans
New Orleans, Louisiana

A new low-temperature calorimeter was built for the study of impurity effects in titanium. In titanium important effects on physical properties, such as the superconducting transition temperature, occur with only small amounts of impurities. Thus an accurate instrument is needed, which we describe below, and apply to measurement of low-temperature specific heats of high-purity titanium, which we prepared for this experiment by electron-beam zone-refining.

The arrangement of the apparatus is shown in Fig. 1. A piece of pure copper was machined so as to include the vapor pressure bulb, the heat shield surrounding the specimen, and the lowest radiation baffle in the vacuum pumping tube, in a single unit. The pressures were read on 1-in. bore mercury and oil manometers, and simultaneous readings of a Cryo-cal germanium thermometer, mounted on the specimen within the enclosure, led to a calibration of the thermometer with respect to the 1958 ^{4}He vapor pressure temperature scale. An indium-tipped mechanical heat switch was effective in coupling the sample to the liquid helium bath below 10°K and helium exchange gas was used only in the later calibration stages of the experiment below this temperature.

Pure titanium for this experiment was prepared by electron-beam zone-refining at 10^{-10} Torr in a baked sputter-ion pumped vacuum system. Electrolytically reduced crystals of 99.9% purity were arc-cast, swaged into 6-mm-diameter rod, and then zone-refined with multiple passes. The metal was soft and ductile and appears to have much less shear resistance than the previous pure titanium samples made from the iodide thermal dissociation process.

The results of the specific heat measurements are shown in Fig. 2, and the electronic specific heat coefficient γ and the Debye temperature θ_D obtained from the intercept and the slope of the linear curve (C/T vs T^2) are compared to the prior results on iodide titanium in Table I.

Note that a significant decrease is observed in the value of electronic specific heat coefficient γ, in agreement with the removal of impurities by zone-refining. Practically all alloying elements which have been added to titanium are observed to increase γ. The zone-refining would be expected to reduce these impurities significantly and thus reduce γ relative to the values for the iodide material previously reported.

The present value of γ may be compared with the values of 2.22–3.97 mJ/g-at deg^2 obtained by Falge[6] from the slopes of the superconducting transition temperature

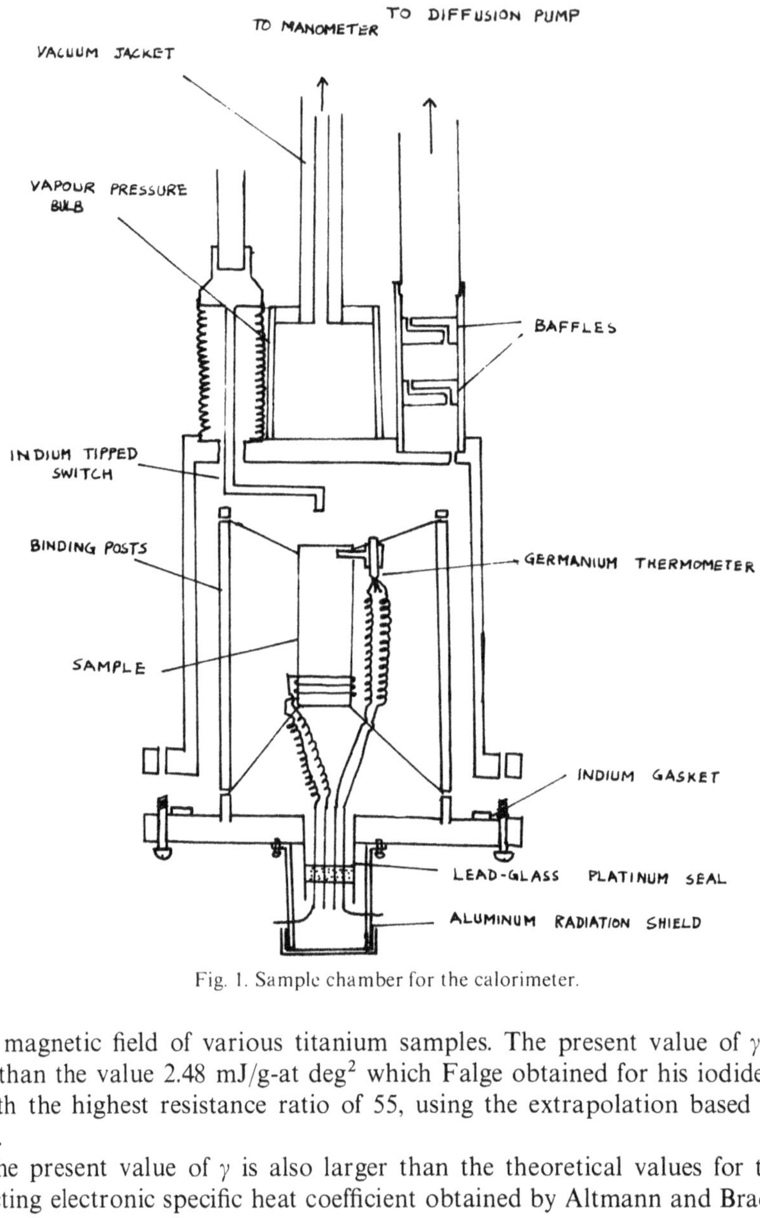

Fig. 1. Sample chamber for the calorimeter.

versus magnetic field of various titanium samples. The present value of γ is 25% larger than the value 2.48 mJ/g-at deg^2 which Falge obtained for his iodide sample D6 with the highest resistance ratio of 55, using the extrapolation based on BCS theory.

The present value of γ is also larger than the theoretical values for the noninteracting electronic specific heat coefficient obtained by Altmann and Bradley[7] by cellular band structure calculations. Altmann and Bradley give noninteracting electronic specific heat coefficients ranging from 1.6 to 6.8 mJ/g-at deg^2, with their preferred value 1.63 mJ/g-at deg^2 based on their most reliable Ti3 potential. This theoretical value of γ involves considerable uncertainty, but at least part of the difference when compared to the present experiment is related to the phonon enhancement of the electronic specific heat in titanium.

The value of the Debye temperature, 423°K, obtained for the zone-refined titanium has not changed appreciably from the iodide value. It is in good agreement

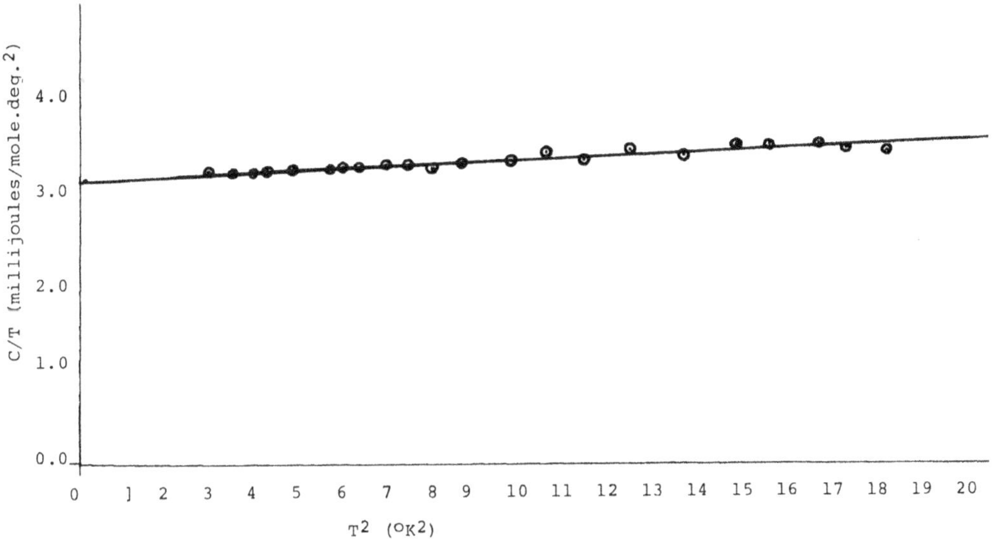

Fig. 2. Specific heat of high-purity titanium between 1.4 and 4.2°K.

Table I. Low-Temperature Specific Heats of Titanium from the Present Experiment (Zone-Refined Metal) and Earlier Measurements (Iodide Metal)

Reference	Specimen resistance ratio, $R_{300}/R_{4.2}$	DPH hardness	Gamma, mJ/g-at deg^2	Debye temperature, °K
Present work	110–145	55*	3.11	423
Kneip *et al.*[1]	56	63	3.346	427
			3.351	430
Aven *et al.*[2]	20–30†	70–80†	3.38	421
Hake and Cape[3]	31	70–80†	3.305	429
Heinger and Muller[4]	20–30†	70–80†	3.3	415
Wolcott[5]	20–30†	70–80†	3.38	430

* Converted from Rockwell 15N hardness.
† Estimated from typical values for iodide titanium.

with the value 425.6°K obtained by Fisher and Renken[8] from low-temperature elastic constants on titanium.

Acknowledgments

The authors wish to thank Louis Cazenavette and D. Apparao for operation of the zone-refiner.

References

1. G.D.Kneip, Jr., J.O. Betterton, Jr., and J.O. Scarbrough, *Phys. Rev.* **130**, 1687 (1963).
2. M.H. Aven, R.S. Craig, T.R. Waite, and W.E. Wallace, *Phys. Rev.* **102**, 1263 (1956).
3. R.R. Cape and J.A. Hake, *Phys. Rev.* **135**, A1151 (1964).
4. F. Heinger and J. Muller, *Phys. Rev.* **134**, A1407 (1964).
5. N.M. Wolcott, *Phil. Mag.* **2**(22), 1256 (1957).
6. R.L. Falge Jr., Thesis, Catholic University of America, Washington, D.C., (1966).
7. S.L. Altmann and C.J. Bradley, in *Proc. Conf. Soft X-ray Band Structure, Univ. Strathclyde, Scotland, 1967*, Academic Press, London and New York (1968), pp. 265–278.
8. E.S. Fisher and C.J. Renken, *Phys. Rev.* **135**, A482 (1964).

A Review of the Effects of High Magnetic Fields on Low-Temperature Thermometers

H. H. Sample*†, L. J. Neuringer, and L. G. Rubin

Francis Bitter National Magnet Laboratory†
Massachusetts Institute of Technology
Cambridge, Massachusetts

Introduction

In this paper we extend our earlier review[1] of the magnetic field dependence of low-temperature thermometers by presenting new high-field data on carbon resistors and thermocouples. In particular, we discuss the magnetoresistance of Speer and Allen-Bradley resistors from 0.5 to 4.2°K, and the magnetic field dependence of Au + 0.07 at.% Fe versus Chromel P and constantan versus Chromel P thermocouples between 4.2 and 45°K in dc magnetic fields up to 140 kG.

Carbon Resistors

We have measured the magnetoresistance of three 1/2-W Speer resistors with nominal room-temperature resistances of 100, 220, and 470 Ω, and one 1/8-W, 10-Ω Allen-Bradley resistor. The experiments were carried out using ^{4}He and ^{3}He cryostats with the specimens immersed directly in the liquid; the temperature in the magnetic field was determined from the vapor pressure of the liquid. The transverse ($\mathbf{I} \perp \mathbf{H}$) and longitudinal ($\mathbf{I} \parallel \mathbf{H}$) magnetoresistances were studied and were found to be the same. However, for practical reasons we generally measured only the transverse magnetoresistance.

The results for the Allen-Bradley resistor are shown in Fig. 1. The magnetoresistance is seen to be positive and increases with decreasing temperature, rising to 35% at 0.5°K and 140 kG, which corresponds to a temperature error of 6%. The data points generally fall within 0.5% of these curves and no effects of thermal cycling on the magnetoresistance are observed.

Representative data for a Speer resistor are shown in Fig. 2 for a 220-Ω unit. In contrast to the data of Fig. 1, all of the Speer resistors exhibit a more complex behavior. The data points of more than 15 different temperature runs have been omitted but they generally lie within 0.2% of the constant-field curves. At 140 kG and 2°K (i.e., near the maximum) the magnetoresistance corresponds to a temperature error of 9%. The results shown in Fig. 2 for the 220-Ω resistor are similar to those for the 100- and 470-Ω units. The key features to be noted are: (i) a negative magne-

* Also at Physics Department, Tufts University, Medford, Massachusetts.
† Supported by the National Science Foundation.

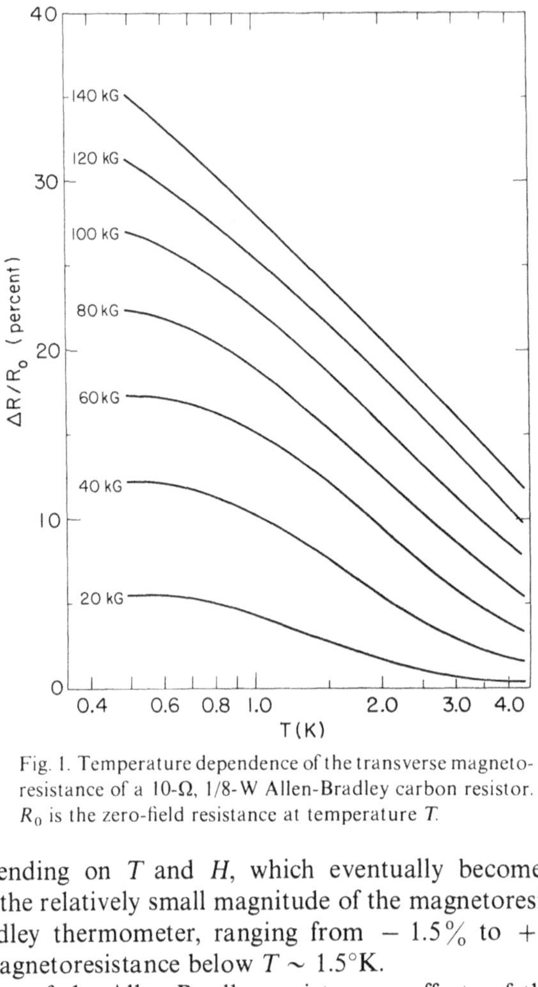

Fig. 1. Temperature dependence of the transverse magnetoresistance of a 10-Ω, 1/8-W Allen-Bradley carbon resistor. R_0 is the zero-field resistance at temperature T.

toresistance, depending on T and H, which eventually becomes positive at the highest fields; (ii) the relatively small magnitude of the magnetoresistance, compared to the Allen-Bradley thermometer, ranging from -1.5% to $+4.5\%$, and (iii) a decrease in the magnetoresistance below $T \sim 1.5°K$.

As in the case of the Allen-Bradley resistor, no effects of thermal cycling on the magnetoresistance were observed. It should be noted that while the magnitude of the magnetoresistance is, over most of the range, much smaller for the Speer resistors than for the Allen-Bradley resistor, nevertheless the temperature errors for the two types are comparable. These errors are summarized in Table I.

We also investigated a second 220-Ω Speer resistor which was ground down parallel to its length so as to have a room-temperature resistance of 440 Ω. The magnetoresistance of this flat resistor was measured at four temperatures between 0.6 and 1°K and was indistinguishable from the one in Fig. 2.

Thermocouples

We present here initial measurements on Au + 0.07 at. % Fe* versus Chromel P and Chromel P versus constantan thermocouples, the two combinations recom-

* Sigmund Cohn Corporation, 0.13 mm diameter.

Table I. Field-Dependent Temperature Errors for Various Low-Temperature Thermometers As a Function of H and T

		Magnetic field*		
Type of sensor	T, °K	Range I $\Delta T/T$, %	Range II $\Delta T/T$, %	Range III $\Delta T/T$, %
Speer resistor	0.5	2	2	6
	1.0	2	4	9
	2.0	4	5	13
	4.2	9	9	13
Allen–Bradley resistor	0.5	1	4	6
	1.0	1	6	9
	2.0	1	6	10
	4.2	< 1	5	10
	10	< 1	3	5
	20	< 1	1	2
Platinum	20	20	100	200
	40	< 1	5	10
	80	< 0.5	1	2
Germanium	4.2–20	5–30†	NR	NR
$SrTiO_3$ capacitor	1.5–20	NM	NM	≦ 0.05
GaAs diode	4.2	2–3	30–50	100–250
	10	1.5–2	25–40	75–200
	20	0.5–1	20–30	60–150
	40	0.2–0.3	4–6	15–30
	80	0.1–0.2	0.5–1	2–5
Au + 0.07 at.% Fe vs. Chromel P thermocouple	5	15	55	85
	10	5	35	50
	20	3	15	25
	40	1	7	10
Constantan vs. Chromel P thermocouple	10	~ 5	~ 10	~ 15
	20	~ 1	~ 4	~ 8
	30	< 1	~ 2	~ 6
Thermistors‡	4.2	< 0.05	1	3
	10	< 0.05	0.3	1
	20	< 0.05	0.1	0.5
	40	< 0.05	0.1	0.5
	60	< 0.05	0.1	0.3

* Range I covers 0–25 kG, range II covers 25–80 kG, and range III corresponds to $H > 80$ kG. NR indicates "not recommended." NM denotes not measurable (too small). For the measurement techniques, see Ref. 3.

† The 30% change corresponds to the lower temperature of 4.2°K.

‡ Includes five different types of Keystone Carbon Company Units.

mended for low-temperature use. (see Ref. 2). The reference junction of the thermocouple was thermally anchored to a copper post in contact with the liquid helium bath at 4.2°K while the measuring junction (hot junction) was anchored to a copper block whose temperature could be maintained above 4.2°K by a heater. A calibrated germanium thermometer was used to measure the block temperature in zero magnetic field. The block was maintained at this temperature, as the field was increased, by a $SrTiO_3$ capacitance thermometer[3] acting as a null detector in a closed-loop system controlling the power input to the heater. A carbon thermometer of known magnetoresistance was also mounted on the block to ensure that the block temperature remained constant in the field. A second calibrated carbon resistor attached to the reference post monitored changes in the reference temperature. A critical design feature of the apparatus is the fact that over the region which included both thermocouple junctions the magnetic field was constant to 1%.

Some typical data for a Au + 0.07 at.% Fe versus Chromel P thermocouple at various fixed temperatures between 5.3 and 45°K are shown in Fig. 3. The field-induced change in the thermocouple emf at a field H and block temperature T is defined as

$$\Delta V(H, T) = V(H, T) - V(0, T) \tag{1}$$

for a constant reference temperature. It is important to recognize that the magnitude of $\Delta V(H, T)$ depends on the particular choice of reference temperature. Small corrections were applied to our data so as to refer the measured $\Delta V(H, T)$ to a constant reference temperature of 4.22°K. In Fig. 3 we also show some results of

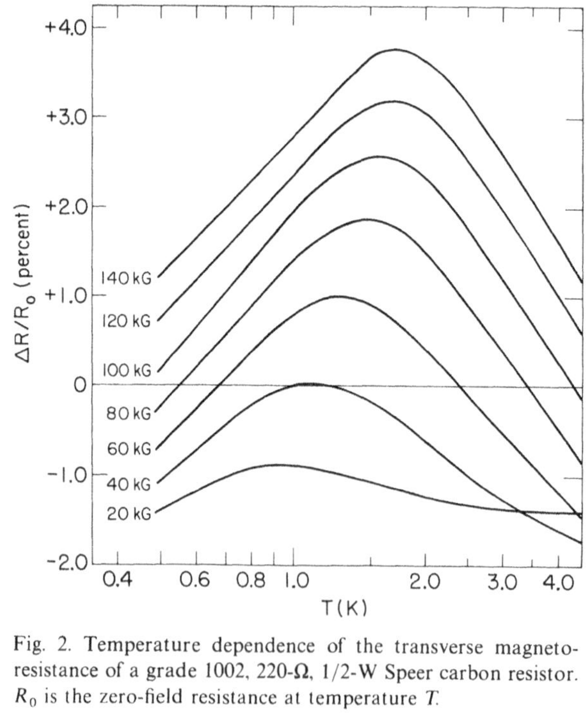

Fig. 2. Temperature dependence of the transverse magnetoresistance of a grade 1002, 220-Ω, 1/2-W Speer carbon resistor. R_0 is the zero-field resistance at temperature T.

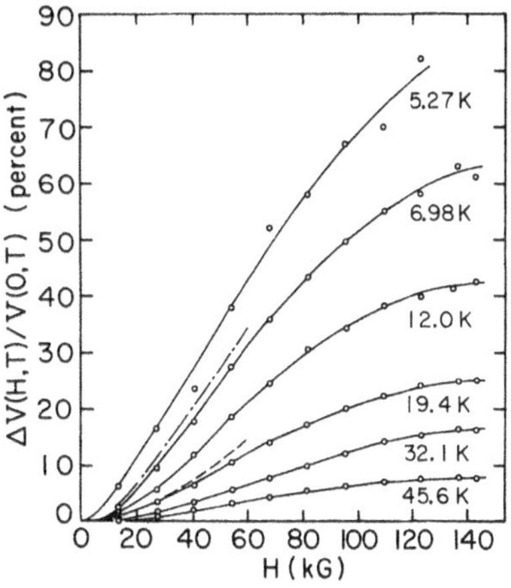

Fig. 3. Percentage change of the emf for a Au + 0.07 at. % Fe vs. Chromel P thermocouple as a function of applied magnetic field, relative to a reference junction temperature of 4.22°K. The hot junction temperatures are listed next to each curve. (——) Present work : (— - —) $T = 5.27$ and 6.98°K (Ref. 4); (– – –) $T = 20$°K (Ref. 4).

von Middendorff[4] on the same thermocouple combination. To make this comparison, it was necessary to integrate his thermocouple sensitivity data and to use values of zero-field sensitivity listed in tables prepared by the NBS. One possible source of the discrepancy between von Middendorff's data and the present work may be due to the fact that his thermocouples did not agree well[5] in zero field with the values in the NBS tables, particularly below 15°K. The zero-field data for our own thermocouple material agreed with the NBS listing to better than 6% at all temperatures. It has been established[2] that variation in thermocouple emf's of up to 6% can exist for different Au + 0.07 at.% Fe specimens in the 4–20°K range.

Measurements are now in progress at our laboratory on a constantan versus Chromel P thermocouple. These experiments indicate that the percentage change of the emf for this combination is three to four times smaller than for the Au–Fe versus Chromel P unit and that it is negative (i.e., the thermocouple becomes less sensitive in high magnetic fields). Von Middendorff has also tested[4] a constantan versus Chromel P thermocouple up to 60 kG and has observed somewhat smaller, positive changes of the emf. This discrepancy is not understood at present. Studies now underway on this type of thermocouple are directed at elucidating the effects of such things as material inhomogeneity and improper thermal tempering on the magnetic-field-induced change in thermocouple emf. An indication as to the maximum temperature errors which might be incurred by using the two types of thermocouples in strong magnetic fields is given in Table I.

Summary of Temperature Errors

The magnetic-field-induced temperature errors for various low-temperature sensors are summarized in Table I. The entries represent the maximum temperature errors in the various field ranges.

Below 4.2°K both the Speer and Allen-Bradley resistors can be used in all three field ranges. However, at 0.5°K the resistance of a 10-Ω Allen-Bradley unit is already 20 kΩ and is increasing rapidly, so that its use much below this temperature might be restricted. This is not true for the Speer resistors. In the 4.2–20°K region the Allen-Bradley resistors are a good choice over the entire magnetic field range, for reasons discussed previously[6]. Another attractive possibility is the thermistor, although it should be noted that a given unit is useful over only a very narrow temperature range. While the germanium thermometer is useful in range I, its large, orientation-dependent magnetoresistance[7] obviates its use at higher fields. Generally, this thermometer should be confined to use in zero field. If experimental conditions require that only a single sensor be used for the entire 4–300°K interval, then the GaAs diode may be recommended in ranges I and II, with proper calibration procedures. As can be seen, there is a spread to the values of $\Delta T/T$ in Table I for the GaAs diodes, reflecting differences in design and operational characteristics of various units. Platinum resistance thermometers are an excellent choice above 30°K in all field ranges because they are so well characterized and highly reproducible. The measurement of small temperature differences, in certain experiments, is often most conveniently accomplished in zero field with thermocouples. The data in Table I indicate that even in strong magnetic fields, at temperatures above 10°K it is practical to correct for magnetic-field-induced temperature errors when using thermocouples. The $SrTiO_3$ glass-ceramic capacitance thermometer has been described in the recent literature.[3] It is still too early to make definite statements about its future in high-accuracy measurements of the absolute temperature. As indicated in Table I it is the preeminent choice for use as a transfer standard of the temperature in high magnetic fields; that is, as a comparison or null-detector thermometer, as used in the present work on thermocouples.

References

1. L.J. Neuringer and L.G. Rubin, in *Temperature, Its Measurement and Control in Science and Industry*, Instrument Society of America, Pittsburgh, Pennsylvania (1972), Vol. IV, Part 2.
2. L.L. Sparks and R.R. Powell, *J. Res. Nat. Bur. Std.* **76A**, 263 (1972), and references therein.
3. W.N. Lawless, *Rev. Sci. Instr.* **42**, 561 (1971); W.N. Lawless, R. Radebaugh, and R.J. Soulen, *Rev. Sci. Instr.* **42**, 567 (1971); L.G. Rubin and W.N. Lawless, *Rev. Sci. Instr.* **42**, 571 (1971).
4. A. von Middendorff, *Cryogenics* **11**, 318 (1971).
5. A. von Middendorff, private communication.
6. L.J. Neuringer and Y. Shapira, *Rev. Sci. Instr.* **40**, 1314 (1969).
7. L.J. Neuringer, A.J. Perlman, L.G. Rubin, and Y. Shapira, *Rev. Sci. Instr.* **42**, 9 (1971).

Hydrodynamic Theory Correlation for 72 Helium Evaporators*

Stephen G. Sydoriak

Los Alamos Scientific Laboratory, University of California
Los Alamos, New Mexico

Introduction

Critical channel boiling heat flux data suitable for use in the design of cryogenic magnets immersed in liquid helium have been published by several laboratories[1–7] but have not yet been fully correlated. It was shown by Wilson[1] that an approximation equation of a 1957 hydrodynamic theory[8] based on the assumptions of frictionless and homogeneous flow of the two phases could be fitted to most of his results (excepting data on large channel gaps and heights) with an "apparent critical exit quality" x_{cr} (the mass fraction of vapor in the effluent) of 0.3. At the same conference Sydoriak and Roberts[2] showed that the efficiency E, which they defined as the ratio of the observed flux Φ_{ideal} for the "ideal evaporator" ($x_{cr} = 1$) of the theory,† was nearly independent of the temperature in their "tall chimney" experiments and of channel height. Subsequently Johannes and Mollard[7] measured x_{cr}, found it to be somewhat higher than Wilson's "apparent" (i.e., fitted) value, and concluded that the theory assumption of homogeneity was questionable for large gaps. They found their own data to agree to $\pm$ 20% with the empirical correlation equation of Lehongre *et al.*,[4] $\Phi_L = [a_1 + a_2(Z/D_e)^{a_3}]^{-1}$, where Z is the height of the channel and D_e is four times the ratio of channel cross-sectional area A to the heated perimeter. The constants a_1, a_2, and a_3 had been fitted by Lehongre *et al.* to data on various channel shapes whose dimensions, unfortunately, cannot be determined from the published figure of Φ_L vs. Z/D_e. More recently Ogata *et al.*[5] fitted a portion of their extensive data to an equation like that of Lehongre *et al.* but with $a_3 = 1$. In none of these correlations was all of the available data used and in general the unused data are a very poor fit to the published equation.

In the present report we shall show how the hydrodynamic theory can be used as the basis for a more elaborate correlation than that of Wilson and one which fits all of the available data,‡ with an average deviation of $\pm 10\%$. The method is identical in detail with that used recently[9] to correlate the critical natural convection boiling data for N_2 and H_2.

* Work performed under the auspices of the U.S. Atomic Energy Commission.

† To a first approximation E is equal to the square root of Wilson's "apparent critical quality."

‡ The data of Ref. 5 were not available at the time the calculations were made, and as explained, data of Ref. 4 were not sufficiently complete to be usable.

The Hydrodynamic Theory of Channel Boiling

The hydrodynamic channel boiling theory[8] consists in a calculation of the rate at which the available hydrostatic forces can accelerate and remove the vapor being produced by the heat power input, thereby allowing more liquid to approach and be vaporized. The flow of the mixture is assumed to be frictionless and homogeneous. It was shown that with these assumptions Bernoulli's differential equation for one-dimensional flow could be integrated to yield the result

$$\dot{Q}_{\text{theo}}^2(\rho_L - \rho_V)(\lambda^2 A^2 \rho_L \rho_V \,\Delta x)^{-1} - \rho_L g Z_e + \rho_L g Z_m = 0 \quad (1)$$

where $\dot{Q}_{\text{theo}}$ is the heat power input, assumed to be uniformly distributed along the axis, A is the (constant) cross-sectional area of the channel, $\Delta x = x_2 - x_1$ is the increase in quality (mass fraction of vapor) between inlet and exit, Z_e is the hydrostatic head of liquid equivalent to the pressure drop in the channel, and Z_m is a complicated term representing liquid head equivalent to the mixture within the channel. The liquid density, vapor density, and latent heat of vaporization are given by ρ_L, ρ_V, and λ, respectively, and g is the gravitational acceleration. Operating pressures for optimizing the heat input or the magnetic field were discussed in Ref. 8. More recently[2] the advantages of an unheated "chimney" above the heated part of a channel were demonstrated and the concepts of an ideal evaporator ($\Delta x = 1$) and of evaporator efficiency $E = Q_{\text{obs}}/Q_{\text{ideal}}$ were introduced.

Agreement between Experiments and the Theory

As an aid to the understanding of a real evaporator and to the choice of appropriate correlation equations, we shall briefly review the results of direct comparisons of Eq. (1) and forced flow N_2 and H_2 data and with critical heat flux natural convection ^{4}He data. These comparisons necessarily exclude experiments in which mass flow rates and hence Δx were not determined.

For nitrogen and hydrogen it is now known[9] that there are areas of excellent agreement (at high flow velocities and large channel cross sections) between the observed heat input $\dot{Q}$ and the Eq. (1) value $\dot{Q}_{\text{theo}}$ in the work of three laboratories[8,10,11] on forced flow evaporators undergoing both nucleate and film boiling. Where agreement of the forced flow data was not excellent the qualitative behavior was such as to suggest that neither friction nor inhomogeneity was an important cause of the deviations, but that deviations appeared to be due primarily to a reduction in A caused by a static layer of bubbles or film attached to the heated wall.

For helium [Fig. 1, calculated from Eq. (1) and the data of Johannes and Mollard[7]] the situation is qualitatively different in that (1) $\dot{Q}/\dot{Q}_{\text{theo}}$ *decreases* with increasing gap, and (2) there is no evidence of presence of a static film, since this would tend to cause $\dot{Q}/\dot{Q}_{\text{theo}}$ to level off at low values of S. We are therefore led to anticipate the need in any correlation equation representative of helium evaporators for a term which will account for the effects of inhomogeneity, but see no need to allow for a term representative of a static film.

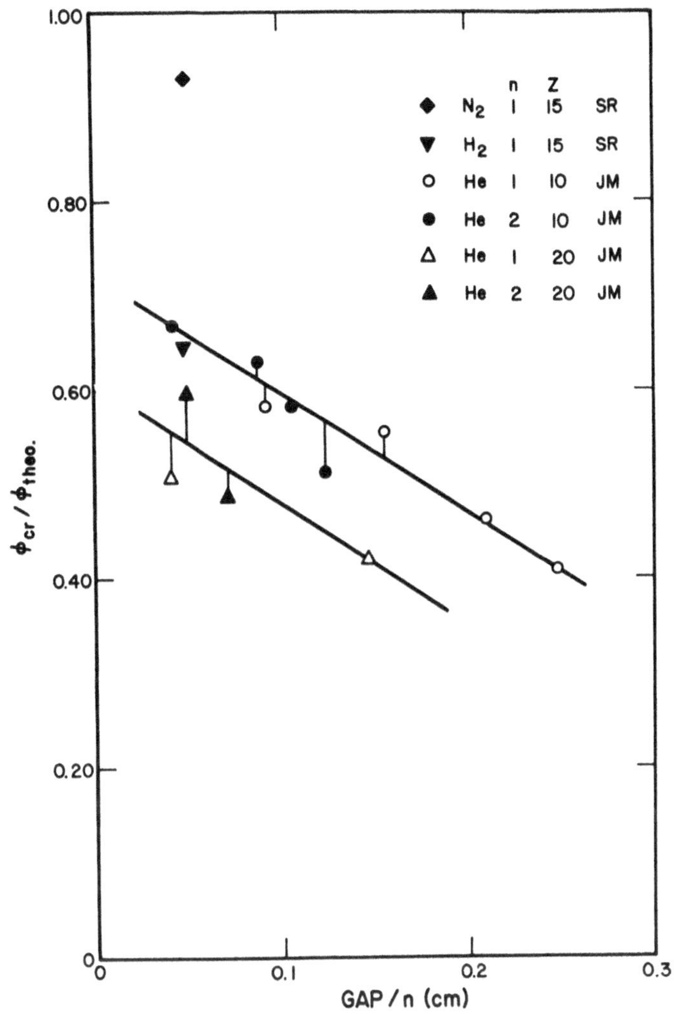

Fig. 1. The ratio of observed to theoretical [Eq. (1)] heat flux to parallel plates of height Z, n of which are heated, versus the gap between the plates divided by n. SR and JM data are from Refs. 8 and 7, respectively.

Analytical Correlation Equations

The hydrodynamic theory is especially useful for the understanding and correlation of critical natural convection heat flux data *not* accompanied by Δx measurements. If Z is the height of a vertical natural convection evaporator, then $Z_e = Z$. Assuming $\rho_L \gg \rho_V$, which is usually the case, and giving Δx its maximum value of unity, we find the Z_m term to be small. Equation (1) applied to an ideal evaporator therefore reduces to $\dot{Q}_{\text{ideal}} = A\lambda(\rho_V \rho_L g Z)^{1/2}$ or, expressed as the heat flux from n (= 1 or 2) heated walls of a parallel plate ideal evaporator,

$$\Phi_{\text{ideal}} = S\lambda(\rho_V \rho_L g/Z)^{1/2}/n \tag{2}$$

We have made a least-square fit of the data of five laboratories[1-3,6,7] on 72 different parallel plate evaporators to the empirical equation

$$E = \Phi/\Phi_{\text{ideal}} = \frac{a}{(S + b_1 + b_2 n)} \tag{3}$$

obtaining $a = 0.0587$ cm, $b_1 = 0.111$ cm, and $b_2 = 0.037$ cm, with S also in cm. The mean percentage deviation of the 26 "$n = 1$" points was 8.5%, while that of the 46 "$n = 2$" points was 10.2%, for an overall average deviation of 9.6%. Deviations are seen in Figs. 2 and 3 to be quite random, in most cases. We have compared the fit of the same body of data to the Ogata correlation, $\Phi_0 = (1.67 + 0.0239nZ/S)^{-1}$, and find the average deviation to be 12.9% for $n = 1$ data and 22% for $n = 2$ data.

Temperatures at which the correlated data of Eq. (3) were taken ranged from 3.95°K to the normal boiling point, 4.2°K. At 4.2°K Eq. (3) reduces to $\Phi = 1.80(S + 0.111 + 0.037n)^{-1}Z^{-1/2}n^{-1}$, with S in cm and the critical flux in W/cm^2.

The data of Ogata *et al.* for $Z = 5$ and $Z = 40$ are also shown in Fig. 2. Although they were not available to us at the time of the fitting calculations, they are seen to be in general agreement with the correlation equation. Their data for $Z = 10$ and $Z = 20$ have not as yet been analyzed.

Since S appears in both numerator and denominator, the fitted equation can be solved for the case of a vertical wall in bulk liquid, $S \to \infty$. It is interesting to note that for $Z > 2$ the five relevant measurements of three laboratories[6,8,9] average 80% of our correlation equation values and appear to follow a $Z^{-0.5}$ dependence. On the other hand, the data of Ref. 4 have a slope of $Z^{-0.2}$ and for $Z > 2$ they average about a factor of two below our correlation equation. Such large variations

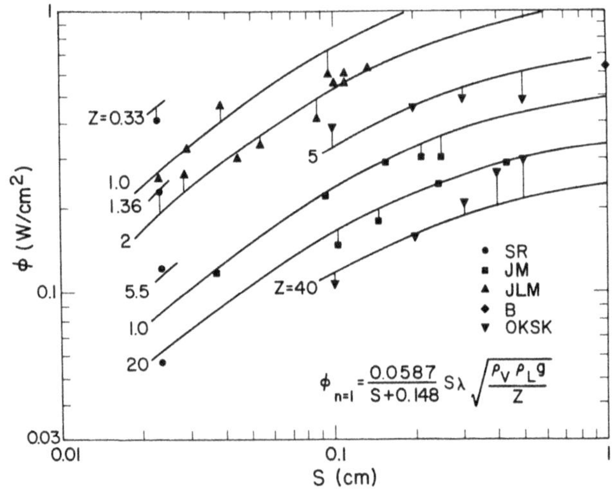

Fig. 2. Critical nucleate boiling heat flux for helium in a parallel plate evaporator. The plates are S cm apart and only one is heated ($n = 1$). Authors SR, JM, JLM, B, and OKSK are given in Refs. 2, 7, 6, 3, and 5, respectively. OKSK data were not available at the time the calculations were made but are seen to be in general agreement with the correlation equation.

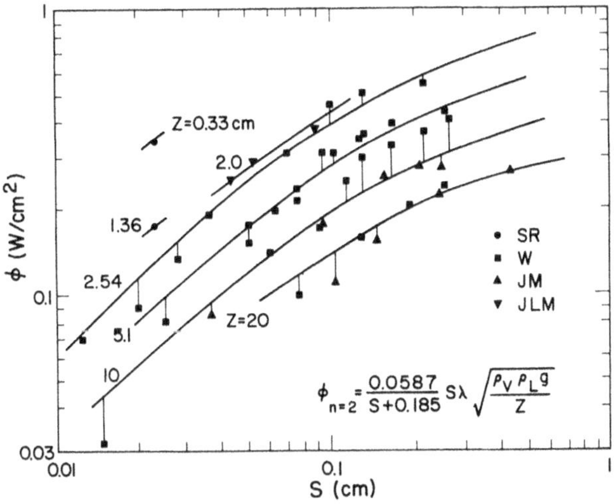

Fig. 3. Critical boiling heat flux to helium between plates S cm apart, both plates being uniformly heated. Authors SR, W, JM, and JLM are given in Refs. 2, 1, 7, and 6, respectively.

in the work of different laboratories are typical of variations also found in pool boiling data, for which values ranging from 0.8 W/cm^2 (Ref. 12) to 1.8 W/cm^2 (Ref. 13) have been reported for helium.

References

1. M.N. Wilson, IIR Comm. I Meeting, *Annexe 1966–5 Bulletin IIF/IIR*, (1966) pp. 109–114.
2. S.G. Sydoriak and T.R. Roberts, IIR Comm. I Meeting, *Annexe 1966-5 Bulletin IIF/IIR*, (1966) pp. 115–123.
3. H.F. Bowman, Thesis, Massachusetts Institute of Technology, Cambridge, Massachusetts (1966).
4. S. Lehongre, J.C. Baissin, C. Johannes, and A. de la Harpe, in *Proc. 2nd Intern. Cryogenics Engrg. Conf.* (1969), p. 274.
5. H. Ogata, M. Kenmochi, S. Sato, and H. Kimura, *Cryogenic Engineering (Tokyo)* **4**, 219–225 (1969).
6. G.B. James, K.G. Lewis, and B.J. Maddock, *Cryogenics* **10**, 480 (1970).
7. C. Johannes and J. Mollard, *Adv. Cryogenic Engrg.* **17**, 332 (1972).
8. S.G. Sydoriak and T.R. Roberts, *J. Appl. Phys.* **28**, 143–148 (1957).
9. S.G. Sydoriak, Cryogenic Engrg. Conf., August 1972, paper S-5.3; LA-DC-72-260.
10. R.J. Richards, R.F. Robbins, R.B. Jacobs, and D.C. Holten, *Adv. Cryogenic Engrg.* **3**, 375 (1960).
11. R.C. Hendricks, R.W. Graham, Y.Y. Hsu, and R. Friedman, NASA TN D-765 (1961) and D-3095, (1966).
12. R.D. Cummings and J.L. Smith, *Intern. Inst. of Ref. Annex 5*, (1966) p. 85.
13. D.N. Lyon, *Adv. Cryogenic Engrg.* **10**, 371 (1965).

Indium–Barrier–Indium Tunneling Junctions

M. D. Jack and G. I. Rochlin

Department of Physics, University of California
and Inorganic Materials Research Division
Lawrence Berkeley Laboratory, Berkeley, California

Introduction

We have successfully produced high-quality In–insulator–In tunnel junctions with resistance at 4°K varying over 12 orders of magnitude—from 0.01 Ω to 0.01 TΩ.

Junction preparation involved the following new technique. An indium film was deposited onto a glass substrate held at liquid nitrogen temperature in a vacuum of 10^{-5}–10^{-6} Torr. After deposition the chamber was valved off and the sample gradually allowed to warm overnight to room temperature. The film was then placed in a reaction chamber consisting of a copper tube with two high-voltage copper electrodes at one end and a roughing line and gas bleeding line at the other. The position of the sample relative to the electrodes was the critical factor in varying the resistance of these junctions. High-impedance junctions faced toward the electrodes, while low-impedance junctions faced away. The vapor from a high-molecular-weight fluorocarbon liquid was continuously pumped through the chamber while a dc plasma discharge was maintained at pressures ranging from 300 to 750 μm for periods of 5 sec to a few minutes. Five cross strips of indium were then deposited at room temperature.

Junctions of the following three types were produced: (1) intermediate junctions, in the range 1–100 Ω, which exhibited pronounced structure in their d^2V/dI^2 characteristic, reflecting the In phonon spectrum; (2) low-impedance junctions, in the range 0.01–1 Ω, which exhibited Josephson junction characteristics; (3) high and ultrahigh junctions in the range 10,000 Ω to 10,000 MΩ.

Intermediate-Resistance Junctions

In addition to reproducing the general features of the phonon structure observed for aluminum–aluminum oxide–indium junctions by various authors,[1-3] we have resolved additional fine structure throughout the region of indium acoustic phonon energies.

In Fig. 1 we compare our raw second derivative data for an In–I–In junction at 0.9°K with the data taken by Rowell *et al.*[3] for an Al–I–In junction taken at 0.3°K. The main differences are fairly evident: a peak instead of a plateau at 7 meV (as measured from the energy gap $2\Delta_0 = 1.05$ meV); complex structure in the range 8–11 meV with a sharp dip at 9 meV; broad humps beyond the longitudinal phonon

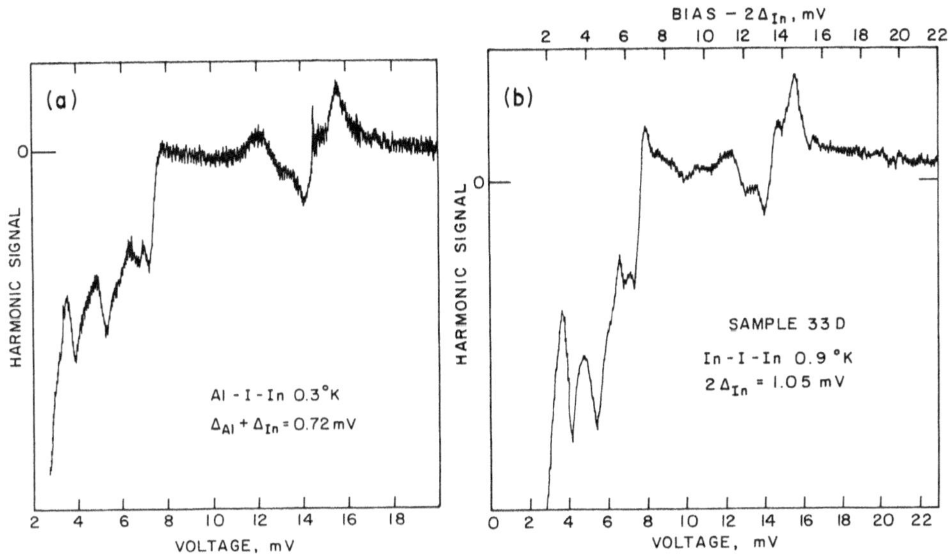

Fig. 1. (a) Second derivative trace for an Al–*I*–In junction at 0.3°K (Ref. 3). (b) Second derivative trace for an In–*I*–In junction at 0.9°K (this work).

peak at 14 meV. In addition, we observed a small peak at 1.6 meV not shown in Fig. 1.

Most of these additional features can be related to structure in the phonon density of states. Recent neutron data by Smith and Reichardt[4] was interpreted using a Born–von Karman (BvK) model to obtain the phonon spectrum for In. Their data indicate a saddle point at approximately 7 meV and a local maximum at 9 meV, corresponding to the peak at 7 meV and the dip at 9 meV in Fig. 1. The broad spectrum beyond the longitudinal peak is most likely attributable to sums and harmonics of the transverse and longitudinal peaks, reflecting the phonon density of states at these higher energies.[5,6] The low-energy peak at 1.5 meV does not reflect any appreciable structure in the constructed BvK spectrum.

These same intermediate junctions also clearly exhibit inelastic tunneling processes via excitation of vibrational modes of molecules in the barrier (Fig. 2). Peaks in the d^2T/dV^2 vs. V characteristic indicate the onset of new tunneling processes at $V = 2\Delta_0 + h\nu$, where $2\Delta_0$ is the superconducting energy gap, 1.05 mV for In, and $h\nu$ is the energy of a vibrational or rotational mode of a barrier molecule. Peaks at 35 and 64 mV were recently observed by Yanson,[7] who attributed them to phonons in INO, which he indicated was the major component of the barrier for his In–I–In junctions. While we do not deny the possibility of some InO in our barrier, our evidence suggests that the major constituent of our barrier is a polymer structure of the type $(C_2F)_n$ bonded to the indium film. Hence we attribute the peak centered at 62 mV not to InO but to In–C stretching modes which lie in the range 58–71 mV for different molecular weight compounds.[8] We also note that InF has a sharp mode at 66 mV.[9] In addition, peaks corresponding to strong vibrational modes of C–F bonds are observed at 54, 79, and 113 mV.[10]

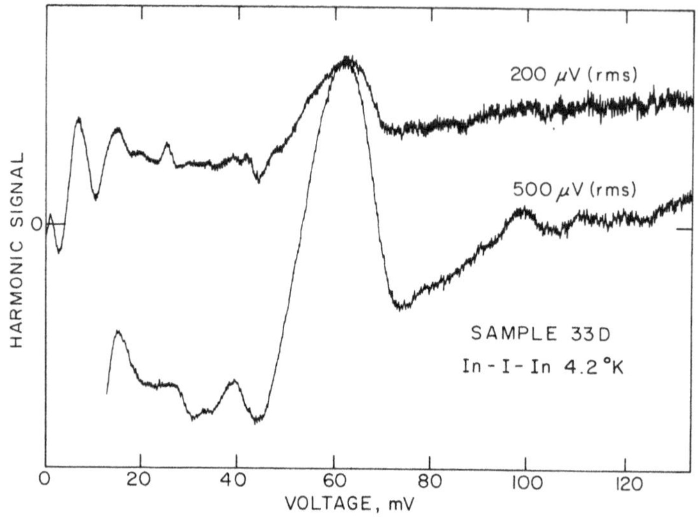

Fig. 2. Second derivative trace for the same junction as in Fig. 1, taken at 4.2°K, above T_c for In. Though there is some residual indium phonon structure for voltages below 20 mV, the higher voltage structure is due to vibrational or rotational excitations of molecules comprising the barrier.

Low-Impedance Junctions

Junctions were also made which exhibited a Josephson current which could be nulled with a field of about 5 G. In zero field the maximum Josephson current for the best of these junctions was 87% of the theoretical result of Ambegaokar and Baratoff,

$$I_{\max} = \frac{\pi \Delta_0}{2R_{nn}}$$

R_{nn} is the normal-state junction resistance. Fulton and McCumber[11] predict a reduced maximum Josephson current of 78.8% and 91.1% of $I_{\max}$ for Pb and Sn junctions, respectively, when strong coupling effects are taken into account. An intermediate value of 87% for In junctions appears quite reasonable since the deviation of the tunneling density of states from the BCS theory is roughly 5%, 1%, and 0.5% for Pb, In, and Sn, respectively.[12]

Fisk steps characteristic of self-resonant junction modes were observed for many low-impedance junctions. We can obtain an approximate value of the ratio of the barrier thickness to the dielectric constant using the expression for the step spacing[13] $\Delta V = hc/4eL(t/d\varepsilon)^{1/2}$ (here ε is the dielectric constant, L is the width of junction, t is the barrier thickness, and $d = 2\lambda + t$, with λ the penetration depth). Typical low-resistance junctions of width 0.3 mm showed a Fisk step spacing of 50 μV, corresponding to $t/\varepsilon = 3$ Å.

High-Impedance Junctions

At the other extreme we have made junctions whose resistance at zero volts and 4°K was at least 10^{10} Ω. A typical trace of the I–V characteristics of one of these junctions is shown in Fig. 3. Gap structure could not be resolved in this case because of possible leakage currents or noise. Note that in the region of the energy gap the total normal current is only 10^{-13} A. Junctions of 1 MΩ did show some gap structure and some phonon structure at higher biases. Capacitance measurements were made and the capacitance of these junctions varied as the junction area. For a typical area of 3.3×10^{-4} cm^2 we measured 100 pF. This amounts to approximately 0.3 μF/cm^2. Interestingly enough, the combination of high impedance and reasonable capacitance necessitated a very slow sweep rate in determining the I–V characteristics. Junctions with an intrinsic RC relaxation time of up to 15 sec have been produced. As determined by capacitance bridge measurements, typical high-impedance diodes had a ratio of $t/\varepsilon = 30$ Å, an order of magnitude larger than the value for the low-impedance junctions. We have attempted to determine the barrier height using the two independent measurements of t/ε. The difference between barrier thicknesses should scale as the logarithm of the ratio of the corresponding junctions resistances, assuming the WKB approximation is valid in both cases, i.e., $R \cong C \exp(\phi_B^{1/2} t)$, where C is assumed independent of dielectric thickness, ϕ_B is the barrier height expressed in eV, and t is the thickness in Å. Using a dielectric constant of about two appropriate for a polymer like polytetrafluoroetylene, we obtain a barrier height of approximately 0.25 eV. This is probably too low because of the questionable validity of the WKB approximation for the very low-resistance Fisk mode junctions, and because of the additional conductance due to nontunneling processes for the high-resistance junctions.

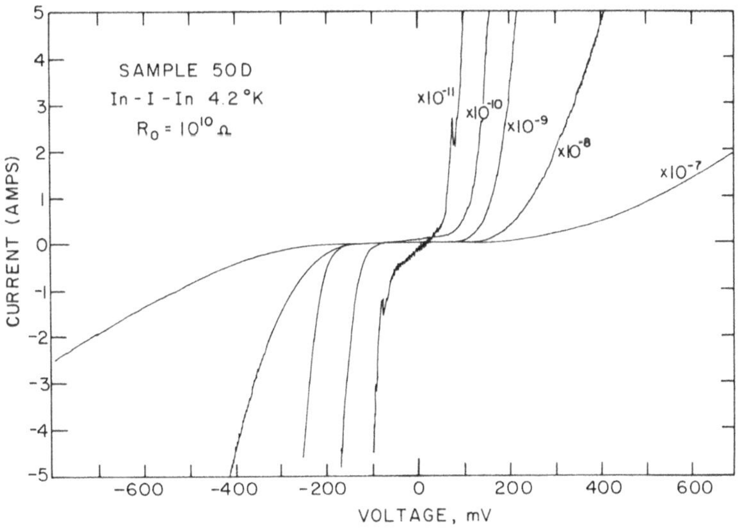

Fig. 3. I–V characteristics of a very high-impedance In–I–In junction taken at 4.2 K. The resistance at 0 V is 10,000 MΩ.

We note that unlike most high-resistance junctions formed from oxides (and usually four orders of magnitude lower resistance), these junctions are very insensitive to transients and rather callous treatment.

Conclusion

We have succeeded in producing tunneling junctions using a material, In, which is a "bad actor" with regard to forming oxide barrier junctions.[2,7] We have observed much of the typical good tunneling junction characteristics and in the process obtained additional information concerning the In phonon density of states and the effect of strong coupling on the magnitude of the Josephson current. In addition, we have produced anomalously high-resistance junctions which are physically very interesting and may very well have thin-film circuitry applications. Studies are now under way applying our barrier production technique to the formation of junctions composed of other materials which cannot readily be utilized with conventional techniques.

Acknowledgments

We wish to acknowledge helpful discussions with Drs. P. K. Hansma, J. M. Rowell, H. G. Smith, F. Szalkowski, G. Somorjai, and G. C. Pimentel.

References

1. J.G. Adler and J.S. Rogers, *Phys. Rev. Lett.* **10**, 217 (1963).
2. J.M. Rowell and L. Kopf, *Phys. Rev.* **137**, 907 (1964).
3. J.M. Rowell, W.L. McMillan, and R.C. Dynes (to be published).
4. H.G. Smith and W. Reichardt (to be published).
5. P. Morel and P.W. Anderson, *Phys. Rev.* **125**, 1263 (1962).
6. J.M. Rowell, private communication.
7. J.K. Yanson, *Soviet Phys.—JETP* **33**, 951 (1971).
8. F. Oswald, *Z. Anal. Chem.* **197**, 309 (1963).
9. R.F. Barrow, D.V. Glaser, and P.B. Zeeman, *Proc. Royal Soc.* **A68**, 962 (1955).
10. P.J.H.Woltz and A.H. Nielsen, *J. Chem. Phys.* **20**, 307 (1951).
11. T.A. Fulton and D.E. McCumber, *Phys. Rev.* **175**, 585 (1968).
12. W.L. McMillan and J.M. Rowell, *Superconductivity*, Vol. I, R.D. Parks, ed., Marcel-Dekker, New York (1969).
13. D.N. Langenberg, D.J. Scalapino, and B.N. Taylor, *Proc. IEEE* **54**, 560 (1966).

Thermal Expansion Coefficient of $La_2Mg_3(NO_3)_{12} \cdot 24H_2O$ and $Ce_2Mg_3(NO_3)_{12} \cdot 24H_2O$ between 10 and 300°K

J. Carp-Kappen, C. W. de Boom, H. J. M. Lebesque, and B. S. Blaisse

Laboratory of Applied Physics
Delft University of Technology, Delft, The Netherlands

Experimental Technique

The linear expansion of the crystal is measured with a Fizeau interferometer. The reflecting planes of the interferometer have a diameter of 20 mm. Three pieces of crystal determine the distance of the planes. These pieces are about 6 mm high and their difference in length is such that an interference pattern with three to five dark lines is formed. A helium–neon gas laser with a wavelength of 632.8 nm is used as a light source.

Prior to each measurement the temperature of the interferometer is stabilized to within 0.01°K, after which the interferometer is either cooled, or heated by about 15°K, and the temperature is stabilized again. The number of interference fringes passing a certain mark as a result of the temperature changes is counted. The temperatures are measured by a platinum resistance thermometer, calibrated at the Kamerlingh Onnes Laboratory in Leiden.

Results

Crystals of these double nitrates grow as thick trigonal or hexagonal plates. Crystals with a length of at least 5 mm are used for measurements in the direction of the *c* axis. Single crystals of this thickness often have a diameter of 60 mm.

The quality of these large, single crystals was usually imperfect: completely clear regions with dimensions of about 8 mm^3 were interspersed with opalescent regions, or with thin layers of a micaceous structure. These thin layers often had an orientation perpendicular to the trigonal axis. Sometimes the position of these imperfect parts in the crystal seemed arbitrary.

It is more than likely that in the less clear regions of the crystal the correct crystal structure is interspersed with holes filled with a saturated solution of the salt.

Measurements of LMN* have been carried out on seven sets of pieces. For the measurements of CMN† there was only one set available for each direction.

In Figs. 1 and 2 experimental data of $\alpha_{\parallel}$ and $\alpha_{\perp}$ are represented. From these

* Lanthanum magnesium nitrate hydrate.
† Cerium magnesium nitrate hydrate.

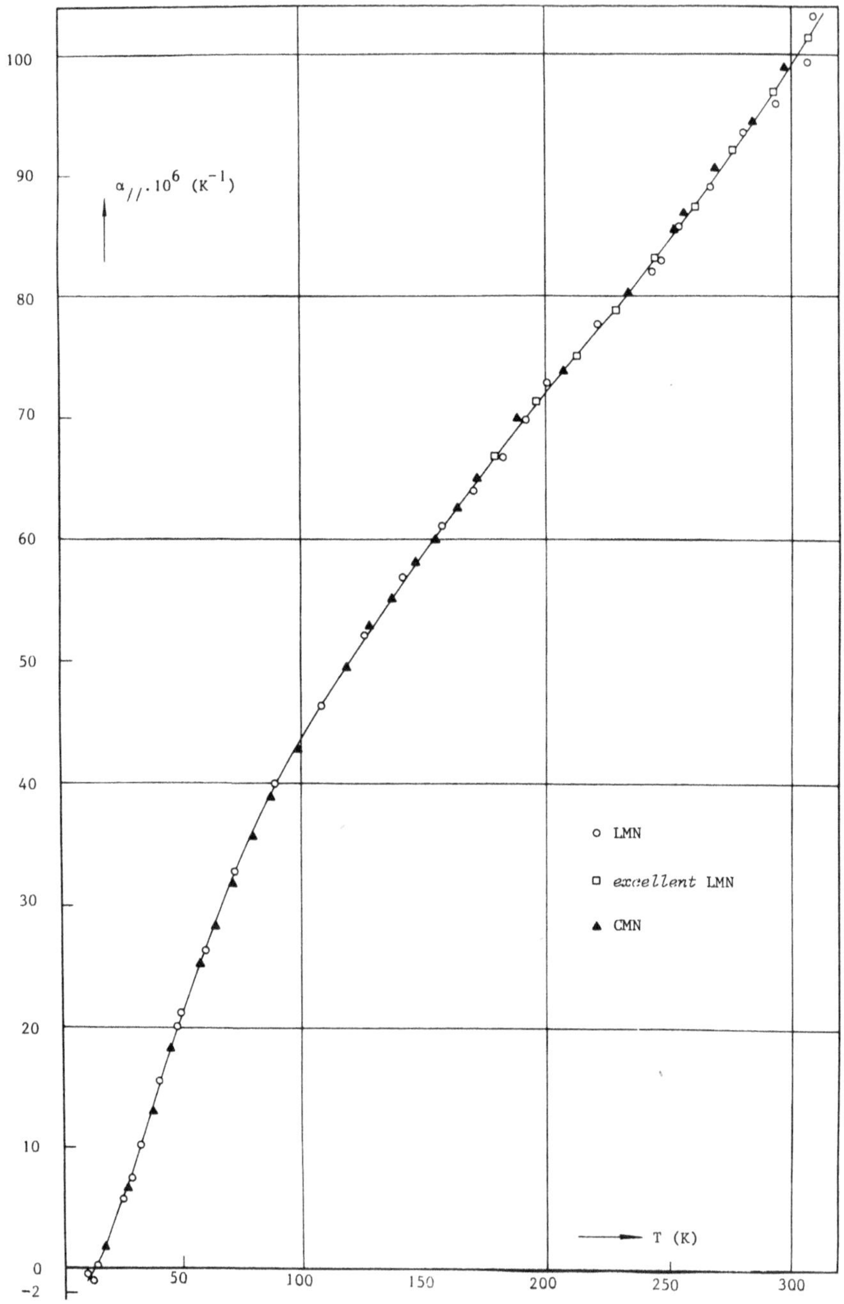

Fig. 1. Linear thermal expansion in the direction of the trigonal axis for LMN and CMN.

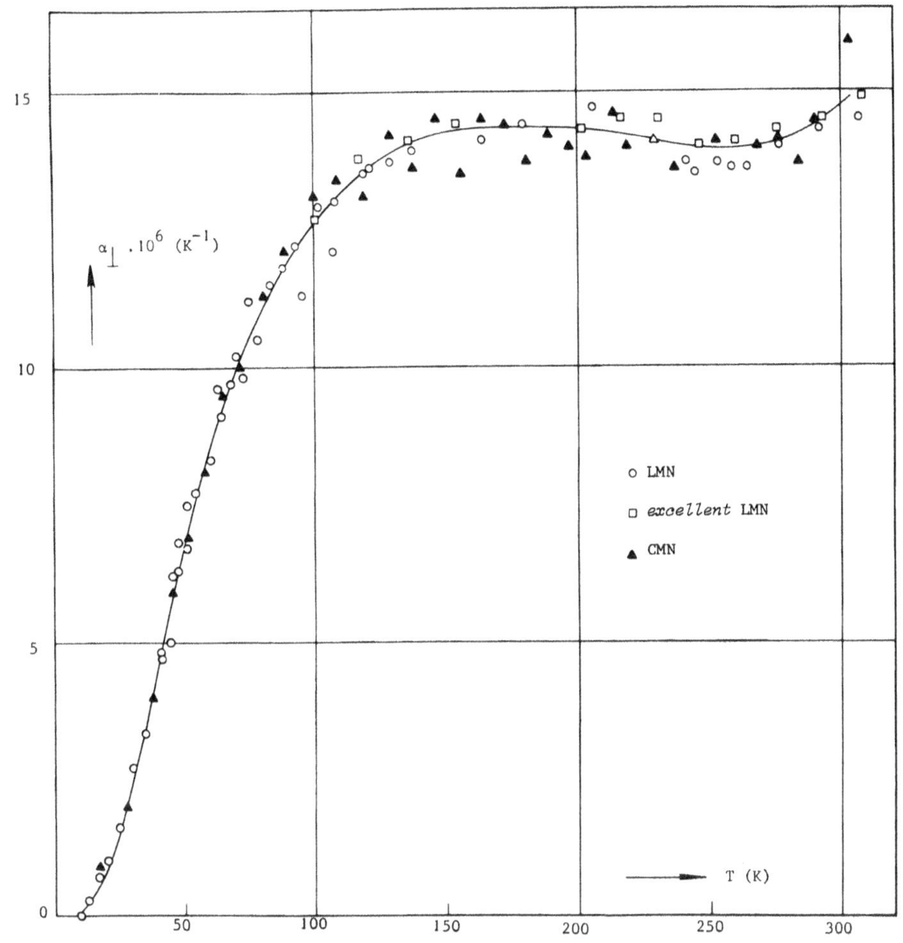

Fig. 2. Linear thermal expansion in a direction perpendicular to the trigonal axis for LMN and CMN.

figures one can see that the thermal expansion coefficient is the same for LMN and CMN, within the accuracy of the measurements.

One completely clear LMN crystal became available after the thermal expansion coefficient had been determined by means of the less clear crystals with a reasonable precision over the whole temperature region. In this $9 \times 25 \times 40$ mm^3 crystal no faults were visible to the naked eye.

For each of the two directions some measurements have been carried out with this clear crystal in that part of the temperature region where the measurement points were spread far apart. The points found with the perfect crystal show a smaller scatter, but they do not differ significantly from the points found with imperfect crystals. (See Figs. 1 and 2.)

It is remarkable that at a temperature of 300°K the thermal expansion coefficient $\alpha_{\parallel}$ is seven times larger than $\alpha_{\perp}$. At 10°K it was found that $\alpha_{\parallel}$ is negative, while $\alpha_{\perp}$

is almost zero. Therefore the volume thermal expansion is negative at that temperature.

Negative thermal expansion coefficients have also been found in other crystalline materials, such as ZnS.[1] Two conditions must be fulfilled for a negative thermal expansion coefficient far away from a phase transition: The temperature must be much lower than the Debye temperature and certain types of transverse waves must be present in the crystal.[2]

Calculations

The thermal expansion coefficients α and the relative expansions $\Delta l/l_0 = \int_0^T \alpha d\tau$ at 10, 20, ..., 290, 300°K are calculated from the experimental data, where l_0 is the length at room temperature. The results are given in Table I.

The lengths of the c and a axes of CMN have been measured by several investigators.[3,4] One can calculate the lengths of these axes at other temperatures, using the relative expansion.

The maximum error in the thermal expansion coefficient, as given in Table I, depends on the magnitude of the thermal expansion coefficient. Values for this maximum error are listed in Table I next to the values of the thermal expansion coefficient.

Measurements of a single crystal of rock salt were carried out with the same apparatus as was used to measure the thermal expansion of the double nitrates. The results that we obtained at 25 different temperatures (between 9 and 290°K) were compared with results contained in the literature.[5,6] A very good agreement was found, showing that no systematic errors were made in $\alpha_\parallel$ and $\alpha_\perp$.

Discussion

The lengths of the crystal axes of CMN at 4.2, 78, and 302°K have been measured by Schiferl.[4] His results can be compared with the thermal expansion coefficients

Table I. Linear Thermal Expansion Coefficients α, the Maximum Error in α, and the Relative Expansions at Various Temperatures for LMN and CMN

T, °K	$\alpha_\parallel \times 10^6$, °K^{-1}	$\alpha_\perp \times 10^6$, °K^{-1}	$(\Delta l/l_0)_\parallel \times 10^4$	$(\Delta l/l_0)_\perp \times 10^4$
10	− 0.7(± 0.3)	0.1(± 0.1)	− 0.1	0.0
20	3.2(± 0.3)	1.0(± 0.3)	0.0	0.0
30	8.5(± 0.4)	2.6(± 0.2)	0.6	0.2
40	14.8(± 0.6)	4.6(± 0.3)	1.8	0.6
50	20.4(± 0.6)	6.8(± 0.4)	3.5	1.1
60	25.9(± 0.7)	9.6(± 0.5)	5.8	1.9
80	35.8(± 0.8)	11.0(± 0.5)	12.0	3.9
100	43.8(± 0.9)	12.6(± 0.5)	20.0	6.3
120	50.2(± 1.0)	13.6(± 0.6)	29.4	8.9
160	61.2(± 1.1)	14.3(± 0.6)	51.7	14.5
200	72.2(± 1.3)	14.3(± 0.6)	78.5	20.3
250	84.6(± 1.4)	14.0(± 0.6)	117.6	27.3
300	99.0(± 1.5)	14.7(± 0.6)	163.4	34.4

that we found: The changes in the length of the c axis measured by Schiferl are in exact agreement with our $\alpha_{\parallel}$.

The changes in length of the a axis over the whole temperature region are much larger than one can calculate from our $\alpha_{\perp}$. The difference for the temperature region 78–302°K is 28%; for the region 4.2–78°K it is −30%.

The results obtained have a bearing on the method of mounting the crystals for low-temperature research. In a number of cases the only possibility is that of gluing a crystal to a rod. In this case a surface perpendicular to the c axis must be glued, since the shrinkage in this plane due to the lowering of the temperature is uniform and much smaller than in a plane parallel to the axis. Moreover, materials exist with approximately the same linear expansion coefficient as $\alpha_{\perp}$.

It is not surprising that no difference has been found between the thermal expansion coefficients of LMN and CMN. Lanthanum and cerium are both rare earth metals. Their atomic weight is 139 and 140, respectively, so one can expect them to be much alike. It will be necessary to carry out some measurements on a few other double nitrates in order to gain some information concerning the variations within the group of double nitrates. Therefore we intend to continue our investigations of the double nitrates taking, for example, bismuth zinc nitrate hydrate.

Acknowledgments

The authors wish to thank Dr. P. Bennema for his suggestions about the crystal faults, Mr. W. H. van der Linden for preparing the crystals, Mr. A. de Vreede for making a reliable thermostat, and Mr. H. Kats for his help in the experimental work.

References

1. R.R. Reeber and G.W. Powell, *J. Appl. Phys.* **38**, 1531 (1967).
2. M. Blackman, *Phil. Mag.* **3**, 831 (1958).
3. A. Zalkin, J.D. Forrester, and D.H. Templeton, *J. Chem. Phys.* **39**, 2881 (1963).
4. D. Schiferl, *J. Chem. Phys.* **52**, 3234 (1970).
5. P.P.M. Meincke and G.M. Graham, *Can. J. Phys.* **43**, 1853 (1965).
6. R.W. Roberts and R. Ruppin, *Phys. Rev. B* **4**, 2041 (1971).

A ^{3}He Thermal Switch for Ultra-Low Temperatures

L. E. Reinstein

Department of Physics, Boston University
Boston, Massachusetts

and

J. A. Gerber*

Department of Physics, Tufts University
Medford, Massachusetts

Thermal isolation at low temperatures is essential for several types of measurements. In both specific heat and adiabatic demagnetization experiments it is necessary to cool a specimen to a given low temperature, and then to quickly isolate it from its surroundings. At high enough temperatures this is easily done with helium exchange gas, which is pumped out at an appropriate time. But at temperatures below 0.3°K, where this cannot be done, the standard method is the use of the superconducting heat switch. This type of switch is not totally satisfactory since it allows for the possibility that the specimen will be heated by eddy currents produced during fast switching, and that it will be incompletely isolated.

We have developed a new type of heat switch which incorporates most of the advantages of the exchange gas method and none of the disadvantages of the superconducting heat switch method. The switch we have developed is schematically illustrated in Fig. 1. The upper chamber is connected by a strong thermal link to the mixing chamber of a ^{3}He/^{4}He dilution refrigerator. Suspended within this chamber, but thermally isolated from it, is the sample to be cooled and then isolated. This is achieved by forcing liquid ^{3}He into the chamber from a lower reservoir to act as a heat conductor between the chamber walls and the sample. When the sample is in thermal equilibrium with the chamber walls the liquid ^{3}He is quickly dropped back into the lower reservoir, leaving the sample totally isolated and at a temperature equal to that of the mixing chamber.

To move the liquid from the lower to the upper chamber, we rely on the vapor pressure of the ^{3}He itself. For instance, at 0.60°K the vapor pressure of ^{3}He is sufficiently high to support a column of liquid ^{3}He 9.1 cm high. Thus by heating the lower reservoir we can drive the liquid ^{3}He slowly into the upper chamber, where it will thermally link the sample to the mixing chamber and enable the mixing chamber to cool the sample to its ultimate low temperature.

In order to lower the total ^{3}He charge from the top chamber to isolate the

* Supported by a grant from the National Science Foundation.

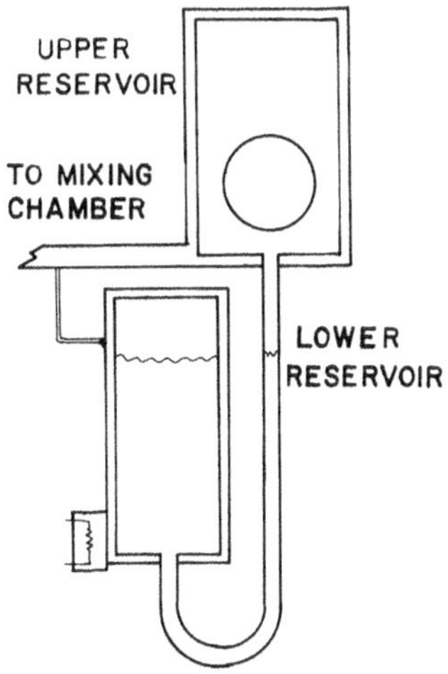

Fig. 1. Schematic diagram of liquid ^{3}He heat switch.

sample, we turn off the heat input to the lower reservoir. The lower reservoir begins to cool slowly through the weak heat link between it and the mixing chamber. At a lower reservoir temperature of 0.6°K a drop of 0.01°K in temperature is sufficient to lower the column of liquid ^{3}He 1 cm and entirely empty the upper chamber. The drop in the liquid level and in the temperature of the lower reservoir is accelerated by the cold liquid from the upper chamber entering and dramatically cooling the lower reservoir. This causes an avalanche which isolates the sample within about 1 min and brings all the liquid into the lower reservoir at a temperature of about 0.2°K. At this temperature the vapor pressure of ^{3}He is approximately 10^{-5} Torr, which guarantees decent thermal isolation.

Figure 2 is a more detailed diagram of the heat switch. We used a cylindrical copper mass as our initial sample to test the operating characteristics of the switch. It is supported by an AGOT graphite tube of 10 mil wall thickness. The upper chamber is connected to the lower reservoir through a 10-mil-I.D. capillary tube. A fine copper wire is the weak thermal link between the lower reservoir and the mixing chamber. This is kept small enough so as not to cause deterioration in the operation of the mixture refrigerator. A copper cylinder and a stainless steel tube in the lower reservoir comprise the two plates of a capacitor used to monitor the level of the liquid helium. Clearly, when the ^{3}He leaves the lower reservoir the capacitance decreases. The temperatures of the lower reservoir, the copper sample, and the mixing chamber are determined using calibrated carbon resistors.

A typical switching sequence is illustrated in Fig. 3. In this sequence we are heating the lower reservoir which initially contains the liquid. At t_1 the liquid reaches the upper chamber and makes thermal contact with the copper mass. This causes the copper mass to quickly cool to the temperature of the mixing chamber, after which they both begin warming because of the influx of warmer ^{3}He into the upper chamber. Also at t_1 we notice that the capacitance in the lower reservoir decreases, indicating that liquid is indeed leaving the lower reservoir.

At t_2 the capacitance levels off, which indicates that all liquid has been pushed out of the lower reservoir. As we see, this enables the lower reservoir to come to

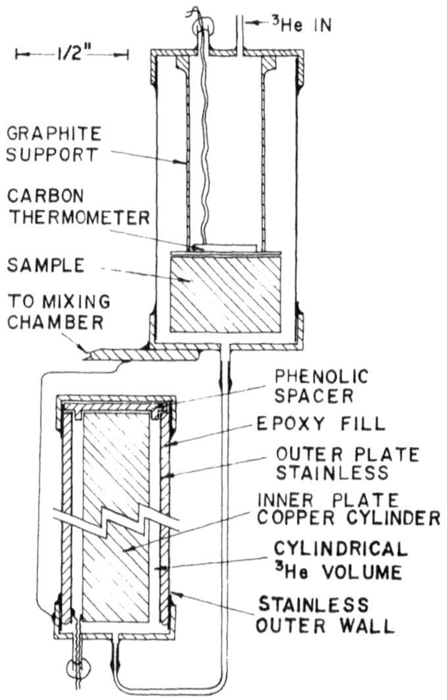

Fig. 2. Detailed drawing of liquid ^{3}He heat switch.

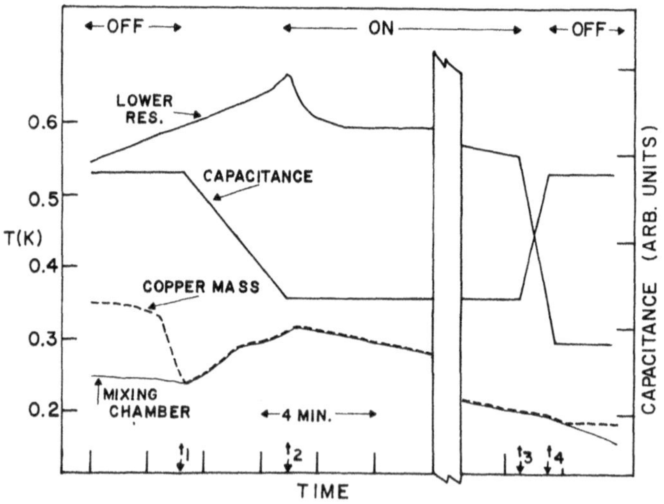

Fig. 3. Graph of the lower reservoir temperature, the copper sample temperature, the mixing chamber temperature, and the capacitance in the lower reservoir all as functions of time, during a typical heat switch operating sequence.

an equilibrium temperature of nearly 0.6°K. This is a very stable operating state of the switch. Since no more warm liquid enters the upper chamber after t_2, the mixing chamber can begin cooling down the copper sample. The ultimate low temperature reached is limited only by the cooling capacity of the mixing chamber.

We turn the switch off at t_3 by ceasing to heat the lower reservoir. Within about 1 min all of the liquid drops to the lower reservoir. The capacitance at t_4 has returned to its original value, the lower reservoir has cooled down, and the mixing chamber becomes isolated from the copper mass. With this switch we have seen thermal conductivity ratios of up to 30 from the on state to the off state. With further improvements we expect to be able to increase this ratio considerably.

In summary, we believe that this heat switch has numerous advantages over the superconducting switch. There is no eddy current heating when it is switched on and off. It can operate unaffected by high magnetic fields. It can be used in experiments in which unwanted trapped magnetic flux could affect the data. In certain ultralow-temperature NMR experiments it can be used for making thermal contact when the varnishes and greases used ordinarily would cause large unwanted NMR peaks. Ultimately, greater thermal isolation of the sample being cooled can be achieved.

Heat Capacity Measurements of Small Samples—Granular Al Films*

C. N. King, R. B. Zubeck, and R. L. Greene†

Department of Applied Physics
Stanford University, Stanford, California

Introduction

In the study of the low-temperature properties of new materials it is often necessary to work with small samples. In this paper we describe a calorimeter in which samples with masses as small as 1 mg and heat capacities as small as 0.5 erg/°K have been measured. Typically, heat capacities of 10 ergs/°K and samples with masses of 50 mg can be routinely measured in an apparatus of simple design. As an example of the capability of these techniques, measurements of the heat capacity of a 1-mg "granular" aluminum film with an enhanced superconducting T_c of 2°K will be reported.

Apparatus Design

The basic components of a calorimeter are a heater, a thermometer, a sample container, and a heat link to the external surroundings. The masses of these components have heat capacity, called the addendum heat capacity, which is of necessity measured along with the sample heat capacity. The addendum heat capacity C_{add} must be minimized in order to measure accurately the heat capacity C_s of a small sample. Ideally, the condition

$$C_{\text{add}} \ll C_s \tag{1}$$

should be met.

To minimize C_{add}, a calorimeter was designed that eliminated all addenda except for the thermometer and its electrical leads. The thermometer used in this design is a silicon resistance device developed by Bachmann *et al.*[1,2] These devices are ideal in the temperature range from 1 to 15°K because of their high sensitivity, small heat capacity, and high thermal conductivity. The silicon thermometers made in our laboratory are surface devices. The advantage of a surface device over a bulk device is that two separate resistive elements may be diffused onto one face of a single silicon chip ($\sim 5 \times 5 \times 1$ mm). One resistive element serves as a thermometer and the other element serves as the sample heater. Contact to the resistive elements is made through diffused degenerate contact pads. Electrical lead wires

* Research supported by U.S. Air Force Office of Scientific Research, Office of Aerospace Research, under Grant No. AFOSR 68-1510C.

† Present address: IBM Research Laboratories, San Jose, California.

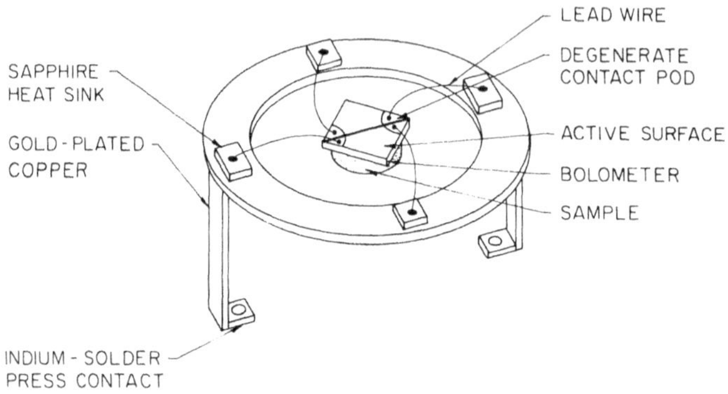

Fig. 1. Ring calorimeter.

are spot-welded to gold films which have been vacuum-deposited on the degenerate contact pads. The other face of the silicon chip is insulating at low temperature and is used as the sample-mounting platform. In Fig. 1 a complete calorimeter is shown which consists of the silicon thermometer–heater, its electrical leads, plus mounting ring and heat sinks.

The electrical lead wires from the thermometer–heater must be thermally anchored to, but at the same time electrically isolated from, the ring. This is accomplished by using either sapphire or quartz heat sinks, which are excellent thermal conductors but are electrical insulators. Gold films are vacuum-deposited on each face of the sapphire or quartz flats. One face of the flat can then be indium-soldered to the copper ring. The wires are indium-soldered to the other face of the crystal heat sink.

An alternate thermometer that works well in the temperature range from 2–50°K is an Au + 0.07 at.% Fe versus Au thermocouple.[3] The thermocouple can be spot-welded or thermal-compression-bonded to an insulating silicon or diamond flat. The sample heater is an evaporated metallic strip of Cr + 5 at.% Ti.

Samples with masses of 50 mg and with one flat face can be easily attached to the silicon with a small quantity (0.2–0.4 mg) of Wakefield grease. The samples are demounted with the aid of trichlorethylene. Metallic films with masses of ~ 1 mg can be attached by sputtering directly onto the silicon.

Measurement Methods

With this single design three different measurement methods can be used: the classic adiabatic pulse method, the ac method, and the relaxation-time method. Which method is most convenient depends on the sample size, temperature range to be covered, and the sample-to-ring relaxation time τ_1, where

$$\tau_1 = (C_s + C_{add})/K_w \tag{2}$$

and K_w is the thermal conductance of the electrical lead wires. In our experiments we tailor τ_1 to our needs by using different alloy wires to vary K_w.

The adiabatic pulse method is the standard large-sample technique and requires thermal isolation of the sample from its surroundings (i.e., τ_1 very long) during the

Table I. Comparison of Methods for Measuring Small-Sample Heat Capacities

Method	Defining relation	Time constant, sec	Useful sample size,* mg	Condition for 1% accuracy
Pulse				
(adiabatic calorimeter)	$C = \Delta Q/\Delta T$	$\tau_1 > 10$	> 100	$\tau_1/50 >$ pulse length $> 5\tau_2$
Relaxation time	$C = k\tau_1$	$\tau_1 > 0.5$	1–100	$\tau_1 > 100\,\tau_2$
AC temperature	$T_{AC} = P/\omega C$	τ_1 short	> 1	$\omega\tau_1/10 > 1 > 10\omega\tau_2$

* By using silicon thermometers the useful sample size can be extended down to 1 mg.

actual measurement to minimize the heat leak correction. The ac method was introduced by Sullivan and Seidel[4] and Handler *et al.*[5] to measure small changes in heat capacity. Previously the relaxation time technique[6] has not been used extensively. It has been found that this method often provides the best way to make absolute measurements of small heat capacities. Table I presents a comparison of the three methods and the criteria which τ_1 must satisfy for accurate results. In the table τ_2 refers to the internal time constant of the sample thermometer system.

In comparing the relaxation time and adiabatic pulse methods one sees that longer time constants (i.e., larger C_s for the same K_w) are necessary in the adiabatic pulse method. For this reason the relaxation time method is better suited for small samples. Also, since the temperature changes much more rapidly in the adiabatic pulse method, τ_2 effects are more of a problem. The advantage of the ac method is that a continuous output proportional to $1/C$ is available. Thus this is an ideal method for sweeping through sharp structure in the specific heat such as phase transitions. The disadvantage of the ac technique is that sometimes it is impossible to satisfy both inequalities in Table I, and only relative, rather than absolute, data can be obtained. This problem is compounded by the fact that τ_2 effects are rather subtle and are difficult to correct for. They tend to show up as a change in phase of the signal. In the relaxation time method τ_2 effects are readily seen as nonexponential decay of the temperature, and corrections can be easily made.[7]

Even though the calorimeter design minimizes the addenda, accurate addenda correction must be made for very small samples. Such addenda as the finite distributed heat capacity of the wires must be considered. It has been shown [6,7] that the addenda correction for the relaxation time method is $C_{add} = C_t + 1/3C_w$ where C_t is the thermometer (silicon) heat capacity and C_w is the total wire heat capacity.

As a check of these methods, we measured the heat capacity of a 50-mg silicon chip between 1.5 and 10°K. The heat capacity of the silicon was as small as 0.5 erg/°K. The data agreed with published data[8] within 2%.

Granular Al Film Measurements

With this calorimeter the specific heat of small grain-size (granular) Al films from 1.3 to 8°K was measured. Whereas pure Al has a T_c of 1.19°K, our films had a T_c of 2°K. Enhancement of T_c in metal films was discovered in resistivity measurements on films evaporated onto a 4°K substrate by Buckel and Hilsch[9] in the 1950's.

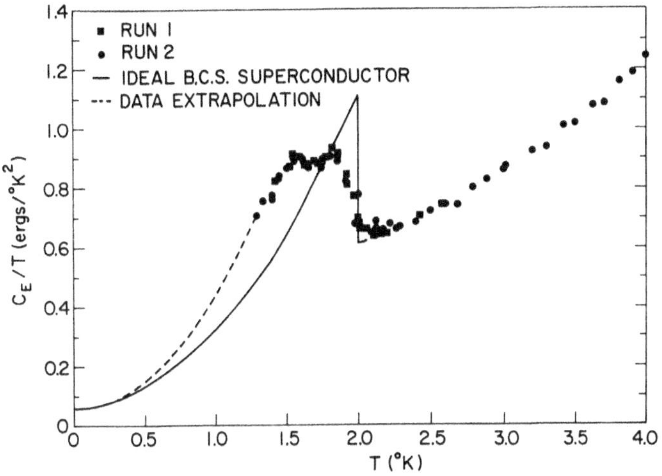

Fig. 2. The heat capacity of a granular Al film and addendum in the superconducting region. The solid line represents the heat capacity of an ideal BCS superconductor with the same electronic heat capacity coefficient as the sample and $T_c = 2.0°$K.

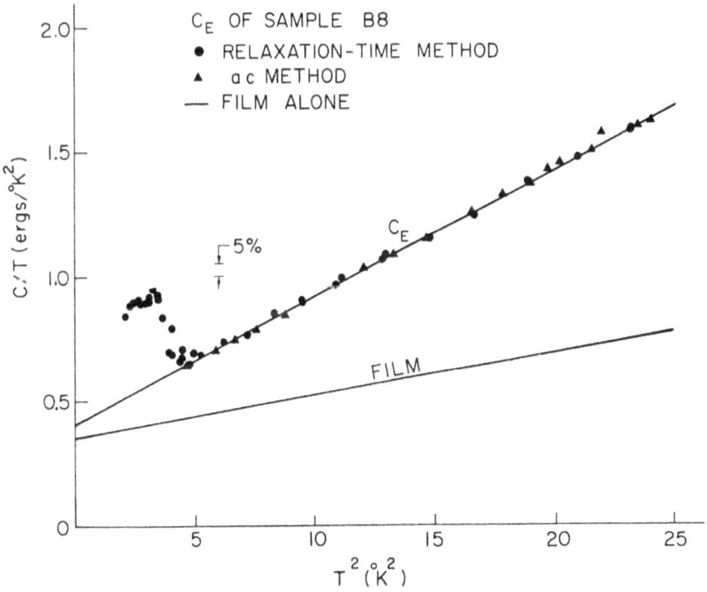

Fig. 3. Heat capacity of a granular Al film and addendum. Correcting for the addendum gives the normal heat capacity of the film alone as shown by the lower line. The electronic specific heat coefficient of the film is $\gamma = 1.37 \pm 1 \times 10^{-3}$ J/°K^2 mole and $\theta_D = 306 \pm 15$°K. These compare with $\gamma = 1.35 \times 10^{-3}$ J/°K^2 mole and $\theta_D = 428$°K for pure Al.[12]

Since then other workers have devised several ways to prepare films with enhanced T_c. It appears that the common characteristic of all of these films is a small grain size (~ 100 Å). The present measurements are the first to determine whether this enhancement is a bulk effect and the first to measure the parameters $N(E_F)$ and θ_D for these films.

The samples were 25-μm-thick Al–Al_2O_3 films with a mass of 1 mg sputtered on a silicon thermometer. The samples were prepared by Hauser.[10] The resistive T_c was 2.06°K and the residual resistivity was 160 $\mu\Omega$-cm. The grain size ranged from 100 to 250 Å, as determined by X-ray diffraction. Electron microprobe and chemical analysis gave the film composition as 83% ± 4% Al.

The results of the specific heat measurements are shown in Figs. 2 and 3. The sample heat capacity was as small as 0.5 erg/°K in these experiments. The entropy under the superconducting transition establishes that the entire volume of the sample is superconducting, although the sample is somewhat inhomogeneous. Within experimental accuracy the electron density of states at the Fermi surface is the same as that of pure Al. The lattice heat capacity is increased in the granular films such that the measured θ_D is 28% less than that of pure Al. This result indicates that the low-frequency phonon spectrum of the granular Al films is significantly different from that of pure Al, in agreement with the tunneling data of Leger and Klein.[11] The data imply that it is a change in the phonons rather than a change in the electron density of states that causes the enhanced T_c of granular Al films.

References

1. R. Bachmann, H.C. Kirsch, and T.H. Geballe, *Rev. Sci. Instr.* **41**, 547 (1970).
2. R. Bachmann, H.C. Kirsch, and T.H. Geballe, *Sol. St. Commun.* **9**, 57 (1971).
3. R.L. Rosenbaum, *Rev. Sci. Instr.* **40**, 557 (1969).
4. P.F. Sullivan and G. Seidel, *Phys. Rev.* **173**, 679 (1968).
5. P. Handler, D.E. Mapother, and M. Rayl, *Phys. Rev. Lett.* **19**, 356 (1967).
6. R. Bachmann, F.J. DiSalvo, Jr., T.H. Geballe, R.L. Greene, R.E. Howard, C.N. King, H.C. Kirsch, K.N. Lee, R.E. Schwall, H.-U. Thomas, and R.B. Zubeck, *Rev. Sci. Instr.* **43**, 205 (1972).
7. C.N. King, Ph.D. Dissertation, Stanford University, Stanford, California (1972).
8. F. Flubacher, A.J. Leadbetter, and J.A. Morrison, *Phil. Mag.* **4**, 273 (1969).
9. W. Buckel and R. Hilsch, *Z. Physik* **138**, 109 (1954).
10. J.J. Hauser, *Phys. Rev. B* **3**, 1611 (1971).
11. A. Leger and J. Klein, *Phys. Lett.* **28A**, 751 (1969).
12. N.E. Phillips, *Phys. Rev.* **114**, 676 (1969).

A 4-MHz Tunnel Diode Oscillator with 0.001 ppm Measurement Sensitivity at Low Temperatures

Craig T. Van Degrift*

University of California, Irvine
Irvine, California

Tunnel diode oscillators have been used for measurements in a wide range of experiments in low-temperature physics since their introduction in 1966 by Boghosian *et al.*[1] Tunnel diodes are one of the few types of commercial semiconductor devices that operate at low temperatures and, in fact, appear to become increasingly temperature independent down to 1°K. These devices can be incorporated into oscillator circuits which act as transducers of changes in position, electric and magnetic susceptibility, and surface impedance to changes in frequency. The changes in frequency may be easily delivered to room temperature and measured with extremely high precision.

Unfortunately, in spite of their widespread use, no systematic study has been published of their performance characteristics at low temperature, such as the dependence of the frequency on temperature, bias voltage, and magnetic field, or even of the intrinsic frequency noise and drift. This paper reports on these parameters and suggests possible areas of further improvement.

Figure 1 is a schematic of our 4-MHz oscillator. A dc bias current of 74 μA enters via the coaxial cable, and biases the diode in its negative resistance region. Since $C_c \ll C_B$ and $R_p \ll |R_n|$, where $|R_n|$ is the magnitude of the negative resistance, at rf frequencies the circuit is essentially an *LC* resonator with negative resistance across a portion of the inductor. The coupling of the diode to the resonator is such that the negative resistance barely maintains the oscillation. R_p suppresses parasitic oscillations due to resonances of diode capacity with the tapped section of the inductor. If the parasitic is not suppressed, all performance characteristics of the oscillator will be seriously degraded. The output coupling capacitor C_c is chosen to allow only the minimal signal required for accurate frequency measurement to pass to room temperature. This electrically isolates the oscillator as much as possible from the outside world, especially from the temperature-dependent impedance of the coaxial cable and from room-temperature rf noise. It should be noticed that we have chosen to vary the temperature of the resonator while maintaining the remainder of the circuit at a fixed bath temperature, the connection being made by leads of poor thermal conductivity.

The choice of the resonator materials must be made with considerable care to minimize the temperature dependence of the frequency. The capacitor and coil

* Supported in part by the National Science Foundation, Grant GP-10699.

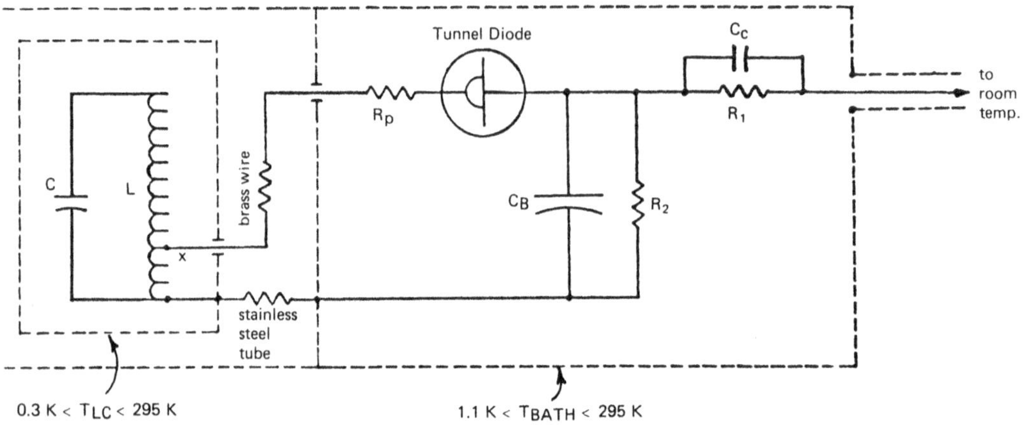

Fig. 1. Schematic of oscillator circuit. Typical values for the components are (in Ω,pF, and μH): $R_1 = 5110$, $R_2,R_p = 1000$, $C_c = 20$, $C_B = 10{,}000$, $C = 200$, $L = 8.0$. The diode is a BD-6 with a negative resistance of $-5600\ \Omega$ at 4°K and the tapping fraction x is 0.23.

housing are fabricated from annealed OFHC copper and the coil from annealed copper wire (0.25 mm diameter) of 99.999% purity. The copper plates of the capacitor are spaced by three small (0.4 mm diameter) sapphire spheres. The coil is wound on a sapphire rod (6 mm diameter) containing less than 3 ppm of magnetic impurity. Sapphire and copper are excellent materials with respect to thermal expansion; also, the surface impedance of copper has a very small temperature dependence. We decided at an early time not to use superconductors in the resonant circuit until we fully understood all aspects of its operation with a normal metal and because we wished to avoid the restriction imposed by a critical temperature and magnetic field.

At room temperature the weak rf signal (0.1 μV into 50 Ω) is separated from the dc current source, amplified, and heterodyned to 20 kHz. The 20 kHz is phase-locked to the reference oscillator of a lock-in amplifier using a time constant of 0.1 sec. The period of the reference oscillator is then measured with an electronic counter.

The frequency dependence on the resonator temperature is shown in Fig. 2. Above 10°K the temperature dependence is dominated by the surface impedance of the copper wire coil. The temperature and frequency are such that the surface impedance must be calculated by the theory of the anomalous skin effect (Ref. 2, p. 313). In that limit the first temperature-dependent correction to the surface impedance is proportional to the dc resistivity and hence to T^5. The thermal expansion of copper contributes a smaller amount which varies as T^2 below 4°K and as T^4 between 4 and 24°K. The thermal expansion of sapphire is not shown but is estimated to be about 3% of the T^4 component of copper. The contribution from the thermal expansion of copper is calculated from the geometry of the LC circuit and the known behavior of copper with no adjustable parameters. The surface impedance was graphically curve-fitted, giving a magnitude about 20% greater than that calculated from the geometry. Higher-order corrections to the anomalous surface impedance will cause the curve to increase less rapidly above 22°K. Between

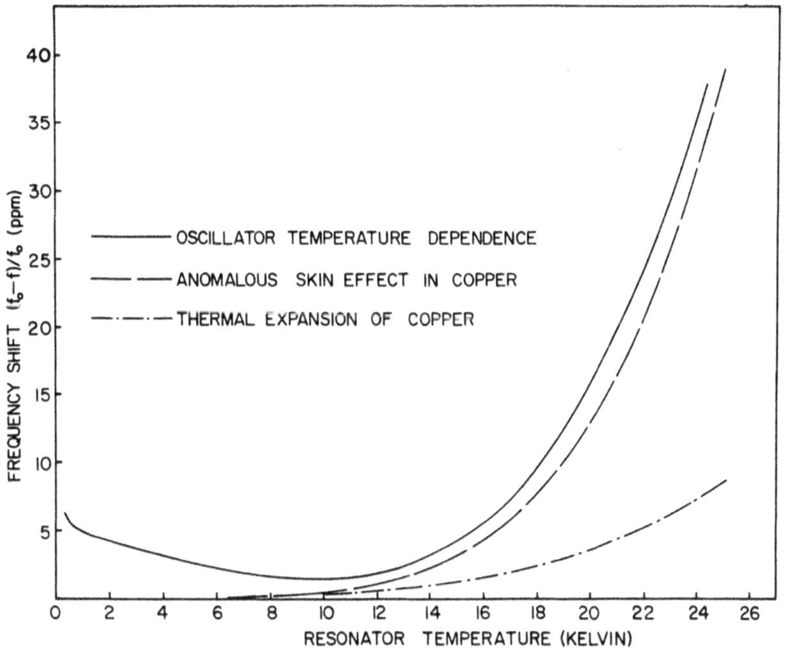

Fig. 2. Fractional frequency change versus resonator temperature. f_0 is a convenient reference frequency.

10 and 0.3°K the frequency decreases from its maximum [the minimum of $(f_0 - f)/f_0$] by 4 ppm. We believe that about 30% of this decrease is the result of a Kondo rise in the copper resistance and the remainder is due to the small amount of magnetic impurity in the sapphire coil form. Unfortunately, since we do not accurately know the concentration of the various magnetic impurities in the sapphire, we cannot claim to understand the temperature dependence below 10°K. More recently we have found that by using a fused quartz tube for the coil form and a different heat treatment of the copper wire the total deviation below 10°K could be reduced to 25% of that shown in Fig. 2.

In Fig. 3 we show the dependence of the frequency on the dc voltage (measured at the coaxial cable) for several different temperatures and several variations of the circuit. The relative positions of these curves have been arbitrarily chosen since it was impractical to remove the much greater temperature dependence due to the thermal expansion and the inductive part of the surface impedance. The curves of Fig. 3(a) correspond to a circuit where the diode drives the whole coil (tapping fraction $x = 1.0$). In this case the circuit oscillates below 265°K. The slope of the frequency vs. bias curve at 262°K may be calculated from the bias voltage dependence of the junction capacity of the diode, which is increasing approximately linearly over this small range of bias voltage. As the temperature is decreased further the Q increases and the diode is driven further into the nonlinear portions of its operating curve; there is a resulting dip in the frequency vs. bias curve. This dip is centered slightly below the bias corresponding to the minimum value of the magnitude of the negative resistance and continues to grow as the temperature decreases to 30°K.

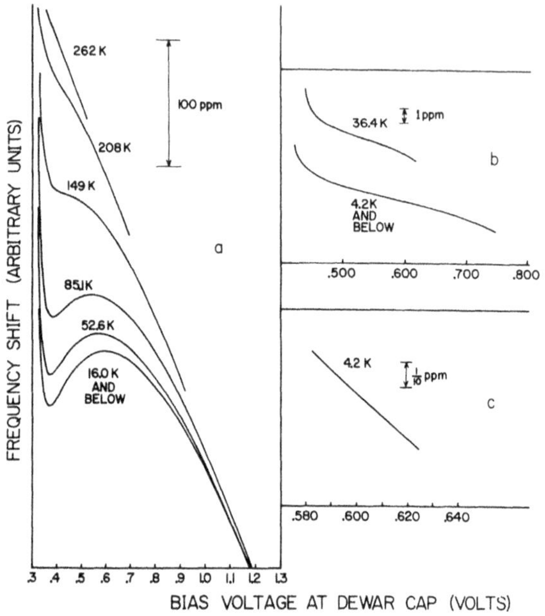

Fig. 3. Frequency versus bias voltage measured at the room-temperature side of R_1. (a) an untapped circuit, (b) the same circuit tapped at $x = 0.23$, and (c) the same circuit tapped at $x = 0.23$ but using a BD-406 diode in place of the BD-6.

Below 30°K the Q becomes nearly temperature independent and hence the bias voltage curves become temperature independent.

Figure 3(b) gives curves at two temperatures for the same circuit as Fig. 3(a) except that the coil has been tapped at $x = 0.23$. In this case the circuit oscillates below 47°K. The frequency scale of these curves is about 20 times smaller than that of the untapped case. Figure 3(c) shows this frequency versus bias curve when a BD-406 diode is substituted for the BD-6. The BD-406 diode is an epoxy packaged diode with considerably reduced case capacity. This particular unit was very well matched to the tuned circuit at 4.2°K, so that no dip was apparent. It should also be noted that tapping the tuned circuit by a factor of x decreases the junction capacitance contribution by x^2. Finally, if the bias current is regulated to 10 ppm, the corresponding frequency deviations are one part in 10^{10} for the characteristics in Fig. 3(b) or 3(c). Hence for 0.001-ppm measurements the bias voltage dependence is no problem.

We briefly investigated the magnetic field dependence of the frequency up to 36 Oe and found that it varies as H^2, with $(f_0 - f)/f_0 H^2 = 3 \times 10^{-12}\,\mathrm{Oe}^{-2}$. This is roughly the expected magnitude and sign of the frequency shift one would expect from the change in surface impedance caused by the magnetoresistance of the copper wire in the coil (Ref. 2, p. 237).

The frequency depends on the bath temperature with a coefficient of about 0.1 ppm/°K from 4 to 1°K, which is sufficiently small that the bath temperature may be easily maintained constant for a 0.001-ppm measurement. It is interesting that this dependence is apparently caused by the metal film resistors and the silver mica capacitors rather than the diode.

We have found that the frequency noise may be quantitatively calculated to within 20% with no adjustable parameters by the use of a formula used in the noise analysis of quartz crystal and atomic frequency standards.[3] It is derived by assuming that the noise in the oscillating feedback loop perturbs the phase of the oscillator, causing it to perform a random walk. This is called "perturbation noise." It is given by

$$\left.\frac{\Delta f}{f}\right|_{\mathrm{rms}} = \frac{1}{Q}\left[\frac{4k(T+T_n)(1/8\tau)}{P}\right]^{1/2}$$

where Q, k, T, T_n, P, and τ are, respectively, the quality factor of the resonator, the Boltzmann constant, the resonator temperature, the noise temperature of the diode, the power dissipated in the resonator, and the averaging time of the counter. The values of these quantities for the circuit of Fig. 1 are $Q = 360$, $T = 4°\mathrm{K}$, $T_n = 290°\mathrm{K}$ (shot noise determined from diode dc $I-V$ curve), and $P = 0.77$ erg/sec (calculated from measured output voltage by linear circuit theory). We find that

$$(\Delta f/f)|_{\mathrm{rms}} = (4.5 \times 10^{-10}\ \mathrm{sec}^{1/2})/\sqrt{\tau}$$

At $\tau = 3.9$ sec this becomes 2.3×10^{-10}, whereas we measure a value of $(2.5 \pm 0.1) \times 10^{-10}$. In addition to "perturbation noise," there is "addition noise"[3] which is caused by amplitude noise added to the signal by the succeeding amplifiers. In our circuits we reduce the effective noise bandwidth of the circuit to 2.5 Hz, which is sufficient to make this noise source unimportant.

The long-term drift of this oscillator is less than three parts in 10^9 per day, provided the bias voltage and bath temperature are held constant. This stability is not degraded by the changing liquid N_2 and liquid He bath levels.

The fundamental limitation of the measurement sensitivity of this oscillator is the perturbation noise and not such problems as temperature dependence, bias voltage dependence, etc. We estimate that a superconducting resonator with a Q of 10^8 would have a noise of about 10^{-15} for a 1-sec averaging time.

A more detailed description of this work will appear elsewhere.

Acknowledgments

I would like to acknowledge the many useful discussions with Profs. J. R. Pellam, W. H. Parker, and D. L. Mills, and the expert technical assistance of Mr. R. T. Horton.

References

1. C. Boghosian, H. Meyer, and J.E. Rives, *Phys. Rev.* **146**, 110 (1966).
2. C. Kittel, *Quantum Theory of Solids*, Wiley, New York (1963).
3. L.S. Cutler and C.L. Searle, *Proc. IEEE* **54**, 136 (1966).

Controllable Electron Sources for Cryogenic Temperatures*

David G. Onn

Physics Department, University of Delaware
Newark, Delaware

and

P. Smejtek and M. Silver

Physics Department, University of North Carolina
Chapel Hill, North Carolina

We have developed criteria for the fabrication and operation of thin-film electron sources (cold cathodes) that permit their reliable use at cryogenic temperatures. Normal energy distributions for the electrons from the devices operating below 200°K are narrower than distributions obtained at room temperature, and are less dependent on applied voltage. Operation above 200°K destroys, or masks, this narrow distribution. We have used these emitters widely in studying electron states in cryogenic fluids.[1,2]

When a sufficient voltage v_d is applied to a pair of metals separated by a thin (~ 100 Å) insulating layer electrons from the base metal enter, either thermally or by tunneling, the conduction band of the insulator. Provided that v_d is greater than the work function of the second metal, and the second metal layer is thin (~ 100 Å), up to 0.1% of the conduction current i_d is emitted from the second metal surface. To be useful the emission current i_e must be stable and constant with time.

A cross section of a typical emitter, as we fabricate it, is shown in Fig. 1. The oxide layer is created by wet electrolysis of the base aluminum layer previously evaporated onto a cleaned glass substrate. Oxide thicknesses from 60 to 200 Å are used. A 100-Å emission layer, generally of gold, is evaporated onto the oxide with thicker electrical contacts to reduce voltage drop. Aluminum, nickel, and silver have also been tried for the emitting layer.

Such techniques have been used in the past[3] for creation of similar emitters, but several innovations have now made success in fabrication almost 100%. During electrolysis a very low current is maintained, up to $\frac{1}{2}$ hr being used to create a 100-Å oxide layer. In addition, frequent reversals of polarity are made. This combination seems to give a very uniform layer free from defects. Further, all the edges of the aluminum base, and any surface areas not required for emission, are electrolyzed to a 25% greater oxide thickness to reduce the spurious emission and edge breakdown effects often reported from these emitters in the past.

* Work Supported by Army Research Office (Durham) and Advanced Research Projects Agency.

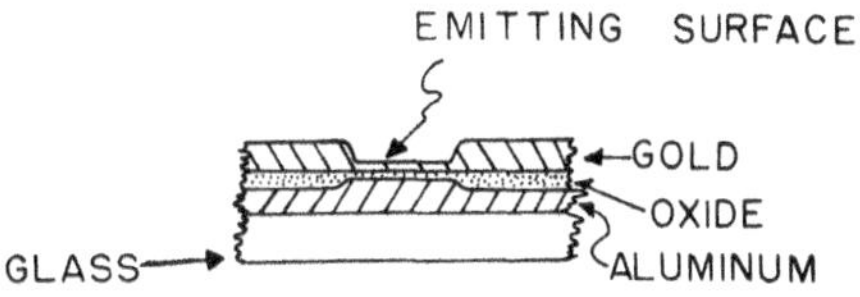

Fig. 1. Cross section of emitting region. Three such emitters are on each glass substrate with a common aluminum base.

However, the main key to successful use of these emitters at cryogenic temperatures appears to reside in their initial electrical treatment, rather than in fabrication. In particular, the emitters must be cooled below 200°K before *any* operating potential is applied. Even a brief test for electrical shorts at room temperature produces some deleterious effects. All testing, for continuity or emission, should be carried out at the anticipated operating temperature or below. Stable operation above 200°K does not seem possible.

A second step to reliable long-term operation consists in "aging" the emitter to its final emission current level. Both i_e and i_d of emitters of this type decay with time after application of v_d. At temperatures above 200°K this decay is followed by a very short period of stable operation, followed by instability and final destruction. In addition, voltage-controlled negative resistance (VCNR) peaks are observed for low voltages. Below 200°K we have established that there is a critical voltage v_c such that as long as v_c is not exceeded the final stage of increasing currents, instability, and destruction does not occur, and VCNR is not observed.

In Fig. 2 we show the variation of i_e with time for various applied voltages for a 100-Å oxide emitter. Above about 8.2 V, i_e increases with time, and soon becomes unstable. At 8.2 V stable operation is possible and has been observed up to 48 hr. Unfortunately, v_c varies by $\pm$ 0.5 V for emitters with the same supposed oxide thickness, probably because the oxide thickness can only be inferred from the electrolyzing voltage, and may depend on other factors. Traces such as those in Fig. 2 can be

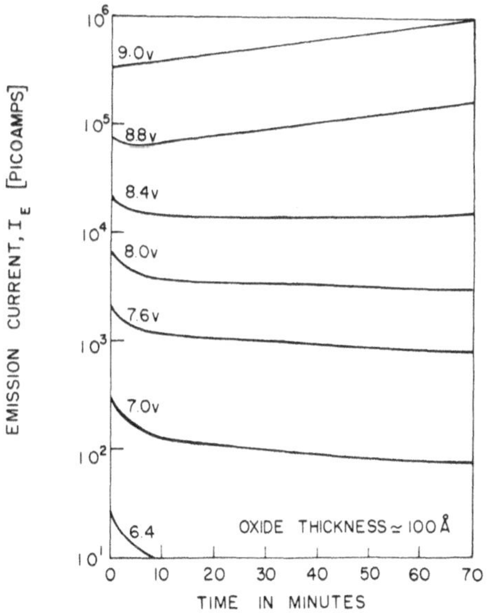

Fig. 2. Emission current i_e against time for increasing values of v_d.

used to determine v_c for one emitter in a set of three: Emitters on the same substrate are consistent.

It is essential that v_d never exceed v_c since this appears to create new conduction mechanisms in the insulator layer, as evidenced by a completely changed set of current–voltage characteristics, higher current levels, broadening of the electron energy distribution, and lower current stability. Furthermore, VCNR appears immediately if the emitter is later operated above 200°K; whereas, if v_c is not exceeded, VCNR appears only after a few minutes of operation at room temperature. The original operating characteristics (i.e., for $v_d < v_c$) cannot be reestablished by lowering v_d, but only by polarity reversal; and even then only a temporary unstable return to the low-voltage operating mode is possible. However, if v_c is not exceeded while at low temperature, the emitter can be cycled to room temperature and back, and the original characteristics reestablished after appropriate aging.

Current levels in the low-temperature operating mode ($v_d < v_c$ and $T < 200$°K) are temperature independent after the necessary aging. Typical emission currents i_e are 10^{-7} A/cm^2 with a typical device area of 0.1 cm^2. Film currents are generally less than about 15 μA at $v_d \simeq v_c$. If the conditions stated are abused, or if a voltage has been applied at room temperature, then higher currents are obtained, with i_e attaining microamp levels, and i_d in the milliamp range. Whether the emission mode for $v_d < v_c$ is masked by these large (and usually unstable) currents or whether it no longer exists is impossible to ascertain. Since current levels quoted by most earlier users of these devices are in the microamp (i_e) and milliamp (i_d) regions, we conclude that the low-temperature emitting mode that we report here has not been clearly identified previously.

We have obtained normal energy distributions from the emitters while operating at cryogenic temperatures, using a planar grid system. Both a retarding potential method and a modulated grid technique were used and gave consistent results. A typical normal energy distribution is shown in Fig. 3 $[dI/dV_R \propto n(E_z)]$. The width at half-height (WHH) is 1.3 eV, compared with more than 2 eV for similar emitters operated at room temperature.[4] We observe the WHH to be independent of temperature for $T < 200$°K and for $v_d < v_c$ and independent of v_d for a given

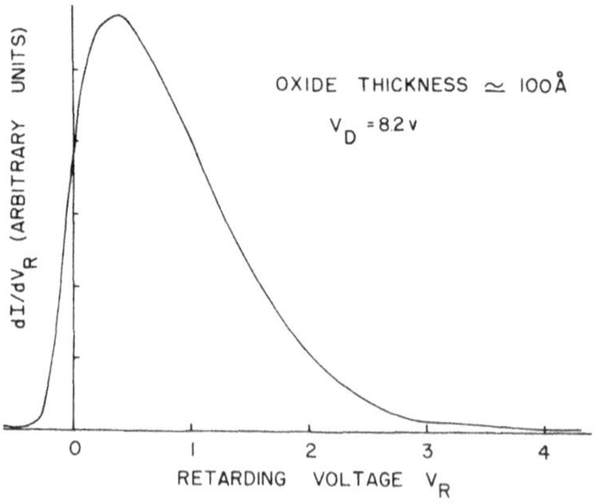

Fig. 3. Normal energy distribution for an aged emitter with $v_d \simeq v_c$.

oxide thickness. For increasing oxide thicknesses the WHH increases, though less rapidly than at room temperature. If v_d is allowed to exceed v_c, then the WHH increases, even below 200°K, to values typical of room-temperature operation, and similar changes are observed if any voltage was applied prior to cooling below 200°K. This observation again emphasizes the different nature of the conduction and emission mechanisms for $v_d < v_c$ and $T < 200$°K.

Various applications of these emitters are possible in cryogenic work, despite the limited range of electron energies obtained. We have used them to study behavior of electrons injected into cryogenic fluids under pressures up to 100 atm. This pressure does not affect the operation of the emitters. Since the operating voltage can be modulated, or a modulating grid technique used, they could provide sources for mobility experiments. However, in all applications in fluids a field high enough to extract an adequate number of electrons from the self-trapping effects of the image potential must be provided.[1,2]

A fuller account of the fabrication and operation techniques and of the energy distributions will be published elsewhere.[5]

References

1. D.G. Onn and M. Silver, *Phys. Rev. A* **3**, 1773 (1971).
2. M. Silver, P. Kumbhare, P. Smejtek, and D.G. Onn, *J. Chem. Phys.* **52**, 5195 (1970).
3. C.R. Crowell and S.M. Sze, *Physics of Thin Films*, G. Haas and R.E. Thun, eds., Academic Press, New York (1967), Vol. 4, p. 325.
4. R.M. Handy, *J. Appl. Phys.* **37**, 4620 (1966).
5. David G. Onn, P. Smejtek, and M. Silver, *J. Appl. Phys.* **45**, 119 (1974).

Improved Heat Exchange in Dilution Refrigerators by Use of Continuous Plastic Exchangers

L. Del Castillo,* G. Frossati,† A. Lacaze, and D. Thoulouze

Centre de Recherches sur les Très Basses Températures
CNRS, Grenoble—Cedex, France

The ultimate performance of a ^3He–^4He dilution refrigerator depends largely upon the efficiency of the heat exchange between the concentrated and the dilute phases. This exchange is limited by the high surface boundary Kapitza resistance. The usual way of overcoming this has been to increase the surface by the use of sintered copper powder or foils.[1,2] Metals were always chosen for their high thermal conductivity, but the exchange can be improved for a given surface by using a material presenting a lower Kapitza resistance.[2,3] Plastics are very useful for this since their Kapitza resistance should be of the order of that of Epibond, that is, about seven times lower than that of pure copper.[3,4] In spite of their low thermal conductivity,[5] the exchange between the two phases remains possible through foils thin enough that the thermal resistance of the bulk material to the exchange is negligible compared to the Kapitza resistances. Furthermore, this low thermal conductivity makes possible the realization of continuous exchangers.

We have made such plastic exchangers providing a large exchange area, in which the heat is exchanged through thin foils of Kapton, 7 μm thick. Their thermal resistance is only about 1% of the Kapitza resistance.

The general arrangement is shown in Fig. 1. The foils are glued on Kapton spacers of a staggered geometry which determines the flow of the ^3He–^4He mixture. The dimensions of the channels punched in the spacers are typically 350 × 2.6 mm, corresponding to an exchange area of 9 cm^2 for each foil. The thickness determines the impedance of the channels. In order to avoid viscous heating, it varies from 0.15 to 0.45 mm on the concentrated side and from 0.30 to 0.75 mm on the dilute side, according to the temperature of the liquid. Elementary exchangers are obtained by gluing in parallel some such circuits in which each phase flows alternately, in counterflow. Then such elements are connected in series to obtain the total surface. The main advantage of this geometry is the possibility of adapting independently the impedance Z and the surface σ. This exchanger is of the continuous type. The parameter Y of Radebaugh[3] is usually smaller than 10^{-2} in all the exchangers we tried, the main contribution being that of the column of liquid.

* CIM, UNAM (Mexico) Fellow.
† CNPq Fellow. On leave from University of Campinas, Brazil.

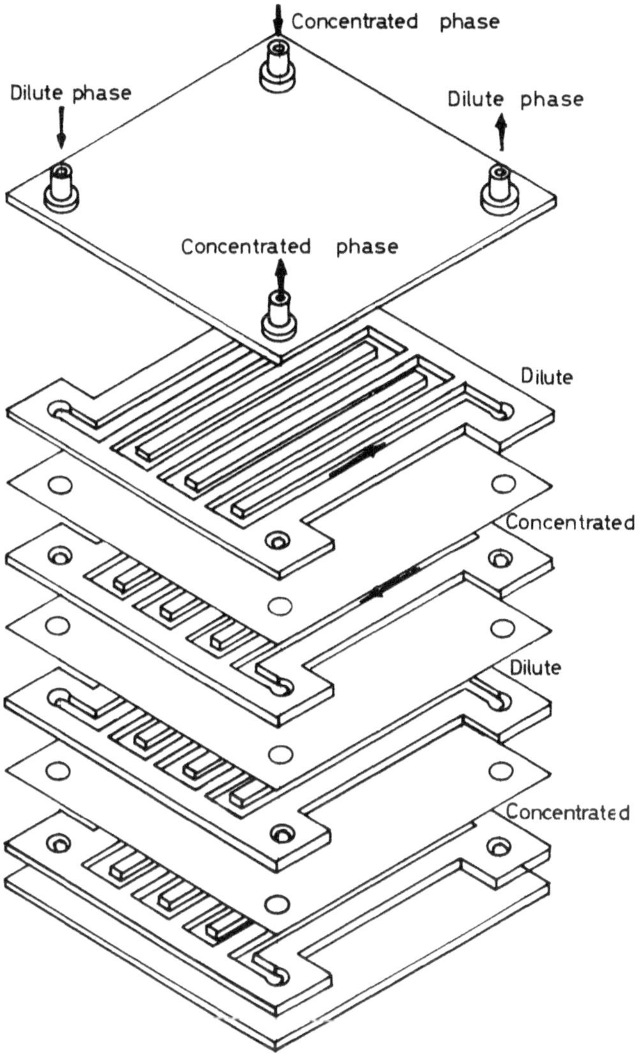

Fig. 1. Experimental arrangement of plastic exchangers.

Experiments

Three different arrangements of exchangers have been studied, and are shown in Table I.

Each time, the plastic exchanger was in series with a classical tubular exchanger consisting of two Cu–Ni tubes 0.22–0.4 mm and 1–1.5 mm in diameter, respectively, and 1.5 m long,[6] which was supposed to stop eventual instabilities at high temperatures.

The temperature T_0 of the concentrated phase was measured at the output of the exchanger with calibrated carbon resistors screwed on a sintered copper block

Table I. Three Different Exchanger Arrangements

1. One exchanger, $\sigma = 36\ cm^2$; five circuits in parallel		
3 for the dilute phase		$Z_d = 2.2 \times 10^7\ cm^{-3}$
2 for the concentrated phase		$Z_c = 2.6 \times 10^8$
2. One exchanger, $\sigma = 200\ cm^2$, 23 circuits in parallel		
12 on the dilute side		$Z_d = 1.7 \times 10^6$
11 on the concentrated side		$Z_c = 6.4 \times 10^6$
3. Three exchangers in series, $\sigma_{total} = 90\ cm^2$		
$\sigma = 18\ cm^2$	2 dilute	$Z_d = 3.3 \times 10^7$
	1 concentrated	$Z_c = 5.2 \times 10^8$
$\sigma = 36\ cm^2$	3 dilute	$Z_d = 2.2 \times 10^7$
	2 concentrated	$Z_c = 2.6 \times 10^8$
$\sigma = 36\ cm^2$	3 dilute	$Z_d = 6.4 \times 10^6$
	2 concentrated	$Z_c = 3.3 \times 10^7$

to ensure a good exchange with the liquid. In the epoxy mixing chamber the minimum temperature T_m^* was obtained by measuring the magnetic susceptibility of a cylinder of powdered cerium magnesium nitrate of height equal to the diameter, calibrated against the ^3He vapor pressure, and with calibrated carbon resistors. The helium flow rate was measured with a heated tube mass flowmeter and the concentration of ^3He was deduced from the ratio of ^3He to ^4He obtained with the mass spectrometer of a leak detector. In all experiments the ^4He concentration was about 1% for a molar flow rate of 26 μmole/sec.

Results

The minimum temperature T_m^* of the liquid in the mixing chamber is reported in Fig. 2 as a function of the ^3He molar flow rate $\dot{n}_3$ for the tubular Cu–Ni exchanger alone and in series with the three different plastic exchangers.

The temperature T_m^* presents a slight minimum for $\dot{n}_3 \simeq 20\ \mu$mole/sec, which may correspond to a spurious heat flux of about 2 ergs/sec for the three arrangements. Thus the performances of the exchangers themselves may be better described by $0.35T_0$, which represents roughly the value of T_m in the absence of heat leaks.

As can be seen from Fig. 2, the results obtained with the first exchanger of 36 cm^2 and the second of 200 cm^2 are the same. This probably results from the fact that in the larger exchanger the liquid circulates only in a few channels, due to some thermal or geometric difference of impedance between the channels. Thus only a small fraction of its surface would be effective. Indeed in the case of the third exchanger, in which the elements are in series instead of being in parallel, it becomes possible to reach much lower temperatures for a smaller total surface and the same impedance.

The significance of the physical parameters, surface, Kapitza resistance, and impedance shows up better by comparing the experimental results to those calcu-

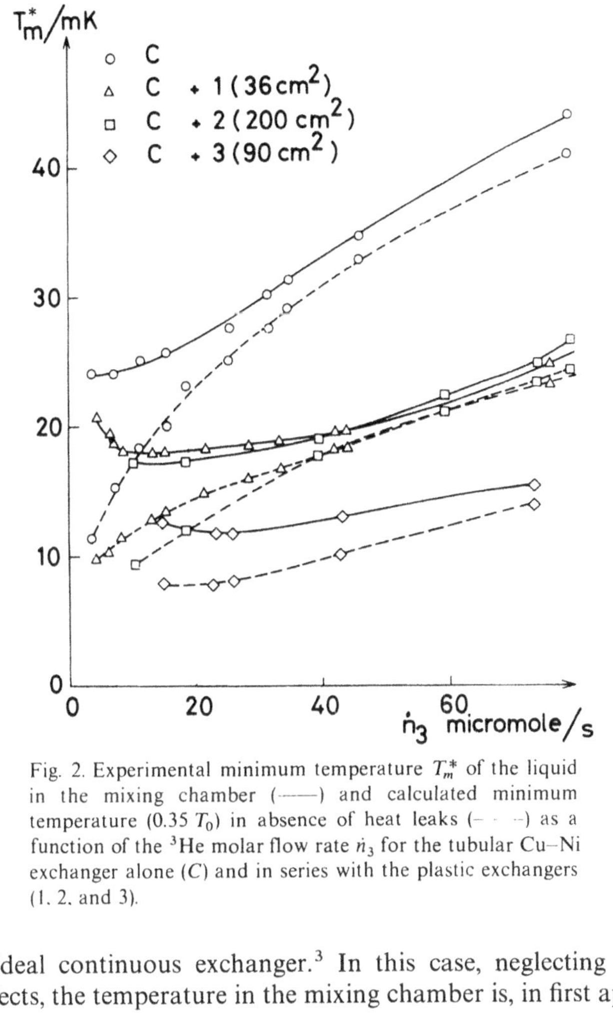

Fig. 2. Experimental minimum temperature T_m^* of the liquid in the mixing chamber (——) and calculated minimum temperature (0.35 T_0) in absence of heat leaks (- - -) as a function of the ^{3}He molar flow rate $\dot{n}_3$ for the tubular Cu–Ni exchanger alone (C) and in series with the plastic exchangers (1, 2, and 3).

lated for an ideal continuous exchanger.[3] In this case, neglecting viscosity and conduction effects, the temperature in the mixing chamber is, in first approximation,

$$T_m^2 \propto \alpha \dot{n}_3/\sigma \tag{1}$$

where σ is the exchange area, and α is the coefficient of the Kapitza resistance between the exchanger and the concentrated phase. In Fig. 3, T_m^* and $0.35T_0$ are reported as a function of the ratio $\sigma/\dot{n}_3$ for the different exchangers. In the calculation of the surface σ of the composite exchangers, the surface of the Cu–Ni exchanger has been reduced to its equivalent in plastic, proportionally to the respective Kapitza resistances, according to formula (1):

$$\sigma_{\text{equiv. Cu-Ni}} = \frac{\alpha_{\text{Epibond}}}{\alpha_{\text{Cu-Ni}}} \sigma_{\text{Cu-Ni}} = \frac{3 \times 10^{-6}}{10^{-5}} \times 10.5 \simeq 3.1 \quad \text{cm}^2$$

As expected, for the same value of $\sigma/\dot{n}_3$ the temperatures which can be reached with the Kapton exchangers are lower than those of the ideal exchanger made of Cu–Ni, assuming for Cu–Ni $\alpha = 10^{-5}$ cm^2 °K^4 sec/erg.[3] However, this value

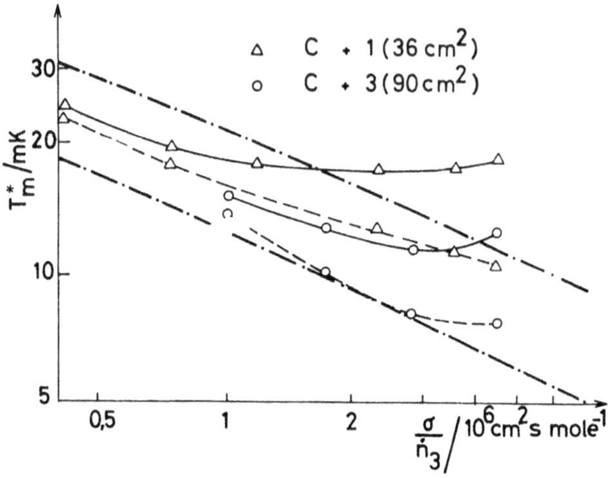

Fig. 3. Experimental (——) and calculated (– – –) minimum mixer temperature as a function of the concentrated stream surface area divided by the ^3He flow rate. The dash-dotted lines are calculated for an ideal continuous heat exchanger: The upper line is for the Kapitza resistivity of Cu–Ni, and the lower line is for that of Epibond.

seems to be rather high, with respect to different results on that type of exchanger. The performance of an ideal exchanger with the Kapitza coefficient of Epibond, $\alpha = 3.10^{-6}$ cm^2 K^4 s/erg, is reached by exchanger 3, which means that Kapton has at most the Kapitza resistance of Epibond. On the other hand, the results on exchanger 1 lie in an intermediate position. As in the case of exchanger 2, this probably results from the fact that only about half of the surface contributes to the exchange. Thus in both cases the fluid would only flow through one or two channels. For exchanger 3 it seems that no phenomenon of this type affects the performance. The curvature of the curves representing T_m^* arises from the heat leaks of about 2 ergs/sec, but it is difficult to explain that of the curves representing $0.35T_0$ since, according to the values of Z and Y, there would be no viscous or conduction effects on the results.

In conclusion, this type of plastic exchanger providing a large exchange area for a required impedance allows one to obtain the performance of the ideal exchanger down to very low temperatures. With a surface of 100 cm^2 we have reached a minimum temperature of 12 m°K in spite of heat leaks amounting to 2 ergs/sec, which would only be possible with surfaces at least ten times larger if using sintered copper exchangers. A reasonable surface of 1000 cm^2 would allow one to reach about 6 m°K with the same heat leak.

References

1. J.C. Wheatley, O.E. Vilches, and W.R. Abel, *Physics* **4**, 1 (1968); B.S. Neganov, Preprint JINR, p. 13, 4014, Dubna, USSR; G.J. Enholm and R.G. Gylling, *Cryogenics* **11**, 40 (1971).

2. P. Roubeau, D. Le Fur, and E.J.A. Varoquaux, *Proc. 3rd Conf. ICEC 3, Berlin, 1970*, p. 315.
3. R. Radebaugh and J.D. Siegwarth, *Cryogenics* **11**, 368 (1971); *Rev. Sci. Instr.* **43**, 197 (1972).
4. A.C. Anderson, J.I. Connolly, and J.C. Wheatley, *Phys. Rev.* **135A**, 910 (1964).
5. R.E. Peterson and A.C. Anderson, *Rev Sci. Instr.* **43**, 834 (1972).
6. J.C. Wheatley, R.E. Rapp, and R.T. Johnson, *J. Low Temp. Phys.* **4**, 1 (1971).

Thermal and Electrical Behavior of a Glass-Ceramic Capacitance Thermometer between 0.004 and 0.040°K*

Donovan Bakalyar, Robert Swinehart, and Walter Weyhmann

School of Physics and Astronomy, University of Minnesota
Minneapolis, Minnesota

and

W. N. Lawless

Research and Development Laboratories
Corning Glass Works, Corning, New York

In a previous paper[1] we reported several findings on the glass-ceramic capacitance thermometer developed at the Research and Development Laboratories of Corning Glass Works.[2-4] In particular, we reported that the capacitance depends linearly on temperature in the 7–25-m°K range with a sensitivity $(1/C)\,dC/dT$ of approximately -10^{-4}/m°K. The temperature measurements were made using the nuclear orientation of ^{60}Co in an iron foil polarized in a 3–4-kG applied field.

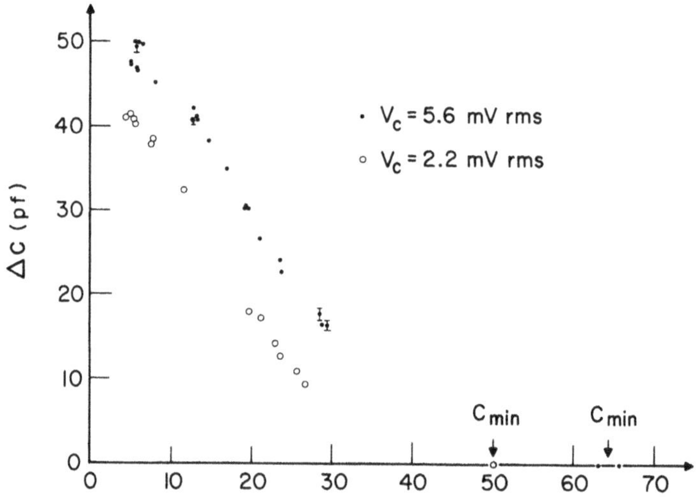

Fig. 1. Voltage and temperature dependences of the incremental capacitance $\Delta C = C(T) - C_{min}$ at very low temperatures. The minima are very broad and therefore not well defined, as shown by the two points for the 5 mV rms curve. $C_{min} \simeq 16{,}400$pF.

* Work supported in part by the U.S. Atomic Energy Commission.

The low temperatures were achieved by enhanced hyperfine nuclear cooling of $PrCu_6$.[5,6]

We have extended these measurements to below 5 m°K and find that the linear temperature dependence continues to at least 5 m°K. A slight evidence of curvature below this cannot be considered significant until the temperature is lowered still further.

More significantly, we find a drastic dependence of the incremental capacitance ΔC (measured with respect to the minimum in the 50–100 m°K region) on the impressed 400-Hz measuring voltage. The results of two such measurements are shown in Fig. 1. The first impression one draws from this figure is that the curves, including the minimum, shift to the left in the figure without changing shape as the voltage is decreased. In any case it is clear that as the voltage is decreased (a) the minimum shifts to lower temperatures, and (b) the incremental capacitance ΔC decreases rapidly for a given temperature.

These results emphasize the need for further research on the origin of this interesting phenomenon and on the need for voltage stability in the use of these devices as thermometers at very low temperatures.

References

1. D. Bakalyar, R. Swinehart, W. Weyhmann, and W.N. Lawless, *Rev. Sci. Instr.* **43**, 1221 (1972).
2. W.N. Lawless, *Rev. Sci. Instr.* **42**, 561 (1971).
3. W.N. Lawless, R. Radebaugh, and R.J. Soulen, *Rev. Sci. Instr.* **42**, 567 (1971).
4. L.G. Rubin and W.N. Lawless, *Rev. Sci. Instr.* **42**, 571 (1971).
5. K. Andres and E. Bucher, *J. Appl. Phys.* **42**, 1522 (1971).
6. E. Bucher, K. Andres, J.P. Maita, A.S. Cooper, and L.D. Longinotti, *J. Phys. (Paris)* (Suppl. 2–3), **32**, Cl-114 (1971).

Electronic Properties of Binary Intermetallic Compounds with Large Nuclear Moments

E. Bucher, E. Ehrenfreund,* A. C. Gossard, K. Andres, J. H. Wernick, J. P. Maita, A. S. Cooper, and L. D. Longinotti

Bell Laboratories
Murray Hill, New Jersey

Recent developments in the generation of millidegree temperatures made it necessary to improve thermometry in this temperature range. Cerium magnesium nitrate becomes unsuitable below 10 m°K, because of its long relaxation time and the thermal contact problem (except in liquid ^3He). Alternatively, the nuclear susceptibility $\chi_N(T)$ has been used, usually taking advantage of the extremely high sensitivity of a SQUID,[1-3] or measuring the spin–lattice relaxation time $T_1 \sim 1/T$ by a pulse method.[4-7] The operation of the SQUID is very often a delicate problem, while the relaxation measurement is discontinuous if operated in a pulsed mode and is not very accurate. Measuring χ_N by NMR in a CW mode[8] suffers from heat input problems and the possibility of erroneous temperature reading.

We wish to point out here that it is possible to measure $\chi_N(T)$ by a conventional dc method, not suffering from the aforementioned difficulties. It has recently been described and found to be very convenient.[9] The important requirement is, however, that χ_N be high enough. In order to have a good signal-to-noise ratio at higher temperatures (e.g., 10–100 m°K), it is necessary to work with large nuclear moments.

The problem, then, is reduced to finding a suitable nuclear thermometric substance meeting the following requirements:

1. Good signal-to-noise ratio: large nuclear moments.
2. Fast response: $T_1 T$ must be small (i.e., < 0.1°K sec).
3. Simple nuclear Curie law: no quadrupole splitting, i.e., small specific heat; cubic point symmetry for $I > 1/2$.
4. No superconductivity or magnetism, or if superconducting, low T_c, type I with low critical field H_c.
5. Easy phase formation (congruent melting very desirable), chemical stability, and ease of forming metallic contacts.
6. High-purity materials; no spurious signals from local moments.

Among the elements, no candidate can be found fulfilling requirements 1–6. Considering 1–5 only, one might think of Van Vleck paramagnetic materials as suitable candidates, since the nuclear Curie constant can be hyperfine-enhanced by orders of magnitude.[10] It has been shown, however, that these materials exhibit

* Present address: Department of Physics, University of Pennsylvania, Philadelphia, Pennsylvania.

nuclear ordering temperatures between 10^{-4} and 10^{-2} °K,[10,11] and therefore their nuclear Curie–Weiss constant will not be negligible. Therefore our search had to be concentrated on intermetallic compounds involving ^{93}Nb, $^{113/115}In$, ^{51}V, ^{45}Sc, ^{59}Co, ^{209}Bi, ^{27}Al, $^{151}Eu^{3+}$, ^{121}Sb, ^{7}Li, $^{185/187}Re$, etc.

Table I summarizes all our results obtained thus far. Along with T_1T, measured by a pulse method, some related electronic properties are listed. However, we will not discuss them further here. No transition metal compounds were found with sufficiently fast relaxation rate to be useful, not even among strongly exchange-enhanced materials (ScCo, $ScCo_2$, $NbCo_2$), due obviously to the fact that χ is enhanced faster than $1/T_1$, leading to magnetic order before T_1 is appreciably reduced. One straightforward conclusion is that the exchange enhancement in these compounds is of itinerant, rather than of local nature (see, e.g., Ref. 12). The slow relaxation rate in transition metal compounds is essentially due to the small d-electron hyperfine coupling constant and the orbital reduction factors.[13] The only alternative was to search for In- and Bi-rich compounds, speculating that these materials would exhibit relaxation rates similar to those of the heavy elements Pt, In, Sn, Pb, Tl, etc. The latter assumption does not turn out to be generally true, but fortunately it worked in the case of $AuIn_2$. This compound provides by far the best compromise in fulfilling the requirements 1–6. Because of its very favorable properties, we will concentrate our discussion on this compound.

$AuIn_2$ becomes superconducting at 0.21°K and has an estimated critical field of 14 Oe (from the BCS theory). The nuclear susceptibility signal (entirely due to ^{113}In and ^{115}In) in a field of 200 Oe, measured with a fluxgate magnetometer, is such that the signal-to-noise ratio below 10 m°K is 30:1. We find that impurity contributions (probably due to about 1 ppm Fe in Au) are present but saturate in 200 Oe below 30 m°K. As demonstrated in Ref. 9, a plot of measured χ_N values vs. $1/T$ in various fields yields perfect agreement with the theoretically calculated straight line from the nuclear Curie law due to the In isotopes, provided the measurements are taken under saturated impurity conditions. We have actually used a single crystal of $AuIn_2$ as a nuclear susceptibility thermometer in nuclear cooling experiments of $PrCu_6$ down to 2.7 m°K.[9] The observed thermal relaxation time of the thermometer at 4 m°K was about 15 sec, which is close to the expected spin–lattice relaxation time of 22 sec at that temperature.[14,15] Thermal contact between $AuIn_2$ and $PrCu_6$ was conveniently made via copper wires by soft-soldering with pure Cd metal.

It should be pointed out that $AuIn_2$ is not only a suitable nuclear magnetic susceptibility thermometer, but it will also be an efficient nuclear cooling material for the temperature range below 1 m°K, provided that it can be prepared in wire form to minimize eddy current heating during the demagnetization process.

For the lowest temperatures the size of the nuclear moment is no longer crucial. The usefulness of a material then only depends on T_1. Unfortunately, Tl metal, with its extremely short T_1,[16] is unsuitable because of its superconductivity and its chemical instability in air. Therefore we included in our study the brasslike Pt–Tl compounds,[17] avoiding both disadvantages. Here T_1 is approximately an order of magnitude longer than in pure Tl but is still shorter than in any other compound studied thus far. The Tl nuclei have the further advantage of zero quadrupole splitting ($I = 1/2$) and having the largest nuclear moment among all metals with nuclear

Table I *

		T_1T, °K sec			
A_nB_m	C_N, 10^{-8} °K units	A site	B site	χ_{300}	χ_{77}, 10^{-6} cm^3/FU
AlCo	12.2	$\gtrsim 200$	5.2	48.5	59.4
$AuIn_2$	6.67	—	$0.089^{14,15}$ (^{115}In)	−98	−98
BiLa	2.62	0.8	—	—	—
Bi_2Pt	3.44	~ 0.2	‡	—	—
Bi_4Rh	3.31	‡	‡	−212	−268
BiSc	5.78	1.5	200	—	—
CaIn†	4.01	—	—	29.6	33.8
Co_2Nb	16.2	3.8	2.9	1224	1352
CoSc	11.3	5.25	67	503	—
Co_2Sc	12.6	~ 2.5	58	3550	3590
Cr_2Nb	6.71	—	1.18	579	588
$Eu^{3+}Pd_3$	0.971	‡	‡	4740	6300
H_2Nb	24.4	> 100	8.74	—	—
InLi	8.48	2.3 (^{115}In)	190 (^7Li)	−25.3	−25.6
In_7Pt_3	7.71	1.4 (^{115}In)	—	−360	−354
In_2Pt	7.19	0.10 (^{115}In)	—	—	—
$InPt_3$	3.73	2.3 (^{115}In)	‡	−16	−16
In_3Sc	9.69	—	—	208	270
$InTi_3$	—	—	—	531	507
NbO	11.5	0.5 ± 0.3	—	29.8	28.3
$NiSc_2$	7.24	—	—	258	306
PdSc	5.08	—	55	82	84
PtSc	5.19	—	—	—	—
PtTl	1.21	‡	0.030	−78	−79
$PtTl_2$	1.53	‡	0.025	−181	−206
ReSi	3.20	1.0	—	26.8	27.7
RhSc	5.44	—	62	29	28

* Listed are the volume nuclear Curie constant C_N (for zero quadrupole splitting), low temperature spin–lattice relaxation constant T_1T, magnetic susceptibility χ_m at 300 and 77°K, electronic specific heat γ, Debye temperature $\theta_D(0)$, and lattice parameters of all compounds explored. The compounds are listed in alphabetical order. FU ≡ formula unit.

† The reported CsCl phases[20] RhIn and PdIn could not be obtained.

‡ No signal observed.

§ Complex[20] not Cu_3Au or $CdMg_3$ as reported.

\# Ordered defect NaCl: $a = 4.21$. 0, 0, 0 and $\frac{1}{2}, \frac{1}{2}, \frac{1}{2}$ positions missing.

spin 1/2. As another alternative, one might consider dropping condition 3 in favor of 2 providing that the quadrupole splitting is known with sufficient accuracy. In this case the nuclear susceptibility may be calculated in a straightforward way. In axial symmetry, e.g., one finds

$$\chi_N(T) = \frac{(N\beta)^2}{RT} \frac{\sum_{-I}^{+I} m^2 \exp[-(\alpha/kT)\, m^2]}{\sum_{-I}^{+I} \exp[-(\alpha/kT)\, m^2]} \tag{1}$$

with $\alpha = 3e^2qQ/4I(2I - 1)$ and $\beta = g_N\mu_N \cos\zeta$, where ζ is the angle between the

γmJ/χ, °K²/FU	$\theta_D(0)$, °K	Structure, Å	Other properties and remarks
—	—	CsCl: $a = 2.860$	—
3.15	187	CaF_2: $a = 6.51$	T_s: 0.21°K, $H_c \sim 14.0$ Oe
0.95	179	NaCl: $a = 6.579$	$\Delta H \sim 200$ G
—	—	Pyrite: $a = 6.69$	Bi site not cubic $T_s = 0.18°$
—	—	bcc: $a = 14.927$	—
—	—	NaCl: $a = 5.95$	—
—	—	CsCl: $a = 3.856$[18]	Unstable in air
18.90	495	$MgCu_2$: $a = 6.777$	Very small Qu pole splitting of Co
2.95	446	CsCl: $a = 3.143$	Sc: $k_N = 0.05\%$, Co: $k_N = 0.8\%$
24.4	457	$MgCu_2$: $a = 6.928$	—
—	—	$MgCu_2$: $a = 6.990$	—
—	—	Cu_3Au: $a = 4.100$	—
—	—	CaF_2: $a = 4.55$[19]	—
—	—	NaTl: $a = 6.789$	Blue-silvery color; $\Delta H \sim 4000$ G
—	—	Ir_3Ge_7: $a = 9.432$	$e^2qQ/h = 65$ MHz
—	—	CaF_2: $a = 6.36$	Phase unstable
8.28	272	Cu_3Au: $a = 3.99$	$\Delta H \sim 300$ G
—	—	Cu_3Au: $a = 4.477$	—
—	—	§	Difficult phase formation
2.47	657	#	$e^2qQ/h = 1.6$ MHz, $T_s = 1.50°$
14.4	352	fcc: 12.120	—
3.68	310	CsCl: 3.283	—
2.64	227	CsCl: 3.270	—
—	—	CoSn: $a = 5.605$[17] $c = 4.639$	—
—	—	$CuAl_2$: $a = 6.822$[17] $c = 5.563$	—
4.28	420	FeSi: $a = 4.775$	Qu pole splitting on Re; Re: $k_N = 4\%$
1.57	409	CsCl: $a = 3.206$	—

crystal axis and the applied field. The inconvenience, however, of a thermometer with a large cooling load still persists unless it is used in the temperature regime where only the lowest nuclear doublet is populated.

References

1. J.H. Bishop, E.C. Hirschkoff, and J.C. Wheatley, *J. Low Temp. Phys.* **5**, 607 (1971), and references therein.
2. J.M. Goodkind and D.L. Stolfa, *Rev. Sci. Instr.* **41**, 799 (1970).
3. R.A. Buhrman, W.P. Halperin, S.W. Schwenterly, J. Reppy, R.C. Richardson, and W.W. Webb, in *Proc. 12th Intern. Conf. Low Temp. Phys. 1970*, Academic Press of Japan, Tokyo (1971), p. 831.
4. R.E. Walstedt, E.L. Hahn, C. Froidevaux, and E. Geissler, *Proc. Roy. Soc.* **A284**, 499 (1965).
5. Yu. D. Anufriev and V.P. Peshkov, *Zh. Eksperm. i Teor. Fiz.* **61**, 345 (1971) [*Soviet Phys.—JETP* **34**, 183 (1971).]
6. O.G. Symko, *J. Low Temp. Phys.* **1**, 451 (1969).
7. P.M. Berglund, G.J. Ehnholm, R.G. Gylling, O.V. Lounasmaa, and R.P. Søvik, in *Proc. 12th Intern. Conf. Low Temp. Phys. 1970*, Academic Press of Japan, Tokyo (1971), p. 719.
8. E.B. Osgood and J.M. Goodkind, *Phys. Rev. Lett.* **18**, 894 (1967).

9. K. Andres and E. Bucher, *J. Low Temp. Phys.* **9**, 267 (1972).
10. K. Andres and E. Bucher, *J. Appl. Phys.* **42**, 1522 (1971).
11. K. Andres, in *Proc. 12th Intern. Conf. Low Temp. Phys., 1970*, Academic Press of Japan, Tokyo (1971), p. 641.
12. A. Narath and H.T. Weaver, *Phys. Rev. B* **3**, 616 (1971).
13. Y. Yafet and V. Jaccarino, *Phys. Rev.* **133**, A1630 (1964).
14. W. Warren, private communication.
15. G.C. Carter, I.D. Weisman, L.H. Bennett, and R.E. Watson, *Phys. Rev. B* **5**, 3621 (1972).
16. N. Bloembergen and T.J. Rowland, *Phys. Rev.* **97**, 1679 (1955).
17. S. Bhan, T. Gödecke, P.K. Panday, and K. Schubert, *J. Less-Common Metals* **16**, 415 (1968).
18. G. Bruzzone and A.F. Ruggiero, *J. Less-Common Metals* **7**, 368 (1964).
19. G. Brauer and H. Müller, *Plansee Proc.* (1958), p. 257.
20. R.P. Elliot, *Constitution of Binary Alloys*, McGraw-Hill, New York (1965), 1st Suppl., p. 555.

Contents
of Volume 1

Quantum Fluids

1. Plenary Topics

2. Quantum Fluids Theory

3. Static Films

4. Flowing Films

5. Superfluid Hydrodynamics

6. Helium Bulk Properties

7. Ions and Electrons

8. Sound Propagation and Scattering Phenomena

9. ^{3}He–^{4}He Mixtures

Contents

of Volume 2

Quantum Crystals

1. Plenary Topics

2. Helium Lattice Dynamics

2.1 Specific Heat and Sound

2.2 Heat Transport in Isotope Mixtures

2.3 Scattering and Vacancies

3. Helium-3 Nuclear Magnetism

4. Helium Monolayers

5. Molecular Solids

9. Low-Dimensional Systems

10. Ferromagnetism

11. Dilute Alloys

12. Theory

13. Magnetism and Superconductivity

14. Small Particles, Heat Capacity, and Paramagnetism

Contents

of Volume 3

Superconductivity

1. Plenary Topics

2. Type II Superconductors

2.1. Structures

2.2. Properties

2.3. Dynamics

3. Josephson Effect and Tunneling

3.1. Josephson Effect

3.2. Tunneling

4. Superconductivity Materials

4.1. Elements and Compounds

4.2. Gases, Films, and Granular Materials

5. Phonons

6. Fluctuations

7. Superconductivity Phenomena

Index to Contributors*

Complete Alphabetical Listing of the Contributors to LT-13, Volumes 1-4

* Volume number appears in parentheses followed by the page number on which the contribution begins.

Subject Index

GPSR Compliance

The European Union's (EU) General Product Safety Regulation (GPSR) is a set of rules that requires consumer products to be safe and our obligations to ensure this.

If you have any concerns about our products, you can contact us on ProductSafety@springernature.com

In case Publisher is established outside the EU, the EU authorized representative is:

Springer Nature Customer Service Center GmbH
Europaplatz 3
69115 Heidelberg, Germany

Batch number: 10838292

Printed by Printforce, the Netherlands